PERSPECTIVES
IN
EXPERIMENTAL BIOLOGY

Volume 2 Botany

PERSPECTIVES IN EXPERIMENTAL BIOLOGY

Volume 2 Botany

Proceedings of the Fiftieth Anniversary Meeting of the Society for Experimental Biology

Edited by N. SUNDERLAND

With the editorial assistance of

E. W. SIMON, J. HESLOP-HARRISON, P. W. BRIAN and D. A. BOULTER

PERGAMON PRESS

OXFORD · NEW YORK · TORONTO · SYDNEY

PARIS · BRAUNSCHWEIG

U.K.	Pergamon Press Ltd., Headington Hill Hall, Oxford OX3 0BW, England
U.S.A.	Pergamon Press Inc., Maxwell House, Fairview Park, Elmsford, New York 10523, U.S.A.
CANADA	Pergamon of Canada, Ltd., 207 Queen's Quay West, Toronto 1, Canada
AUSTRALIA	Pergamon Press (Aust.) Pty. Ltd., 19a Boundary Street, Rushcutters Bay, N.S.W. 2011, Australia
FRANCE	Pergamon Press SARL, 24 rue des Ecoles, 75240 Paris, Cedex 05, France
WEST GERMANY	Pergamon Press GmbH, 3300 Braunschweig, Postfach 2923, Burgplatz 1, West Germany

First edition 1976

Library of Congress Catalog Card No. 75-34711

Printed in Great Britain by A. Wheaton & Co., Exeter

ISBN 0 08 019868 6 (Volume 2)

0 08 018767 6 (Volume 1)

0 08 019939 9 (2 Volume set)

CONTENTS

LIST OF CONTRIBUTORS

ATTRIDGE, T. H., Department of Physiology and Environmental Studies, University of Nottingham School of Agriculture, Sutton Bonington, Loughborough, Leics., U.K.

BALDRY, C. W., Tate & Lyle Limited, Group Research and Development, P.O. Box 68, Reading, U.K.

BATT, T., Department of Plant Sciences, Baines Wing, University of Leeds, Leeds, U.K.

BOWLING, D. J. F., Department of Botany, University of Aberdeen, U.K.

BRADBEER, J. W., Department of Plant Sciences, University of London, King's College, 68 Half Moon Lane, London, U.K.

BRUINSMA, J., Department of Plant Physiology, Agricultural University, Wageningen, The Netherlands.

BUCKE, C., Tate & Lyle Limited, Group Research and Development, P.O. Box 68, Reading, U.K.

BUTT, V. S., Botany School, South Parks Road, University of Oxford, Oxford, U.K.

CALLOW, MAUREEN E., Department of Plant Sciences, University of Leeds, Leeds, U.K.

CLARKSON, D. T., A.R.C. Letcombe Laboratory, Wantage, U.K.

CLOWES, F. A. L., Botany School, South Parks Road, University of Oxford, U.K.

COOMBS, J., Tate and Lyle Limited, Group Research and Development, P.O. Box 68, Reading, Berks, U.K.

DAVIES, D. D., School of Biological Sciences, University of East Anglia, Norwich, U.K.

DOUGHERTY, C. T., Department of Plant Science, Lincoln College, Canterbury, New Zealand.

ELLIS, R. J., Department of Biological Sciences, University of Warwick, Coventry, U.K.

EVANS, L. V., Department of Plant Sciences, University of Leeds, Leeds, U.K.

FOWDEN, L., Rothamsted Experimental Station, Harpenden, Herts, U.K.

GILCHRIST, A. J., Botany School, South Parks Road, University of Oxford, Oxford, U.K.

GORE, J. R., Department of Botany, University of Edinburgh, Mayfield Road, Edinburgh, U.K.

GOULD, A. R., Botanical Laboratories, School of Biological Sciences, University of Leicester, Leicester, U.K.

HEATH, O. V. S., 10 St. Peter's Grove, London, U.K.

HESLOP-HARRISON, YOLANDE, Cell Physiology Laboratory, Royal Botanic Gardens, Kew, London, U.K.

HUMPHREY, T. J., School of Biological Sciences, University of East Anglia, Norwich, U.K.

INGLE, J., Department of Botany, University of Edinburgh, Mayfield Road, Edinburgh, U.K.

JACKSON, S. MARGARET, Department of Biology, University of York, Heslington, York, U.K.

JOHNSON, C. B., Department of Physiology and Environmental Studies, University of Nottingham School of Agriculture, Sutton Bonington, Loughborough, Leics., U.K.

JUNIPER, B. E., Botany School, South Parks Road, University of Oxford, Oxford, U.K.

KING, JOHN, Botanical Laboratories, School of Biological Sciences, University of Leicester, Leicester, U.K.

LANGER, R. H. M., Department of Plant Science, Lincoln College, Canterbury, New Zealand.

LEAVER, C. J., Department of Botany, University of Edinburgh, Mayfield Road, Edinburgh, U.K.

LEECH, RACHEL M., Department of Biology, University of York, Heslington, York, U.K.

LEFTLEY, J. W., Department of Botany and Microbiology, University College, Swansea, U.K.

LEWIS, D. H., Department of Botany, University of Sheffield, Sheffield, U.K.

LILLEY, R. MCC., Department of Botany, University of Sheffield, Sheffield, U.K.

LOUGHMAN, B. C., Department of Agricultural Science, Parks Road, University of Oxford, U.K.

LOVETT, J. S., Department of Biological Sciences, Purdue University, West Lafayette, Indiana, U.S.A.

MACROBBIE, ENID A. C., Botany School, University of Cambridge, U.K.

MANSFIELD, T. A., Department of Biological Sciences, University of Lancaster, Bailrigg, Lancaster, U.K.

MEIDNER, H., Biology Department, University of Stirling, U.K.

MILBORROW, B. V., Shell Research Limited, Milstead Laboratory of Chemical Enzymology, Sittingbourne Research Centre, Sittingbourne, Kent, U.K.

MILTHORPE, F. L., Macquarie University, N. Ryde, 2113, Australia.

OSBORNE, DAPHNE J., Agricultural Research Council Unit of Development Botany, 181A Huntingdon Road, Cambridge, U.K.

RAVEN, J. A., Department of Biological Sciences, University of Dundee, U.K.

ROBARDS, A. W., Department of Biology, University of York, Heslington, York, U.K.

RUSSELL, R. SCOTT, A.R.C. Letcombe Laboratory, Wantage, U.K.

SCHWABE, W. W., Wye College, University of London, Wye, Kent, U.K.

SIMS, A. P., School of Biological Sciences, University of East Anglia, Norwich, U.K.

SMITH, D. C., Department of Botany, University of Bristol, Bristol, U.K.

SMITH, H., Department of Physiology and Environmental Studies, University of Nottingham School of Agriculture, Sutton Bonington, Loughborough, Leics., U.K.

STEGWEE, D., Department of Plant Physiology, University of Amsterdam, The Netherlands.

STEWARD, F. C., Cornell University, Ithaca, N.Y. and Department of Cellular and Comparative Biology, State University of New York at Stony Brook, Stony Brook, N.Y. 11794, U.S.A.

STEWART, W. D. P., Department of Biological Sciences, University of Dundee, Dundee, U.K.

STREET, H. E., Botanical Laboratories, School of Biological Sciences, University of Leicester, Leicester, U.K.

SUTCLIFFE, J. F., School of Biological Sciences, University of Sussex, Falmer, Brighton, U.K.

SYRETT, P. J., Department of Botany and Microbiology, University College, Swansea, U.K.

TIMMIS, J. N., Department of Botany, University of Edinburgh, Mayfield Road, Edinburgh, U.K.

WALKER, D. A., Department of Botany, University of Sheffield, Sheffield, U.K.

WAREING, P. F., University College of Wales, Aberystwyth, U.K.

WILKINS, M. B., Department of Botany, University of Glasgow, U.K.

WIMBLE, R. H., Wye College, University of London, Wye, Kent, U.K.

WOOLHOUSE, H. W., Department of Plant Sciences, Baines Wing, University of Leeds, Leeds, U.K.

PREFACE

The Fiftieth Anniversary of the inauguration of the Society for Experimental Biology was celebrated by a special meeting held at the University of Cambridge from July 16th to 19th, 1974. Prominent members of the Society were each invited to present a paper on their own specialized field of research, providing a historical review, a statement of current thought in that area and, if possible, a predictive element. In the belief that the material thus presented would be of considerable use to other research workers, the speakers were invited to contribute their papers to these two volumes. The chapters therefore are each a personal affirmation, and reflect the approaches, opinions and styles of their individual authors.

For convenience, the papers have been collected into one zoological and one botanical volume. Within each volume papers on a particular discipline have been grouped together, although these are not necessarily the groupings that were used when the papers were read at the meeting. This perhaps underlines the difficulty of classification in such a diversified field as is here presented. However, this diversity gives an indication of the breadth of the Society's interests and of the topics which tend to appear most frequently in its programmes. The diversity too, emphasizes the strength and aims of the Society; to provide a forum for research workers in both animal and plant sciences to discuss current research, to present work which is often still in progress and incomplete, and above all, to encourage the cross-fertilization of ideas between specialized disciplines.

It is hoped that the material assembled in these two volumes will provide for young research workers a perspective of the field of experimental biology and for undergraduates a source of reference to a wide range of biological topics.

Several invited speakers were unable to provide a manuscript and we wish to record our thanks to them for their contribution to the Anniversary meeting. We are also indebted to those distinguished members of the Society who acted as chairmen to the sessions of the meeting. They were: T. Weis-Fogh, R. Brown, G. M. Hughes, A. Punt, P. W. Brian, Sir Vincent Wigglesworth, J. Chayen, R. D. Keynes, F. R. Whatley, E. W. Simon, G. E. Fogg, J. Heslop-Harrison, J. W. S. Pringle, Sir Rutherford Robertson, J. S. Kennedy, L. C. Beadle, D. A. Boulter, G. P. Wells, J. D. Robertson and Helen K. Porter. Finally particular thanks are due to the local organizing team: Helen Skaer, D. A. Hanke, C. G. Gill and the able Local Secretary, D. B. Sattelle.

N. SUNDERLAND,
John Innes Institute,
Colney Lane,
Norwich, U.K.

P. SPENCER DAVIES,
Department of Zoology,
The University,
Glasgow, U.K.

IN PRAISE OF EXPERIMENTS

O. V. S. HEATH

10 St. Peter's Grove, London, U.K.

IN THIS paper I propose to air my prejudices in favour of experimental science as compared with non-experimental, or observational, science. At this 50th Anniversary Meeting of the Society for Experimental Biology I may hope to be, in the main, preaching to the converted. In fact, most of what I have to say will probably be considered trite—and the rest no doubt wrong-minded.

If the Society had been formed say five years earlier I might well now be a zoologist. As it was, just nine months before the birth of the Society in December 1923, I visited my zoology professor at Imperial College, Professor MacBride, to tell him that I wanted to transfer from zoology to botany. He said that he could not conscientiously recommend anyone to take up zoology, as the prospects of employment were almost nil. In fact those prospects were not the reason for my decision, but I did not feel it was necessary to tell him that the main reason was that elementary zoology, as then taught at Imperial College, consisted entirely of anatomy and morphology, with the promise of embryology to come. We did not of course carry out any experiments in practical classes and the only ones I can remember being told about were Kamerer's experiments on the nuptial pads of the midwife toad—alleged to prove the inheritance of acquired characters.

In botany, which we did on days alternating with zoology, Professor Farmer did his best to undo the harm done by MacBride and assured us of the impossibility of acquired characters being inherited, but of more interest to me, the practicals included a course in plant physiology when we carried out a number of intriguing experiments. I do not think we were explicitly called upon to consider *why* we did these experiments, or, for instance, that they tested predictions arising from hypotheses, but they were certainly great fun. Nor did the fact that some of them were not exactly new detract from my enjoyment. For example, we repeated some of the experiments on root pressure devised by Stephen Hales (1727). We used manometers attached to the stumps of rooted plants much as he did. Hales himself knew quite well why *he* did these experiments: 'And since we are assured that the all-wise Creator has observed the most exact proportions, *of number, weight and measure*, in the make of all things; the most likely way therefore, to get any insight into the nature of those parts of the creation, which come within our observation, must in all reason be to number, weigh and measure.' His aim was, therefore, to justify the ways of God to man—an appropriate activity for a clergyman.

Hales also took the view that it was not enough to 'number, weigh and measure' natural objects in a state of nature (that is, natural history studies) but it was also necessary to carry out experiments, for he wrote: 'For the wonderful and secret operations of Nature are so involved and intricate, so far out of the reach of our senses, as they present themselves to us in their natural order, that it is impossible for the most sagacious and penetrating genius to pry into them, unless he will be at the pains of analysing Nature, by a numerous and regular series of Experiments; which are the only solid foundation whence we may reasonably expect to make any advance, in the real knowledge of the nature of things.'

He realized, however, that although honest experimental data are enduring facts, there is always some degree of uncertainty of interpretation: 'I have been careful in making, and faithful in relating the result of these Experiments, and wish I could be as happy in drawing the proper inferences from them.'

In the Botany Department at Imperial College there was a great deal of experimental research going on, both in Plant Pathology under William Brown and in the Research Institute of Plant Physiology under V. H. Blackman, and when I became an Assistant Demonstrator (on a salary of £225 p.a.) I began investigations on my own account. Later I was seconded to work at Rothamsted, to fill the gap between the departure of Maskell for Trinidad and the arrival of W. O. James appointed in his place. I was fortunate to be at Rothamsted at a time of great intellectual excitement, shortly after the publication of R. A. Fisher's (1925) *Statistical Methods for Research Workers.* This soon became my bible—like the fundamentalist's bible everything was there, if only one could understand it.

Fisher was a founder member of the Society and a member of its first Council. He probably, in my view, did more for experimental biology than any other one person in this century, especially by placing experiments with small samples on a sound basis, but also by providing methods for analysis of factorial experiments which provide information on how factors interact in affecting biological material; information that is unobtainable in any other way.

He was a great believer in the experimental approach. Many years later F. G. Gregory and I went to consult him about a problem. We were interested in the possibility that the cells in apical growing points were 14-sided tetrakaidodekahedra, as suggested by D'Arcy Thompson (1917), that being the space-filling solid nearest to a sphere and therefore of minimal surface, implying the predominance of surface tension forces. There are no intercellular spaces in the growing point and therefore the shape of the cells must be one that fills space in close packing. It seemed to me that assuming they were tetrakaidodekahedra it should be possible to predict the frequencies of triangles, quadrilaterals, pentagons, hexagons etc. as seen in sections of the growing point, but of course I was quite unable to do so. Fisher received us with a most benign and gracious condescension, and after brief consideration of the problem, he suggested that I should prepare large numbers of plasticene tetrakaidodekahedra, cut them in random planes and see. He then allowed us to feel that the interview was at an end—we felt we ought really to walk out backwards.

During the War I was privileged to share a room in the Imperial College Laboratory at Rothamsted with F. J. Richards. In one of our many discussions I put forward a view very much like that which I have quoted from Stephen Hales (whose book I had not then read), namely that without experiments it was impossible to arrive at an understanding of how things worked and that any so called science that did not involve experiment was not a science at all but stamp collecting. Richards, who was a keen amateur astronomer, mildly enquired if I thought astronomy was a science and I had to admit reluctantly that perhaps it was. After a good deal of thought I came to the conclusion that in Observational Science there were experiments, but they were provided by Nature, who on the whole designed them very badly, and generally carried them out in a slip-shod manner. A formal scheme for such Observational Science is shown in Fig. 1.

Recently (Heath, 1970) I suggested the following definition of science: 'Science consists essentially in an attempt to understand the *relations* of selected aspects of things and events in the real world, an attempt which should have both intuitive and logical components, and which must be based on observation and tested by further observation.'

This definition of course excludes mathematics, which does not have to be based on observations (data) but only on postulates which need not have any relevance to the real world. It consists of a number of systems of logic, written in shorthand, and it can be used to study relations in isolation, without reference to things or events. To the scientist, mathematics makes up an important part of the tool-kit of techniques that he must use but in my view it is not in itself a science.

I still believe in my definition of science and it implies that all science must begin with observation of things in the real world and what they do (i.e. events)—that is it must begin with natural history (1). For the observational scientist such natural history observations have two functions: first, exciting his interest and suggesting a problem for scientific investigation, and secondly, providing part of the data for one of Nature's experiments. These two functions may of course be filled by different sets of observations.

The next two stages are: (2) to make a hypothesis to account for the feature of the observations that has excited interest and (3) to predict certain consequences of the hypothesis which should be observable under specified natural conditions if it is true but which have not yet been observed. For the next stage (4) it is necessary to wait for the specified natural conditions to occur, and this may involve long delay or even going to another part of the world. For example, one might have to wait for a different season and/or go to a different latitude in order to test a hypothesis that flowering of a plant or mating behaviour of an animal was a response to a particular photoperiod. The contrasting natural conditions for the initial observations (1) and the final ones (4) constitute

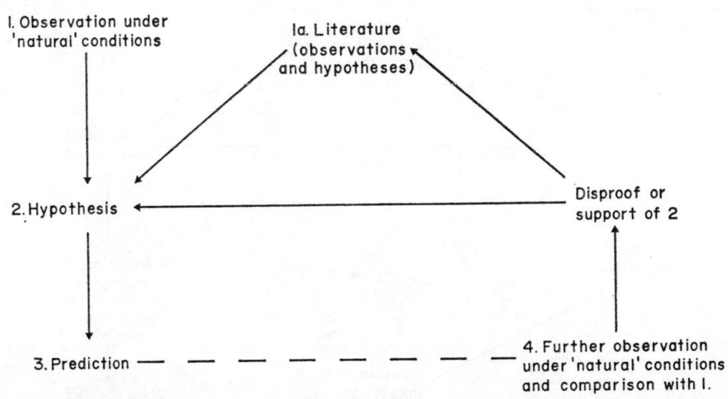

FIG. 1. Formal scheme for the stages in non-experimental science. For explanation see text. (After Heath, 1970).

the experimental treatments provided by Nature, but as (1) and (4) are usually widely separated in time and often in space also, there are manifold changes of other conditions confounded with the one that is intended. Thus at different latitudes or in different seasons, not only daylength but also light intensity and temperature (to mention only two other factors) will be different and it is virtually impossible to separate their effects. The organisms studied will also be of different populations, for all biological material changes with time and usually from place to place also. For a valid and unbiassed experimental comparison, each individual in the population studied must have an equal chance of being subjected to any one of the experimental treatments, but with 'treatment 1' applied initially to one whole group of individuals and 'treatment 4' later to another this cannot hold. The experiment is badly designed.

However, if the further observations agree with the prediction this can be held to give some support to the hypothesis, while if they do not it may (sometimes) provide disproof. Statistical tests of significance are not very meaningful in such natural experiments. To use them would be rather like testing the significance of a difference in mean height between samples of inhabitants of England and France. The test might show that the English were 'really' taller, with a certain level of probability, but we should not know whether to attribute this to the climate, or genetics, or the fact that the French eat frogs.

Fig. 2 shows the corresponding formal scheme for Experimental Science. The observations under natural conditions (1) now have only the one important function of exciting interest and suggesting problems for investigation. The investigator can of course make use of someone else's observations as recorded in the literature (1a) but not only does using his own observations give him more emotional drive to sustain him in the frightful labour of trying to think connectedly and logically, it also enables him to collect data of just the sort he wants—which the literature never provides.

He next has to think of as many as he can of the multitudinous hypotheses (2) that might provide the explanation for the problem raised by the observations, and select the one that most appeals to him. Again, someone else's hypothesis may be taken from the literature, but I think his own hypothesis (by definition a good one) and the hope of

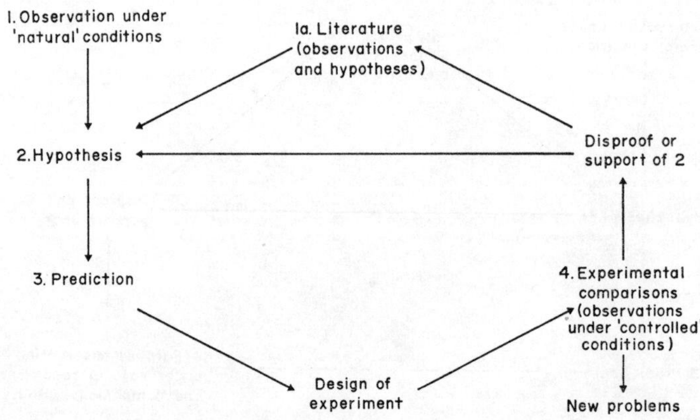

FIG. 2. Formal scheme for the stages in experimental science. For explanation see text. (After Heath, 1970).

supporting it gives him the greatest drive. However, a good second is someone else's hypothesis (by definition a bad one) and the hope of disproving it.

You will note that I am no believer in the unemotional and totally objective scientist of fiction—and if he existed I do not think he would be successful.

The prediction (3) does not now have to be one that can be tested by observation under natural conditions—new conditions can be imposed experimentally. The comparisons can therefore all be *within* stage (4), which becomes 'Experimental Comparisons' instead of between the widely separated (4) and (1).

The prediction and design of experiment should usually be considered together—it is seldom useful to predict effects which could only be observed under conditions that cannot be realized, unless indeed they are effects of such importance that the prediction will stimulate someone else to devise a new technique for the purpose. (Such stimulation is the principal role of the theoretician, who takes his natural history from the literature, carries out stages (2) and (3), and then hands the investigation to any experimental scientist who has the interest and ability to take it further.)

The imposition of experimentally controlled treatments to test the hypothesis has enormous advantages over the method of Observational Science. First and foremost, the treatment comparisons can be made *valid* and *unbiassed* by allotting the treatments randomly among the available material used for the experiment. Secondly, all the experimental treatments can be given *at the same time*, with a great reduction of irrelevant effects (for instance, of climate); alternatively, the treatments can be applied one after another to the *same individuals* with the removal of genetic differences in response to treatment; in either case there is generally a great improvement in *precision*. Thirdly, the treatments can be replicated, which with suitable randomization makes meaningful statistical *tests of significance* possible. Fourthly, the levels of the factors investigated can cover much *wider ranges* than are provided by Nature and can be distributed over those ranges in any way that suits the purpose of the experiment (e.g. in equal steps). Fifthly, different levels of different factors can be combined in an orthogonal *factorial design* to give information on how the factors *interact* in affecting the material, and thus make possible some approach to an understanding of the complex behaviour of plants and animals under natural conditions.*

Factorial experiments are also valuable in counteracting our innate tendency to select experimental treatments the results of which seem to support our pet hypothesis. If such treatments are incorporated in a wide-ranging factorial design the results are seen in context and are less likely to be misleading. The hypothesis is therefore more rigorously tested than in a simple experiment. Unexpected discoveries frequently emerge in experiments and this is especially true of those factorial experiments in which some of the combination treatments are such as never occur in nature—the organism's response is then unpredictable, as there has been no selection pressure to direct it, but may be extremely informative. Such discoveries pose new problems (Fig. 2) and provide 'Stage 1' for new cycles of investigation.

* The large factorial experiment of Heath and Russell (1954), on the interactions of light and carbon dioxide in affecting stomatal behaviour in wheat leaves, was discussed in the lecture as an example of the comprehensive information that such experiments may provide—see Heath (1975) for a condensed account.

I cannot here discuss what constitutes disproof or support of the hypothesis; I have given the matter some consideration in Heath (1970). Following disproof the hypothesis should be replaced by another: following support we should wish to extend it.

I do not wish to suggest that practising scientists often consciously work round cycles of procedure such as those in Figs. 1 and 2. The whole process is normally much less tidy and more intuitive. Nevertheless I believe that all the stages are essential components of a scientific investigation, whether it is all carried out by one investigator or by different investigators for different stages. I have left out of this diagram one very important source of information and inspiration—namely gossip. Ideas arising from gossip (and also from reading) are brought in at *all* stages. Scientific gossip in fairly remote subjects seems to be especially stimulating, which is perhaps why S.E.B. meetings are so extremely valuable in this respect.

I have said that in Observational Science the initial and final sets of observations constituting the 'treatments' are subject to many unwanted differences of conditions confounded with the desired change, but this is also true of the treatments in an experiment. It is, I believe, impossible to apply an experimental treatment intended to change one factor without at the same time altering an indefinitely large number of others. Suppose for example we wished to investigate the effects of supplying tissue cultures with different amounts of K^+ ion. This would also involve adding equivalent anions. We might think it important to hold the amounts of such anions as Cl^-, SO_4^{2-} etc. constant and therefore use potassium hydroxide, but this would change the pH. We could use a buffer to hold the pH constant but then the OH^- ions would alter the proportions of certain ions and undissociated molecules in the buffer mixture. Further, by altering the amount of potassium we should change the total salt concentration and hence the osmotic pressure, unless at the same time we varied some other solute (perhaps a sodium salt) which might then really be responsible for the observed effects of the treatments. Giving different amounts of potassium would also change the ratio of K^+ ion to every other sort of ion in the mixture and in fact balance of two or more kinds of ion might be the most important factor. And so on.

A well designed experiment, with the treatments properly randomized, can give unbiassed estimates of the effects of those treatments *as applied*. It tells us nothing of which of the innumerable components of the treatments are responsible for the effects observed. That is a matter of interpretation, in which the experimenter has to rely upon what he calls his judgement and his rivals call his prejudices. However, the experimentalist is much better off than the observational scientist, who can seldom eliminate *any* of the unwanted changes confounded with the desired one, however important he may think them. (As I mentioned, in Nature changes of daylength are *always* accompanied by changes in light intensity and temperature which usually have important effects on responses to photoperiod.) In the design of an experiment, *some* confounded components thought to be important can be eliminated or reduced, though at the cost of introducing or increasing others. In the same or subsequent experiments *different* changes of conditions can be confounded with the desired components of the treatments and in this way information can be obtained as to the apparent importance of the desired changes over a wider range of conditions. If, for example, additional K^+ ions appear to have consistent effects whether given with Cl^-, Br^-, SO_4^{2-} or OH^- ions, one believes that there is an effect of K^+ ions *per se*, but it *might* still be an osmotic effect—it is still a matter of interpretation.

In fact, the common criticism of other people's experiments, 'this was not a truly controlled comparison', must always be true of our own also.

Finally I might say something about the level of control in different types of experiment. There are two quite different reasons for controlling conditions in an experiment: (i) to impose experimental treatments, as for instance if we wished to compare the effects of high and low day temperature upon the growth of plants; (ii) to reduce variation (in space and time) of circumstances which are extraneous to the intended treatment comparisons, but which may nevertheless interact with the treatments and so reduce precision if not controlled. An example would be growing all our plants at a single controlled day temperature while investigating some other factor, for instance photoperiod.

Such control, applied for either of the two reasons, almost always reduces the size and complexity of the experimental units that can be used. In botany the lowest level of control and the largest and most complex units are found in field experiments—these, whether ecological or agricultural, can yield results which are very directly applicable in an empirical way. The questions of immediate interest to the farmer can often be answered at a purely empirical level by well designed field experiments; for example, we may be able to tell him that the application of a particular fertilizer mixture at a specified rate and in a certain manner may be expected to increase the yield of corn by a stated quantity (plus or minus so much)—for the sort of land and season concerned. Here, as always, the 'treatment effect' is the effect of all the confounded components—it tells us nothing of *how* the treatment works. The *reductio ad absurdum* for this completely empirical approach is, that without theoretical knowledge to make generalizations from experimental results possible, it would be necessary on every field in every year to carry out an elaborate experiment to discover what treatment it ought to have had.

If we leave the purely empirical approach and seek more fundamental information, the results of field experiments are the most difficult to interpret because of the complexity of the system and the very low degree of control of variation in important extraneous circumstances (temperature, soil moisture and so forth). In a properly randomized experiment, these are not confounded with the treatments, whose effects are validly estimated *for that experiment*, but they often do interact with treatment so that different results are obtained in another season or place. We therefore want to control them to increase precision and may also wish to do so to investigate their effects and interactions (as an example, in a factorial experiment on temperature and photoperiod).

We are therefore tempted first into the greenhouse to control night temperature and perhaps watering, then to sand or water culture to eliminate soil heterogeneity, then to the growth room to control light intensity, photoperiod and day temperature, then to organ culture in test tubes to substitute known supplies of organic and inorganic substances for the confusing effects of one plant organ on another, then to single cell culture (perhaps for similar reasons, though it must be said that one of the principal aims of this technique is to persuade the cells to make organs and whole plants), then to isolated organelles, then to enzyme preparations.

Each successive increase of control of external conditions improves precision, and each step in some sense simplifies interpretation as well as making further experimental treatments possible; however, each step also introduces a different set of confounded components for a similar intended treatment and makes the results less directly applicable to what happens in Nature or agriculture.

It seems that we have the choice between obtaining results that we can apply but not understand *or* those which we think we understand but cannot apply. The only way out of this dilemma seems to be to have experiments all up and down the range between field and test tube, and to argue back and forth from one to the other. Fortunately that provides a niche for each of us.

But whether we experiment with corn crops or chloroplasts, leaping locusts or lysosomes, it is salutary to remember the basic uncertainty of our conclusions. Like Stephen Hales we should say: 'I have been careful in making, and faithful in relating the result of these Experiments, and wish I could be as happy in drawing the proper inferences from them.'

REFERENCES

FISHER, R. A. (1925) *Statistical Methods for Research Workers*, Oliver & Boyd, Edinburgh (14th edition, 1970).

HALES, S. (1727) *Vegetable Staticks*, Printed for W. and J. Innys . . . and T. Woodward, London (reprinted 1961, Oldbourne, London).

HEATH, O. V. S. (1970) *Investigation by Experiment*, Edward Arnold, London.

HEATH, O. V. S. (1975) *Stomata*, Oxford University Press.

HEATH, O. V. S. and RUSSELL, J. (1954) An investigation of the light responses of wheat stomata with the attempted elimination of control by the mesophyll. Part 2 Interactions with external carbon dioxide, and general discussion. *J. exp. Bot.* **5**, 269–292.

THOMPSON, SIR D'ARCY W. (1917) *On Growth and Form*, Cambridge University Press (2nd edition, 1942).

MULTIPLE INTERACTIONS BETWEEN FACTORS THAT CONTROL CELLS AND DEVELOPMENT*†

F. C. STEWARD

Cornell University, Ithaca, N.Y. and Department of Cellular and Comparative Biology, State University of New York at Stony Brook, Stony Brook, N.Y. 11794, *U.S.A.*

INTRODUCTION

The life of the Society for Experimental Biology (1924–74) coincides with the period of my own preoccupations with research on plants. This being so, reference to the problems personally investigated and to the methods of approach may be appropriate inasmuch as this experience reflects some changing and continuing trends in plant physiology. An outstanding trend is from the simplistic explanation of biological phenomena, based on their separate investigation in isolation, to the need to recognize their inter-relationships within the complex organization of cells, tissues and organs. Therefore, this essay can draw largely on published material, restricting the presentation of new data to a minimum.*

Hopefully, the choice of *Growth and Development of Higher Plants* for the first plant physiological session of the Fiftieth Anniversary Meeting was deliberate, not fortuitous. If so, this also is a facet of the changing scene. In the mid-1920s the bright hope was that understanding of the properties of matter, especially of organic matter, and the direct application of the then known principles of chemistry and physics, collated in a body of knowledge labelled *General Physiology*, equally applicable to plant and animal cells, would lead to great advances. In a measure, this occurred as knowledge about cell permeability, osmotic relations, respiration and metabolism proliferated. Meanwhile, investigators reduced the complexity of their systems to the minimum, attempted to control all experimental variables except one and to present their data in graphs with two coordinates by points that fell neatly upon a line or a derived curve. The rise of enzymology and descriptive biochemistry soon raised a monument to the talent and skills of chemists in the whole body of knowledge of intermediary metabolism. But soon this subject was to embark upon another adventure through biochemical genetics and the linkage of genes to enzymes and thence to metabolism and biosyntheses. Consequently

* The background of this paper emerged from the work of many years with many associates. This was supported by successive research grants, latterly GM 09609 from the National Institutes of Health of the U.S. Public Health Service and recently by contracts from the U.S. National Aeronautics and Space Administration (NAS2-7846).

† The original extemporary presentation was copiously illustrated. This is here impossible, so the references cited must suffice inasmuch as they will lead to illustrations similar to those that were used. The references, therefore, are to work in which the author participated, and were chosen with this in mind.

text books became thicker, courses of instruction became more detailed and numerous, as information about plant physiological properties and functions proliferated. However, topics were treated in seemingly ever more deliberate isolation, and rarely was there time at the end of it all to discuss how plants integrate their varied attributes and activities, as they grow, develop and reproduce. So hopefully, the trend is now to face these complexities at the outset, to recognize that the current need is for synthesis and a treatment of growth and development of higher organisms, especially plants, as the end to which all other attributes of organisms and their cells are directed. No plant physiological function, or cell physiological problem, is fully comprehended until its operation is seen and understood in the context of the entire system that grows.

Over the years, we have come to know much about the structure and function of biological membranes and freely hypothesize upon the means by which they are traversed, actively or passively, in quiescent cells by electrolytes; nevertheless it is still surprisingly difficult to state how and why the cells of angiosperms regulate their internal solute composition as they grow. Metabolic charts, as on a laboratory wall, have long told in great detail the reactions that may occur, but they do not predict what does happen in any given situation or foretell the different metabolisms of organs or their response to environmental change. Despite all the detailed descriptions of how proteins are made one cannot yet produce a system for protein biosynthesis that is remotely able to rival plant cells and organs as they grow. Today we seem to know so much about the physics and chemistry of photosynthesis, except the most important point of all, namely, why plants and their cells *in situ* are so incredibly better at the task than any physical or chemical alternative that can yet be substituted. Moreover, while the means by which information that perpetuates biological identity is seemingly widely understood, even to-the niceties of structure of the hereditary substance, the means by which it is so efficiently programmed that diversity of form, function and composition emerge has not yet been rendered either generally intelligible or credible. Thus there is an obvious limit to the attempted understanding of the complex by breaking it down, for convenience, to other simpler units or by analogy with much simpler systems.

After 50 years, one must face the obvious conclusion: that the problems of growth and development are paramount and that physiological properties and functions of cells or organs are not fully understood until they can be seen in the context of systems as they grow. Also, in research, one should allow the full complications of active growth, metabolism and development free play, so that multiple interactions both occur and are detected. For too long the simplistic explanations that emerge from single variable experiments or systems at, or near, equilibrium have been either partial or misleading. 'Reductionism' without 'Holism' may become a sterile exercise.

The emergence of these points of view may now be illustrated by reference to such salient problems as (a) solute composition and osmotic regulation of plant cells, (b) interactions of respiration and nitrogen metabolism in growing and quiescent cells, (c) factors that induce growth in quiescent cells, i.e. growth-promoting substances and systems, (d) consequences of balance and imbalance in growth-promoting systems and their interactions with environments, (e) cultures of free cells and/or protoplasts; morphogenesis and somatic embryogenesis and (f) effects of environmental stimuli on morphogenesis in shoot apices.

Solute composition and osmotic regulation in plant cells

The mid 1920's saw, in the work of Hoagland and others, the first major swing away from equilibrium concepts of the unequal concentrations that obtain across cytoplasmic membranes separating the internal fluids of plant cells and their ambient media. Through the effect of the daily duration and intensity of light on the uptake and internal accumulation of ions (principally bromide ions) in *Nitella*, Hoagland saw the protracted accumulation of ions as a non-equilibrium expenditure of metabolic energy, which was originally fixed by light, thus enabling the system to do physicochemical work. In this sense the work on plants antedates by many years the concepts of active transport or dynamic transfer of solutes between cells and their media which are now such familiar features of cell physiology including the physiology of animal cells and micro-organisms. The idea here is not to trace the involvement of metabolism through energy-coupling and ATP, nor to indulge in an analysis of ion fluxes and permeases, for in spite of all the highly specific ideas that have been formulated over the years in this field it has been necessary to maintain that the primary uptake and accumulation of solutes (particularly halides of alkali metals and, latterly, of organic non-electrolytes) is a function also of the growth and development of the cells in question (for refs. see Steward and Sutcliffe, 1959).

It was a fortunate, but by no means accidental, circumstance that in 1929–33 *thin* discs of potato tuber tissue could be used as an experimental system but with a more realistic appreciation of their biological properties and salt absorbing propensities than hitherto. The conditions compatible with *de novo* accumulation of the bromide ion from very dilute solutions into *very thin*, well washed, discs in large volumes of well aerated solutions, proved to be those in which (a) the cells were all metabolically active, (b) they generated antibiotic protection along with a superficial browning of the discs (polyphenol oxidase activity) and (c) they possessed, through starch hydrolysis and a high content of non-protein nitrogen, all the prerequisites for protein synthesis and growth. Under appropriate conditions of oxygen tension etc. the respiratory intensity of the discs became an index of their ability to synthesize protein and to revive the latent activities of the erstwhile dormant cells harnessed to the constructive metabolism of cells re-embarking upon growth and the uptake of solutes (see Fig. 1 in Steward, 1970a). It is a matter of record that, in one long series of papers, the linkage of ion (Br^-, Cl^-, K^+, Rb^+) intake and accumulation with respiration in the actively metabolizing cells of potato was pursued. In another and later series, various aspects of metabolism compatible with salt accumulation were investigated while, in still other papers, the active centres of salt accumulation were located in centres of growth in the developing plant body. While all this was going on in my laboratory, elsewhere, Hoagland was pursuing somewhat similar studies using, principally, excised root systems of barley (for refs. and figs. see Steward and Sutcliffe, 1959). One must by-pass many other highly detailed studies, on carrot and other storage organs, in which the ionic interchanges observed, were all secondary, often short range interchanges between essentially non-growing quiescent cells and their ambient media.

After 6 years of wartime interruption, and with a renewed interest in the problem of salt uptake new approaches were adopted. These sought, first, an aseptic system capable of contrasting highly active, rapidly growing and proliferating cells with their quiescent but surviving counterparts. Out of this search emerged a carrot-phloem-tissue culture system involving small explants stimulated to grow by growth factors, though full exploitation of this system in the area of salt and solute uptake was long delayed. A second objective was to find a means of studying intimate details of nitrogen metabolism and the conversion of soluble non-protein nitrogen compounds to protein in the actively metabolizing and proliferating carrot cells; this was met through the then novel technique of paper chromatography (which incidentally had great and diverting implications in the study of nitrogen metabolism). Still a third objective, contemplated but then only dreamed of, was to study nutrition and solute uptake in zygotes isolated from embryo sacs. Unexpectedly this objective was destined to be met indirectly, for the carrot-phloem-explant system gave rise to cells which proved to be capable of somatic embryogenesis.

To return, however, to the concept that *de novo* intake and accumulation of solutes and ions in the cells of plants is one facet of their growth and development and that devices which seemingly isolate the one from the other, and which lead to simplistic causal explanations, may evade the primary problem (Steward and Mott, 1970; Mott and Steward, 1972 and refs. cited). We found that the heterotrophically growing carrot cells as they multiply may remain small, with small vacuoles, and in this state do not build up the great contrast between the osmotic value of their solutes and the ambient medium. The first solutes which such cells preferentially do absorb are non-electrolytes; any inorganic ions absorbed are mainly potassium (largely unaccompanied by halide but balanced by organic acid anions). Substantial accumulation of inorganic cations and anions (K^+, Cl^-, Br^-) occurs later when the growth of the cells is mainly by enlargement and when organic solutes may be reversibly replaced by salts from the medium, and vice versa. In fact the prime need as the cells vacuolate is *to reduce the internal activity of their water by solutes* so that it becomes a matter of supply and demand, of metabolic use or disuse, whether one solute or another is accumulated. A recent major summary chart (see Figs. 1a–c in Steward and Rao, 1971) showed the very great range of organic solute concentrations and compositions in standard carrot explants (all of the same clone) after exposure to many different conditions for a standard period; these were brought about solely by different combinations of growth factors, media, and environments. In a later paper (Steward *et al.*, 1973) the same theme was developed with special reference to the inorganic ions (K^+, Na^+, Cl^-) in the cells. For these reasons, and in the writer's view, so many modern investigations which restrict cells by their states of development, or by the experimental conditions, to such limited degrees of freedom necessarily fail to comprehend the full range and potentiality of the cells to accumulate, actively, both organic and inorganic (i.e. charged and uncharged) solutes and the many interacting variables by which their internal concentrations may be determined during their growth.

A recent small body of data has far reaching significance here. The small pro-embryonic cells in the carrot culture system seem able to exist, in or near, osmotic equilibrium with their ambient medium (although with potassium accumulated to some degree), while the larger, more highly vacuolated cells preserve their high osmotic values (with ions accumulated to very high ratios) virtually unchanged, despite variations in the osmolarity of the ambient medium (Steward *et al.*, 1975, table 2, a nd fig. 55).

INTERACTIONS OF RESPIRATION AND NITROGEN METABOLISM IN GROWING AND QUIESCENT CELLS

The first novel application of paper chromatography to the qualitative identification and quantitative analysis of compounds in the pool of soluble compounds in cells, referred to above, led to a broad survey of the nitrogenous composition of plants (leaf, stem and root, etc.) as they grow under different environments. It also traced the composition of cultured tissue explants as they synthesize protein and metabolize in full nutrient media under the influence of different levels and combinations of growth-regulating substances (Steward and Durzan, 1965 and refs. there cited). Whereas Schulze and Prianischnikov had spent almost a lifetime identifying protein amino acids free in plants (and by 1906 had listed only six of these) chromatography soon revealed the great diversity of simple, non-protein, nitrogenous plant constituents. Moreover, the bulk composition of soluble pools bore no evident relationship to the composition of the protein synthesized at the pool's expense. Older ideas of the use, and re-use, of nitrogen in plants as it passed into protein and into breakdown products of protein, which are reworked into the nitrogen-rich compounds (such as glutamine, asparagine, arginine) for storage or translocation were consolidated. This was so because, often, protein synthesis occurred in light, as in green leaves, whereas protein breakdown, and storage of breakdown products or their translocation to other organs, often occurred in darkness. Later, use of radioactively labelled nitrogen compounds revealed that their carbon entered into protein, as when synthesis was stimulated by growth-promoting substances, and re-entered soluble pools and even respired carbon dioxide in ways that necessitate that protein breakdown be regarded as a continuing feature of such cells (Steward *et al.*, 1956). In fact, whereas earlier, nitrogen and protein metabolism might have seemed to be a linear sequence involving the entry of nitrogen from inorganic sources into organic combinations and the consequent synthesis of protein for the work of cells or storage, it was to become recognized as a complex network of metabolic relationships. In these relationships nitrogen compounds were cross-linked via keto acids to carbohydrate metabolism and to respiration and sequences of synthesis and breakdown were seen as parts of the metabolic machinery of cells. In this machinery, regulated in green cells by light or darkness or in growing cells by growth-regulating substances, nitrogen and protein metabolism reflect the pace or intensity of metabolic activity and the ability of the cells to accomplish physiological work. All this requires that nitrogen compounds be compartmented; when they are inactive into storage pools and, when actively metabolized, they must be mobilized for passage to more active sites. Such ideas emerged even prior to the current knowledge of the role of tRNAs which may sequester given amino acids and transport them to sites of protein synthesis in the correct linear array.

The point here is that application of the distinctively new techniques (first amino-acid chromatography and later acrylamide-gel electrophoresis) not only gave new dimensions to the descriptive biochemistry of nitrogen compounds in plants, but also involved the reactions of nitrogen metabolism (of protein synthesis, breakdown and re-synthesis) cross-linked to carbohydrate metabolism and respiration, in the network of reactions that represent the working parts of the chemical machinery of plants. Hence few topics, from photosynthesis to ion accumulation and growth regulation, should now be considered without reference to nitrogen and protein metabolism.

FACTORS THAT CONTROL GROWTH IN QUIESCENT CELLS: GROWTH-PROMOTING SUBSTANCES AND SYSTEMS

The life of the Society for Experimental Biology also spans the productive period of our modern knowledge of plant growth-regulating substances i.e. the so-called plant growth hormones, however inappropriate that term may now seem to be in comparison with the hormones of the animal body. Prior to 1924 plant physiologists, notably of the Dutch school, referred to *the* 'growth substance' and confidently declared 'No growth without growth substance' and these ideas preceded the knowledge even of indol-3yl-acetic acid (IAA) as the longest known naturally-occurring auxin. Instead of this simple idea of one growth substance we now have several classes of growth regulators, variously distinguished by operational tests or chemical characteristics and, in each class, there are both naturally-occurring representatives and a steadily increasing array of synthetic equivalents. While a library of books has been written about this store of chemical and plant physiological knowledge, most references here in question may be obtained from Steward and Krikorian (1971) or from Thimann (1972). The relevant point is in the multiplicity of interacting factors that control growth, over and above the organic and inorganic nutrients and vitamins, both exogenous or endogenous, which plant cells utilize and require.

The auxins (IAA in particular) were to be the peculiar outcome of work on the coleoptile of grasses (which for over a quarter of a century kept the attention focused on what is, essentially, the final phase of cell elongation). In fact some still so restrict the term 'growth' as to preclude cell multiplication! Nevertheless, it was the induction of rapid proliferating growth in aseptic plant tissue explants, greatly stimulated in the case of carrot by the addition of coconut milk to the medium (Caplin and Steward, 1948, 1949) that stimulated the study of cell-division factors. Early expectations that the effect of coconut milk (the liquid endosperm of the nut) would be due to a single cell-division hormone-like substance, which was even initially termed coconut-milk factor or CMF, were soon confounded. However, the effectiveness of autoclaved, old samples of nucleic acid on tobacco explants, attributable (Miller *et al.*, 1955) to the substance 6-furylamino-purine, which was isolated, synthesized and called kinetin, again tended to emphasize first one chemical substance and, later, one type of chemical configuration (i.e. N^6-substituted adenyls) as the essential factor in stimulating cell division. Even so this simplistic idea overlooked the fact that often, as in carrot, the exogenous effect of kinetin or its relatives was realized only in the presence of exogenous IAA. In coconut milk and other similar liquid endosperms that nourish immature embryos, there are at least two types of system that will trigger cell division in otherwise quiescent mature carrot cells. In one of these (System I) the motivating cofactors are hexitols (replaceable by myo-inositol) and, among the active substances (AF_1), two known examples are glycosides of IAA. In the other (System II) the known cofactor is IAA and, among the active substances (AF_2) the known natural example is zeatin (6-(4-hydroxy-3-methylbut-*trans*-2-enylamino)purine). Nevertheless, the effect of whole coconut milk on carrot explants still exceeds the best that Systems I and II can do, even when they are accentuated by casein hydrolysate, and the trace elements (Fe and Mn) in appropriate interacting concentrations (Steward, 1970a, b; Steward and Rao, 1971). In fact, the representation of the various effects of the components of these systems, tested singly and in all possible combinations, on such varied parameters as the number and average size of carrot cells, their content of protein, of total nucleic acid, or almost any other important biochemical constituent, only becomes concisely feasible by resort to polygonal diagrams used under certain conventions. However, with all these developments, in 1974 the idea of a single, master, controlling growth substance seems long outmoded. Synergistic combinations of powerful but nevertheless simple molecules, which interact with trace elements and casein hydrolysate to make their maximum impact upon cells, and which also act upon them sequentially in different combinations, are a part of the requirement of the somatic cells (now known to be essentially totipotent) for their full heterotrophic nutrition and growth. This requirement comprises a great range, or network, of exogenous substances over and above the organic and inorganic nutrients and the known vitamins. In short what was belatedly recognized in 1972 as the role of 'multiple hormone systems' (Thimann, 1972) was definitely apparent much earlier. This being so, experiments on growth should surely be so designed that effects arising from interaction of known (and still undiscovered) control factors may be revealed. The mind should remain open, even in 1974, to consider all that is required to push growth and metabolism of angiosperm cells to the upper limit of their genetic potentialities.

BALANCE AND IMBALANCE IN GROWTH-PROMOTING SUBSTANCES: INTERACTIONS WITH ENVIRONMENT

The early lesson of studies on the induction of growth in carrot phloem explants was that no single substance, or even class of substance, 'unlocked the door of cell division'. This still holds true despite much current stress on the so-called cytokinins which are N-6 substituted adenyl compounds. The demonstrable effects of naturally-occurring leucoanthyocyanins extractable from endosperms and of other phenolic compounds, of diphenylurea, and even of the synthetic substance benzothiazolyl-2-oxyacetic acid (BTOA) were all unexpected (Steward, 1968; cf. pp. 132–206 therein). None of the active fractions from coconut milk, immature corn grains or fluids from *Aesculus* fruits (all of which have strong activity in the carrot-phloem-assay system) has yet owed its activity to substances that have assayed as gibberellins. Nevertheless the powerful inhibitor abscisic acid (ABA), like many extracts from natural sources, is inhibitory to the growth induced by coconut milk and by other sources of similar stimuli and, in the carrot-phloem system, this inhibition is reversible by gibberellic acid (GA_3). Thus, the whole system is open to regulatory control, through balanced or imbalanced actions of the component parts of growth-promoting systems, and these in turn are still accessible to the action of both synthetic and naturally-occurring substances that act as inhibitors and still others that reverse these effects. Even the unorganized, proliferative growth of carrot explants (which occurs at 21°C, in continuous diffuse light, in a full nutrient medium with coconut milk as a balanced source of growth regulatory stimuli) is still responsive to the daily duration of the light period, interacting with temperature (Steward, 1970a; cf. Fig. 9). The consequences of these interactions are very evident in the total growth (fresh weight) and colour of the explants, but these criteria could be extended to other useful indices of their growth and metabolism. Inevitably, the more restrictive are the experimental controls imposed upon the cells, and the more incomplete the growth-promoting systems being investigated, the more chances there are that unitary, simplistic, conclusions may emerge. Such conclusions seem unrealistic in relation to the unlimited activity of the whole.

MORPHOGENESIS AND EMBRYOGENESIS IN CULTURES OF FREE CELLS AND/OR PROTOPLASTS

It was a fortuitous, and fortunate, circumstance that the aseptic system of cultured carrot phloem explants that so effectively provided a system to contrast vigorously growing and mature quiescent cells, should also have yielded somatic cells capable of a simulated embryogenesis. In 1974, one may note the great contrast between the new procedures and the old. In the 1920's Stiles and his colleagues used relatively thick slices of carrot root shaken mechanically in limited volumes of unaerated solutions of single salts, i.e. conditions effectively designed to suppress vital activity in the system (except, albeit, of anaerobic bacteria). The more recent conditions are those under which small aseptic carrot explants proliferate, absorb solutes, and grow heterotrophically and also yield free cells as capable of morphogenesis as zygotes. It is not necessary to recapitulate the first observations of freely suspended viable carrot cells, released into the ambient medium, or of their division and growth, nor of the steps by which it was demonstrated that they could, in very large numbers, give rise to embryoids and then to plantlets and to clones of plants (Steward, 1968, pp. 466 *et seq.*). That tale has been told and retold. One may, however, reconsider here, the totipotency of free somatic cells of angiosperms, foretold by Haberlandt in 1902, because of its wide biological implications.

Explants of all the Umbelliferae so far tested (Steward *et al.*, 1970) proliferate when aerated in liquid media, whether they are isolated from mature organs or from seedlings, and the proliferating explants in turn yield suspension cultures of cells and small derived units from which plantlets arise in large numbers. The particular exogenous stimuli which initiate cell division have varied with the species, cultivar, or clone in question and with the site in the plant body from which the isolates were made. Also the useful devices, involving synergistic combinations of growth regulators, often applied in prescribed sequences (Steward *et al.*, 1967) that have fostered embryogenesis, have also varied with the circumstances. Whenever these devices have succeeded the conclusion has been that the genetic information was maintained intact in somatic cells throughout their prior development, together with a cytoplasmic machinery effective to transcribe it. Thus, when liberated from the restrictions of their location in the tissue of origin, and suitably nourished and stimulated, the isolated free cells of angiosperms may function again like zygotes. But zygotic embryogenesis proceeds virtually infallibly over its prescribed course whereas the somatic embryonic parallel, though often close, does not reproduce exactly the performance of zygotes *in situ*. Moreover, there are still many examples in which cell and tissue cultures of angiosperms merely proliferate and are still, morphogenetically, recalcitrant (Steward and Krikorian, 1974).

Recent work on the conditions most conducive to somatic embryogenesis in carrot has tested those conditions most able to foster development from the smallest units (i.e. free cells or small derived clusters) which have been propagated as pro-embryonic cultures, graded by filtration, and spread sparsely upon, or in, the surfaces of agar media in small disposable petri dishes. The smallest and most morphogenetically competent units contain relatively unvacuolated cells, as free of extraneous cytoplasmic inclusions (Wilson *et al.*, 1974) as they are of secondary reserve substances: in this they resemble meristematic cells. Moreover, imbalance in the cultural conditions that divert the cells prematurely to accumulate extraneous cytoplasmic inclusions seem to be unfavourable to somatic embryogenesis. In fact the presumption now is that the media and conditions that best favour somatic embryogenesis simulate, or replace, those that apply to zygotes in ovules and these, empirically detected or imitated, are as follows.

Like the fluids of embryo sacs, culture media conducive to somatic embryogenesis are, osmotically, relatively concentrated. Metabolically neutral solutes (sorbitol, mannitol) and reduced amounts of sucrose in the medium keep the cells reversibly in an arrested, 'poised' state. The two ways of consistently triggering organized and uniform development of the smallest carrot units on agar are (a) to include in the medium samples of cell-free 'conditioned' media taken from densely-inoculated and vigorously-grown suspension cultures of carrot cells heterogeneous as to their unit size and (b) to interpose a period of total darkness with cultures whose morphogenetic ability has been suppressed by light. The length of the period of darkness and light and the degree to which media are preconditioned for embryonic growth interact, positively, to contribute to isolated small pro-embryonic carrot units whatever inductive stimuli they otherwise may receive in mass heterogeneous cultures. Such heterogeneous cultures *en masse*, continuously subcultured in light, tend to become morphogenetically less effective with time, and they may need to be periodically 'revived' by re-establishing new lines of cultured cells from subdivided cultured embryos (see Steward *et al.*, 1975 for relevant figures).

All of this morphogenetic 'lore' is hard to pinpoint, as yet, in precise physical and chemical terms. But it emphasizes how subtle are the exogenous messages that otherwise totipotent somatic cells of angiosperms need in order that they may express their innate genetic information in the manner of embryos. In fact, for the most subtle of these expressions in terms of particular biosyntheses, one has to conclude that the 'information' alone is, as yet, not enough without the morphology which creates the internal environment in which it can be expressed (Israel *et al.*, 1969; Steward, 1970b).

Nineteen years after the clear laboratory evidence of totipotency and somatic embryogenesis in carrot it is curious to reflect upon the attention that has been given to this question in the botanical literature. The obvious needs were (a) to obtain such knowledge that populations of somatic cells, not only of carrot but of any angiosperm, could routinely, and in high yield, be caused to develop into plants and (b) to understand how, in development, cells which carry all the information to preserve identity in their progeny yet produce *in situ* the required diversity of cells, tissues, organs and composition. The problems in (a) have been pursued and some limited and continuing advances documented (cf. Steward *et al.*, 1975) but those in (b) are still central to the mystery of development.

Despite the great challenges in (a) and (b) above, an inordinate amount of attention has been paid, in the literature, to the question whether single isolated cells, or only small preformed clusters, form embryos and plantlets. This discussion has been sterile and unimaginative since no single cell, whether zygotic or somatic, ever becomes an 'embryo' without passing through a multicellular, often globular, stage. It was the frequency with which free carrot cells could divide, whether in liquid or on agar media and could assume multicellular patterns and display somatically the eventual sequence of globular, heart-shaped, torpedo and cotyledonary embryos in cultures, that vindicated Haberlandt's prophecy of 1902. This was not concluded without the knowledge that, under the special culture conditions of their origin, free and single carrot cells could divide and their progeny remain attached, that filtered preparations consisting predominately of viable single cells could be obtained and these used to inoculate, sparsely and successively, strains of suspension cultures. The enormous numbers of embryonic forms, and eventual plantlets, that have developed in sparsely-inoculated cultures placed beyond all reasonable doubt that the original carrot somatic cells and their progeny had all the intact genetic information to form, and reform, plants by events which (more or less closely depending on the conditions) simulate zygotic embryogeny. As techniques and knowledge improve, it may well become possible, even though impractical, to introduce one single, pro-embryonic cell into a facsimile of an embryo sac and expect it, without fail, to develop. But at present, it is better to pursue this objective, with the statistical advantage of numbers, knowing that the evidence justifies the theoretical possibility. Therefore, faced with the larger questions in (a) and (b) above, it becomes trivial to dwell unduly upon the relative frequency of development from single cells or small multicellular clusters. The latter, being further advanced, present fewer problems, and the farther along the developmental route, whether somatic or zygotic, that one intervenes, the fewer the problems encountered in continuing that development. The latest study (Steward *et al.*, 1975) of the conditions that best permit very small pro-embryonic globules to develop profusely, and on agar, to form very compact embryos in large numbers has re-established all the essentials of somatic embryogenesis and has brought new knowledge of the environmental factors that are best conducive to the parallelism with zygotes. Furthermore this has been done with different personnel, with different cultivars, in a different laboratory setting in a different decade! But perhaps, in the perspective of 50 years, one may say here that, too often in plant physiology, we examine the brush work and the composition of the pigments without seeing, from afar, the design and message of the picture; to the frequent detriment be it said, of successive generations of students.

I should digress too far to discuss here the even closer zygotic parallel that would obtain if free protoplasts, cells divested enzymically of their walls, could constitute the effective starting point in lieu of free cell cultures. Strangely, however, viable protoplasts prepared from the very pro-embryonic globules which will develop into plantlets with great ease, are far more reluctant than the cells from which they came to develop into plants. Thus the particular 'dialogue' between the nucleus and cytoplasm which prescribes what cells do in any particular circumstance is obviously modulated by other exogenous signals, not yet deciphered, that they may receive. To pursue this thought further, other avenues of investigation, which involve cells in the organized growing regions of angiosperms, need to be touched upon.

EFFECTS OF ENVIRONMENTAL STIMULI ON MORPHOGENESIS IN ANGIOSPERM SHOOT APICES

The apical growing points of angiosperms, capable of repetitive organ reproduction, have no counterparts in higher animals; hence it is not surprising that totipotency of somatic cells is as yet a more distinctive plant than animal characteristic (Steward, 1970a, b). The formation of whole plants from cells that have originated in both shoots and roots, reinforces the view that the differences in behaviour of the apices of shoots and roots results from their contrasted organization, not from any innate differences between their cells. Moreover, it is a feature of shoot tips that the forms to which they give rise are far more responsive to environmental factors than are those of roots.

The period now under review has seen conspicuous advances to knowledge of plant responses to diurnal cycles of light and darkness and of day and night temperatures; nevertheless, thinking has tended to dwell on those examples in which a single factor may seemingly trigger a single observed response. (Such examples are flowering in the cocklebur which is stimulated by a single short day and long night and reversed by a single light flash in a long dark period; or the light-induced germination of a particular strain of lettuce seed which is alternatively promoted and reversed by red and far-red radiation respectively.) These examples suggest that the train of events so triggered in the growing regions, or in the subsequent metabolism, are more simple and unitary than is generally the case. In fact, if one scans the voluminous literature on classes of growth-regulating substances and the varied responses they affect, and the literature on biological clocks and rhythms and on the interactions between the effects of environmental factors it becomes apparent that situations are rare in which all may be simplistically stated as due to a unit response to a single causal factor. It is, in fact, more realistic that each of the living cells in the organized shoot apex of angiosperms is conditioned by its location to a certain range of response, and that organs, when they are induced to form from primordial cells, express their metabolism in characteristic ways which are as much a consequence of organogenesis as are their forms.

The limitations of current 'molecular biology' are here most apparent. In any real biological situation complex organization above the molecular level is involved whether at the levels of organelles, cells in isolation or *in situ*, of organs and at the levels of their correlated actions in 'divisions of labour' within the plant body. Thus, it is one thing to invoke the inherited, innate, information in DNA to describe the way in which the repeating patterns of large molecules are directed; but it is quite another to understand how the complex organization of which they eventually form a part actually works, or how great blocks of gene- and enzyme-controlled reactions are selectively and simultaneously called into play during development. If the events of active transfers of solutes, organic and inorganic, have not yet yielded to simple causal explanations and the regulations of cell multiplication and enlargement involve so many factors it is not surprising that plant morphogenesis with its basis in the origin and behaviour of apical growing regions, is still in its descriptive, not causally interpreted, phases.

The 'molecular biology of development' seems, therefore, to be at present a contradiction in terms and it is a salutory exercise to recognize the range of morphological and metabolic events that are triggered by the morphogenetic stimuli that affect angiosperm shoots. While examples are innumerable, references are here made to but a few, especially to perennial plants in which both flowering and vegetative organs of perennation are induced by interacting environmental stimuli which ultimately control localized events in the shoot apices; these events are, no doubt, locally due to the activity of various substances which can induce the target cells to multiply, to enlarge or metabolize.

For me the first 'case history' of this kind was the peppermint plant of commerce (*Mentha piperita* L.) (see Steward *et al.*, 1962). Peppermint flowers under long days, albeit without setting seed so that it is vegetatively and clonally reproduced; it also forms stolons under short days and then exhibits distinctive metabolism (especially nitrogen metabolism) in its leaves, stems and roots and the detailed differences in the metabolism of these organs are as strikingly regulated by environmental factors as are their forms. Long and short days interact with high and low night temperatures and low night temperatures tend to offset the effect of long days. But all these events, recently re-investigated, are traceable to the growing regions where they obviously originate.

The potato plant, and its various hybrids, provides an even more dramatic case history (Moreno, 1970; Roca-Pizzini, 1972). Particular species, or cultivars, are strikingly photoperiodic as flowering is induced under long days (aided by high night temperatures) while the initiation of stolons (which later may form tubers), fostered by short days is aided by low night temperatures. Thus, whereas the vegetative apical-growing point of potato plants may be switched over to flowering, with a profoundly changed set of forms and developing organs in its apex, axillary buds, *even on the same axis,* may adopt the stoloniferous habit with a very different set of organizational changes at their shoot tips. Thus, if one contrasts the terminal growing points and their corresponding stoloniferous buds under potentially flowering conditions (long days and relatively high temperatures) with those of non-flowering, potentially tuberizing conditions (short days and relatively low temperatures) one sees a set of four sharply contrasting states of apices. In all of these states in apices (composed of essentially totipotent cells, which make up the organization) cells behave in very different ways in response to the signals that they receive (for brief summaries see Steward, 1971; Steward and Krikorian, 1972 cf. fig. 5 and pp. 608–9). Moreover, the consequences of the applied environments, with their interlocking effects of day length and temperature, so modify the growth of shoot tips that they may produce such contrasting organs as flowers and tubers. However, in particular locations, these variables also produce great contrast in the forms of cells and in their metabolism. In the procambial region immediately below the shoot apex of stolon- and tuber-forming branches (i.e. under short days and low temperatures) there may be tremendous activity while the cells of the pith, or medulla, become large and open and full of conspicuous starch grains. Immediately below the *terminal* vegetative shoot apex there is far less secondary growth in the cambial region while the cells of the pith, or medulla, elongate greatly and divide transversely into longitudinal files of starch-free cells. However, under maintained long-day conditions (which transform the main apex into a floral shoot) the lower axillary branches on the same plant remain, not only vegetative, but their development into stolons is suppressed.

Innumerable other examples could be cited. Certain cultivars of peas unpredictably form the substance *O*-acetyl-homoserine which accumulates in the ovary wall but, in a strange way, their soluble nitrogen pools emphasize either the *O*-acetyl homoserine together with homoserine, or other nitrogen compounds, e.g. asparagine, according to the particular temperature conditions that obtain at night during development (Grobbelaar and Steward, 1969). (High temperatures during long nights favour asparagine over *O*-acetyl-homoserine.) Many plants, e.g. the cultivated gros Michel banana (Steward, 1968, pp. 337–370), manage to make tremendous morphogenetic shifts at the vegetative terminal shoot apex, seemingly without the aid of exogenous environmental stimuli such as daylength or temperature. This shift in the banana transforms the vegetative habit of their pseudostems, which bear large green leaves and no flowers, to the production of a true stem with its tremendous production of flowers and fruit and reduced leafy bracts. This transition, which transforms the shoot apex, seems to occur under endogenous stimuli which are the consequence of the prior formation of a given number of leaves. In a tulip bulb, another monocotyledon, the capacity to flower, is in part a function of the size of the bulb but, in this case, temperature also exerts a powerful morphogenetic effect as the Dutch have long known (Steward, 1968, figs. 8–14; Steward *et al.*, 1971).

Concluding remarks

As problems of organization, growth and development loom larger in biology the limitations of the time-honoured approaches inherited from physics and chemistry become ever more apparent, despite their evident successes in areas of genetics and intermediary metabolism (these approaches best deal with systems which, reduced to their simplest terms, behave as equilibria and respond predictably to factors that may be varied singly). Moreover, it is now often regarded as a scientific virtue (even a necessity) to seem to know beforehand what to expect, to prescribe the methods of approach to the last detail and even to predict the results. Carried to extremes these fashionable restrictions (beloved of research-granting agencies) stifle originality and eliminate respect for the essential mystery of the *bio* part of biochemistry and of biology which derives from the often unpredictable consequences of complexity and organization. This mystery often entails that, even when we know *what* the living system does we must recognize that, in so doing, its complexity may be increased and its entropy is locally and inexplicably reduced.

In these circumstances the single-factor physiological experiment of my youth, with its unitary simplistic interpretations, needs to give way to others which are designed to reveal and to represent complex multiple interactions: interactions between many different factors operating at all levels, from the purely physical to the morphogenetic, triggering effects which may occur in cells, organs, or whole plants. To follow this effectively and comprehensively, requires that individual investigators, who can only deal with one facet of the problem, should be part of a team. All this being so, one should not shy away and shun the study of the seemingly insoluble, as in the physiology of cells as they multiply, grow and differentiate. Plant physiologists should recognize anew the distinctive importance of the growing regions where ultimately growth and development are regulated. Finally, a major goal is still to know how the specialized behaviour of genetically totipotent plant cells is programmed *in situ*, throughout development, so that, when isolated, otherwise mature cells can again recapitulate the behaviour morphogenetically and biochemically, of the entire organism even down to such insignificant detail as in the last few pink flowers in the complex umbel of a Queen Anne's Lace, reared from somatic cells (Steward, 1970a, cf. fig. 21).

REFERENCES

CAPLIN, S. M. and STEWARD, F. C. (1948) Effect of coconut-milk on the growth of explants from carrot root. *Science N.Y.* **108**, 655–657.

CAPLIN, S. M. and STEWARD, F. C. (1949) A technique for the controlled growth of excised plant tissue in liquid media under aseptic conditions. *Nature Lond.* **163**, 920–923.

GROBBELAAR, N. and STEWARD, F. C. (1969) The isolation of amino acids from *Pisum sativum*. Identifications of L(—)-homo-serine and L(+)-*O*-acetylhomoserine and certain effects of environment upon their formation. *Phytochemistry* **8**, 553–559.

ISRAEL, H. W., MAPES, M. O. and STEWARD, F. C. (1969) Pigments and plastids in cultures of totipotent carrot cells. *Am. J. Bot.* **56**, 910–917.

MILLER, C. O., SKOOG, F., VON SALTA, M. H. and STRONG, F. M. (1955) Kinetin, a cell division factor from deoxynucleic acid. *J. Am. chem. Soc.* **77**, 1392.

MORENO, U. (1970) Physiological investigations on the potato plant. Ph.D. Thesis, Cornell University, Ithaca, New York.

MOTT, R. L. and STEWARD, F. C. (1972) Solute accumulation in plant cells: V. An aspect of nutrition and development. *Ann. Bot.* N.S. **36**, 915–937.

ROCA-PIZZINI, W. M. (1972) The development of certain crop plants as affected by environments. Ph.D. Thesis, Cornell University, Ithaca, New York.

STEWARD, F. C. (1968) *Growth and Organization in Plants*. Addison-Wesley, Reading, Mass.

STEWARD, F. C. (1970a) From cultured cells to whole plants: The induction and control of their growth and morphogenesis. The Croonian Lecture, 1969 *Proc. R. Soc.* B **175**, 1–30.

STEWARD, F. C. (1970b) Totipotency, variation and clonal development of cultured cells. *Endeavour* **29**, 117–124.

STEWARD, F. C. (1971) Plant physiology: The changing problems, the continuing quest. *A. Rev. Pl. Physiol.* **22**, 1–21.

STEWARD, F. C. and DURZAN, D. J. (1965) Metabolism of nitrogenous compounds. In *Plant Physiology: A Treatise*, ed. STEWARD, F. C., vol. IVA, pp. 379–686. Academic Press, New York.

STEWARD, F. C. and KRIKORIAN, A. D. (1971) *Plants, Chemicals and Growth*, 232 pp. Academic Press, New York.

STEWARD, F. C. and KRIKORIAN, A. D. (1972) Problems of integration and organisation. In *Plant Physiology: A Treatise*, ed. STEWARD, F. C., Vol. VIC, pp. 367–419, Academic Press, New York.

STEWARD, F. C. and KRIKORIAN, A. D. (1974) The culturing of higher plant cells: Its status, problems and potentialities. In *Form, Structure, and Function in Plants*, pp. 1–27, Prof. B. M. Johri Commemoration Volume, University of Delhi.

STEWARD, F. C. and MOTT, R. L. (1970) Cells, solutes and growth: Salt accumulation in plants reexamined. *Int. Rev. Cytol.* **28**, 275–370.

STEWARD, F. C. and RAO, K. V. N. (1971) Investigations on the growth and metabolism of cultured explants of *Daucus carota*. IV. Effects of iron, molybdenum and the components of growth promoting systems and their interactions. *Planta* **99**, 240–264.

STEWARD, F. C. and SUTCLIFFE, J. F. (1959) Plants in relation to inorganic salts. In *Plant Physiology: A Treatise*, ed. STEWARD, F. C., Vol. II, pp. 253–478, Academic Press, New York.

STEWARD, F. C., BIDWELL, R. G. S. and YEMM, E. W. (1956) Protein metabolism, respiration and growth. A synthesis of results from the use of C^{14}-labelled substrates and tissue cultures. *Nature Lond.* **178**, 734–738; 789–792.

STEWARD, F. C., HOWE, K. J., CRANE, F. A. and RABSON, R. (1962) Growth, nutrition and metabolism of *Mentha piperita* L. *Mem. Cornell Univ. agric. Exp. Stn* **379**, 1–144.

STEWARD, F. C., KENT, A. E. and MAPES, M. O. (1967) Growth and organization in cultured cells: Sequential and synergistic effects of growth-regulating substances. *Ann. N.Y. Acad. Sci.* **144**, 326–334.

STEWARD, F. C., AMMIRATO, P. V. and MAPES, M. O. (1970) Growth and development of totipotent cells: Some problems, procedures and perspectives. *Ann. Bot.* N.S. **34,** 761–788.

STEWARD, F. C., BARBER, J. T., BLEICHERT, E. F. and ROCA, W. M. (1971) The behavior of shoot apices of *Tulipa* in relation to floral induction. *Devl Biol.* **25,** 310–335.

STEWARD, F. C., MOTT, R. L. and RAO, K. V. N. (1973) Investigations on the growth and metabolism of cultured explants of *Daucus carota.* V. Effects of trace elements and growth factors on the solutes accumulated. *Planta* **111,** 219–243.

STEWARD, F. C., ISRAEL, H. W., MOTT, R. L., WILSON, H. J. and KRIKORIAN, A. D. (1975) Observations on growth and morphogenesis in cultured cells of carrot (*Daucus carota*). *Phil. Trans. R. Soc.* (In press).

THIMANN, K. V. (1972) The natural plant hormones. In *Plant Physiology: A Treatise*. ed. STEWARD, F. C., vol. VIC, pp. 3–365, Academic Press, New York.

WILSON, H. J., ISRAEL, H. W. and STEWARD, F. C. (1974) Morphogenesis and the fine structure of cultured carrot cells. *J. Cell Sci.* **15,** 57–73.

MERISTEMS

F. A. L. CLOWES

Botany School, South Parks Road, University of Oxford, U.K.

INTRODUCTION

In complexly constructed organisms cell cycling becomes more or less confined to special regions in the adult, the meristems. In the few examples where there is any quantitative information about cell cycling it is usually confined to the overall rate of mitosis for the whole meristem and this gives a poor idea of the complex organization of cell kinetics. In meristems that have been adequately investigated we find regional diversity in cell cycles; in particular, there seems always to be a region with a low rate of mitosis. For example, in erythropoiesis the cell cycle lasts for 9 or 10 hr at the pro-erythroblast and basophil erythroblast stages, but the pluripotent stem cells have cycles of about 30 hr. In the intestinal crypts the overall rate of mitosis may be once in 10 hr, but the cells at the base of the crypts divide only once in 45 hr.

The apical meristems of plant roots are similar in the possession of great regional diversity in the duration of the cell cycle and, for several technical reasons, are much better characterized than the meristems of animals. So the rest of this article will be confined to root tips and the quantitative data will come from *Zea mays* because the organization of the cells is easier to follow in root meristems of grasses than in most other kinds of plants.

OVERALL RATES OF DIVISION IN ROOT MERISTEMS

Average rates of cell division in root meristems were determined by cell counting procedures about 25 years ago (Brown and Rickless, 1949; Gray and Scholes, 1951; Erickson and Sax, 1956) and were found to be about one cell per cell per day at normal growing temperatures. Unlike mammalian meristems, root tips produce new cells at a fairly wide range of temperature and rates of proliferation can vary by a factor of ten with temperature (Evans and Savage, 1959). Because of the widespread use of root tips in studies related to radiotherapy much of the early research concerned the influence of X-rays on cell proliferation, for, although ionizing radiations had been in use for some time in the treatment of cancer, there was no clear reason why they should have a beneficial effect. Their use was based upon the empirical law of Bergonié and Tribondeau (1906) that their deleterious effect was greatest in those cells with the highest rates of proliferation and long reproductive futures. Thus a dose of X-irradiation should damage cells of a tumour more than the normal cells of the human body which have negligible rates of division. The meristems of the human body, particularly the bone marrow and gut epithelia, are badly damaged by X-rays, but can be shielded in the treatment of tumours in some sites.

Unfortunately the plants chosen for these studies, usually *Vicia faba*, have root apices whose cell patterns were difficult to analyse and so what actually happens when meristems are irradiated was obscured. Work on animals is hampered by the mobility of their cells and again the complexity of the reaction to irradiation of a meristem was not discovered. A better analysis of root tip organization has become available recently and there is now some chance of understanding the behaviour of meristems.

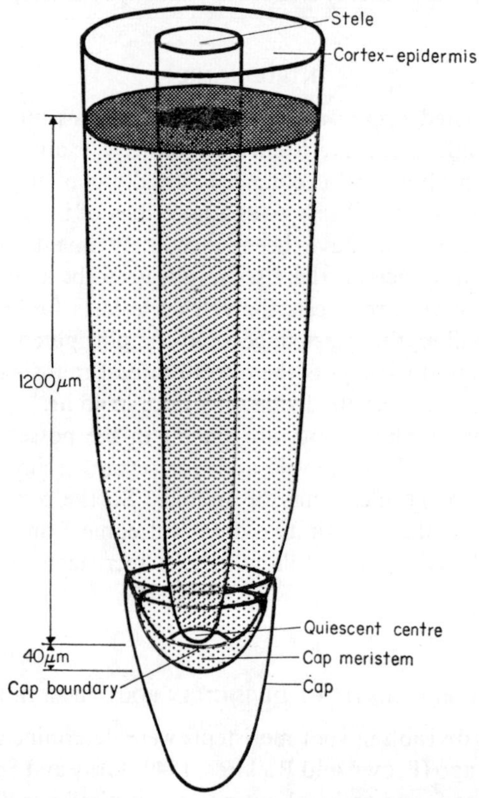

FIG. 1. Diagram of a root tip of *Zea mays* showing the extent of the meristem having measurable rates of mitosis.

Roots of *Zea* have a discrete cap at the tip with its own meristem normally separate from that of the rest of the root (Fig. 1). They also have a broad columella within the cap i.e. a region whose cells divide only transversely to the axis. A purely geometrical analysis of the cell pattern showed that the meristem as a whole could not function if all its cells divided at similar rates. There had to be a region at the pole of the stele and cortex having a negligible rate of mitosis (Clowes, 1954). This was confirmed by the use of radioactive precursors of DNA which also showed that all root tips, even those with quite different cellular organizations, have quiescent centres to the meristem (Clowes, 1956a, b; D'Amato and Avanzi, 1965).

As more sophisticated methods for measuring rates of mitosis became available it was shown that the quiescent centre has an elemental rate of cell formation some ten to fifteen times lower than in the surrounding cells (Clowes, 1961). More recently, further quantitative information has been acquired about cell kinetics of *Zea* root meristems and this has provided a better idea of how meristems in general react to perturbation.

Measurement of cell cycles

The new methods of measuring mitotic cycles in asynchronous meristems involve either (i) stopping the cycle at a specific phase and then observing the accumulation of that phase over a short period or (ii) labelling cells at a specific phase momentarily and then either watching the progress of the labelled fraction of the population through the mitotic cycle or relabelling after an interval to determine the rate of entry into the phase. With plants we commonly use colchicine to block the cycle at metaphase under conditions that produce no change in the rate of entry into prophase and exclude restitution. Tritiated thymidine is often used to label the DNA-synthetic phase (S phase), but caffeine can also be used to label telophase by inhibiting cytokinesis and producing cells with two nuclei.

These methods provide three different values for the rate of cycling and it is important to be clear exactly what is being measured. Treatments that block the cycle give an average duration of the cycle for all cells in the meristem or region of the meristem i.e. a cell-doubling time. Pulse-labelling with precursors of DNA as usually practised gives an average duration of the cycle for only that part of the cell population displayed by the curve of labelled mitoses (Quastler and Sherman, 1959). This curve is of damped sine wave form and clearly excludes non-cycling cells from the average duration of the cycle. The duration of the cycle is measured between peaks of the curve and also excludes cells whose cycles differ widely from the average. It is also possible to calculate the average duration of the cycle of all cycling cells as the ratio of the duration of the S phase to the fraction of labelled mitoses integrated over one cycle. The calculations involved are given in detail by Clowes (1971, 1975a) and De la Torre and Clowes (1974).

In many meristems we find that cycle duration times based on all cycling cells are less than the cell-doubling time as some of the cells are not contributing to the increase in cell number. These cells are not necessarily permanently inert. The size of the proliferating cell fraction, usually called the growth fraction, can be estimated as the ratio of the fraction of all labelled cells to the fraction of labelled mitoses integrated over one cycle or some derived equation (Clowes, 1971). These estimates can be misleading if the meristem contains cells that synthesize DNA for some purpose other than normal cycling. Endoreduplication prior to the formation of polytene or endopolyploid nuclei may occur at some period or in some parts of a meristem and, in such instances, the growth fraction should be calculated from independent estimates of the cycle duration of cycling cells and the cell-doubling time.

The root apical meristem of *Zea mays*

Root meristems have two margins at which cell division peters out (Fig. 1). In primary roots of *Zea mays* cv. Golden Bantam, the proximal margin at which a rate of mitosis can still be measured occurs at 1200 μm from the cap boundary though some roots have meristematic cells higher up in both the stele and cortex. The distal margin of the cap meristem having a measurable rate of mitosis occurs only 40 μm below the cap boundary in the fourth tier of cells though some roots have meristematic cells as far as the eighth tier.

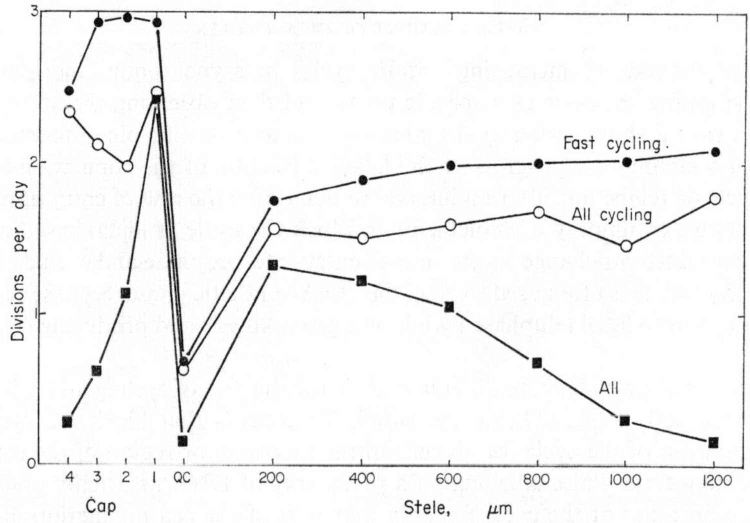

FIG. 2. Rates of mitosis in the first four tiers of cap cells, in the quiescent centre and in the stele between 200 and 1200 μm above the cap boundary in root tips of *Zea mays* at 23°C. Rates are plotted as averages for all cells, all cycling cells and fast cycling cells displayed by the pulse-labelled mitosis curve. Data from Clowes (1975a).

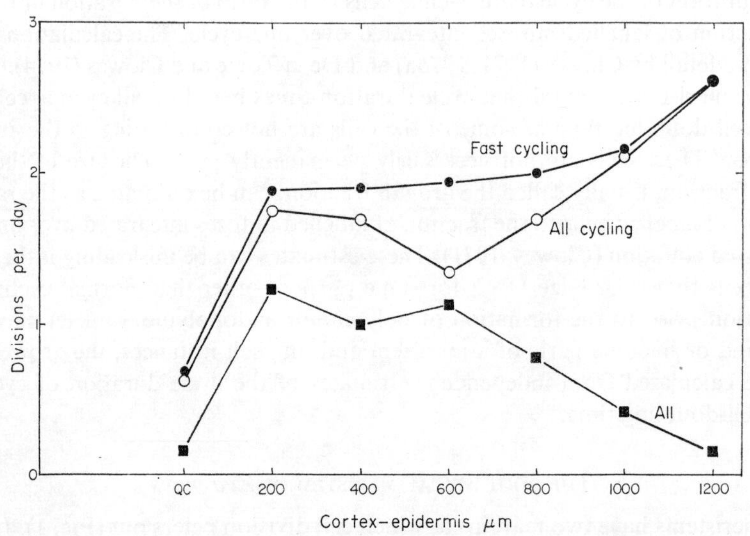

FIG. 3. As Fig. 2, but for the quiescent centre and the cortex–epidermis complex of cells.

Rates of mitosis as averages for all cells, all cycling cells and fast cycling cells displayed by the pulse-labelled mitosis curve are given in Figs. 2 and 3 at intervals along the meristem in the cap and in the stele and cortex separately. It will be seen that the quiescent centre has the lowest rate of cell formation and the adjacent cells of the cap meristem have the highest rate. The cap meristem includes cells with the shortest cycles even where meristematic activity is rapidly diminishing. Also, in stele and cortex, rates of mitosis of cycling cells increase towards the proximal margin of the meristem although meristematic activity is diminishing (Figs. 2, 3). Thus variation in cycle duration makes no contribution to the contraction of cell production as the cells enlarge and differentiate. The contraction must therefore be due to a change in growth fraction, whose levels are shown in Fig. 4. This is not wholly true however of the quiescent centre. Its diminished contribution of new cells is due both to a lengthening of the cell cycle and to a reduction in its growth fraction.

An analysis of the duration of each phase in cycling cells (Fig. 5) shows that the main difference between the quiescent centre and other regions of the meristem is in the duration of G_1, i.e. the period between the end of mitosis and the onset of DNA synthesis. This phase is lengthened greatly in the quiescent centre. It is eliminated in the dividing cells of the cap so that the S phase overlaps the latter part of the previous mitosis, M. In the stele and cortex the cell cycle becomes shortened towards the proximal margin of the meristem largely by a reduction in the duration of G_1; this phase is eliminated at 1000 or 1200 μm. The consequent overlap of DNA synthesis and mitosis is confirmed by the presence of labelled telophases immediately after a pulse of [^{3}H]thymidine.

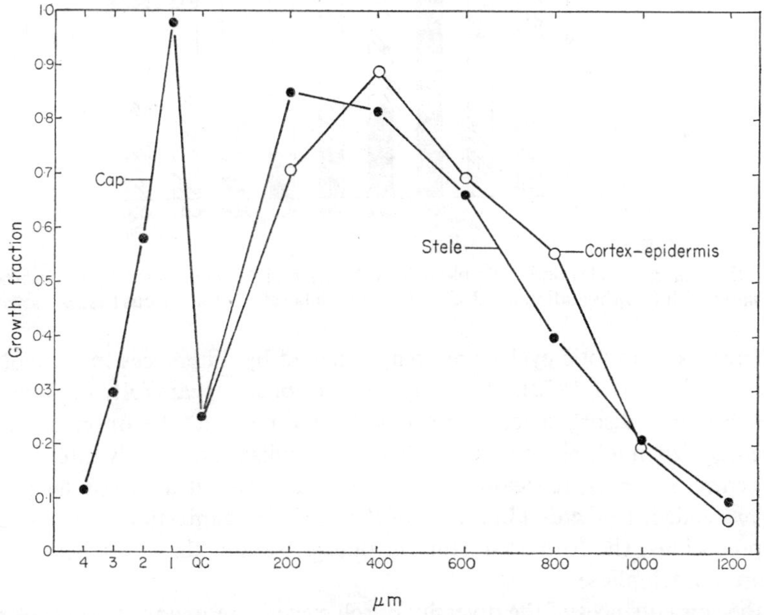

FIG. 4. Growth fractions in the first four tiers of cap cells, in the quiescent centre and in the stele and cortex-epidermis between 200 and 1200 μm above the cap boundary. Data calculated from the ratio of the cell-doubling time to the average duration of the mitotic cycle of all cycling cells in root tips of *Zea mays* at 23°C.

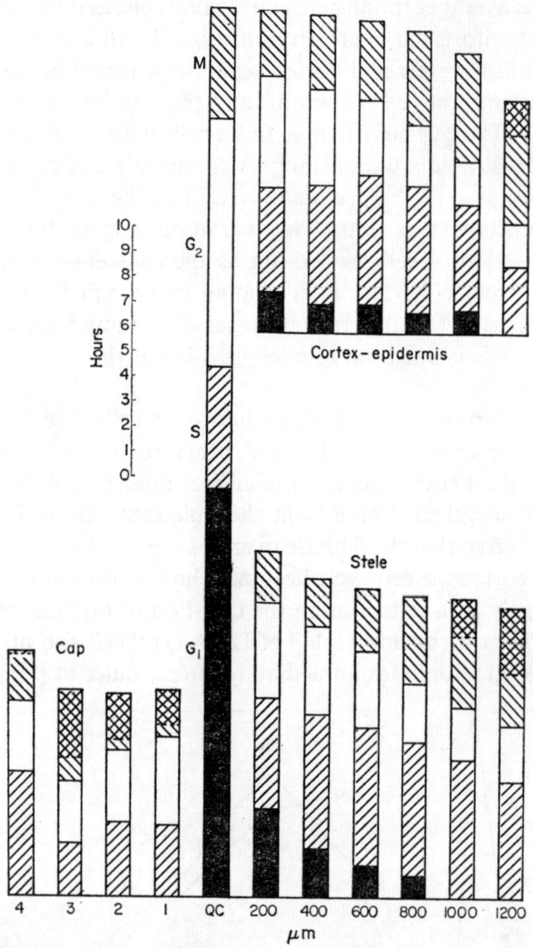

FIG. 5. Durations of cell cycles and their phases G_1, S, G_2 and M determined by pulse-labelling root tips of *Zea mays* with [^{3}H]thymidine at 23°C. Measurements refer to the various tissues shown in Fig. 4.

The differences in mitotic cycles are complemented by differences in nucleolar cycles (De la Torre and Clowes, 1972). Normally the nucleolus appears fully organized during interphase, becomes disorganized in prophase, disappears at the onset of metaphase, starts to reorganize in telophase and completes reorganization in early interphase. In the quiescent centre, however, reorganization is delayed so that it does not start until after telophase has ended. In the dividing cells of the cap, disorganization and reorganization are both precocious, the former starting in interphase and the latter being completed halfway through telophase.

One of the consequences of the diversity of cell cycles is reorganization of the meristem after perturbation. This has been investigated most thoroughly after X-irradiation, but other experimental treatments or even small changes in the conditions of growth lead to reorganization of the constituent cells (Clowes, 1975b). The reorganization is seen as a stimulation of cycling in the quiescent centre and an inhibition elsewhere in the meristem. After heavy doses of X-rays the normally dividing cells stop dividing and the meristem is renewed by proliferation of the quiescent centre (Clowes, 1959). When the perturbation is monitored in *Zea*, after an acute dose of 1700 rads, it is seen that the rate of cycling in the quiescent centre can reach three times the equilibrium rate within an hour (Clowes, 1972), whereas in other parts of the meristem the cycling rate falls to 0.3–0.6 times the normal rate with the cap cells reacting most violently. The rate of cycling may drop even further before recovery, which occurs at about 100 hr in the stele and rather later in the cap. In the quiescent centre, on the other hand, the rate of cycling may fall after the initial stimulation and then rise again to a second peak of seven times normal at 20 hr, after which the rate again falls and quiescence is regained beyond 100 hr.

This information concerns average rates of mitosis for all cells and therefore reflects total cell production. There are no methods available for monitoring rapid changes in cycling rates of cycling cells only because these take too long to determine. So we cannot tell directly whether the changes are brought about by changes in cycle duration or by changes in growth fraction. But we can see that the nucleolar cycles are altered; fully organized nucleoli become available in telophase briefly in the quiescent centre and are lost in the cap cells. This suggests that the cycles are changed, possibly by elimination of G_1 and advancement of S into M of the previous cycle as in the most rapidly cycling cells of a root tip in equilibrium. In these circumstances the quiescent centre builds a virtually new meristem from a small number of its cells and it is likely that similar differential effects apply in the irradiation of a tumour or in the intra- or extra-corporeal irradiation of peripheral blood, though it is more difficult to acquire the basic information here. The reason for the differences in radiosensitivity must be sought in the differences in the cell cycles. An obvious candidate is the proportion of time spent in G_1 and hence the size of the DNA target. In plant tissues a small advantage here could be amplified by the superior chance of regeneration from a group of contiguous undamaged cells.

A further look at the various cell cycles throughout the meristem (Fig. 5) shows that G_1 and G_2 in both stele and cortex become shorter with increasing distance from the quiescent centre towards the proximal margin. This suggests that quiescence might be maintained by starvation or competition for nutrients coming down the root from the shoot or endosperm and which are required for cells to pass the control points of entry into S and M. Or it could be that there is inhibition exerted on cycling, diminishing from the quiescent centre upwards. There does indeed appear to be attenuation of the supply of substances coming down the root, but this in no way accounts for the maintenance of high cycling rates in cap initials on the distal side of the quiescent centre (Clowes, 1970). We also know from work on excised pea roots that cells arrested in cycle by carbohydrate starvation are prevented by γ-irradiation from proceeding into S and M after release from starvation (Van't Hof and Kovacs, 1970) whereas, after irradiation, cells of the quiescent centre are stimulated into S and M. Starvation is therefore an unlikely cause of quiescence at the pole of the root meristem.

In animals one of the phenomena associated with control of cycling is contact inhibition. This is a complex series of events not fully understood and clearly there can be nothing like this in plants whose cells share common walls. However, in plants restraint on cell growth may serve the same function as contact. Pressure of the surrounding cells could account for the inability of the cells of the quiescent centre to expand and therefore divide since there is probably some critical mass phenomenon associated with cycling (Barlow, 1973). Pressure could also account for the variation in shape of the quiescent centre in roots having different cell patterns (Clowes, 1975c) and the reorganization that occurs after perturbation could stem from relief of pressure. Pressure could not account for cessation of cycling at the margins of the meristem; this must be under quite different control. Although the margins and the pole of the meristem share similar low rates of cell production, the rates are brought about differently as we have seen.

By attributing restriction of cell cycling in the quiescent centre to constraints exerted by surrounding cells one does not end the search for control mechanisms. Nor of course, does this interpretation deny a hormonal control theory. The kind of explanation we seek depends on us. This is a morphologist's approach, but the results reported here must change the way all of us look at meristems. One example of an inevitable change is in the concept of stem cells or initial cells in development.

References

Barlow, P. W. (1973) Mitotic cycles in root meristems. In *The Cell Cycle in Development and Differentiation*, ed. Balls, M. and Billett, F. S. Cambridge University Press.

Bergonié, J. and Tribondeau, L. (1906) Interpretation de quelques résultats de la radiothérapie et essai de fixation d'une technique rationelle. *C.r. hebd. Séanc. Acad. Sci. Paris* **143,** 983–985.

Brown, R. and Rickless, P. (1949) A new method for the study of cell division and cell extension with some preliminary observations on the effect of temperature and of nutrients. *Proc. R. Soc.* B **136,** 110–125.

Clowes, F. A. L. (1954) The promeristem and the minimal constructional centre in grass root apices. *New Phytol.* **53,** 108–116.

Clowes, F. A. L. (1956a) Nucleic acids in root apical meristems of *Zea. New Phytol.* **55,** 29–34.

Clowes, F. A. L. (1956b) Localization of nucleic acid synthesis in root meristems. *J. exp. Bot.* **7,** 307–312.

Clowes, F. A. L. (1959) Reorganization of root apices after irradiation. *Ann. Bot.* **23,** 205–210.

Clowes, F. A. L. (1961) Duration of the mitotic cycle in a meristem. *J. exp. Bot.* **12,** 283–293.

Clowes, F. A. L. (1970) Nutrition and the quiescent centre of root meristems. *Planta* **90,** 340–348.

Clowes, F. A. L. (1971) The proportion of cells that divide in root meristems of *Zea mays. Ann. Bot.* **35,** 249–261.

Clowes, F. A. L. (1972) Cell cycles in a complex meristem after X-irradiation. *New Phytol.* **71,** 891–897.

Clowes, F. A. L. (1975a) The cessation of mitosis at the margin of a root meristem. *New Phytol.* **74,** 263-271.

Clowes, F. A. L. (1975b) Control of cell cycles in plant meristems. *Phytomorphology* in press.

Clowes, F. A. L. (1975c) The quiescent centre. In *Development and Function of Roots* ed. Torrey, J. G. and Clarkson, D. T. Academic Press, London, in press.

D'Amato, F. and Avanzi, S. (1965) DNA content, DNA synthesis and mitosis in the root apical cell of *Marsilea strigosa. Caryologia* **18,** 383–394.

De la Torre, C. and Clowes, F. A. L. (1972) Timing of nucleolar activity in meristems. *J. Cell Sci.* **11,** 713–721.

De la Torre, C. and Clowes, F. A. L. (1974) Thymidine and the measurement of rates of mitosis. *New Phytol.* **73,** 919–925.

Erickson, R. O. and Sax, K. (1956) Rates of cell division and cell elongation in the growth of the primary root of *Zea mays. Proc. Am. Phil. Soc.* **100,** 499–514.

Evans, H. J. and Savage, J. R. K. (1959) The effect of temperature on mitosis and on the action of colchicine in root meristem cells of *Vicia faba. Expl Cell Res.* **18,** 51–61.

Gray, L. and Scholes, M. E. (1951) The effect of ionizing radiations on the broad bean root. Part viii growth rate studies and histological analysis. *Br. J. Radiol.* **24,** 82–92, 176–180, 228–236, 285–291, 348–352.

Quastler, H. and Sherman, F. G. (1959) Cell population kinetics in the intestinal epithelium of the mouse. *Expl Cell Res.* **17,** 420–438.

Van't Hof, J. and Kovacs, C. J. (1970) Mitotic delay in two biochemically different G_1 cell populations in cultured roots of pea. *Radiat. Res.* **44,** 700–712.

QUANTITATIVE ASPECTS OF LEAF GROWTH

F. L. MILTHORPE

Macquarie University, N. Ryde, 2113, Australia

INTRODUCTION

Just three years before the Society's inauguration, F. G. Gregory (1921) wrote '. . . it is surprising that no quantitative studies have yet appeared dealing with the increase in area of leaves of a plant. . . .' Gregory's classic studies were extended by F. J. Richards and many others so that, some fifty years later, it is possible to predict—with reasonable precision but often inadequate understanding—the size, form and physiology of successive leaves of a plant axis over a range of environments. In this paper, I will confine discussion to a current assessment of events on a single stem axis, leaving aside the important issue of generation of new axes and, reluctantly, abandoning any attempt to present a historical perspective.

STRUCTURE

The development of the leaf surface depends on meristems (cf. Clowes, this volume)—first, the apical meristem, which persists until flowers are ultimately differentiated and which gives rise to the stem tissue and, secondly, transient meristems of the leaf primordia. Leaf primordia arise amid a mass of dividing cells some short distance from the apex at a point which seems to be the most remote from all existing primordia (Fig. 1a). The cells first divide periclinally at much the same rate as those of the apical meristem but soon assume a rate some two to four times greater. Much still remains unknown about the details of this region despite the basic contributions of Sunderland and Brown (1956) and Sunderland *et al.* (1957). In a constant environment, the rate of leaf production, dN_L/dt leaves d^{-1}—and hence its reciprocal the plastochron—is constant with time (Fig. 1b), but it is influenced by the irradiance on the expanded leaves, the data of Fig. 1 being described by $dN_L/dt = 0.40 + Q_V/(0.8\ Q_V + 1.95)$ where Q_V is the daily visible radiation in MJ m^{-2} d^{-1} (Milthorpe and Newton, 1963); by temperature, following an optimum relationship with the cardinal points varying between species (Milthorpe, 1959); and by water supply (see below).

Soon after initiation, the young leaf primordium is visible as a protuberance, the relative rate of increase in volume with time thereafter sometimes decreasing (Hannam, 1968) but frequently remaining constant during part of the pre-unfolding phase (Sunderland, 1960; Williams, 1960; Williams and Bouma, 1970). On the other hand, the relative rate of increase in cell number is usually constant with time from initiation to unfolding and is influenced by irradiance and temperature in much the same way as the rate of leaf

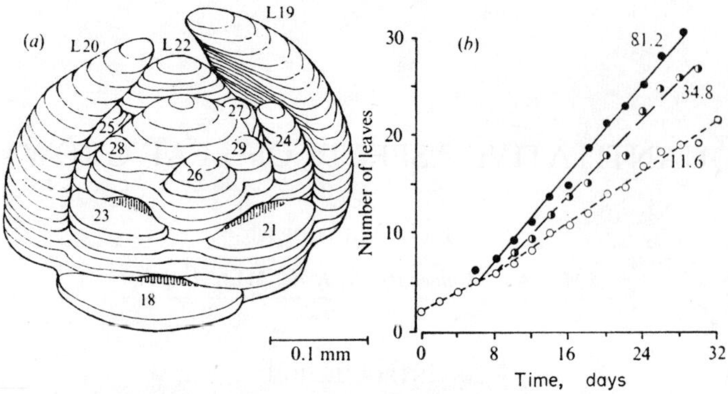

FIG. 1. (a) Perspective drawing of a shoot apex of an 11-day-old seedling of flax (*Linum usitatissimum* L.). The projection is built up from outline drawings of successive 10-μm sections. Leaves 18, 21 and 23 have been removed to allow the apex and younger primordia to be seen (after Williams, 1970). (b) Changes in total numbers of leaves produced with time by cucumber plants kept in darkness until day 4 and then exposed to the irradiance (W m^{-2}) shown for 15 hr daily. Data from Milthorpe and Newton, 1963.

production (Milthorpe, 1959; Newton, 1963). (Cell division in leaves of the plants described in Fig. 1b, for example, is given by $(1/N)dN/dt = (0.86 + 0.465/Q_V)^{-1}$ d^{-1}.) Over the same period there is a three- or four-fold increase in average cell volume. From around the time of unfolding, the rate of cell division in the leaf as a whole falls rapidly and that of cell expansion increases (perhaps, up to twenty-fold), leading to rapid expansion of the leaf. Over this later phase, the course of leaf area or weight, W, can often be described by the logistic equation $W = W_f/\{1 + b \exp(-kt)\}$ where $W_f/(1 + b) \leq W \leq W_f$ when $0 \leq t \leq \infty$.

The decrease in the proportion of mitotic figures found with time before unfolding (Fig. 2) indicates either a continually decreasing proportion of cells dividing or a shorter time being spent in mitosis than in interphase or a combination of both. Although I know of no rigorous substantiating data, the development of the leaf as a whole and its constituent tissues (or at least the palisade as shown by the data of Clough (1971)) is consistent with the assumptions that cells divide at a given constant size, that the proportion of cells going into another cycle of division falls with time—following either a negative logistic (Milthorpe and Newton, 1963; Clough and Milthorpe, unpublished) or a negative exponential (Dale, 1970) function—, that the rate of division of those cells which do divide increases throughout and that expanding cells of any one tissue follow the same time course of expansion irrespective of when they are formed. These criteria can only be accepted as tentative; they allow the growth of the leaf to be simulated adequately but are likely to cloak more complex compensating effects. It is known that the rate of expansion of area of the leaf varies both in time and space (Avery, 1933), differential growth, of course, leading to differences in shape (Ashby, 1948). Cell division and expansion in the component tissues are also displaced in time but are described by the same functions (with different values for the parameters). Most of the first-formed cells form vascular tissue (Hannam, 1968) and cell division possibly ceases first in the epidermis and vascular tissues, then in the spongy mesophyll and finally in the palisade.

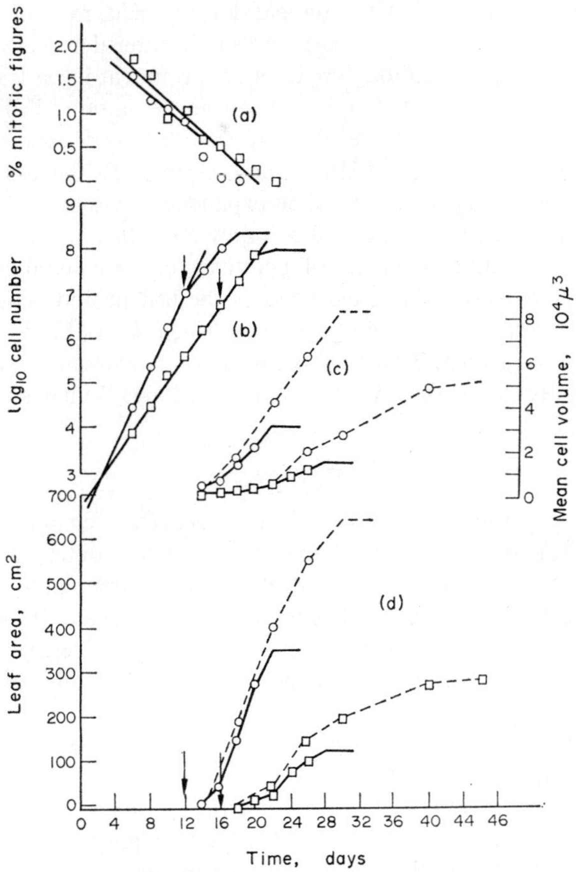

FIG. 2. Changes with time in (a) percentage of cells in meta-, ana- and telophases, (b) $\log_{10}$ cell numbers, (c) mean cell volume, and (d) area of the third leaf of cucumber exposed to radiation of 0.5 (squares) and 3.2 (circles) MJ m^{-2} d^{-1} with the apices above the leaf present (continuous lines) or absent (broken lines). Arrows show times of unfolding of the leaf. Data from Wilson, 1966.

Whereas the potential size of any leaf (determined by the rate and duration of cell division) appears to be set by the irradiance on the older expanded leaves and by temperature, the realization of that potential depends on cell expansion which appears to be primarily related to the supply of mineral elements, temperature (with an optimum lower than that for cell division), and irradiance on the expanding leaf which encourages cell expansion in the dimension normal to the epidermis (Newton, 1963; Wilson, 1966; Milthorpe and Moorby, 1969).

Monocotyledon leaves (Williams, 1960) follow similar general functions to those used above to describe the dicotyledon leaf but the temporal and spatial relations differ—the cessation of cell division and expansion proceeding basipetally, there being no distinct palisade mesophyll, and all expansion occurring within the portion of the leaf still enclosed by surrounding sheaths.

Shortage of water has marked effects on leaf development, rates of leaf initiation and of cell expansion being particularly sensitive to small water deficits (Slatyer, 1967). The rate of cell division in the leaf, although reduced, continues at lower leaf water potentials than do the two other processes (Clough and Milthorpe, unpublished). The reasons for the different responses of the apex and the leaf require further exploration especially if there is confirmation of the finding of Husain and Aspinall (1970) that the water potential of the apex still remains very high even when expanded leaves are wilting.

Leaves produced at successive nodes also follow a predictable pattern. For example, the growth of each leaf on the main axis of a cultivar of wheat could be described by the logistic equation cited above where, taking B as the leaf position counted acropetally, the two coefficients were given by (Morgan, unpublished) $k = (0.758 + 0.117\,B)^{-1}\ \mathrm{d}^{-1}$; $W_f = \exp\,(7.64 - 4.4\,k)$ mg. That is, successive leaves grew more slowly but reached successively larger sizes (cf. Hannam, 1968; Williams, 1960; Williams and Bouma, 1970).

FUNCTION

The changes in physical structure described above are accompanied by changes in chemical composition and physiological activity. There is a continued increase in RNA, protein, chlorophyll and other nitrogen and phosphorus compounds, on both a per leaf and per cell basis, until the leaf is almost fully grown. There are usually allometric relationships between these substances, indicating *inter alia* increasing productions of RNA per unit DNA and increasing protein per unit RNA until after unfolding (Williams and Rijven, 1965; Williams and Bouma, 1970). Cell wall material increases at a proportionately accelerated rate during much of development and continues to be deposited after the leaf has ceased expansion.

For much of its life the leaf is very much a heterotrophic organ depending on imports of substrates from older leaves. The dicotyledon leaf commences to be a net exporter of carbohydrate soon after unfolding (Fig. 3) although some is still imported until it is about one-third its final size. Phosphorus and potassium are imported over most of its growth, that which enters in the early stages coming from older leaves rather than directly from the roots (cf. Milthorpe and Moorby, 1969). Here, older senescing leaves play a significant role (Hopkinson, 1966). The wheat leaf depends much more on import of carbohydrates than does the more exposed dicotyledon leaf, all being imported (mainly from the second leaf and to a less extent from the first leaf below) until the tip of the lamina emerges, about 80 per cent is being imported when the lamina is half emerged and still 25 per cent when it is fully emerged. Even half of the dry weight of the sheath, which develops after much of the lamina has emerged, is imported from the next two oldest leaves (Patrick, 1972).

The major physiological process in the leaf is, of course, photosynthesis. For the purpose of this general discussion, we may use the approximate relationship of photosynthesis, P, to flux density of visible radiation, I, in the form $P \approx (a + b/I)^{-1}$ where a represents a series of resistances describing transport of carbon dioxide to the leaf surface r'_a, its diffusion into the intercellular spaces, r'_l, and movement to the chloroplast and the activity of the carboxylation processes r'_x, and b a resistance reflecting the activity of the photochemical reactions. Generally, b does not seem to vary very widely during

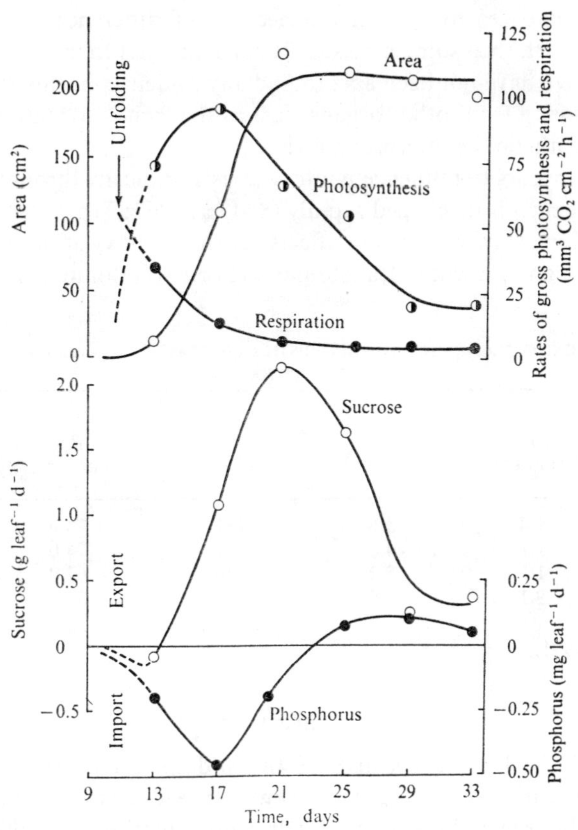

FIG. 3. Change in leaf area, rates of photosynthesis and respiration, and import and export of sucrose and phosphorus during ontogeny of the second leaf of cucumber. Data from Hopkinson, 1964.

expansion, under different environmental conditions or even between species. Most of the variation in rates of photosynthesis is attributable to changes in r_l' and r_x.

Changes in photosynthesis and respiration during expansion of a leaf are shown in Fig. 3. Here, changes in the rate of photosynthesis appeared to be attributable mainly to stomatal responses (r_l'), a response also found in citrus leaves by Kriedemann (1971). However, a comparable pattern of change in the rate of photosynthesis with age of wheat leaves was associated mainly with changes in r_x' (Osman and Milthorpe, 1971). There is still too little information to generalize with confidence, over a wide range of conditions, concerning the components which lead to the decline in rate of photosynthesis. Nevertheless, there is a growing body of evidence which suggests that it changes rapidly, as shown in Fig. 3, during phases of leaf production but stays remarkably constant over a long period of time while fruits are growing and leaf growth is suspended, as in *Capsicum*. Many experiments now show that the rate of photosynthesis is closely related to the demand by 'sinks'; removal of the fruits of *Capsicum*, for example, leads to the decrease usually found and this effect is exerted almost entirely through r_l' (Hall, unpublished). The responses found are such as to suggest that control resides within the leaf itself; if photosynthate commences to accumulate because of differences between the rate of production and export, then some message is soon conveyed to the stomata causing them to close partially. We have not been able to find any evidence, in this system, of controlling substances coming from other organs such as the fruit—no matter how tenuous the evidence nor sympathetic the interpretation!

A water deficit appears to influence photosynthesis primarily through stomatal closure particularly if the deficit is developed rapidly (Slatyer, 1967; Troughton, 1969). However, if the deficit develops slowly there are effects on the carboxylation and photochemical components, although in a wilted leaf stomatal effects still dominate (Table 1).

TABLE 1. CARBON DIOXIDE EXCHANGE CHARACTERISTICS OF TOBACCO LEAVES OVER A RANGE OF WATER POTENTIAL

Leaf water potential (J kg^{-1})	r' (s cm^{-1})	r'_x (s cm^{-1})	b (kJ g^{-1})	P_{max} (ng cm^{-2} s^{-1})	Dark respiration (ng cm^{-2} s^{-1})
− 500	1.4	3.6	157	74.6	18.7
− 660	1.1	4.0	211	74.0	18.1
− 850	1.1	5.2	246	67.0	10.0
−1190	9.1	6.6	—	32.7	11.2

Data from Clough, 1971.

SIGNIFICANCE

Gregory voiced his lament because of the widespread direct extrapolation from measurements of the intensity of the photosynthetic reactions to the growth of the plant as a whole. He did a tremendous service by drawing attention to a self-evident dictum—that the output of a system must be equal to the product of the *number* of components and the *intensity* of the activity of those components of that system. Many investigations since have also emphasized the obvious—that variations in size far exceed those in intensity. Analysis of growth in terms of leaf area and net assimilation rate has made important contributions (cf. Watson, 1952) although it is now appreciated that the net assimilation rate is a far more complex concept than was originally envisaged, including a number of internal influences some of which have been mentioned above. However, it is sad to relate that simplistic extrapolation from observations on a detailed component to behaviour of the whole plant is still widely current—possibly it is an enforced legacy of a dominant Cartesian philosophy. The dampening effects involved in moving through the different levels of organization in a plant are well illustrated by the comparison of C_3 and C_4 plants (Table 2).

As the leaf surface plays such a central role in plant growth, efforts must be directed towards a complete understanding of its development and functioning and integrating this with other aspects to give an adequate concept of the whole plant. Many efforts are now being directed towards achieving an acceptable degree of cohesion; this involves the establishment of quantitative functions to describe the several processes, knowledge of the values of the parameters of these functions in the different species and how they vary with environment, and the integration of the whole so as to simulate the growth of the plant in the range of variable environments it experiences. Some of these functions have been mentioned above; others and approaches that are being made to achieve these aims have been discussed by Milthorpe and Moorby (1974).

TABLE 2. RELATIVE ACTIVITY OF CARBON DIOXIDE EXCHANGE PROPERTIES OF C_3 AND C_4 SPECIES AT SEVEN LEVELS OF ORGANIZATION

Level of organization	Relative activity C_4/C_3
In vitro primary carboxylation	60 – 70
Carboxylation pathway	5
Net photosynthesis of intact leaf	1.5–3
Net photosynthesis of crop stand	1.5
Short-term crop growth rate	1
Forage yield (annual total)	2.5–3.5
Grain yield (seasonal total)	1

Ratios are based on the highest values quoted in the literature.
Data from Gifford, 1974.

REFERENCES

AVERY, G. S. (1933) Structure and development of the tobacco leaf. *Am. J. Bot.* **20,** 565–592.

ASHBY, E. (1948) Studies in morphogenesis of leaves. I. An essay on leaf shape. *New Phytol.* **47,** 153–176.

CLOUGH, B. F. (1971) The effects of water stress on photosynthesis and translocation in tobacco. Ph.D. Thesis, University of Sydney.

DALE, J. E. (1970) Models of cell number increase in developing leaves. *Ann. Bot. NS* **34,** 267–273.

GIFFORD, R. M. (1974) A comparison of potential photosynthesis, productivity and yield of plant species with differing photosynthetic metabolism. *Aust. J. Pl. Physiol.* **1,** 107–117.

GREGORY, F. G. (1921) Studies in the energy relations of plants. I. The increase in area of leaves and leaf surface of *Cucumis sativus*. *Ann. Bot.* OS **35,** 93–123.

HANNAM, R. V. (1968) Leaf growth and development in the young tobacco plant. *Aust. J. biol. Sci.* **21,** 855–870.

HOPKINSON, J. M. (1964) Studies on the expansion of the leaf surface. IV. The carbon and phosphorus economy of a leaf. *J. exp. Bot.* **15,** 125–137.

HOPKINSON, J. M. (1966) Studies on the expansion of the leaf surface. VI. Senescence and the usefulness of old leaves. *J. exp. Bot.* **17,** 762–770.

HUSAIN, I. and ASPINALL, D. (1970) Water stress and apical morphogenesis in barley. *Ann. Bot.* NS **34,** 393–407.

KRIEDEMANN, P. E. (1971) Photosynthesis and transpiration as a function of gaseous diffusive resistances in orange leaves. *Physiologia Pl.* **24,** 218–225.

MILTHORPE, F. L. (1959) Studies on the expansion of the leaf surface. I. The influence of temperature. *J. exp. Bot.* **10,** 233–249.

MILTHORPE, F. L. and MOORBY, J. (1969) Vascular transport and its significance in plant growth. *A. Rev. Pl. Physiol.* **20,** 117–138.

MILTHORPE, F. L. and MOORBY, J. (1974) *An Introduction to Crop Physiology*. University Press, Cambridge.

MILTHORPE, F. L. and NEWTON, P. (1963) Studies on the expansion of the leaf surface. III. The influence of radiation on cell division and leaf expansion. *J. exp. Bot.* **14,** 483–495.

NEWTON, P. (1963) Studies on the expansion of the leaf surface. II. The influence of light intensity and day length. *J. exp. Bot.* **14,** 458–482.

OSMAN, A. M. and MILTHORPE, F. L. (1971) Photosynthesis of wheat leaves in relation to age, illuminance and nutrient supply. II. Results. *Photosynthetica* **5,** 61–70.

PATRICK, J. W. (1972) Distribution of assimilate during stem elongation in wheat. *Aust. J. biol. Sci.* **25,** 455–467.

SLATYER, R. O. (1967) *Plant-water Relationships*. Academic Press, New York.
SUNDERLAND, N. (1960) Cell division and expansion in the growth of the leaf. *J. exp. Bot.* **11,** 68–80.
SUNDERLAND, N. and BROWN, R. (1956) Distribution of growth in the apical region of the shoot of *Lupinus albus. J. exp. Bot.* **7,** 127–145.
SUNDERLAND, N., HEYES, J. K. and BROWN, R. (1957) Protein and respiration in the apical region of the shoot of *Lupinus albus. J. exp. Bot.* **8,** 55–70.
TROUGHTON, J. H. (1969) Plant water status and carbon dioxide exchange of cotton leaves. *Aust. J. biol. Sci.* **22,** 289–302.
WATSON, D. J. (1952) The physiological basis of variation in yield. *Adv. Agron.* **4,** 101–145.
WILLIAMS, R. F. (1960) The physiology of growth in the wheat plant. I. Seedling growth and the pattern of growth at the shoot apex. *Aust. J. biol. Sci.* **13,** 401–428.
WILLIAMS, R. F. (1970) The genesis of form in flax and lupin as shown by scale drawings of the shoot apex. *Aust. J. Bot.* **18,** 167–173.
WILLIAMS, R. F. and BOUMA, D. (1970) The physiology of growth in subterranean clover. I. Seedling growth and the pattern of growth at the shoot apex. *Aust. J. Bot.* **18,** 127–148.
WILLIAMS, R. F. and RIJVEN, A. H. G. C. (1965) The physiology of growth in the wheat plant. II. The dynamics of leaf growth. *Aust. J. biol. Sci.* **18,** 721–743.
WILSON, G. L. (1966) Studies on the expansion of the leaf surface. V. Cell division and expansion in a developing leaf as influenced by light and upper leaves. *J. exp. Bot.* **17,** 440–451.

CONTROL OF FLOWER INITIATION IN LONG- AND SHORT-DAY PLANTS—A COMMON MODEL APPROACH

W. W. SCHWABE and R. H. WIMBLE*

Wye College, University of London, Wye, Kent, U.K.

INTRODUCTION

Photoperiodic induction of flowering in higher plants includes the extreme opposite types of obligate short-day (SD) and obligate long-day (LD) requiring plants, with several intermediate types, the quantitative long-day (QLD), quantitative short-day (QSD) and day-neutral (DN) plants. In addition there are a few species which have dual requirements for either short day followed by long day (S–LD) or vice versa (L–SD). Nevertheless, the taxonomic relationships of these different response types are often quite close and extreme opposites may be found within the same family or genus. Even different cultivars within the same species may differ in their requirements for flower induction. Some examples of such behaviour taken from the literature are given in Table 1. The implication of these facts is that the step leading to a specific photoperiodic requirement is probably a late one in the course of evolution and must have arisen several times, or even many times. In metabolic terms, such a step is unlikely to have resulted in far-reaching biochemical differences. It would seem plausible therefore for such a difference to be quantitative rather than qualitative in nature. Indirectly this suggestion is also supported by the frequent occurrence of photoperiodic ecotypes. The many successful graft transmissions of the final flowering stimulus between opposite response types and between different species, genera and even families, seem to suggest that the final end product stimulating transition from the vegetative to the floral condition may be identical for most if not all species (Zeevaart, 1958; Okuda, 1954; Melchers and Lang, 1941; Wellensiek, 1970).

Some years ago Schwabe (1959) proposed a Scheme (Scheme I) to account for the photoperiodic behaviour of a number of SD and QSD plants such as *Kalanchoe blossfeldiana*, *Glycine max*, *Soja* (var. Biloxi) and *Perilla ocymoides*. It is now proposed to put forward an elaboration of the original scheme which incorporates LD and QLD plants and other response types (Scheme II). It has been possible to express the extended reaction steps of this hypothetical scheme by appropriate mathematical formulae (simulation model); and these with the aid of a computerized programme have permitted the revised scheme to be tested for discrepancies and for predictions to be made about the flowering response in given experimental situations.

* Present address: Rothamsted Experimental Station, Harpenden, Herts.

TABLE 1. CLOSELY RELATED SPECIES OR CULTIVARS HAVING DIVERGENT DAYLENGTH REQUIREMENTS

Species or cultivar		Daylength requirement
Campanula alliariaefolia	(S)*	QLD plant (low temp. reqd.)
medium		S-LD plant (no direct temp. effect; SD replaced by low temp.)
persicaefolia		QLD plant (low temp. reqd.)
primulaefolia		QLD plant (low temp. reqd.)
Fragaria chiloensis	(S)	SD plant (no causative temp. effect; critical dark period inversely proportional to temp.)
do. cv.		QLD plant (no direct temp. effect)
do. cv.		DN plant (no causative temp. effect)
Iberis durandii	(S)	QS-LD plant (SD effect replaced by low temp.; no direct temp. effect)
intermedia		QLD plant (low temp. reqd.)
Lespedeza cuneata Don.		Requires 13 hr day
stipulacea		SD plant (qualitative or absolute); (no direct temp. effect)
thunbergii		LD plant
Nicotiana sylvestris	(S)	DN plant (no causative temp. effect)
tabacum cv.		SD plant (at high temp.; DN at low temp.; no direct effect)
tabacum cv.		DN plant (no causative temp. effect)
tabacum cv.		QLD plant (no direct temp. effect)
Poa pratensis cv.	(S)	QLD plant (at high temp.; DN at low temp.; no causative temp. effect)
pratensis cv.		Absolute-S-QLD plant (which reqs. low temp.)
Solanum nigrum	(S)	QLD plant (at high temp. DN at low temp.; no causative temp. effect)
tuberosum cv.		DN plant (no causative temp. effect)
tuberosum cv.		QSD plant (no causative temp. effect)
tuberosum cv.		QLD plant (no direct temp. effect)
Trifolium pratense		LD plant (no direct temp. effect)
repens L.		S-LD plant.

*(S) = quoted from Salisbury (1963).

SCHEME I

The original scheme which accounted for numerous quantitative responses observed in *Kalanchoe* (Table 2) is reproduced here to indicate the evolution of the fuller scheme. The principal features of Scheme I were as follows.

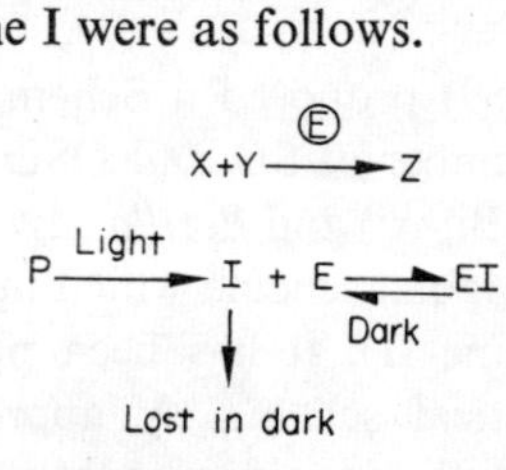

1. An enzyme *E*, catalyses a reaction between *X* and *Y* (where *Y* might be a photosynthetic product) to produce the flowering hormone (*Z*).
2. *E* is limiting in amount and increases as the induction reaction proceeds, i.e. with increasing numbers of SDs or dark hours.

3. E can be blocked by competitive action of an inhibitor I, formed from a precursor P by the action of light, which forms a reversible bond (EI) with the enzyme.
4. Inhibitor I accumulates only up to a certain maximum during the light periods, and is gradually lost during the dark periods.
5. It is assumed that phytochrome catalyses the rapid light reaction ($P \rightarrow I$) (perhaps acting as an enzyme in the P_{fr} state).

Scheme I served a useful purpose in bringing together a whole series of observations relating to SD plants e.g. accelerated flowering (approximate doubling of flower numbers for each additional SD), the proportionality of inhibition by intercalated LDs and the eventual continuation of flowering in continuous light. In view of the often close taxonomic relationships between LD and SD plants it seemed highly desirable to explore the possibility of extending Scheme I so as to include the numerous phenomena related to LD plants and other response types.

TABLE 2. QUANTITATIVE OBSERVATIONS ON THE PHOTOPERIODIC BEHAVIOUR OF *Kalanchoe blossfeldiana* ACCOUNTED FOR BY SCHEME I.

1.	Cumulative induction by SD.
2.	Linear relation between number of days of induction and log flower number (under optimum conditions each SD roughly doubles flower number).
3.	Inhibition of flowering by LD's intercalated between inducing SDs.
4.	Proportionality of inhibition to induction (i.e. each LD destroys effect of approximately 2 SDs).
5.	Light breaks as effective as LDs in inhibiting flowering.
6.	There is an upper limit to the amount of inhibition which is reached after 1–2 LDs. No accumulation of inhibition beyond this level.
7.	The inhibition acts in a forward direction and does not destroy previous induction.
8.	The need for at least a very short light period during SD induction (1 sec per day—'Harder' effect).
9.	Need for CO_2 in the light.

SCHEME II

In the case of LD plants the scheme must account for (i) the requirement for a light period in excess of a limiting duration, and (ii) an increase of flowering with further extension of daylengths beyond this limiting value. While one might almost say that in general the behaviour of LD plants represents the *mirror image* of SD plants, there are also some significant deviations from this pattern and two such differences may be referred to here. Thus the response to light-breaks in the night, which is qualitatively similar to the response of SD plants by reversing the dark effect of a long night, differs quantitatively, in that the light-break usually needs to be of several hours' duration, or several light-breaks must be given in a long dark period to be effective. In numerous LD plants the rate of flower induction increases with increasing daylength right up to 24 hr light. This is seen in *Epilobium adenocaulon* in which the effectiveness of different durations of LDs can best be assessed as the minimum number of inductive cycles needed, to get the first signs of flowering (Fig. 1). It was found, for instance, that compared with continuous illumination, even a very short dark period, inserted in the light, could have

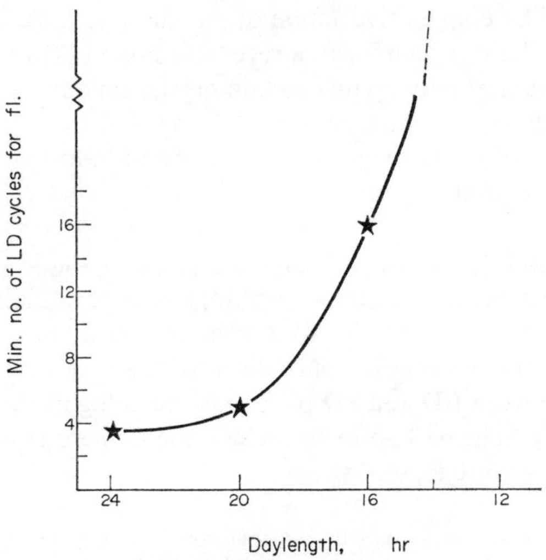

FIG. 1. Minimum number of photoperiodic cycles of different lengths required to induce flowering in the long-day plant *Epilobium adenocaulon.*

a seriously detrimental effect. One might almost speak of a dark-break effect or 'dark-flash' when plants given as little as 20 min darkness per 24-hr period required *more* inductive cycles than those held in continuous light. The response of *Viscaria candida* to such conditions was very similar, e.g. four cycles of continuous light caused 100 per cent flowering, whereas four cycles of 22 hr light gave no flowering.

Another distinct difference in behaviour is that in LD plants interruption of a number of inductive LD cycles by non-inductive SD cycles appears not to diminish the inductive effect of the favourable cycles appreciably, unless the interruption lasts for several, if not many, cycles.

How then can Scheme I be extended to include the response of LD requiring species and other types? Clearly, limitations at other stages in the reaction chain must be involved and yet a common end product for all response types and the diametrically opposite effects of light-breaks in LD and SD plants must be accounted for by way of quantitative differences. Two additional assumptions only need to be made:

(i) the prime limitation for LD plants is in the production of substrate for enzyme *E* or its precursor, and

(ii) in contrast to SD plants, LD plants have adequate amounts of *E* available at all times (hence there is no dark requirement).

The first assumption can be incorporated into Scheme I by postulating that the limiting substrate *S* is produced from the precursor *P* in the reaction chain leading from *P* to inhibitor *I*. Given one specific limiting step for SD plants (amount of *E*) and another for LD plants (amount of *S*) it follows that the flowering process in S–LD and L–SD plants could be subject to both limiting steps. The revised scheme, therefore, which would be applicable to all response types and could account for their differences by merely quantitative changes can be summarized as follows:

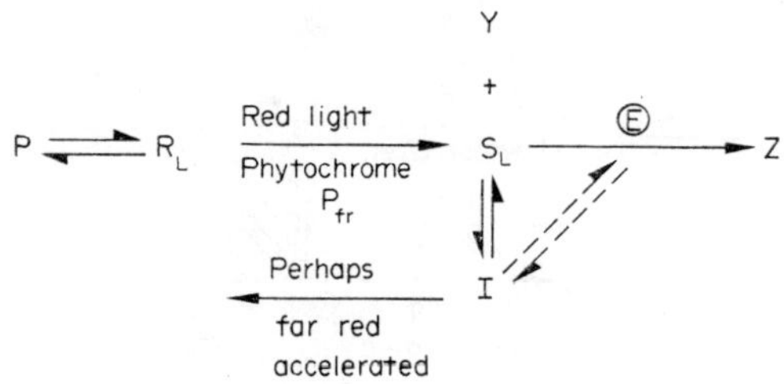

Three important assumptions are made as to the hypothetical chemical steps which basically make up the reaction system.

1. An intermediate substance R is produced at an accelerating rate from the precursor, $P \rightarrow R$ is then transformed by a light reaction (red light) to serve as substrate S for an enzyme reaction which ultimately results in the production of the flowering promoter (Z). The light reaction is probably phytochrome (P_{fr})-mediated. There is also an upper limit (R_L) of precursor which can be accumulated and when this is reached production falls to zero, which also results in a maximum level for S, i.e. S_L.
2. Enzyme E is believed to be adaptive or inductive—increasing in amount with activity.
3. The inhibition required is produced from S either by a rapid but reversible reaction to the inhibitor I which blocks E (equilibrium being tilted heavily in favour of I), or more simply by excess S acting through substrate inhibition of E. If a molecular change of S to I is adopted, then inhibitor I would competitively block enzyme E and thus stop it from operating. Also, inhibitor I will be lost again at a steady rate not specifically dependent on light or dark or any other condition.

Conditions for all the steps in Scheme II as well as the rates of reactions and limits in product accumulation are shown in Table 3.

The light and dark reactions would be as follows:

(a) *In light*, the accumulated maximum of R (R_L) would be transformed to S and I. R production from P would start with the onset of light, the rate accelerating gradually with time. All R would immediately go to form S, etc. Depending on the amount of E, thc enzyme would either all be blocked by I as in non-induced SD plants, or E would operate to give some Z from S, as in LD plants if he light period is long enough to produce adequate amounts of R.

(b) *In darkness*, R production would go on until R_L was reached and then stop. Inhibitor I would gradually disappear as no new I would be made from S, and eventually EI would dissociate and some free E be formed, which could then react with any S to give Z, i.e. in SD plants given adequate dark periods.

In formulating the scheme quantitatively it is easy to allow for the requirement that R_L, the maximum amount of R, will be important in LD plants, i.e. R_L is likely to be small for LD plants—this means also that light-breaks are less effective and the 'boost' effect of R_L conversion is small.

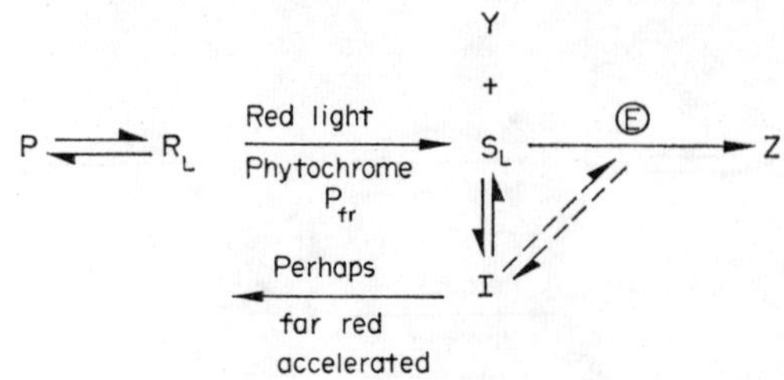

TABLE 3. CONDITIONS FOR EACH REACTION STEP, RATES OF EACH REACTION AND LIMITATIONS ON PRODUCT ACCUMULATION, IN SCHEME II

Reaction	Condition for reaction	Rate of reaction	Limitation on product accumulation
$P \rightleftharpoons R_L$	none	Sigmoid	Up to maximum R_L
$R \rightarrow S$	Red light, phytochrome P_{fr}	Generally very rapid	S/I equilibrium with I predominating heavily
$S \rightleftharpoons I$	none	Very rapid	S_L and I_L
Loss of I	Perhaps far red light accelerated	Roughly constant	Until I exhausted
$Y+S \rightarrow Z$	Presence of free enzyme E	Dependent on amount of E	As E increases Z will exceed threshold level and accumulate until full flowering
$E+I \rightleftharpoons EI$	nonc	Rapid, constant	Competitive combination and reversal as level of I declines in dark
E increase	Free E functioning	Dependent on activity	Increasing with activity until full induction
Y	Perhaps dependent on photosynthesis	Unspecified	May become limiting in very long dark periods

The deceleration of R production as R_L is approached in darkness will determine the length of darkness tolerated by LD plants. The rate may then drop gently or very sharply, e.g. the 20 min dark effect in *Epilobium*.

In SD plants the rate of E increase and the loss rate of I are of crucial importance. This will determine the sensitivity of SD plants.

For brevity's sake, the numerous phenomena which need to be accounted for in each response type have been listed in Table 4 together with the appropriate steps concerned in Scheme II.

THE MATHEMATICAL FORMULATION OF SCHEME II

It seemed imperative to test the scheme for internal inconsistencies and discrepancies and this prompted us to formulate the scheme as a mathematical model which could be programmed for computer testing of the implications of changes in parameters such as reaction rates, constants, limiting amounts, light/dark conditions and initial levels of the hypothetical enzyme, etc. The computer results would also permit comparisons with actual data obtained experimentally.

TABLE 4. FACTS WHICH CAN BE ACCOUNTED FOR BY INDIVIDUAL STEPS OF SCHEME II

Effect	Step in reaction scheme
(A) *General effects not confined to a particular photoperiodic category*	
Photosynthetic requirement (High light intensity reaction)	Y production
Transmission from receptive organ (usually the leaf) to growing points	Final product Z is mobile: S might also be translocated in some species
Graft transmission of flowering stimulus	Direct Z transfer, or transfer of tissue making Z
Timing reaction in the dark for both LD and SD species	Loss of I
Light-break effect in both LD and SD plants	$R \rightarrow S$ $\upharpoonleft\downharpoonright$ I
Far-red light reversal of red light effect	Loss of I accelerated or loss of P_{fr}, inhibiting $R \rightarrow S$ conversion
Maximum efficiency of light-break after some hours (usually middle of dark period)	High level of R built up yielding large amount of I on red irradiation
Summation of effects of inductive cycles	Accumulation of Z to threshold level
Involvement of circadian rhythms in response to light and dark	In the absence of conclusive information it is suggested that endogenous rhythms are directly linked to phytochrome levels and activity, affecting $R \rightarrow S$
(B) *Obligate short-day plants*: Only small amount of native E present	
SD requirement	Adequate time for loss of I; multiplication of E
Greater induction caused by small part of single leaf given several SD cycles than by numerous leaves given only one or few cycles	Multiplication of E with repeated SD cycles
LD inhibition	Insufficient time for loss of I $\therefore$ excess of I.
Light-break inhibition	Re-formation of I, when its level had dropped in darkness
Increased efficiency of successive SDs ('development of response mechanism' cf. above)	Adaptive increase of E
'Harder effect': 1 sec light per day promoting flowering in *Kalanchoe*	$R \rightarrow S$, S being substrate in reaction $S + Y \rightarrow Z$
Quantitative response with sub-optimal induction	Amount of adaptively formed E remains less than maximum amount of I, see below
Continued flowering after full induction, even when in LD for long period	Amount of E formed exceeds possible maximum of I
Inhibitory effect of intercalated LDs on following 1–2 SDs	Excess of I (maximum) built up which needs more than one long dark period for removal
Upper limit to LD inhibition (i.e. one LD approx. as inhibitory as several LDs), no 'counter-induction'	Maximum level of I and also of R
Proportionality of inhibition to previous induction, in *Kalanchoe* at least (flower number divided by constant factor for each symmetrically intercalated LD, cf. Schwabe, 1955)	Competitive combination of I with E prevents multiplication of E as well as Z production
Beneficial effect of 24-hr dark period following intercalated LD, and preceding SD, but no effect if 24 hr dark follows SD	Removal of excess I during long dark allows next SD to be effective, some free E available during dark period of SD
Reduced SD induction if temperature lowered during dark period	Reaction causing loss of I slowed down

continued

TABLE 4—*continued*

Effect	Step in reaction scheme
Somewhat reduced SD induction if temperature lowered during short photoperiod throughout induction treatment	Reaction $P \rightarrow R$ slowed by low temperature, hence less substrate for $R \rightarrow S$.
(C) *Obligate long-day plants*: Adequate amounts of native E	
LD requirement (critical length of photoperiod)	Rate of $P \rightarrow R$ increasing slowly after lag, i.e. light period must be long enough for rate to accelerate to high level
SD inhibition of flowering	Rate of $P \rightarrow R$ cannot accelerate sufficiently before being reduced again to very low level by accumulation of R_L
Favourable effects of light-break	Removal of R allows rate of $P \rightarrow R$ to increase to effective level provided R_L is not reached
Shorter critical daylength at lower temperature at least in *Hyoscyamus*	Rate of loss of I in the dark reduced at low temperature, hence more I left to serve as reservoir for $S + Y \rightarrow Z$
(D) *Other categories*:	
Quantitative SD plants	Native E exceeds maximum I only slightly, thus SD beneficial to speed E increase
Day-neutral plants	Excess of native E, no limitation of $P \rightarrow R$
Quantitative LD plants	Shortage of R from $P \rightarrow R$, but not severe enough to stop all Z production
S–LD plants	Low level of native E as well as low initial rate of $P \rightarrow R$, hence, after E level built up, LD needed for greater R production
L–SD plants	Possibly low synthesis of R from P has to be stepped up before E-formation can be stimulated by presence of S.

As far as inhibitor I is concerned, its functional relations could be such that one is dealing merely with an 'excess substrate' effect, in which event $S = I$. While it is felt that this is probably not the case there is no overriding reason to assume that I is greatly different from S. Hence in devising the mathematical model S was taken as equal to I.

Y as the reactant with S in the hypothetical reaction catalysed by E, has been assumed to represent a product of photosynthesis which normally has no limiting effect on Z production. Y can account for the high light-intensity reaction and may become limiting in conditions involving very long dark periods. Since the availability of Y does not normally impose a restraint on the system it was unnecessary to include it specifically in the mathematical model.

The notation and formulation used in the model for the various reaction steps of Scheme II were as follows:

R = the precursor-based substance, measured in the same units of molar concentration as the substrate S

t = time in hours

R_L = maximum level of R

$(\mathrm{d}R/\mathrm{d}t)_L$ = maximum level of $\mathrm{d}R/\mathrm{d}t$

S = the substrate for the enzyme reaction catalysed by E

S_L = the upper limit to the concentration of S which the system can accept
G = a constant determined by the initial and maximum levels of $\mathrm{d}R/\mathrm{d}t$
B = a constant determining the maximum rate of increase of $\mathrm{d}R/\mathrm{d}t$
A = the rate of loss of S per hour from the system
K = a constant affecting the rate of Z production (Eqn. 4)
a = a constant affecting the rate of exponential increase of E
M and J = constants chosen to give large and equivalent inhibition when S has a high or a low concentration
S_1 and S_2 = limits of S between which the inductive increase of E occurs.

In light, the rate of R production normally climbs from time zero to its maximum in a sigmoid fashion represented by

$$\frac{\mathrm{d}R}{\mathrm{d}t} \Big/ \left(\frac{\mathrm{d}R}{\mathrm{d}t}\right)_L = \frac{G \exp (Bt)}{1 + G \exp (Bt)} \tag{1}$$

In darkness, this basic relation is modified by the inclusion of a factor $(1 - R^2/R_L{}^2)$ to provide for the virtual cessation of R production as the level of R accumulated approaches its upper limit, giving

$$\frac{\mathrm{d}R}{\mathrm{d}t} \Big/ \left(\frac{\mathrm{d}R}{\mathrm{d}t}\right)_L = \frac{G\,(1 - R^2/R_L{}^2) \exp (Bt)}{1 + G \exp (Bt)} \tag{2}$$

Small values of G delay R production, large values advance it. In light, the maximum rate of increase of $\mathrm{d}R/\mathrm{d}t$ occurs at a time $(-\ 1/B) \log_e G$ hours after daybreak. The maximum rate of increase itself is $(B/4)(\mathrm{d}R/\mathrm{d}t)_L$.

Substrate S *and hormone* Z

Change in the level of S may be due to conversion to Z, to augmentation in light from the precursor, or to loss from the system. Symbolically, the dynamic equation is

$$\frac{\mathrm{d}S}{\mathrm{d}t} = -\frac{\mathrm{d}Z}{\mathrm{d}t} + \frac{\mathrm{d}R}{\mathrm{d}t} - A. \tag{3}$$

In darkness, the term $\mathrm{d}R/\mathrm{d}t$ is omitted from the right-hand side of the equation since the substance R can only be transformed to S in light.

Production of hormone Z

The rate of production of Z is determined by the amount of enzyme E present, the amount of substrate to be acted on and the level of inhibitor I. Since for the mathematical model, inhibition is thought of as the result of excess substrate, the most satisfactory formulation is obtained by using the substrate concentration squared, reflecting possibly a reaction with two molecules of substrate. The dynamic relation is

$$\frac{dZ}{dt} = \frac{KE}{1 + (M/S^2) + (S^2/J)}. \tag{4}$$

In Eqn. 4, both low and high concentrations of substrate are in different ways unfavourable to the reaction. It can be shown that the maximum rate of hormone production for given E occurs when the level of S satisfies the equation $S^4 = MJ$. For a fixed value of this product, larger values of M make the reaction more sensitive to inhibition at both high and low concentrations, and smaller values, less.

Inductive increase of enzyme E

The enzyme concentration is represented as increasing at a rate proportional to itself when it is very active and remaining at a constant concentration otherwise. Symbolically,

$$\frac{dE}{dt} = aE, \quad S_1 \leqslant S \leqslant S_2 \tag{5}$$

and

$$\frac{dE}{dt} = 0, \quad S < S_1 \quad \text{or} \quad S_2 < S \leqslant S_L \tag{6}$$

Choice of values for the constants and of the initial values of the variables

The first step in choosing values is to define the range of permitted variation in substrate concentration, S_L. As an arbitrary choice the figure of 20 units of molar concentration was used.

As a next step the loss rate of S was considered; large values, such as 1 unit per hr, will lead to rapid induction of a SD plant in SDs, since the substrate inhibition will be broken down in the first 16-hr dark period. With small values, induction may be delayed to a greater or lesser extent, even in repeated SDs, i.e. before inhibition will fall to a level at which enzyme activity can occur. This is illustrated in Table 5; values of 0.75 and 0.80 are not inductive.

TABLE 5. EFFECT OF DIFFERENT VALUES OF THE LOSS-RATE CONSTANT A ON PRODUCTION OF THE FLOWERING HORMONE Z IN A HYPOTHETICAL SD PLANT GIVEN TEN INDUCTIVE CYCLES EACH OF 8-HR DAYLENGTH

Value of A (units/hr)	0.75	0.80	0.85	1.0
Z production (units)	5	6	60	99

The system is very sensitive to the loss rate and flowering may be much reduced or accelerated with slight variations in A.

The constant, B, determines the rapidity with which R production builds up in light. Figure 2 shows that a value of 0.80 leads to an increase in the rate of production from 2 units per hr to 6 units per hr in about 3 hr, whereas with the value of 0.48 this increase takes nearly 5 hr. Large values of B make the simulated plant very sensitive to photoperiod. The timing of the onset of this increase in the rate of R production will depend on the rate at time zero.

Figure 3 shows the increase in the rate of R production over time for two initial rates; the greater one leads to substantial R production being achieved about 4 hr earlier.

This initial rate is used with the maximum rate (dR/dt_L) in the calculation of the constant, G. Specifically, if the ratio of the initial to the maximum rate is denoted by r, then G is $r/(1 - r)$.

The maximum rate of R production $(dR/dt)_L$ must be large enough to lead to substrate inhibition in a SD plant given LDs, and small enough not to lead to inhibition in a LD plant; a value of 8 units has been used in deriving the graphs in Fig. 2 and 7 units in Fig. 3.

Values of L and J were chosen as 100 units and 1 unit respectively, so as to give minimum inhibition when S lies between 4 and 3 units. The rapidity of induction increases as the interval between the levels S_1 and S_2 on either side of the value of S giving minimum inhibition widens; values of 1 and 6 were found satisfactory.

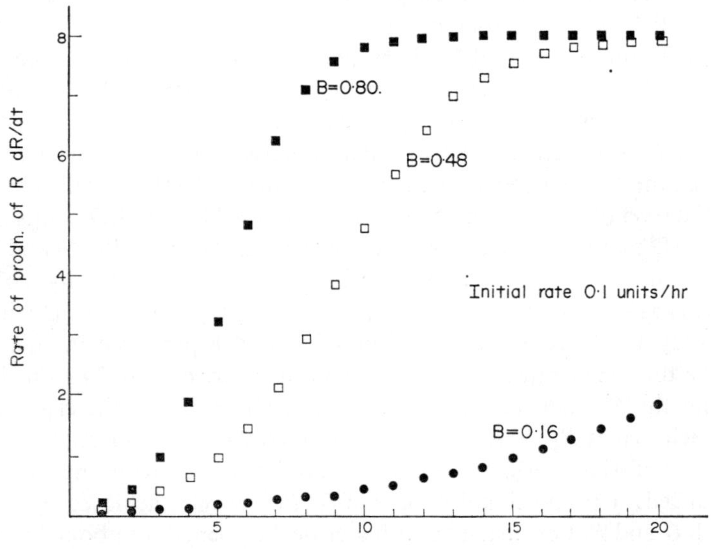

FIG. 2. Changes in the rate of R production with time for three values of constant B at an initial rate of 0.1 units per hr.

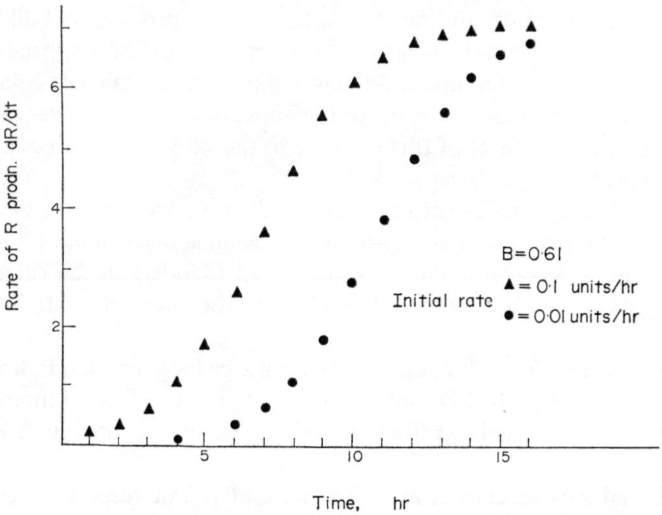

FIG. 3. Changes in the rate of R production with time for two initial rates at a value of 0.61 for constant B. A faster rate, 0.1 units/hr would apply to SD plants, a slower rate, 0.01 units/hr to LD plants.

A large value of K will lead to rapid onset of induction but, because S will rapidly become exhausted, the increase in enzyme concentration will occur relatively slowly. Values of 1 and 0.25 have been used in this work.

The three constants R_L, G and 'a' determine together with the loss rate A, the day-length status of the simulated plant. LD plants are characterized by small values of G or equivalently low initial rates of R production; furthermore, they must have very small values of R_L since otherwise they would eventually flower in short days. Finally, the value of 'a', the rate of exponential increase in enzyme level, must also be very small, since the initial level of enzyme concentration is supposed high in LD plants and close to its maximum. SD plants, on the other hand, have larger values of R_L, G, and 'a', and the larger R_L is, the more effectively inhibitory will a light-break be in a long dark period.

Where there has been no prior induction, critical daylength in both LD and SD plants is determined by a balance between the sum of R_L and R produced in light on the one hand, and the total substrate lost to the system in 24 hr, namely $24A$ on the other. If $R(t)$ represents the R production in t hours after daybreak, then the critical daylength, t_c, is that which makes $R_L + R(t_c)$ slightly less than $24A$. A proportional reduction of R_L, $R(t)$ and A will slow down the onset of induction while preserving the critical day-length unchanged; a proportional reduction in $R(t)$ may be achieved by manipulating either or both G and B. For instance, with a critical daylength of about 11 hr, values of $R_L = 6$, $B = 0.40$ and $A = 0.84$ give very rapid induction, whereas values of $R_L = 3$, $B = 0.27$ and $A = 0.42$ give slower induction.

The mathematical model posits that a SD plant, kept in continuous light before induction, will have very little enzyme present and complete substrate inhibition. Thus initially an arbitrary small value of 4 units is taken for E, and S is taken as 20 units. R and Z are assumed zero, and a minimum rate of R production, 0.1 units per hr, is chosen: this minimum rate together with a maximum $(dR/dt)_L$ of 7 units per hr gives a value of G of 0.0145.

For a LD plant kept in SDs, flowering is inhibited, as has been stated earlier, by the slow rate of build-up of R production. This initial minimum is taken as 0.01 units per hour. The large enzyme concentration is taken as 150 units and, with this amount present, the substrate concentration will have been reduced to a very low level, taken as 0.5 units in a regime of SDs.

COMPUTER SIMULATION AND PREDICTION OF FLOWERING BEHAVIOUR WITH THE MODEL

The effect of various daylength environments on the production of flowering hormone Z was predicted in a series of hypothetical situations by means of the mathematical model and an IBM application programme called System/360 Continuous System Modelling Programme (CSMP). This provides an input language for systems of ordinary differential equations and permits these to be linked into a single structure by means of FORTRAN statements. When the initial values of all constants and variables are specified, the programme advances time by a small interval, calculates new values of variables and evaluates integrals over the interval using the Runge–Kutta technique. Time is then advanced and the process repeated. After the end of a light period, the programme executes the equations of the model for the dark period until 24 hr have elapsed. Thereafter, the sequence may be stopped or repeated or altered by re-setting some constants; for instance, if the daylength is altered from 8 hr to 16 hr, a cycle of SD/LD is given which itself may be repeated. More generally, light-breaks may be simulated in dark periods, short dark periods inserted in light periods, or periods of continuous light given. A 'day' of more or less than 24 hr may be simulated and its effects studied.

In the following Tables and Figures further illustrating the performance of the model, values for the production of the flowering hormone, Z, are given. Values within the same table are comparable, and they are intended to show that the model simulates some particular aspect of known flowering behaviour. Values in different tables are, however, not usually comparable, because they are derived from different sets of values of certain constants. It would not have been difficult to achieve full comparability between tables, but to economize in the use of computer time, no attempt was made to do so.

Using the programme described, a considerable number of runs of the model were carried out and simulations of SD and LD behaviour were achieved by modification of the several constants. A selection of these is shown below in Tables 6–9 and Figs. 4 and 5. Details are shown in each legend. No internal inconsistencies have been revealed by the simulation trials.

While it has been shown that the simulation can quantitatively reproduce patterns of flowering behaviour seen in actual physiological experiments, predictions from the model should be tested experimentally. One such prediction is the increase in level and function of enzyme E in SD plants under progressive inductive treatment. Thus the critical daylength of partly induced SD plants should increase with inductive treatment, i.e. such plants should benefit from longer light and shorter dark periods than controls without previous induction.

TABLE 6. EFFECT OF TEN CYCLES OF 10, 12, 14, 16, 18, 20, 22, 23 AND 24 HR DAYLENGTHS ON PRODUCTION OF THE FLOWERING HORMONE Z IN A HYPOTHETICAL LD PLANT

Daylength (hours)	10	12	14	16	18	20	22	23	24
Z production (units)	16	44	135	240	359	481	603	661	1350

Flowering is virtually inhibited when the light period falls below about 12 hr. Z production steadily increases as the light period increases. In continuous light for 10 days more than 1,350 units of Z are produced, over twice as much as is produced in 10 × 23 hr days, i.e. the 1 hr of darkness every 24 hr is sufficient to reduce Z production by 50 per cent.

TABLE 7. EFFECT OF TEN CYCLES OF 6, 8, 10 AND 12 HR DAYLENGTHS ON PRODUCTION OF FLOWERING HORMONE Z IN A HYPOTHETICAL SD PLANT

Daylength (hours)	6	8	10	12
Z production (units)	42.9	50.6	62.8	4.1

The optimum daylength is around 10 hr. Shorter daylengths give lower Z production, light periods longer than 12 hr are completely inhibitory.

TABLE 8. PRODUCTION OF FLOWERING HORMONE Z IN A HYPOTHETICAL SD PLANT GIVEN 12 SDs WITH VARIOUS NUMBERS OF INTERCALATED LDs

Number of intercalated long days	0	1	2	3	5	11
Z production (units)	15.0	14.2	5.5	2.2	Nil	Nil
Percentage flowering nodes in Biloxi soybean	86	83	50	2	2	Nil

Long days were intercalated in a symmetrical fashion; e.g. a single LD occurred after 6 SDs, 2 LDs divide the 12 SDs into three sets of four, etc. With this hypothetical SD plant, the simulated model showed that 1 LD can annul 1½ SDs approximately, a value found experimentally in *Kalanchoe*. For comparison the percentage flowering nodes in Biloxi soybean given such treatment are also shown.

TABLE 9. ACCUMULATION OF THE FLOWERING HORMONE Z AND INCREASE IN LEVEL OF ENZYME E IN A HYPOTHETICAL SD PLANT GIVEN TEN CONSECUTIVE SD CYCLES OF 10 HR LIGHT AND 14 HR DARK

Day	1	2	3	4	5	6	7	8	9	10
Z production (units)	0.3	1.2	3.0	9.4	17.8	26.5	35.3	44.3	53.4	62.8
Enzyme level, E	4	4	7	22	57	89	129	165	198	236
Times of enzyme activity, from dawn*	None	23–24	18–24	1–14	1–11	1–11	0–3 7–11	0–2 7–11	0–2 8–10	0–2 8–11

Flowering hormone accumulation is shown for each day and the enzyme level as it would be at the end of the dark period. Towards the end of the induction period there would also be enzyme activity during the light hours. Thus, as is known, a SD plant can be induced with longer days once it has had some induction and eventually it becomes effectively a LD plant and LDs are then flower promoting.

* The daylight period is 0–10 hr.

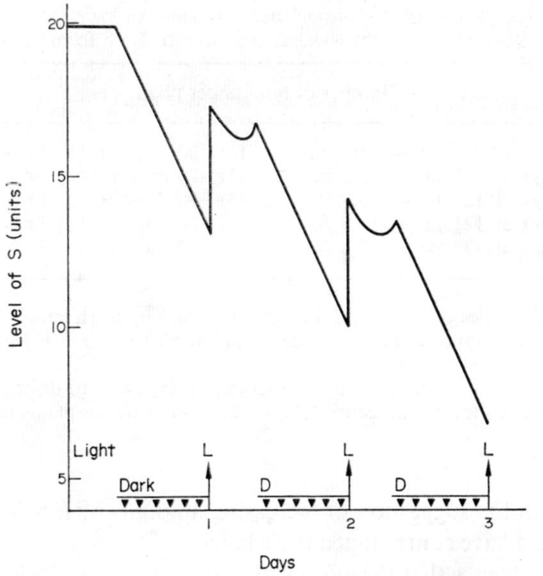

FIG. 4. Level of S during light and dark periods of three successive short days, in a SD plant.

This was done in an experiment (Schwabe, 1959), the design and results of which are shown in Table 10. From treatments nos. 4 and 5 it is seen that daylengths of 12½ and 13 hr can still be inductive, provided that they are preceded by treatment with shorter daylengths. When the same daylengths are given for 20 days (controls) with no prior induction, then they are clearly non-inductive (treatments 9 and 10). This result is there-

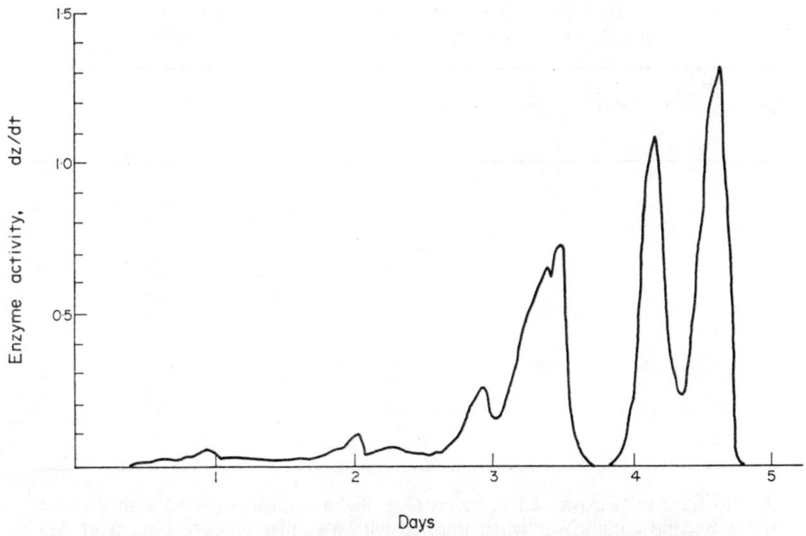

FIG. 5. Activity of enzyme E with progressive induction in a SD plant given four SD cycles.

TABLE 10. EFFECT ON FLOWERING OF *Kalanchoe* GIVEN A SERIES OF DAYLENGTH TREATMENTS IN WHICH CYCLES OF INCREASING LENGTH WERE ACCUMULATED (Data from Schwabe, 1959)

Number of flowers per plant	
(1) 4 days at 11 hr — 0 fl.	(6) 20 days at 11 hr — 81.3 fl.
(2) as (1) plus 4 days at 11½ hr — 2.6 fl.	(7) 20 days at 11½ hr — 66.5 fl.
(3) as (2) plus 4 days at 12 hr — 5.9 fl.	(8) 20 days at 12 hr — 14.2 fl.
(4) as (3) plus 4 days at 12½ hr — 13.0 fl.	(9) 20 days at 12½ hr — 0.4 fl.
(5) as (4) plus 4 days at 13 hr — 17.8 fl.	(10) 20 days at 13 hr — 0 fl.

The controls received 20 cycles of 11, 11½, 12, 12½, or 13 hr daylengths (treatments 6–10). Treated plants (treatments 1–5) received 4 days at 11 hr (1) or 4 days at 11 hr plus 4 days at 11½ hr, and so on, as shown in the Table.

Note that partially induced plants are further induced by exposure to daylengths of 12½ and 13 hr, while these daylengths given for 20 consecutive days are not inductive in plants having received no prior SD.

fore compatible with the suggestion of increasing amounts of *E* being formed, while the opposite result would have contradicted the scheme.

There is further evidence that the effect of LD conditions results in the production of a specific inhibitory substance. This is based on the effects on flowering by injecting crude extracts of leaves into single leaves of experimental plants while these leaves were being induced by SD inductive treatments. The extracts compared were prepared from leaves subjected to either prolonged LD or prolonged SD treatment. The degree to which flowering was reduced by the inhibitory effect of LD extracts is shown in Table 11 in which the results of ten experiments are shown.

TABLE 11. EFFECT ON FLOWER NUMBERS OF TWO INJECTIONS OF 'CRUDE LD EXTRACT' INTO SINGLE LEAVES OF *Kalanchoe blossfeldiana* WHILE THESE WERE BEING SUBJECTED TO 12 SD CYCLES. (INJECTIONS ON FIFTH AND TENTH SD)

Experiment No.	As % water control	As % SD extract
1	57	63
2	58	58
3	41	54
4	18	29
5	52	26
6	12	13
7	44	43
8	6	73
9	17	23
10	18	38
	Mean 32	42

Results have been expressed as percentage of the controls injected with distilled water (second column) or when injected with a similar crude extract from SD treated leaves (third column).

In each case the inhibition of flowering by a LD extract is shown as a percentage of the flower numbers produced in two sets of control plants, those injected with distilled water and those injected with similar crude extracts from SD treated leaves. In the ten experiments quoted it is quite clear that LD sap injections given on 2 out of 14 SDs are inhibitory to flowering to a very large extent (Schwabe, 1972). The magnitude of this effect almost equals the inhibition due to photoperiodic LD treatment given on 2 days intercalated among 12 SDs, as seen in other experiments. SD sap injection is only very slightly detrimental to flowering in much the same way as injection of water when compared with non-injected controls.

These results give considerable credence to the reality of an inhibitory substance being produced in LD conditions. However, a recent publication by Pryce (1972) claiming to have identified this inhibitor as gallic acid could not be confirmed by assay with *Kalanchoe*. Injection of up to 2000 ppm of gallic acid gave no consistent inhibitory effects, and if applied together with crude LD sap the effect was not greater than the crude extract alone.

Clearly Scheme II retains validity as long as it can account for the major observations of effects or as long as it can conveniently be adjusted to do so without the introduction of a large number of subsidiary hypotheses.

The incorporation of effects arising from circadian rhythms has not been attempted since it is quite uncertain where the effects might operate. However, if, as seems most likely, rhythmic effects interact or affect phytochrome levels or phytochrome 'setting', then these would easily incorporate into the present scheme. It is hoped that it will be possible to make predictions from this scheme which would be open to test in the future.

References

Melchers, G. and Lang, A. (1941) Weitere Untersuchungen zur Frage der Blühhormone. *Biol. Zbl.* **61,** 16–39.

Okuda, M. (1954) Flower formation of *Xanthium canadense* under long-day conditions induced by grafting with long-day plants. *Bot. Mag. Tokyo* **66,** 247–255.

Pryce, R. J. (1972) Gallic acid as a natural inhibitor of flowering in *Kalanchoe blossfeldiana. Phytochemistry* **11,** 1911–1918.

Salisbury, F. B. (1963) *The Flowering Process*, Pergamon Press, Oxford.

Schwabe, W. W. (1959) Studies of long-day inhibition in short-day plants. *J. exp. Bot.* **10,** 317–329.

Schwabe, W. W. (1972) Flower inhibition in *Kalanchoe blossfeldiana.* Bioassay of an endogenous long-day inhibitor and inhibition by (±) abscisic acid and xanthoxin. *Planta* **103,** 18–23.

Wellensiek, S. T. (1970) The floral hormones in *Silene armeria* L. and *Xanthium strumarium. Z. Pflphysiol.* **63,** 25–30.

Zeevaart, J. A. D. (1958) Flower formation as studied by grafting. *Meded. LandbHoogesch. Wageningen* **53,** 1–88.

PHYSIOLOGY OF GRAIN YIELD IN WHEAT

R. H. M. LANGER and C. T. DOUGHERTY

Department of Plant Science, Lincoln College, Canterbury, New Zealand

INTRODUCTION

The physiology of grain yield in wheat and other temperate cereals has been a subject of active discussion over the last 50 years, ever since Engledow and Wadham (1923) published their classical work on field populations of wheat. Through what were essentially census methods they analysed grain yield in terms of its component fractions, in the hope that it would be possible to identify the most limiting component and to improve it. The letters *p e n g* are no doubt indelibly inscribed in the memory of many students of agricultural science. These studies were invaluable by breaking up the cumbersome complex of yield, but the individual fractions turned out to be too highly compensating to allow any single one to be isolated as the most limiting. Such inter-relationships, together with considerable environmental effects, imposed severe limits on the physiological significance of these components of yield.

The first really important step in our understanding of the physiology of grain yield was taken by Porter (Porter *et al.*, 1950) who demonstrated convincingly that grain carbohydrates in barley are derived predominantly from current photosynthesis following anthesis. This work was soon extended by several workers, notably by Buttrose and May (1959) and Thorne (1965) who were able to estimate the relative contributions of the various assimilating surfaces—the flag leaf, peduncle and the ear itself (including awns if present). In wheat, the ear was shown to contribute some 17 per cent of the carbohydrates entering the grain, the shoot system about 83 per cent. The apparently heavy dependence of grain yield on flag leaf photosynthesis prompted Watson and his colleagues (Watson *et al.*, 1963) to calculate flag leaf area duration, a measure of flag leaf area from ear emergence to leaf senescence, and to relate this with some degree of success to grain yield. The average efficiency of the green surface between anthesis and maturity was estimated by *G*, the grain : leaf ratio, computed by dividing final grain yield by leaf area duration (Welbank *et al.*, 1966).

These studies were concerned with measuring the extent to which grain yield was source-limited, and the high correlation of *G* with mean daily radiation gave support to the belief that size and efficiency of the assimilating surfaces largely determined grain yield. There were, however, hints in the results of Watson and his colleagues that the positive effect of temperature on *G* could be explained in terms of grain growth and thus sink capacity. More recent work, for example that by Evans and Rawson (1970) and by Bremner and Rawson (1972), also suggests that sink-limitation may, at least under certain circumstances, be equally if not more important than source-limitation in determining grain yield. It is in relation to this argument that we wish to present some of our own data.

It will generally be accepted that the main objective in obtaining high yield is to produce the maximum number of grains per unit area, on the assumption that mean weight per grain plays a less important part in yield determination. Grain numbers depend on the number of ears, and this in turn is governed by the extent of tillering and the survival of these tillers to become fertile. However, grain number per unit area is also determined by how many grains each ear contains, and this depends on the number of spikelets and the number of grains in each of them.

In this paper we wish to pay particular attention to the factors determining spikelet numbers per ear and the number of grains within the spikelet. There is, of course, plenty of evidence that the size of the ear population is highly important in relation to yield, and also that there are inter-relationships between this component and the size of each ear. Our own experiments would not be complete without taking into account all yield components, but for the present we wish to focus attention on the performance of the ear, in terms of both spikelet production and grain set.

SPIKELET PRODUCTION

The vegatative shoot apex in wheat initiates primordia at intervals, and these expand to form leaves. However, in general the rate of primordium initiation is greater than the rate of leaf production, so that primordia accumulate on the apex. This process has been referred to as stacking. Given appropriate conditions, floral initiation will occur and its first sign is the appearance of spikelet initials in the axil of leaf primordia of the lower-central region of the apex. Development of further spikelet initials now occurs in both an upward and downward direction, and the shoot apex elongates much more rapidly than it did while leaf primordia were being laid down. Kirby (1974) has shown that spikelet accumulation proceeds at a linear rate which is faster than the rate of stacking of leaf primordia. When the terminal spikelet has been determined, the process is complete and no further spikelets can now be added.

What are the factors that determine the number of spikelets per ear? Table 1, based on a field experiment with New Zealand spring wheat sown in autumn and winter, shows that cultivars differ in spikelet number and that the environment plays a large part. Cultivar Arawa tended to have more spikelets per ear than the cultivar Hilgendorf. In

TABLE 1. MEAN NUMBER OF SPIKELETS PER EAR IN SPRING WHEAT CV. ARAWA AND CV. HILGENDORF SOWN AT DIFFERENT TIMES

Month of sowing (southern hemisphere)	Arawa	Hilgendorf
March	18.2	15.0
April	18.4	16.1
May	16.3	14.9
June	14.8	13.0
	S.E. ± 0.39	

both, a prolonged vegetative period achieved by early sowing and the day-length conditions in which they differentiated had a large effect. For example, sowing in April was attended by a long period of vegetative growth followed by a long interval between differentiation and ear emergence. Sowing in June reduced the duration of vegetative growth only slightly but severely curtailed the length of the reproductive phase.

It would appear that a long vegetative period provides time for stacking of primordia to occur, although the rate of accumulation is presumably too slow and the number of uncommitted primordia not large enough to cause any appreciable increase in final spikelet numbers. What is probably more important is the duration of spikelet initiation and this is influenced by photoperiod and, in cold-requiring wheats, by vernalization. There are numerous records showing that the number of spikelets declines with increases in photoperiod. Rawson (1971) has shown that the rate of spikelet initiation increased as the photoperiod was extended, but the duration of the process was drastically reduced, thus leading to lower spikelet numbers. Vernalization also hastened development and decreased spikelet numbers (Rawson, 1970). Nitrogen supply also has a significant effect. Raising nitrogen levels at the beginning of the reproductive process increases the rate of primordium production and extends the duration of primordium formation (i.e. delays determination of the terminal spikelet), and consequently more spikelets tend to be formed (Table 2).

TABLE 2. NUMBER OF SPIKELETS PER EAR IN WHEAT IN RESPONSE TO 15 PPM OR 150 PPM OF NITROGEN SUPPLIED BETWEEN DEVELOPMENTAL STAGES OF DOUBLE RIDGES AND FLORET INITIATION

Nitrogen ppm	Mean number of spikelets
15	9.8
150	11.3
	S.E. ± 0.25

Data from Langer and Liew (1973).

However, not all spikelets necessarily contain ear-bearing florets. Those at the base of the ear may be quite rudimentary, depending on genotype and environment. This is shown in Table 3 taken from the same experiment as Table 1 in which wheat cultivars sown at different times in autumn and winter were compared.

It will be noted that the advantages of early sowing were somewhat less, if grain-bearing spikelets are taken as a criterion, and the superiority of cv. Arawa was also less distinct. We have so far not studied the physiological factors contributing to the appearance of barren spikelets, athough it is significant that they are either absent or reduced in number in long days and with high nitrogen. Both treatments are known to increase the rate of primordium formation, but, whereas long days reduce the length of the spikelet formation phase, high nitrogen extends it. The rate of the process would thus seem to be the most important consideration in this respect, especially as the spikelet primordia involved are at the base of the spike, directly above those primordia destined to become leaves.

TABLE 3. NUMBER OF BARREN AND FERTILE SPIKELETS PER EAR IN SPRING WHEAT CV. ARAWA AND CV. HILGENDORF SOWN AT DIFFERENT TIMES

Month of sowing (southern hemisphere)	Arawa		Hilgendorf	
	barren	fertile	barren	fertile
March	4.2	14.0	3.2	12.8
April	4.4	14.0	3.4	12.7
May	2.7	13.6	2.1	12.8
June	1.9	12.9	2.0	11.0
		barren:	S.E. ± 0.30	
		fertile:	S.E. ± 0.45	

FLORET DEVELOPMENT

The eight or nine florets in each spikelet are initiated over a period of several days, and according to Williams (1966) the interval between successive floret primordia can be of the order of 1.5 days at 17°C in long days. However, florets which are later going to produce a grain reach the stage of anthesis at approximately the same time, so that rates of floret development must be assumed to increase acropetally within the spikelet. That this is in fact the case has been shown by Langer and Hanif (1973) (Table 4). Distal florets which fail to set grain have much slower rates of development and, presumably fail to reach a minimum size.

TABLE 4. MEAN RATE OF FLORET DEVELOPMENT (DEVELOPMENTAL UNITS/DAY) IN WHEAT OVER A 9-day PERIOD

Floret	Spikelet 2	Spikelet 6
1	0.19	0.25
2	0.20	0.25
3	0.25	0.31
4	0.26	0.28
5	0.17	0.27
6	0.05	0.17
Mean	0.19	0.25
	S.E. ± 0.016	

Data from Langer and Hanif (1973).

Table 4 also shows that the rate of floret development is slower in basal spikelets, and this may well be a factor contributing to their potential barrenness. A long photoperiod was found to speed up development of potentially fertile florets. As a result of these different rates of development within the spikelet, only the basal florets had the chance to become fertilized and set grain, while the late-formed and slow-developing distal florets degenerated at or before ear emergence. This now raises the question of what determines the number of grains set among the florets available for fertilization.

Grain number

In discussing the number of grains per spikelet we will consider only those florets which continue to develop until anthesis and which do not abort prematurely, i.e. potentially fertile florets. How many of these become fertilized and set grain appears to depend on physiological events during the period from their initiation until pollination occurs, or possibly even later. We believe the early part of this period to be highly important, and in this respect it seems to us that the evidence that has been procured so far suggests that one or several of at least the following four physiological factors could be involved in preventing all potentially fertile florets from setting grain.

1. Transport difficulties;
2. shortage of nitrogen;
3. correlative inhibition;
4. shortage of assimilates.

We will now examine these possibilities in turn.

1. *Transport difficulties*

Hanif and Langer (1972) have shown that in common New Zealand cultivars of wheat only the lowest three florets are linked directly with the main vascular supply of the spikelet. If this has any physiological significance, it might be assumed that these florets occupy the most favourable positions to receive supplies of assimilates and minerals. However, under optimal conditions, especially with high nitrogen nutrition, more distal florets may also set grain, even though they are connected by only sub-vascular elements. By contrast, one or two of the directly linked florets may fail to set grain, if environmental conditions are unfavourable. We are thus forced to conclude that the vascular system does not provide the main limitation to the setting of grain, although it may possibly do so under certain limiting conditions. Perhaps we should also stress that we have not examined, from this point of view, those cultivars which commonly have 4–5 grains per spikelet.

2. *Shortage of nitrogen*

Pot experiments with de-tillered single plants reduced to the main stem have shown that a high level of nitrogen in the early stages of floral organ development has a significant effect on grain set (Langer and Liew, 1973). Table 5 shows that this effect was particularly striking if nitrogen supply was raised between the appearance of double ridges and floret initials, although there was still evidence of a response in the next phase of development between floret initial formation and ear emergence. From other work we know that the rate of floret development should have been increased by high nitrogen supply, but we are not sure how long this boost would have lasted. Following ear emergence, nitrogen had no further effect on grain set.

These results refer to single plants reduced to the main shoot, and furthermore the observed spikelet was in the middle region of the ear. In field populations, and taking all ears and spikelets into account, we have obtained rather different results. For example in

TABLE 5. EFFECT OF NITROGEN SUPPLY ON NUMBER OF GRAINS PER SPIKELET IN WHEAT

Stage of development	nitrogen supply ppm 15	150
Double ridges to floret initiation	2.6	4.3 ***
Floret initiation to ear emergence	3.2	3.9 ***
After ear emergence	3.5	3.6 N.S.

*** Difference significant ($p < 0.001$). N.S. Not significant.

Data from Langer and Liew (1973).

an irrigated crop of cv. Kopara we measured 1.14 grains per spikelet with no nitrogen fertilizer added but only 0.83 with 200 kg/ha of nitrogen (Dougherty and Langer, 1974). Since there were no differences in the number of ears per unit area, we were not dealing with ears from late-formed tillers induced by nitrogen. We must therefore assume that it is possible for other factors to interact with nitrogen and to override its effect.

3. *Correlative inhibition*

The supposition that grain set may be affected by growth regulators comes from experiments in which one or more of the basal florets within a spikelet have been sterilized. Evans *et al.* (1972) as well as Walpole and Morgan (1973), have shown that such treatment is often accompanied by compensatory grain formation in more distal florets. Although it is possible to interpret these responses in terms of sink strength in relation to assimilate supply, a number of reasons have been advanced to suggest that this was unlikely. For example preventing ear photosynthesis by 3-(3,4-dichlorophenyl)-1,1-dimethylurea (DCMU) did not alter grain set, and furthermore the rate of assimilate supply appeared to be quite adequate for the initial growth of grains (King *et al.*, 1967). On the other hand, there was at least circumstantial evidence of possible hormonal influences. Thus grain set in floret 3 occurred at the same time as more basal ovaries lost viability, delayed hand-pollination of basal florets gave increasingly positive responses in floret 3, and removal of anthers appears to have an effect on grain set in other spikelets, whereas ovary removal caused a reaction within the same spikelet (Evans *et al.*, 1972). These results relate to events following ear emergence, but other evidence suggests that the same considerations could also apply at an earlier stage. For example, the different rates of floret development and the abortion of distal florets could be interpreted in similar fashion as indicating correlative inhibition (Langer and Hanif, 1973). Furthermore, morphological studies by Fisher (1973) in various genotypes have shown differences in spikelet development, in that in Norin 10 derivatives glumes and lemmas grew to a greater extent before stamen primordia developed than in standard Canadian cultivars. This was interpreted as stronger apical dominance preventing precocious development of basal florets and consequently leading to a greater number of fertile florets in the Mexican genotypes.

We may be inclined to accept the weight of this evidence as indicating the presence of hormonal control, but it will undoubtedly be difficult to prove it. The first time that the developing ear is readily accessible to treatment is following its emergence, but by then the fate of most florets is determined. Certainly abortion of distal florets is then inevitable, as our own morphological studies have shown. We have investigated the possible effects of growth regulators straight after ear emergence by applying successively droplets of cytokinin, indol-3yl-acetic acid (IAA) or gibberellin to floret 3, but all we observed was reduced grain set, even though final dry weight of the grain was apparently increased by IAA. It will require some ingenuity to devise experiments to demonstrate the action of growth regulators in floret development and grain set.

4. *Shortage of assimilates*

Although the effect on grain numbers caused by removal of basal grains has been interpreted as indicating a hormonal correlation, the possibility of competition for a limited supply of assimilates cannot be entirely dismissed. There is other evidence to show that following anthesis, grain set may be reduced at low light intensities of 20 K lx or less (Wardlaw, 1970). In the field we would hardly expect light intensities to be so low as to have this effect, and yet recent results lead us to suspect that assimilates could be limiting for grain numbers, and that this may occur at an early stage during reproductive development. One line of evidence comes from an experiment in which we thinned small areas 30 × 30 cm within a crop to leave a single main tiller. This was done on fresh areas every week so as to cover the whole reproductive phase until anthesis. As Table 6 shows, early thinning was attended by an increase in the number of grains contained in the middle six spikelets. Weight per grain was not greatly affected.

TABLE 6. EFFECT OF THINNING AT DIFFERENT TIMES ON GRAIN NUMBER AND WEIGHT PER GRAIN IN WHEAT

Time of thinning (southern hemisphere)	No. grains/spikelet	Weight/grain (mg)
5 Nov.	3.51	42
12 Nov.	3.42	42
19 Nov.	3.26	42
26 Nov.	3.21	42
3 Dec.	3.21	44
10 Dec.	3.18	44
Unthinned	2.87	40

It seems that the early part of the reproductive phase was responsive to what we assume to have been an improvement in light conditions, since the plots were both irrigated and given a nitrogen fertilizer. Supporting evidence comes from other field experiments for which we calculated correlation coefficients between the mean number of grains per spikelet at harvest and crop growth parameters during the early part of the reproductive process. The negative coefficients in the case of crop growth rate, leaf area index and tiller density, but the positive association with net assimilation rate, suggest that in all probability we were dealing with competition for assimilates between vegetative and reproductive structures, to the detriment of grain numbers (Table 7).

It is this apparent competition for assimilates which, we think, is responsible for depressions of yield which we have obtained with irrigation and nitrogen fertilizer. Bingham (1972) has also stressed the importance of assimilate supply during ear development, while Lupton *et al.* (1974) found flag leaf area duration and rate of ear growth to be the most important variables determining grain yield.

TABLE 7. SIMPLE CORRELATION COEFFICIENTS BETWEEN SOME CROP PARAMETERS DURING EARLY REPRODUCTIVE DEVELOPMENT IN SPRING AND MEAN NUMBER OF GRAINS PER SPIKELET AT HARVEST IN WHEAT CV. AOTEA

Parameter	Date	Correlation coefficient
Crop growth rate	3–25 Oct.	−0.24
Net assimilation rate	3–25 Oct.	+0.21
Leaf area index	3 Oct.	−0.50
	25 Oct.	−0.47
Tiller density	3 Oct.	−0.32
	25 Oct.	−0.27

All coefficients significant ($p < 0.01$)

CONCLUSION

We do not suggest that the factors we have discussed work independently of one another, but rather that they indicate several interrelated facets of the same phenomenon. If we are correct in our interpretation, then it also follows that we are dealing here with a source limitation that prevents full development of potential sink size. At a later stage, when sink size has been determined we may however be justified in concluding that, under Australian and New Zealand conditions, wheat crops could be sink-limited during grain filling. This is probably not the case in the United Kingdom where lower levels of radiation may be responsible for a source-limiting situation. Whether or not these generalizations are correct, we intend to continue our studies of the crucial events before ear emergence when grain number per ear is determined.

REFERENCES

BINGHAM, J. (1972) Physiological objectives in breeding for grain yield in wheat. *Proceedings of the Sixth Congress of Eucapria, Cambridge 1971*, pp. 15–29.

BREMNER, P. M. and RAWSON, H. M. (1972) Fixation of $^{14}CO_2$ by flowering and non-flowering glumes of the wheat ear, and the pattern of transport of label to individual grains. *Aust. J. biol. Sci.* **25**, 921–930.

BUTTROSE, M. S. and May, L. H. (1959) Physiology of cereal grain. 2. The source of carbon for the developing barley kernel. *Aust. J. biol. Sci.* **12**, 40–52.

DOUGHERTY, C. T. and LANGER, R. H. M. (1974) An analysis of a nitrogen-induced depression of yield in irrigated Kopara wheat. *N.Z. J. agric. Res.* **17**, 325–331.

ENGLEDOW, F. L. and WADHAM, S. M. (1923) Investigations of yield in the cereals, *J. agric. Sci. Camb.* **13**, 390–439.

EVANS, L. T., BINGHAM, J. and ROSKAMS, M. A. (1972) The pattern of grain set within ears of wheat. *Aust. J. biol. Sci.* **25**, 1–8.

EVANS, L. T. and RAWSON, H. M. (1970) Photosynthesis and respiration by the flag leaf and components of the ear during grain development in wheat. *Aust. J. biol. Sci.* **23,** 245–254.

FISHER, J. E. (1973) Developmental morphology of the inflorescence in hexaploid wheat cultivars with or without the cultivar Norin 10 in their ancestry. *Can. J. Pl. Sci.* **53,** 7–15.

HANIF, M. and LANGER, R. H. M. (1972) The vascular supply of the spikelet in wheat (*Triticum aestivum*). *Ann. Bot.* NS **36,** 721–727.

KING, R. W., WARDLAW, I. F. and EVANS, L. T. (1967) Effect of assimilate utilization on photosynthetic rate in wheat. *Planta* **77,** 261–276.

KIRBY, E. J. M. (1974) Ear development in spring wheat. *J. agric. Sci. Camb.* **82,** 437–447.

LANGER, R. H. M. and HANIF, M. (1973) A study of floret development in wheat (*Triticum aestivum*). *Ann. Bot.* NS **37,** 743–751.

LANGER, R. H. M. and LIEW, F. K. Y. (1973) Effects of varying nitrogen supply at different stages of the reproductive phase on spikelet and grain production and on grain nitrogen content in wheat. *Aust. J. agric. Res.* **24,** 647–656.

LUPTON, F. G. H., OLIVER, R. H. and RUCKENBAUER, P. (1974) An analysis of the factors determining yield in crosses between semi-dwarf and taller wheat varieties. *J. agric. Sci. Camb.* **82,** 483–496.

PORTER, H. K., PAL, N. and MARTIN, R. V. (1950) Physiological studies in plant nutrition. 15. Assimilation of carbon by the ear of barley and its relation to the accumulation of dry matter in the grain. *Ann. Bot.* NS **14,** 55–68.

RAWSON, H. M. (1970) Spikelet number, its control and relation to yield per ear in wheat. *Aust. J. biol. Sci.* **23,** 1–15.

RAWSON, H. M. (1971) An upper limit for spikelet number per ear in wheat, as controlled by photoperiod. *Aust. J. agric. Res.* **22,** 537–546.

THORNE, G. N. (1965) Photosynthesis of ears and flag leaves of wheat and barley. *Ann. Bot.* NS **29,** 317–329.

WALPOLE, P. R. and MORGAN, D. G. (1973) The effects of floret sterilization on grain number and grain weight in wheat ears. *Ann. Bot.* NS **37,** 1041–1048.

WARDLAW, I. F. (1970) The early stages of grain development in wheat: response to light and temperature in a single variety. *Aust. J. biol Sci.* **23,** 765–774.

WATSON, D. J., THORNE, G. N. and FRENCH, S. A. W. (1963) Analysis of growth and yield of winter and spring wheats. *Ann. Bot.* NS **27,** 1–22.

WELBANK, P. J., FRENCH, S. A. W. and WITTS, K. J. (1966) Dependence of yield of wheat varieties on their leaf area durations. *Ann. Bot.* NS **30,** 291–299.

WILLIAMS, R. F. (1966) Development of the inflorescence in Gramineae. In *The Growth of Cereals and Grasses*, ed. MILTHORPE, F. L. and IVINS, J. D., Butterworths, London.

EXOGENOUS AND ENDOGENOUS GROWTH-REGULATING SUBSTANCES IN STUDIES ON PLANT MORPHOGENESIS

J. BRUINSMA

Department of Plant Physiology, Agricultural University, Wageningen, Netherlands

INTRODUCTION

The role of hormones in plant morphogenesis can be studied in two ways. On the one hand, growth-regulating substances can be administered directly to plants or plant parts and their effects on the development of the organs studied. On the other hand, effects of hormones can be deduced indirectly from the way in which internal hormone levels change during the development of a particular organ. These two ways complement each other, each having its own possibilities and limitations which will be discussed in this paper. The term *growth regulators* will be used in a general way to indicate all substances, both natural and synthetic, which in low concentrations affect the growth and development of plant parts, whereas the term *hormones* is confined to the naturally occurring auxins, cytokinins, gibberellins, abscisins, and ethylene.

EXOGENOUS GROWTH REGULATORS

The administration of growth-regulating substances, both natural and synthetic, is often the more rapid and easier way to evaluate which groups of hormones may be involved in the regulation of a particular developmental process. In order to avoid interference by hormones from other plant parts, the organ under study is preferably isolated from the rest of the plant and cultivated *in vitro* on media of known compositions. The various growth substances can then be added and their effects studied, thus indicating which hormones may participate in the regulation of the development of the organ. Influences exerted by other plant parts from which these hormones are derived can then be analysed in further studies *in vivo*.

An example is the analysis of pistil abortion in flower buds of the spider flower, *Cleome spinosa*. The flowers are normally bisexual, but they develop into males by pistil abortion when fruits and flowers develop simultaneously on the same inflorescence. The developing seeds in these fruits actively produce indol-3yl-acetic acid (IAA), which causes the carpels to grow (Fig. 1). A considerable sink activity is thereby created at the cost of the flower buds, and the pistils fail to grow out (de Jong and Bruinsma, 1974a).

In order to determine which factor in the sap stream is so essential for pistil development, young flower buds were cultured *in vitro*. For this purpose, a smaller-flowered

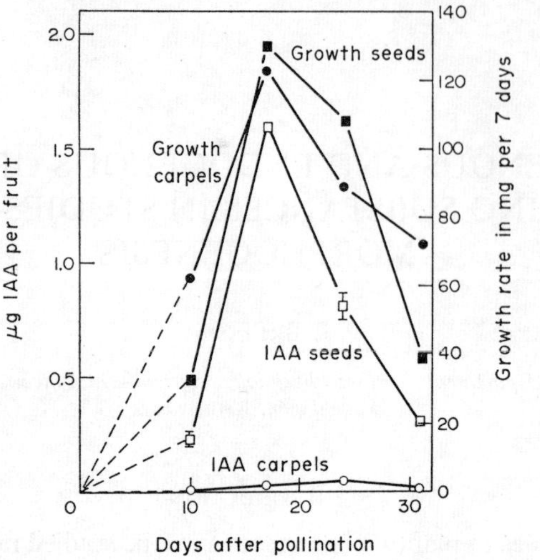

FIG. 1. Growth rate and IAA content of *Cleome spinosa* seeds and carpels (data from de Jong and Bruinsma, 1974a).

species of *Cleome*, *C. iberidella*, was used. *C. iberidella* shows the same pistil abortion as *C. spinosa.* In these studies *in vitro* pistils did not develop unless specific cytokinins were added, viz. zeatin or benzyladenine. Other cytokinins were more or less ineffective (Fig. 2). Moreover, lack of nutrients in the medium was always found to limit petal growth rather than pistil development, so that it is improbable that nutrients were responsible for the occurrence of male flowers (de Jong *et al.*, 1974). It was presumed, therefore, that developing fruits withdraw cytokinins from the sap stream which are essential for pistil development.

Because cytokinins are mainly derived from roots, we returned to studies with intact *Cleome spinosa* plants. These were grown in pots of different sizes in order to limit to varying degrees the development of roots, and, hence the cytokinin supply. Under conditions of limited root growth pistil abortion could be reduced by zeatin injected into the base of the inflorescence. The results, presented in Table 1, would have been more

TABLE 1. EFFECT OF 10^{-4} M ZEATIN, INJECTED DAILY DURING 5 WEEKS AT THE BASE OF THE INFLORESCENCE OF *Cleome spinosa* PLANTS, GROWN IN POTS OF DIFFERENT SIZES TO REDUCE ROOT DEVELOPMENT

Pot volume litre	% female abortion		LSD 5%	No. of replicates
	control	zeatin		
0.5	56.3	48.5	6.8	24
0.5	67.2	52.5	7.1	18
1.8	49.5	39.6	5.8	12
4.0	51.0	48.0	n.s.	12

Data from de Jong and Bruinsma, 1974b.

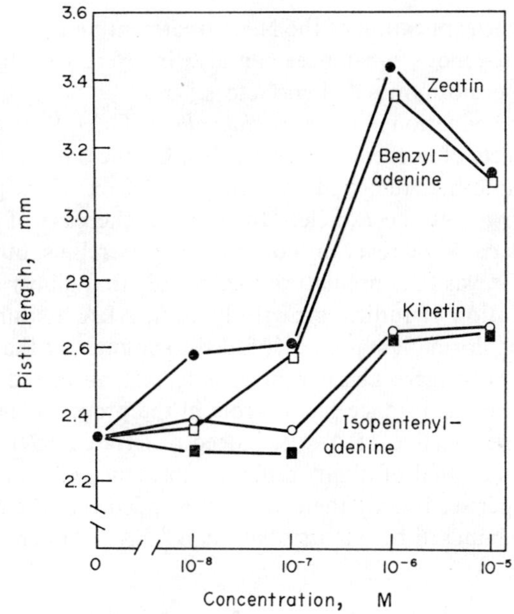

FIG. 2. Effect of different cytokinins on pistil development in *Cleome iberidella* flower buds grown *in vitro* during 14 days.

pronounced if the temperature in the greenhouse had not risen to levels adverse to pistil development.

Experiments *in vitro* are suited not only for analysing effects of individual hormonal and nutritional factors, but are also informative about interactions between these various factors. For instance, it is generally agreed that ethylene promotes pistil development, whereas gibberellins suppress it. From our experiments on *Cleome* flower buds grown *in vitro* it became obvious that ethylene alone is as adverse to pistil development as gibberellins. However, the adverse effect of gibberellins is counteracted by ethylene, so that in the presence of gibberellins, which apparently is the normal situation *in vivo*, ethylene does promote pistil growth (de Jong and Bruinsma, 1974b).

Endogenous cytokinin contents have not yet been determined in *Cleome*, but in tomato it has been demonstrated that fruits can, indeed, be strong sinks for root-produced cytokinins (Varga and Bruinsma, 1974).

It is always dangerous to jump to conclusions from experiments with exogenous growth regulators only. These administered substances cannot simply be taken as a substitute for the corresponding internal regulators, because they may interfere with the hormonal pattern in other ways than the endogenous substances would do. For instance, if the level of the corresponding internal component is already optimal, then the addition of an exogenous component may fail to exert an effect or may cause artifacts because of overdosing. Moreover, the exogenous component need not merely act additively to the corresponding endogenous component. Thus, in the afore-mentioned study on *Cleome spinosa*, spraying plants with 1-naphthylacetic acid (NAA) caused the internal IAA level in the seeds to drop, and shifted the IAA peak towards an earlier stage of development.

Such changes render the explication of the NAA treatment on the basis of an auxin effect extremely difficult. Exogenous substances can also interfere with biosynthesis, activity, and breakdown of other hormones and co-factors.

In a study on seed germination in *Chenopodium album*, C. M. Karssen in my laboratory found two distinct effects of exogenous gibberellins. On the one hand, gibberellins broke dormancy and, on the other, removed the inhibiting effect of exogenous abscisic acid (ABA) on the outgrowth of the radicle. However, in the case of radicle growth, the adverse effect of ABA could be reversed not only by gibberellins, but also by cytokinins, ethylene or red light. It was thus premature to conclude that gibberellins are involved in the endogenous regulation of radicle growth. In turn, ABA was unable to prevent the seed escaping from the dormant state, i.e. it failed to counteract the dormancy-breaking effect of gibberellins. The mere absence of an effect of an added substance does not imply, of course, that the substance has no role in the process. Therefore, it had to be established first, that radioactive ABA does indeed penetrate into the dormant seed and, secondly, that samples of seed of highly different dormancy contain equally little ABA, only a few picogram per seed. Only then was it concluded that the dormancy of *Chenopodium album* seeds is imposed by factors other than ABA (Karssen, 1975).

ENDOGENOUS GROWTH REGULATORS

In experiments using added growth regulators, the effect of the already present endogenous substances can never be ignored, even if the chemical nature of the latter is unknown. Quantitative determination of the endogenous components of the hormonal pattern which regulates the developmental process under study, is essential. However, it must be questioned whether the level of a hormone is always a reliable indicator of its activity. Apart from supra-optimal levels, it is generally agreed that the effect of a hormone is positively related to its concentration, the relationship being mostly a logarithmic one.

In the tomato fruit, as in *Cleome* fruit, IAA is practically confined to the seeds. After the first week the tomato fruit grows by cell enlargement only and the ultimate size of the fruit is proportional to the number of seeds. It is plausible therefore that the growth of the fruit is related to the IAA content of the seeds. However, the amount of IAA increases only after the growth rate has reached its optimal level (Fig. 3). It looks rather as if IAA is utilized during growth at about the rate at which it is produced and that it accumulates only when the rate of growth declines. The sharpness of the peak indicates that a fairly efficient feedback mechanism exists between the sites of utilization and production of IAA. This may well be a rather common phenomenon: if a hormone is used up when exerting its effect, the resulting hormone concentration will be low when the effect is large, and a negative relationship will be found between concentration and effect of the hormone.

This point may be illustrated by reference to a study on abscission of pedicels of *Begonia* flower buds. In these pedicels, the abscission zone is already present at a very young stage, and in the still elongating pedicel the ethylene level is high enough to induce the production of cell-wall solubilizing enzymes (Hänisch ten Cate and Bruinsma, 1973). However, the IAA produced by the flower bud keeps the pedicel tissue insensitive to

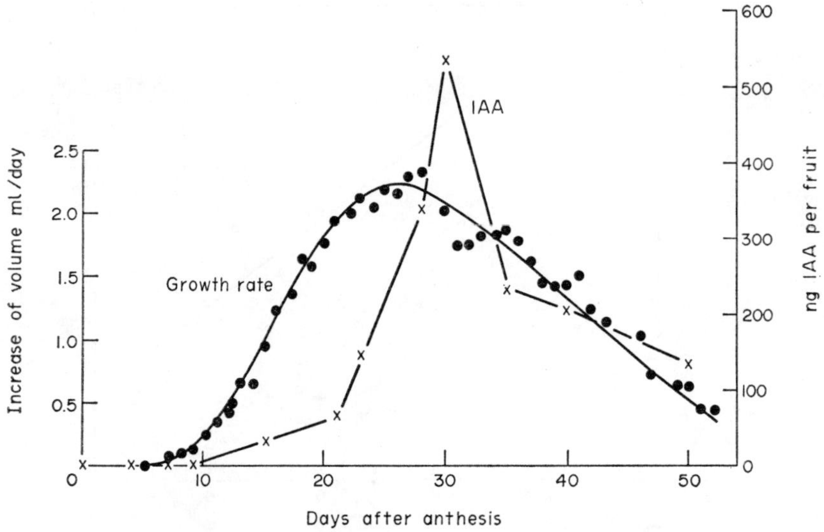

FIG. 3. Growth rate and IAA content during the development of the tomato fruit *cv.* Moneymaker (Data from Varga and Bruinsma, in press).

ethylene and thus prevents abscission. In a monoecious *Begonia fuchsioides* hybrid, the inflorescence consists of a central male flower and two female side flowers. The male flowers drop before anthesis whereas the females reach full bloom; IAA production of the females is 100 times that of the males so that the abscission zones of the latter are no longer prevented from responding to the prevailing ethylene concentration (Table 2).

TABLE 2. IAA CONTENT AND PERCENTAGE BUD ABSCISSION OF FEMALE AND MALE FLOWERS OF *Begonia fuchsioides F* × *B. foliosa* HYBRID

	female	male
% flower bud abscission	0	99
IAA content (ng/bud)	10.7 ± 0.7	0.12 ± 0.02

Data from Hänisch ten Cate *et al.*, 1975.

However, in another *Begonia* hybrid, having male flowers only, no positive relationship is found between the auxin content of the buds and their prevention from abscission. On the contrary, the relationship is a negative one, because the auxin content of the bud is the result, not only of the production of IAA, but also of its release to the pedicel. This release is apparently blocked (Fig. 4): the higher the auxin content of the flower (bud) the lower is the diffusion of IAA into the pedicels, causing premature shedding by cell-wall solubilization in the abscission zone of the pedicels.

These examples demonstrate that it is often hazardous to relate hormone effects to their levels and, moreover, these effects are frequently influenced by other factors which alter the turnover rates of the hormones while they exert their influence. It is these turn-

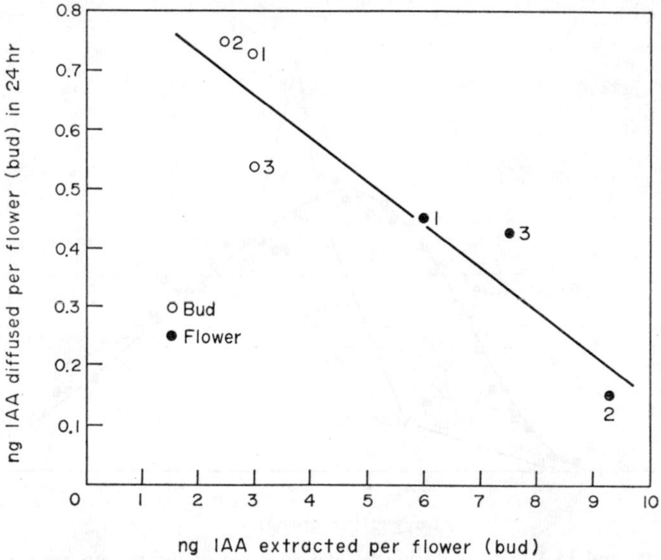

FIG. 4. Reverse relation between the IAA content of, and IAA release from, buds and flowers of (*Begonia cinnabarina* × *B. micranthera*) × *B. davisii* hybrid. Large buds and open flowers of three clones were either extracted or allowed to diffuse during 24 hr (data from Hänisch ten Cate *et al.*, 1975).

over rates which need to be determined, rather than the levels, but this of course is extremely difficult. One possibility is to determine the activity of key enzymes and co-factors of the biosynthetic pathways of hormones. Quantitative determination of absolute hormone concentrations is already very intricate. In the early days of hormone research one was restricted to bioassays which were highly sensitive but not accurate owing both to unknown losses that occur during purification and to the logarithmic nature of the dose–response relationship. Moreover, their specificity is questionable: bioassays respond differently to different members of a group of hormones and are generally sensitive to several groups of promoting and inhibiting substances. Physico-chemical assays are therefore to be preferred. Further progress in hormone research will depend upon the elaboration of suitable methods for the rapid extraction, purification and determination of the hormonal substances. Losses sustained during extraction and purification must also be taken into account in the estimation of absolute concentrations.

In our laboratory, for instance, a method has been worked out for extraction and determination of IAA in nanogram quantities (Knegt and Bruinsma, 1973). No elaborate purification is required since IAA can be easily converted into indole-α-pyrone which can be determined spectrofluorometrically. The addition of a small amount of radio-active IAA to the extract permits measurement of the amount lost and thus the absolute amount of IAA in the original extract can be calculated. The method is rapid and from six to twelve samples can be estimated in a few hours.

For determinations of ABA and ethylene good gas–liquid chromatographic methods are available but determinations of cytokinins and, particularly, the numerous gibber-ellins are much more difficult. High pressure liquid chromatography and gas–liquid chromatography combined with mass spectrometry give the best prospects, provided that the purification procedures can be speeded up and recovery calculations included.

These physicochemical methods, however, do not render the bioassay superfluous; the bioassay is still the principal method of checking whether a substance determined physicochemically has physiological activity. Moreover, it is the bioassay which informs us about the possible occurrence of still unknown hormonal substances in the extracts and diffusates. This is the case, for example, with hormonal regulation of phototropic curvature in light-grown seedlings of sunflower (*Helianthus annuus*). According to the Cholodny–Went theory, auxin is produced in the top of the seedling, the cotyledons. This auxin diffuses into the hypocotyl in which it is laterally transported upon unilateral illumination. The resulting asymmetric distribution causes hypocotyl bending by inducing different rates of cell elongation on opposite sides of the organ.

In testing this theory by resort to Knegt's method for determination of endogenous IAA, we found that the seedling does indeed contain appreciable amounts of IAA, of the order of 1–3 $\times 10^{-7}$ M. An erect hypocotyl has an even auxin distribution, with equal amounts of IAA in each longitudinal half (Table 3). However, this distribution is not changed in hypocotyls induced to bend by unilateral illumination (Table 4). Further-

TABLE 3. DISTRIBUTION OF IAA IN LONGITUDINAL HALVES OF ERECT SUNFLOWER HYPOCOTYLS

Experiment No.	Half I		Half II		No. of plants
	ng IAA/g	%	ng IAA/g	%	
I	44.2	*49.5*	45.4	*50.5*	55
II	63.9	*48.5*	67.8	*51.5*	60

Data from Bruinsma *et al.*, 1975.

Seedlings were grown under 16 hr per day of fluorescent light from above. After 4 days, aerial parts were halved longitudinally in a plane perpendicular to the cotyledonary axis. Cotyledons and apex were removed in order to determine the IAA content of the hypocotyl halves.

TABLE 4. DISTRIBUTION OF IAA IN LIGHT AND SHADED HALVES OF SUNFLOWER HYPOCOTYLS

Experiment No.	Light half		Shaded half		Curvature (°)	No. of plants
	ng IAA/g	%	ng IAA/g	%		
I	53.7	*52*	48.8	*48*	21	44
II	63.2	*51*	61.5	*49*	23	45

Data from Bruinsma *et al.*, 1975.

Seedlings were treated as in Table 3, but horizontal illumination was included for 4 hr prior to dissection to obtain phototropic curvature.

more, the diffusate from sunflower hypocotyls hardly contains any endogenous IAA; the very small amounts present are of the order of the amount of added radioactive IAA (Table 5). Thus although IAA, as required by the Cholodny–Went theory, is indeed present in the hypocotyl, it nevertheless does not seem to be involved in phototropic curvature. IAA is a rather immobile substance and may be confined, for example, to the vacuoles (Dela Fuente and Leopold, 1972). As yet unidentified auxin-like substances other than IAA are probably involved in curvature, but at present they can only be demonstrated by means of the *Avena* coleoptile straight growth test (Table 5). We have

Table 5. Auxin activity and occurrence of IAA in diffusates from sunflower hypocotyls after 24 hr diffusion in light

Experiment No.	*Avena* bio assay IAA equivalents ng/plant	IAA ng/plant
I	0.85	0.07
II	1.93	0.04

Data from Bruinsma *et al.*, 1975.

also indications from other sources that the substance released from the cotyledons into the hypocotyl, and which is required in the curvature response, is not auxin. It is, on the contrary, a growth-inhibiting substance, released from the cotyledons upon illumination. The chemical nature of this inhibitor is under investigation (Bruinsma *et al.*, 1975).

References

Bruinsma, J., Karssen, C. M., Benschop, M. and Dort, J. B. van (1975) Hormonal regulation of phototropism in the light-grown sunflower seedling (*Helianthus annuus*): immobility of endogenous indoleacetic acid and inhibition of hypocotyl growth by illuminated cotyledons. *J. exp. Bot.* **92,** 411–418.

Dela Fuente, R. K. and Leopold, A. C. (1972) Two components of auxin transport. *Pl. Physiol. Lancaster* **50,** 491–495.

Hänisch ten Cate, Ch.H. and Bruinsma, J. (1973) Abscission of flower bud pedicels in *Begonia.* I. Effects of growth-regulating substances on the abscission with intact plants and with explants. *Acta bot. neerl.* **22,** 666–674.

Hänisch ten Cate, Ch.H., Berghoef, J., Van der Hoorn, A. M. H. and Bruinsma, J. (1975) Hormonal regulation of abscission of flower bud pedicels in *Begonia. Physiologia Pl.*, in press.

Jong, A. W. de and Bruinsma, J. (1974a) Pistil development in *Cleome* flowers. I. Effects of mineral nutrition and the presence of leaves and fruits on female abortion in *Cleome spinosa* Jacq. *Z. Pfl.-physiol.* **72,** 220–226.

Jong, A. W. de and Bruinsma, J. (1974b) Pistil development in *Cleome* flowers. IV. Effects of growth-regulating substances on female abortion in *Cleome spinosa* Jacq. *Z. Pfl.physiol.* **73,** 152–159.

Jong, A. W. de, Smit, A. L. and Bruinsma, J. (1974) Pistil development in *Cleome* flowers. II. Effects of nutrients on flower buds of *Cleome iberidella* Welw. ex Oliv. grown *in vitro. Z. Pfl.physiol.* **72,** 227–236.

Karssen, C. M. (1975) Uptake and effect of abscisic acid during induction and progress of radicle growth in seeds of *Chenopodium album* L. In press.

Knegt, E. and Bruinsma, J. (1973) A rapid, sensitive and accurate determination of indole-3-acetic acid. *Phytochemistry* **12,** 753–756.

Varga, A. and Bruinsma, J. (1974) The growth and ripening of tomato fruits at different levels of endogenous cytokinins. *J. hort. Sci.* **49,** 135–142.

IDENTIFICATION OF AUXIN IN GROWING PLANT ORGANS AND ITS ROLE IN GEOTROPISM

M. B. WILKINS

Department of Botany, University of Glasgow, U.K.

INTRODUCTION

The way in which a plant organ regulates its direction of growth has attracted the attention of plant physiologists for more than 150 years. There are essentially two major aspects of this problem. The first concerns the mechanism by which the organ detects that it is deviating from the normal direction of growth which has been established during the course of evolution. The second concerns the mechanism which regulates the relative rates of cell extension on the two sides of an organ that has been displaced from its preferred orientation and thus brings about the curvature necessary to regain the normal angle. This paper deals only with the latter aspect of the problem. Most of the work I shall describe relates to geotropism because this phenomenon has been studied in more detail than phototropism, and is of greater overall significance in the growth and development of a plant. Virtually all the work described in this paper has been published elsewhere; only a summary of the present position is therefore justified.

THE CHOLODNY–WENT HYPOTHESIS

The Cholodny–Went hypothesis has dominated thought in the field of tropic responses ever since it was propounded in 1926. Broadly speaking, Cholodny and Went suggested independently that auxin, a growth regulating substance, was produced at the apices of roots and shoots, that it was normally symmetrically transported basipetally and thus controlled the rates of extension of cells in the sub-apical region of the organ. Upon geotropic stimulation, however, they proposed that auxin underwent a lateral displacement towards the lower side of the organ. In the shoot, this led to an enhancement of growth in the lower half of the organ and a diminution of growth in the upper half with the result that an upward curvature occurred. In the horizontal root, accumulation of auxin in the lower half was supposed to inhibit growth, and hence give rise to downward curvature.

Attempts to ascertain the validity of the Cholodny–Went theory have been made repeatedly over the past 46 years but many of these have proved to be of little or no value because of the lack of critical thought in experimental design. In addition, very little experimental attention has been given to responses in intact, non-mutilated organs—the plant physiologist's love of segmentology seems always to have dominated the approach.

The validity of the Cholodny–Went theory depends essentially upon two main facts being established. First, a growth regulating substance (auxin) must be shown to be present in the tips of growing roots and shoots and unequivocally identified by the methods of analytical chemistry. Secondly, the substance must be shown to undergo lateral transport towards the lower side of a horizontal root and shoot. Over the past ten years a number of my associates and I have investigated these questions and this paper gives a brief overall summary of our findings. My remarks must be regarded as referring specifically to the coleoptile and primary root of *Zea mays*, since it is by no means clear that the results we have obtained with this species are also true for others.

Geotropic response of coleoptiles

To deal first with the geotropic response of coleoptiles, the first question which must be answered is whether or not an auxin-like substance arises from the apex of the organ and, if so, its chemical identity. The presence of a factor which promotes cell extension was established by Went (1928) in his classical experiments, and a certain amount of subsequent circumstantial evidence, based on chromatographic separative techniques and bioassay procedures, suggested that this substance might be indol-3yl-acetic acid (IAA). Astonishing as it might seem in view of the vast amount of work which has been carried out on the substance, and the multitudinous roles attributed to it, over the past 40 to 50 years, there has been no unequivocal identification of IAA in *growing* plant tissues until recently. Attempts in 1944 by Berger and Avery to identify IAA in dormant maize kernels revealed a substance that had a melting point of 166–167°C, but the infra-red spectrum indicated that up to 30 per cent impurities were present in the crystals that were thought to be IAA. Haagen-Smit *et al.*, (1942, 1946) also obtained from corn meal crystals melting at 165°C. Notwithstanding that this evidence is not very satisfactory from the chemical standpoint, it has not been sufficiently recognized that the identification of IAA in the storage tissue (endosperm) of *Zea mays* fruits is quite irrelevant to the question of its existence in the apices of *growing* plant organs where it has been held to control growth and many developmental phenomena.

Greenwood *et al.* (1972) showed IAA to be present in the apices of *Zea* coleoptiles and to diffuse out into agar blocks. Fifteen thousand 2-mm coleoptile apices were placed on agar blocks for 2 hr. The blocks were then extracted and the substances present separated by sequential chromatography in three different solvent systems. The growth active region at the *Rf* of authentic IAA on the final chromatogram was eluted and introduced directly into an AEI MS30 mass spectrometer. The molecular fractionation pattern of the sample was identical with that of authentic IAA, while a volume of agar equal to that used for the diffusion revealed no peaks at *m/e* 175, 130 and 103 (Fig. 1). Another sample was subjected to high resolution mass spectrometry to determine accurately the mass of the molecular ion. The calculated mass of IAA is 175.0633 and the experimentally determined mass of our sample was 175.0628. There is thus no doubt whatsoever that IAA diffuses out of *Zea* coleoptile tips. Extraction of *Zea* coleoptile tips in cold methanol immediately after excision also revealed the presence of IAA in the tissue with the methodology outlined above. Recent investigations by my colleagues Hall and Medlow (1974) have shown IAA to be present also in the ploem and xylem sap of *Ricinus com-*

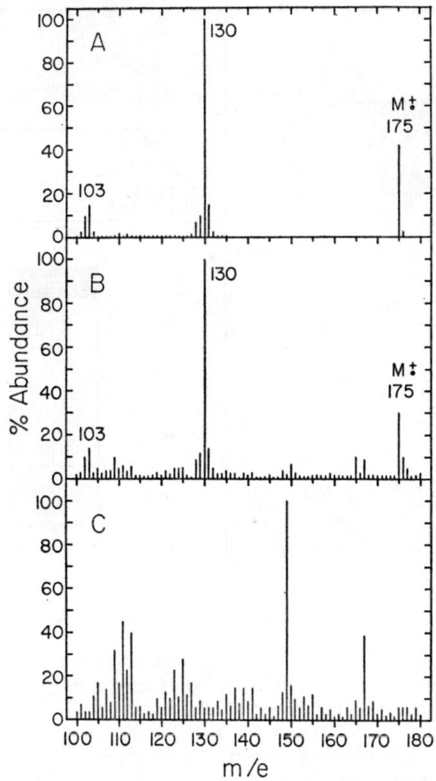

FIG. 1. Mass spectra of (A) authentic IAA at a sample concentration of 50 μg in 0.05 ml of methanol and a probe temperature of 280°C; (B) the violet fluorescent spot from the IAA *Rf* obtained after the chromatographic purification of the diffusate from 15,000 *Zea* coleoptile apices, probe temperature 270°C; and (C) the same zone of a chromatogram as used in (B) except that the extracts were made of blank agar blocks; probe temperature 220°C.

munis plants, and White *et al.* (1975) have shown IAA to be present in the apical buds of *Phaseolus vulgaris* plants.

Having established the presence of IAA in the coleoptile apex of *Zea*, it is next necessary to ascertain whether or not it undergoes lateral transport in a geotropically stimulated organ. The early work of Dolk (1930) showed that a *gradient* of growth promoting activity is established between the upper and lower agar receiver blocks in contact with the basal end of a detached coleoptile tip following geotropic stimulation (Fig. 2A). This kind of experiment, however, shows *no more* and *no less* than that a *gradient* is established; it cannot under any circumstances indicate *by what mechanism* the gradient is established. The claim made even as recently as 1974 (Went, 1974) that this gradient demonstrates a downward lateral transport of auxin completely ignores the possibility that the same result could be achieved by the upward movement of a growth inhibitory substance, or that geotropic stimulation may change the rate of synthesis, longitudinal transport or secretion of auxin (and/or an inhibitor) in the upper and lower halves of the organ! For exactly the same reasons, experiments with horizontal sub-apical segments of coleop-

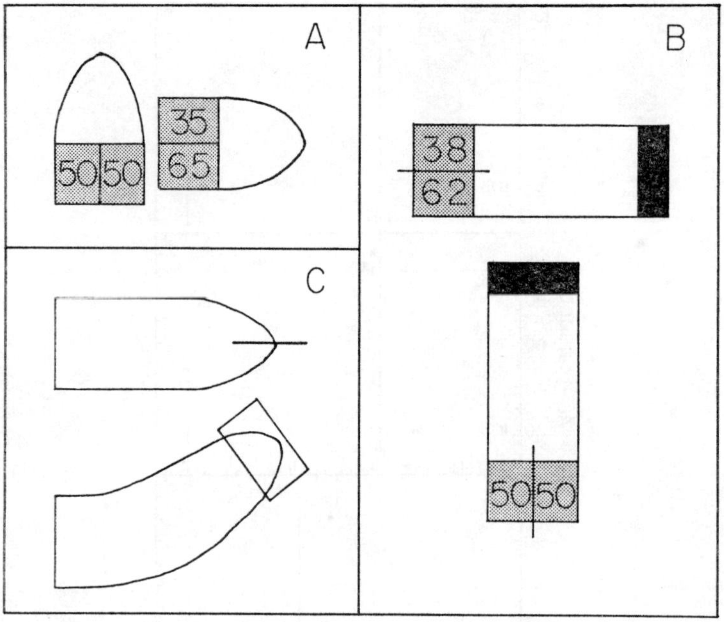

FIG. 2. Diagram illustrating various techniques that have been used on apical and sub-apical segments of coleoptiles to study the validity of the Cholodny–Went theory of geotropism. Tissue is shown in white, agar receiver blocks are shaded and agar donor blocks are shown in black. The numbers refer to the percentage distribution of growth activity (A and B) in the two agar receiver blocks at the basal end of the tissues. The curvature of the coleoptiles in C reflects the interference with lateral transport of a growth regulator on insertion of an impermeable barrier into the end of the coleoptile. Data of (A and B) from Dolk (1930) and (C) from Brauner and Appel (1960). For further explanation see text.

tiles supplied with a symmetrical donor block covering their entire apical surfaces and two receiver blocks at their basal ends (Fig. 2B), can reveal only whether or not a gradient of growth promoting activity is established (or of radioactivity when [^{14}C]IAA is used), and nothing else. That the lateral movement of a growth regulating molecule is taking place in geotropically stimulated coleoptiles is revealed by the simple—but critical—experiment of Brauner and Appel (1960) in which impermeable barriers to lateral movement were shown to abolish or greatly reduce the geotropic curvature (Fig. 2C). In this experiment, of course, a distinction could not be made between downward movement of a promoter and upward movement of an inhibitor, but at least the experiment established that lateral movement of something was involved in the geotropic response mechanism.

The first unequivocal demonstration that a downward lateral transport of IAA takes place in horizontal *Zea* coleoptiles was made by Goldsmith and Wilkins (1964). They used the simple technique of supplying [^{14}C]IAA *asymmetrically* to the apical end of coleoptile segments (Fig. 3A). The logic behind these experiments was that any IAA found in the non-donated half of the coleoptile *must have moved* laterally in the tissue from the donated half. Thus, if more IAA was found to be present in the non-donated half of a horizontal segment with an upper donor block, than in the non-donated half

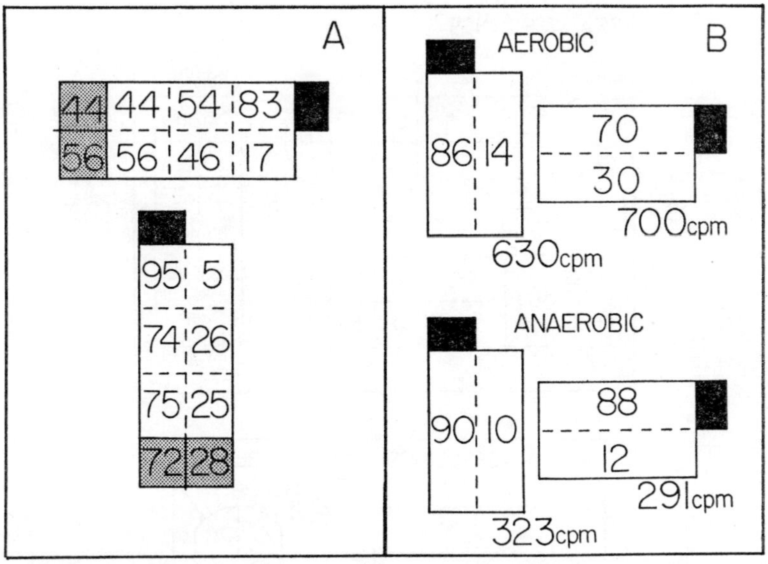

FIG. 3. Diagram illustrating lateral transport of IAA towards the lower side of a horizontal sub-apical segment of *Zea mays* coleoptile (A) and its dependence upon aerobic metabolism (B). Agar donor blocks containing [1-^{14}C]IAA are shown in black and agar receiver blocks are shaded. The numbers refer to the percentage distribution of IAA across the three portions of the 15-mm segment and the receiver blocks (A) or between the two halves of the 10-mm segment (B). Total cpm in the segments is indicated in B for both aerobic and anaerobic conditions. Data of (A) from Goldsmith and Wilkins (1964) and (B) from Wilkins and Whyte (1968).

of a vertical segment, a downward lateral transport of IAA must have been induced by geotropic stimulation. This was found to be so (Fig. 3A). This polarized downward lateral transport of IAA in horizontal *Zea* coleoptile segments was shown to be dependent upon metabolic energy (Wilkins and Whyte, 1968) (Fig. 3B).

The use of segments is open to criticism to the extent that the tissue is damaged and isolated, and its pattern of behaviour might substantially differ from that of intact tissue. In addition, the assay of substances in the agar receiver blocks is really quite irrelevant to the pattern of hormone transport in the tissue because of the numerous complicating factors involved in the transfer of hormone from the living tissue to the agar across an interface of mutilated cells. The critical information required is the pattern of hormone transport in tissues of an intact, undamaged plant organ. Shaw and Wilkins (1974) recently developed a micro-application technique to facilitate the study of hormone movement in relatively undamaged, intact plant organs. Micro-electrodes, of the type used in neurophysiology were employed as micro-pipettes to apply radioactive hormones of high specific activity to plant organs. The pipette acts as a point source of diffusion and allows a small pulse of hormone to be introduced into a highly localized zone of the tissue with the absolute minimum of tissue damage. No detectable change occurred in the volume of solution in the pipette after the tip had been introduced into the tissue for a short time, often as little as 5 secs. By means of this technique, Shaw *et al.* (1973) have shown that downward lateral transport of IAA occurs in geotropically stimulated coleoptiles of *Zea* and *Avena*, and within 10 min of the onset of geotropic stimulation (Fig. 4).

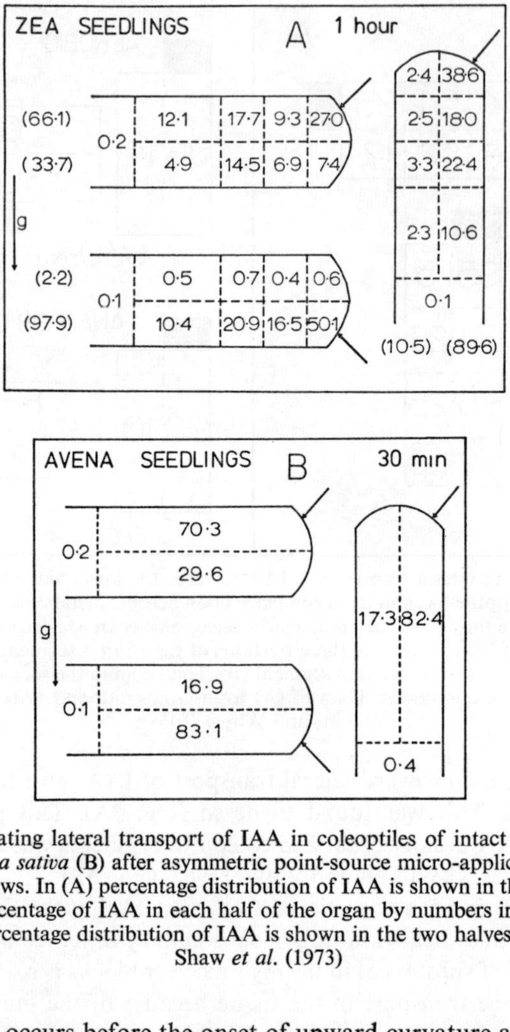

FIG. 4. Diagram illustrating lateral transport of IAA in coleoptiles of intact undamaged seedlings of *Zea mays* (A) and *Avena sativa* (B) after asymmetric point-source micro-application of [^{3}H]IAA at the point shown by the arrows. In (A) percentage distribution of IAA is shown in the several portions of the tissue, and the total percentage of IAA in each half of the organ by numbers in brackets at the bases of the diagrams. In (B) percentage distribution of IAA is shown in the two halves of the tissue. Data from Shaw *et al.* (1973)

The movement thus occurs before the onset of upward curvature and at about the same time as the geoelectric effect develops (Grahm and Hertz, 1962, 1964; Wilkins and Woodcock, 1965; Woodcock and Wilkins, 1969a). The latter effect is attributable entirely to the asymmetric distribution of IAA in *Zea* coleoptile tissues following the onset of geotropic stimulation (Grahm and Hertz, 1964; Wilkins and Woodcock, 1965; Woodcock and Wilkins, 1969a, b, 1970, 1971; Bridges and Wilkins, 1971).

Is the Cholodny–Went theory therefore correct in the case of coleoptiles? The answer is that it is correct, as far as it goes—but it accounts for only a part of the overall mechanism. An indication that another phenomenon might be involved in the geotropic response of *Zea* coleoptiles came first from the work of Naqvi and Gordon (1966). They found that slightly more IAA moved basipetally in the physically separated lower half of a horizontal *Zea* coleoptile than in the upper half. This point was taken up and confirmed by Cane and Wilkins (1969) who used coleoptiles that had been opened out into flat pieces of tissue. The value of this procedure is that in the flattened coleoptile *every cell has an identical orientation with respect to gravity*, something which is impossible to achieve in a solid plant organ. Cane and Wilkins (1969) were able to show that gravity can modify the longitudinal basipetal transport of IAA in coleoptile tissue, and induce lateral transport of IAA, *quite independently* of each other. Both processes can, however, contribute to the gradient of IAA established in the growing zone of the normal, horizontally-placed coleoptiles.

In addition to these two independent processes, others may also be involved in the geotropic response mechanism. For example, differences between the upper and lower halves of the horizontal coleoptile may occur in the synthesis, release, and inactivation of IAA, in the sensitivity of the elongating cells, and in the production, movement and effectiveness of other growth regulating molecules. That these possibilities must be taken seriously is shown by the recent demonstration by Webster and Wilkins (unpublished) of an apparent *upward* lateral movement of GA_3 in horizontal *Zea* coleoptiles. Thus, although substantial progress has been made towards understanding the mechanism of the geotropic response of coleoptiles many questions remain unanswered, and the mechanism involved is certainly much more complex than that envisaged by Cholodny and Went.

Geotropic response of roots

Elucidation of the geotropic response of roots presents a more formidable task than that of shoots primarily because virtually nothing is known about the natural hormonal regulation of the growth and development of intact roots. Only in the past year or so has the presence of IAA been established beyond doubt in primary roots of *Zea mays*. The presence of an auxin with the chromatographic properties of IAA was detected in steles but not in cortices of *Zea* roots and very little was detectable in the extreme 2–3 mm at the tip (Greenwood *et al.*, 1973). The growth active substance has been extracted and its trimethylsilyl (TMS) derivative shown to be identical with the TMS derivative of authentic IAA by combined gas-chromatography and mass spectrometry (GCMS) (Bridges *et al.*, 1973) (Fig. 5). The *quantitative distribution* of the IAA determined by a GCMS single ion monitoring technique closely paralleled the distribution of the substance found by Greenwood *et al.* (1973) to be growth active. Independent confirmation of the presence of IAA in *Zea* roots has been provided by Elliott and Greenwood (1974) also with GCMS techniques.

In the primary root of *Zea mays* seedlings most of the naturally occurring IAA is thus located in the stele, and it is only in this tissue that IAA undergoes polar transport (Bowen *et al.*, 1972; Shaw and Wilkins, 1974). This transport takes place only towards the root tip (see Scott and Wilkins, 1968; and Wilkins *et al.*, 1972, for list of references relating to this topic). Only a little IAA has been detected in 2–3 mm root tips and none in the root cap (Kundu and Audus, 1974a,b). Extraction of root caps on a large scale has also failed to reveal IAA (Hillman, unpublished). Wilkins and Wain (1974) make no report of growth promoting substances in the root cap of *Zea*.

Physiological studies have clearly established that the geotropic response of *Zea* and *Pisum* roots is brought about by the *basipetal* movement of a growth regulator from the

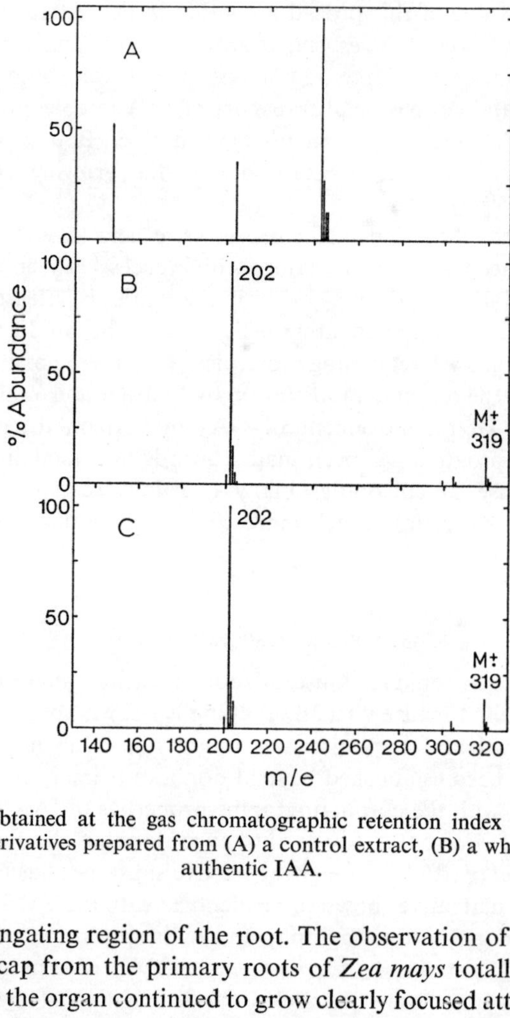

FIG. 5. Mass spectra obtained at the gas chromatographic retention index of the bis-trimethylsilyl derivative of IAA for derivatives prepared from (A) a control extract, (B) a whole root extract and (C) authentic IAA.

root cap into the elongating region of the root. The observation of Juniper *et al.* (1966) that removal of the cap from the primary roots of *Zea mays* totally abolished the geotropic response while the organ continued to grow clearly focused attention upon the root cap as the site of at least a part of the geotropic response mechanism. Gibbons and Wilkins (1970) confirmed this observation and extended it by removing only one half of the root cap. Roots so treated always developed substantial curvature towards the side of the root upon which the remaining half-cap was located regardless of the orientation of the root with respect to gravity. The implication of these results is that at least one inhibitor of cell extension arises from the cap and moves basipetally into the elongating zone of the root. Shaw and Wilkins (1973) made a detailed study of the involvement of the cap in the geotropic response of roots of both *Zea* and *Pisum* which gave essentially similar results. The experiment involved (i) removal of one half of the root caps, and (ii) insertion of minute impermeable barriers in various parts of the root when the cap was intact and when it had been totally removed. Their principal findings may be summarized as follows. Nothing that is in any way involved in the geotropic response of the roots moves *acropetally* into the growing zone. At least one inhibitor of growth arises in the root cap and moves basipetally into the extending zone of the root. A cap inhibitor undergoes downward lateral movement towards the lower side of the geotropically stimulated root. The root tip proper (as distinct from the cap) is not the source of any growth regulating substances when the cap is removed. The results obtained by Shaw and Wilkins (1973) cannot be ascribed to injury effects, because experimental injury did not induce curvature, and Pilet (1973) has shown that half-caps can be detached and re-applied without inducing significant curvature in the roots. Whether or not processes other than lateral movement of a growth inhibitor are involved in the geotropic response of roots has yet to be resolved, but it seems likely that they may be in view of what is known about the response mechanism in coleoptiles. The possibility that there is a differential production, release and longitudinal transport of inhibiting substances in the upper and lower halves of horizontal roots has yet to be investigated.

Physiological evidence for the production of at least one inhibitor in the root cap has recently been supported by extraction and bioassay studies. Those of Kundu and Audus (1974a, b) have revealed the presence of two, possibly three inhibitors, and Wilkins and Wain (1974) have reported similar results. Which of the inhibitors undergoes lateral displacement as a result of geotropic stimulation has not yet been established but the field is now at a stage where substantial advances are being made.

To return, then, to the original question of the validity of the Cholodny–Went theory in relation to roots. The answer, on present evidence, is that *in principle* the theory is correct in that at least one growth regulator produced in the root cap undergoes lateral displacement into the lower half of the horizontal organ, and thereby elicits different rates of growth in the upper and lower halves of the root. In detail, however, present evidence suggests that, in *Zea*, the growth regulator is not IAA since IAA cannot be detected in the root cap. Longitudinal transport of IAA is polarized in the wrong direction for the auxin to move basipetally in the tissue, unless polarization of the transport system is reversed in the extreme 0.5 mm of the root. In this connection, it is pertinent that we have been unable to find any lateral transport of IAA in *Zea* roots by means of micro-injection procedures (Shaw and Wilkins, unpublished).

Another relevant question is whether or not IAA has any regulatory influence on cell extension in roots when applied in low concentrations. I believe this to be an open question at the present time. Experiments in which roots, or worse root segments, are immersed in hormone solutions are virtually impossible to interpret and of little value except in metabolic studies. Uptake and accumulation of IAA, and its rapid metabolism to numerous products which are *different in different root tissues* (Greenwood *et al.*, 1973) are processes which render the evaluation of most such experiments very difficult indeed from the standpoint of growth regulation. The approach of immersing plant organs in solutions of hormones is one which is frequently used, because of convenience, but is one which I believe should be heavily criticized on two counts, first, because neither the *amount* of hormone entering the tissue nor the location of the hormone in the tissue is known, and secondly, because *plants patently do not normally derive their hormone* supply from the surrounding environment. Plants possess sophisticated control mechanisms for the synthesis, transport and metabolism of hormones which are often localized in specific tissues of an organ (Bridges *et al.*, 1973), and perhaps in specific organelles within particular cells (Railton *et al.*, 1974). We are unlikely to advance our understanding of the mechanisms controlling growth and development by indiscriminate external application of massive and unknown doses of hormones to the entire surfaces of roots, shoots or both either by immersion or spraying techniques! To emphasize the point, just how much attention would be paid to an endocrinologist who immersed his experimental animals in a hormone solution for 5 sec to test its activity? For these reasons Shaw and I are attempting to develop a microapplication technique to permit highly localized application of very small amounts of labelled hormones to intact organs. We believe that the development and exploitation of such a technique will lead to experimental results that will provide a firmer basis for elucidation of the hormonal mechanisms controlling the natural regulation of growth and development in intact plants.

REFERENCES

BERGER, J. and AVERY, G. S. (1944) Isolation of an auxin precursor and an auxin (indoleacetic acid) from maize. *Am. J. Bot.* **31,** 199–203.

BOWEN, M. R., WILKINS, M. B., CANE, A. R. and MCCORQUODALE, I. (1972) Auxin transport in roots. VIII. The distribution of radioactivity in the tissues of *Zea* root segments. *Planta* **105,** 273–292.

BRAUNER, L. and APPEL, E. (1960) Zur Problem der Wuchsstoff-Querverschiebung bei der geotropischen Induktion. *Planta* **55,** 226–234.

BRIDGES, I. G. and WILKINS, M. B. (1971) The effects of electrolyte and non-electrolyte solutions on the tropic responses of *Avena* coleoptiles. *J. exp. Bot.* **22,** 208–212.

BRIDGES, I. G., HILLMAN, J. R. and WILKINS, M. B. (1973) Identification and localization of auxin in primary roots of *Zea mays* by mass spectrometry. *Planta* **115,** 189–192.

CANE, A. R. and WILKINS, M. B. (1969) Independence of lateral and differential longitudinal movement of indoleacetic acid in geotropically stimulated coleoptiles of *Zea mays. Pl. Physiol. Lancaster* **44,** 1481–1487.

CHOLODNY, N. (1926) Beiträge zur Analyse der geotropischen Reaktion. *Jb. wiss. Bot.* **65,** 447–459.

DOLK, H. E. (1930) Geotropic en groeistof. Dissertation of the University of Utrecht. (English translation in *Recl. Trav. bot. néerl.* **33,** 509–585, 1936).

ELLIOTT, M. C. and GREENWOOD, M. S. (1974) Indol-3yl-acetic acid in roots of *Zea mays. Phytochemistry* **13,** 239–241.

GIBBONS, G. S. B. and WILKINS, M. B. (1970) Growth inhibitor production by root caps in relation to geotropic responses. *Nature Lond.* **226,** 558–559.

GOLDSMITH, M. H. M. and WILKINS, M. B. (1964) Movement of auxin in coleoptiles of *Zea mays* L. during geotropic stimulation. *Pl. Physiol. Lancaster* **39,** 151–162.

GRAHM, L. and HERTZ, C. R. (1962) Measurement of geoelectric effect in coleoptiles by a new technique. *Physiologia Pl.* **15,** 96–114.

GRAHM, L. and HERTZ, C. R. (1964) Measurement of the geoelectric effect in coleoptiles. *Physiologia Pl.* **17,** 231–261.

GREENWOOD, M. S., SHAW, S., HILLMAN, J. R., RICHIE, A. and WILKINS, M. B. (1972) Identification of auxin from *Zea* coleoptile tips by mass spectrometry. *Planta* **108,** 179–183.

GREENWOOD, M. S., HILLMAN, J. R., SHAW, S. and WILKINS, M. B. (1973) Localization and identification of auxin in roots of *Zea mays. Planta* **109,** 369–374.

HAAGEN-SMIT, A. J., LEECH, W. D. and BERGREN, W. R. (1942) The estimation, isolation and identification of auxins in plant materials. *Am. J. Bot.* **29,** 500–506.

HAAGEN-SMIT, A. J., DANDLIKER, W. B., WITTWER, S. H. and MURNEEK, A. E. (1946) Isolation of 3-indoleacetic acid from immature corn kernels. *Am. J. Bot.* **33,** 118–120.

HALL, S. M. and MEDLOW, G. C. (1974) Identification of IAA in phloem and root pressure saps of *Ricinus communis* L. by mass spectrometry. *Planta* (In Press).

JUNIPER, B. E., GROVES, S., LANDAU-SCHACHAR, B. and AUDUS, L. J. (1966) Root cap and the perception of gravity. *Nature Lond.* **209,** 93–94.

KUNDU, K. K. and AUDUS, L. J. (1974a) Root-growth inhibitors from root tips of *Zea mays* L. *Planta* **117,** 183–186.

KUNDU, K. K. and AUDUS, L. J. (1974b) Root growth inhibitors from root cap and root meristem of *Zea mays* L. *J. exp. Bot.* **25,** 479–489.

NAQVI, S. M. and GORDON, S. A. (1966) Auxin transport in *Zea mays* L. coleoptiles. I. Influence of gravity on the transport of indoleacetic acid-2-^{14}C. *Pl. Physiol. Lancaster* **41,** 1113–1118.

PILET, P. E. (1973) Growth inhibitor from the root cap of *Zea mays. Planta* **111,** 275–278.

RAILTON, I. D., REID, D. M., GASKIN, P. and MACMILLAN, J. (1974) Characterization of abscisic acid in chloroplasts of *Pisum sativum* L. cv. Alaska by combined gas chromatography and mass spectrometry. *Planta* **117,** 179–182.

SCOTT, T. K. and WILKINS, M. B. (1968) Auxin transport in roots. II. Polar flux of IAA in *Zea* roots. *Planta* **83,** 323–334.

SHAW, S. and WILKINS, M. B. (1973) The source and lateral transport of growth inhibitors in geotropically stimulated roots of *Zea mays* and *Pisum sativum. Planta* **109,** 11–26.

SHAW, S. and WILKINS, M. B. (1974) Auxin transport in roots. X. Relative movement of radioactivity from IAA in the stele and cortex of *Zea* root segments. *J. exp. Bot.* **25,** 199–207.

SHAW, S., GARDNER, G. and WILKINS, M. B. (1973) The lateral transport of IAA in intact coleoptiles of *Avena sativa* L. and *Zea mays* L. during geotropic stimulation. *Planta* **115,** 97–111.

WENT, F. W. (1926) On growth accelerating substances in the coleoptile of *Avena sativa. Proc. K. ned. Akad. Wet.* **30,** 10–19.

WENT, F. W. (1928) Wuchstoff und Wachstum. *Recl. Trav. bot. néerl.* **25,** 1–116.

WENT, F. W. (1974) Reflections and speculations. *A. Rev. Pl. Physiol.* **25,** 1–26.

WHITE, J. C., MEDLOW, G. C., HILLMAN, J. R. and WILKINS, M. B. (1975) Correlative inhibition of lateral bud growth in *Phaseolus vulgaris* L. Isolation of indoleacetic acid from the inhibitory region. *Planta* (in press).

WILKINS, M. B. and WHYTE, P. (1968) Relationship between metabolism and the lateral transport of IAA in corn coleoptiles. *Pl. Physiol. Lancaster* **43,** 1435–1442.

WILKINS, M. B., CANE, A. R. and MCCORQUODALE, I. (1972) Auxin transport in roots. IX. Movement, export, resorption and loss of radioactivity from IAA by *Zea* root segments. *Planta* **106,** 291–310.

WILKINS, M. B. and WOODCOCK, A. E. R. (1965) The origin of the geoelectric effect in plants. *Nature Lond.* **208,** 990–992.

WILKINS, H. and WAIN, R. L. (1974) The root cap and control of root elongation in *Zea mays* L. seedlings exposed to white light. *Planta* (in press).

WOODCOCK, A. E. R. and WILKINS, M. B. (1969a) The geoelectric effect in plant shoots. I. The characteristics of the effect. *J. exp. Bot.* **20,** 156–169.

WOODCOCK, A. E. R. and WILKINS, M. B. (1969b) The geoelectric effect in plant shoots. II. Sensitivity of concentration 'chain' electrodes to reorientation. *J. exp. Bot.* **20,** 687–697.

WOODCOCK, A. E. R. and WILKINS, M. B. (1970) The geoelectric effect in plant shoots. III. Dependence upon auxin concentration gradients. *J. exp. Bot.* **21,** 985–996.

WOODCOCK, A. E. R. and WILKINS, M. B. (1971) The geoelectric effect in plant shoots. IV. Interrelationship between growth, auxin concentration and electrical potentials in *Zea* coleoptiles. *J. exp. Bot.* **22,** 512–525.

CONTROL OF CELL SHAPE AND CELL SIZE BY THE DUAL REGULATION OF AUXIN AND ETHYLENE

DAPHNE J. OSBORNE

Agricultural Research Council Unit of Development Botany, 181A Huntingdon Road, Cambridge, U.K.

INTRODUCTION

This paper describes a model for the control of cell growth based on opposing actions of the two hormones, auxin and ethylene. The scheme in no way excludes involvement of other hormones, nor does it imply that the quantitative relationships presented here will be similar in all situations, but it does propose a central role for auxin and ethylene in determining the eventual shape and size that a cell may achieve.

The model has been developed from physiological and biochemical investigations of the rate, extent and orientation of cell growth that follows external application of auxin and ethylene to shoots of intact, etiolated pea plants. It is proposed that such a model could be operative during the course of normal growth and development, and that the variety of cell shapes and sizes produced in expanding and elongating shoots could be controlled by endogenous levels of auxin and ethylene.

Two facts are central to the model. The first, established by Morgan and Hall (1962, 1964) in cotton, is that production of ethylene by a tissue is closely linked to the level of auxin *supplied* to it: the higher the auxin level, the higher is the rate of ethylene production. Later this was extended to relate *endogenous* levels of auxin to basal rates of ethylene production by tissues (Burg and Burg, 1968; Kang *et al.*, 1971). The second central fact is that handling, tactile stimulus, pressure or wounding of a plant at once induces a short period of enhanced ethylene production. This is the 'wound' ethylene response (Goeschl *et al.*, 1966).

In etiolated pea plants, ethylene production is highest in those regions that are highest in endogenous auxin, i.e. the hook and the bud (Burg and Burg, 1968). For the model, it is envisaged that auxin transported from the apex determines the 'basal' level of ethylene production of cells through which it passes. Further ethylene may be produced by cells above this 'basal' level when any kind of contact or internal pressures are exerted within the tissues. This can be considered as *endogenously* induced 'wound' ethylene production. Certainly, if growing tissues make contact with any kind of mechanical impedance [e.g., when roots of young shoots grow through the soil or when pea seedlings are grown through glass beads or press up against the cover of their growth chamber (Goeschl *et al.*, 1966)], then *exogenously* induced 'wound' ethylene is produced. Subsequent growth of the shoot then exhibits characteristics normally associated with those of a plant that has been exposed to ethylene treatment.

It has become clear from our studies that plants can possess cells with very different sensitivities and responses to auxin and ethylene. The juxtaposition of these different cells within defined tissues represent precise examples of positional differentiation, and one example will be outlined for abscission zones of *Phaseolus vulgaris* (Wright and Osborne, 1974).

STUDIES WITH INTACT PEA PLANTS

The peas (either Alaska or Meteor) are planted in sand and grown in darkness for a period of 6 days, seedlings receiving green light during watering. By then, the second internode is fully expanded and the third is about 1.5 cm long. Two spots of a lanolin–charcoal mixture are placed 1 cm apart on the shoot, the upper spot being just below the hook. These marks span the most rapidly elongating tissues of the shoot.

When shoots of 6-day-old etiolated plants are given a single spray of an aqueous solution of the sodium salt of indol-3yl-acetic acid (IAA) at 10^{-3} M or are enclosed in 10–15 l glass or perspex containers with ethylene (50 μl/l), differences in growth are clearly seen between treated plants and controls within 24 hr (Fig. 1). After 3 or 6 days, when cell expansion in the marked areas is complete, both auxin-and ethylene treated plants are stunted and at first sight they appear similarly swollen in the marked region. Critical investigation shows, however, that auxin-induced and ethylene-induced swellings are quite distinct.

The effects of auxin and ethylene on elongation growth are both rapid; for pea segments exposed to solutions of IAA at a concentration of 2.8×10^{-5} M, elongation is enhanced in about 7 min (Uhrström, 1974) and for intact pea plants exposed to 50 μl/l ethylene elongation growth is also arrested within 5–6 min (A.V. Atack, unpublished).

From changes that take place in the length and breadth of cortical cells in a precisely defined region of the shoot immediately below the upper of the two lanolin–charcoal marks, it is possible to determine the kinetics of cell expansion until mature cell size. This has been done after the shoots are treated with either a single spray of IAA solution (10^{-5}–10^{-3} M) or with ethylene (1–50 μl/l). At intervals, from 0 to 72 hr, longitudinal sections of the marked region of the shoot have been cut, photographed under the light microscope, and cell length and cell breadth determined from the photographs. In this way, the pattern of cell enlargement of many hundreds of control cells has been compared with that of cells receiving enhanced levels of either auxin or ethylene.

Values for the rate, extent and orientation of cell enlargement at the highest concentrations of IAA (10^{-3} M) and ethylene (50 μl/l.) are presented diagrammatically in Fig. 2. In control and ethylene treatments, cell volumes at 72 hr are 30×10^3 and 25×10^3 μm^3 respectively, but control cells are nearly twice the length of those in ethylene. Cells of ethylene treated plants expand more slowly, but at final cell size their volumes do not differ significantly from those of the controls (see also Ridge and Osborne, 1969, using different techniques). After the IAA treatment, however, cell volume at 9 hr is 100 per cent greater than that of the controls, and enhanced expansion is predominantly in the longitudinal direction. As cortical cells of auxin treated plants continue to enlarge (at a reduced rate), the orientation of expansion is altered to predominantly the lateral direction, so that although the final cell volume is still nearly 50 per cent greater than in

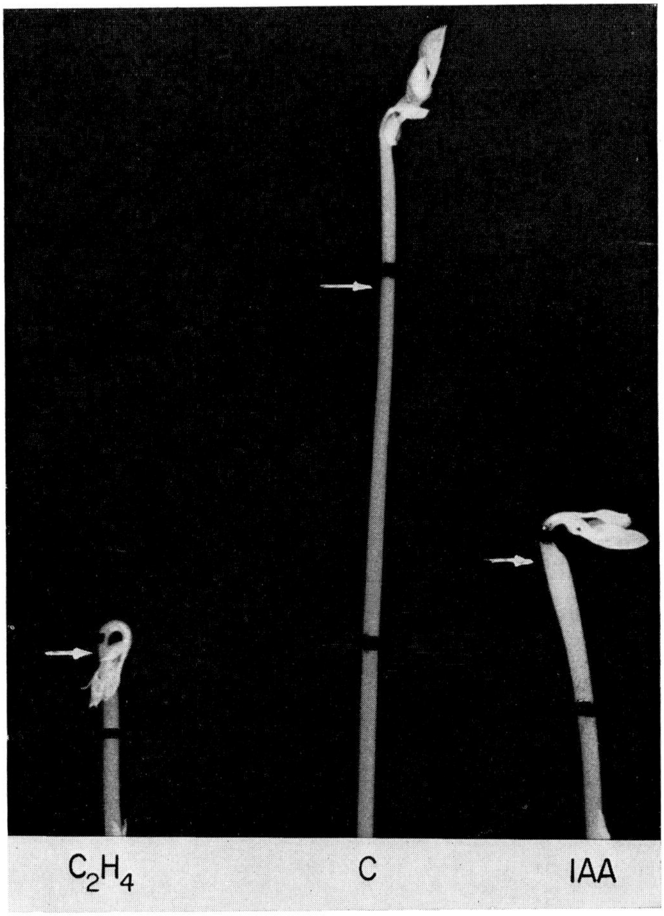

FIG. 1. Shoots of etiolated pea plants (var. Alaska) (6 days old) after 24 hr treatment in ethylene (100 μl/l.) or a single spray of IAA solution (10^{-3} M). Charcoal bands indicate the growth of a 1 cm marked region and arrows indicate the position from which tissues were sampled for cell size determinations and for electron microscopy.

Facing p. 90

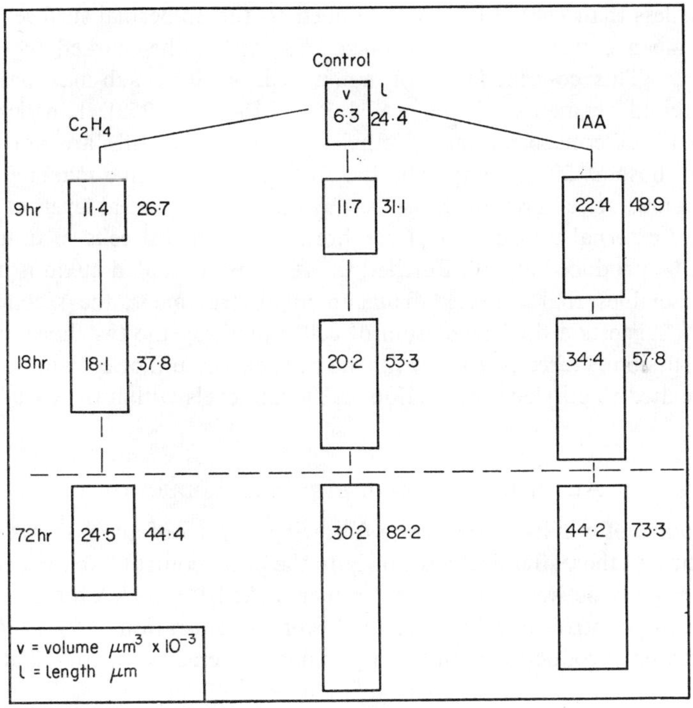

FIG. 2. Kinetics of growth in cell length and cell volume in the 'arrowed' position (Fig. 1) following continuous treatment of etiolated pea plants (var. Meteor) with ethylene (50 μl/l.) or a single spray of IAA solution (10^{-3} M). Cell lengths and volumes IAA vs. controls, and cell lengths ethylene vs. controls, are significantly different at 95 per cent confidence limits (Data from Atack and Osborne, in Osborne, 1974).

the controls (44×10^3 μm³ compared with 30×10^3 μm³) final cell length is less (73 compared with 82 μm).

There are, therefore, two distinct phases recognizable in the growth response to applied auxin. The first is distinguishable as a stimulation in the rates of both expansion and elongation, the second is characterized by a slower rate of expansion compared with controls, but with an enhanced lateral growth component. The change between these two phases occurs between 9 and 18 hr. The first phase has characteristics similar to those for auxin-induced cell expansion in excised segments of the pea shoot taken from the same marked position; the second phase has characteristics of expansion of such segments in ethylene alone (Burg and Burg, 1968).

When lower concentrations of auxin (IAA 10^{-4} and 10^{-5} M) are supplied, smaller but longer cells develop; *final* cell lengths and volumes are respectively 76 μm and 40×10^3 μm³ and 80 μm and 34×10^3 μm³ compared with 82 μm and 30×10^3 μm³ for controls. In fact, the lowest concentration of auxin tested (10^{-5} M) causes little modification in final cell shape and size (differences between control and IAA 10^{-5} M being non-significant at 95 per cent confidence limits). This is important because using [¹⁴C]IAA, experiments showed that the amount of IAA that will penetrate the epidermis from a single spray is less than 1/40 of the IAA supplied to the epidermal surface. For IAA at 10^{-3} M, the mean concentration of applied IAA within the marked region does not exceed 10^{-6} M. This concentration of auxin will elicit a sub-maximal elongation growth in excised segments of the pea shoot (Osborne, 1958). Kinetics of growth expansion in lower concentrations of ethylene (10 and 1 μl/l) are not significantly different from those at 50 μl/l, implying that all three levels elicit a maximal response.

This time course of cell growth shows that by modifying internal levels of either auxin or ethylene by external application of the hormones, cortical cells of different shapes and sizes can be produced at will. Further, the response to added auxin is biphasic; the first phase resembles that expected from an auxin treatment, the second that from ethylene. This suggests a dual regulation of cell growth by the two hormones, and the following discussion is directed to how far the rate, extent and orientation of cell expansion can be related to ethylene production and auxin levels within the tissue.

AUXIN REGULATION OF ETHYLENE PRODUCTION

The time course of ethylene production following a spray of auxin shows a maximum between 5–9 hr and thereafter declines slowly to the basal control level. Exact timing and maxima vary slightly between different experiments. At 10^{-3} M IAA there is a thirty-fold increase over the control at 7 hr (Fig. 3); lower concentrations of IAA (not shown) induce less ethylene production with lower maxima. Whereas with 10^{-3} and 10^{-4} M, the

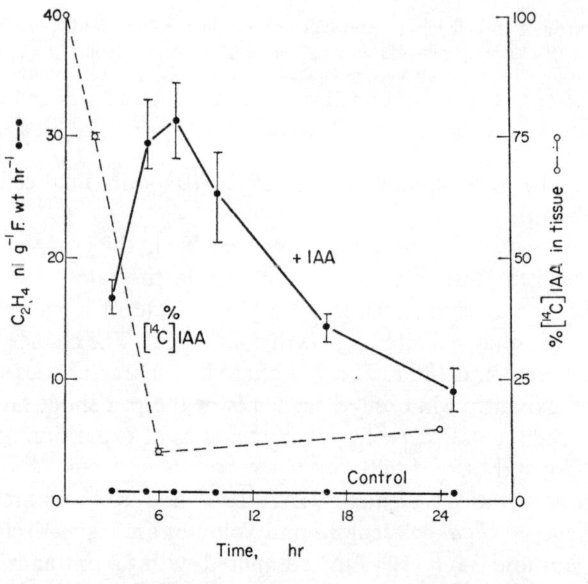

FIG. 3. Time course of ethylene production (—) by 10 cm 'apical' regions of etiolated pea shoots (var. Meteor) given a single spray with water (control) or IAA solution (10^{-3} M), and the percentage of ^{14}C remaining as [^{14}C]IAA (– – –) in tissue of the marked segment (Fig. 1) following decapitation and subsequent application of 5 μl [^{14}C]IAA (4.5 × 10^{-4} M) to the stump 30 min after a non-radioactive spray of IAA. For total levels of ^{14}C in the marked segment see Fig. 4. (Atack and Osborne, unpublished.)

elevated ethylene production continues for longer than 24 hr, with 10^{-5} M the stimulation is less than two-fold at maximum and returns to the control level within 9 hr.

Since so much ethylene is produced by auxin treated tissue in the first 9 hr after a spray, one must ask why the pattern of cell growth for this period (Fig. 2) represents a growth enhancement rather than a growth inhibition. An explanation becomes apparent from a study of the levels of the applied IAA that are present in the tissue after different lengths of time.

METABOLISM OF APPLIED IAA

The rate of metabolism of applied IAA within intact tissues has been determined under conditions comparable with those of the experiments already described.

Whole plants are first sprayed with IAA (10^{-3} M) as before, and when dry (30 min) the bud is removed and a 5 μl drop of 2-[^{14}C]IAA at 4.5 $\times$ 10^{-4} M applied to the cut surface. Some of this IAA is transported basipetally to cells below in the marked region of the shoot (Fig. 1), and in all tissues it is converted to a number of metabolites. A balance sheet of the total counts present above, below and in, the marked region, at different times, shows that 80 per cent of the radioactivity present in the marked region at 24 hr is already transported to that tissue by 4.5 hr. The proportion of this radioactivity that represents [^{14}C]IAA in the marked region at the different times was determined in the following way. The segments were extracted in methanol at -20°C, concentrated and chromatographed on thin layer plates of silica gel G in isobutanol:methanol:water (80:5:15). Results show that 2 hr after application of the radioactive drop to the apex 75 per cent of counts present represent extractible [^{14}C]IAA. By 6 hr, only 12 per cent can be recovered as [^{14}C]IAA, and at 24 hr the value is 15 per cent (Fig. 3). This rate of metabolism resembles that reported for *Phaseolus* shoots by Patrick and Woolley (1973).

In other experiments designed to check the level of [^{14}C]IAA in its freely diffusible form, the marked segments have been excised after different intervals and receiver agar blocks attached to both the apical and the basal ends for a period of 3 hr. In samples taken later than 7.5 hr after application of the drop of [^{14}C]IAA (Fig. 4), no radioactivity was found in receiver blocks above that recovered in blocks placed in contact with either end of the radioactive segments for 5 sec only (i.e. 'contact' contamination control blocks). Radioactivity transported to receiver blocks before 7.5 hr has been identified as [^{14}C]IAA by extraction and co-chromatography with authentic IAA.

Conversion of applied IAA to other products is therefore rapid since little [^{14}C]IAA remains unmetabolized after 7–9 hr. However, auxin-enhanced ethylene production continues long after the [^{14}C]IAA level has fallen to nearly zero. We can envisage therefore that at all times after 9 hr, enlarging cells of auxin treated pea shoots are developing under the influence of additional ethylene. Before 9 hr (0–6 hr probably) cells are primarily responding to the enhanced level of IAA; hence the biphasic pattern of growth that follows the treatment with auxin. As long as a sufficiently high level of auxin remains in the tissues, the cells either do not recognize or show no response to the levels of ethylene produced. Only when the auxin level falls below a critical value for that cell does the ethylene begin to elicit an ethylene-type growth response. This concept

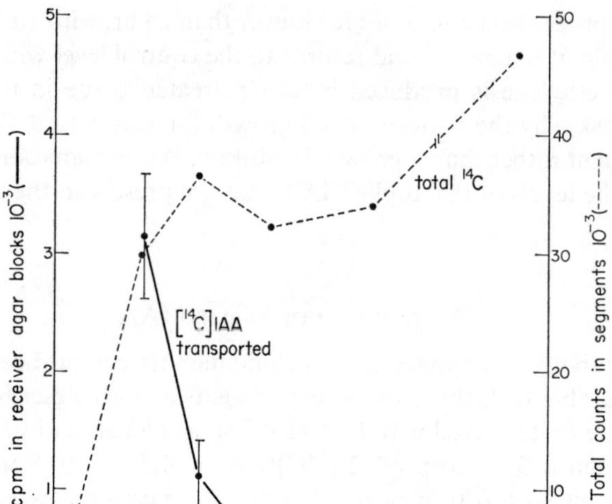

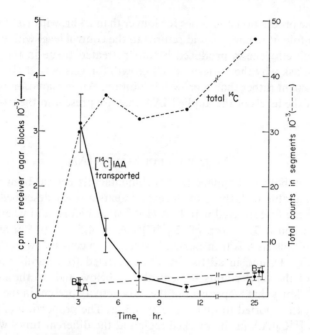

FIG. 4. Time course of diffusible [^{14}C]IAA from the marked segment of etiolated pea shoots (var. Meteor) transported during 3 hr into basally applied agar receiver blocks (no radioactivity transported into apically applied receivers). Counts in apical or basal agar blocks in contact with the segment for 5 sec only are denoted as A and B. Also shown are total counts of ^{14}C in the marked segment following decapitation and subsequent application of [^{14}C]IAA (4.5×10^{-4} M) to the stump 30 min after a non-radioactive spray of IAA (Atack and Osborne, unpublished).

of a dual regulation by auxin and ethylene is the basis of the present model for the control of cell growth.

NATURE OF THE AUXIN AND ETHYLENE RESPONSES

Why should the rate, extent and orientation of cell growth be altered by auxin and ethylene? Clearly, a cell enlarges if the water potential increases, and the water potential can increase by virtue of (a) an increased osmotic concentration within the cell or (b) a relaxation of the cell wall resulting from a lowering of the restriction to expansion. With respect to the former possibility, cell growth is not dependent upon an increase in internal osmotic concentration as was shown many years ago by Ordin *et al.* (1956); expansion must therefore depend upon an increased extensibility of the wall. Burström *et al.* (1970) pointed out that a decrease in the elastic modulus of the wall could result from an increase in permeability of a cell to water as well as to an alteration in the properties of the cell wall itself. However, an increase in the permeability of pea segments to tritiated water following auxin treatment has not been demonstrated (Dowler *et al.*, 1974) and as the permeability of cells to water is always high, it is doubtful whether this factor can limit sustained growth. Extension of a cell must therefore involve changes in the physical and chemical properties of its wall.

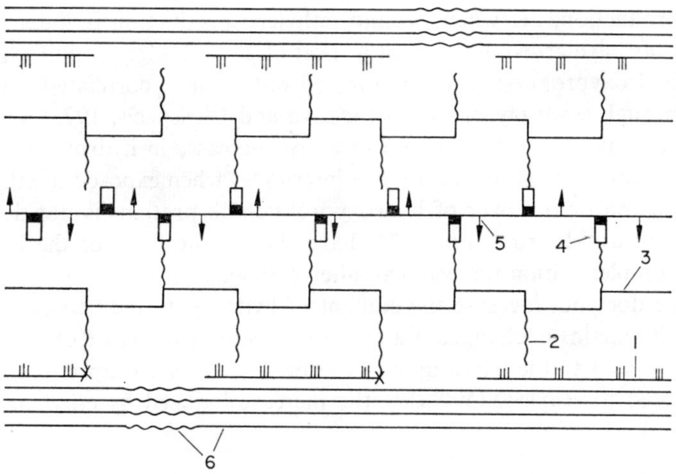

FIG. 5. Model of the higher plant cell wall (reproduced from Keegstra *et al.*, 1973). 1. Hemicellulose (xyloglucan). 2. Pectin (arabinan-galactan side chain). 3. Pectin (rhamnogalacturon main chain). 4. Protein (arabino-galactosyl serine). 5. Protein (hydroxyproline-*O*-glycosidyl arabinoside). 6. Cellulose (crystalline and amorphous fibril).

THE CELL WALL

Cell walls comprise a complex of microfibrils of long chain polymers of α-cellulose with alternating crystalline and amorphous regions, orientated in a precise manner within a matrix of pectic and hemicellulosic polysaccharides and protein: the composition of a mature wall is of the order of 50 per cent cellulose, 5 per cent pectin, 25 per cent hemicellulose and 20 per cent protein. Minor constituents such as ions (Ca^{2+}, Mg^{2+} and K^{+}) and phenolic compounds (e.g. ferulic acid) may be important in determining secondary characteristics. Cellulose microfibrils can be considered as structural reinforcement of the wall and the protein–polysaccharide matrix as the component which determines the major constraints on extensibility. In the model of a plant cell wall (Fig. 5) proposed by Keegstra *et al.* (1973) two loci may be pinpointed as being of particular potential importance in constraining extension. The first is a proposed hydrogen bonding of hemicellulose to cellulose [Fig. 5(1)], and the second is the crosslinking of pectins to hydroxylated amino acids, e.g. serine (4), or hydroxyproline (5) within the matrix protein. Although the precise nature of these latter covalent bondings is still uncertain, the extent of crosslinking within the wall is considered by both Lamport (1970) and Keegstra *et al.* (1973) to constitute a constraint to extension. Preclusion of the formation of these linkages, or their subsequent breakage should therefore lead to enhanced wall lossening and an enhanced rate and extent of cell expansion.

It has long been known that the content of hydroxyproline-rich protein in walls of pea shoots increases as the tissue becomes fully expanded (Cleland and Karlsnes, 1967). Also, exposing plants to ethylene increases the level of hydroxyproline in the walls while markedly reducing the rate of cell elongation (Ridge and Osborne, 1970). These increases are concentration-dependent and are detectable after treatment with as little as 0.1 μl/l ethylene in CO_2-free air. An increase can also be detected following an *in vivo* increase in ethylene production, i.e. when 'wound' ethylene production is induced by pressure when pea shoots have grown up against a sheet of glass (Ridge, unpublished). Although increases in hydroxyproline-rich protein in cell walls can be correlated with cessation of elongation through a variety of causes (Sadava and Chrispeels, 1973) an association is not always clear (Winter *et al.*, 1971). Certainly, increases in hydroxyproline also occur in the walls of non-elongating mature pea internodes when exposed to ethylene, indicating that deposition or turnover of hydroxyproline-rich proteins is not dependent upon cell growth. But as Monro *et al.* (1974) have shown, the form of the hydroxyproline-rich protein complex within the wall can alter with age.

Auxin alone does not increase the content of hydroxyproline-rich protein in pea cell walls; the level remains unchanged if a continuous supply of IAA (0.5 per cent in petroleum jelly) is applied to the shoot apex. Further, if this IAA is applied to plants before they are exposed to ethylene (500 μl/l), the increase induced by ethylene is suppressed (Osborne, 1973).

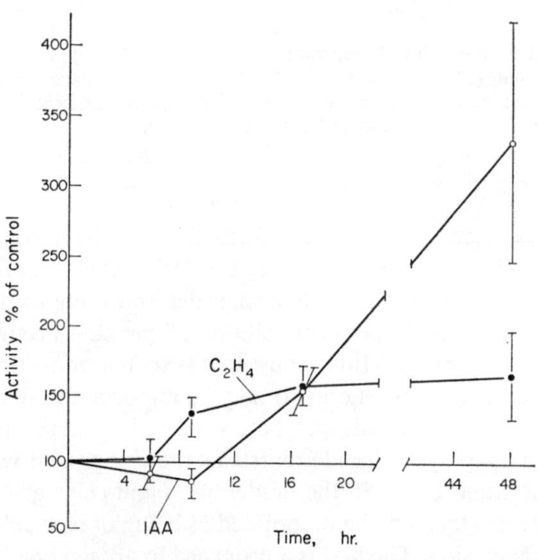

FIG. 6. Cytoplasmic peroxidase activity expressed as percentage of control values in marked segments of etiolated pea shoots (var. Meteor) following continuous treatment in ethylene (50 μl/l) or a single spray of IAA solution (10^{-3} M). For methods see Ridge and Osborne (1970). (Atack and Osborne, unpublished.)

The rise in hydroxyproline content in pea cell walls is positively correlated with an increase in activity of cytoplasmic and wall peroxidases (Ridge and Osborne, 1971). A time course of this activity in the marked region following ethylene and auxin treatment of whole plants is shown in Fig. 6. Ethylene increases peroxidase activity at all times. In contrast, the response to applied auxin is biphasic, with an initial decrease followed by a rise, the change occurring at approximately 9 hr. If IAA (0.5 per cent) is applied to the shoot apex before plants are given ethylene (500 μl/l), the three-fold increase in peroxidase activity induced by ethylene alone is completely suppressed (Osborne, 1973). The concept that auxin can 'buffer' cells against the effects of ethylene is now developed further.

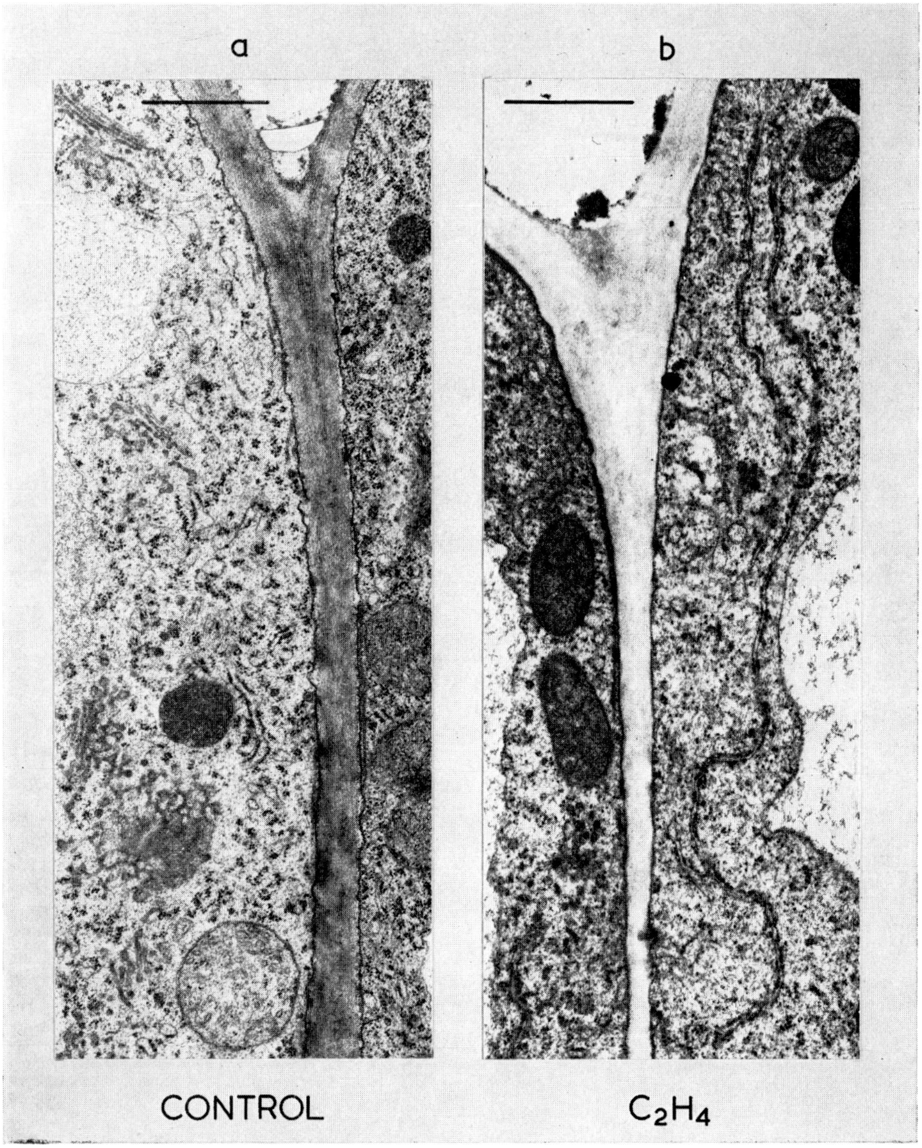

FIG. 7. Ultrastructural differences in the endoplasmic reticulum of cortical cells in the marked region (Fig. 1) of etiolated pea shoots (var. Alaska). Intact plants in air (control) or ethylene 10 μl/l for 18 hr (Sargent, unpublished).

Ultrastructural changes

During the growth responses induced in pea shoots by ethylene, major structural changes occur in (a) cell walls and (b) the endoplasmic reticulum.

Cell walls

In cortical cells, ethylene causes a 30 per cent increase in the thickness of the walls within 24 hr, and a 250 per cent increase in 4 days. After IAA treatment no increase is at first observed, but from 24 hr wall thickening slowly takes place and by 6 days reaches the value attained in ethylene at 24 hr (Sargent *et al.*, 1973). More precise measurements of changes in thickness can be made on the outer walls of epidermal cells in which significant ethylene-enhanced thickening (40 per cent) is detectable after 18 hr though not before (Sargent *et al.*, 1974). With auxin, however, the cuticle of these epidermal cells appears stretched and from 6 to 12 hr the wall becomes thinner. Subsequently the wall increases in thickness at a rate comparable with that in ethylene alone and by 24 hr is nearly 40 per cent thicker than in the controls. These results suggest that relaxation and stretching of the wall, and enhanced rate of cell expansion, both occur for only as long as some of the applied auxin remains in the tissues as unmetabolized IAA (see Figs. 2, 3, 4). A second, and wall-thickening, phase is the response to auxin-induced ethylene production and is reflected in the declining rate of elongation and enhanced lateral cell expansion.

Membranes

An early effect of ethylene in cells of the pea shoot is the development of elongated profiles of endoplasmic reticulum (e.r.). This is statistically significant by 6 hr (Sargent and Osborne, 1975) and becomes more extensive with time; it is clearly visible by 18 hr (Fig. 7). No such changes are discernible in an IAA-treated plant at 12 hr though some enlargement of the e.r. is visible at 18 hr. Involvement of both e.r. and golgi in secretion of cell wall proteins and polysaccharides has been implicated by a number of studies including those of Mühlethaler (1967), Chrispeels (1970) and Northcote (1972). Evidence suggests that peptides synthesized on polysomes of rough e.r. are hydroxylated and subsequently transported within the lumen of the reticulum, later to be secreted in vesicles of the smooth reticulum. Glycosylation may then occur in association with golgi vesicles, glycoproteins being liberated to the wall on fusion of vesicle membranes with the plasmalemma.

As part of the study of membrane changes induced by ethylene, the rate of incorporation of [^{14}C]glycerol into total phospholipids has been determined in the elongating region of etiolated pea epicotyls. Exposure of plants to 10 μl/l ethylene for 2.5–3 hr reduces the level of incorporation by subsequently excised segments by as much as 30–50 per cent (Fig. 8a) during a 45 min incubation in [^{14}C]glycerol at all concentrations up to pool saturation (Irvine and Osborne, 1973). Similar decreases are observed when [^{14}C]choline is the precursor.

Reduction in incorporation into phospholipids occurs at very low levels of ethylene and can be measured 2.5 hr after the burst of endogenous 'wound' ethylene that follows

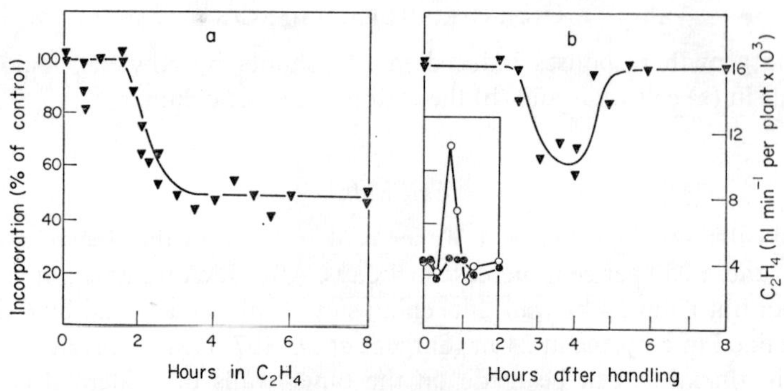

FIG. 8. Incorporation of [^{14}C]glycerol into phospholipids after (a) treatment of intact etiolated pea plants (var. Meteor) with ethylene (10 μl/l), and (b) handling of plants to induce 'wound' ethylene. Data from Irvine and Osborne (1973).

the normal laboratory handling of plants (Fig. 8b). No reduction in incorporation of [^{14}C]glycerol is detectable at these times in peas sprayed with IAA (10^{-3} M) so that, as with other effects reported here, reduction attributable to ethylene can be precluded by an auxin treatment.

If synthesis and secretion of hydroxyproline-rich protein and the composition of the cell wall matrix are associated with the ethylene induced changes in membrane phospholipids, then the changes are presumably associated with long term cell expansion rather than with the cessation of growth that occurs after 6 min. Some other function of ethylene must perforce determine the rapid inhibition of elongation in the pea shoot.

Rapid cell extension responses induced by auxin are associated in many tissues with an equally rapid extrusion of protons through the plasmalemma to the external solution (Evans, 1974). By inserting microelectrodes into cortical cells of pea segments, Marrè *et al.* (1974) measured an increased negative transmembrane potential associated with enhanced growth and a decrease in the pH of the medium, 60 min after treatment with IAA. The plasmalemma has been implicated as a site of specific proteins having a high affinity for auxin molecules (Hertel *et al.*, 1972), and Hager *et al.* (1971) propose that a membrane-bound Ca^{2+}-dependent ATPase may function as the auxin-activated proton pump. An attraction of protons (the so-called *second messengers*) in regulating rapid cell wall extensibility lies in their possible function in the weakening of hydrogen-bonding between xylan-glucan hemicelluloses and cellulose microfibrils (crosslinking (1), Fig. 5). However, doubt concerning the direct effect of protons in the initial auxin responses has been cast by Ilan (1973). No involvement of ethylene in proton fluxes has yet been reported, but the possibility of active binding sites for both IAA and ethylene at the plasma membrane cannot be excluded.

ORIENTATION OF CELL GROWTH

Probine (1964), following the work of Castle and van Iterson (quoted in Probine, 1964), pointed out that the orientation of deposition of cellulose microfibrils in the wall could determine the direction of expansion of a cell: transverse deposition restricting expansion in the lateral direction and vice-versa. Using polarizing optics, Veen (1970) showed that concentrations of IAA that enhance elongation growth of pea segments (10^{-5} M) cause deposition of transverse microfibrils at the inner surface of the wall. Higher concentrations (10^{-3} M) in the presence of 2 per cent sucrose (which result in considerable ethylene production and lateral swelling of stem segments) lead to deposition of longitudinal microfibrils. Following this, Ridge (1973) has shown that fibrils are deposited longitudinally within 3–6 hr in plants treated with ethylene (1–10 μl/l) and this can again be correlated with lateral cell expansion. From electron micrographs, Sargent *et al.* (1974) have confirmed that thickening of cell walls seen at 18 hr in laterally-expanding ethylene-treated peas comprises an inner layer of longitudinally deposited microfibrils, and have shown that this longitudinal deposition is absent from the corresponding auxin treatment and controls.

A model for control of cell shape and size

From the information discussed so far it may be concluded (a) that auxin and ethylene regulate cell expansion in opposing ways and (b) that if the concentration of auxin within the tissue remains sufficiently high the cell will be insensitive or unresponsive to ethylene up to, but not above, a certain threshold level. In other words, the presence of auxin 'buffers' the cell against ethylene action, but the cell responds to levels of ethylene that are above this 'buffered' threshold.

In an intact plant it is proposed that an expanding cell at the shoot apex is nicely balanced with respect to the ethylene it produces via endogenous auxin levels, but any additional ethylene will be perceived and evoke an ethylene-type response. It is presumed that ethylene moves freely in all directions (Fig. 9), but auxin (particularly in elongating tissues) moves predominantly in the basipetal direction in a gradient of free IAA from

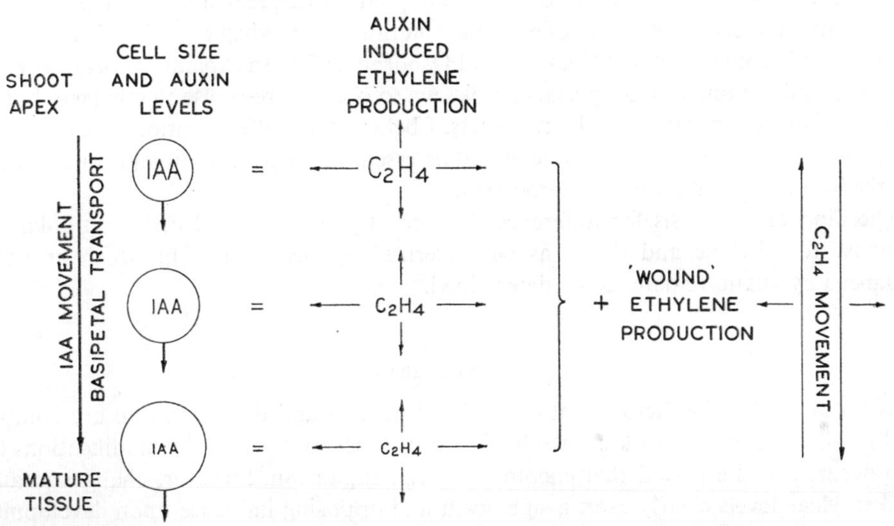

FIG. 9. Proposed movement of IAA, and of ethylene, in apical regions of etiolated pea shoots in a model for auxin and ethylene control of cell expansion. Concentrations indicated by the size of print.

the apex. Any local increase in ethylene production, therefore, could evoke an altered growth response in surrounding cells and particularly in the cells below where the auxin level is lower (Fig. 9). Such a locally elevated production of ethylene could be induced by an external wound stimulus as demonstrated by Goeschl *et al.* (1966) for pea shoots growing against an impedance, or by internal perturbations and pressures exerted by developing cells one upon the other. The model envisages that cells are continuously monitoring and responding to levels of auxin and ethylene which they are producing or to which they are exposed, auxin then determining the extent of cell expansion and ethylene regulating the orientation of that expansion.

TWO CLASSES OF CELLS

It would seem however, that not all cells in a plant are regulated in the same fashion by auxin and ethylene. Certain specialized cells respond in the reverse way, growth being enhanced by ethylene and arrested by auxin. I believe that they may constitute a second class of cell having possibly different receptor sites for auxin and ethylene and also, perhaps, different cell wall characteristics.

These specialized cells were probably first observed in fruits, e.g. grape (Hale *et al.*, 1970). The initial phase of enlargement of the berry is stimulated by auxin and retarded by ethylene. A period of slow growth rate is then succeeded by a second phase of enlargement associated with ripening of the berry. This phase is stimulated by ethylene but suppressed by auxin. Harris *et al.* (1968) showed that the volume of the cells increases two-fold during this ripening period.

A few of these specialized cells may be present in other localized regions of plants, and we have found such cells in abscission zones. In bean (Wright and Osborne, 1974) a single row is differentiated in the petiole immediately adjacent to the most proximal cells of the pulvinus (Fig. 10). Their enlargement (and subsequent abscission of the leaf) depends upon a local production of ethylene and this occurs when the cells of the pulvinus above them become senescent (Jackson and Osborne, 1970). In explants of bean (Fig. 10), expansion of this one row of special cells is seen to occur in response to ethylene (but not to auxin) before separation of the two parts. I believe that differentiation of these special cells, precisely at the abscission zone, could be essential for separation and could account for the localization of abscission processes.

The biochemical basis for differences between these specialized cells that enlarge in response to ethylene and those, as characterized by pea shoots, in which growth is enhanced by auxin, remains to be determined.

CONCLUSION

Without doubt, the factors that control cell growth and differentiation are complex, and no simple system invoking two hormones will account for all the modifications that can occur. For the period that encompasses cell expansion, however, changes in auxin and ethylene levels clearly exert a significant and opposing influence upon development. The regulatory role that has been portrayed for auxin and ethylene in the model presented is novel for biological systems in that one component is volatile and is readily

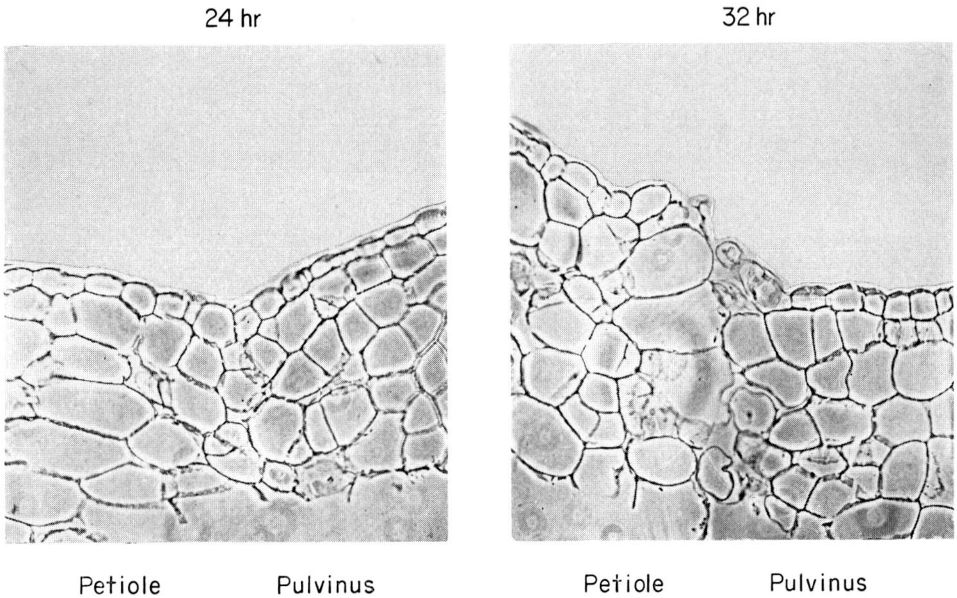

(a)

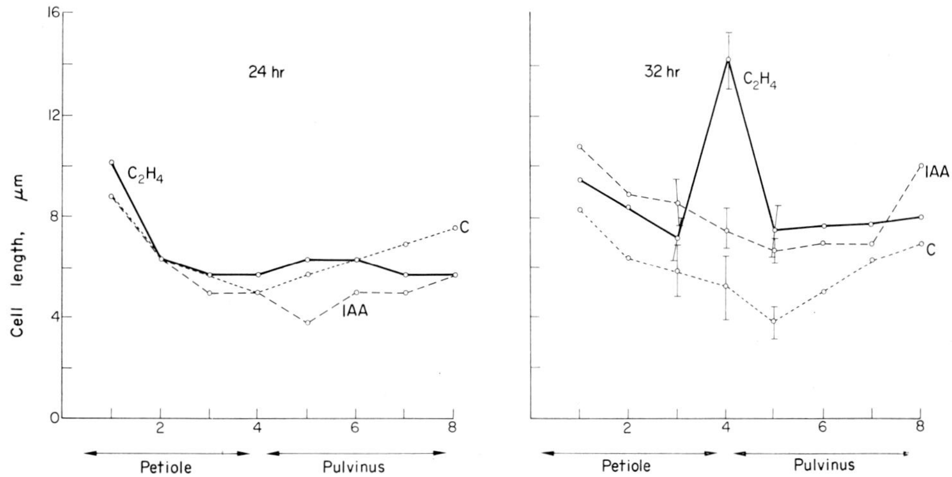

(b)

FIG. 10 (a) Light microscopy. Section across the abscission zone of *Phaseolus vulgaris* showing cells after 24 and 32 hr in ethylene (10 μl/l). (b) Lengths of cells across the abscission zone at 24 and 32 hr in explants exposed to (i) ethylene (10μl/l), (ii) air + mercuric perchlorate to absorb ambient ethylene, C, and (iii) in explants treated with a 2 μl drop of IAA (10^{-3} M) at the pulvinar end.

No abscission in any treatment before 32 hr, nor in (ii) and (iii) at 48 hr. 100 per cent abscission in (i) at 48 hr. Data from Wright and Osborne (1974).

lost to the external environment. One hormone (auxin) not only controls production of the other (ethylene) but auxin also buffers cells against the levels of ethylene that are produced. It seems possible that gradients of these hormones, set up between cells, could account for many aspects of the control of rate, extent and orientation of cell growth observed *in vivo*.

References

BURG, S. P. and BURG, E. A. (1968) Auxin stimulated ethylene formation: its relationship to auxin inhibited growth, root geotropism and other plant processes. In *Biochemistry and Physiology of Plant Growth Substances*, ed. WIGHTMAN, F. and SETTERFIELD, G. pp. 1275–1294. Runge Press, Ottawa.

BURSTRÖM, H. G., UHRSTRÖM, I. and OLAUSSON, B. (1970) Influence of auxin on Young's modulus in stems and roots of *Pisum* and the theory of changing the modulus in tissues. *Physiologia Pl.* **23,** 1223–1233.

CHRISPEELS, M. J. (1970) Biosynthesis of cell wall protein: sequential hydroxylation of proline, glycosylation of hydroxyproline and secretion of the glycoprotein. *Biochem. biophys. Res. Commun.* **39,** 732–737.

CLELAND, R. and KARLSNES, A. M. (1967) A possible role of hydroxyproline-containing protein in the cessation of cell elongation. *Pl. Physiol. Lancaster* **42,** 669–671.

DOWLER, M. J., RAYLE, D. L., CANDE, W. Z., RAY, P. M., DURAND, H. and ZENK, M. H. (1974) Auxin does not alter the permeability of pea segments to tritium-labelled water. *Pl. Physiol. Lancaster* **53,** 229–232.

EVANS, M. L. (1974) Rapid responses to plant hormones. *A. Rev. Pl. Physiol.* **25,** 195–223.

GOESCHL, J. D., RAPPAPORT, L. and PRATT, H. K. (1966) Ethylene as a factor regulating the growth of pea epicotyls subjected to physical stress. *Pl. Physiol. Lancaster* **41,** 877–884.

HAGER, A., MENZEL, H. and KRAUSS, A. (1971) Versuche und Hypothese zur Primarwirkung des Auxins beim Strechungswachstum. *Planta* **100,** 47–75.

HALE, C. R., COOMBE, B. G. and HAWKER, J. S. (1970) Effects of ethylene and 2-chloroethanephosphonic acid on the ripening of grapes. *Pl. Physiol. Lancaster* **45,** 620–623.

HARRIS, J. M., KRIEDEMANN, P. E. and POSSINGHAM, J. V. (1968) Anatomical aspects of grape berry development. *Vitis* **7,** 106–119.

HERTEL, R., THOMSON, ST. K. and RUSSO, V. E. A. (1972) *In vitro* auxin binding to particulate cell fractions from corn coleoptiles. *Planta* **107,** 325–340.

ILAN, I. (1973) On auxin-induced pH drop and on the improbability of its involvement in the primary mechanism of auxin induced growth promotion. *Physiologia Pl.* **28,** 146–148.

IRVINE, R. F. and OSBORNE, D. J. (1973) The effect of ethylene on 1-[^{14}C]glycerol incorporation into phospholipids of etiolated pea stems. *Biochem. J.* **136,** 1133–1135.

JACKSON, M. B. and OSBORNE, D. J. (1970) Ethylene, the natural regulator of leaf abscission. *Nature Lond.* **225,** 1019–1022.

KANG, B. G., NEWCOMB, W. and BURG, S. F. (1971) Mechanism of auxin-induced ethylene production. *Pl. Physiol. Lancaster* **47,** 504–509.

KEEGSTRA, K., TALMADGE, K. W., BAUER, W. D. and ALBERSHEIM, P. (1973) The structure of plant cell walls. III. A model of the walls of suspension-cultured sycamore cells based on the inter-connections of the macro-molecular components. *Pl. Physiol. Lancaster* **51,** 188–196.

LAMPORT, D. T. A. (1970) Cell wall metabolism. *A. Rev. Pl. Physiol.* **21,** 235–270.

MARRÈ, E., LADO, P., FERRONI, A. and DENTI, A. B. (1974) Transmembrane potential increase induced by auxin, benzyladenine and fusicoccin. Correlation with proton extrusion and cell enlargement. *Pl. Sci. Lett.* **2,** 257–265.

MONRO, J. A., BAILEY, R. W. and PENNY, D. (1974) Cell wall hydroxyproline polysaccharide associations in *Lupinus* hypocotyls. *Phytochemistry* **13,** 375–382.

MORGAN, P. W. and HALL, W. C. (1962) Effect of 2,4-dichlorophenoxyacetic acid on the production of ethylene by cotton and grain sorghum. *Physiologia Pl.* **15,** 420–427.

MORGAN, P. W. and HALL, W. C. (1964) Accelerated release of ethylene by cotton following application of indolyl-3-acetic acid. *Nature Lond.* **201,** 99.

MÜHLETHALER, K. (1967) Ultrastructure and formation of plant cell walls. *A. Rev. Pl. Physiol.* **18,** 1–24.

NORTHCOTE, D. H. (1972) Chemistry of the plant cell wall. *A. Rev. Pl. Physiol.* **23,** 113–132.

ORDIN, L., APPLEWHITE, T. H. and BONNER, J. (1956) Auxin-induced water uptake by *Avena* coleoptile sections. *Pl. Physiol. Lancaster* **31,** 44–53.

OSBORNE, D. J. (1958) Growth of etiolated sections of pea internode following exposures to indole-3-acetic acid, 2,4-dichlorophenoxyacetic acid and 2,5-dichlorobenzoic acid. *Pl. Physiol. Lancaster* **33,** 46–57.

OSBORNE, D. J. (1973) Ethylene and protein synthesis. In *Biosynthesis and its Control in Plants,* ed. MILBORROW, B. V. pp. 127–142. Academic Press, London and New York.

OSBORNE, D. J. (1974) Auxin, ethylene and the growth of cells. In *Mechanisms of Regulation of Plant Growth.* ed. CRESSWELL, M., FERGUSON, A. R. and BIELESKI, R. L. pp. 645–654, Bulletin 12, Royal Society of New Zealand.

PATRICK, J. W. and WOOLLEY, D. J. (1973) Auxin physiology of decapitated stems of *Phaseolus vulgaris* L. treated with indol-3yl-acetic acid. *J. exp. Bot.* **24,** 949–957.

PROBINE, M. (1964) Chemical control of plant cell wall structure and of cell shape. *Proc. R. Soc. B.* **161,** 526–537.

RIDGE, I. (1973) The control of cell shape and rate of cell expansion by ethylene: effects on microfibril orientation and cell wall extensibility in etiolated peas. *Acta bot. neerl.* **22,** 144–158.

RIDGE, I. and OSBORNE, D. J. (1969) Cell growth and cellulases: regulation by ethylene and indole-3-acetic acid in shoots of *Pisum sativum. Nature Lond.* **223,** 318–319.

RIDGE, I. and OSBORNE, D. J. (1970) Hydroxyproline and peroxidase in cell walls of *Pisum sativum*: regulation by ethylene. *J. exp. Bot.* **21,** 843–856.

RIDGE, I. and OSBORNE, D. J. (1971) Role of peroxidase when hydroxyproline rich protein in plant cell walls is increased by ethylene. *Nature New Biol.* **229,** 205–208.

SADAVA, D. and CHRISPEELS, M. J. (1973) Hydroxyproline rich cell wall protein (extensin): role in the cessation of elongation in excised pea epicotyls. *Devl Biol.* **30,** 49–55.

SARGENT, J. A., ATACK, A. and OSBORNE, D. J. (1973) Orientation of cell growth in etiolated pea stem. Effect of ethylene and auxin on cell wall deposition. *Planta* **109,** 185–192.

SARGENT, J. A., ATACK, A. and OSBORNE, D. J. (1974) Auxin and ethylene control of growth in epidermal cells of *Pisum sativum*: a biphasic response to auxin. *Planta* **115,** 213–225.

SARGENT, J. A. and OSBORNE, D. J. (1975) An effect of ethylene on the endoplasmic reticulum of expanding cells of etiolated shoots of *Pisum sativum. Planta* (in press).

UHRSTRÖM, I. (1974) The effect of auxin and low pH on Young's modulus in *Pisum* stems and on water permeability in potato parenchyma. *Physiologia Pl.* **30,** 97–102.

VEEN, B. W. (1970) Control of plant cell shape by cell wall structure. *Proc. K. ned. Akad. Wet.* **73,** 118–121.

WRIGHT, M. and OSBORNE, D. J. (1974) Abscission in *Phaseolus vulgaris.* The positional differentiation and ethylene-induced expansion growth of specialized cells. *Planta* **120,** 163–170.

WINTER, H., MEYER, L., HENGEVELD, E. and WIERSMA, P. K. (1971) The role of wall-bound hydroxy proline-rich protein in cell extension. *Acta bot. neerl.* **20,** 489–497.

ENDOGENOUS CYTOKININS AS GROWTH REGULATORS

P. F. Wareing

University College of Wales, Aberystwyth, U.K.

Introduction

Cytokinins are now fully recognized as one of the major groups of endogenous plant hormones. The discovery that one of the breakdown products of DNA, 6-furfuryl aminopurine (kinetin) has high activity as a cell-division factor immediately raised the question as to whether similar substances occur naturally in plant tissues, and the search for these substances resulted in the isolation by Letham (1963) of a substance, zeatin, from immature maize kernels, which proved to be a substituted purine with an isoprenoid side chain. Since then a number of other biologically active substances have been isolated which (with one exception) are either bases closely related to zeatin, or are the nucleosides (ribosides) or nucleotides (ribotides) of these bases.

The primary criterion of a cytokinin is its ability to stimulate cell division in certain types of callus tissue, such as that derived from tobacco pith or soybean cotyledons. However, most cytokinins have a number of other physiological effects such as the stimulation of growth of lateral buds, the delay of senescence of detached leaves and the stimulation of germination of certain seeds.

As well as kinetin, a considerable number of other non-natural substances have been found to possess cytokinin activity of which one of the best known is 6-benzyl aminopurine. Hitherto all the known natural cytokinins were substituted adenine derivatives with an isoprenoid side chain, but recently we isolated a fraction from poplar leaves which we identified as 6-hydroxy benzyl adenosine (Horgan *et al.*, 1973a), the first example of a naturally-occurring cytokinin with an aromatic side chain.

The discovery that certain species of transfer-RNA (namely those corresponding to codons with the initial letter U) contain 2-isopentenyladenosine (2iPA) (Armstrong *et al.*, 1969), in a position adjacent to the anti-codon of the tRNA immediately suggests that the biological activity of endogenous plant cytokinins may be related to their occurrence in tRNA. However the demonstration that the 2iPA moiety is formed *in situ* by attachment of the side chain to adenine already incorporated into the tRNA (Chen and Hall, 1969), suggests that free cytokinins are not incorporated directly into tRNA, despite some evidence to the contrary (Fox and Chen, 1967; Burrows *et al.*, 1971).

On the other hand, the fact that the cytokinin group in tRNA is formed *in situ* raises the question as to whether the free cytokinins which can be extracted from plant tissues are derived from tRNA breakdown, a question that remains unresolved. However, certain observations suggest that it is unlikely that all free cytokinins in the plant are derived from tRNA breakdown. Thus, Chen and Hall (1969) found that even cytokinin-dependent tissues i.e. those which require cytokinin in the medium in order to grow, possess tRNA molecules which include cytokinin moieties in their structure, and that the supplied exogenous cytokinin is *not* the precursor of the cytokinin in the tRNA. On this and other evidence it would seem possible that there is a separate route for cytokinin biosynthesis other than through the breakdown of tRNA.

These studies on the molecular and cellular role of cytokinins have given interesting and important results. However, there is also the question of whether endogenous cytokinins are involved in the regulation of growth and development within the whole plant, as appears to be the case with auxins and gibberellins.

This latter problem leads on to the question of where the sites of cytokinin biosynthesis are in the plant. The observation that kinetin will delay the senescence of detached leaves (Richmond and Lang, 1957), suggests that endogenous cytokinins may be identified with the hypothetical 'root factor' postulated by Chibnall (1939) to be produced by roots and transported to the leaves, where its presence is essential for the maintenance of a healthy, functional condition; the rapid protein breakdown and senescence occurring in detached leaves was attributed to lack of this root factor. Much evidence supports this hypothesis. Thus, it is known that cytokinins occur in relatively high amounts in root tips (Weiss and Vaadia, 1965) and in bleeding sap from roots (Kende, 1965). Moreover, this bleeding sap is capable of delaying senescence of leaf tissue from the same species (Kulaeva, 1962). It would appear, therefore, that cytokinins are produced in roots and transported thence in the transpiration stream to the leaves and other parts of the shoot.

The question arises as to whether buds also synthesize free cytokinins. Some evidence suggests that buds, as well as expanded leaves, require a supply of cytokinins from the roots. Thus if shoots of *Solanum andigena* are severed from the roots and placed in a nutrient solution for 12 days and then cut into two-node sections, the axillary buds are not capable of growing out unless supplied with exogenous cytokinin such as zeatin or benzyladenine (Woolley and Wareing, 1972). If buds do indeed require a supply of cytokinins from the roots, it is possible that inhibited axillary buds fail to grow partly because they are deprived of cytokinins as a result of competition with the main apex.

It is clear that there is a lack of information on some of the most basic aspects of cytokinin production and distribution in the plant and that we need to be able to answer the following questions.

1. What is the chemical identity of the endogenous cytokinins in various parts of the plant?
2. What are the sites of cytokinin biosynthesis? Is biosynthesis limited to the roots?
3. In what form are cytokinins transported within the plant, and what is the pattern of their distribution?

Several years ago we set ourselves the task of trying to solve some of these problems, but it was clear that we would not achieve much progress so long as we were dependent upon the current methods of isolation and assay of endogenous cytokinins.

Methods of study for endogenous cytokinins

Hitherto, physiological studies on endogenous cytokinins have largely involved standard methods of partial purification, such as paper chromatography and partition on Sephadex columns, followed by detection and estimation by various types of bioassay, e.g. growth of callus cultures of tobacco or soybean, growth of discs from radish cotyledons (Letham, 1968), or development of anthocyanin in *Amaranthus caudatus* (Kohler and Conrad, 1966). The results obtained with these techniques have been valuable but they do not permit positive identification of the cytokinins present in the active fractions and the most reliable of the assay methods, the growth of callus cultures, is slow in giving results and is too laborious to permit sufficient replication for satisfactory statistical analysis. Studies with other types of plant hormone have been considerably advanced by the application of a number of physical techniques, such as optical rotatory dispersion, spectrofluorimetry and especially gas–liquid chromatography (GLC). GLC not only gives much better separation of hormones such as gibberellins, in plant extracts, but in combination with mass-spectrometry (MS) can give positive identification and the value of combined gas chromatography and mass spectrometry (GC.MS) has been amply demonstrated by J. MacMillan and others.

Although it was demonstrated several years ago that GLC can be applied readily to synthetic cytokinins (Most *et al.*, 1968) and to endogenous cytokinins in extracts of microorganisms (Upper *et al.*, 1972) there have been few successful attempts, in the past, to use this technique for the study of endogenous cytokinins in higher plants. The special problems associated with endogenous cytokinins arise partly from their relatively polar nature which makes purification difficult, and partly from the small quantities present in extracts, which makes it difficult to obtain the minimal amount required for identification. However, during the past few years we have devoted considerable effort towards overcoming these difficulties and as a result of the work of J. and R. Horgan and J. Purse, we have successfully developed techniques for the application of GLC and GC.MS to endogenous cytokinins in plant extracts. The purification procedure (Fig. 1) involves fairly standard techniques, but a valuable innovation has been the operation of columns of Sephadex LH 20 under a pressure of 50 p.s.i. (3.6×10^4 kg m^{-1}) which speeds up the procedure and reduces peak-spreading. However, although the purification problems have largely been mastered, there tend to be new problems with each new plant species to be examined.

For gas chromatography, 3 per cent OV-1 columns are used, with Gaschrome Q as solid support. For GC.MS the columns are conditioned for at least 2 weeks before use; in this way very low bleed rates are obtained. Derivatization is carried out in pure BSA (*N*,*o*-bis(trimethylsilyl)acetamide), the sample being heated with the BSA at 80°C for 1 hr.

The presence of known cytokinins can easily be determined by means of GC.MS of their trimethylsilyl (TMS) derivatives. To determine the structure of unknown cytokinins a more complex procedure has to be used, since the TMS derivatives of cytokinins give a very limited amount of structural information. Biological activity is located by means of preparative GLC and bioassay, and the GC peak then checked for homogeneity by GC.MS. If the peak is homogeneous it is trapped, the TMS groups removed and a mass spectrum obtained by use of the heated probe. Samples of as little as 0.1 μg can be dealt with in this way.

We have so far successfully identified zeatin riboside (Horgan *et al.*, 1973b) and dihydrozeatin (J. Purse and R. Horgan, unpublished results) from sycamore sap as well as the new cytokinin, *o*-hydroxy benzyl adenosine. Where the identity of an endogenous

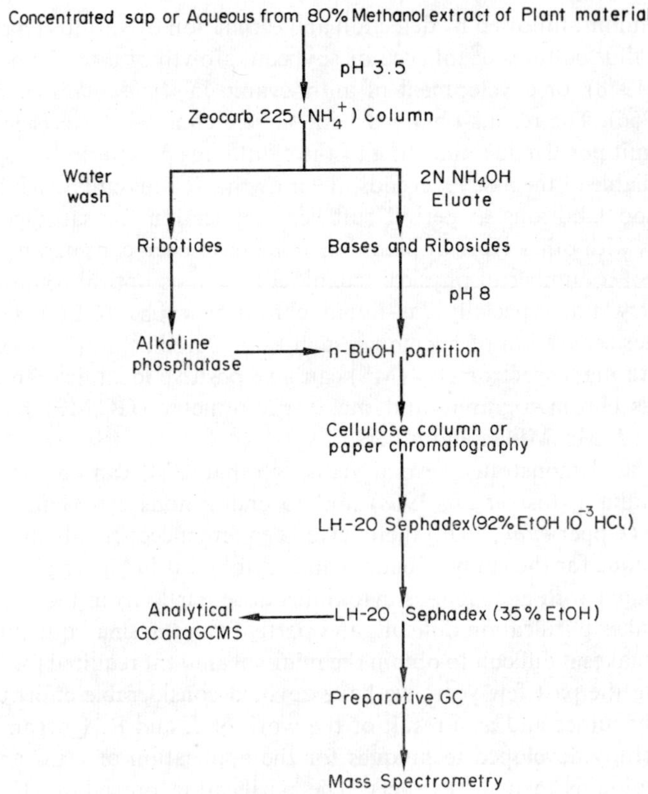

FIG. 1.

cytokinin fraction has been established it is possible to use 'single peak monitoring' in which the mass spectrometer is focused on a single ion unique to the cytokinin in question, to measure the amounts of the latter present in plant extracts, even when they occur only in nanogram amounts.

EFFECTS OF EXTERNAL FACTORS ON LEVELS OF ENDOGENOUS CYTOKININS

In the Introduction the question was raised as to whether endogenous cytokinins may play a co-ordinating role within the plant as a whole—for example, in shoot–root interactions, and in apical dominance. Apart from this possible function, there is increasing evidence that endogenous cytokinins, in common with other types of plant hormone, may be involved in growth responses to various environmental factors, as is indicated by the following examples.

Effects of photoperiod and of red light

There is increasing evidence that phytochrome control of plant growth and development may be partly mediated through effects on endogenous plant hormones. There have been a number of reports of the effects of long photoperiods in green plants, and of short periods of red light in etiolated tissue, on endogenous gibberellin levels (Reid *et al.*, 1967; Beevers *et al.*, 1970). Reports of similar effects on cytokinin levels have been less numerous, but we have reported several examples of these effects recently. Thus, van Staden and Wareing (1972a) found that exposure of the light-requiring seeds of *Rumex obtusifolius* to only 10 min of red light resulted in an immediate increase in the extractable cytokinin levels. Similar effects have been reported for seeds of *Spergula arvensis* in which there is an interaction between the effects of red light and of ethylene on the endogenous cytokinin levels (van Staden *et al.*, 1973).

A further example of the effects of red irradiation on cytokinin levels is seen in poplar (*Populus robustan*) leaves (Hewett and Wareing, 1973). It was found that whereas leaves collected from plants growing in the open contained quite high cytokinin levels, leaves from plants maintained in growth rooms under fluorescent lamps supplemented by tungsten filament lamps contained very much lower levels of cytokinins. It was thought that possibly this effect might be due to a difference in light quality, since light from fluorescent lamps tends to be deficient in the red and far-red regions of the spectrum, in comparison with natural daylight. To test this hypothesis, plants were exposed to fluorescent lamps at an intensity of 0.3 cal cm^{-2} mm^{-1} for 16 hr daily and then for a further 4 hr either to white fluorescent lamps or to radiation from red fluorescent lamps at an intensity of 4.2×10^{-3} cal cm^{-2} mm^{-1}. After 21 days of these treatments the cytokinin levels in the leaves were determined and it was found that the levels in the leaves of plants exposed to 4 hr of additional red light per day were twenty times greater than in those exposed daily to white fluorescent lamps.

The effects of short periods of red light were then investigated. Leaves were collected from plants growing outdoors and exposed to various periods, ranging from 0 to 30 min of red light. It was found that the maximum effect on cytokinin levels was obtained after 5 min of red light and that the levels declined with longer periods of irradiation. When leaves were exposed to 5 min of red light, followed by varying periods of darkness before extraction, at either the beginning or at the end of the dark period, the greatest effect was obtained by irradiation at the end of the dark period, followed by a further 30 min of darkness before extraction. The cytokinin fraction, which varied with exposure to red light, was later identified as hydroxy-benzyl adenosine (Horgan *et al.*, 1973a).

Treatments with short periods of red light are, of course, highly artificial and under natural conditions plants are exposed to long periods of light of mixed spectral composition. Preliminary studies carried out by the GC.MS 'single peak monitoring' technique suggest that there is a diurnal cycle in the cytokinin levels in poplar leaves, the peak level occurring shortly after dawn (Thompson and Horgan, unpublished results). We are currently studying this diurnal cycle in greater detail.

In the classical short-day plant, *Xanthium strumarium*, which will flower in response to a single long dark period, the levels of cytokinin in the leaves, buds and root sap fall quite dramatically when the plants are transferred from long days to short days (van Staden and Wareing, 1972b). We have subsequently found that the cytokinin levels decline markedly after the plants have been exposed to a single long dark period (Henson and Wareing, 1974). Moreover, this effect of a long dark period can be nullified if it is interrupted by a short period (60 min) of red irradiation; under these latter conditions the cytokinin levels in the leaves and the sap remain at about the same level as in plants kept under long day conditions. Thus, although we have not yet shown red/far-red reversibility in this system, it seems likely that the cytokinin levels in *Xanthium* plants is under phytochrome control.

Effects of mineral nutrition on cytokinin levels

It is well known that plants differ considerably in their responses to varying levels of mineral nutrients and we have reason to think that the differences in response in tree species, such as birch (*Betula pubescens*) and sycamore (*Acer pseudoplatanus*), may not be due simply to differences in their requirements for 'substrate levels' of nitrogen, phosphorus, calcium etc. for synthesis of proteins, nucleic acids and cell walls, but to disturbances of the hormonal balance in the plants. Thus, when seedlings of birch and sycamore are maintained under conditions of nutrient deficiency, the birch seedlings continue to grow slowly whereas the sycamore seedlings rapidly form a terminal resting bud and cease growth altogether. However, tissue analysis indicates that there are equally significant differences between the two species in their overall demands for various mineral nutrients (van den Driessche and Wareing, 1966). Since it appears that photoperiodic control of extension growth and resting bud formation in these species is mediated through hormonal factors, it seems likely that hormones are also involved in their responses to varying levels of mineral nutrients. We have, therefore, been studying the effects of nitrogen and phosphorus nutrition on cytokinin and gibberellin levels in seedlings of the two species.

We have found (J. M. Horgan and P. F. Wareing, unpublished results) that when birch seedlings are transferred from an adequate to a deficient supply of nitrogen (supplied as ammonium nitrate), within a few days there is a dramatic decline in the levels of endogenous cytokinins in the leaves. Of course, the decrease in cytokinin levels does not necessarily imply that growth is limited by a deficiency of these hormones. However, if exogenous benzyladenine is applied to the shoot apices or axillary buds of birch seedlings growing under nitrogen-deficient conditions, growth is stimulated within 2–3 days, suggesting that the low levels of cytokinins, rather than a deficiency of 'substrate' levels of nitrogen for protein synthesis, may well have been the limiting factor.

In sycamore, nitrogen deficiency has a less marked effect on cytokinin levels in the leaves, and it appears that there is a greater effect on gibberellin levels. On the other hand, effects of nitrogen nutrition on cytokinin levels similar to those observed in birch have been observed in plants of sunflower (Salama and Wareing, unpublished results). There is a rapid decline in the hormone levels not only in the leaves, but also in extracts of roots, and of the bleeding sap, when sunflower plants are transferred to a nitrogen-deficient regime.

It is apparent from the various examples we have considered that environmental factors such as light, temperature and mineral nutrients can have a profound effect on the endogenous cytokinin levels of the plant and hence certain plant responses to such factors may be mediated through variations in these growth hormones. However, extensive further studies will be required before such an intermediary role for cytokinins can be firmly established.

References

ARMSTRONG, D. J., BURROWS, W. J., SKOOG, F., ROY, K. L. and STOLL, D. (1969) Cytokinins: distribution in transfer RNA species of *Escherichia coli*. *Proc. natn. Acad. Sci. U.S.A.* **63,** 834–841.

BEEVERS, L., LOVEYS, B., PEARSON, J. A. and WAREING, P. F. (1970) Phytochrome and hormonal control of expansion and greening of etiolated wheat leaves. *Planta* **90,** 286–294.

BURROWS, W. J., SKOOG, F. and LEONARD, N. J. (1971) Isolation and identification of cytokinins located in the transfer ribonucleic acid of tobacco callus grown in the presence of 6-benzylaminopurine. *Biochemistry* **10,** 2189–2194.

CHEN, C. and HALL, R. (1969) Biosynthesis of $N^6(\Delta^2$-isopentenyl) adenosine in the transfer ribonucleic acid of cultured tobacco pith tissue. *Phytochemistry* **8,** 1687–1695.

CHIBNALL, A. C. (1939) *Protein Metabolism in the Plant*. Yale University Press, New Haven, Connecticut.

DRIESSCHE, R. VAN DEN and WAREING, P. F. (1966) Nutrient supply, dry matter production and nutrient uptake of forest tree seedlings. *Ann. Bot.* NS **30,** 657–672.

FOX, J. E. and CHEN, C.-M. (1967) Characterisation of labelled ribonucleic acid from tissue grown on C^{14} containing cytokinins. *J. biol. Chem.* **242,** 4490–4494.

HENSON, I. and WAREING, P. F. (1974) Cytokinins in *Xanthium strumarium*: a rapid response to short day treatment. *Physiologia Pl.* **32,** 185–187.

HEWETT, E. W. and WAREING, P. F. (1973) Cytokinins in *Populus* × *robusta* (Schneid): Light effects on endogenous levels. *Planta* **114,** 119–129.

HORGAN, R., HEWETT, E. W., PURSE, J. G., and WAREING, P. F. (1973a) A new cytokinin from *Populus robusta*. *Tetrahedron Lett.* **30,** 2827–2828.

HORGAN, R., HEWETT, E. W., PURSE, J. G., HORGAN, J. M. and WAREING, P. F. (1973b) Identification of a cytokinin in sycamore sap by gas chromatography and mass spectrometry. *Plant Sci. Lett.* **1,** 321–324.

KENDE, H. (1965) Kinetin-like factors in the root exudate of sunflowers. *Proc. natn. Acad. Sci. U.S.A.* **53,** 1302–1307.

KOHLER, K. H. and CONRAD, K. (1966) Ein quantitativer phytokinin-test. *Biol. Rundschau* **4,** 36–37.

KULAEVA, O. (1962) The effect of roots on leaf metabolism in relation to the action of kinetin on leaves. *Sov. Pl. Physiol.* **9,** 182–189.

LETHAM, D. S. (1963) Zeatin, a factor inducing cell division isolated from *Zea mays*. *Life Sci.* **8,** 569–573.

LETHAM, D. S. (1968) A new cytokinin bioassay and the naturally-occurring cytokinin complex. In *Biochemistry and Physiology of Plant Growth Substances* ed. WIGHTMAN, F. and SETTERFIELD, G. Runge Press, Ottawa, Canada.

MOST, B. H., WILLIAMS, J. C. and PARKER, K. J. (1968) Gas chromatography of cytokinins. *J. Chromatography* **38,** 136–138.

REID, D. M., CLEMENTS, J. B. and CARR, D. J. (1967) Red light induction of gibberellin synthesis in leaves. *Nature Lond.* **217,** 580–582.

RICHMOND, A. E. and LANG, A. (1957) Effect of kinetin on protein content and survival of detached *Xanthium* leaves. *Science N. Y.* **125,** 650–651.

STADEN, J. VAN and WAREING, P. F. (1972a) The effect of light on endogenous cytokinin levels in seeds of *Rumex obtusifolius*. *Planta* **104,** 126–133.

STADEN, J. VAN and WAREING, P. F. (1972b) The effect of photoperiod on levels of endogenous cytokinins in *Xanthium strumarium*. *Physiologia Pl.* **27,** 331–337.

STADEN, J. VAN, OLATOYE, S. T. and HALL, M. A. (1973) Effect of light and ethylene upon cytokinin levels in seed of *Spergula arvensis*. *J. exp. Bot.* **24,** 662–666.

UPPER, C. D., HELGESON, J. P. and SCHMIDT, C. J. (1972) Identification of cytokinins by gas–liquid chromatography and gas–liquid chromatography–mass spectrometry. In *Plant Growth Substances 1970* ed. CARR, D. J., Ed. Springer-Verlag, Berlin.

WEISS, C. and VAADIA, Y. (1965) Kinetin-like activity in root apices of sunflower plants. *Life Sci.* **4,** 1323–1326.

WOOLLEY, D. J. and WAREING, P. F. (1972) The role of roots, cytokinins and apical dominance in the control of lateral shoot form in *Solanum andigena*. *Planta* **105,** 33–42.

RECENT STUDIES ON ABSCISIC AND PHASEIC ACIDS

B. V. Milborrow

Shell Research Limited,
Milstead Laboratory of Chemical Enzymology,
Sittingbourne Research Centre, Sittingbourne, Kent, U.K.

Introduction

The structure (I) of the growth-inhibitory plant hormone known as (+)-abscisic acid (ABA) was published in 1965 (Ohkuma *et al.*,), and confirmed by chemical synthesis a few weeks later (Cornforth *et al.*). Availability of the synthetic material stimulated research into many aspects of its action on plants, and some 700 papers have now appeared on the subject. The role of ABA in the regulation of plant growth has been reviewed recently (Milborrow, 1974). ABA is probably the only active member of this most recently discovered group of plant growth regulators and, in bioassays, it antagonizes the effect of other growth regulators, such as the auxins, gibberellins and cytokinins. Time/concentration studies indicate that ABA plays a part in the abortion of young fruit (van Steveninck, 1959; Davis and Addicott, 1972), and it may also be implicated in leaf and fruit abscission. The highest natural concentrations are found in ripe fruit (Milborrow, 1967) and (±)-ABA has been found to affect ripening of grapes when applied to the skin of the fruit (Coombe and Hale, 1973). Although concentrations in seeds are usually about one tenth of those in fruit, ABA is probably the main chemical agent controlling seed dormancy and has been found to inhibit synthesis of hydrolytic enzymes during germination (Chrispeels and Varner, 1967). Perhaps the clearest evidence for one of its functions has come from the discovery that it causes stomata to close (Mittelheuser and van Steveninck, 1969; Jones and Mansfield, 1970).

During the last few years we have investigated the routes by which ABA is biosynthesized and degraded in an attempt to discover the regulatory mechanisms that control its concentration within plant tissues. The work has been carried out with the help of R. Mallaby, R. C. Noddle, D. R. Robinson, G. Ryback and J. A. D. Zeevaart whose contributions are gratefully acknowledged.

Parts of the plant in which ABA is synthesized

Wright (1969) and Wright and Hiron (1969) found that when wheat (*Triticum aestivum* cv Eclipse) plants wilt their ABA content increases about forty-fold. This important observation represents a clear, measurable response of a plant's regulatory apparatus to an environmental stimulus. We have tried to discover how the control mechanisms that adjust the (+)-ABA concentration are activated by wilting (Milborrow and Noddle, 1970; Milborrow and Robinson, 1973). To do this we have studied incorporation of [^{14}C]mevalonate into ABA in a range of tissues. (The carbon skeleton of ABA is made up of three 5-carbon isoprene units, each derived from one molecule of mevalonate.) Wilted wheat leaves when treated with [^{14}C]mevalonate incorporated more of the precursor into ABA than turgid leaves (Milborrow and Noddle, 1970), indicating that the 'extra ABA' in wilted leaves is formed *de novo* from mevalonate rather than by release from a conjugate or by elaboration of a close precursor.

A favourable material for these incorporation studies has proved to be ripening mesocarp of the avocado fruit (*Persea gratissima*), and techniques for the isolation of ABA in a radiochemically pure state have been developed with, and extensively used on, this tissue (Noddle and Robinson, 1969). The procedures are adequately described elsewhere (Milborrow, 1972; Milborrow and Robinson, 1973) and will not be discussed further, except to say that the percentage recovery of ^{3}H or ^{14}C label from mevalonate incorporated into ABA can be determined by first adding to the extraction medium, synthetic (±)-ABA, labelled with the alternate isotope, and then counting both isotopes in the final radiochemically-pure sample; the total amount of (+)-ABA is measured by its optical activity and the value corrected for losses incurred during extraction (Milborrow and Robinson, 1973). Apart from the fruit of avocado, cotyledons, stems and leaves of this plant also convert mevalonate into ABA. Furthermore, the (+)-ABA content of isolated roots of avocado increases as they dry out. Hence it appears that ABA can be synthesized throughout most of the plant (Table 1).

TABLE 1. INCORPORATION OF [2-^{14}C]MEVALONATE (ADDED AS LACTONE) INTO ABSCISIC ACID BY DIFFERENT ORGANS OF AVOCADO

	Duration of incorporation, hr	% recovery of ABA	(+)-ABA corrected for % recovery, μg/kg	dpm	Sample fresh weight, g
Leaves (wilted) (A)	41	69.5	2670	1370	20
Fruit mesocarp (B)	24	79.5	4050	3650	20
Cotyledon (C)	24	70.0	500	40	2.5
Stem (D)	39	42.0	930	3640	59
	Increase in (+)-ABA content of excised avocado roots				
Excised roots (no label added)					
Wet, turgid	8	24.5	27	—	400
Dry, wilted	8	52.5	85	—	400

A = 5.2 × 10^{6} dpm; B = 24 × 10^{6} dpm; C = 6.4 × 10^{6} dpm; D = 19.8 × 10^{6} dpm.

This observation makes the criteria of *synthesis in a restricted site* and *action at a distance*, two of the basic tenets in the definition of a hormone, inapplicable to ABA. In this respect ABA resembles other plant growth regulators.

STEREOCHEMISTRY OF ABA BIOSYNTHESIS

Investigations of the stereochemistry of ABA biosynthesis (Robinson and Ryback, 1969; Milborrow, 1972) were carried out in an attempt to discover if there is a qualitative difference between the biosynthetic pathways of ABA and carotenoids. ABA could be a degraded carotenoid fragment (cf. I, II, III) but its carbon skeleton might be synthesized by a unique pathway. The procedure adopted was to feed to avocado mesocarp, mixtures of stereospecifically ^{14}C- and ^{3}H-labelled mevalonic acids in defined ratios, then to extract and purify the ABA synthesized, and to measure its $^{14}C:^{3}H$ ratio. This permitted conclusions to be drawn about whether or not a given hydrogen atom of mevalonate was retained in ABA.

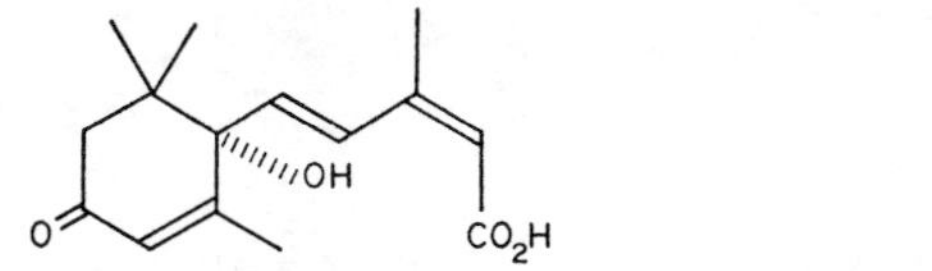

I (+)-Abscisic acid (revised absolute configuration)

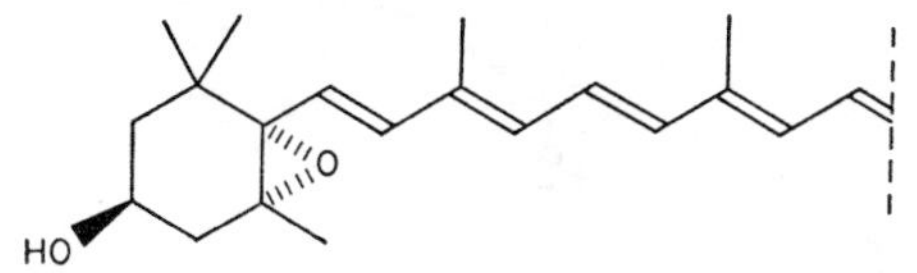

II Violaxanthin

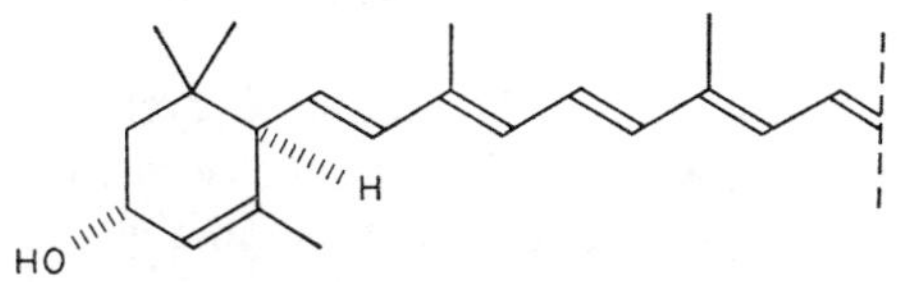

III Lutein

Mevalonic acid (IV, Fig. 2) contains three pairs of methylene hydrogen atoms and six stereospecific syntheses have been carried out to replace the hydrogen at each position with tritium (Popjak and Cornforth, 1966). We used these materials to define the origin of the hydrogen atoms of ABA and our results are summarized in Fig. 1. The derivation of each of the analogous hydrogen atoms proved to be identical in both carotenoids and ABA, and this indicated that a pathway involving carotenoids (40 carbon atoms) could not be distinguished from one in which no precursor exceeded 15 carbon atoms.

A stereochemical feature of ABA not hitherto investigated is the origin of the two *geminal* methyl groups. From the known sequence in terpenoid biosynthesis (Fig. 2) it seems highly probable that one of these groups is derived from the C-2 methyl group, and the other from the C-3 methyl group of mevalonic acid (Popjak and Cornforth, 1966). If the individual methyl groups could be identified the direction of cyclization in the precursor of ABA could then be deduced. In our studies we took advantage of the fact that (+)-ABA is broken down in the plant to the unstable *Metabolite C* (V, Fig. 3)

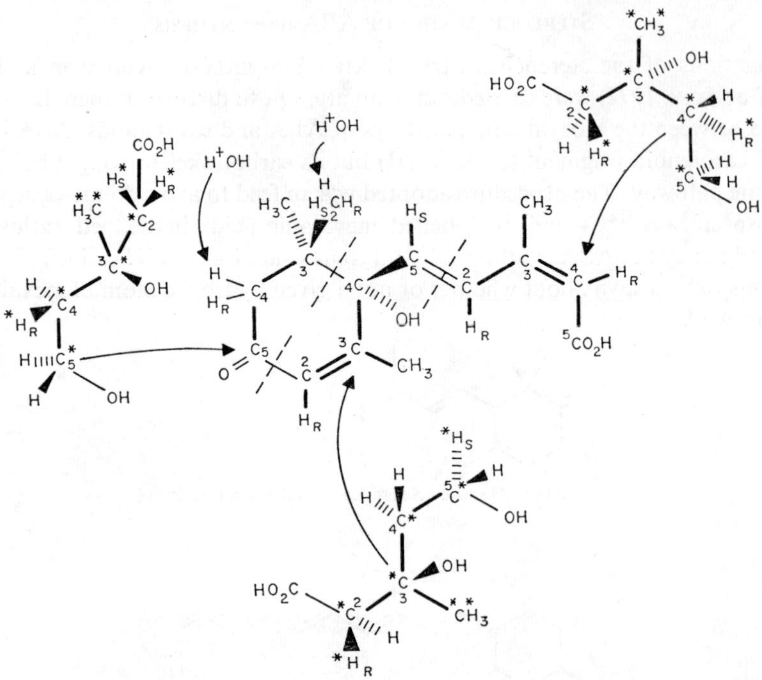

FIG. 1. Derivation of the carbon and hydrogen atoms of (+)-(*S*)-abscisic acid from those of three molecules of 3-(*R*)-mevalonic acid. Carbon atoms of the three isoprene residues are joined by thick bonds. The C and H atoms of mevalonate which are retained in abscisic acid are marked thus: * and their numbering and chirality in the mevalonate are shown.

by hydroxylation of one 6′-methyl group. Metabolite C cyclizes spontaneously to form phaseic acid (VI, Fig. 3) (Milborrow, 1969b). Only one isomer of phaseic acid is formed indicating that hydroxylation is stereospecific and one methyl group only is attacked.

We first prepared labelled ABA biosynthetically from a mixture of [2-^{14}C]- and [2-^{3}H]mevalonate as already described, and after measuring the ^{14}C:^{3}H ratio, fed the ABA to tomato shoots (*Lycopersicon esculentum* cv Arastor). Phaseic acid produced by hydroxylation of the ABA was then isolated and counted. Labelling of the ABA was as shown in Fig. 4. Theoretically, if the *geminal* methyl group carrying two tritium atoms, (CH$_3$) in Fig. 4, was the one hydroxylated, then the ^{14}C:^{3}H ratio would be expected to change from 3.0:3.0 (abscisic acid) to 3.0:2.3 (phaseic acid)—a loss of about 23 per cent of tritium, but if the other methyl group, shown as [CH$_3$] in Fig. 4, was the one hydroxylated then the ratio would not change. However, the carbon–tritium bond is shorter and stronger than a carbon–protium (H, light hydrogen) bond, and it is possible that tritium might be preferentially rejected by the hydroxylating enzyme, in other words, a methyl group containing a tritium atom might not show a significant loss of label on hydroxylation. The specific activity of the [2-^{3}H]mevalonate used in the experiments was 91 μCi/μM, i.e. less than one in a thousand of the methyl groups derived from C-2 carried two tritium atoms. Thus, although the C-2 atom of mevalonate was nominally tritiated

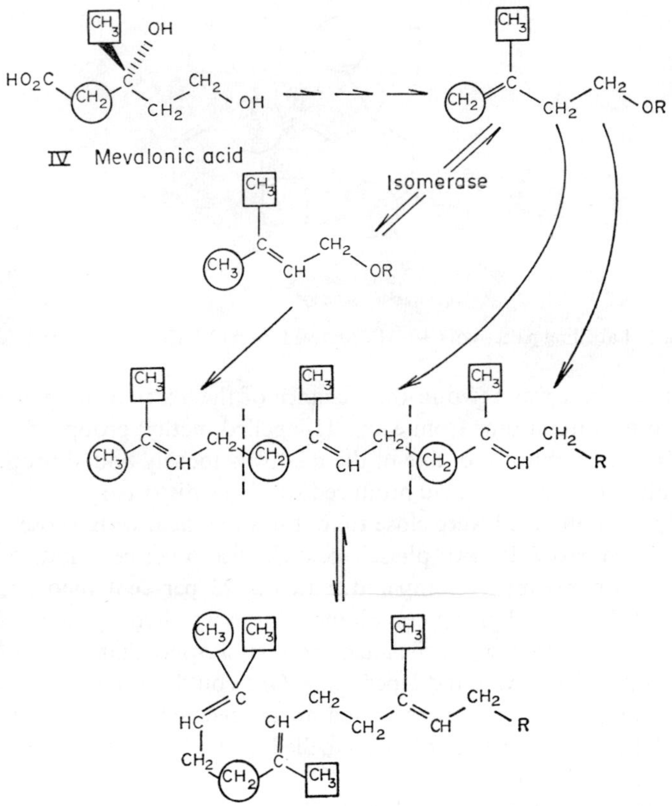

FIG. 2. Positions of C-2 ◯ and the C-3 methyl groups ☐ of mevalonic acid in an uncyclized precursor of ABA.

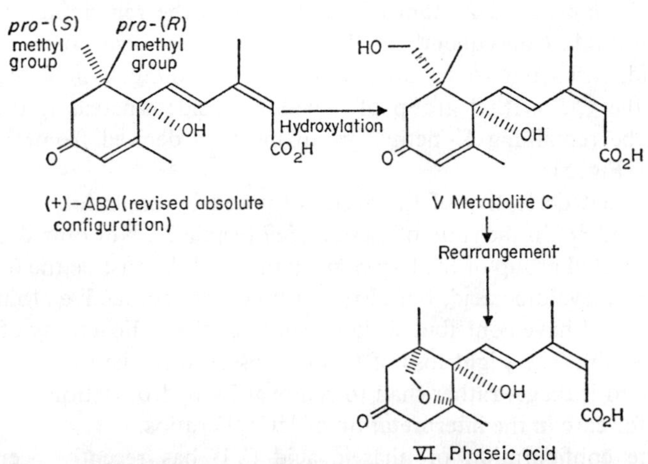

FIG. 3. Stereochemistry of the formation of phaseic acid from abscisic acid.

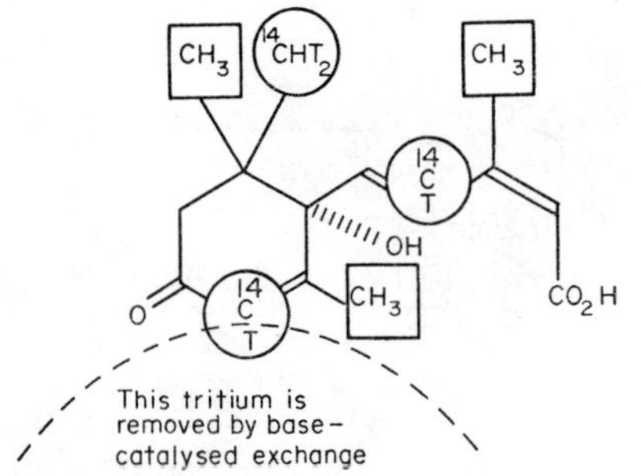

FIG. 4. Labelling pattern of (+)-ABA derived from (2-^{14}C, 2-$^{3}H_2$) mevalonic acid.

in both positions, there was a two-in-three chance of the hydroxylating enzyme replacing one of the two protium atoms from each ^{3}H-labelled methyl group. If an isotope effect operated in favour of removal of one of the relatively loosely bound protium atoms then the ^{14}C:^{3}H ratio of the phaseic acid produced would be distorted.

The actual ratios obtained were close to, but not identical with, those of the ABA fed to the plants (Milborrow, 1975a): phaseic acid showed 6 per cent and 15 per cent losses of tritium in two experiments compared with the 23 per cent theoretical loss on the assumption that the methyl group containing tritium was hydroxylated (Table 2). However, when the ABA itself was reisolated, counts showed that it also had lost small amounts of tritium (1 per cent and 5 per cent). One could not be certain, therefore, that the observed change in the ratio represented an unspecific leakage of tritium or whether it was caused by hydroxylation of the labelled methyl group with an isotope effect favouring retention of tritium.

A second experiment was designed to overcome the difficulty. [^{14}C, ^{3}H]phaseic acid was prepared as before, isolated and subjected to Kuhn–Roth oxidation to convert the three methyl groups, and the carbon atoms to which they are attached, into acetic acid. In phaseic acid, the only methyl group, or carbon to which it is attached, that can be derived from the labelled C-2 atom of mevalonate is the remaining methyl of the 6′-*geminal* pair of ABA. Consequently, if the resulting acetic acid, formed after oxidation of phaseic acid, turns out to be unlabelled, the remaining 6′-methyl group must be derived from the C-3 methyl group of mevalonic acid; whereas if the acetic acid is labelled then the remaining 6′-methyl group must be derived from the C-2 atom of mevalonic acid (Fig. 5).

A highly purified derivative of the acetate formed by oxidation of the phaseic acid contained ^{14}C and ^{3}H in the ratio of 1.000:1.637 (Table 3). Not only did this show that the remaining methyl group of the 6′-*geminal* pair of ABA must be the one derived from the C-2 atom of mevalonic acid, but also that no other radioactive atom of the phaseic acid molecule could have contributed significantly to the radioactivity of the acetic acid (Milborrow, 1975b). The slight loss of tritium observed in the earlier experiments was thus attributed to leakage, rather than to removal by hydroxylation. This result underlines the need for care in the interpretation of ^{14}C:^{3}H ratios.

The absolute configuration of phaseic acid (VI) has recently been characterized (Milborrow, 1975b), and so the absolute stereochemistry of the *geminal* pair of 6′-methyl

TABLE 2. ^{14}C:^{3}H RATIOS IN ABA FORMED FROM (±)-[2-^{14}C, 2-^{3}H]MEVALONIC ACID AND IN PHASEIC ACID FORMED BY TOMATO PLANTS FROM THIS ABA. DATA FROM TWO EXPERIMENTS

	Experiment no.	[^{14}C]dpm	[^{3}H]dpm	Normalized ^{14}C:^{3}H ratio	% ^{3}H lost in tomato plants and during extraction	
					from ABA	from phaseic acid
Mevalonolactone supplied to avocado fruit	1	9632	24,548	(3:6.000)	—	—
	2	2293	17,817	(3:6.000)	—	—
ABA after removal of exchangeable ^{3}H 10% counted	1	1623	1782	3:2.484	—	—
	2	1207	3398	3:2.174	—	—
ABA recovered from tomato plants (total sample)	1	868	943	3:2.458	1.1	—
	2	243	648	3:2.063	5.2	—
Phaseic acid isolated from tomato plants (total sample)	1	5338	5524	3:2.341	—	5.8
	2	1258	3014	3:1.850	—	15.0

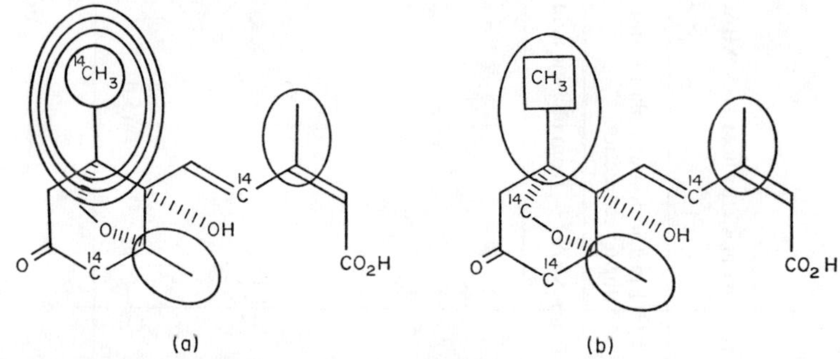

FIG. 5. Phaseic acid showing the possible positions of ^{14}C-label derived from [2-^{14}C]mevalonate. Kuhn–Roth oxidation oxidizes methyl groups, and the carbon atoms to which they are attached (shown ringed), to acetic acid. Only the 6′-methyl of (a) (triple ringed) can give ^{14}C-labelled acetic acid.

TABLE 3. OCCURRENCE OF ^{14}C AND ^{3}H IN ACETATE DERIVED BY KUHN–ROTH OXIDATION FROM THE 6′-METHYL GROUP OF PHASEIC ACID

	[^{14}C]dpm	[^{3}H]dpm	^{14}C:^{3}H	Normalized to 3:3
Methyl phaseate before oxidation	3786	7401	1:1.955	(3:3)
Expected ratio in acetate if 6′-methyl were derived from C-2 of mevalonate	—	—	1:3.910	(1:2)
p-bromophenacyl acetate after 2 recrystallizations	295	943	1:3.200	1:1.637

The abscisic acid carried ^{14}C and ^{3}H in the 6′-methyl group derived from C-2 of mevalonic acid.

groups of ABA can be deduced (Fig. 6). It is the same as that of the analogous groups in β-carotene found by Bu'Lock *et al.* (1970). In this feature also the stereochemistry of biosynthesis of ABA is identical to that of the carotenoids. Arrangement of the methyl groups is determined during cyclization, but now that their stereochemistry in ABA is known, the arrangement of the precursor before cyclization can also be deduced. This is as shown in Fig. 7(a), and not as in Fig. 7(b).

COLD TRAP EXPERIMENTS

In early attempts to define the precursors of ABA we prepared an equal mixture of the two isomeric diols (VII and VIII) by reduction of ABA with sodium borohydride

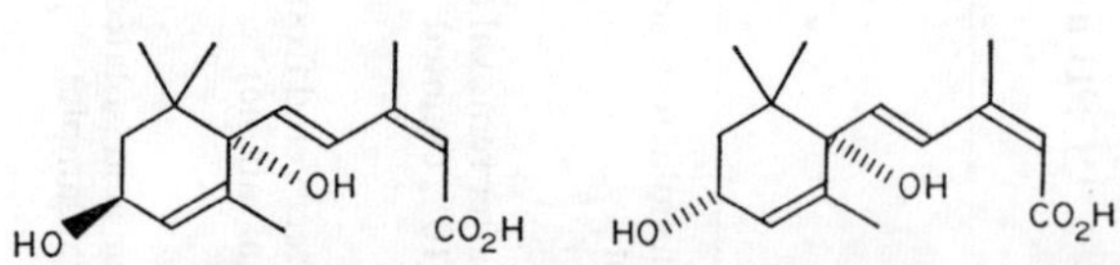

VII 1′,4′-*trans*-diol of (+)-ABA

VIII 1′,4′-*cis*-diol of (+)-ABA

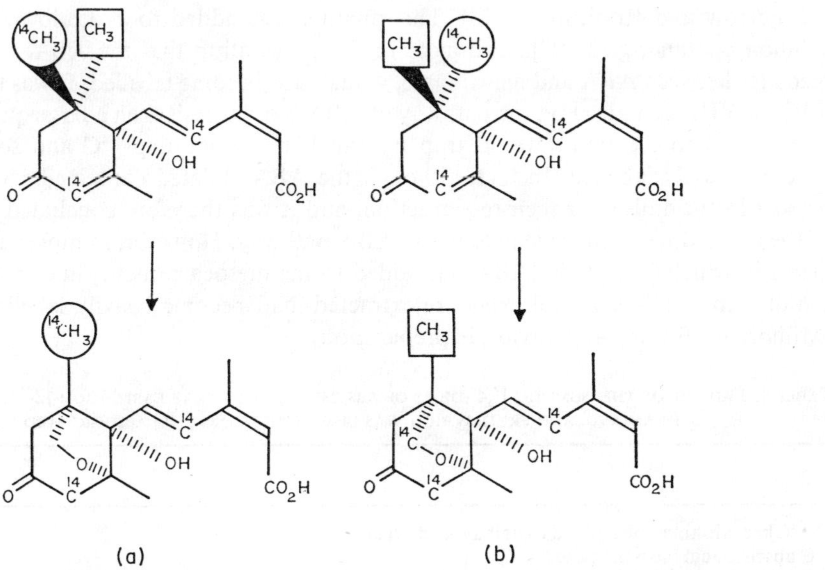

FIG. 6. The two possible cyclization mechanisms (Fig. 7(a) and (b)) which could give rise to abscisic acid (upper) produce two alternative labelling patterns in phaseic acid (lower). The 6′-methyl group of abscisic acid derived from the ^{14}C-labelled C-2 of mevalonic acid is shown (CH$_3$), the other 6′-methyl group derived from the C-3 methyl of mevalonic acid is shown [CH$_3$].

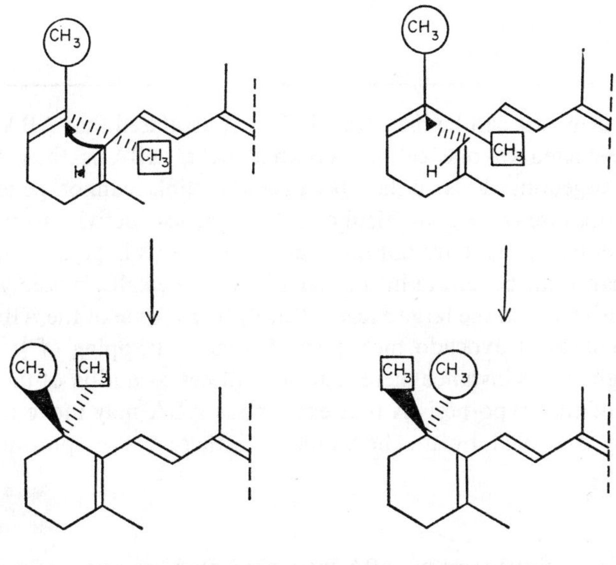

FIG. 7. Arrangements during cyclization of the 6′-*geminal* methyl groups of abscisic acid showing their derivation from C-2 (CH$_3$), and the C-3 methyl [CH$_3$], of mevalonic acid. The experimental results show that ABA undergoes cyclization by mechanism (a).

(Milborrow and Robinson, 1973). This mixture was added to avocado mesocarp in a solution containing [2-^{14}C]mevalonate in the expectation that the ABA, and all intermediates between ABA and mevalonate, would also become labelled. It was thought that if VII or VIII are indeed on the pathway of ABA biosynthesis then a large quantity of the diols added to the tissue would trap intermediates containing ^{14}C and so themselves become heavily labelled. In fact, although the ABA isolated contained ^{14}C, none was present in the diols after their re-extraction, and it was therefore concluded that neither of them could be an intermediate in the ABA pathway. However, in more recent experiments in which (±)-[2-^{14}C]ABA was added to the mesocarp tissue, in conjunction with a diol trap, 1′,4′-*trans*-diol when re-extracted had become heavily labelled (Table 4) (Milborrow, Burden and Taylor, in preparation).

TABLE 4. FAILURE OF THE ISOMERIC 1′,4′-DIOLS OF ABSCISIC ACID TO TRAP LABEL FROM [2-^{14}C]MEVALONATE IN AN AVOCADO FRUIT WHICH WAS BIOSYNTHESIZING [^{14}C]ABSCISIC ACID

	dpm
[2-^{14}C]mevalonolactone added to half avocado fruit	4×10^6
^{14}C abscisic acid biosynthesized	50,000
1′,4′-*cis*-diol of abscisic acid (1 mg) added as 'cold trap'	
1′,4′-*trans*-diol of abscisic acid (1 mg) added as 'cold trap'	
^{14}C in 'cold trap' diols after reisolation	2.0
(±)-[2-^{14}C]abscisic acid added to half avocado fruit	1×10^6
1′,4′-*cis*-diol of abscisic acid (1 mg) added as 'cold trap'	
1′,4′-*trans*-diol of abscisic acid (1 mg) added as 'cold trap'	
^{14}C in 1′,4′-*cis*-diol after reisolation	296
^{14}C in 1′,4′-*trans*-diol after reisolation	10,300

The earlier experiments can be interpreted if it is postulated that ABA synthesis from mevalonate is restricted to one cellular compartment. ABA is then prevented from exchange with exogenous diols, either because the diols cannot penetrate into this compartment or because there is no alcohol dehydrogenase activity to reduce the ABA. Results of the later experiment are not inconsistent with this hypothesis since exogenous ABA and diols can both penetrate into other parts of the cells, possibly where there is dehydrogenase activity, and the large excess of diols traps some of the ABA on reduction (cell free homogenates of avocado mesocarp also cause trapping of [^{14}C]ABA by the *trans*-diol although in this instance there is an overall net oxidation of the diols to ABA). The significance of this hypothesis is that exogenous ABA may move into parts of the cell not normally occupied by the hormone and there cause spurious physiological responses.

SYNTHESIS OF ABA BY A CELL-FREE SYSTEM

As already indicated, incorporation of labelled mevalonate into ABA is slight in turgid leaves (Milborrow and Noddle, 1970). This is reminiscent of a similar low incorporation of labelled mevalonate into carotenoids observed in 1958 by Goodwin. This low incorporation, taken in conjunction with the similarities in carbon skeletons and stereochemistry of biosynthesis of the two types of compound, suggest a common site of synthesis. Carotenoid synthesis is known to occur in plastids (Goodwin, 1958). We therefore tested the ability of isolated chloroplasts of avocado mesocarp (a tissue that synthesizes ABA rapidly) to incorporate mevalonate into ABA.

Chloroplasts and leucoplasts were isolated from tissue homogenates; samples were then lysed in a hypotonic solution and [^{3}H]mevalonate added. Unlabelled ABA was added 2–16 hr later, and ABA then isolated as its radiochemically pure methyl ester. This compound contained tritium label (Table 5). The presence of ^{3}H in the ester was confirmed by reduction to its constituent diols by means of sodium borohydride. ^{3}H was distributed equally between the two diols both of which co-chromatographed with authentic markers. In contrast, the non-chloroplast fractions failed to incorporate mevalonate into ABA. On the other hand extremely small amounts of labelled mevalonate were incorporated into ABA by excised tomato shoots.

TABLE 5. INCORPORATION OF [2-$^3H_2$]MEVALONOLACTONE BY INTACT AND LYSED CHLOROPLASTS ISOLATED FROM RIPENING AVOCADO FRUIT

	ABA isolated as methyl ester, μg	dpm in methyl ABA	dpm in 1′,4′-*cis*-diol of methyl abscisate	dpm in 1′,4′-*trans*-diol of methyl abscisate
Intact chloroplasts	23.6	546	238	201
Intact chloroplasts with acetone 0.5 ml/3.5 ml and Tween 20	21.7	910	300	280
Lysed chloroplasts	24.2	2860	1376	1197

17 hr incubation 2.0 × 10^7 dpm per sample, 0.1 M phosphate buffer. Cold carrier (±)-ABA (28 μg) was added when the preparation was killed. Incubation was carried out with laboratory light 480 lx. The methyl ABA was reduced with sodium borohydride to give equal amounts of isomeric 1′,4′-diols.

The question that at once arises is whether the ABA synthesized in plastids passes out into the rest of the cell. The presence of ABA in nectar, taken from flowers of the silky oak (*Grevillia robusta*) (Milborrow, 1969a), shows that it does because nectar is devoid of plastids whereas the presence of ABA in bleeding phloem and xylem sap, aphid exudate and honey could possibly be accounted for by the presence of a few plastids in these liquids.

The discovery of plastids as the site of synthesis of ABA and its ability to cause rapid closure of stomata (Jones and Mansfield, 1970; Mittelheuser and van Steveninck, 1969), together with the ubiquitous occurrence of chloroplasts in the guard cells of active stomata, suggests that these chloroplasts might synthesize ABA and so play a part in the regulation of stomatal apertures.

EFFECTS OF INHIBITORS ON ABA BIOSYNTHESIS

Many of the inhibitors of gibberellin and carotenoid biosynthesis, such as Phosphon D, CCC, diphenylamine and 3-amino, 1,2,4-triazole, have been tested in the avocado mesocarp system but without effect on incorporation of mevalonate into ABA. Two further inhibitors have been tested recently in the same system, namely AMO 1618 and Sandoz 6706; the latter is claimed to inhibit desaturation in carotenoid precursors (Bartels and McCullough, 1972). At 10^{-5} M, AMO 1618 has a negligible effect, but at 10^{-3} M, incorporation is enhanced to about three times that of the control (Table 6). Sandoz 6706 at 10^{-3} M on the other hand, strongly inhibits ABA synthesis, has little effect at 10^{-4} M and causes a three times enhancement at 10^{-5} M. This enhancement gradually declines with decreasing concentration of the inhibitor, and the inhibitor becomes ineffective at 10^{-8} M (Fig. 8).

TABLE 6. EFFECT OF AMO 1618 ON THE INCORPORATION OF [2-^{14}C]MEVALONATE INTO ABSCISIC ACID BY AVOCADO FRUIT

	Experiment 1 dpm in ABA/kg fruit	Experiment 2 dpm in ABA/kg fruit
Control	33,200	148,000
AMO 1618, 10^{-5} M	47,000	—
AMO 1618, 10^{-4} M	70,000	—
AMO 1618, 10^{-3} M	94,200	169,000
AMO 1618, 10^{-2} M	—	235,000

Approximately 40 g mesocarp was supplied with 2.5×10^6 dpm [2-^{14}C]-mevalonolactone in water (10 ml) containing the appropriate concentration of AMO 1618.

These results can be best interpreted in relation to a model in which most of the exogenous mevalonate supplied is utilized in the cytoplasm outside the chloroplasts for the synthesis of triterpenes and steroids. The small amount of mevalonate not utilized in the cytoplasm penetrates into the chloroplasts and becomes available for synthesis of carotenoids and ABA. If synthesis in the cytoplasm is inhibited to any extent by AMO 1618, as shown for other tissues by Paleg and Seamark (1968), more labelled mevalonate might be expected to reach and enter the chloroplasts and thus become incorporated into carotenoids and ABA. Separation of the neutral extract, followed by saponification and silica gel thin layer chromatography did indeed reveal considerably less radioactivity in steroids and triterpenes, and more radioactivity in carotenoids and ABA, in tissues treated with the inhibitor. It must be borne in mind that incorporation of more ^{14}C into ABA in the presence of the inhibitor does not necessarily imply that the rate of ABA synthesis is increased; on the contrary, it is likely that the rate is inhibited, but the degree of inhibition is outweighed by the high specific activity of the precursors.

In the case of Sandoz 6706, however, analysis of the neutral fractions showed no effect of the inhibitor on steroid synthesis. It is possible that this inhibitor exerts a sparing action within the chloroplast by reducing carotenoid synthesis, and thereby directing more labelled isoprenoid precursor into the ABA pathway. So far no specific inhibitor of abscisic acid biosynthesis has been found and consequently the investigation of the role of ABA *in vivo* by the prevention of its synthesis has been unattainable.

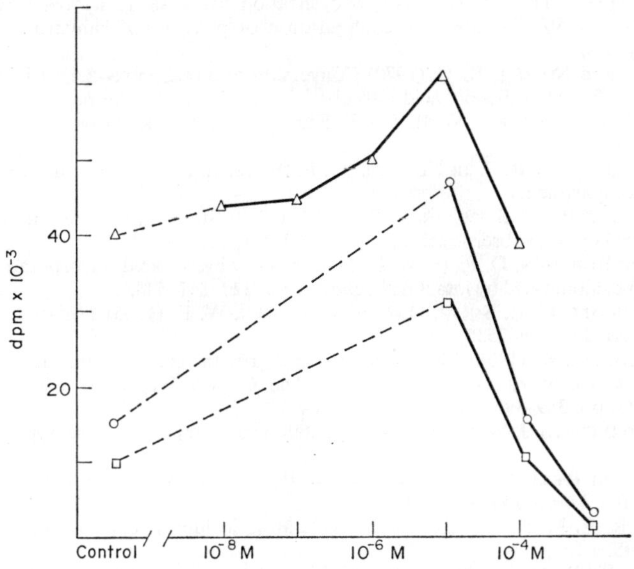

FIG. 8. Effect of Sandoz 6706 on the incorporation of [2-^{14}C]mevalonate into abscisic acid by avocado slices. Four or six segments of one fruit were used in each experiment and the results expressed as dpm of ^{14}C in ABA from 1 kg mesocarp. 2.5×10^6 dpm (±)-mevalonolactone (10.5 mCi/mM) were added to each segment.

REFERENCES

BARTELS, P. G. and MCCULLOUGH, C. (1972) A new inhibitor of carotenoid synthesis in higher plants: 4-chloro-5-(dimethylamino-)2-α,α,α, (trifluoro-m-tolyl)-3(2H)-pyridazinone (Sandoz 6706). *Biochem. biophys. Res. Commun.* **48,** 16–22.

BU'LOCK, J. D., AUSTIN, D. J., SNATZKE, G. and HRUBAN, L. (1970) Absolute configuration of trisporic acids and the stereochemistry of cyclization in β-carotene biosynthesis. *Chem. Commun.* 255–256.

CHRISPEELS, M. J. and VARNER, J. E. (1967) Hormonal control of enzyme synthesis: on the mode of action of GA and abscisin in aleurone layers of barley. *Pl. Physiol. Lancaster* **42,** 1008–1016.

COOMBE, B. G. and HALE, C. R. (1973) The hormone content of ripening grape berries and the effects of growth substance treatments. *Pl. Physiol. Lancaster* **51,** 629–634.

CORNFORTH, J. W., MILBORROW, B. V. and RYBACK, G. (1965) Synthesis of (±)-abscisin II. *Nature Lond.* **206,** 715.

DAVIS, L. A. and ADDICOTT, F. T. (1972) Abscisic acid: correlations with abscission and with development in the cotton fruit. *Pl. Physiol. Lancaster* **49,** 644–648.

GOODWIN, T. W. (1958) The incorporation of $^{14}CO_2$,[2-^{14}C]acetate and [2-^{14}C]mevalonate into β-carotene by illuminated etiolated maize seedlings. *Biochem. J.* **70,** 612.

JONES, R. J. and MANSFIELD, T. A. (1970) Suppression of stomatal opening in leaves treated with abscisic acid. *J. exp. Bot.* **21,** 714–719.

MILBORROW, B. V. (1967) The identification of (+)-abscisin II ((+)-dormin) in plants and measurement of its concentrations. *Planta* **76,** 93–113.

MILBORROW, B. V. (1969a) The occurrence and function of abscisic acid in plants. *Sci. Prog. Oxf.* **57,** 533–558.

MILBORROW, B. V. (1969b) Identification of "Metabolite C" from abscisic acid and a new structure for phaseic acid. *Chem. Commun.* 966–967.

MILBORROW, B. V. (1972) Stereochemical aspects of the formation of double bonds in abscisic acid. *Biochem. J.* **128,** 1135–1146.

MILBORROW, B. V. (1974) The chemistry and physiology of abscisic acid. *A. Rev. Pl. Physiol.* **25,** 259–307.

MILBORROW, B. V. (1975a) The stereochemistry of cyclization in abscisic acid. *Phytochemistry* **14,** 123–128.
MILBORROW, B. V. (1975b) The absolute configuration of phaseic and dihydrophaseic acids. *Phytochemistry* (in press).
MILBORROW, B. V. and NODDLE, R. C. (1970) Conversion of 5-(1,2-epoxy-2,6,6-trimethylcyclohexyl)-3-methylpenta-*cis*-2-*trans*-4-dienoic acid into abscisic acid in plants. *Biochem. J.* **119,** 727–734.
MILBORROW, B. V. and ROBINSON, D. R. (1973) Factors affecting the biosynthesis of abscisic acid. *J. exp. Bot.* **24,** 537–548.
MILBORROW, B. V., BURDEN, R. S. and TAYLOR, H. F. In preparation. The occurrence of abscisic acid in two cellular compartments.
MITTELHEUSER, C. J. and VAN STEVENINCK, R. F. M. (1969) Stomatal closure and inhibition of transpiration induced by (*RS*)-abscisic acid. *Nature Lond.* **221,** 281–282.
NODDLE, R. C. and ROBINSON, D. R. (1969) Biosynthesis of abscisic acid: incorporation of radioactivity from [2-^{14}C]mevalonic acid by intact fruit. *Biochem. J.* **112,** 547–548.
OHKUMA, K., ADDICOTT, F. T., SMITH, O. E. and THIESSEN, W. E. (1965) The structure of abscisin II. *Tetrahedron Lett.* **29,** 2529–2535.
PALEG, L. and SEAMARK, R. (1968) The effects of plant growth retardants on rat liver cholesterol biosynthesis. In *Biochemistry and Physiology of Plant Growth Substances.* eds. WIGHTMAN, F. and SETTERFIELD, G. pp. 389–398, Runge Press, Ottawa.
POPJAK, G. and CORNFORTH, J. W. (1966) Substrate stereochemistry in squalene biosynthesis. *Biochem. J.* **101,** 553–568.
ROBINSON, D. R. and RYBACK, G. (1969) Incorporation of tritium from [(4*R*)-4-^{3}H]mevalonate into abscisic acid. *Biochem. J.* **113,** 895–897.
VAN STEVENINCK, R. F. M. (1959) Abscission accelerators in lupins (*Lupinus luteus* L.). *Nature Lond.* **183,** 1246–1248.
WRIGHT, S. T. C. (1969) An increase in the 'inhibitor-β' content of detached wheat leaves following a period of wilting. *Planta* **86,** 10–20.
WRIGHT, S. T. C. and HIRON, R. W. P. (1969) (+)-Abscisic acid, the growth inhibitor induced in detached wheat leaves by a period of wilting. *Nature Lond.* **224,** 719–720.

PHYSIOLOGY OF SEXUAL REPRODUCTION IN MUCORALES

D. STEGWEE

Department of Plant Physiology, University of Amsterdam, The Netherlands

INTRODUCTION

Fifty years ago Burgeff (1924) published a lengthy paper entitled 'Untersuchungen ueber Sexualitaet und Parasitismus bei Mucorineen'. This event marked the beginning of a period of gradually intensified research on the physiology of sexual reproduction in the group of lower fungi which we now prefer to designate as the order of the Mucorales. This order is one of the class Zygomycetes, implying that characteristically sexual zygospores are formed by complete fusion of two gametangia. Familiar representatives are species of *Mucor*, *Blakeslea*, *Rhizopus*.

The aim of the present paper is to review briefly the early work and then to present more extensively results of recent investigations in which a group in my laboratory actively participated. Finally an attempt will be made to evaluate the truly fascinating results in terms of possible lines of future research. For extensive coverage of the literature on the subject up till 1970 the reader is referred to a previous review article by van den Ende and Stegwee (1971).

EARLY HISTORY

Sexual reproduction in the Mucorales was discovered by Blakeslee (1904). He observed that zygospores are formed by fusion of sexual hyphae which originate either from a single thallus or from two morphologically indistinguishable but physiologically different (i.e. compatible) thalli. Thus the distinction between homothallic and heterothallic species was introduced. Compatible strains of a heterothallic species were labelled by Blakeslee as + and −. The first assignment, of course, was rather arbitrary. From a particular pair of compatible strains of *Mucor mucedo* the one showing the most luxuriant growth was termed +. Thanks to the fact that very often incomplete matings occur between different heterothallic species a more or less consistent distinction between + and − strains of heterothallic Mucorales could be made (Blakeslee and Cartledge, 1927; Blakeslee *et al.*, 1927).

At this point it should be noted that so-called illegitimate matings also occur between heterothallic and homothallic species. According to Satina and Blakeslee (1930), the latter may exhibit either a + or a − tendency, or something in between. For the present, however, the discussion will be limited to heterothallic species.

Blakeslee and Cartledge (1927) concluded that '. . . the strict sexual dimorphism in the forms studied, as well as their morphological simplicity or sex differentiation, renders the group peculiarly adapted for use in an investigation of the fundamental differences between sexes'. Obviously Burgeff, whom I mentioned in the introduction, had also been aware of this. From Blakeslee's work it had already become evident that sexual hyphae or progametangia as they were named later by Cutter (1942) never occurred on separate thalli of single mating-types. They were formed only when compatible strains grew together. Burgeff (1924) demonstrated that no true physical contact between the partners is needed for progametangia to be formed. Even separation of a + and a − mycelium of *Mucor mucedo* by a collodion membrane did not prevent the initiation of sexual activity (the outgrowth of progametangia), although naturally it was a definite obstacle for achieving fulfillment (copulation of gametangia, the terminal cells formed in the progametangia by deposition of a cross-wall). Burgeff's conclusion was that there had to be a mutual interaction-at-a-distance between the partners ('telemorphosis') that could only be due to diffusion of a specific sexual compound. In present-day terminology we would for such a compound use the word 'sex-hormone' or preferably 'sex-pheromone'.

The existence of such pheromones was made all but certain when it was shown that filtered cell-free media from mated cultures of *Mucor mucedo* were able to induce the formation of progametangia in cultures of a single mating-type, but this confirmation of Burgeff's original hypothesis did not come until 30 years later (Banbury, 1954). Koehler (1935) still felt that instead of chemicals some kind of mysterious 'mitogenic rays' was involved. This merely serves to illustrate that for years and years little or nothing was added to our knowledge. Banbury's work and subsequently that of Plempel and his associates during the period 1957–1963 called for a drastic change of this situation. Plempel (1957, 1963) and Plempel and Braunitzer (1958) succeeded in the isolation, partial purification and characterization of the active principle from mated cultures of *Mucor mucedo*. Because the substance induced the formation of progametangia in both + and − mating-types, the active principle was considered to be a mixture of at least two sex-specific compounds which were termed 'gamones'. Because 'gamones' were produced in mated cultures, Plempel speculated that individual mating-types would secrete so-called 'progamones' which in a sex-specific way would induce the formation and secretion of sex-specific 'gamones' in the partners. This speculation was borne out by his observation that when a − strain was grown in culture medium from a + culture it apparently formed a 'gamone' active towards + only. The other way round the experiment worked equally well. However, attempts to separate the two 'gamones' produced in mated cultures remained unsuccessful, while a further identification also failed (Plempel, 1963).

THE PAST TEN YEARS

About 1965 van den Ende in Amsterdam entered the field. In his first paper (van den Ende, 1967) he described the isolation of a substance from a mated culture of *Blakeslea trispora*, capable of induction of progametangia in + and − *Mucor mucedo*. It was suggested that this sexual factor (the term 'gamone' was rejected) might be identical with trisporic acid C, a metabolite discovered at about the same time in mated cultures of *Blakeslea trispora* by Caglioti *et al.* (1967). Van den Ende (1968, 1969) readily proved his suggestion to be correct, but also showed that the sexual factor comprised two substances: trisporic acid C and trisporic acid B that had been isolated from the same source by Cainelli *et al.* (1967). The Italian workers had recognized the trisporic acids as potent stimulators of the biosynthesis of carotenes. Now these substances appeared to act as sex-pheromones as well. Significantly, the pheromones produced by mated cultures of *Mucro mucedo* appeared to be identical to those secreted by *Blakeslea trispora.* Austin *et al.* (1969) independently reached similar conclusions.

Important as the discovery of these pheromones undoubtedly was, neither trisporic acid B nor C met the requirements of sex-specificity set by Burgeff and claimed by Plempel for his 'gamones'. Either one or the other trisporic acid induced the formation of progametangia in + and − *Mucor mucedo.* Moreover, it was clearly demonstrated by van den Ende and Stegwee (1971) and by van den Ende *et al.* (1972) that trisporic acids are produced by both partners in a combined culture. Combination of a radioactive mycelium (obtained by growing + or − *Blakeslea trispora* on [^{14}C]glucose) with a non-radioactive counterpart always resulted in the production of [^{14}C]trisporic acids, irrespective of whether the + or − strain had been labelled.

Originally we hypothesized (van den Ende and Stegwee, 1971) that cultures of single mating-types would produce identical trisporic acid precursors. In mated or combined cultures then a mechanism for the conversion of precursors into trisporic acids would have to be induced and here eventually sex-specificity could come into play. However, in a note added at the last moment to our paper we indicated that Sutter (1970) had provided an alternative by showing that a + culture of *Blakeslea trispora* contained a substance that was converted into trisporic acid by the − strain only. Indeed, this strongly suggests that the trisporic acid precursors synthesized by the partners are not identical and, therefore, sex-specific. Confirmation of Sutter's (1970) suggestion came from two sides. Bu'Lock *et al.* (1972, 1973) discovered that in mated cultures of *Blakeslea trispora,* besides trisporic acids, a large number of related compounds was formed. Some of these might be considered as intermediates in the biosynthesis of trisporic acids. Significantly, the intermediates were partly also found in single + cultures, whereas others were encountered in single − cultures and mated ones.

Van den Ende *et al.* (1972) confirmed and extended Sutter's (1970) findings by demonstrating in culture media of both single + and − *Blakeslea trispora* the presence of different trisporic acid precursors which were converted into trisporic acids by mycelium of the opposite mating-type only. Sutter *et al.* (1973) reached the same conclusion. Bu'Lock *et al.* (1973), on the basis of his own results and of those obtained by the other groups, formulated a 'metabolic grid hypothesis' accounting for a complementary interaction between mating-types, whereby the complete biosynthesis of trisporic acids is only achieved in a mixed culture.

Werkman and van den Ende (1974) in the meantime turned their attention to *Mucor mucedo.* With this species they encountered a situation similar to that found in *Blakeslea trispora.* Nieuwenhuis and van den Ende (1974) identified the sex-specific precursors of trisporic acid in *Mucor mucedo.* The + strain produces 4-hydroxyl-methyltrisporates B and C, which were shown to be converted by the − strain into trisporic acids B and C. The conversion probably proceeds via methyltrisporates, traces of which were found in the incubation medium. These findings fully agree with those of Bu'Lock *et al.* (1973) for *Blakeslea.* The − strain of *Mucor mucedo* produces compounds which we propose to call trisporins B and C. Upon incubation with + mycelium these are converted into

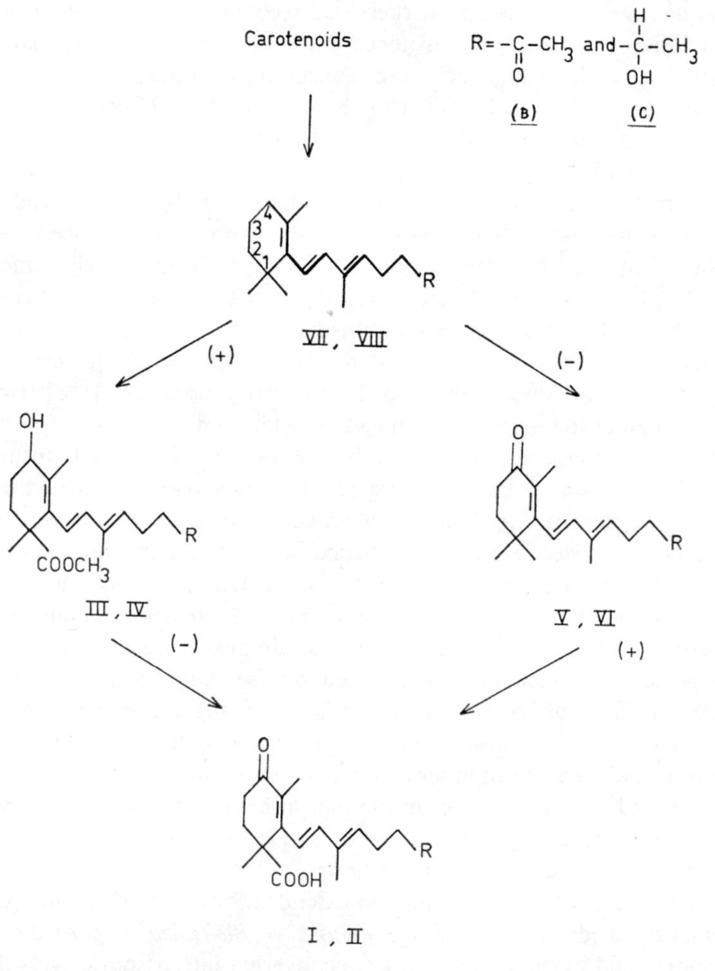

FIG. 1. Dual pathway for the biosynthesis of trisporic acids B and C in mated or combined cultures of *Mucor mucedo* Brefeld. I, II: trisporic acids B and C; III, IV: 4-OH-methyltrisporates B and C; V, VI: trisporins B and C; VII, VIII: β-C_{18}-ketone and -alcohol; (+) and (−) refer to (enzymatic) transformations carried out by the + and − mating-types respectively. (After Nieuwenhuis and van den Ende, 1974).

trisporic acids, probably by way of trisporols, which according to Bu'Lock *et al.* (1973) are the precursors produced by the − strain of *Blakeslea trispora.* Apparently there is after all—not surprisingly—a difference between *Mucor* and *Blakeslea* in that in the former the reaction trisporin→trisporol is carried out more efficiently by the + strain, and in the latter by the − strain.

The results discussed so far are summarized in Fig. 1, which depicts the situation in *Mucor mucedo.* Starting from a hypothetical β-C_{18}-ketone and a similar alcohol (both thought to be derived from β-carotene for reasons not to be discussed here) trisporic acids are synthesized along two different pathways. Because in + and − strains different reactions of these pathways are repressed, trisporic acids are only synthesized when both mating-types can complement each other.

One puzzling observation remains to be explained, viz. that the amounts of precursor compounds produced by single + or − cultures are grossly insufficient to account for the relatively high production of trisporic acids in mated cultures. Sutter *et al.* (1973) invoked hypothetical factors, named π and μ, produced by + and − strains respectively, which somehow stimulated the synthesis of trisporic acid precursors in the opposite mating-type. Here we recognize the old concept of Plempel's 'progamones'. Bu'Lock *et al.* (1973) on the other hand believed that the precursors themselves would act as de-repressors of precursor synthesis in the partners. Werkman and van den Ende (1974) demonstrated unequivocally that the trisporic acids are the substances promoting the production of their own precursors in both mating-types. Thus the complementary system of trisporic acid synthesis in mated cultures becomes de-repressed when and while it works.

Prospects for the future

In retrospect, so far the study of the physiology of sexual reproduction in the Mucorales has been highly rewarding. Referring to the above-mentioned quotation from Blakeslee and Cartledge (1927) one may state that at present the differences between the two sexes can largely be explained in simple biochemical terms. In the very near future undoubtedly the enzymology of the complementary biosynthesis of trisporic acids will be elucidated.

But then still only part of the differences are explained. It is all very well to be able to show how trisporic acids are formed in mated cultures. Next we can demonstrate that these acids induce the formation of progametangia, but nothing is known about how they do this. Yet this might well be a very fundamental question, part of the general problem of the mode of action of hormones and the like.

Once progametangia are formed, those from opposite mating-types appear to exert a distinct mutual attraction upon one another. This phenomenon had already been discovered by Burgeff (1924) who named it 'zygotropism'. Mesland *et al.* (1973) found that both the induction of the formation of progametangia and the subsequent zygotropism are brought about by volatile sex-specific pheromones which are probably closely related to the known precursors of trisporic acids. It is felt that a further study of the initiation of progametangia by trisporic acids and of the zygotropic behaviour may reveal some of the secrets of the fundamental processes behind the growth of fungal hyphae.

Zygotropism could possibly be explained by assuming a growth-promoting action of trisporic acids formed from certain (volatile) sex-specific precursors by equally sex-specific reactions. But then, when two progametangia meet and make contact they will only react when they are of opposite sex. This means that there has to be a mechanism for mutual recognition, enabling fusion, formation of gametangia and, finally, of the zygote. A study of the problem of how one cell is able to recognize another is obviously of more than limited importance.

In the second section of this communication I mentioned briefly the occurrence of illegitimate matings between homo- and hetero-thallic species. Werkman and van den

Ende (1974) were able to show that a number of homothallic Mucorales can convert precursors from single cultures of *Mucor mucedo* into trisporic acids. This suggests that in homothallic species a similar if not identical pheromonal system is operative. It seems that exploration of this possibility will lead to a better understanding of the essence of homo- and hetero-thallism in this interesting group of fungi.

REFERENCES

AUSTIN, D. J., BU'LOCK, J. D. and GOODAY, G. W. (1969) Trisporic acids: sexual hormones from *Mucor mucedo* and *Blakeslea trispora*. *Nature Lond.* **223,** 1178–1179.

BANBURY, G. H. (1954) Processes controlling zygophore formation and zygotropism in *Mucor mucedo* Brefeld. *Nature Lond.* **173,** 499–500.

BLAKESLEE, A. F. (1904) Sexual reproduction in the Mucorineae. *Proc. natn. Acad. Sci. U.S.A.* **40,** 205–319.

BLAKESLEE, A. F. and CARTLEDGE, J. L. (1927) Sexual dimorphism in Mucorales. II. *Bot. Gaz.* **84,** 51–58.

BLAKESLEE, A. F., WELCH, D. S. and BERGNER, A. D. (1927) Sexual dimorphism in Mucorales. I. *Bot. Gaz.* **84,** 27–50.

BU'LOCK, J. D., DRAKE, D. and WINSTANLEY, D. J. (1972) Specificity and transformations of the trisporic acid series of fungal sex hormones. *Phytochemistry* **11,** 2011–2018.

BU'LOCK, J. D., JONES, B. E., QUARRIE, S. A. and WINDSKILL, N. (1973) The biochemical basis of sexuality in the Mucorales. *Naturwiss.* **60,** 550–553.

BURGEFF, H. (1924) Untersuchungen ueber Sexualitaet und Parasitismus bei Mucorineen. *Bot. Abh.* **4,** 1–135.

CAGLIOTI, L., CAINELLI, G., CAMERINO, B., MONDELLI, R., PRIETO, A., QUILICO, A., SALVATORI, T. and SELVA, A. (1967) The structure of trisporic-C acid. *Tetrahedron* Suppl. **7,** 175–187.

CAINELLI, G., GRASSELLI, P. and SELVA, A. (1967) Struttura dell' acido trisporico B. *Chim. è l'Ind.* (Milano) **7,** 628–629.

CUTTER, V. M., JR. (1942) Nuclear behavior in the Mucorales. *Bull. Torrey Botan. Club.* **69,** 480–508.

ENDE, H. VAN DEN (1967) Sexual factor of the Mucorales. *Nature Lond.* **215,** 211–212.

ENDE, H. VAN DEN (1968) Relationship between sexuality and carotene synthesis in *Blakeslea trispora*. *J. Bact.* **96,** 1298–1303.

ENDE, H. VAN DEN (1969) Over de sexualiteit van enkele Mucorales. Ph.D. thesis, University of Amsterdam.

ENDE, H. VAN DEN and STEGWEE, D. (1971) Physiology of sex in Mucorales. *Bot. Rev.* **37,** 22–36.

ENDE, H. VAN DEN, WERKMAN, B. A. and BRIEL, M. L. VAN DEN (1972) Trisporic acid synthesis in mated cultures of the fungus *Blakeslea trispora*. *Arch. Mikrobiol.* **86,** 175–184.

KOEHLER, F. (1935) Genetische Studien an *Mucor mucedo* Brefeld. *Z. indukt. Abstamm. und Vererb.-Lehre* **70,** 1–54.

MESLAND, D. A. M., HUISMAN, J. G. and ENDE, H. VAN DEN (1973) Volatile sexual hormones in *Mucor mucedo*. *J. gen. Microbiol.* **80,** 111–117.

NIEUWENHUIS, M. E. J. and ENDE, H. VAN DEN (1975) Sex-specificity of hormone synthesis in *Mucor mucedo*. *Arch. Mikrobiol.* **102,** 167–169.

PLEMPEL, M. (1957) Die Sexualstoffe der Mucoraceae, ihre Abtrennung und die Erklaerung ihrer Funktion. *Arch. Mikrobiol.* **26,** 151–174.

PLEMPEL, M. (1963) Die chemischen Grundlagen der Sexualreaktion bei Zygomyceten. *Planta* **59,** 492–508.

PLEMPEL, M. and BRAUNITZER, G. (1958) Die Isolierung der Mucorineen-Sexualstoffe. I. *Z. Naturf.* **13,** 302–305.

SATINA, S. and BLAKESLEE, A. F. (1930) Imperfect sexual reactions in homothallic and heterothallic Mucors. *Bot. Gaz.* **90,** 299–311.

SUTTER, R. P. (1970) Trisporic acid synthesis in *Blakeslea trispora*. *Science N.Y.* **168,** 1590–1592.

SUTTER, R. P., CAPAGE, D., HARRISON, T. L. and KEEN, W. A. (1973) Trisporic acid biosynthesis in separate plus and minus cultures of *Blakeslea trispora*: identification *Mucor* by assay of two mating-type specific components. *J. Bact.* **114,** 1074–1090.

WERKMAN, B. A. and ENDE, H. VAN DEN (1974) Trisporic acid synthesis in homothallic and heterothallic Mucorales. *J. gen. Microbiol.* **82,** 273–278.

CHLOROPLAST DEVELOPMENT IN GREENING LEAVES

J. W. Bradbeer

Department of Plant Sciences, University of London King's College, 68 Half Moon Lane, London, U.K.

Introduction

It was appropriate that the Fiftieth Anniversary Meeting of the Society for Experimental Biology should have been held in Cambridge where Miss A. A. Irving made what may be regarded as the first precise investigation of the development of photosynthesis in greening seedlings (Irving, 1910). Irving measured the output of CO_2 from etiolated seedlings subsequent to their transfer from continuous darkness to diurnal conditions of light and dark. When the CO_2 output, measured by the Pettenkofer method, became consistently lower in the light than in the dark it was inferred that photosynthetic activity had commenced. Irving found that the onset of photosynthesis lagged considerably behind the appearance of chlorophyll, there often being a rapid development of photosynthesis when the leaves had almost attained a full green colour.

Contrary results were obtained by Willstätter and Stoll (1918) who reported detectable photosynthesis in greening leaves after only a small amount of chlorophyll had been formed. The controversy was elegantly resolved by Briggs (1920) who pointed out that Willstätter and Stoll's results had been obtained with older etiolated seedlings than those used by Irving. For the same species (*Phaseolus vulgaris*), Briggs showed that the illumination of leaves cut from 7-day-old etiolated seedlings gave results like those of Irving, while illumination of leaves from 12-day-old seedlings gave results like those of Willstätter and Stoll. Briggs deduced that the activity of the photochemical part of the photosynthetic mechanism in young seedling leaves as compared with that of more mature seedling leaves was in some way limited and that the photochemical part depended for its intensity not only upon chlorophyll but also upon some other factor. This factor increased with the age of the leaf. Results obtained in my laboratory are in full agreement with these conclusions and suggest that the other photochemical factor might be the thylakoid membrane system.

Development of the membrane system of etioplasts

In our work we have made a quantitative study of the development of the internal membrane system of the plastids in the primary leaves of *Phaseolus vulgaris* cv. Alabaster seedlings during growth in the dark (Bradbeer *et al.*, 1974a). Measurements were based on both phase-contrast and electron microscopy and the results, expressed on a per leaf basis, are summarized in Table 1. During the first 14 days of growth a large number of

TABLE 1. ETIOPLAST DEVELOPMENT IN DEVELOPING MESOPHYLL CELLS OF PRIMARY LEAVES OF *Phaseolus vulgaris* L. CV. ALABASTER SEEDLINGS DURING GERMINATION AND GROWTH IN THE DARK AT 23°C.

Days from sowing	Number of plastids/leaf $\times 10^{-8}$	Membrane in cm^2/leaf		
		Thylakoid membrane	Prolamellar body membrane	Total internal plastid membrane
4	0.11	—	—	—
6	0.73	3	0	3
7	1.49	—	—	—
9	1.74	52	6	58
10	2.04	94	52	146
12	2.68	154	84	238
14	2.88	134	139	273
30	3.19	105	145	250

cell divisions and plastid divisions occurred. Between 4 days (the earliest stage at which plastids became clearly distinguished under phase-contrast) and 14 days the total number of cells in the primary leaf increased by eleven times while the number of plastids in the developing mesophyll cells increased by twenty-six times. Most of this increase in plastid number occurred between days 4 and 7; the mechanism by which the plastids increased was not clear. However, between 7 and 14 days there was a further doubling in plastid number and it was then evident from electron micrographs that the increase resulted from plastid constriction. Figure 1b shows an oblique section through a constricted plastid at 9 days.

The earliest stage at which the amount of thylakoid membrane could be measured was at 6 days (see Fig. 1a), when the plastids were little more than proplastids, somewhat irregular in shape with an average diameter of about 1.5 μm. Only a few porous lamellar sheets (thylakoids) could be discerned at this stage, as invaginations of the inner membrane of the proplastid. The sheets were associated with prominent starch grains. Formation of porous thylakoid sheets continued to 9 days, after which condensation of the sheets and their assembly to form the prolamellar body commenced (Table 1, Fig. 1b). For the next 3 days both the thylakoid and prolamellar body membranes increased in amount, there being a fourfold increase in the total internal membrane system. Subsequently there was only a small increase in the total membrane, most of the further development of the prolamellar body being at the expense of the thylakoid membrane. At 14 days the prolamellar body had developed its characteristic structure of a regularly branched system of tubules (Fig. 1c) (Weier and Brown, 1970). The time course of prolamellar body formation was longer in our studies than in those of either Weier and Brown (1970) or Klein and Schiff (1972). This difference was probably due to the use of a different genotype and a lower temperature for germination.

It would seem likely that plastids in the young leaves studied by Irving (1910) were similar to those shown in Fig. 1a whereas the plastids in the older etiolated leaves studied by Willstätter and Stoll (1918) were similar to those in Fig. 1c. The data of Table 1 indicate that the total membrane content of the plastids in Fig. 1a is only about 1 per cent of the content of the plastids in Fig. 1c. The difference could account for the second factor postulated by Briggs (1920).

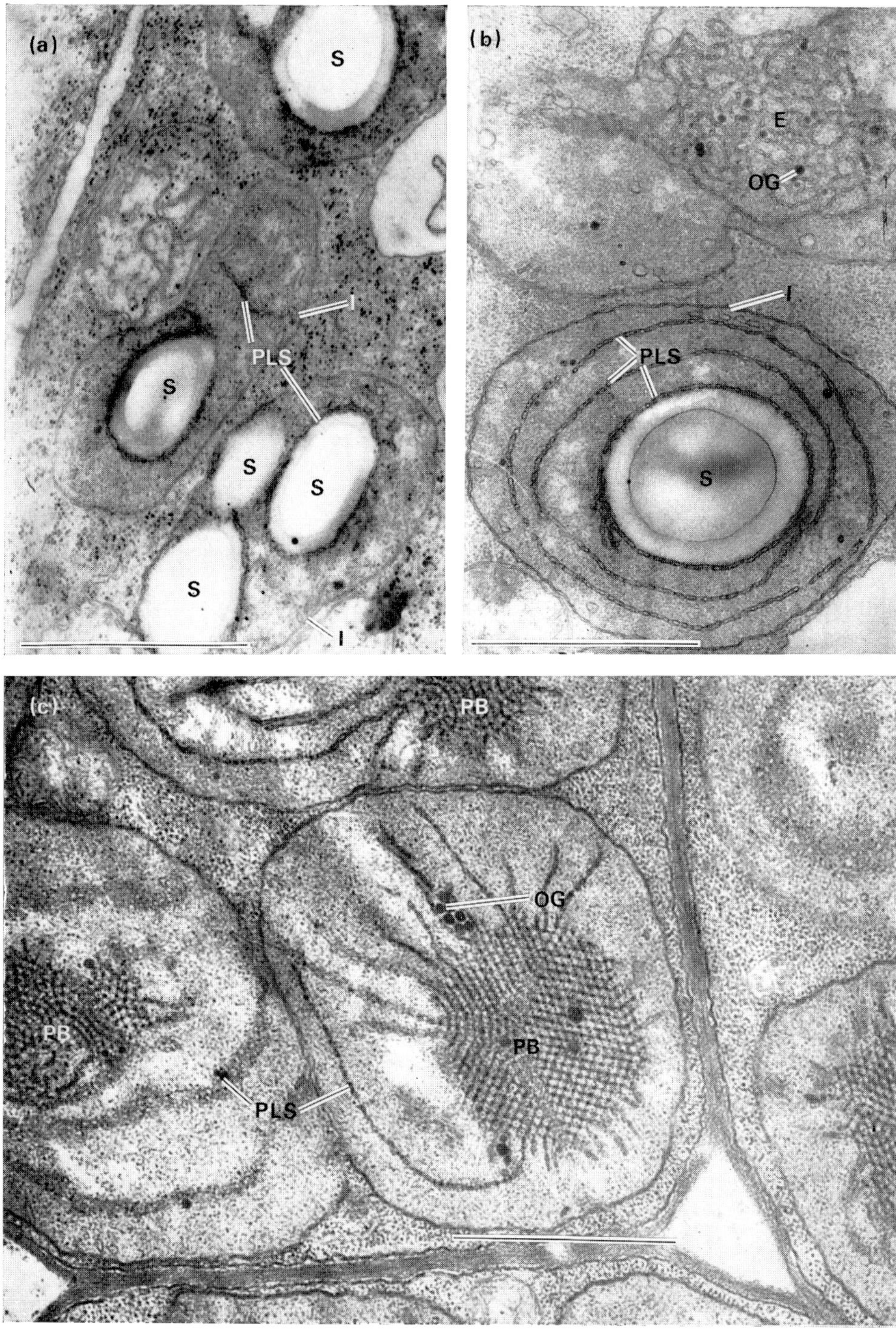

FIG. 1. Stages of etioplast development during growth in darkness of the primary leaves of *Phaseolus vulgaris*. The magnification is ×33,000, the scale lines representing 1 μm. Key to lettering: E, early stage of prolamellar body formation; I, invagination of plastid inner membrane; OG, osmiophilic globules; PB, prolamellar body; PLS, porous lamellar sheet (thylakoid); S, starch grain. (a), 6 days' dark growth; (b), 9 days' dark growth; (c), 14 days' dark growth.

LIGHT-INDUCED CHANGES IN ETIOPLASTS

So far our studies on the conversion of etioplasts into chloroplasts have been mainly restricted to 14-day-old plants of *Phaseolus*, which have well developed prolamellar bodies in the etioplast (Fig. 1c). A full reinvestigation of the Irving, Willstätter and Stoll and Briggs work on younger leaves has still to be realized.

Ultrastructural changes

Ultrastructural changes during the light-induced conversion of etioplasts have been described by many workers. In contrast to the earlier descriptive reports (see e.g. von Wettstein, 1958; Weier *et al.*, 1970) we have emphasized the quantitative aspects of these changes.

During the first few hours of illumination the prolamellar body is dispersed so that all of the internal plastid membrane appears as thylakoids (Bradbeer *et al.*, 1974b). Under our experimental conditions, *de novo* formation of thylakoids did not commence until between 10 and 15 hr after the beginning of illumination (Fig. 2). The formation of grana, evident at first as 'overlaps' between adjacent thylakoids, also became evident between 10 and 15 hr and by 15 hr the samples averaged 45 per cent of their total thylakoid membrane in grana. Thylakoid and grana formation continued throughout the rest of the 160 hr of the experimental period and, under the light intensity used (1.6 mW/ cm^2), almost 80 per cent of the thylakoid membrane became located in the grana.

Chlorophyll formation

Fourteen-day-old etiolated leaves of *Phaseolus* contained no chlorophyll and about 0.6 nmoles of protochlorophyllide *a*/leaf. After less than 1 min of illumination the protochlorophyllide *a* was converted into chlorophyllide *a* which was then phytylated to give chlorophyll *a* within about 20 minutes. *De novo* synthesis of chlorophyll occurred slowly at first but the rate of formation increased with time and reached a maximum at 15 hr, the maximal rate being maintained up to 72 hr, after which a slower accumulation of chlorophyll occurred up to the end of the experimental period. Chlorophyll formation in greening leaves has been reviewed by Kirk (1970).

Development of photosynthetic activity

Figure 2 also shows the development of CO_2 fixation during greening, the values having been obtained from differences between CO_2 efflux in darkness and CO_2 efflux or influx in light. Measurements were obtained by an infra-red gas analyser for leaves held in an atmosphere of 300 ppm CO_2 in air. The first significant measurements of CO_2 fixation were obtained after 10 hr; thereafter the rate of fixation increased with time up to 96 hr and then a slight decline set in. Extrapolation of the curve to the *y* axis indicated that fixation began after about 5 hr, a somewhat earlier time than that obtained previously by less sensitive methods (Bradbeer, 1969).

The data for CO_2 fixation and chlorophyll formation shown in Fig. 2 are in accord with those of Willstätter and Stoll (1918) and those obtained from 12-day seedlings by Briggs (1920) Chlorophyll formation begins before photosynthetic CO_2 fixation, as expected, since chlorophyll is an essential part of the photosynthetic energy-trapping system, but the data go further in showing that CO_2 fixation begins before *de novo* synthesis of thylakoids or formation of grana. Photosynthetic CO_2 fixation was found in leaves which did not possess conventional grana (see also Butler *et al.*, 1972) but in no case was a photosynthesizing leaf found to lack overlapping and appression of the thylakoids. Before illumination of the seedlings, the etioplasts had 3 per cent of their total internal membrane involved in such overlaps (Bradbeer *et al.*, 1974b), but after illumination additional overlapping was not detectable until about 7.5 hr. Robertson and Laetsch (1974) have likewise concluded that the overlapping thylakoids formed in darkness are responsible for the initial CO_2 fixation in the light. The data of Butler *et al.* (1972) are also consistent with this interpretation since their electron micrographs (De Greef *et al.*, 1971) clearly show the presence of overlapping and appressed thylakoids.

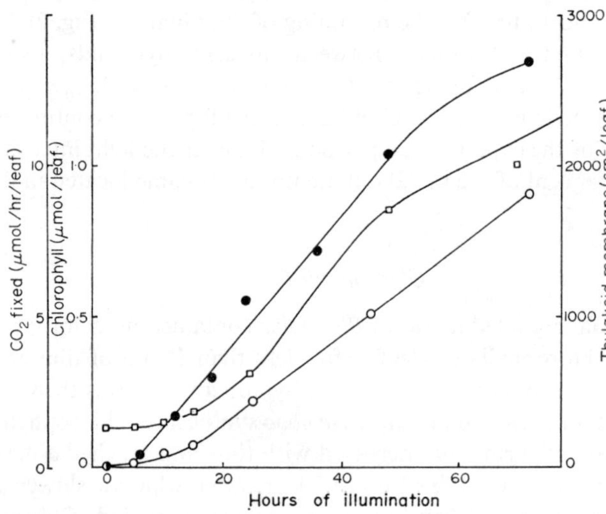

FIG. 2. Development of thylakoids and onset of photosynthesis during light-induced greening of primary leaves of 14-day-old dark-grown seedlings of *Phaseolus*. The illumination treatment amounted to 1.6 mW/cm² at 25°C. ○, chlorophyll content; □, area of thylakoid membrane; ●, rate of CO_2 fixation.

ACTIVATION OF ENZYMES OF THE CARBON REDUCTION CYCLE

In the context of CO_2 fixation by leaves during the light-dependent development of etioplasts into chloroplasts, it is desirable to know whether this fixation is mediated by enzymes synthesized during the dark period of growth or after the onset of illumination. In addition there is now considerable evidence to show that there is a light-induced activation of certain enzymes, which is a process which does not involve protein synthesis. An insight into the mechanism of this light activation of enzymes has come from experiments on crude extracts of plastids with which the light-induced activation can be simulated if NADPH and ATP are supplied in the incubation medium (Müller *et al.*, 1969; Müller, 1970). In my laboratory we have investigated three enzymes of the photosynthetic carbon reduction cycle which are subject to allosteric regulation. The enzymes are triosephosphate dehydrogenase ($NADP^+$) (EC 1.2.1.13), phosphoribulokinase (EC 2.7.1.19) and ribulosebisphosphate carboxylase (EC 4.1.1.39). The first two enzymes will be considered together because they show very similar activation properties after illumination of either intact leaves or isolated chloroplasts (Ziegler and Ziegler, 1965; Latzko and Gibbs, 1969; Steiger *et al.*, 1971).

The triosephosphate dehydrogenase reduces 1,3-diphosphoglycerate to glyceraldehyde-3-phosphate (Eqn. (1)), and uses either NADH or NADPH as reductant, while phosphoribulokinase catalyses the synthesis of ribulosebisphosphate and requires ATP (Eqn. (2)).

$$\text{3-phospho-D-glyceroyl phosphate} + \text{NADPH (NADH)} = \text{D-glyceraldehyde-3-phosphate} + \text{orthophosphate} + NADP^+ (NAD^+) \quad (1)$$

$$\text{ATP} + \text{D-ribulose 5-phosphate} = \text{ADP} + \text{D-ribulose 1,5-bisphosphate} \quad (2)$$

Extracts of etioplasts isolated from 14-day-old etiolated seedlings of *Phaseolus* can carry out these reactions without light and indeed light has no effect on these extracts. Pretreatment of these extracts with ATP, NADPH or dithiothreitol (DTT) results in substantial activation of these enzymes, as shown in Fig. 3 (Wara-Aswapati, 1973). Extracts were incubated with appropriate concentrations of either ATP or NADPH for 30 min before assay or with DTT for 72 min. 6 mM ATP gave a 550 per cent increase in activity of the triosephosphate dehydrogenase and 4 mM ATP increased the activity of phosphoribulokinase by 1000 per cent. 3 mM NADPH gave a 430 per cent increase of triosephosphate dehydrogenase activity and a 230 per cent increase of phosphoribulokinase activity. 7 mM DTT gave a 200 per cent increase of triosephosphate dehydrogenase activity and a 1150 per cent increase in phosphoribulokinase activity. It may be noted that the curves in Figs. 3a and 3b are sigmoidal in form.

Pupillo and Piccari (1973) found that pretreatment of crude extracts of spinach chloroplasts in the presence of $NADP^+$ resulted in a dissociation of the triosephosphate dehydrogenase molecule and released a more active moiety having a different molecular weight and $NAD^+/NADP^+$ activity ratio. Wara-Aswapati consequently examined the dissociation of triosephosphate dehydrogenase and phosphoribulokinase in extracts of *Phaseolus* etioplasts pretreated with ATP, NADPH and DTT. Extracts before and after pretreatment were fractionated on a Sepharose 6B column. The data of Fig. 4 show that pretreatment causes a shift in the elution profiles of both enzyme activities (Wara-Aswapati, 1973; Wara-Aswapati *et al.*, in preparation). Before pretreatment, triosephosphate dehydrogenase activity was recovered solely in one peak corresponding to a molecular weight of about 8×10^5 and having an NADH/NADPH activity ratio of 6. Phosphoribulokinase, on the other hand, dissociated into two active fractions having molecular weights of about 7×10^5 and 1.7×10^5. After pretreatment in 6 mM ATP for 24 hr (Fig. 4b) all of the triosephosphate dehydrogenase activity was again recovered in one peak, but this now corresponded to a lower molecular weight of about 2.2×10^5 and an NADH/NADPH activity ratio of about 0.3. Phosphoribulokinase activity was also recovered in only one peak, namely that corresponding to the lower molecular

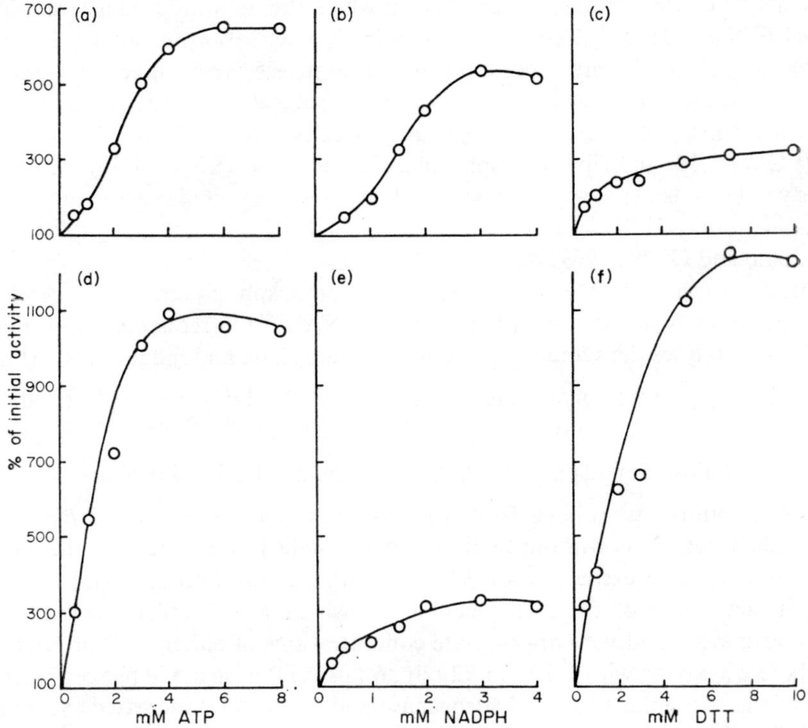

FIG. 3. Activation of triosephosphate dehydrogenase and phosphoribulokinase in an extract of etioplasts of primary leaves of 14-day-old dark-grown seedlings of *Phaseolus*. Extracts were preincubated at 25°C with either the appropriate concentration of activator or with water for 30 min in the case of both ATP and NADPH and 72 min for DTT. Assays were then performed and results expressed as the percentage of the initial activity (△ O.D. at 340 nm/min) of the water control, which was constant during the preincubation period. a, b and c: triosephosphate dehydrogenase; d, e and f: phosphoribulokinase. Data from Wara-Aswapati (1973).

weight. A similar elution profile of phosphoribulokinase activity was obtained after pretreatment with 7 mM DTT (Fig. 4c); however, the elution profile of triosephosphate dehydrogenase activity was again changed, activity now corresponding to a molecular weight of 3×10^4 and an NADH/NADPH activity ratio of 0.2. Pretreatment with DTT caused a considerable reduction in the recovery of triosephosphate dehydrogenase activity from the column.

Pawlizki and Latzko (1974) have performed similar experiments on the triosephosphate dehydrogenase of spinach chloroplasts after fractionation on a Biogel A-1.5 column. They found that the enzyme dissociated into two active fractions having molecular weights of 2.4×10^5 and 7.9×10^4 respectively and NADH/NADPH activity ratios of 2.5 and 0.36. Pretreatment with 0.03 mM NADP increased activity in the fraction of lower molecular weight while pretreatment with 10 mM DTT resulted in the appearance of all of the activity in this fraction. These workers also found that some of the activity lost during fractionation of the enzyme could be recovered by recombining the higher

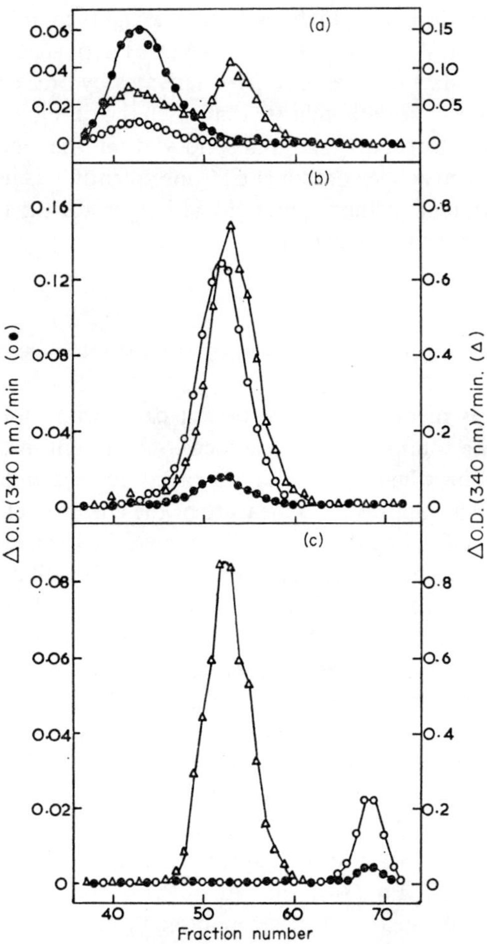

FIG. 4. Sepharose 6B-elution profiles of triosephosphate dehydrogenase and phosphoribulokinase activities of a partially purified extract from isolated chloroplasts of *Phaseolus*. 5 ml of the enzyme solution was preincubated for 24 hr with either (a) buffer (0.1 M Tris-HCl, pH 8; 0.01 M EDTA; 0.01% 2-mercaptoethanol), or (b) buffer plus 6 mM ATP, or (c) buffer plus 7 mM DTT, before being applied to a 2.5 × 37 cm column of Sepharose 6B. The Sepharose was preincubated and eluted with the appropriate solution. 2.5 ml fractions were collected and assayed for activity. ○, triosephosphate dehydrogenase with NADPH; ●, triosephosphate dehydrogenase with NADH; △, phosphoribulokinase. Data from Wara-Aswapati (1973).

and lower molecular weight fractions and that a further increase in activity could be obtained by the addition of a lower molecular weight fraction. Sodium dodecylsulphate (SDS)-polyacrylamide gel electrophoresis of the spinach enzyme revealed the presence of two different polypeptide subunits having molecular weights of 42,000 and 39,000. The higher and lower molecular weight fractions contained both subunits but in different relative proportions.

The discrepancies between the data of Wara-Aswapati and of Pawlizki and Latzko show that the mechanism of enzyme activation is still not fully understood. Activation clearly involves dissociation of a high molecular weight complex into smaller components. It is possible that ATP, $NADP^+$ and NADPH activate through reactions close to the active site of the enzymes, whereas DTT activates by breaking down S-S linkages between enzyme subunits. The data indicate that when the developing plastids commence photosynthetic electron transport and phosphorylation the above enzymes will be dissociated into their more active components. Consequently it is necessary to determine the responsibility of protein synthesis and light activation for the light-induced increase in the activity of these chloroplast enzymes.

ASSAY OF ENZYMES DURING THE GREENING PROCESS

The assay procedures mentioned above permit calculation not only of the enzyme activities but also of the total amount of each enzyme present in the extracts. For each enzyme the activity found when the extract is assayed under standard conditions before pretreatment with ATP is regarded as a measure of the enzyme activity in the leaf and it is expressed as μmoles of substrate consumed or product formed/minute/leaf at 25°C. After pretreatment the enzyme is then considered to show its full potential activity which is expressed as enzyme units/leaf (1 enzyme unit consumes 1 μmole of substrate or forms 1 μmole of product/minute at 25°C under specified standard conditions). By definition the full potential activity of the enzyme is directly proportional to the amount of an enzyme protein present. Hence changes in the amount of an enzyme during greening can be assessed from activation experiments.

To give an example of the determination of both enzyme activity and the amount of enzyme that this activity represents the assay of phosphoribulokinase will be considered (Eqn. (2)). We measured ADP in the conventional manner by its reaction with pyruvate kinase (EC 2.7.1.40, Eqn. (3)) and subsequent reduction of the pyruvate by lactate dehydrogenase (EC 1.1.1.27, Eqn. (4)), the rate of the overall reaction being determined by the oxidation of NADH measured spectrophotometrically at 340 nm.

$$\text{ADP} + \text{phosphoenolpyruvate} = \text{ATP} + \text{pyruvate} \tag{3}$$

$$\text{Pyruvate} + \text{NADH} = \text{lactate} + \text{NAD}^+ \tag{4}$$

For determination of enzyme activity the reaction mixture contained 50 μmol of Tris-HCl buffer pH 7.8, 10 μmol $MgCl_2$, 2 μmol DTT, 5 μmol of a ribulose-5-phosphate preparation, 40 μmol KCl, 2 μmol ATP, 2 μmol phosphoenolpyruvate, 100 μg NADH and 5 μg each of pyruvate kinase and lactate dehydrogenase (Racker, 1957; Bradbeer, 1969). The reaction was started by the addition of 0.10 ml of appropriately diluted enzyme extract to give a total volume of 1 ml. Since ATPase and NADH-oxidase interfere with this reaction it was necessary to run a control without ribulose-5-phosphate. The activity was determined between 2 and 4 min after addition of the enzyme and the control value was subtracted to give phosphoribulokinase activity. To determine the potential activity as enzyme units the assay was performed in a similar manner but the reaction was started with enzyme extract which had been preincubated for 12 min at 25°C in the presence of Tris-HCl, $MgCl_2$, KCl, ATP, and DTT.

Figure 5 shows changes in the actual and potential activities of both triosephosphate dehydrogenase and phosphoribulokinase with time after illumination of 14-day-old etiolated *Phaseolus* seedlings. Over an experimental period of 120 hr the potential activity of triosephosphate dehydrogenase showed a five and a half fold increase and there was no evident lag in its development. This increase in potential activity presumably resulted from *de novo* synthesis of the protein which seems to have commenced very soon after the onset of illumination. In contrast the actual enzyme activity increased by about fifty fold over the same period and there was a clear lag period before the increase in activity commenced. Before the onset of illumination the actual activity of triosephosphate dehydrogenase in the etiolated leaves amounted to only 10 per cent of the potential activity in the extract while after 90 hours of illumination the actual activity amounted to over 90 per cent of the potential activity. The increase in actual activity resulted in part from *de novo* synthesis of the enzyme and in part from *in vivo* activation of the enzyme by ATP, NADPH and possibly sulphydryl compounds formed as photosynthetic products. The lag period probably resulted from delay in the development of photophosphorylation and photosynthetic electron transport in the greening leaves.

A similar explanation can be offered for the ten fold increase in potential activity and the sixty fold increase in actual activity of phosphoribulokinase over the same period.

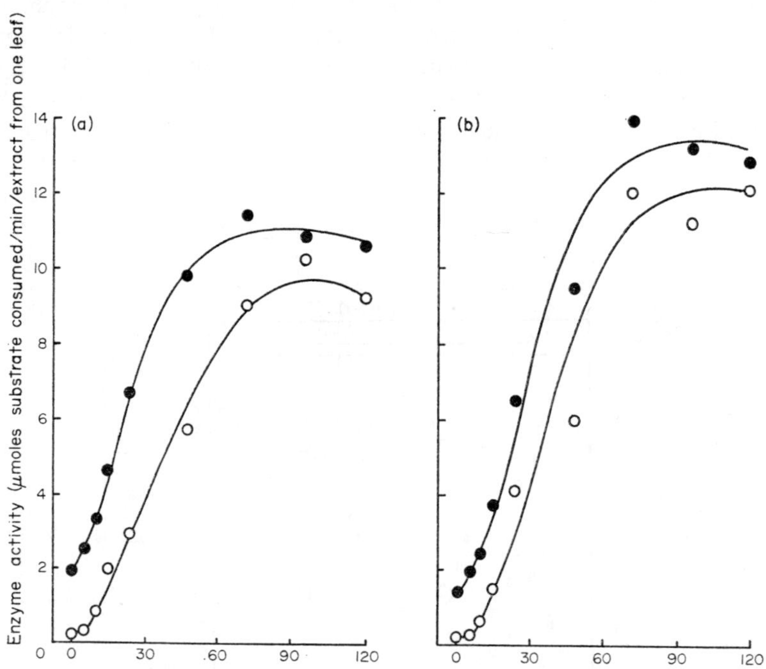

FIG. 5. Effects of the transfer of 14-day-old dark-grown seedlings of *Phaseolus* to continuous illumination (1.6 mW/cm^2) on the triosephosphate dehydrogenase (NADPH) and phosphoribulokinase activities of the primary leaves. (a) triosephosphate dehydrogenase; (b) phosphoribulokinase, ○, actual activity; ●, potential activity determined after 12 min activation as described in the text. Data from Wara-Aswapati (1973).

ANALYSIS OF PLASTID POLYPEPTIDES

The use of potential activity as a measure of the amount of an enzyme protein is subject to two objections, firstly that it is not a direct measure of absolute amounts of protein, but only of relative amounts, and secondly that the calculation is based on the assumption that activation of the enzyme is complete. These objections will be overcome when the mechanism of activation is fully understood. Direct measurement of protein however has its pitfalls as the protein in question needs to be freed from all other proteins and interfering substances and none of the protein must be lost during the purification process. One-dimensional polyacrylamide gel electrophoresis has proved to give satisfactory resolutions of the components of some fairly simple mixtures of polypeptides obtained in studies of organelle biogenesis (Blair and Ellis, 1973; Tzagoloff and Akai, 1972), but for the separation of more complex mixtures a technique such as two-dimensional polyacrylamide gel electrophoresis is necessary. This technique, developed by Martini and Gould (1971) for the separation of ribosomal polypeptides, has been adapted for the separation of stroma polypeptides in developing plastids (Arron and Bradbeer, 1974).

For the investigation of stroma polypeptides the soluble protein fraction of whole leaves was used. The fraction was first subjected to conventional polyacrylamide disc electrophoresis in buffered gel to allow the proteins to migrate in their undissociated state. The gel was then transferred through a number of solutions containing increasing concentrations of SDS, β-mercaptoethanol and urea and finally embedded as the starting line near the top of a 200 × 200 × 4 mm plate of SDS-polyacrylamide gel. A typical separation of polypeptides by the method is illustrated in Fig. 6. Electrophoresis in SDS-polyacrylamide gel causes the polypeptides to migrate as dissociated subunits. Figure 6A indicates the numerous subunits obtained from the supernatant of a whole leaf extract of *Phaseolus* while Fig. 6B shows those present in the supernatant of an

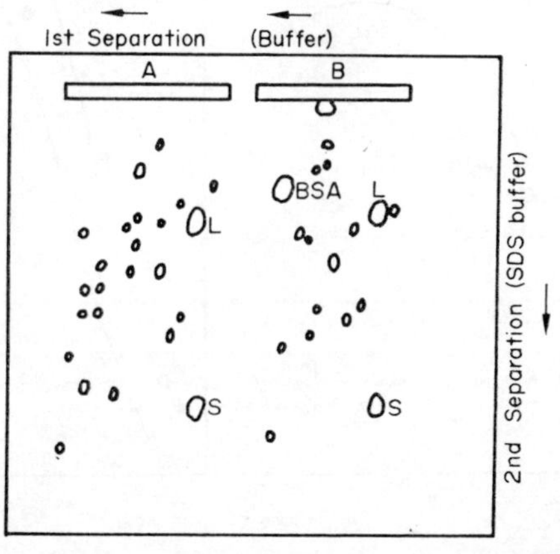

FIG. 6. Diagram of two-dimensional polyacrylamide gel separations of (A) a leaf extract and (B) a chloroplast extract obtained from primary leaves of greenhouse-grown *Phaseolus*. Key: L, large subunit of ribulosebisphosphate carboxylase; S, small subunit of ribulosebisphosphate carboxylase; BSA, bovine serum albumin used in the chloroplast preparation. After Arron and Bradbeer (1974).

osmotically ruptured preparation of chloroplasts isolated from the leaves. In both cases the most prominent components are the large and small subunits of ribulosebisphosphate carboxylase, having molecular weights of about 5.2×10^4 and 1.3×10^4 respectively. The identification of the other polypeptides has yet to be made.

To determine the amounts of the large and small subunits of ribulosebisphosphate carboxylase, after separation by gel electrophoresis, wc followed the normal procedure of treating the gel with Coomassie brilliant blue. The stained polypeptides were each eluted by dimethyl sulphoxide and the intensity of the stain then determined by spectrophotometry at 635 nm. Appropriate standards and blanks were included. Changes in the amount of each subunit after illumination of 10-day-old etiolated plants of *Phaseolus* are shown in Fig. 7. The amount of each subunit increased over the first 50 hr of illumination. There was no evidence of a delay in the synthesis of either subunit; both commenced to increase immediately after illumination. The large subunit increased in amount by about fifteen fold and the small one by about twenty-five fold over the first 50 hr. The discrepancy between the rates of increase is probably experimental error. Between 50 and 100 hr the amount of each subunit fell. No extract contained any detectable amount of free subunits. In contrast the enzyme activity increased throughout the whole of the experimental period, there being a forty-six fold increase in 100 hr and a substantial rise between 50 and 100 hr when the amount of each subunit decreased. It is evident that the increase in activity of ribulosebisphosphate carboxylase after illumination of etiolated leaves was to a large extent due to *de novo* protein synthesis but enzyme activation was also involved. Although *de novo* synthesis seemed to commence promptly in response to illumination the enzyme activity increased only slowly at first. The rapid increase in activity between 25 and 50 hr ot illumination may have involved some light activation of newly synthesized enzyme while the increase between 50 and 100 hr must have resulted wholly from activation. The enzyme is known to be under allosteric control

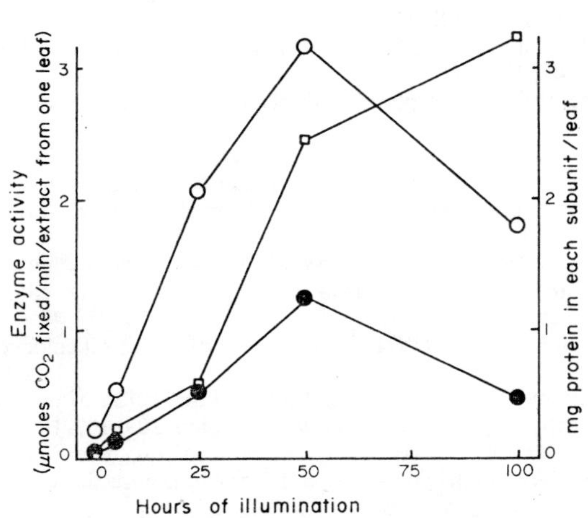

Fig. 7. Effects of the transfer of 10-day-old dark-grown seedlings of *Phaseolus* to continuous illumination (1.6 mW/cm^2), on the activity and amounts of the two subunits of ribulosebisphosphate carboxylase in the primary leaves. □, enzyme activity in μmoles CO_2 fixed/minute/leaf at 25°C, ○, large subunit in mg/leaf; ●, small subunit in mg/leaf. After Arron and Bradbeer (1974).

(Wildner and Criddle, 1969) but we have not so far had consistent results of activation of the enzyme *in vitro*. It is of interest to compare these results with those reported for the formation of ribulosebisphosphate carboxylase in greening barley leaves, in which the newly synthesized ribulosebisphosphate carboxylase was labelled during its biosynthesis and subsequently precipitated from extracts with the specific antibody (Smith *et al.*, 1974). As for bean, enzyme biosynthesis occurred early in greening while activation occurred later, but unlike bean no clear increase in the amount of enzyme protein was apparent.

Acknowledgements: I should like to acknowledge the excellent electron microscopy carried out by H. J. W. Edge and L. Neville and to thank Elsevier for permission to reproduce Figs. 6 and 7.

REFERENCES

ARRON, G. and BRADBEER, J. W. (1974) Use of two-dimensional polyacrylamide gel electrophoresis in the study of the biosynthesis of ribulosebisphosphate carboxylase. In *Proceedings of the 3rd International Congress on Photosynthesis Research*, ed. AVRON, M. pp. 2081–2088, Elsevier, Amsterdam.

BLAIR, G. E. and ELLIS, R. J. (1973) Protein synthesis in chloroplasts. I. Light-driven synthesis of the large subunit of fraction I protein by isolated pea chloroplasts. *Biochim. biophys. Acta* **319,** 223–234.

BRADBEER, J. W. (1969) The activities of the photosynthetic carbon cycle enzymes of greening bean leaves. *New Phytol.* **68,** 233–245.

BRADBEER, J. W., IRELAND, H. M. M., SMITH, J. W., REST, J. and EDGE, H. J. W. (1974a) Plastid development in primary leaves of *Phaseolus vulgaris*. VII. Development during growth in continuous darkness. *New Phytol.* **73,** 263–270.

BRADBEER, J. W., GYLDENHOLM, A. O., IRELAND, H. M. M., SMITH, J. W., REST, J. and EDGE, H. J. W. (1974b) Plastid development in primary leaves of *Phaseolus vulgaris*. VIII. The effects of the transfer of dark-grown plants to continuous illumination. *New Phytol.* **73,** 271–279.

BRIGGS, G. E. (1920) Experimental researches on vegetable assimilation and respiration. XIII. The development of photosynthetic activity during germination. *Proc. R. Soc.* B **91,** 249–268.

BUTLER, W. L., DE GREEF, J., ROTH, T. F. and OELZE-KAROW, H. (1972) The influence of carbonyl-cyanide-*m*-chlorophenylhydrazone and 3-(3,4-dichlorophenyl)-1,1-dimethylurea on the fusion of primary thylakoids and the formation of crystalline fibrils in bean leaves partially greened in far red light. *Pl. Physiol. Lancaster* **49,** 102–104.

DE GREEF, J., BUTLER, W. L. and ROTH, T. F. (1971) Greening of etiolated bean leaves in far-red light. *Pl. Physiol. Lancaster* **47,** 457–464.

IRVING, A. A. (1910) The beginning of photosynthesis and the development of chlorophyll. *Ann. Bot.* OS **24,** 805–818.

KIRK, J. T. O. (1970) Biochemical aspects of chloroplast development. *A. Rev. Pl. Physiol.* **21,** 11–42.

KLEIN, S. and SCHIFF, J. A. (1972) The correlated appearance of prolamellar bodies, protochlorophyll(ide) species, and the Shibata shift during development of bean etioplasts in the dark. *Pl. Physiol. Lancaster* **49,** 619–626.

LATZKO, E. and GIBBS, M. (1969) Effect of O_2, arsenate, sulfhydryl compounds and light on the activity of ribulose-5-phosphate kinase. In *Progress in Photosynthesis Research*, ed. METZNER, H., Volume 3, pp. 1624–1630, Tübingen.

MARTINI, O. H. W. and GOULD, H. J. (1971) Enumeration of rabbit reticulocyte ribosomal proteins. *J. molec. Biol.* **62,** 403–405.

MÜLLER, B. (1970) On the mechanism of the light-induced activation of the NADP dependent glyceraldehyde phosphate dehydrogenase. *Biochim. biophys. Acta* **205,** 102–109.

MÜLLER, B., ZIEGLER, I. and ZIEGLER, H. (1969) Lichtinduzierte, reversible Aktivitatssteigerung der NADP-abhängigen Glycerinaldehyd-3-phosphat-Dehydrogenase in Chloroplasten. *Eur. J. Biochem.* **9,** 101–106.

PAWLIZKI, K. and LATZKO, E. (1974) Partial separation and interconversion of NADH- and NADPH-linked activities of purified glyceraldehyde-3-phosphate dehydrogenase from spinach chloroplasts. *FEBS Lett.* **42,** 285–288.

PUPILLO, P. and PICCARI, G. (1973) The effect of NADP on the subunit structure and activity of spinach chloroplast glyceraldehyde-3-phosphate dehydrogenase. *Archs Biochem. Biophys.* **154,** 324–331.

RACKER, E. (1957) The reductive pentose phosphate cycle. I. Phosphoribulokinase and ribulose diphosphate carboxylase. *Archs Biochem. Biophys.* **69,** 300–310.

ROBERTSON, D. and LAETSCH, W. M. (1974) Structure and function of developing barley plastids. *Pl. Physiol. Lancaster* **54,** 148–159.

SMITH, M. A., CRIDDLE, R. S., PETERSON, L. and HUFFAKER, R. C. (1974) Synthesis and assembly of ribulosebisphosphate carboxylase during greening of barley plants. *Archs Biochem. Biophys.* **165,** 494–504.

STEIGER, E., ZIEGLER, I. and ZIEGLER, H. (1971) Unterschiede in der Lichtaktivierung der NADP-abhängigen Glycerinaldehyd-3-phosphat-Dehydrogenase und der Ribulose-5-phosphat-Kinase bei Pflanzen des Calvin- und des C_4-Dicarbonsäure-Fixierungstypus. *Planta* **96,** 109–118.

TZAGOLOFF, A. and AKAI, A. (1972) Assembly of the mitochondrial membrane system. VIII. Properties of the products of mitochondrial protein synthesis in yeast. *J. biol. Chem.* **247,** 6517–6523.

WARA-ASWAPATI, O. (1973) The development of photosynthetic activity and of some photosynthetic enzymes in greening leaves. Ph.D. Thesis. University of London.

WARA-ASWAPATI, O., KEMBLE, R. J. and BRADBEER, J. W. The activation of triosephosphate dehydrogenase and phosphoribulokinase in chloroplast extracts. In preparation.

WEIER, T. E. and BROWN, D. L. (1970) Formation of the prolamellar body in 8-day, dark-grown seedlings. *Am. J. Bot.* **57,** 267–275.

WEIER, T. E., SJOLAND, R. D. and BROWN, D. L. (1970) Changes induced by low light intensities on the prolamellar body of 8-day, dark-grown seedlings. *Am. J. Bot.* **57,** 276–284.

WETTSTEIN, D. VON (1958) The formation of plastid structures. *Brookhaven Symp. Biol.* **11,** 138–159.

WILDNER, G. F. and CRIDDLE, R. S. (1969) Ribulosediphosphate carboxylase. 1. A factor involved in light activation of the enzyme. *Biochem. biophys. Res. Commun.* **37,** 946–951.

WILLSTÄTTER, R. and STOLL, A. (1918) *Untersuchungen über die Assimilation der Kohlensäure*, pp. 127–135, Springer, Berlin.

ZIEGLER, H. and ZIEGLER, I. (1965) Der Einfluss der Belichtung auf die $NADP^+$-abhangige Glycerinaldehyd-3-phosphat-dehydrogenase. *Planta* **65,** 369–380.

PLASTID DEVELOPMENT IN ISOLATED ETIOCHLOROPLASTS AND ISOLATED ETIOPLASTS

RACHEL M. LEECH

Department of Biology, University of York, Heslington, York, U.K.

INTRODUCTION

The availability of techniques for the isolation of functional chloroplasts from mature leaf cells and the investigation of their biochemical activities *in vitro* have been most significant contributory factors in the progress of our understanding of the biochemistry of photosynthesis (Walker, 1970). In the study of plastid development, on the other hand, most investigations have been carried out on intact leaf tissue (for recent reviews see papers in Boardman *et al.*, 1971), partly for the obvious reason that the control of plastid development is a cellular function (Smillie and Steele-Scott, 1969; Kirk, 1972) but also because procedures for the isolation of developing plastids have only become available during the past few years. The study of plastid development *in vitro* is therefore relatively new and the potential of the use of isolated organelles in studies of chloroplast differentiation has still to be fully exploited. This paper reviews results of recent *in vitro* studies of differentiating plastids and considers their relevance to the understanding of chloroplast development *in vivo*.

Two experimental approaches (Fig. 1) have been used in many laboratories to study isolated developing plastids and should be carefully distinguished since the interpretation of the results from each method depends directly on knowledge of the characteristics of the starting material. In both cases leaves of etiolated plants are used. In the most frequently used system (A), etiolated (dark grown) leaves greening in the light are sampled at timed intervals after illumination begins, i.e. the plants are illuminated *before* isolation of the plastids. In the alternative procedure (B) the plastids are first isolated (in darkness), and only illuminated during incubation *in vitro*. While the first approach has considerable value in the study of light-induced plastid changes once they are initiated, the second has greater potential for the study of the initiation processes themselves and has been used to study pigment conversions and membrane changes. Partially differentiated plastids in etiolated leaves have been known as *etioplasts* since Kirk and Tilney-Bassett coined the term in 1967. To distinguish greening etioplasts from their precursors they will be referred to as *etiochloroplasts*.

CHARACTERISTICS OF PLASTIDS IN ETIOLATED LEAVES

A brief description of the current knowledge of the morphological characteristics of plastids in etiolated leaf tissue before greening (Weier and Brown, 1970) will be helpful

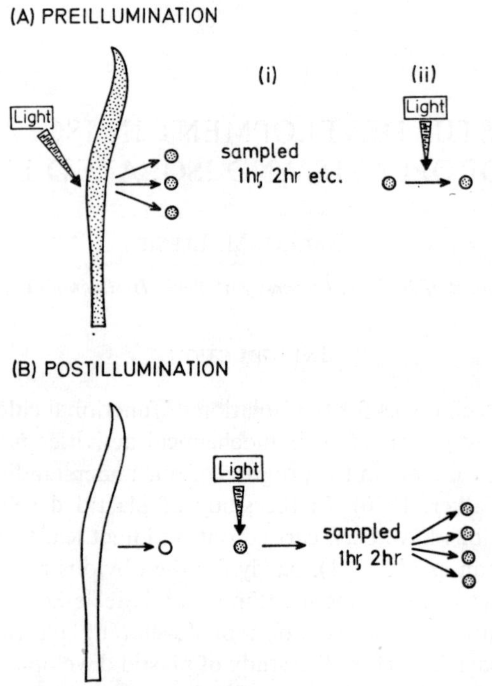

FIG. 1. Diagram illustrating the two experimental approaches used to investigate plastid development in isolated systems. (A) Etiolated plants are illuminated before isolation of the etiochloroplasts (pre-illumination). The isolated etiochloroplasts may be examined immediately or after further periods of illumination in the suspension medium. (B) Etioplasts are isolated from etiolated plants in the dark and then illuminated in the suspension medium (post-illumination).

in a consideration of the results of experiments on isolated etioplasts. Etioplasts develop from simple proplastids, c. 1 μm in diameter, which possess only a few internal membranes (Fig. 2a). Etioplasts may be up to 5 μm in diameter and are characterized by the presence of a central regular paracrystalline assembly of tubules, termed the *prolamellar body* (Fig. 2b, c, d). Etioplasts contain DNA (Jacobson *et al.*, 1963; Gunning, 1965) and ribosomal RNA (Jacobson *et al.*, 1963; Brown and Gunning, 1966), and synthesize 70 *s* ribosomes as is evidenced by the presence of large numbers of ribosomes (Boardman, 1966; Jacobson and Williams, 1968) and polyribosomes (Brown and Gunning, 1966). Etioplasts from different plant species are not necessarily at equivalent stages in development. The distinction between monocotyledons and dicotyledons is particularly clear: in general after the same growing period, etioplasts of grasses such as maize or barley reach a more sophisticated level of development than do etioplasts in dicotyledonous leaves. Development of photochemical activity may occur more rapidly in monocotyledons. After illumination (Whatley *et al.* (1972) found substantial increases in all cytochromes of *Phaseolus* leaves whereas in maize these components are synthesized in the dark (Plesnicar and Bendall, 1973). Robertson and Laetsch (1974) have recently pointed out that plastids in a variety of developmental stages from proplastid to etioplast

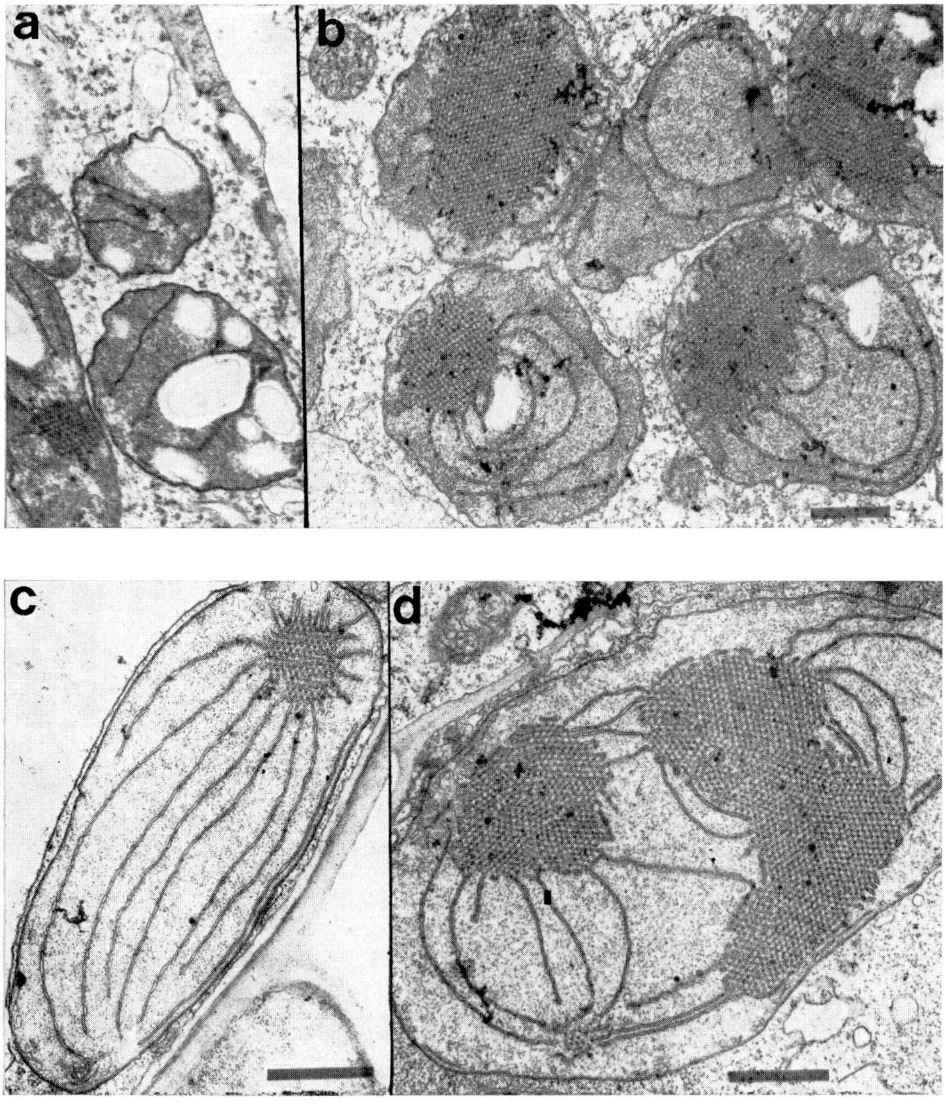

FIG. 2. Electronmicrographs of proplastids and etioplasts in cells of a 7 day-old etiolated maize (*Zea mays*) leaf (Leech and Mackender, unpublished). (a) Proplastids in cells 1 cm from the leaf base (× 14000); (b) young etioplasts in cells 2 cm from the leaf base (× 14000); (c) etioplast in a bundle sheath cell 4 cm from the leaf base (× 14000); (d) etioplast in a mesophyll cell 4 cm from the leaf base (× 14000). The scale represents 1μ.

Facing page 146

are found in the cells of a single etiolated leaf. In linear monocotyledonous leaves, such as barley (Robertson and Laetsch, 1974) and maize (Fig. 2), proplastids are found in the youngest cells at the base of the leaf, close to the intercalary meristem, and successively larger and more highly developed etioplasts are present in cells nearer the leaf tip. The developmental sequence found in a 7-day-old etiolated maize (*Zea mays*) leaf is shown in Fig. 2 and the system has been successfully used in our laboratory to provide a series of suspensions of plastids of increasing age.

CHARACTERISTICS OF ISOLATED ETIOCHLOROPLASTS

The photosystems

Illumination of an etiolated leaf initiates an immediate and extensive rearrangement of the membranes of the prolamellar body and chlorophyll synthesis begins. These early changes are followed by sequential functional development of the two photosystems. Fluorescence induction studies first indicated that Photosystem I becomes active before Photosystem II (Butler, 1965), and cyclic photophosphorylation can be detected in isolated etiochloroplasts before non-cyclic photophosphorylation (Gyldenholm and Whatley, 1968). Indeed the barley etiochloroplast is capable of extensive photochemical activity associated with Photosystem I after only a few minutes illumination, i.e. as soon as a trace of chlorophyll is detectable (Plesnicar and Bendall, 1973), whereas Photosystem II activity, measured as ferricyanide reduction or non-cyclic photophosphorylation, is not detectable until at least $2\frac{1}{2}$ hr after greening commences. CO_2 fixation begins later still (Bradbeer, 1969). The time course of greening seems to be independent of the light intensity since the pattern of development of the photosystems is similar whether 60 lx (Nadler *et al.*, 1972), 600 lx (Plesnicar and Bendall, 1973) or 8000 lx (Phung Nhu Hung *et al.*, 1970) is used. Recently it has become clear that relatively few qualitative molecular changes are induced by illumination of etiolated leaves. Synthesis of chlorophyll *a*, chlorophyll *b* and possibly cytochrome b_{559} are among the essential light-induced reactions. Before illumination, several components of the electron transport system, plastocyanin, cytochrome f, cytochrome $b_{559_{LP}}$ and cytochrome b_{563} are already present in sufficient concentration in etioplasts (Plesnicar and Bendall, 1973). The rate-limiting step in the development of Photosystem I activity appears to be the quantity of accessory pigment synthesized on illumination and Photosystem II activity depends on the prior light-dependent synthesis of chlorophyll *b* (Thorne and Boardman, 1971). Etioplasts also possess a Ca^{2+}-dependent ATPase (Lockshin *et al.*, 1971; Horak and Hill, 1972) which functions as a coupling factor for photosynthetic phosphorylation when added to chloroplast membranes deprived of their native coupling factor by washing with EDTA. Again, functional activity of the components already synthesized depends on the prior formation of the photosynthetic pigments.

Nucleic acids, proteins and lipids

In order to study the characteristics of the photosystems, preparations of isolated plastid membranes only are needed. Considerable synthesis of proteins, nucleic acids and lipids occurs in greening leaves at the same time as photochemical functions develop and membrane assembly is taking place (Kirk, 1970), but for detailed analysis of these changes in non-membrane plastid components, isolated intact etiochloroplasts are required. Several methods are available for the isolation of intact etioplasts (see later) but etiochloroplasts are very fragile in suspension and no leading worker in the field has so far claimed successful isolation of well preserved etiochloroplasts. In several studies, centrifuged fractions have been isolated from greening leaves and analysed as 'plastids'. In the absence of electron microscopical examination of such 'plastids', the analytical results should be treated with some caution. Isolation of 'plastid fractions' at intervals over 4 days after illumination of etiolated plants has permitted changes in size and various components to be quantified. In maize there is a threefold increase in volume and a doubling in dry weight; insoluble protein increases about threefold and total lipid more than doubles (Lürssen, 1970). The composition of the lipid does not change, in accordance with similar findings from the analysis of greening leaf tissue (Tevini, 1971; Tremolieres and Lepage, 1971; Roughan and Boardman, 1972) and in leaf tissue analysed during the development of proplastids directly into chloroplasts (Leech *et al.*, 1972, 1973). Isolated 'plastid' fractions have also been used to investigate changes in plastid proteins following illumination. Major quantitative differences in polypeptide pattern are not detectable until between 12 and 20 hr after illumination begins (Cobb and Wellburn, 1973) and light may also be a factor in controlling the type of protein synthesized since the relative incorporation into different polypeptide bands as determined by sodium dodecyl sulphate (SDS) polyacrylamide gel electrophoresis changes with increasing illumination (Hearing, 1973).

In order to establish the quantitative importance of such changes, data on the percentage plastid recovery and on the extent of protein loss from the plastids during isolation are essential. Only leaf tissue analyses are available describing the slow accumulation of chloroplast ribosomal RNA after illumination (Ingle *et al.*, 1970, 1971); the increase presumably reflects the increase in plastid size and also plastid number (Possingham, 1973; Boasson *et al.*, 1972).

In essence the pre-illuminated leaf system is an induced tissue system from which sequential samples of only one cellular component, the plastid, are taken and examined. The synthetic capacities of the etiochloroplasts after isolation can then be investigated by standard biochemical techniques. Recently, Rebeiz and Castelfranco and their co-workers have successfully demonstrated chlorophyll biosynthesis *in vitro* using homogenates isolated from pre-illuminated leaves. After $2\frac{1}{2}$ hr pre-illumination, incorporation of ^{14}C from δ-aminolaevulinic acid (ALA) into chlorophyll *a*, and after $4\frac{1}{2}$ hr into chlorophyll *b* (Rebeiz and Castelfranco, 1971b) could be demonstrated.

In isolated etiochloroplasts, [^{14}C]leucine is incorporated into lamellar and soluble proteins in about equal amounts (Parenti and Margulies, 1967; Margulies and Parenti, 1968), and as in chloroplasts from green leaves (Ireland and Bradbeer, 1971; Blair and Ellis, 1973; Ellis *et al.*, 1973), the major products of protein synthesis in etioplasts appear to be the large subunit of fraction I protein and a limited number of lamellar and envelope proteins (Siddell and Ellis, 1975). Pre-illumination also stimulates lipid synthesis by etiochloroplasts and has a marked effect upon the type of fatty acid into which [1-^{14}C]acetate is incorporated *in vitro* (Panter and Boardman, 1973). The proportion of unsaturated fatty acids synthesized is lower in darkness and in the presence of 3-(3,4 dichlorophenyl)-1,1-dimethylurea (DCMU). In pea and bean seedlings there is a delay of 6–8 hr before illumination causes an increase in lipid synthesis (Roughan and Boardman, 1972).

Characteristics of isolated etioplasts

Many stimulating questions are posed by the observations made on intact leaf tissue and isolated etiochloroplasts, not least those concerning the exact nature of the interactions between the plastids and other cellular components during development. For this reason attempts have recently been made to follow the development of etioplasts after isolation. This has proved a particularly useful approach in studying the induction of pigment and membrane changes by light. Because of the complexity of the spectral changes and the rapidity of the membrane reorganization processes, it is often instructive to isolate the etioplasts in the dark and then initiate and control the developmental process *in vitro*.

The situation has been neatly summarized by Woodcock and Bogorad (1971):

> Fundamental understanding of chlorophyll synthesis and membrane assembly is not yet possible because of our inability to study the etioplast in the presence of extra-chloroplast effects. Plastids capable of doing *in vitro* what they do *in vivo* need to be prepared. It will then be possible to dissect the system so that separate processes can be followed separately.

New problems are raised by this approach and isolation and incubation procedures are needed which preserve the integrity and function of the etioplast.

Etioplast isolation

The first suspensions of etioplasts were isolated in the early 1960s by Klein and Poljakoff (1960, 1961) and Boardman and Wildman (1962). The etioplasts were shown to be morphologically well preserved and capable of limited development after isolation, as manifested by prolamellar body proliferation and photoconversion of protochlorophyllide into chlorophyllide. Since then techniques have been considerably refined, and two, based on different physical principles, are currently in favour. The first relies on differential centrifugation to separate intact etioplasts from other cell components (Jacobson, 1968), whereas in the second, this separation is achieved by means of a macromolecular sieve such as Sephadex G25 (Wellburn and Wellburn, 1971a). In my laboratory we have modified the Jacobson procedure by inclusion of (i) an initial rapid spin (as used in isolation of chloroplasts) and (ii) a discontinuous sucrose density gradient. These steps help to clean up the preparations (Leese *et al.*, 1972). The entire procedure is rapid and can be completed within about 10 min. The etioplasts are remarkably well preserved as can be seen from Fig. 3. The only contaminants are mitochondria and bacteria, both at less than 1 per cent of the total profiles. The Sephadex method also gives suspensions of well preserved etioplasts frcc of contaminants. The method works excellently on *Avena* leaves (for which it was developed) but we have been less successful using the method for leaves of barley and maize.

TABLE 1. COMPOSITION OF MEDIA USED IN THE ISOLATION AND INCUBATION OF MAIZE ETIOPLASTS

Component	Isolation	Incubation
Sucrose	0.5 M	0.5 M
BSA	0.2%	0.4%
Ficoll	—	2.5%
PVP	1.0%	—
Cysteine	5 mM	5 mM
Mg	1 mM	0.5 mM
Fe	—	0.1 mM
Mn	—	1 mM
EDTA	—	1.5 mM
SO_4	—	0.5 mM
Cl	1 mM	1 mM
Acetate	—	0.5 mM
Glutamate	—	1 mM
HCO_3	—	0.5 M
Na/KPO_4	67 mM	67 mM
pH	7.5	7.5

In the case of etioplasts all manipulations must be carried out in complete darkness (or exceptionally under a dim green safelight). A few seconds exposure of the etioplasts to low intensity white light is sufficient to initiate both the photoreduction of protochlorophyllide to chlorophyllide and the associated transformation of the prolamellar body. This proviso does not of course apply to isolation of etiochloroplasts.

Membrane development in isolated etioplasts

Avena etioplasts, isolated by the Sephadex method, can be maintained in isolation for about 1 hour. On illumination, prolamellar body transformation and dispersion occur and membrane rearrangement continues at least to the membrane association stage (Wellburn and Wellburn, 1973). Further development is arrested. We have observed the same sequential changes in etioplasts from maize isolated by differential centrifugation (incubation medium detailed in Table 1). As in the leaf (Fig. 4) the first observable change on illumination of the isolated etioplasts is a change in the prolamellar body from a *tight* to a *loose* configuration (Figs. 3 and 5a). Etioplasts containing portions of prolamellar body in both tight and loose configuration ('mixed') are also present. Dispersal of the prolamellar body membranes first becomes visible around the periphery of the prolamellar body and eventually a network of continuous perforated membranes appears. This is similar to the dispersal in etioplasts in leaves described by Weier *et al.* (1970). The first stage of grana formation, i.e. the overlapping of lamellae, also takes place in isolated etioplasts but overlapping beyond the bithylakoid stage has not so far been observed *in vitro*.

Under the conditions of incubation currently in use, plastid development is far from synchronous. Wellburn and Wellburn (1973) have therefore introduced the concept of a *developmental index* to quantify the proportions of plastids in different morphological

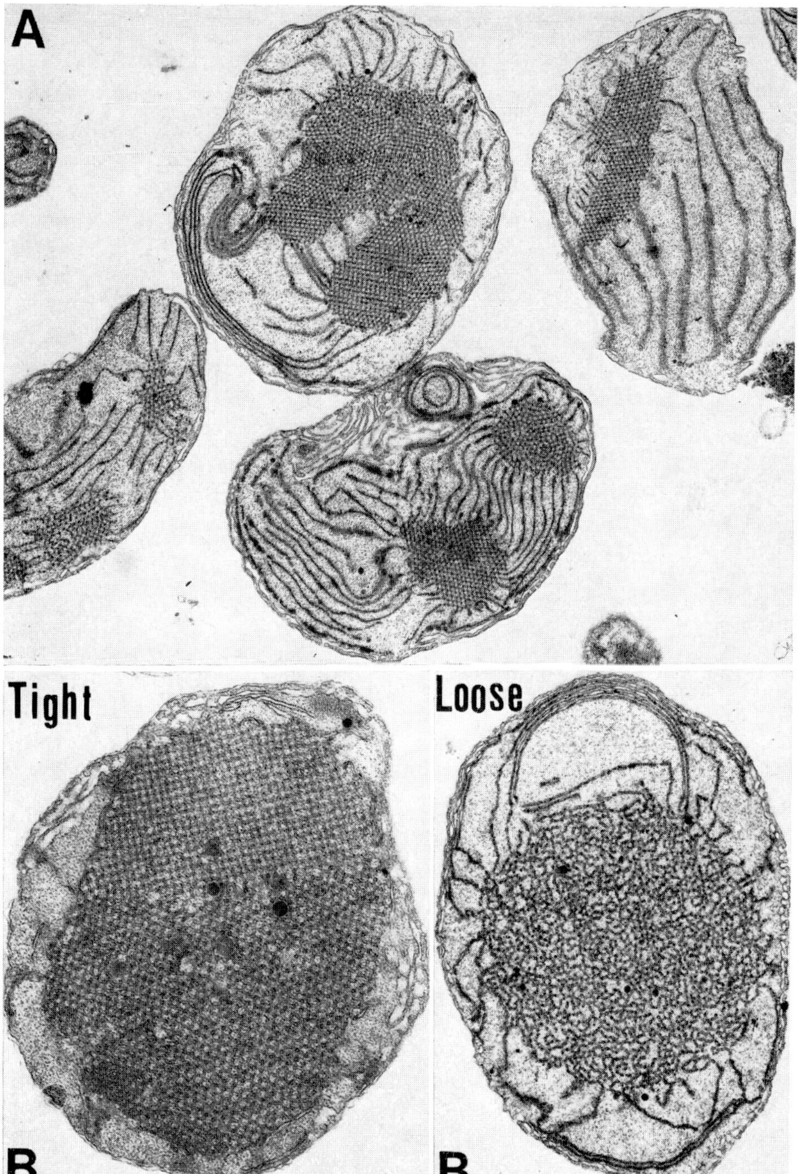

FIG. 3. Electronmicrographs of isolated etioplasts from 7 day-old etiolated maize leaves (Leese *et al.*, 1972). (A) Before illumination (× 10,000); (B) after illumination, showing examples of tight (untransformed) and loose (transformed prolamellar) bodies.

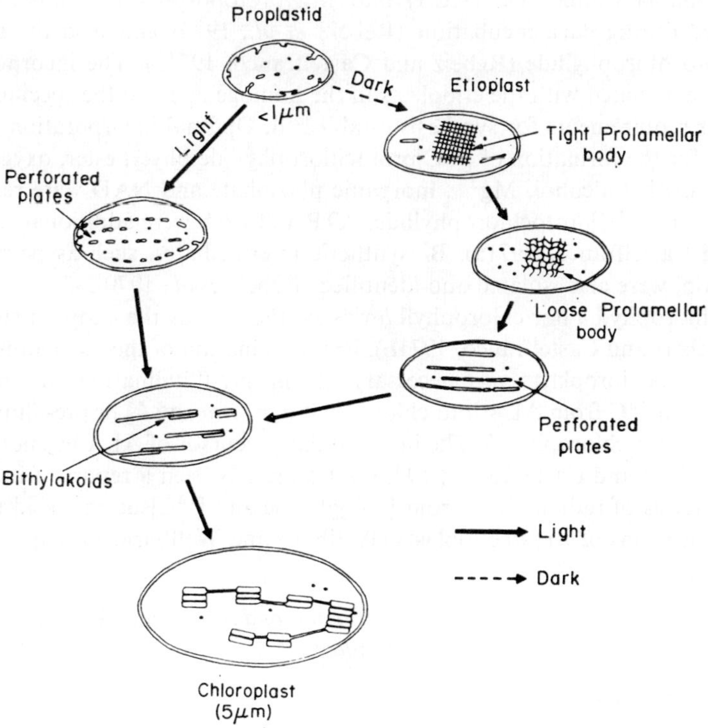

FIG. 4. Diagram illustrating the development of a proplastid into a chloroplast and the development of a proplastid into an etioplast which can then, after illumination, develop into a chloroplast.

states at different times after isolation. Extensive dispersion of the prolamellar body or extensive membrane proliferation (which the authors regarded as an indication of advanced development) were both given high ratings whereas the presence of tight prolamellar bodies or the absence of lamellate membranes were given low ratings. Until knowledge of the details of membrane development becomes more complete the choice of criteria must be somewhat subjective. The index has been used to study the effect of gibberellic acid (GA) on etioplast behaviour (Wellburn *et al.*, 1973). No change in fraction I protein was found in the presence of GA. It may be unwise to regard a negative finding at this stage as final since it is possible that the developmental index in its present form may be too imprecise to pick up small changes in structure and composition.

Chlorophyll formation in isolated etioplasts

The biochemistry of chlorophyll formation has been greatly extended by studies of suspensions of etioplasts isolated from cucumber cotyledons. Many of the studies have been carried out by Rebeiz and Castelfranco and their co-workers and a comprehensive review of the subject has been published (Reibez and Castelfranco, 1973). Incorporation of [^{14}C]δ-aminolaevulinic acid (ALA) into Mg protoporphyrin monoester has been demonstrated during dark incubation (Rebeiz *et al.*, 1970) and also the synthesis of labelled protochlorophyllide (Rebeiz and Castelfranco, 1971a). The incorporation was shown to be associated with the etioplasts in the homogenate and the specific conditions and cofactor requirements for synthesis established. Optimal incorporation occurred at pH 7.7, and for the formation of [^{14}C]protochlorophyllide phytyl ester, oxygen, reduced glutathione, methyl alcohol, Mg^{2+}, inorganic phosphate and NAD were required. For the formation of [^{14}C]protochlorophyllide, ATP and CoA were additional requirements (Rebeiz and Castelfranco, 1971a). Biosynthetic intermediates, such as porphyrins and Mg porphyrin, were also isolated and identified (Rebeiz *et al.*, 1970).

Neither chlorophyll *a* nor chlorophyll *b* was synthesized by the etioplast suspension in the dark (Rebeiz and Castelfranco, 1971b). Pre-illumination of the cotyledons and isolation of the etiochloroplasts was necessary. 2½ hr pre-illumination was required for incorporation of ^{14}C from ALA into chlorophyll *a in vitro* and 4½ hr pre-illumination for incorporation into chlorophyll *b*. The intracellular precursor of ALA has not so far been identified (Rebeiz and Castelfranco, 1973) but there has been a report of incorporation of very low levels of radioactivity from [^{14}C]glycine and [^{14}C]succinic acid into chlorophyll by suspensions of *Avena* etioplasts (Wellburn and Wellburn, 1971b).

Relationship between pigment photoconversion and membrane changes in isolated etioplasts

Spectral changes in vivo

In etiolated leaves, accumulation of chlorophyll precursors and changes in membrane configuration occur so rapidly after illumination that the relationships between them are difficult to determine. However in suspensions of isolated etioplasts held at 20°C in a defined medium under low illumination, the membrane and spectral changes take place more slowly and their characteristics are thus more easily investigated. We have used such suspensions of isolated etioplasts to investigate a problem which had been unresolved in studies of etiolated leaf tissue. The question to be decided is whether the pigment transformations *initiate* rearrangement of the membranes or whether the pigment synthesis is triggered by the initial membrane change in the prolamellar body itself.

Two major post-illumination spectral shifts have been recorded from studies on leaf tissue and are illustrated in Fig. 6.

*Photoconversion of protochlorophyll(ide) (PChl/e) to chlorophyll(ide) (Chl/e)**: This photoreduction is identified by a shift in absorption maximum from c. 650 nm to c. 680

* The following abbrevations have been adopted: PChle: protochlorophyllide; PChl: protochlorophyll; Chle: chlorophyllide; Chl: chlorophyll.

For identification purposes the spectral forms of protochlorophyll(ide) and chlorophyll(ide) are referred to as follows:

PChl/e	630	—	Absorption maximum at 628–630 nm
PChl/e	650	—	Absorption maximum at 650 nm
Chl/e	670	—	Absorption maximum at 669–674 nm
Chle	680	—	Absorption maximum at 680–685 nm

The range of absorption maxima refers to previously published values.

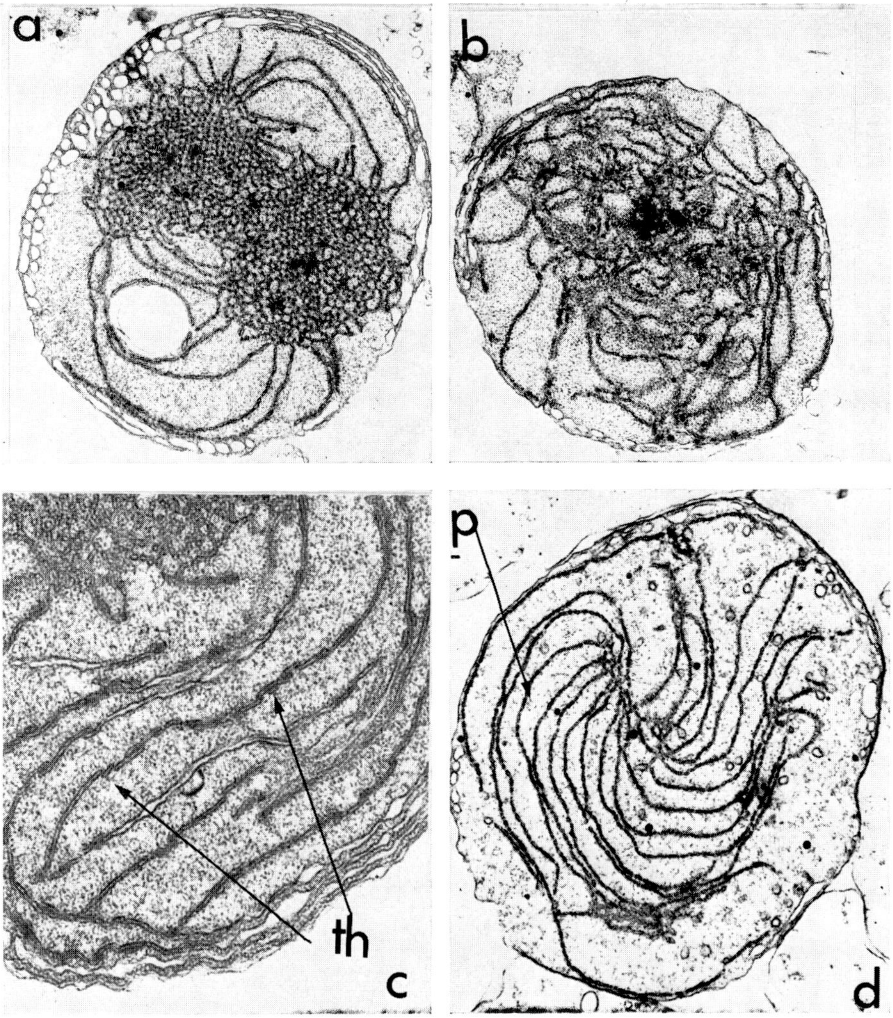

FIG. 5. Electronmicrographs of isolated etioplasts from maize after illumination (1.6×10^3 erg cm^{-2} sec^{-1}) for 5 hr. (a) An etioplast with partially-transformed prolamellar bodies and numerous lamellar performations ($\times$ 15000); (b) an etioplast showing dispersed prolamellar body and perforated lamellae ($\times$ 15000); (c) part of an etioplast at high magnification ($\times$ 40000) showing extensive bithylakoid formation(th); (d) an etioplast showing only peripheral transformation of the prolamellar body but extensive bithylakoid formation ($\times$ 15000) (p = perforated lamella).

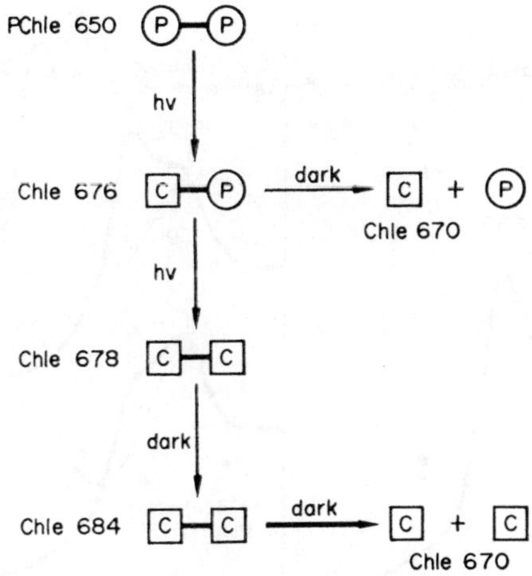

FIG. 6. Scheme proposed to account for spectroscopic changes observed in etiolated leaves after illumination (adapted from Mathis and Sauer). P and C represent protochlorophyllide and chlorophyllide monomers respectively. Dimeric species are represented as ○—○ and □—□. Numbers refer to wavelengths of absorption maxima.

nm. On illumination of etiolated leaves the 650 nm peak disappears and a subsidiary absorption peak between 630 and 635 nm remains (non-convertible PChl/e). Chlorophyll synthesis requires the continuous formation of PChl/e 630 and its conversion to PCh/le 650.

The Shibata shift: Photoreduction (above) is succeeded by a series of spectral changes collectively known as the *Shibata shift* (Shibata, 1957). These changes result in the shift of the absorption maximum from c. 680 nm, immediately after photoreduction, to c. 670 nm (the absorption maximum of the stable forms of chlorophyll). A portion of the pigment absorbing maximally at 670 nm is certainly phytylated (Treffry, 1970; Horton and Leech, 1975b) although phytylation itself does not give rise to the Shibata shift.

Spectral and membrane changes in vitro

In suitably prepared suspensions of maize etioplasts we have been able to study each of the spectral changes independently and new facts about the inter-relation of the spectral shifts and the membrane transformations have been revealed, particularly by analysing the effects of exogenously added ATP.

Photoreduction in isolated etioplasts: Before illumination the main PChl/e peak in isolated etioplasts is at 650 nm with a subsidiary peak or shoulder between 630 and 635 nm (Fig. 7a). After illumination the chlorophyllide absorption peak appears at 680 nm, the 650 nm peak disappears (convertible PChle) and non-convertible PChl/e remains absorbing maximally around 630 nm. Flashing-light treatment results in the conversion of a part of the remaining PChl/e 630 to PChl/e 650 (Horton and Leech, 1972, 1975b).

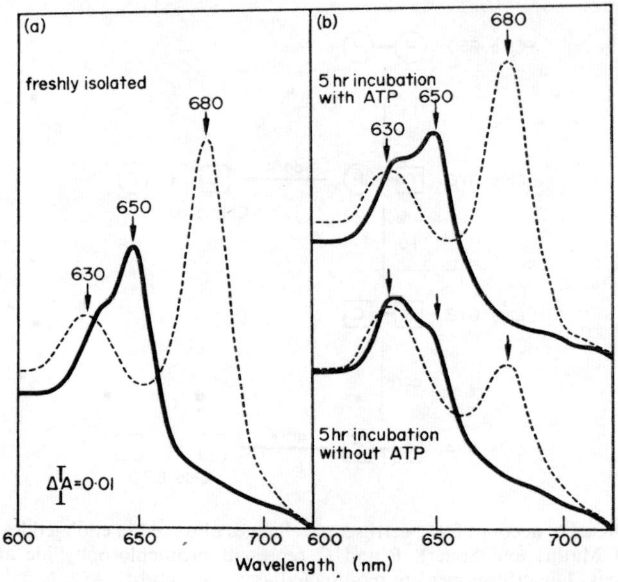

FIG. 7. (a) Absorption spectra of freshly-isolated maize etioplasts before and after illumination. Data from Horton and Leech (1975b). ———, Spectrum of etioplasts at zero time (dark); ------, spectrum after 40 sec illumination. (b) Absorption spectra of maize etioplasts after 5 hr incubation. Incubations with and without 1.5 mM ATP as indicated. Data from Horton and Leech (1975b). ———, Spectra of etioplasts before illumination; ------, spectra of etioplasts after 40 sec illumination.

In freshly-isolated etioplasts, more than 70 per cent of the PChl/e is in the photo-convertible form but 5 hr after isolation, a much larger proportion of the protochloro-phyll(ide) is now in the non-convertible (630) form (Table 2, Fig. 7(b)). Photoconvert-

TABLE 2. EFFECT OF ATP ON PHOTOCONVERTIBILITY OF PChl

Treatment	% Photoconvertibility	$\frac{A_{650}}{A_{630}}$
Intact etioplasts		
(a) 0 hr	100	1.95
5 hr	33	1.02
5 hr + ATP	75	1.56
Etioplasts without envelopes		
(b) 0 hr	100	1.71
5 hr	37	0.78
5 hr + ATP	76	1.27

(a) Effect of addition of 1.5 mM ATP (final conc.) to intact etioplasts; (b) effect of addition of 1.5 mM ATP to etioplasts prepared without envelopes. Incubation conditions as described in the text. The degree of photoconvertibility of PChl was measured by expressing the Chl peak height at 680 nm produced by illumination as a percentage of that obtained by illuminating freshly-isolated etioplasts. (Horton and Leech, 1972) ibility is preserved, however, if ATP (optimally at 1.5 mM) is added to the incubation medium. ADP has a smaller effect, and AMP no effect on photoconvertibility (Horton and Leech, 1975b). Analysis of the isolated pigments revealed that only unphytylated PChle is involved in the ATP effect (Table 3). Specific association between holochrome and protochlorophyllide is required for photoconversion (Kahn, 1968; Kahn *et al.*, 1970) and ATP appears to facilitate this binding as suggested in the model in Fig. 8. The proposal that ATP induces specific binding between a porphyrin and a protein is not unique. Recently, Haddock and Schairer (1973) have described an ATP-induced reconstitution of cytochrome *b* oxido-reduction in *Escherichia coli* homogenates in the presence of added haematin. They interpret their observation as indicating that ATP induces binding of the haematin to the apoprotein to form a functional cytochrome.

The Shibata shift in isolated etioplasts: The Shibata shift can be studied in isolated etioplasts but takes longer for completion than in the leaf. The molecular basis for the Shibata shift is less well-understood than is photoconversion and it seemed possible that ATP, since it appears to facilitate pigment-protein binding may also be a useful probe in the further characterization of the shift. ATP (1.5 mM) inhibits the completion of the shift (Fig. 9) in isolated etioplasts where the absorption maximum after 3½ hr is 657 nm. The proportion of phytylated to non-phytylated chlorophyll(ide) is unchanged in the presence of ATP (Table 3), confirming the suggestion from intact leaf studies that the Shibata shift is not a manifestation of phytylation. If it is assumed that ATP is also affecting the conformation of the holochrome during the Shibata shift (as it does in the maintenance of photoconvertibility), then the inhibitory effect of ATP on the shift can be interpreted as an *inhibition* of the *dissociation* of the individual holochrome subunits. The effect is thus analogous to the ATP effect on photoconvertibility where ATP apparently increases the *association* between PChle and the holochrome. Such an inter-

TABLE 3. CONCENTRATIONS OF PHYTYLATED AND UNPHYTYLATED PIGMENTS AFTER INCUBATION UNDER CONTINUOUS ILLUMINATION WITH AND WITHOUT 1.5 mM ATP

Pigment	Concentration (nmol/ml)	
	+ATP	−ATP
Phytylated:		
Chl	0.12	0.09
PChl	0.14	0.16
Unphytylated:		
Chle	0.68	0.61
PChle	0.08	0.17
Total Chl/e	0.80	0.70
Chl/e Phytylation	15%	13%
Total Chl/e + PChl/e	1.02	1.03

Phytylation (%) is expressed as the percentage of the total Chl/e which is phytylated (Chl). Each value is a mean of duplicate extractions and separations. (Horton and Leech, 1975a)

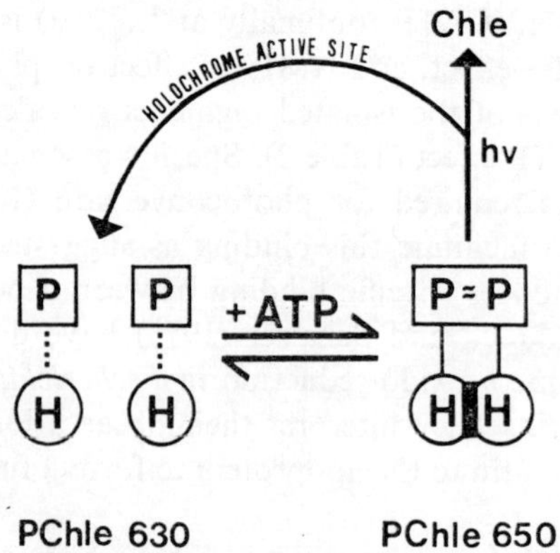

FIG. 8. Model to explain the effect of ATP on photoconvertibility. Illumination (hv) induces Chle formation and allows further PChle 650 synthesis under the influence of ATP. (Horton and Leech, 1975b) ○, Holochrome; □, protochlorophyllide; ..., feeble association between P and H; P≈P, pigment–pigment interaction; ——, photoreactive binding between P and H induced by ATP; ○○, association between two holochrome molecules.

pretation would support proposals that the Shibata shift results from changes in interactions between porphyrin rings (Kirk, 1970), perhaps as a result of dissociation of pigment-protein dimers (Mathis and Sauer, 1972; Schultz and Sauer, 1972) or alternatively conformational relaxation of the individual holochrome subunits (Henningsen *et al.* (1974).

Morphological configurational changes in tubules of the prolamellar body: Quantitative analyses of electron micrographs of the illuminated etioplasts showed that the frequency of untransformed prolamellar bodies was higher in the presence of ATP than in its absence. The values for tight and loose prolamellar bodies are given in Table 4 and

TABLE 4. QUANTITATIVE COMPARISON BETWEEN THE DEGREE OF THE SHIBATA SHIFT AND THE DEGREE OF TRANSFORMATION OF THE PROLAMELLAR BODY

Treatment	Chle 680 (as % total Chl/e)	Tight prolamellar bodies (as % total area of prolamellar body)
a) plus ATP (1.5 mM)	36	73
minus ATP	6	22
b) washed etioplasts (minus ATP)	55	66
crude etioplasts (minus ATP)	30	26

Curve analyses of spectra taken from the same samples as used for electron microscopy. (a) Comparison in the presence and absence of ATP after 3½ hr illumination. (b) Comparison of etioplasts and washed etioplasts after 5 hr illumination. (Horton and Leech, 1975a)

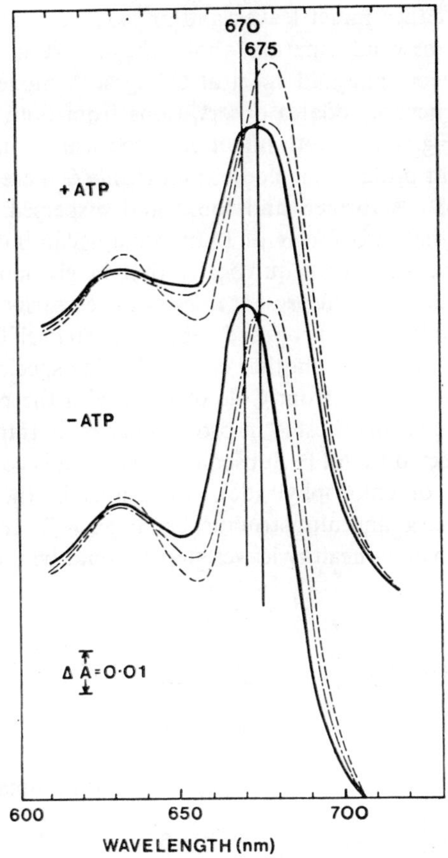

FIG. 9. Effect of ATP on the Shibata shift: absorption spectra during dark incubation. Crude etioplasts were incubated in darkness after photoconversion of PChle. ATP (1.5 mM) was added just prior to illumination. Data from Horton and Leech (1975a).
------, 0 hr (after photoconversion); –·–·, $1\frac{1}{2}$ hr; ———, $4\frac{1}{2}$ hr.

electron micrographs showing the appearance of representative etioplasts in Fig. 10. This distinct and dramatic effect of ATP on the morphological conformation of the prolamellar body, in addition to its effect on pigment spectral changes, suggested that further analysis of the two types of effect of ATP might help to resolve the nature of the normal relationship between the two kinds of changes. The extent of both the Shibata shift and the membrane transformation was therefore measured simultaneously in the presence and absence of ATP. The results are shown in Table 4. It can be seen that quantitatively the changes are of a similar order. In contrast there is no correlation between the membrane changes and the degree of photoconvertibility of PChle. These results imply that the Shibata shift and prolamellar body transformation may be related events and that both can be inhibited by the presence of ATP. ATP may therefore have an important role in regulating the early stages of plastid development *in vivo*.

The results of the experiments on isolated etioplasts extend considerably the earlier observations made on either intact leaves and/or isolated membrane and holochrome preparations, and add new information about the nature and characteristics of the relationship between membrane and spectral changes. A model proposing an indirect two stage process that accommodates observations from both leaf tissue and isolated etioplasts is shown in Fig. 11. In both intact etiolated leaf tissue and isolated etioplasts before illumination, tight prolamellar bodies and PChle 650 are present: after illumination chlorophyll(ide) 680 is formed and loose and dispersed prolamellar bodies are found. It would seem that two forms of tight prolamellar bodies must co-exist since Weier *et al.* (1970) have shown unequivocally that in greening etiolated leaves total photoconversion of all the pigment present does not necessarily lead to total conversion of all the prolamellar bodies. It is proposed that Chle 680 itself causes no ultrastructural change but can be converted into another chlorophyllide species specified here as Chle 680^x, spectrally indistinguishable from Chle 680 but with the potential to undergo the Shibata shift. This change is manifested in prolamellar body transformation to the loose configuration and is affected by ATP. Subsequent dispersal is correlated with the formation of the 670 forms of chlorophyll(ide). Variations in the pattern of observable correlation between spectral and ultrastructural changes will occur if the rates of steps 1 and 2 vary (Fig. 11). In young healthy leaves step 1 would be rapid, as virtually instant-

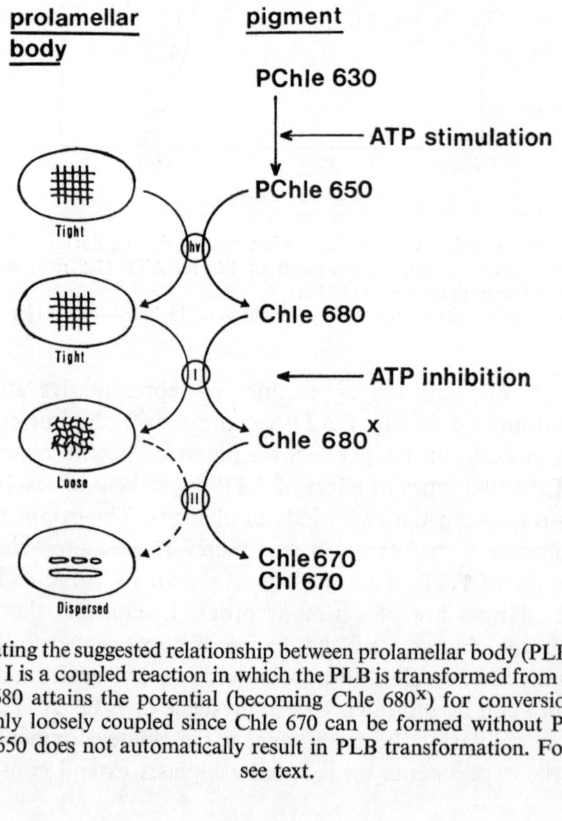

FIG. 11. Model illustrating the suggested relationship between prolamellar body (PLB) transformation and the Shibata shift. Step I is a coupled reaction in which the PLB is transformed from a tight to a loose configuration and Chle 680 attains the potential (becoming Chle 680^X) for conversion into Chle 670 and Chl 670. Step II is only loosely coupled since Chle 670 can be formed without PLB dispersal. Photoconversion of PChle 650 does not automatically result in PLB transformation. For further explanation see text.

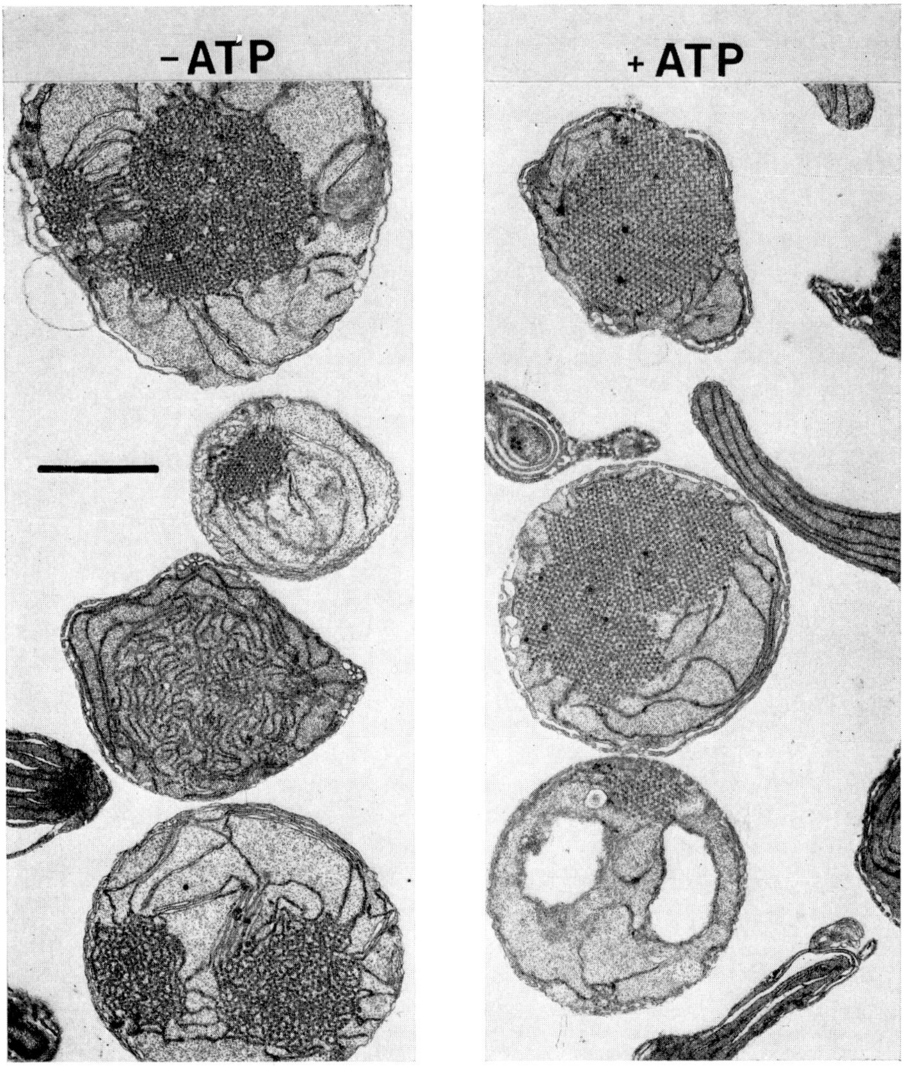

FIG. 10. Electron micrographs of isolated etioplasts from 7 day-old etiolated maize leaves after incubation for 5 hr in the suspension medium (−ATP) and after incubation for 5 hr in the presence of 1.5 mM ATP (+ATP). Note that all the prolamellar bodies show the tight configuration. After initial illumination of 1.1×10^4 erg cm^{-2} sec^{-1} for 2 min the samples were continuously illuminated at 1.6×10^3 erg cm^{-2} sec^{-1} and rotated. The scale represents 1 μ.

aneous prolamellar body transformation is observed correlated to PChle photoconversion. In older leaves the slowing down of step 1 would cause much slower transformation so the ensuing Shibata shift would then be better correlated to transformation.

IN PROSPECT

The experimental potential of suspensions of isolated etioplasts for the study of discrete developmental changes over short time periods is clear. Although adequate techniques are not yet available to promote long term 'survival' of etioplasts *in vitro*, the potential value of etioplast and proplastid suspensions in the study of longer term developmental processes is probably also considerable. For extended studies of development of the two photosystems *in vitro*, etioplasts need to be maintained for periods of several hours or days in isolation and we can look for exciting developments here in the future. Problems of synchrony, bacterial contamination and etioplast decay need to be solved first before experiments over extended time periods are possible. Plastid development is a collaborative cellular process and clearly long term 'culture' of etioplasts and proplastids will require the addition of a large variety of cofactors and even other organelles. Plastid differentiation depends on the continuing interaction of plastid and nuclear genomes so further work on isolated systems will rely heavily on information about the nature and characteristics of nuclear as well as messenger RNA and the movement of large and small molecules across the plastid envelope membranes. Although not yet directly demonstrated, the movement of molecules such as proteins across the chloroplast envelope has to be considered as a realistic possibility since older plastids seem to be able to synthesize relatively few of their own proteins. So far only the large subunit of fraction I protein, and a few (but not all) of the envelope and lamellar proteins have been shown to be synthesized on chloroplast ribosomes (Ellis *et al.*, 1973). The development of methods for maintaining plastids in isolation in a state of functional efficiency will enable organelle interactions involved in plastid development to be studied under controlled conditions.

It should not be forgotten that under natural conditions etioplasts are rarely encountered and that, in normal diurnal light regimes, chloroplasts develop directly from proplastids. The etioplast is clearly at a more sophisticated level of plastid development than the proplastid. Occasionally very small prolamellar bodies consisting of a few tubules are found in developing chloroplasts, often when grana are already present, but these never show the extensive development characteristic of the prolamellar body of an etioplast. Methods for the isolation of proplastids are now available (Thomson *et al.*, 1972; Leese *et al.*, 1972) so it is possible to begin to study their biochemical characteristics.

It is to be hoped that with increased knowledge of the effective manipulation of both isolated etioplasts and proplastids, the full potential of these systems can be exploited to increase our understanding of plastid development in the same way as the study of chloroplasts in isolation so greatly increased our knowledge of photosynthesis.

Acknowledgements: I am most grateful to P. Horton for many enlightening discussions and for permission to use some unpublished results. We acknowledge the expert technical skill of Mrs. W. Crosby in the preparation of the electron micrographs. Permission to reproduce Tables 2, 3 and 4 and Figures 8 and 9 is acknowledged.

REFERENCES

BLAIR, G. E. and ELLIS, R. J. (1973) Light-driven synthesis of the large subunit of Fraction 1 protein by isolated pea chloroplasts. *Biochim. biophys. Acta* **319,** 223–234.

BOARDMAN, N. K. (1966) Ribosome composition and chloroplast development in *Phaseolus vulgaris. Expl Cell Res.* **43,** 474–482.

BOARDMAN, N. K. and WILDMAN, S. G. (1962) Identification of proplastids by fluorescence microscopy and their isolation and purification. *Biochim. biophys. Acta* **59,** 222–224.

BOARDMAN, N. K., LINNANE, A. and SMILLIE, R. M. (1971) *Autonomy and Biogenesis of Mitochondria and Chloroplasts,* Elsevier, New York.

BOASSON, R., LAETSCH, W. W. and PRICE, I. (1972) The etioplast-chloroplast transformation in tobacco: correlation of ultra-structure, replication and chlorophyll synthesis. *Am. J. Bot.* **59,** 217–223.

BRADBEER, J. W. (1969) The activities of the photosynthetic carbon cycle enzymes of greening bean leaves. *New Phytol.* **68,** 233–245.

BROWN, F. A. M. and GUNNING, B. E. S. (1966) Distribution of ribosome-like particles in *Avena* plastids. In *Biochemistry of Chloroplasts,* ed. GOODWIN, T. W., Vol. I, pp. 365–373, Academic Press, New York.

BUTLER, W. L. (1965) Development of photosystems 1 and 2 in the greening leaf. *Biochim. biophys. Acta* **102,** 1–8.

COBB, A. H. and WELLBURN, A. R. (1973) Developmental changes in the levels of SDS-extractable polypeptides during plastid morphogenesis. *Planta* **114,** 131–142.

ELLIS, R. J., BLAIR, G. E. and HARTLEY, M. R. (1973) The nature and function of chloroplast protein synthesis. *Biochem. Soc. Symp.* **38,** 137–162.

GUNNING, B. E. S. (1965) The fine structure of chloroplast stroma following aldehyde osmium-tetroxide fixation. *J. Cell Biol.* **24,** 79–93.

GYLDENHOLM, A. O. and WHATLEY, F. R. (1968) The onset of photophosphorylation in chloroplasts isolated from developing bean leaves. *New Phytol.* **67,** 461–468.

HADDOCK, B. A. and SCHAIRER, H. U. (1973) Electron transport chains of *Escherichia coli*: reconstitution of respiration in a δ-amino-laevulinic acid-requiring mutant. *Eur. J. Biochem.* **35,** 34–45.

HEARING, V. J. (1973) Protein synthesis in isolated etioplasts after light stimulation. *Phytochemistry* **12,** 277–282.

HENNINGSEN, K. W., THORNE, S. W. and BOARDMAN, N. K. (1974) Properties of protochlorophyllide and chlorophyllide holochromes from etiolated and greening leaves. *Pl. Physiol. Lancaster* **53,** 419–425.

HORAK, A. and HILL, R. D. (1972) Adenosine triphosphatase in plastids. *Pl. Physiol. Lancaster* **49,** 365–370.

HORTON, P. and LEECH, R. M. (1972) The effect of ATP on photoconversion of protochlorophyllide into chlorophyllide in isolated etioplasts. *FEBS Lett.* **26,** 277–280.

HORTON, P. and LEECH, R. M. (1975a) The effect of ATP on the Shibata shift and on associated structural changes in the conformation of the prolamellar body in isolated maize etioplasts. *Pl. Physiol. Lancaster* **55,** 393–400.

HORTON, P. and LEECH, R. M. (1975b) The effect of ATP on the photoconversion of protochlorophyllide in isolated etioplasts of *Zea mays. Pl. Physiol. Lancaster* (in press).

INGLE, J., POSSINGHAM, J. V., WELLS, R., LEAVER, C. J. and LOENING, U. E. (1970) The properties of chloroplast ribosomal-RNA. In *Control of Organelle Development,* ed. Miller, P. L. pp. 303–325. Proceedings of the 24th Symposium of the Society for Experimental Biology.

INGLE, J., WELLS, R., POSSINGHAM, J. V. and LEAVER, C. J. (1971) The origins of chloroplast ribosomal-RNA. In *Autonomy and Biogenesis of Mitochondria and Chloroplasts,* eds. BOARDMAN, N. K., LINNANE, A. and SMILLIE, R. M., pp. 393–401. North Holland, Amsterdam.

IRELAND, H. M. and BRADBEER, J. W. (1971) The effect of D-threo and L-threo chloramphenicol on the light-induced formation of enzymes of the photosynthetic carbon pathway. *Planta* **96,** 254–261.

JACOBSON, A. B. (1968) A procedure for isolation of proplastids from etiolated maize leaves. *J. Cell Biol.* **38,** 238–244.

JACOBSON, A. B. and WILLIAMS, R. W. (1968) Sedimentation studies on RNA from proplastids of *Zea mays. Biochim. biophys. Acta* **169,** 7–13.

JACOBSON, A. B., SWIFT, H. and BOGORAD, L. (1963) Cytochemical studies concerning the occurrence and distribution of RNA in plastids of *Zea mays. J. Cell Biol.* **17,** 557–570.

KAHN, A. (1968) Developmental physiology of bean leaf plastids. II. Negative contrast electron microscopy of tubular membranes in prolamellar bodies. *Pl. Physiol. Lancaster* **43,** 1781–1785.

KAHN, A., BOARDMAN, N. K. and THORNE, S. W. (1970) Energy transfer between photochlorophyllide molecules: evidence for multiple chromophore in the photoactive protochlorophyllide-protein complex *in vivo* and *in vitro*. *J. molec. Biol.* **48,** 85–101.

KIRK, J. T. O. (1970) Biochemical aspects of chloroplast development. *A. Rev. Pl. Physiol.* **21,** 11–42.

KIRK, J. T. O. (1972) The genetic control of plastid formation: recent advances and strategies for the future. *Subcell. Biochem.* **1,** 333–361.

KIRK, J. T. O. and TILNEY-BASSETT, R. A. E. (1967) *The Plastids: Their Chemistry, Structure, Growth and Inheritance.* Freeman, London and San Francisco.

KLEIN, S. and POLJAKOFF-MAYBER, A. (1960) Isolation of proplastids from etiolated bean leaves. *Expl. Cell Res.* **24,** 143–145.

KLEIN, S. and POLJAKOFF-MAYBER, A. (1961) Fine structure and pigment conversion in isolated etiolated proplastids. *J. biophys. biochem. Cytol.* **11,** 433–440.

LEECH, R. M., RUMBSY, M. G., THOMSON, W. W., CROSBY, W. and WOOD, P. (1972) Lipid changes during plastid development in developing maize leaves. In *IInd International Congress of Photosynthesis,* eds. FORTI, G., AVRON, M. and MELANDRI, A. Vol. II, Dr. Junk, The Hague.

LEECH, R. M., RUMSBY, M. G. and THOMSON, W. W. (1973) Plastid differentiation, acyl lipid and fatty acid changes in developing green maize leaves. *Pl. Physiol. Lancaster* **52,** 240–245.

LEESE, B. M., LEECH, R. M. and THOMSON, W. W. (1972) Isolation of plastids from different regions of developing maize leaves. In *IInd International Congress of Photosynthesis,* Vol. III, Dr. Junk, The Hague.

LOCKSHIN, A., FALK, R. H., BOGORAD, L. and WOODCOCK, C. L. F. (1971) A coupling factor for photophosphorylation from plastids of light and dark grown maize. *Biochim. biophys. Acta* **226,** 366–382.

LÜRSSEN, K. (1970) Volumen, Trockengewicht und Zusammensetzung von Etioplasten und ihren Ergrünungsstadien. *Z. Naturf.* **25b,** 1113–1119.

MATHIS, P. and SAUER, K. (1972) Circular dichroism studies on the structure and the photochemistry of protochlorophyllide and chlorophyllide holochrome. *Biochim. biophys. Acta* **267,** 498–510.

MARGULIES, M. M. and PARENTI, F. (1968) *In vitro* protein synthesis by plastids of *Phaseolus vulgaris.* III. Formation of lamellar and soluble chloroplast protein. *Pl. Physiol. Lancaster* **43,** 504–514.

NADLER, K. D., HERRON, H. A. and GRANICK, S. (1972) Development of chlorophyll and Hill activity. *Pl. Physiol. Lancaster* **49,** 388–392.

PANTER, R. A. and BOARDMAN, N. K. (1973) Lipid biosynthesis by isolated plastids from greening pea, *Pisum sativum. J. Lipid Res.* **14,** 664–671.

PARENTI, F. and MARGULIES, M. M. (1967) *In vitro* protein synthesis by plastids of *Phaseolus vulgaris.* 1. Localization of activity in the chloroplasts of a chloroplast containing fraction from developing leaves. *Pl. Physiol. Lancaster* **42,** 1179–1186.

PHUNG NHU HUNG, S., HOARAU, A. and MOYSE, A. (1970) Etude de l'evolution en chloroplastes des plastes étoiles d'orge. II. Photophosphorylation et photoreduction du NADP. Formation de ferredoxine, en eclairement continue et par l'action d'eclairs. *Z. Pfl.physiol.* **62,** 245–258.

PLESNICAR, M. and BENDALL, D. S. (1973) The photochemical activities and electron carriers of developing barley leaves. *Biochem. J.* **136,** 803–812.

POSSINGHAM, J. V. (1973) Chloroplast growth and division during the greening of spinach leaf discs. *Nature New Biol.* **245,** 93–94.

REBEIZ, C. A. and CASTELFRANCO, P. A. (1971a) Protochlorophyll biosynthesis in a cell-free system from higher plants. *Pl. Physiol. Lancaster* **47,** 24–32.

REBEIZ, C. A. and CASTELFRANCO, P. A. (1971b) Chlorophyll biosynthesis in a cell-free system from higher plants. *Pl. Physiol. Lancaster* **47,** 33–37.

REBEIZ, C. A. and CASTELFRANCO, P. A. (1973) Protochlorophyll and chlorophyll biosynthesis in cell-free systems of higher plants. *A. Rev. Pl. Physiol.* **24,** 129–172.

REBEIZ, C. A., HAIDAR, M. A., YAGHI, M. and CASTELFANCO, P. A. (1970) Porphyrin biosynthesis in cell-free homogenates from higher plants. *Pl. Physiol. Lancaster* **46,** 543–549.

ROBERTSON, D. and LAETSCH, W. M. (1974) Structure and function of developing barley plastids. *Pl. Physiol. Lancaster* **54,** 148–160.

ROUGHAN, P. G. and BOARDMAN, N. K. (1972) Lipid composition of pea and bean leaves during chloroplast development. *Pl. Physiol. Lancaster* **50,** 30–34.

SCHULTZ, A. and SAUER, K. (1972) Circular dichroism and fluorescence changes accompanying the protochlorophyllide to chlorophyllide transformation in greening leaves and holochrome preparations. *Biochim. biophys. Acta* **267,** 320–340.

SHIBATA, K. (1957) Spectroscopic studies on chlorophyll formation in intact leaves. *J. Biochem. Tokyo* **44,** 147–173.

SIDDELL, S. G. and ELLIS, R. J. (1975) Characteristics and products of *in vitro* protein synthesis in etioplasts and developing chloroplasts from pea leaves. (in press).
SMILLIE, R. M. and STEELE-SCOTT, N. (1969) Organelle biosynthesis: the chloroplast. *Prog. molec. subcell. Biol.* **1,** 136–202.
TEVINI, M. (1971) The formation of lipids following illumination of etiolated seedlings. *Z. Pfl.physiol.* **65,** 266–272.
THOMSON, W. W., FOSTER, P. and LEECH, R. M. (1972) The isolation of proplastids from roots of *Vicia faba* L. *Pl. Physiol. Lancaster* **49,** 270–272.
THORNE, S. W. and BOARDMAN, N. K. (1971) Formation of chlorophyll *b*, and the fluorescence properties and photochemical activities of isolated plastids from greening pea seedlings. *Pl. Physiol. Lancaster* **47,** 252–261.
TREFFRY, T. E. (1970) Phytylation of chlorophyllide and prolamellar body transformation in etiolated peas. *Planta* **91,** 279–284.
TREMOLIERES, A. and LEPAGE, M. (1971) Changes in lipid composition during greening in etiolated pea seedlings. *Pl. Physiol. Lancaster* **47,** 329–334.
WALKER, D. A. (1970) Three phases of chloroplast research. *Nature Lond.* **226,** 1204–1208.
WELLBURN, A. R. and WELLBURN, F. A. M. (1971a) A new method for the isolation of etioplasts with intact envelopes. *J. exp. Bot.* **22,** 972–979.
WELLBURN, F. A. M. and WELLBURN, A. R. (1971b) Chlorophyll synthesis by isolated intact etioplasts. *Biochem. biophys. Res. Commun.* **45,** 747–750.
WELLBURN, A. R. and WELLBURN, F. A. M. (1973) Developmental changes of etioplasts in isolated suspensions and *in situ*. *Ann. Bot.* NS **37,** 11–19.
WELLBURN, F. A. M., WELLBURN, A. R., STODDART, J. L. and TREHARNE, K. J. (1973) Influence of gibberellic and abscisic acid and the growth retardant CCC upon plastid development. *Planta* **111,** 337–346.
WEIER, T. E. and BROWN, D. L. (1970) Formation of the prolamellar body in 8-day dark-grown seedlings. *Am. J. Bot.* **57,** 267–275.
WEIER, T. E., SJOLAND, R. D., BROWN, D. L. (1970) Changes induced by low light intensities on the prolamellar body of 8-day dark-grown seedlings. *Am. J. Bot.* **57,** 276–284.
WHATLEY, F. R., GREGORY, P., HASLETT, B. G. and BRADBEER, J. W. (1972) Development of electron transport intermediates in greening chloroplasts. In *IInd International Congress of Photosynthesis*, eds. FORTI, G., AVRON, M. and MELANDRI, A. Vol. III, Dr. Junk, The Hague.
WOODCOCK, C. L. F. and BOGORAD, L. (1971) Nucleic acids and information processing in chloroplasts. In *Structure and Function of Chloroplasts*, ed. GIBBS, M. Springer-Verlag, Berlin.

THE NATURE AND REGULATION OF SENESCENCE IN PLASTIDS

H. W. WOOLHOUSE and T. BATT

Department of Plant Sciences, Baines Wing, University of Leeds, Leeds, U.K.

INTRODUCTION

Perhaps the greatest difference between plant and animal cells is that the former possess plastids. Plastids are so named because they can exist in several different forms which are, in some measure at least, interconvertible or plastic, although generally speaking one form is found within a given cell at a particular stage of its development. Schimper (1885) distinguished three groups of plastids primarily on the basis of their colour: leucoplasts (white), chloroplasts (green), and the chromoplasts (yellow).

It has been known for about a hundred years that plastids may enjoy some measure of autonomy and continuity in plant cells. This view developed from observations of division in the chloroplasts of dividing cells of the green alga, *Spirogyra* (Strasburger, 1882), and the bryophyte, *Anthoceros* (Davis, 1899). Higher plants differ in that it is not the mature chloroplast which divides at cell division but the undifferentiated precursor, the proplastid. These earliest observations stimulated work on chloroplasts so that through the present century there have been a succession of findings pointing to some measure of chloroplast autonomy within the cell. The following findings are of particular significance in this respect: (i) chloroplasts mutate and mutant forms frequently show exclusively maternal inheritance (Baur, 1909); (ii) isolated chloroplasts are capable of synthesizing proteins (Stephenson *et al.*, 1956); (iii) chloroplasts contain RNA (Jagendorf, 1955), ribosomes and polyribosomes (Lyttleton, 1962); and (iv) chloroplasts contain DNA and the enzymes necessary for replication (DNA polymerase) and transcription (RNA polymerase) (Kirk, 1964). The chloroplast would thus appear to possess the basic elements to sustain a self-replicating system.

It must be noted, however, that there has been a steady accumulation of evidence during the past 75 years to show that the nucleus of the plant cell also controls a wide variety of events occurring in the plastids. Firstly, in the study of variegated plants, large numbers of nuclear genes have been identified in such species as *Avena sativa*, *Epilobium parviflorum*, *Pharbitis nil*, *Ballota nigra*, *Zea mays*, and *Oryza sativa* all of which control various aspects of pigment synthesis and chloroplast development. Moreover, hybridization annealing studies with chloroplast rRNA of tobacco suggest homology with both chloroplast and nuclear DNA (Tewari and Wildman, 1970). Secondly, recent investigations involving application of translational inhibitors of ribosome function to greening cells of higher plants and algae have shown that only a restricted number of components of chloroplasts are actually encoded in the plastid and translated on 70S ribosomes. The majority of enzymes of the photosynthetic carbon reduction cycle together with both structural and electron transport proteins of chloroplast membranes are coded for in the nucleus and synthesized on cytoplasmic 80S ribosomes (Boulter *et al.*, 1972; Ellis, 1974; Ellis *et al.*, 1973).

Finally, to emphasize the increasing complexity of the situation, Bourque and Wildman (1973) using fertile crosses between two closely related species of *Nicotiana* have shown that certain of the proteins of the 50S subunit of the 70s chloroplast ribosomes are different between the species, are inherited in a Mendelian fashion, and are probably synthesized on cytoplasmic ribosomes before passing into the chloroplast to complete the assembly of the plastid ribosomes.

This thumbnail sketch of current knowledge concerning the partial autonomy of plastids and their interdependence with the rest of the cell, must now be borne in mind in discussing the question of plastid senescence.

PLASTID SENESCENCE

If for reasons of time and convenience we confine ourselves to chloroplasts in leaves, then attention to plastid senescence is drawn from two directions: the visible yellowing of a leaf in the later stages of its senescence; and the measurement of physiological performance, such as photosynthetic and respiratory rates, throughout the life of the leaf. At one extreme some leaves live but 5 or 6 weeks on the plant whilst in others, e.g. the Bristle Cone Pine, *Pinus aristata*, individual leaves may live up to 30 years. Senescence of leaves may occur in a variety of circumstances. It may be caused by environmental signals such as a dry season, short days, or low temperatures; by internal factors such as sequential senescence which is a consequence of factors associated with the changing

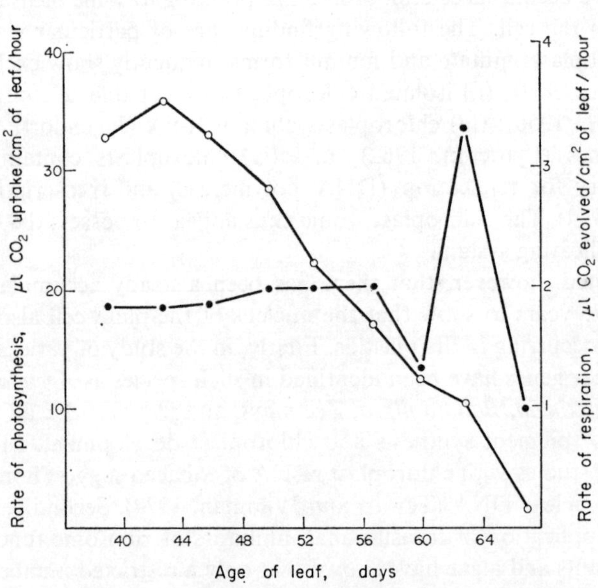

FIG. 1. Changes in photosynthetic (○) and respiratory (●) rates of the third pair of leaves of *Perilla frutescens* during leaf ageing. Data from Hardwick *et al.* (1968).

relative position of the leaf along a shoot; or senescence may be associated with the development of flowers and fruits.

To take a specific example, if the rates of photosynthesis and of respiration are monitored during the sequential senescence of a particular leaf in a vegetative specimen of the monocarpic, short-day plant, *Perilla frutescens*, then the relationship shown in Fig. 1 is found. In this experiment, the completion of leaf expansion occurs at about 35 days. It is seen that from this point onwards there is a progressive decline in the rate of photosynthesis until about 4 days prior to leaf abscission when the rate of decline increases. The pattern of respiration on the other hand stays relatively steady after completion of leaf expansion and shows marked changes only during advanced senescence. The decline in photosynthetic activity can be analyzed by means of simultaneous transpiration studies so that stomatal resistance is measured. When this is done, it is found that the decline in the rate of photosynthesis is due to increased internal (mesophyll) resistance rather than to stomatal closure (Woolhouse, 1968).

Biochemical and anatomical examination of the leaf over this period reveals a complex of changes of which the most noteworthy are a decline in the protein, chlorophyll, and RNA contents of the leaf from the time of completion of leaf expansion onwards (Hardwick and Woolhouse, 1967; Hardwick *et al.*, 1968). The decline in protein content is due primarily to breakdown of soluble proteins, most notably Fraction 1 protein (Kannangara and Woolhouse, 1968) (Fig. 2), the homogeneous chloroplast protein which comprises the enzyme, ribulose-1,5-diphosphate carboxylase. Secondly, there are concurrent changes in the membranes of the chloroplast as revealed by fine structural studies (Haber *et al.*, 1969). In the newly developed chloroplast, thylakoids forming the grana are tightly stacked and the stromal matrix containing the ribosomes is very dense. After completion of expansion, when there has as yet been little loss of chlorophyll but a substantial decline in the rate of photosynthesis, the chloroplasts swell as the granal

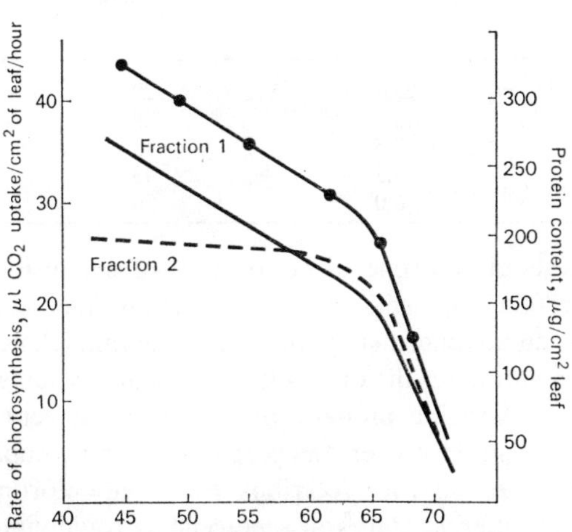

FIG. 2. Decline of photosynthetic rate (—) and concomitant decrease of Fraction 1 protein (●) with leaf age in *Perilla*. The heterogeneous protein fraction other than Fraction 1 protein is designated as Fraction 2 protein (_ _ _). Data from Woolhouse (1967).

thylakoids become distended and move apart. There is a weakening of the granular appearance of the stroma apparently associated with a loss of ribosomes and osmiophilic bodies are developed. These bodies appear in most species as the plastid ages, and the favoured speculation is that they are composed of lipids which are released in the course of breakdown of the chloroplast membranes. As leaf senescence proceeds and loss of chlorophyll commences, there is a further separation of thylakoid membranes, loss of stroma matrix and ribosomes and partial degradation of the chloroplast envelope. Lipid granules may also be lost at this stage, but in other species the granules may be more persistent (Butler and Simon, 1971).

CHLOROPLAST-PROTEIN SYNTHESIZING CAPACITY DURING LEAF AGEING

The combined evidence of decline in the photosynthetic activity, the decrease in amount of a major chloroplast protein (which may comprise up to 50 per cent of the total soluble protein of a leaf), and the fine structural evidence of degradative changes involving membrane disorganization and loss of ribosomes, led us to investigate the relative protein-synthesizing capacities of the chloroplastic and cytoplasmic fractions of the ageing leaf. A lengthy and detailed study was made of the isolation of *Perilla* chloroplasts and of a cytoplasmic ribosome fraction with respect to the optimal conditions of buffer, pH, metal ions and co-factors and these were then used to measure the extent of protein synthesis of the two fractions throughout the life of the leaf (Callow *et al.*, 1972) (Table 1). It is seen that there occurs a significant loss of capacity of chloroplasts to

TABLE 1. INCORPORATION OF [2-^{14}C]GLYCINE BY CHLOROPLASTIC AND CYTOPLASMIC RIBOSOMES ISOLATED FROM LEAVES OF *Perilla frutescens* OF DIFFERENT AGE. Data from Callow *et al.* (1972)

Chloroplast ribosomes		*Cytoplasmic ribosomes*	
Leaf age days	dpm/mg RNA $\times 10^{-3}$	Leaf age days	dpm/mg RNA $\times 10^{-3}$
22	23.1	20	13.3
30	13.2	28	13.9
37	7.5	35	13.1
44	8.3	43	10.4
51	6.0	56	6.4

synthesize protein at about the time of completion of expansion of the leaf, whereas synthesis mediated by 80S cytoplasmic ribosomes is maintained at a significant level even in the yellowed leaf at an advanced stage of senescence immediately prior to abscission. There is a worry in presenting results of this kind since one is acutely aware that even the best cell-free protein-synthesizing preparations show but a very small fraction of the activity of the *in vivo* system. However, the preparation and comparison of the 80S ribosome-based cytoplasmic profiles with 70S ribosome-based chloroplast polysome profiles prepared from leaves at different stages of senescence, provide interesting corroborative evidence (Fig. 3). It is seen that the young leaves contain polymers of both cytoplasmic and chloroplastic ribosomes but subsequently the chloroplast polyribosomes are preferentially lost.

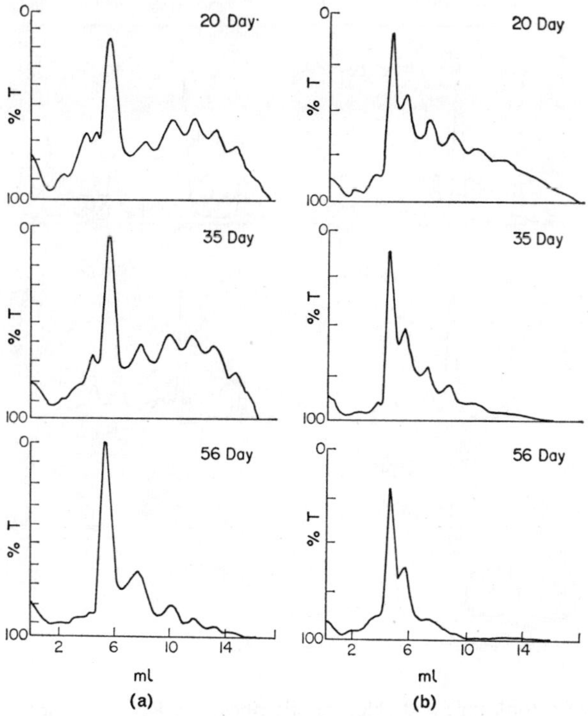

FIG. 3. Extinction profiles of cytoplasmic polyribosomes (a) and chloroplast ribosomes and polyribosomes (b) from young (20 day), fully expanded (35 day), and senescing (56 day) *Perilla* leaves. In (a), the 80S monomer peak is followed by a series of peaks of approximate sedimentation coefficients 120S, 150S, 177S, 200S, and 220S. In (b), the major peak is the 70S monomer followed by contaminating 80S ribosomes and a polysome series of 100S, 140S, 160S, and 184S. Percentages of polyribosomes in each profile were, (a): 60, 62, and 42 per cent, respectively; (b): 62, 45 and 20 per cent, respectively. Data from Callow *et al.* (1972).

The possibility was explored that this loss of chloroplast polyribosome formation might be associated with a cessation of synthesis of chloroplast ribosomes. An experiment was performed in which leaves of *Perilla* in different stages of senescence were given a pulse of $^{32}PO_4$ (200 μCi) for 2 hr prior to extraction of RNA by a phenol procedure (Callow *et al.*, 1972). The gels were frozen, sliced at 0.5 mm intervals, and the segments counted in a liquid scintillation counter. The results are summarized in Fig. 4 which shows that there is a cessation of incorporation of $^{32}PO_4$ into chloroplast rRNA at a relatively early stage in the life of the leaf but incorporation into the cytoplasmic rRNA continues, albeit at a somewhat diminished rate.

With this information in mind we now arrive at the question of what proteins would normally be synthesized by these chloroplast ribosomes (cf. our introductory discussion). This question becomes important, because most cell proteins are normally turned over, that is to say they are generally inactivated or broken down and replaced by synthesis of more of the same proteins. Thus if, as our evidence suggests, the chloroplast ribosomes cease to function the chloroplast will lose the capacity to replace these proteins which it is

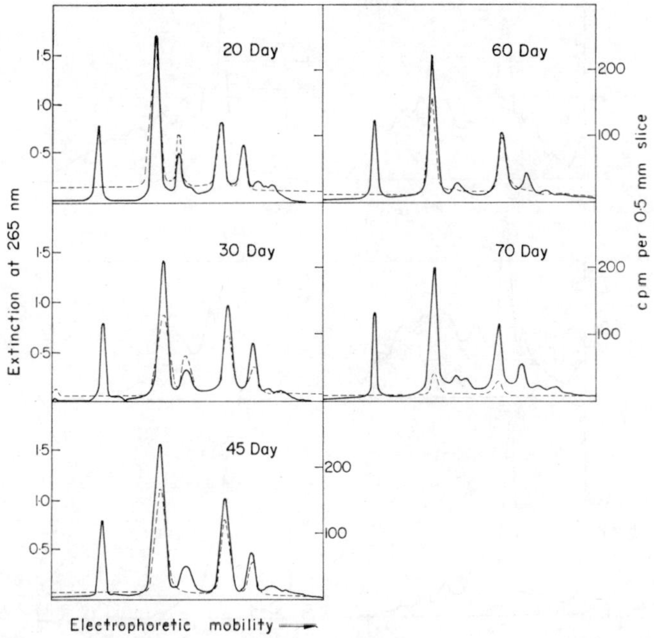

FIG. 4. Separations of cytoplasmic and chloroplastic ribosomal RNAs isolated from *Perilla* leaves of indicated ages, on 2.4 per cent acrylamide gels. The leaves were labelled for 2 hr as described in the text. In order of increasing mobility, the main peaks represent DNA, 1.3×10^6 M Wt cytoplasmic rRNA, 1.1×10^6 chloroplastic rRNA, 0.7×10^6 cytoplasmic rRNA, and 0.56×10^6 chloroplastic rRNA. Gels were scanned at 265 nm (solid line) and radioactivity assessed in 0.5 mm slices (discontinuous line). Data derived from Callow *et al.* (1972).

responsible for making. Accordingly, since the rate of photosynthesis had previously been shown to be declining with a concomitant rise in the mesophyll resistance, the activity of enzymes which play a part in the regulation of photosynthesis was examined at intervals throughout the life of the leaf.

CHANGING ACTIVITIES OF PHOTOSYNTHETIC ENZYMES DURING THE COURSE OF FOLIAR SENESCENCE

Figure 5 shows the profiles of changing activities of six photosynthetic enzymes involved in the carbon reduction cycle during leaf ageing in *Perilla*, selected because such enzymes have known regulatory functions in the Calvin cycle (Bassham, 1971; Preiss and Kosuge, 1970). Enzymes were partially purified and individually characterized to provide optimal conditions to be described elsewhere (Batt and Woolhouse, 1975). The activities of ribose-5-phosphate isomerase, phosphoribulokinase, ribulose-1,5-diphosphate carboxylase and NADPH-linked glyceraldehyde-3-phosphate dehydrogenase (expressed as units per cm^2 leaf area) all declined before or immediately prior to the attainment of full leaf expansion, achieved in this experiment at around day 30. In contrast, high activities

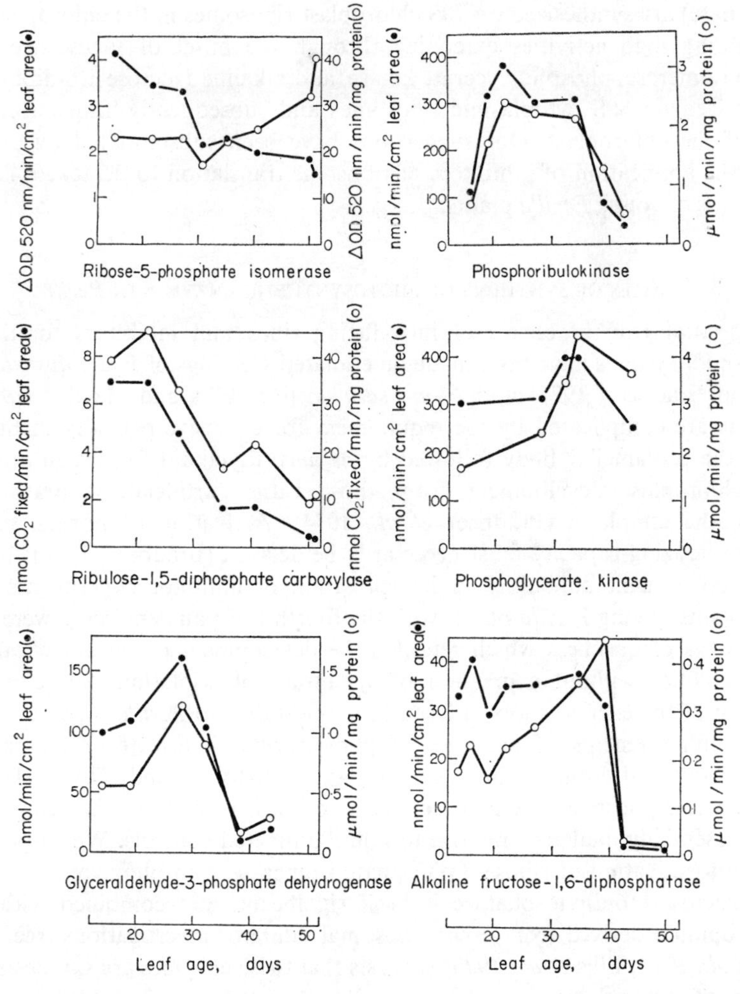

FIG. 5. Changing activities of several photosynthetic enzymes accompanying leaf ageing in the third pair of leaves of *Perilla frutescens*. Activities are expressed both on cm² leaf area and mg protein bases. The complication of a progressively decreasing protein content (180 and 40 μg/cm² leaf area at days 15 and 50, respectively) necessitates description of activity in the present text on a leaf area basis. Data from Batt and Woolhouse (1975).

of phosphoglycerate kinase and alkaline fructose-1,6-diphosphatase were maintained until the lamina began to turn yellow between days 40 and 45. Ribose-5-phosphate isomerase showed some early decline in activity but resembled the phosphoglycerate kinase and the diphosphatase in that it remained relatively active (40 per cent of original activity remained) even at the time of abscission of the leaf.

Such changes during leaf ageing in *Perilla* suggested the hypothesis that those photosynthetic enzymes showing early and sustained deterioration of activities (phosphoribulokinase, ribulose-1,5-diphosphate carboxylase, and NADPH-glyceraldehyde-3-phosphate dehydrogenase) are synthesized on 70S chloroplast ribosomes in the chloroplast whereas those retaining high activities extending through the onset of senescence (ribose-5-phosphate isomerase, phosphoglycerate kinase, and alkaline fructose-1,6-diphosphatase) are synthesized on 80S cytoplasmic ribosomes and subsequently transported into the confines of the chloroplast. Our suggestions have been substantiated by experiments involving the application of inhibitors of ribosome translation to detached, illuminated shoot systems of young *Perilla* plants.

SITES OF SYNTHESIS OF PHOTOSYNTHETIC ENZYMES IN *Perilla*

Previous studies of the effects of introducing ribosomal inhibitors to illuminated, detached shoot systems have been made on etiolated seedlings of *Phaseolus vulgaris* (e.g. Ireland and Bradbeer, 1971) and *Pisum sativum* (e.g. Ellis and Hartley, 1971). Such experiments are complicated by the requirement for extensive rearrangement of membranes of the prolamellar body followed by *de novo* thylakoid formation during light-induced chloroplast development from those rather artificial, laboratory-induced organelles, the etioplasts (Bradbeer *et al.*, 1974). Activation of normal chloroplast protein synthesizing capacity must necessarily be delayed (Bradbeer, 1970). In order to avoid the consequent difficulties in interpretation of inhibitor experiments involving etiolated plants, young *Perilla* plants with the fourth leaf pairs emerging were subjected to only 2 days of darkness which retarded the development of certain photosynthetic enzymic activities without expression of symptoms of etiolation. Subsequent 24 hr illumination of the excised shoots placed in ten-fold diluted *Perilla* root sap resulted in significant enhancements of activities of phosphoribulokinase (four and a half-fold increase), ribulose-1,5-diphosphate carboxylase (five-fold), and NADPH-dependent glyceraldehyde-3-phosphate dehydrogenase (two-fold), whilst all other enzymes tested remained essentially unaltered relative to non-illuminated controls. We suspect that the lack of photoactivation of ribose-5-phosphate isomerase, phosphoglycerate kinase, and alkaline fructose-1:6-diphosphatase is itself significant, and combined with the consensus of opinion derived from other ribosomal inhibitor investigations (see reviews of Boulter *et al.*, 1972; Ellis *et al.*, 1973) suggests that these enzymes are synthesized on 80s ribosomes in the cytoplasm.

Introduction of the D-threo isomer of chloramphenicol, a translational inhibitor blocking 70S ribosome-mediated protein synthesis (Ellis, 1969), to illuminated *Perilla* shoots restricted the increases of the kinase, carboxylase, and dehydrogenase whereas the L-threo isomer had virtually no effect (Table 2).

The stereospecific inhibition by D-threo chloramphenicol suggests that the inhibitor acts specifically at the ribosomal level with little or no disruption of ion uptake and phosphorylation with associated electron transport (Wara-Aswapati and Bradbeer, 1974). The inability of chloramphenicol to inhibit the activities of these photosynthetic enzymes to levels encountered in dark controls, is probably due to light re-activation of enzymes already synthesized before commencement of the 2 day dark pretreatment. Recent confirmatory experiments using the 80S ribosomal inhibitor, 2-(4-methyl-2,6-dinitroanilino)-N-methylpropionamide (MDMP) (Weeks and Baxter, 1972; Baxter *et al.*, 1973), have demonstrated stereo-specific inhibition by the active D-form of MDMP

TABLE 2. EFFECTS OF D- AND L-THREO CHLORAMPHENICOL (CAP) (10 μg/ml) AND D- AND L-MDMP 2-(4-METHYL-2,6-DINITROANILINO)-N-METHYLPROPIONAMIDE ON ACTIVITIES OF PHOSPHORIBULOKINASE, RIBULOSE-1,5-DIPHOSPHATE CARBOXYLASE, AND NADPH-LINKED GLYCERALDEHYDE-3-PHOSPHATE DEHYDROGENASE DURING 24 HR ILLUMINATION OF DETACHED *Perilla* SHOOTS

Enzyme	Per cent illuminated control activities			
	D-threo CAP	L-threo CAP	D-MDMP	L-MDMP
Phosphoribulokinase	64.2 ± 2.6	98.2 ± 12.9	79.8 ± 7.3	61.5 ± 2.9
Ribulose-1,5-diphosphate carboxylase	68.3 ± 6.2	91.8 ± 3.0	64.0 ± 7.4	93.6 ± 6.5
Glyceraldehyde-3-phosphate dehydrogenase	62.5 ± 13.7	106.2 ± 8.5	119.9 ± 5.1	104.6 ± 8.7

Activities were initially calculated as μmol/min/g fresh weight and expressed here as percent of values recorded for illuminated control shoots placed in diluted sap alone. The variation from the average in duplicate treatments is indicated.

only for ribulose-1,5-diphosphate carboxylase (Table 2). Similarly, the carboxylase has been shown to be inhibited by cycloheximide supplied at 25 μg/ml to levels of 31 per cent of illuminated control values. Recent preliminary experimentation using lincomycin, a 70S ribosome inhibitor (Ellis and Hartley, 1971), has revealed inhibition of both ribulose-1,5-diphosphate carboxylase and NADPH-glyceraldehyde-3-phosphate dehydrogenase (74 and 72 per cent of illuminated control values, respectively) whilst phosphoribulokinase activity was enhanced slightly (110 per cent controls) in lincomycin-treated shoots.

All three enzymes are considered on this accumulating evidence to be synthesized, in part at least, on 70S chloroplast ribosomes. The inhibition of the carboxylase by both D-threo chloramphenicol and D-MDMP agrees with combined results of inhibitor studies, genetic investigations, and *in vitro* chloroplast protein-synthesizing systems (see reviews of Ellis, 1973, 1974; Ellis *et al.*, 1973) that the catalytically active large subunit of Fraction 1 protein is encoded by chloroplast DNA and synthesized on 70S chloroplast ribosomes, whilst the small subunit is encoded by nuclear genes and synthesized on 80S cytoplasmic ribosomes. The small subunit is necessary for either the transcription or translation resulting in large subunit synthesis. We suggest that in addition to synthesis of the large subunit of ribulose-1,5-diphosphate carboxylase and chloroplast ribosomal and membrane proteins, chloroplast ribosomes may also mediate the synthesis of phosphoribulokinase and NADPH-dependent glyceraldehyde-3-phosphate dehydrogenase (see also Berger and Feierabend, 1967; Graham *et al.*, 1970; Ireland and Bradbeer, 1971). Our present view of the co-operation between nucleus, cytoplasm and chloroplast in the synthesis of chloroplast components is summarized in Fig. 6.

The interpretation of changing activities of photosynthetic enzymes accompanying foliar senescence in *Perilla* (Fig. 5), based upon suggested chloroplastic- or cytoplasmic-located synthesis are thus substantiated by these inhibitor studies. This approach is currently being extended to determine whether we may generalize to the point of stating that all or most of the chloroplast components synthesized on the plastid ribosomes decline from an early stage of leaf development whereas nuclear controlled constituents do not. Evidence is available showing deterioration of cyclic phosphorylation capacity of

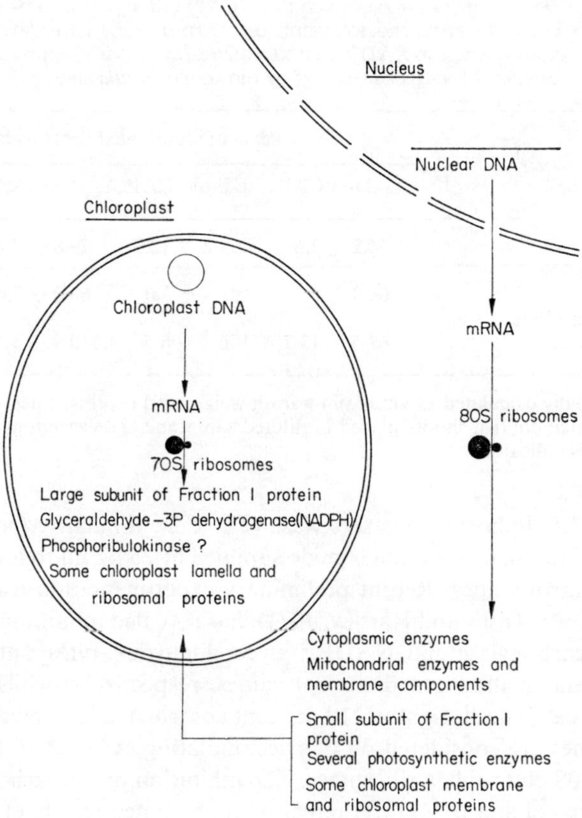

FIG. 6. Suggested interdependence of nucleus, cytoplasmic ribosomes, and chloroplast in synthesis of chloroplast photosynthetic enzymes, and ribosomal and membrane proteins.

Perilla chloroplasts throughout the life of the leaf (Callow, 1969) and we would on this hypothesis expect to recognize 70S-synthesized components in this system. Fractionation of the structural and electron-transporting proteins of chloroplasts of *Perilla* leaves is now proceeding.

HORMONAL CONTROL OF SENESCENCE

It has been known since the time of the gardener Knight in the 1830s that if leaves are taken from plants such as *Begonia rex* and are induced to form roots upon the petiole, then the senescence of the lamina may be arrested. A similar observation on rooted leaves of *Phaseolus* led Chibnall (1954) to suggest that hormones derived from the roots may be involved in the regulation of leaf metabolism. Richmond and Lang (1957) showed that the synthetic cytokinin, kinetin (6-furfurylamino purine), applied to detached leaves of *Xanthium pensylvanicum* delayed the onset of senescence, prompting the conclusion that similar naturally-occurring compounds produced in roots could be involved in the control of leaf senescence *in vivo*. Kulaeva (1962) demonstrated senescence-retarding activity in exudates from cut stems of tobacco. There followed a gamut of reports of hormones delaying the senescence of leaf discs and the presence of gibberellins and cytokinins in xylem sap was confirmed (e.g. Kende, 1971). It must be said, however, that the results of this multitude of leaf-disc experiments have been disappointing. The experimental procedures have frequently been unsound in that unphysiologically high concentrations of applied hormones are involved and little account has been taken of bacterial effects.

Members of each of the five groups of plant hormones have been implicated, either individually or in combination, with the regulation of both senescence and the subsequent abscission of the leaf. An alternative approach has been to measure the endogenous hormone content or diffusible hormone production from leaves. Such studies carried out during leaf senescence have shown increased auxin production, diminished gibberellin content, decline of cytokinins, and increases in the production of ethylene and the abscisic acid (ABA) content.

The difficulty surrounding almost all of these studies is that the amount of hormone extracted from the leaves at a given moment gives no indication of its relative rates of synthesis and degradation, which may be the critical factor in regulating leaf senescence. It seems very probable that hormone turnover rates do change with leaf age and that complex interactions between the hormones influence this process. Thus the application of cytokinins to leaf discs may counteract the senescence-accelerating effect of exogenously applied ABA and it has been shown that cytokinins may be rapidly transformed when applied to leaves (Itai and Vaadia, 1971; Back *et al.*, 1972). However, in leaves of *Rumex pulcher*, GA_3, while delaying senescence did not have any effect on kinetin metabolism although ABA, which accelerates senescence, slowed down the rate of conversion of [^{14}C]kinetin to adenine (Back *et al.*, 1972). At the present time one can go no further than to suggest that all the main plant hormones are probably involved in the regulation of leaf senescence in processes which involve their interacting with one another in a complex manner.

It has been claimed that the retarding of senescence in leaf discs by application of hormones is associated with a stimulation of RNA and protein synthesis (Osborne, 1962), although this work has been strongly criticized in that changes in pool sizes were not considered (Kende, 1971). Ribonucleases, proteases, and β-1,3-glucanases have been claimed to be synthesized *de novo* during leaf senescence (Udvardy *et al.*, 1969; Martin and Thimann, 1972; Moore and Stone, 1972; Lontai *et al.*, 1973).

The fourteen-fold increase in β-1,3-glucanase observed during senescence of *Nicotiana* leaves (Moore and Stone, 1972) was influenced by IAA and ABA, but no judgement can be made concerning the specificity of these effects. Similar difficulties arise in connection with the observed effects of serine on protease activity in senescing wheat leaves (Martin and Thimann, 1972), of IAA on RNAase activity in ageing barley leaves (Moore and Stone, 1972), or of ABA on RNA polymerase in maize coleoptiles (Bex, 1972). The general conclusion to be reached from the multitude of observed effects of plant hormones on RNA metabolism is that a great many are relatively non-specific. The techniques for handling plant mRNAs are not yet sufficiently developed for it to be possible to say whether particular hormone treatments act first through influencing mRNA or more generally by stimulating rRNA synthesis. There are suggestions of the involvement of specific hormone-binding proteins (Matthysse and Phillips, 1969) and of activity initiated at the cell surface and mediated through an adenyl cyclase system. These studies are not, however, at a sufficiently critical level to give any clear indication of whether the initial targets of hormone action in regulating senescence changes are located at the cell surfaces or at sites within the cell (Woolhouse, 1974). We are currently endeavouring to study the effects of endogenous hormone changes following decapitation to cause re-greening, involving stimulation of chloroplast ribosomal RNA synthesis (Callow and Woolhouse, 1973). It is our hope to characterize the plasmalemma and chloroplast membranes following such treatment and to explore whether the changing hormone levels act first at the surface membranes and whether these in turn mediate the activation of the RNA polymerases needed to restore the ribosome synthesis.

This is of course a long term programme but we hope that the present article has indicated that the study of plastid senescence is not a fringe topic to be included at the end of a symposium, rather it has all the ingredient excitements of modern developmental biology.

REFERENCES

BACK, A., BITTNER, S. and RICHMOND, A. E. (1972) The effect of abscisic acid on the metabolism of kinetin in detached leaves of *Rumex pulcher*. *J. exp. Bot.* **23,** 744–750.

BASSHAM, J. A. (1971) The control of photosynthetic carbon metabolism. *Science N.Y.* **172,** 526–534.

BATT, T. and WOOLHOUSE, H. W. (1975) Changing activities during senescence and sites of synthesis of photosynthetic enzymes in leaves of the Labiate, *Perilla frutescens* (L.) Britt. *J. Exp. Bot.* (in press).

BAUR, E. (1909) Das wesen und die Erbslichtkeitsverhältnisse der "Varietates albomarginatae hort" von *Pelargonicum zonale*. *Z. indukt. Abstam.-u. VererbLehre* **1,** 330–351.

BAXTER, R., KNELL, V. C., SOMERVILLE, H. J., SWAIN, H. M. and WEEKS, D. P. (1973) Effect of MDMP on protein synthesis in wheat and bacteria. *Nature New Biol.* **243,** 139–142.

BERGER, C. and FEIERABEND, J. (1967) Plastidenentwicklung und Bildung von Photosynthese-Enzymen in etiolierten Roggenkeimlingen. *Physiol. Végét.* **5,** 109–122.

BEX, J. H. M. (1972) Effects of abscisic acid on the soluble RNA polymerase activity in maize coleoptiles. *Planta* **103,** 11–17.

BOULTER, D., ELLIS, R. J. and YARWOOD, A. (1972) Biochemistry of protein synthesis in plants. *Biol. Rev.* **47,** 113–175.

BOURQUE, D. P. and WILDMAN, S. G. (1973) Evidence that nuclear genes code for several chloroplast ribosomal proteins. *Biochem. biophys. Res. Commun.* **50,** 532–537.

BRADBEER, J. W. (1970) Plastid development in primary leaves of *Phaseolus vulgaris*. An initial lag phase in light-induced chloroplast development. *New Phytol.* **69,** 635–637.

BRADBEER, J. W., GYLDENHOLM, A. O., IRELAND, H. M. M., SMITH, J. W., REST, J. and EDGE, H. J. W. (1974) Plastid development in primary leaves of *Phaseolus vulgaris*. VIII. The effects of the transfer of dark-grown plants to continuous illumination. *New Phytol.* **73,** 271–279.

BUTLER, R. D. and SIMON, E. W. (1971) Ultrastructural aspects of senescence in plants. *Advances in Gerontological Research* **3,** 73–129.

CALLOW, J. A. (1969) The control of chloroplast metabolism during foliar senescence in *Perilla frutescens*. Ph.D. Thesis, University of Sheffield.

CALLOW, M. E. and WOOLHOUSE, H. W. (1973) Changes in nucleic-acid metabolism in regreening leaves of *Perilla*. *J. exp. Bot.* **24,** 285–294.

CALLOW, J. A., CALLOW, M. E. and WOOLHOUSE, H. W. (1972) *In vitro* protein synthesis, ribosomal RNA synthesis, and polyribosomes in senescing leaves of *Perilla*. *Cell Differ.* **1,** 79–90.

CHIBNALL, A. C. (1954) Protein metabolism in rooted runner-bean leaves. *New Phytol.* **53,** 31–37.

DAVIS, B. M. (1899) The spore-mother cell of *Anthoceros*. *Bot. Gaz.* **28,** 89–109.

ELLIS, R. J. (1969) Chloroplast ribosomes: stereo-specificity of inhibition by chloramphenicol. *Science N.Y.* **163,** 477–478.

ELLIS, R. J. (1973) Fraction 1 protein. Commentaries in Plant Science, in *Curr. Adv. Pl. Sci.* **3,** 29–38.

ELLIS, R. J. (1974) The biogenesis of chloroplasts: protein synthesis by isolated chloroplasts. *Biochem. Soc. Trans.* 544*th Meet. Lond.* **2,** 179–182.

Ellis, R. J. and Hartley, M. R. (1971) Sites of synthesis of chloroplast proteins. *Nature New Biol.* **233,** 193–196.

Ellis, R. J., Blair, G. E. and Hartley, M. R. (1973) The nature and function of chloroplast protein synthesis. *Biochem. Soc. Symp.* **38,** 137–162.

Graham, D., Hatch, M. D., Slack, C. R. and Smillie, R. M. (1970) Light-induced formation of enzymes of the C_4-dicarboxylic acid pathway of photosynthesis in detached leaves. *Phytochemistry* **9,** 521–532.

Haber, A. H., Thompson, P. J., Walne, P. L. and Triplett, L. L. (1969) Nonphotosynthetic retardation of chloroplast senescence by light. *Pl. Physiol. Lancaster* **44,** 1619–1628.

Hardwick, K. and Woolhouse, H. W. (1967) Foliar senescence in *Perilla frutescens* (L.) Britt. *New Phytol.* **66,** 545–552.

Hardwick, K., Wood, M. and Woolhouse, H. W. (1968) Photosynthesis and respiration in relation to leaf age in *Perilla frutescens* (L.) Britt. *New Phytol.* **67,** 79–86.

Ireland, H. M. M. and Bradbeer, J. W. (1971) Plastid development in primary leaves of *Phaseolus vulgaris*: the effects of D-threo and L-threo chloramphenicol on the light-induced formation of enzymes of the photosynthetic carbon pathway. *Planta* **96,** 254–261.

Itai, C. and Vaadia, Y. (1971) Cytokinin activity in water-stressed shoots. *Pl. Physiol. Lancaster* **47,** 87–90.

Jagendorf, A. T. (1955) Purification of chloroplasts by a density technique. *Pl. Physiol. Lancaster* **30,** 138–143.

Kannangara, C. G. and Woolhouse, H. W. (1968) Changes in the enzyme activity of soluble protein fractions in the course of foliar senescence in *Perilla frutescens* (L.) Britt. *New Phytol.* **67,** 533–542.

Kende, H. (1971) The cytokinins. *Int. Rev. Cytol.* **31,** 301–338.

Kirk, J. T. O. (1964) DNA-dependent RNA synthesis in chloroplast preparations. *Biochem. biophys. Res. Commun,* **14,** 393–397.

Kulaeva, O. N. (1962) Translated from *Fiziologiya Rast.* **9,** 229.

Lontai, I., Van Loon, L. C. and Bruinsma, J. (1973) *Z. Pfl.physiol.* **67,** 146.

Lyttleton, J. W. (1962) Isolation of ribosomes from spinach chloroplasts. *Exp. Cell Res.* **26,** 312–317.

Martin, C. and Thimann, K. V. (1972) The role of protein synthesis in the senescence of leaves. 1. The formation of protease. *Pl. Physiol. Lancaster* **49,** 64–71.

Matthysse, A. G. and Phillips, C. P. (1969) A protein intermediary in the interaction of a hormone with the genome. *Proc. natn. Acad. Sci. U.S.A.* **63,** 897–903.

Moore, A. E. and Stone, B. A. (1972) Effect of senescence and hormone treatment on the activity of a β-1,3-glucan hydrolase in *Nicotiana glutinosa* leaves. *Planta* **104,** 93–109.

Osborne, D. J. (1962) Effect of kinetin on protein and nucleic acid metabolism in *Xanthium* leaves during senescence. *Pl. Physiol. Lancaster* **37,** 595-602.

Preiss, J. and Kosuge, T. (1970) Regulation of enzyme activity in photosynthetic systems. *A. Rev. Pl. Physiol.* **21,** 433–466.

Richmond, A. E. and Lang, A. (1957) Effect of kinetin on protein content and survival of detached *Xanthium* leaves. *Science N.Y.* **125,** 650–651.

Schimper, A. F. W. (1885) Untersuchungen über die Chlorophyllkörner und ihnen homologen Gebilde. *Jb. wiss. Bot.* **16,** 1–27.

Stephenson, M. L., Thimann, K. V. and Zamecnik, P. C. (1956) Incorporation of C^{14}-amino acids into proteins of leaf discs and cell-free fractions of tobacco leaves. *Archs. Biochem. Biophys.* **65,** 194–209.

Strasburger (1882) Quoted in Guilliermond, L. (1941) *The Cytoplasm of the Plant Cell.* Translated by L. R. Atkinson. Chronica Botanica, Waltham, Mass.

Tewari, K. K. and Wildman, S. G. (1970) Information content in the chloroplast DNA. In *Control of Organelle Development.* Society for Experimental Biology Symposium ed. Miller, P. L. Vol. 24, pp. 147–180. Cambridge University Press.

Udvardy, J., Farkas, G. L. and Marré, E. (1969) On RNAase and other hydrolytic enzymes in excised *Avena* leaf tissues. *Pl. Cell Physiol. Tokyo* **10,** 375–386.

Wara-Aswapati, O. and Bradbeer, J. W. (1974) Chloramphenicol as an energy transfer inhibitor in spinach chloroplasts. *Pl. Physiol. Lancaster* **53,** 691–693.

Weeks, D. P. and Baxter, R. (1972) Specific inhibition of peptide-chain initiation by 2-(4-methyl-2,6-dinitroanilino)-N-methylpropionamide. *Biochemistry* **11,** 3060–3064.

Woolhouse, H. W. (1967) The nature of senescence in plants. In *Aspects of the Biology of Ageing.* Symposium of the Society for Experimental Biology, ed. Woolhouse, H. W. Vol. 21, pp. 179–214. Cambridge University Press.

Woolhouse, H. W. (1968) Leaf age and mesophyll resistance as factors in the rate of photosynthesis. *Hilger J.* **11,** 7–12.

Woolhouse, H. W. (1974) Longevity and senescence in plants. *Sci. Prog. Oxf.* **61,** 123–147.

C4 PHOTOSYNTHESIS

J. Coombs, C. W. Baldry and C. Bucke

Tate and Lyle Limited, Group Research and Development, P.O. Box 68, Reading, Berkshire, U.K.

Introduction

The story of events leading to the concept of C4 photosynthesis began about 25 years after the inauguration of the Society for Experimental Biology when the long-lived radioactive isotope of carbon (^{14}C) became available for use as a tracer in investigations of the physiology and biochemistry of plants. At the same time the development of paper chromatography made it possible to separate and identify the large number of metabolic intermediates which became radioactive when cultures of the unicellular alga *Chlorella* were exposed to $^{14}CO_2$ for even a few seconds. Details of some of these early studies were reported at the fifth symposium of the Society (see Brown, 1950), although at this time the interpretation of results was mainly speculative. However, within a few years the complete sequence of reactions which are now known to comprise the photosynthetic carbon reduction (PCR) cycle had been worked out (Bassham and Calvin, 1957).

The primary chemical reaction of the PCR cycle is that catalysed by an enzyme (E.C.4.1.1.39) which is located exclusively within chloroplasts. This enzyme has been previously known as carboxydismutase or ribulose *di*phosphate carboxylase*. Recently it has been renamed ribulose *bis*phosphate carboxylase (RBPC). This enzyme catalyses the formation of two molecules of the three-carbon compound phosphoglyceric acid (PGA) from one molecule of CO_2 and one molecule of ribulose *bis*phosphate (RBP). The PGA is then reduced to sugar phosphates in a sequence of reactions which are essentially the reverse of those which occur during the glycolytic breakdown of hexose in respiration. Some of the sugar phosphates are used in the cyclic regeneration of the CO_2 acceptor (RBP), the rest constitute the net product of photosynthesis, and as such are metabolized further to cellular constituents.

Recently higher plants have been classified on the basis of the number of carbon atoms in the first compound to become radioactive when their leaves are exposed to $^{14}CO_2$ in the light. For this reason those species in which the major route of fixation of atmospheric CO_2 is through the PCR cycle are known as C3 plants.

Elucidation of the mechanism of the PCR cycle stimulated intensive research on photosynthetic carbon metabolism and led to the subsequent discovery of two further pathways related metabolically to the PCR cycle. The first of these is the C2 or glycollate pathway (Tolbert, 1963) and the second the C4 dicarboxylic acid cycle (Kortschak *et al.*, 1965; Hatch and Slack, 1966). As discussed below in more detail the production of glycollate and its further metabolism through the C2 pathway results in a decrease in the

* Used by several contributors to this volume. [Ed.]

ability of C3 plants to fix and retain CO_2 under the present atmospheric composition of low CO_2 concentration and high oxygen partial pressure. On the other hand, the C4 cycle appears to have evolved as a mechanism to overcome these deficiencies.

GLYCOLLATE PRODUCTION AND THE C2 PATHWAY

A high concentration of $^{14}CO_2$ (1 to 5 per cent) was used in the early studies of photosynthesis in *Chlorella*. When similar experiments were carried out with the alga illuminated at the normal atmospheric concentration (0.03 per cent) of CO_2 it was found that a considerable proportion of the fixed carbon was recovered in the form of a two-carbon compound glycollic acid. It is now apparent that production of glycollate under conditions of low CO_2 concentration is an ubiquitous feature of plants with the PCR cycle. Glycollate has its origins in the chloroplasts where it is derived from sugar phosphates. Two separate mechanisms, both of which involve the participation of oxygen, are believed to contribute to glycollate production.

In the first-proposed mechanism of glycollate formation (Coombs and Whittingham, 1966) the interaction of sugar phosphates with O_2 is indirect. It is suggested that O_2 reacts initially with reduced ferredoxin (a product of the light reactions of photosynthesis) to form H_2O_2. This peroxide then oxidizes a two-carbon fragment from the dihydroxyethylthiamine pyrophosphate intermediate transketolase complex formed during the regeneration of RBP in the PCR cycle. More recently it has been shown (Bowes *et al.*, 1971) that RBP can be oxidized directly to form one molecule of PGA and one molecule of phosphoglycollate. The phosphoglycollate is then hydrolysed to the free acid in a reaction catalysed by a specific phosphatase located in the chloroplasts. The formation of phosphoglycollate from RBP has been studied in detail (Andrews *et al.*, 1973) and shown to be catalysed by the enzyme RBPC acting as an oxygenase.

Production of glycollate by either of these mechanisms will decrease the ability of plants with the PCR cycle to assimilate CO_2. Apart from the direct inhibition of RBPC by oxygen (Ogren and Bowes, 1971), loss of carbon from the chloroplasts as glycollate depletes the pools of the carboxylation substrate (RBP) and other intermediates of the cycle. Net photosynthesis is also decreased since CO_2 is produced during the conversion of glycollate back to sugar through the C2 pathway. In the C2 pathway, glycollate is metabolized through glyoxylate, glycine, serine, hydroxypyruvate, glycerate and PGA; most of the steps are catalysed by enzymes located within specific microbodies—peroxisomes (Tolbert and Yamazaki, 1969; Tolbert, 1971). However, the step in which CO_2 is produced, i.e. the condensation of two molecules of glycine to one of serine, occurs in the mitochondria; ATP is also produced in this reaction (Bird *et al.*, 1972).

The facts that plants produce CO_2 and consume O_2 in light-dependent reactions are now well established (Jackson and Volk, 1970). The exact magnitude of this gas exchange (photorespiration) is difficult to establish since both CO_2 evolution in 'dark' respiration and CO_2 assimilation in photosynthesis interfere with its measurement. However, in spite of these difficulties it is generally agreed that in C3 plants this loss is of sufficient magnitude seriously to decrease productivity (Zelitch, 1971) in many important crop plants.

Over the last 10 years or so considerable information has accumulated which suggests that certain species of higher plants are resistant to inhibition by oxygen and lack the manifestation of photorespiration. These plants, which are typified by tropical grasses such as sugar cane, maize and sorghum, but include a wide variety of both monocotyledons and dicotyledons (see Hatch *et al.*, 1971), have an additional carboxylation reaction sequence which is ancillary to the PCR cycle. In these plants the first compound to become radioactive when they are exposed to $^{14}CO_2$ in the light is the four-carbon compound oxaloacetic acid. For this reason these plants are known as C4 plants, and the carboxylation sequence as the C4 dicarboxylic acid cycle.

The C4 cycle

The occurrence of four-carbon organic acids as initial products of $^{14}CO_2$ assimilation in the light was first noted during attempts to demonstrate the presence of the PCR cycle in leaves of sugar cane. These observations were made in the early 1950s but did not receive wide publicity prior to the publication by Kortschak *et al.* (1965). Russian workers made similar observations on maize in the early 1960s (see Karpilov, 1974).

The incorporation of radioactivity into four-carbon acids was not of itself a new observation. However, as reported by Kortschak (1968) 'Finally the facts became too clear to be denied. When radioactive CO_2 is fed to a sugar cane leaf the first stable compounds to become radioactive are malic and aspartic acids. PGA is an intermediate product appearing after these and before hexose phosphate . . . essentially all of the carbon gained by the leaf passed through malic acid then through PGA. At this point, having spent years on the problem I urged Dr. Hatch to continue this study'.

It was left to Hatch and Slack to make the critical biochemical observations which led to the elucidation of the individual reactions of the pathway. Extensive radioactive-tracer experiments, including time-course studies and investigations of the distribution of label in the products (Hatch and Slack, 1966), were complemented by identification of the enzymes involved (Slack and Hatch, 1967; Andrews and Hatch, 1969; Johnson and Hatch, 1970). The primary carboxylation reaction was recognized as that catalysed by phosphoenol pyruvate (PEP) carboxylase, and the first stable product as oxaloacetic acid (OAA).

Although the literature on C4 photosynthesis contains many different complex diagrams of suggested routes of carbon flow in the C4 cycle the essential reactions are in fact very simple as shown in Fig. 1. The initial carboxylation reaction is followed by reduction of the product (OAA) to malic acid. Several alternative routes of reduction have been suggested; these include reduction by NADP-specific malate dehydrogenase

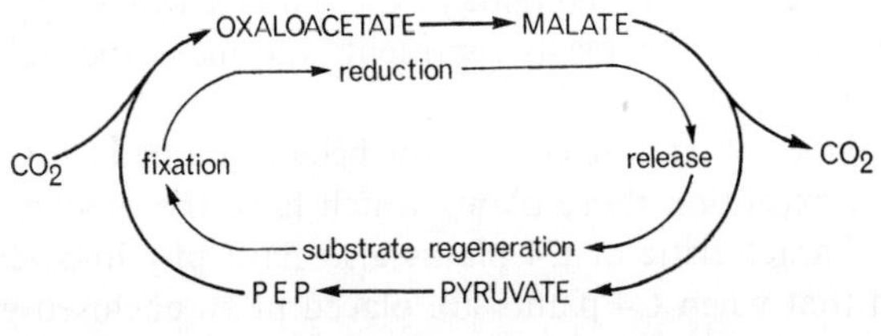

Fig. 1. The C4 cycle.

in the chloroplasts (Johnson and Hatch, 1970), by a NAD-specific enzyme in the cytoplasm (Coombs *et al.*, 1973b) or directly by ferredoxin in the chloroplasts (Baldry *et al.*, 1969). OAA may also be transaminated to form aspartic acid. The relative amounts of the two acids produced vary in different species of C4 plants (Downton, 1970).

In the next step of the cycle the four-carbon acids are decarboxylated. For malic acid this decarboxylation may be catalysed directly, either by a specific NADP malic enzyme (Johnson and Hatch, 1970) or by a NAD malic enzyme (Hatch *et al.*, 1974). For aspartic acid the decarboxylation, catalysed by PEP carboxykinase follows a transamination reaction in which the aspartate is converted back to OAA (Hatch and Mau, 1973). The relative importance of these different mechanisms of decarboxylation in various species of C4 plants has been discussed in detail by Gutierrez *et al.*, 1974.

The carbon dioxide liberated in the C4 pathway is subsequently refixed by RBPC and metabolized further through the PCR cycle to cellular constituents. The residual three-carbon fragment (pyruvate) is used to regenerate the carboxylation substrate (PEP) in a reaction catalysed by pyruvate P_i dikinase (Andrews and Hatch, 1969).

It is obvious that the sequence of reactions which constitute the C4 pathway does not result in a net assimilation of CO_2 into organic matter. It is now generally accepted that the C4 pathway functions as a shuttle mechanism moving CO_2 from the atmosphere to the site of reductive re-assimilation through the PCR cycle. Inherent in this suggestion is the assumption that there is little, if any, direct assimilation of atmospheric CO_2 through the reaction catalysed by RBPC in C4 plants.

The possession of the additional (PEP) carboxylation reaction in some way overcomes the deleterious effects of oxygen on photosynthesis and prevents the output of CO_2 in photorespiration. In order to understand how photorespiration is overcome it is necessary to consider the anatomical and physiological properties of C4 plants in more detail.

CHARACTERISTICS OF C4 PLANTS

In addition to their distinctive pathway of CO_2 assimilation, C4 plants also differ from C3 plants in a number of anatomical and physiological characteristics. In all C4 plants the photosynthetic tissue consists of two distinct cell layers arranged concentrically around the vascular tissue. The inner cell layer is generally known as the bundle sheath, whereas the outer layer has been termed the mesophyll tissue.

In many of the large panicoid grasses, such as sugar cane, maize and sorghum there is a marked dimorphism in chloroplast structure (Laetsch, 1974). Chloroplasts of the mesophyll cells appear normal whereas those of the bundle sheath are larger and contain few grana. These bundle-sheath plastids preferentially accumulate starch. Although this dimorphism is not an essential characteristic of C4 plants it would appear that the loss of grana in the bundle-sheath chloroplasts represents the most specialized and advanced (on an evolutionary basis) situation.

A wide variety of higher plant species has now been examined for C3 or C4 characteristics. Almost without exception those plants which have the biochemical pathway and anatomical features characteristic of C4 plants also have physiological peculiarities. In particular it is found that when C4 plants are placed in an enclosed gas-proof chamber in the light the concentration of CO_2 within the chamber is reduced to near zero (0 to 10 ppm CO_2). This is in contrast to C3 plants, which under similar conditions reach a point of equilibrium at which the rate of uptake of CO_2 by photosynthesis is equal to the rate of output of CO_2 in 'dark' respiration and photorespiration. The concentration of CO_2 (about 100 ppm) obtained at equilibrium is known as the *CO_2 compensation point.* In a similar manner a *light compensation point* can be reached by decreasing the light intensity until assimilation and respiration again compensate one another with no net gas exchange. Under these conditions, light compensation points of about 107 lx are obtained for C4 plants in contrast to values of about 1076–5242 lx for C3 plants.

The low compensation points observed in C4 plants reflect the fact that leaves of C4 plants do not photorespire. As a result C4 plants are capable of assimilating a greater proportion of the atmospheric CO_2; effects on production of dry matter and hence total crop yields are obvious. Since rates of photorespiration increase under high temperatures and high light intensities the C4 pathway is of particular advantage in the tropics, where such plants have in fact evolved.

The long term interest in C4 photosynthesis, therefore, arises from the practical possibility of increasing production of food and other plant-based commodities. However, much of the work carried out on C4 plants has been concerned with developing a theoretical basis to account for the observed effects in biochemical terms. In particular the distribution of enzymes of the C3 and C4 pathways has been studied in detail.

Enzyme distribution studies

Explanations of C4 photosynthesis have been influenced considerably by the scheme proposed by Hatch and Slack (1971) which makes the following assumptions:

(i) the carboxylation reaction catalysed by PEP carboxylase is associated with the mesophyll layer,

(ii) the carboxylation reaction catalysed by RBPC is associated with the bundle-sheath cells,

(iii) the C4 organic acids produced in mesophyll cells function as a carrier moving fixed carbon from one cell layer to the other and

(iv) CO_2 released on decarboxylation of the C4 acids is refixed by RBPC in the bundle sheath.

On the other hand, we have suggested (Baldry *et al.*, 1971; Bucke and Long, 1971; Coombs, 1971, 1973) that mesophyll cells do in fact possess PCR cycle activity, and that the C4 cycle is located in the cytoplasm of mesophyll cells. In this case the four-carbon acids malate and aspartate would function as carbon carriers moving CO_2 to the PCR cycle in both mesophyll and bundle-sheath chloroplasts.

The question of the location of PEP carboxylase would now appear to be resolved. Early suggestions of Hatch, Slack and others indicated that PEP carboxylase occurs within the mesophyll chloroplasts or in association with the chloroplast outer envelope (Hatch and Mau, 1973). However, this conclusion has since been revised (Kagawa and Hatch, 1974). There is now general agreement with our suggestions (Baldry *et al.*, 1971; Bucke and Long, 1971; Coombs and Baldry, 1972) that PEP carboxylase is cytoplasmic.

Pyruvate P_i dikinase would appear to be the only enzyme of the C4 cycle located exclusively in the mesophyll chloroplasts.

On the other hand, the question of the exact location of PCR cycle activity has yet to be resolved. It has now been shown that PGA-kinase, glyceraldehyde-3-phosphate dehydrogenase, triosephosphate dehydrogenase and certain enzymes of carbohydrate metabolism are located within the mesophyll cells as well as in the bundle sheath cells (Hatch and Kagawa, 1973; Kagawa and Hatch, 1974; Chen *et al.*, 1974; Bucke and Coombs, 1974). Hence the question may be simplified to—is there RBPC activity in mesophyll cells? An unequivocal answer is not easily obtained owing to difficulties which arise in methods of isolating chloroplasts, in separating the contents of the two cell types, and in overcoming the effects of inhibitory compounds released during disruption of the leaves.

Initial studies revealed little RBPC activity in C4 plants (Slack and Hatch, 1967), but improved techniques have enabled high rates of RBPC activity to be demonstrated in bundle-sheath extracts. Using a differential grinding technique, we have demonstrated RBPC in mesophyll-cell extracts, although the interpretation of our results has been questioned (Black, 1973).

A wide range of techniques has been used in attempts to resolve this problem. Chloroplasts have been isolated by aqueous (Baldry *et al.*, 1968; O'Neal *et al.*, 1972), or non-aqueous (Slack *et al.*, 1969) methods. Leaf tissue has been disrupted by differential grinding procedures (Bjorkman and Gauhl, 1969; Baldry *et al.*, 1971; Bucke and Long, 1971; Hatch and Mau, 1973; Hatch and Kagawa, 1973), or the two cell types separated by filtration or density gradients following mechanical (Edwards and Black, 1971) or enzymic (Kanai and Edwards, 1973; Chen *et al.*, 1974) disruption. The distribution of RBPC within these fractions has then been estimated. However, an unqualified proof of the presence of RBPC in mesophyll cells has yet to be obtained.

In our opinion one of the main problems yet to be overcome is the inhibitory effect of oxidation products of phenolic compounds liberated during disruption of the leaf tissue. We have shown that leaves of sugar cane contain high concentrations of chlorogenic acid (Baldry *et al.*, 1970a) and a specific *o*-diphenol oxidase (Coombs *et al.*, 1974a). When leaves are disrupted the substrate and enzyme come together resulting in the production of highly reactive *o*-diquinones which can preferentially inhibit RBPC.

By means of differential grinding procedures we established that the concentration of both phenol oxidase and phenolic substrate are higher in the mesophyll than in the bundle-sheath cells (Baldry *et al.*, 1971; Bucke and Long, 1971). This conclusion is supported by results of a histochemical test for phenol oxidase. Fresh hand-cut sections of leaves were incubated in a 10 mM solution of dihydroxyphenyl alanine (DOPA) plus catalytic amounts (1 mM) of chlorogenic acid. The resulting melanin pigmentation developed only in the mesophyll cell layer (Fig. 2). If this layer does represent the locality of phenol oxidase the inhibitors would preferentially affect enzymes located in the mesophyll layer; hence, the low level of RBPC activity observed could be an artefact. This problem is recognized by most plant biochemists. However, attention paid to it in studies of C4 photosynthesis has been slight and at times confused (Huang and Beevers, 1972). A limited range of reagents has been used in attempts to protect isolated enzymes or organelles from effects of phenol oxidation products. We have shown that, with most of

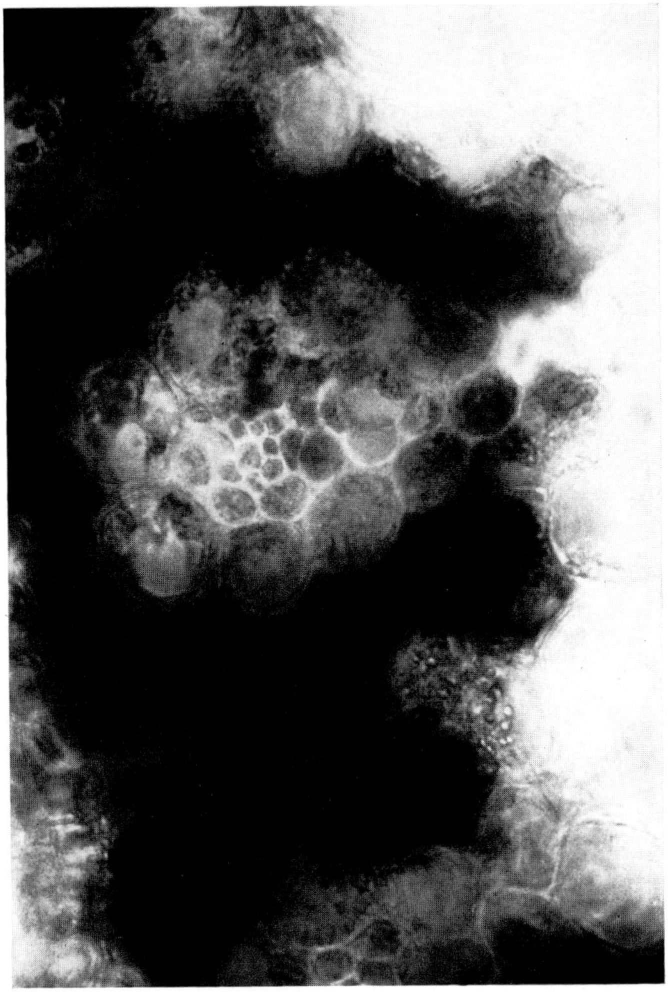

FIG. 2. Transverse section of leaf of sugar cane stained for *o*-diphenol oxidase activity. Note the formation of black (melanin) pigmentation in the mesophyll cell layer.

Facing p. 182

the protective reagents used, several minutes may elapse before activity of the phenol oxidase is inhibited to a significant extent (Baldry *et al.*, 1970b).

In spite of the case for rigid compartmentation of the carboxylation reactions proposed by Hatch and Slack (1971) and by Black (1973) recent reviews of the subject (Laetsch, 1974; Heber, 1974) favour a more cautious interpretation. To emphasize the nature of the argument it should be recalled that attempts to isolate active chloroplasts or RBPC from leaves of many C3 plants, other than peas and spinach, have failed. However, one does not conclude from this failure that temperate species do not carry out photosynthesis.

Biochemical basis of physiological phenomena

Although there is still some disagreement on the details of the biochemistry it is clear that the transport of CO_2 to the PCR cycle by the C4 pathway overcomes photorespiration. The apparent lack of photorespiration in C4 plants cannot be explained in terms of a total absence of the photorespiratory mechanism since C4 plants do show a consumption of oxygen in the light (Volk and Jackson, 1972), presence of peroxisomes and also enzymes of photorespiration (Frederick and Newcomb, 1971; Hilliard *et al.*, 1971; Osmond and Harris, 1971; Huang and Beevers, 1972; Lui and Black, 1972).

It is possible that PEP carboxylase serves to refix CO_2 produced by photorespiration before it can be lost by diffusion from the leaf. If this is so, not only would refixation decrease the total CO_2 loss, but by keeping the concentration of CO_2 in the substomatal cavity low (near zero) it would also maintain a diffusion gradient into the leaf. At the same time the inhibitory effects of O_2 would be overcome if the C4 cycle increases the local concentration of CO_2 in the vicinity of RBPC (Chollet *et al.*, 1974).

Regulation of C4 photosynthesis

Recently we have turned our attention to the mechanisms by which assimilation of CO_2 in C4 plants is regulated at the metabolic level. The various partial reactions found in both the C4 cycle and the PCR cycle parallel one another as shown in Fig. 3. Both cycles are initiated by a carboxylation step, the respective products are reduced by pyridine nucleotides and the carboxylation substrates for both PEP and RBPC are regenerated in phosphorylation reactions which require ATP. As a result the two systems may be in competition, not only for CO_2, but also for NAD(P)H and ATP. For the system to work efficiently one might expect the activity of the initial (C4) carboxylation step to be regulated by both light and the products or metabolites associated with the second (PCR cycle) carboxylation reaction. We have therefore, investigated the possibility that PEP carboxylase acts as a regulatory enzyme.

Initial observations suggesting that PEP carboxylase is a regulatory enzyme in C4 plants were made independently in three laboratories (Coombs and Baldry, 1972; Wong and Davies, 1973; Ting and Osmond, 1973a, b). In general, these reports and others (Lowe and Slack, 1971; Coombs *et al.*, 1973a, 1974b; Nishikido and Takanashi, 1973) indicate activation of the enzyme by sugar phosphates and inhibition by organic acids.

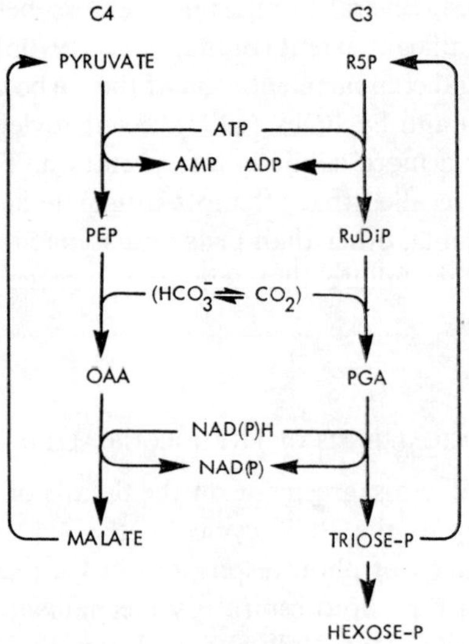

FIG. 3. Interaction between metabolites and co-factors common to the C4 and the C3 pathways.

The two effects may combine to give close control of the activity of PEP carboxylase (Fig. 4).

Although such feedback control may be of great importance a number of other factors may also serve to modulate the rate of carboxylation. For instance the two enzymes, PEPC and RBPC, preferentially use different forms of inorganic carbon. It is generally accepted that the substrate for RBPC is CO_2 (Cooper *et al.*, 1969) and we have recently confirmed that the C4 PEP carboxylase uses bicarbonate as substrate (Coombs *et al.*, 1975). Competition between the two pathways for inorganic carbon could thus be regulated by changes in the size of pools of bicarbonate or CO_2. The size of these pools could in turn be influenced by the activity of carbonic anhydrase, or by light-dependent shifts in the pH of cytoplasm or chloroplast stroma.

In vitro the activity of the PEP carboxylase from *Pennisetum purpureum* can be modified by the inclusion of both pyridine nucleotides and adenylates in experimental reaction mixtures. In general carboxylation rates are increased by NADH and decreased by NADP, NAD and NADPH. The increase observed with NADH may be due in part to the reduction of the end product inhibitor OAA, to the less inhibitory malate, by malate dehydrogenase present as an impurity in the enzyme preparations.

Effects of adenylates (ATP, ADP and AMP) may be direct or complicated by interactions of components of the experimental system with Mg^{2+} ions (Coombs *et al.*, 1974b). *In vivo* changes in the level of one adenylate form must lead to changes in the ratio between other forms since, in the short term at least, the total remains constant. Such changes may be defined in terms of energy charge (ATP + 0.5 ADP/ATP+ADP+AMP). Reaction mixtures can be prepared so as to contain a varying charge. It has been reported (Wong and Davies, 1973) that the activity of PEP carboxylase decreases as the

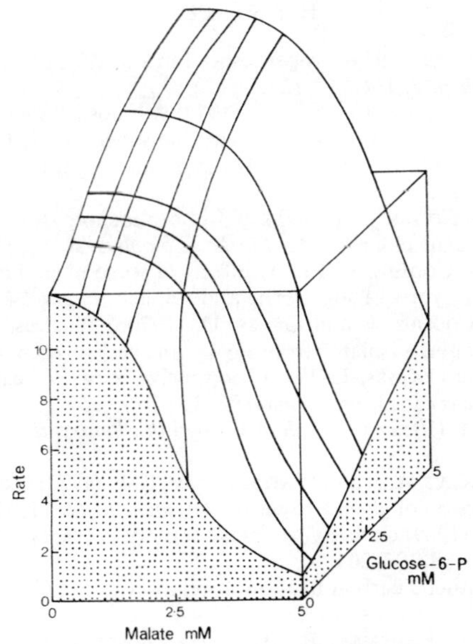

FIG. 4. Interaction of a feedback inhibitor (malate) and a feedback activator (glucose-6-phosphate) on the activity of PEP carboxylase. The activity of PEP carboxylase was determined by measuring the incorporation of $^{14}CO_2$ into OAA in reaction mixtures containing saturating concentrations (5 mM) of PEP, HCO_3^- and Mg^{2+} plus enzyme purified from *Pennisetum purpureum*. Activator (glucose-6-phosphate) and inhibitor (malate) were added at varying concentrations up to 5 mM.

energy charge is increased. However, such a result is inconsistent with the role of PEP carboxylase in C4 photosynthesis since charge increases in the light. We found that inhibition at high energy charge can be overcome by increasing the concentration of Mg^{2+} or by addition of the activator glucose-6-phosphate.

The magnitude and direction of several of the effects described above depends on both the pH of the reaction mixture and the concentration of available Mg^{2+}. Illumination is known to affect the concentration of both hydrogen and Mg^{2+} ions within various metabolic compartments of leaf cells. These effects coupled to feedback control by metabolites could account for the rapid light-activation of the C4 pathway.

CONCLUSIONS

Extensive investigations have clarified many aspects of C4 photosynthesis. The practical value of these results has yet to be realized. The selection of varieties with low photorespiration, and the introduction of C4 characteristics into temperate plants are possibilities which can be explored. It is probably in these areas that major advances will be made.

REFERENCES

ANDREWS, T. J. and HATCH, M. D. (1969) Properties and mechanism of action of pyruvate, phosphate dikinase from leaves. *Biochem. J.* **114,** 117–125.

ANDREWS, T. J., LORIMER, G. H. and TOLBERT, N. E. (1973) Ribulose diphosphate oxygenase I. Synthesis of phosphoglycollate by fraction I protein of leaves. *Biochemistry* **12,** 11–18.

BALDRY, C. W., COOMBS, J. and GROSS, D. (1968) Isolation and separation of chloroplasts from sugar cane. *Z. Pfl.physiol.* **60,** 78–81.

BALDRY, C. W., BUCKE, C. and COOMBS, J. (1969) Light/phosphoenolpyruvate dependent carbon dioxide fixation by isolated sugar cane chloroplasts. *Biochem. biophys. Res. Commun.* **37,** 828–832.

BALDRY, C. W., BUCKE, C. and COOMBS, J. (1970a) Effects of some phenol oxidase inhibitors on chloroplasts and carboxylating enzymes of sugar cane and spinach. *Planta* **94,** 124–133.

BALDRY, C. W., BUCKE, C., COOMBS, J. and GROSS, D. (1970b) Phenols, phenol oxidase and photosynthetic activity of chloroplasts isolated from sugar cane and spinach. *Planta* **94,** 107–123.

BALDRY, C. W., BUCKE, C. and COOMBS, J. (1971) Progressive release of carboxylating enzymes during mechanical grinding of sugar cane leaves. *Planta* **97,** 310–319.

BASSHAM, J. A. and CALVIN, M. (1957) *The Path of Carbon in Photosynthesis,* Prentice Hall, Englewood Cliffs, N.J.

BIRD, I. F., CORNELIUS, M. J., KEYS, A. J. and WHITTINGHAM, C. P. (1972) Oxidation and phosphorylation associated with the conversion of glycine to serine. *Phytochemistry* **11,** 1587–1594.

BJORKMAN, O. and GAUHL, E. (1969) Carboxydismutase activity in plants with and without β-carboxylation photosynthesis. *Planta* **88,** 197–203.

BLACK, C. C. (1973) Photosynthetic carbon fixation in relation to net CO_2 uptake. *A. Rev. Pl. Physiol.* **24,** 253–286.

BOWES, G., OGREN, W. L. and HAGEMAN, R. H. (1971) Phosphoglycollate production catalysed by ribulose diphosphate carboxylase. *Biochem. biophys. Res. Commun.* **45,** 716–722.

BROWN, R. (ed.) (1950) *Carbon Dioxide Fixation and Photosynthesis.* Society for Experimental Biology, Symposium V. Cambridge University Press, Cambridge.

BUCKE, C. and COOMBS, J. (1974) Regulation of CO_2 assimilation in *Pennisetum purpureum.* Proceedings 3rd International Congress of Photosynthesis, Israel. In press.

BUCKE, C. and LONG, S. P. (1971) Release of carboxylating enzymes from maize and sugar cane leaf tissue during progressive grinding. *Planta* **99,** 199–210.

CHEN, T. M., DITTRICH, P., CAMPBELL, W. H. and BLACK, C. C. (1974) Metabolism of epidermal tissue, mesophyll cells, and bundle sheath strands resolved from mature nutsedge leaves. *Archs Biochem. Biophys.* **163,** 246–262.

CHOLLET, R., OGREN, W. L. and BOWES, G. (1974) Oxygen effects in C_3 and C_4 photosynthesis. *What's New in Plant Physiology* **6,** 1–5.

COOMBS, J. (1971) The potential of higher plants with the phosphopyruvic acid cycle. *Proc. R. Soc.* B, **179,** 221–235.

COOMBS, J. (1973) β-Carboxylation, photorespiration and photosynthetic carbon assimilation in C_4 plants. *Curr. Adv. Pl. Sci.* **2** (3), 1–10.

COOMBS, J. and BALDRY, C. W. (1972) C-4 pathway in *Pennisetum purpureum. Nature New Biol.* **238,** 268–270.

COOMBS, J. and WHITTINGHAM, C. P. (1966) The mechanism of inhibition of photosynthesis by high partial pressures of oxygen in *Chlorella. Proc. R. Soc.* B **164,** 511–520.

COOMBS, J., BALDRY, C. W. and BUCKE, C. (1973a) The C-4 pathway in *Pennisetum purpureum.* I. The allosteric nature of PEP carboxylase. *Planta* **110,** 95–107.

COOMBS, J., BALDRY, C. W. and BUCKE, C. (1973b) The C-4 pathway in *Pennisetum purpureum* II. Malate dehydrogenase and malic enzyme. *Planta* **110,** 109–120.

COOMBS, J., BALDRY, C. W., BUCKE, C. and LONG, S. P. (1974a) *o*-Diphenol O_2 oxidoreductase from leaves of sugar cane. *Phytochemistry* **13,** 2703–2708.

COOMBS, J., MAW, S. L. and BALDRY, C. W. (1974b) Metabolic regulation in C_4 photosynthesis; PEP carboxylase and energy charge. *Planta* **117,** 279–292.

COOMBS, J., MAW, S. L. and BALDRY, C. W. (1975) Metabolic regulation in C_4 photosynthesis; the inorganic substrate for PEP carboxylase. *Pl. Sci. Lett.* **4,** 97–102.

COOPER, T. G., FILMER, D., WISHNICK, M. and LANE, M. D. (1969) The active species of 'CO_2' utilized by ribulose diphosphate carboxylase. *J. biol. Chem.* **244,** 1081–1083.

DOWNTON, W. J. S. (1970) Preferential C_4 dicarboxylic acid synthesis, the post-illumination CO_2 burst, carboxyl transfer step and grana configuration in plants with C_4 photosynthesis. *Can. J. Bot.* **48,** 1795–1800.

EDWARDS, G. E. and BLACK, C. C. (1971) Isolation of mesophyll cells and bundle sheath cells from *Digitaria sanguinalis* (L) Scop. leaves and a scanning microscopy study of the internal leaf cell morphology. *Pl. Physiol. Lancaster* **47**, 149–156.

FREDERICK, S. E. and NEWCOMB, E. H. (1971) Ultrastructure and distribution of microbodies in leaves of grasses with and without CO_2 photorespiration. *Planta* **96**, 152–174.

GUTIERREZ, M., GRACEN, V. E. and EDWARDS, G. E. (1974) Biochemical and cytological relationships in C4 plants. *Planta* **119**, 279–300.

HATCH, M. D. and SLACK, C. R. (1966) Photosynthesis by sugar cane leaves. A new carboxylation reaction and the pathway of sugar formation. *Biochem. J.* **101**, 103–111.

HATCH, M. D. and SLACK, C. R. (1971) Photosynthetic CO_2 fixation pathways. *A. Rev. Pl. Physiol.* **21**, 141–162.

HATCH, M. D. and KAGAWA, T. (1973) Enzymes and functional capacities of mesophyll chloroplasts from plants with C_4 pathway photosynthesis. *Archs Biochem. Biophys.* **159**, 842–853.

HATCH, M. D. and MAU, S. (1973) Activity, location, and role of aspartate amino transferase and alanine amino transferase isoenzymes in leaves with C_4 pathway photosynthesis. *Archs Biochem. Biophys.* **156**, 195–206.

HATCH, M. D., MAU, S. and KAGAWA, T. (1974) Properties of leaf NAD malic enzyme from plants with C4 pathway photosynthesis. *Archs Biochem. Biophys.* **165**, 188–200.

HATCH, M. D., OSMOND, C. B. and SLATYER, R. O. (ed.) (1971) *Photosynthesis and Photorespiration.* Wiley–Interscience, London, New York.

HEBER, U. (1974) Metabolite exchange between chloroplasts and cytoplasm. *A. Rev. Pl. Physiol.* **25**, 393–421.

HILLIARD, J. H., GRACEN, V. E. and WEST, S. H. (1971) Leaf microbodies (peroxisomes) and catalase localization in plants differing in their photosynthetic carbon pathways. *Planta* **97**, 93–105.

HUANG, A. H. C. and BEEVERS, H. (1972) Microbody enzymes and carboxylase in sequential extracts from C_4 and C_3 leaves. *Pl. Physiol. Lancaster* **50**, 242–248.

JACKSON, W. A. and VOLK, R. J. (1970) Photorespiration. *A. Rev. Pl. Physiol.* **21**, 385–432.

JOHNSON, H. S. and HATCH, M. D. (1970) Properties and regulation of leaf nicotinamide–adenine dinucleotide phosphate malate dehydrogenase and malic enzyme in plants with the C_4-dicarboxylic acid pathway of photosynthesis. *Biochem. J.* **119**, 273–280.

KAGAWA, T. and HATCH, M. D. (1974) Light dependent metabolism of carbon compounds by mesophyll chloroplasts from plants with the C_4 pathway of photosynthesis. *Aust. J. Pl. Physiol.* **1**, 51–64.

KANAI, R. and EDWARDS, G. E. (1973) Separation of mesophyll protoplasts and bundle sheath cells from maize leaves for photosynthetic studies. *Pl. Physiol. Lancaster* **51**, 1133–1137.

KARPILOV, Y. S. (ed.) (1974) *Photosynthesis in Maize. Structural and Functional Properties of Photosynthetic Apparatus.* (In Russian) Academy of Sciences of the USSR, Puschino-on-Oka.

KORTSCHAK, H. P., HARTT, C. E. and BURR, G. O. (1965) Carbon dioxide fixation in sugar cane leaves. *Pl. Physiol. Lancaster* **40**, 209–213.

KORTSCHAK, H. P. (1968) Photosynthesis in sugar cane and related species. In *Photosynthesis in Sugar Cane*, ed. COOMBS, J. Tate & Lyle, London.

LAETSCH, W. M. (1974) The C_4 syndrome: A structural analysis. *A. Rev. Pl. Physiol.* **25**, 27–52.

LUI, A. Y. and BLACK, C. C. (1972) Glycollate metabolism in mesophyll cells and bundle sheath cells isolated from crab grass *Digitaria sanguinalis* (L) Scop. leaves. *Archs Biochem. Biophys.* **149**, 269–280.

LOWE, J. and SLACK, C. R. (1971) Inhibition of maize leaf phosphopyruvate carboxylase by oxaloacetate. *Biochim. biophys. Acta* **235**, 207–209.

NISHIKIDO, T. and TAKANASHI, H. (1973) Glycine activation of PEP carboxylase from monocotyledenous C4 plants. *Biochem. biophys. Res. Commun.* **45**, 716–722.

OGREN, W. L. and BOWES, G. (1971) Ribulose diphosphate carboxylase regulates soybean photorespiration. *Nature New Biology* **230**, 159–160.

O'NEAL, D., HEW, C. S., LATZKO, E. and GIBBS, M. (1972) Photosynthetic carbon metabolism of isolated corn chloroplasts. *Pl. Physiol. Lancaster* **49**, 607–614.

OSMOND, C. B. and HARRIS, B. (1971) Photorespiration during C_4 photosynthesis. *Biochim. biophys. Acta* **234**, 270–282.

SLACK, C. R. and HATCH, M. D. (1967) Comparative studies on the activity of carboxylase and other enzymes in relation to the new pathway of photosynthetic CO_2 fixation in tropical grasses. *Biochem. J.* **103**, 660–665.

SLACK, C. R., HATCH, M. D. and GOODCHILD, D. J. (1969) Distribution of enzymes in mesophyll and parenchyma sheath chloroplasts of maize leaves in relation to the C_4-dicarboxylic acid pathway of photosynthesis. *Biochem. J.* **114**, 489–498.

TING, I. P. and OSMOND, C. B. (1973a) Activation of plant P-enolpyruvate carboxylases by glucose-6 phosphate: a particular role in crassulacean acid metabolism. *Pl. Sci. Lett.* **1,** 123–128.

TING, I. P. and OSMOND, C. B. (1973b) Multiple forms of plant phosphoenolpyruvate carboxylase associated with different metabolic pathways. *Pl. Physiol. Lancaster* **51,** 448–453.

TOLBERT, N. E. (1963) Glycolate pathway. In *Photosynthetic Mechanisms in Green Plants,* pp. 648–662, National Academy of Sciences – National Research Council Publication 1145, Washington, D.C.

TOLBERT, N. E. (1971) Microbodies, peroxisomes and glyoxysomes. *A. Rev. Pl. Physiol.* **22,** 45–74.

TOLBERT, N. E. and YAMAZAKI, R. K. (1969) Leaf peroxisomes and their relation to photorespiration and photosynthesis. *Ann. N.Y. Acad. Sci.* **168,** 325–341.

VOLK, R. J. and JACKSON, W. A. (1972) Photorespiratory phenomena in maize. Oxygen uptake, isotope discrimination and carbon dioxide efflux. *Pl. Physiol. Lancaster* **49,** 218–223.

WONG, K. F. and DAVIES, D. D. (1973) Regulation of phosphoenolpyruvate carboxylase of *Zea mays* by metabolites. *Biochem. J.* **131,** 451–458.

ZELITCH, I. (1971) *Photosynthesis, Photorespiration and Plant Productivity*. Academic Press, New York.

RIBULOSE BISPHOSPHATE CARBOXYLASE—AN ENIGMA RESOLVED?

D. A. WALKER and R. McC. LILLEY*

Department of Botany, University of Sheffield, Sheffield, U.K.

RIBULOSE bisphosphate carboxylase is a central feature of the Benson–Calvin cycle which, in turn, plays the major role in photosynthetic carbon metabolism. For a brief period it seemed that the Benson–Calvin cycle might be displaced from this major role, at least in C4 species, but more recently it has been accepted that phosphoenolpyruvate carboxylation is an adjunct to the cycle rather than a substitute for it (Hatch, 1970). Thus, in C4 plants as in C3 plants, virtually all the carbon assimilation in photosynthesis must involve ribulose bisphosphate carboxylase at one stage or another (Fig. 1). For many years, however, the carboxylase has been something of an enigma because its characteristics have seemed inadequate for the task which it is believed to perform. Pure air contains only about 330 ppm of CO_2 and yet at first it appeared that the carboxylase would

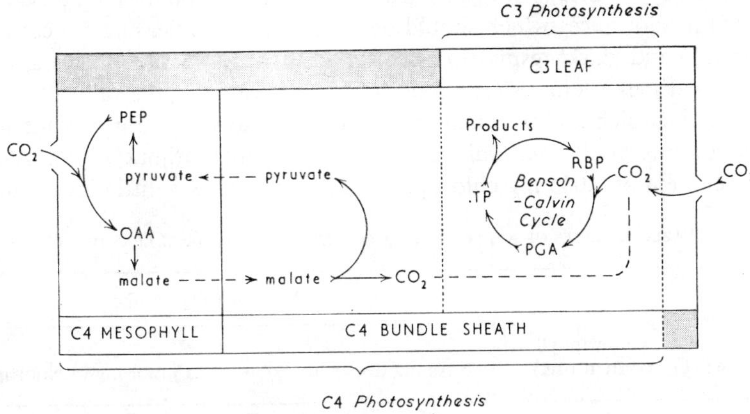

FIG. 1. The relationship between C3 photosynthesis and C4 photosynthesis. This simplified scheme shows β-carboxylation as an adjunct to the Benson–Calvin cycle in C4 plants. In both C3 and C4 species, the Benson–Calvin cycle (which alone is capable of autocatalysis—see e.g. Walker, 1974) is central to photosynthetic carbon assimilation. In the C3 leaf, the CO_2 entering the cycle is derived directly from the external atmosphere (or from CO_2 released in the cytoplasm as the result of respiratory processes). In C4 photosynthesis there is a preliminary β-carboxylation of phosphoenolpyruvate which probably occurs in the cytoplasm of the mesophyll cells. Malate (or aspartate) derived from the end-product (oxaloacetate) of this carboxylation is transported to the bundle-sheath where it undergoes decarboxylation, releasing CO_2 which enters the cycle as before. Abbreviations: OAA, oxaloacetate; PEP, phosphoenolpyruvate; PGA, 3-phosphoglycerate; RBP, ribulose-1,5-bisphosphate; TP, triosephosphate.

*Department of Biology, Wollongong University College, University of New South Wales, Northfields Lane, North Wollongong 2500, P.O. Box 1144, Wollongong, Australia.

require about four or five hundred times this concentration in order to attain maximal velocity (Weissbach *et al.*, 1956). For example, in 1961, Peterkofsky and Racker found more ribulose bisphosphate carboxylase activity in spinach than in four other photosynthetic organisms which they investigated, but at 20°C and in the presence of excess bicarbonate, the rate of carboxylation was only 150 μmol mg^{-1} chlorophyll hr^{-1}. For maximal rates in these experiments, the crude enzyme extract required 3×10^{-2} M bicarbonate.

Because the concentration of CO_2 within the chloroplast stroma is still a matter for conjecture, it is not an easy matter to say precisely what the equivalent rate of carboxylation would be inside the chloroplast for an enzyme with these characteristics. However, in the absence of active transport or facilitated diffusion a difference in partial pressure of 100–200 ppm would be needed, between source and sink, to enable atmospheric CO_2 to diffuse into the leaf at rates commensurate with whole leaf photosynthesis (see e.g. Heath, 1969).

Even if the concentration of CO_2 at the carboxylation site were as high as 200 ppm the 1961 values of Peterkofsky and Racker would not permit rates of fixation faster than about 0.2 μmol mg^{-1} chlorophyll hr^{-1}. They were therefore entirely justified in their conclusion that there was a considerable discrepancy between the performance of the isolated enzyme and that of the intact organism. Again it is difficult to say with any certainty precisely how large this discrepancy really was because of the problems involved in the determination of 'true' photosynthesis. Net CO_2 uptake can now be measured with great accuracy, by means of infra-red analysis, but uncertainties still arise in deciding the extent of the allowances which should be made for CO_2-releasing processes such as photorespiration and 'dark' respiration (see e.g. Zelitch, 1971). Nevertheless, although a few species sometimes attain very rapid rates of photosynthesis, the value of about 20 mg CO_2 dm^{-2} hr^{-1} (which Rabinowitch (1956) ascribed to the average plant under favourable conditions in its natural environment) is not without merit. Assuming a chlorophyll content of 0.045 mg chlorophyll cm^{-2}, which is a little lower than that in

TABLE 1. RATES OF APPARENT PHOTOSYNTHESIS BY SPINACH LEAF DISCS

Disc	Fresh wt (g)	Chlorophyll content (mg)	Rate of CO_2 uptake: As mg dm^{-2} hr^{-1}		As μmol mg^{-1} chlorophyll hr^{-1}	
(1)	0.574	0.56	20.5	17.8	83.4	72.2
(2)	0.497	0.62	21.6	17.1	79.2	62.8
(3)	0.460	0.55	18.5	16.4	76.4	67.7
light			White	Red	White	Red

Circular discs, of 10 cm² area, were cut from spinach leaves with a sharp punch and illuminated with a 150 W quartz iodine slide projector in an open differential infra-red gas analysis system. Light from the projector was passed through 15 cm of water and a Balzers Calflex C interference filter to remove infra-red (Delieu and Walker, 1972) and, where indicated, through red perspex (ICI.400) to cut off light below 625 nm so that the intensity and quality was then virtually the same as that used in the chloroplast experiments (Fig. 4). Air at 20–21°C, containing 0.0305 per cent CO_2, was passed over the discs at 1 l. min^{-1} and rates of apparent photosynthesis were calculated from the observed decrease in CO_2 concentration in the steady state.

spinach as it is grown in Sheffield, the rate of 20 mg CO_2 dm^{-2} hr^{-1} is equivalent to 100 μmol mg^{-1} chlorophyll hr^{-1} and very close to the rate of true photosynthesis in spinach at 20°C and 300 ppm CO_2 (Table 1). On this basis, the activity of the organism would be 500-fold greater than that of the enzyme on which it is believed to depend.

This then was the enigma and it came to be centred on the affinity of the extracted enzyme for CO_2. Regardless of the fact that a great part of leaf protein is ribulose bisphosphate carboxylase (Kawashima and Wildman, 1970), its affinity was simply too low to be acceptable. Even if on a chlorophyll basis the *activity* had been 500 times higher (which would, in theory, have made the enzyme equal to its task), its low *affinity* would have remained an oddity. The only credible explanation which was forthcoming was that the enzyme might owe its low affinity to the fact that it evolved at a time when there was much more CO_2 in the atmosphere (Stiller, 1962). If low affinity was combined with low activity it seemed inescapable that the characteristics of the enzyme must be altered on extraction or that the effective concentration of CO_2 within the chloroplast must be very much larger than that in the external atmosphere (Gibbs *et al.*, 1967; Walker and Crofts, 1970).

The first step in the resolution of this enigma came with the demonstration that ribulose bisphosphate carboxylase utilizes CO_2 rather than bicarbonate (Cooper *et al.*, 1969). This was an important observation but, contrary to some comments made at the time, it only called for a moderate downward revision of the CO_2 requirement from about 6 per cent to about 1.8 per cent. The original arguments had been based on the concentration of CO_2 which would be required in the gas phase in order to maintain the reaction at half maximal velocity under steady state conditions. At pH 7.0 the 1956 value of 11 mM bicarbonate (Weissbach *et al.*, 1956) was equivalent to 6 per cent CO_2 in the gas phase, and if all the various ionic forms of CO_2 in solution were also in equilibrium it would have been immaterial whether the enzyme used CO_2 or bicarbonate (Walker, 1973). However, the concentration of CO_2 in solution is independent of pH, whereas the distribution of total dissolved CO_2 amongst its ionic forms, including bicarbonate, is not. Following the observations of Cooper *et al.* (1969) it therefore became necessary to recalculate the CO_2 in the gas phase in relation to the bicarbonate required for half maximal velocity *at the pH of the actual determination.* Using the (then) current value for the K_m of 22 mM at pH 7.8 this became 1.8 per cent CO_2. This value then applied over the whole pH range under consideration and did not rise to 12 per cent CO_2 at pH 7.0 as it would if it had been necessary to ask how much gaseous CO_2 would be required to maintain 22 mM bicarbonate at neutrality (Walker, 1973).

The second advance came with the growing recognition (following earlier and perhaps neglected observations of Pon (1959) that the apparent K_m varied with the Mg content of the assay medium. In this connection there was a succession of reports in which the K_m CO_2 edged gradually downwards and in which various schemes of Mg activation *in vivo* were proposed (see e.g. Pon *et al.*, 1963; Jensen and Bassham, 1968; Sugiyama *et al.*, 1968; Jensen, 1971; Lin and Nobel, 1971; Stokes *et al.*, 1972; Walker, 1972; Jensen and Bahr, 1972; Hind *et al.*, 1974; Lilley *et al.*, 1974; Walker, 1974). Finally it seemed that the ultimate had been reached when Bahr and Jensen (1974) described a kinetic form of ribulose bisphosphate carboxylase with an apparent K_m CO_2 equivalent of approximately 300 ppm in the gas phase. However this particular form also had a very low V_{max}

so that even though the gap between enzyme and organism had narrowed enormously, perhaps from 500 to about 5, the problem was not entirely resolved.

At Sheffield, two developments (Lilley and Walker, 1974, 1975) have occurred recently which we believe will narrow the gap still further if not eliminate it entirely. The first involves an improved assay for ribulose bisphosphate carboxylase (Lilley and Walker, 1974) and the second a method of relating carboxylation to chlorophyll content in a way which permits a more realistic evaluation of overall enzyme activity. Ruptured chloroplasts lose their stromal protein but retain their chlorophyll. If it were possible to determine accurately the proportion of ruptured chloroplasts in a given preparation it would obviously be possible to express ribulose bisphosphate carboxylase activity on a chlorophyll basis in a much more meaningful manner. The chloroplast envelope is thought to be largely impermeable to ferricyanide (Cockburn *et al.*, 1967; Heber and Santarius, 1970) and therefore the rate of ferricyanide reduction (or ferricyanide dependent O_2 evolution) may provide an accurate measure of envelope integrity. On the other hand, we suspect that envelopes can rupture and then reseal, following protein loss, so that the 20–30 per cent correction which we presently apply on this basis is very probably an underestimate of the actual *in vivo* protein/chlorophyll ratio. Moreover, the fact that

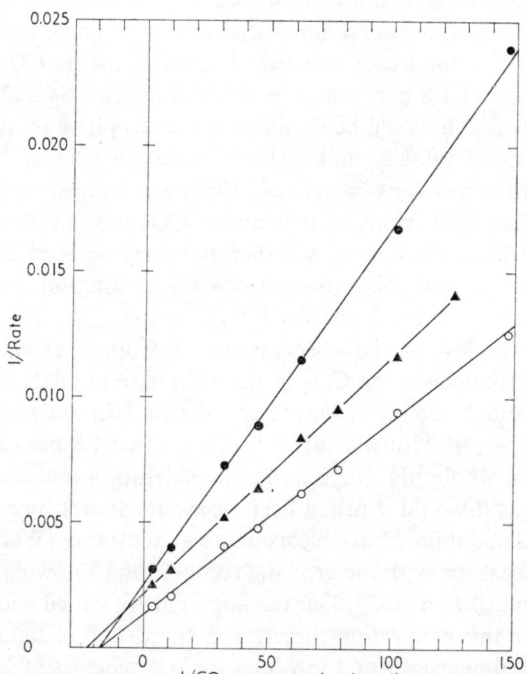

FIG. 2. Reciprocal plots of rates of CO_2 fixation by ribulose bisphosphate carboxylase in spinach chloroplast extracts against CO_2 concentration (data from Lilley and Walker, 1975). Data for enzyme activity in extracts from three chloroplast preparations containing the following proportions of intact chloroplasts: ●, 80 per cent; ○, 69 per cent; △, 76 per cent. Ribulose bisphosphate carboxylase activity was determined by a spectrophotometric assay (Lilley and Walker, 1974) coupled to NADH oxidation. The assay mixture contained chloroplast extract equivalent to 2.5 μg chlorophyll, 0.33 M glucitol, 50 mM N-2-hydroxyethylpiperazine-N′-ethanesulphonic acid, 10 mM KCl, 1 mM EDTA.

the various characteristics of the carboxylase appear to vary considerably according to the prevailing pH and Mg concentration (see e.g. Sugiyama *et al.*, 1968) makes it unlikely that the present assay, although greatly improved, will guarantee the fastest possible rates. In short, the true activity within the chloroplast is almost certainly somewhat better than the present measurements might suggest.

Our improved assay for the carboxylase yields an apparent K_m CO_2 (Fig. 2) which is still four to five times as high as atmospheric CO_2 but which, when taken together with improved maximal velocities approaching 1000 μmol mg^{-1} chlorophyll hr^{-1} (see Table 2 and Fig. 3 and cf. the 1961 figure of 150 reported by Peterkofsky and Racker) is just about sufficient to account for the performance of the parent tissue (cf. Tables 1, 2).

TABLE 2. RELATIONSHIP BETWEEN CARBOXYLATION RATE AND CO_2 CONCENTRATION AT THE SITE OF CARBOXYLATION FOR RIBULOSE BIPHOSPHATE CARBOXYLASE WITH CHARACTERISTICS SIMILAR TO THAT FROM SPINACH

Maximal velocity (μmol mg^{-1} chlorophyll hr^{-1})	Rates in CO_2 concentrations of 50 ppm	100 ppm	150 ppm	200 ppm	CO_2 conc (ppm) to give fixation rate of 100 μmol mg^{-1} chlorophyll hr^{-1}
800	27	52	75	98	206
940	*23*	*61*	*88*	*114*	*172*
1000	33	64	93	120	161
1250	42	80	116	150	127
1600	53	102	148	192	98
2000	67	130	188	244	82

In Fig. 2 ribulose bisphosphate carboxylase displays K_m values of approximately 46.5 μM CO_2 and maximal velocities as high as 940 μmol mg^{-1} chlorophyll hr^{-1}. This table shows the rates which would be obtained in reaction mixtures in equilibrium with partial pressures of CO_2 between 50–200 ppm assuming a K_m of 46.5 μM CO_2 and maximal velocities in the range of 800–2000 μmol mg^{-1} chlorophyll hr^{-1}. In addition, in the last column, the CO_2 concentration required to give a rate of 100 μmol mg^{-1} chlorophyll hr^{-1} is listed. It will be seen that, for the highest value of V_{max} actually recorded (figures in italics), a partial pressure of 100 ppm would have yielded a rate of 61 or that for a rate of 100 (which is taken as that of the average plant in its natural environment) a partial pressure of 172 ppm CO_2 would have been required at the carboxylation site.

It should be stressed, of course, that Table 2 is almost entirely a theoretical exercise and there is no certainty that behaviour within the chloroplast can be confidently predicted on the basis of results obtained with the soluble enzyme. However, if Peterkofsky and Racker (1961) could conclude that the levels of ribulose bisphosphate carboxylase which they measured were inadequate to satisfy the rate of CO_2 fixation by the whole organism it is not unreasonable to ask if the position is in any way changed as a consequence of the new measurements. The values in Table 2 show that the situation is indeed changed and they also draw attention to the fact that the maximum velocity is no less important than the K_m. In a sense this is self-evident but, because the fastest rates which were first measured were not in themselves inadequate, attention became focused on affinity rather than V_{max} (e.g. Peterkofsky and Racker's value of 150 μmol mg^{-1} chlorophyll hr^{-1} compares very favourably with that of the average plant and is only unsatisfactory in relation to the massive bicarbonate concentration required to attain

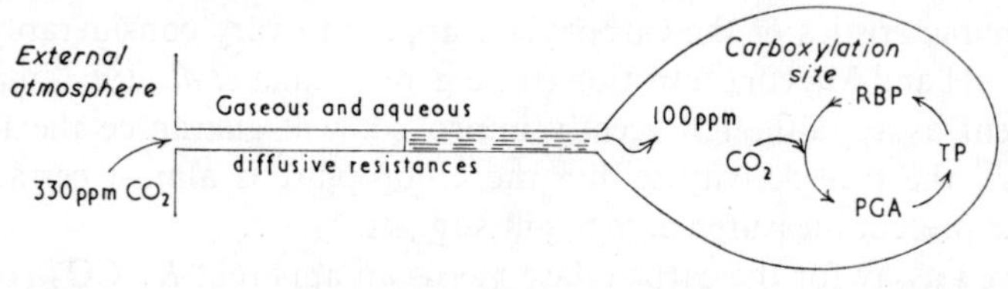

FIG. 3. The relationship between carboxylation and diffusion. Under optimal conditions, the rate of carboxylation will be governed by the concentration of CO_2 within the chloroplast and the activity and characteristics of the carboxylase. The rate of entry of CO_2 into the chloroplasts will depend upon the diffusive resistances (in the gas and liquid phases) and upon the difference in partial pressure of CO_2 between the exterior and the site of carboxylation. A low internal concentration would favour entry but limit carboxylation (and vice versa) so that the steady state concentration of CO_2 at the carboxylation site needed to maintain a given rate could be calculated if the external concentration and diffusive resistances were accurately known. Applying values for the diffusive resistance derived by Heath and others (Heath, 1969), a concentration of 100 ppm at the carboxylation site would allow CO_2 to enter sufficiently fast to maintain a rate of photosynthesis of 100 μmol mg^{-1} chlorophyll hr^{-1} (which approximates to that displayed by the average plant in air). If the carboxylase works at half maximal velocity in 46.5 μM CO_2 this would call for a maximal velocity of approximately 1500 (see also Table 2). Abbreviations as in Fig. 1.

this rate). The present affinity (say 0.13 per cent CO_2 for half maximal velocity), although a marked improvement on 6 per cent CO_2, is still appreciably higher than the probable CO_2 concentration at the carboxylation site during steady state photosynthesis, but, as Table 2 shows this would be immaterial given a sufficiently high V_{max} (cf. Raven, 1970). It would also account for the marked increases in photosynthetic carbon fixation which occur as a consequence of relatively small increases in CO_2 concentration (Heath, 1969; Milthorpe and Moorby, 1974). Furthermore, it would be possible for species such as sunflower to exhibit greatly superior rates of photosynthesis, not because of any innate superiority in their carboxylation capacity but simply because they have leaves which present a smaller diffusive resistance to CO_2 than the average plant (cf. Hesketh, 1963).

It may also be asked why we believe that the affinity which we measure, accurately reflects that of the enzyme within parent tissue. The term 'K_m' (used in the sense of the substrate concentration required for half maximal velocity) has been applied to both chloroplast (see e.g. Jensen, 1971) and whole leaves (Goldsworthy, 1968) and, in terms of CO_2 in the gas phase is probably about 0.03 per cent. Whether or not it is reasonable to apply terms borrowed from enzyme kinetics to whole organisms (or organelles) is a matter for argument. It can be misleading, because under appropriate conditions the maximal velocity of an enzyme-catalysed reaction is determined by the characteristics of that enzyme, whereas in the intact organism a variety of other factors could be involved. This is also true of chloroplasts where much emphasis has been placed on the fact that half maximal velocity was reached at the same concentration of CO_2 as the parent organism. In this respect therefore the organelle was apparently very much superior to the soluble enzyme. In fact new determinations (Lilley and Walker, 1974, 1975) of the rate of photosynthesis (by chloroplasts) as a function of CO_2 concentration show that there is a marked departure from linearity in a double reciprocal plot (Fig. 4). This occurs at high CO_2 concentrations and it is a consequence of a ceiling imposed by electron transport (Fig. 5). Clearly if the maximal rate of electron transport (Table 3) and associated phosphorylation is insufficient to regenerate ribulose bisphosphate as rapidly as it

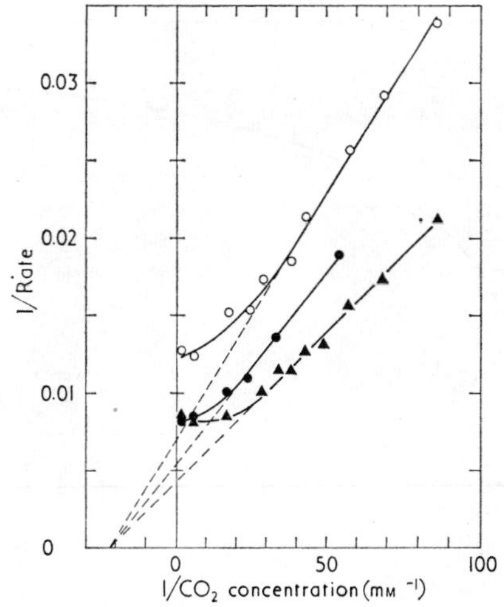

FIG. 4. Reciprocal plots of rates of CO_2-dependent O_2 evolution against CO_2 concentration for three spinach chloroplast preparations (Lilley and Walker, unpublished data). Proportions of intact chloroplasts: ●, 62 per cent; ○, 69 per cent; ▲, 66 per cent. Oxygen evolution by the chloroplasts was measured using apparatus described previously (Delieu and Walker, 1972). The reaction mixture contained chloroplasts (200 μg chlorophyll), 0.33 M glucitol, 1 mM $MgCl_2$, 1 mM $MnCl_2$, 2 mM EDTA, 50 mM N-2-hydroxyethylpiperazine-N′-ethanesulphonic acid, 5 mM $Na_4P_2O_7$, 0.5 mM Pi, pH 7.6 in a 2.0 ml volume, and was illuminated with red light at 20°C. A linear rate of O_2 evolution was recorded when an appropriate amount of $NaHCO_3$ was added to the reaction mixture following 3 min preillumination.

TABLE 3. COMPARISON OF RIBULOSE BISPHOSPHATE CARBOXYLASE ACTIVITY WITH CHLOROPLAST PHOTOSYNTHETIC ACTIVITY

	Mean ± S.E.M. (14 determinations)	Mean corrected to 100% intact
	As μmol CO_2 fixed or μmol O_2 evolved mg^{-1} chlorophyll hr^{-1}	
Ribulose bisphosphate carboxylase activity in chloroplast extracts (in 10 mM HCO_3^-/302 μM CO_2 at pH 7.9 and 20°C)	426 ± 22	573*
Whole chloroplasts—Rate of CO_2-dependent O_2 evolution (in 10 mM HCO_3^-/584 μM CO_2 at pH 7.6)	134 ± 4	180
Chloroplasts ruptured by osmotic shock and uncoupled—Rate of ferricyanide-dependent O_2 evolution	266 ± 11**	

* These rates were determined at 20°C whereas Peterkofsky and Racker (1961) made their measurements at 30°C and corrected them to 20°C by assuming a Q_{10} of 2. Applying a similar correction the 1969 values of Bjorkman and Gauhl (for a variety of species) and the 1967 value of Johnson and Bruff (for spinach chloroplasts) would be 60–252 and 165 respectively. By interpolation to infinite substrate concentration the highest maximum velocity recorded (Fig. 2, lower curve) in the present work (Lilley and Walker, 1975) was 940 μmol CO_2 fixed mg^{-1} chlorophyll hr^{-1}, in respect to intact chloroplasts.
** It should be noted that our case that carboxylation in intact chloroplasts is limited by electron transport does not rest on abnormally low rates in the Hill reaction. The values given here are substantially higher than those summarized by Rabinowitch (1956) or those determined more recently by several workers (see e.g. Reeves and Hall, 1973; West and Wiskich, 1973).

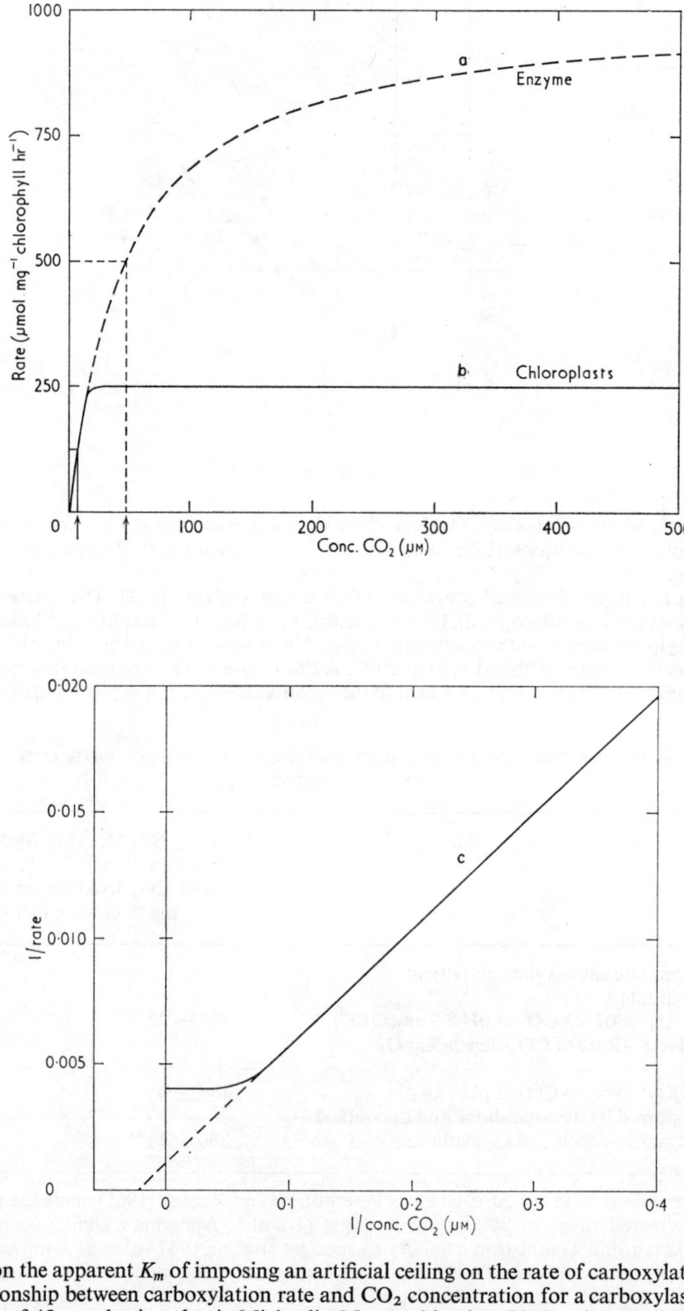

FIG. 5. Effect on the apparent K_m of imposing an artificial ceiling on the rate of carboxylation. (a) Broken line. The relationship between carboxylation rate and CO_2 concentration for a carboxylase with a V_{max} of 1000 and a K_m of 45 μM obeying classic Michaelis–Menton kinetics. (b) Continuous line. The apparent relationship which would emerge (e.g. with chloroplasts) if an artificial ceiling of 250 were applied. The CO_2 required for half maximal velocity is correspondingly diminished so that the apparent affinity of the system for CO_2 is less than the real affinity. (c) A double reciprocal plot of (b) in which true kinetic constants can be derived only if the departure from linearity imposed by the artificial ceiling is disregarded.

is used then the true maximal rate of carboxylation will not be displayed. Likewise, if the *apparent* maximum is less than the *real* maximum the CO_2 concentration required to give half of this rate will also be correspondingly smaller. For this reason the departure from linearity has been ignored in Fig. 4 and the straight line portion of each curve extrapolated to a common intercept equivalent to 45 μM CO_2. If this intercept is accepted as the true value of the CO_2 concentration required for half maximal velocity it would follow that the affinity of ribulose bisphosphate carboxylase for CO_2 is the same *in situ* as it is when the stromal protein is released from the chloroplast by osmotic shock.

These results do not in themselves provide any clues about active transport or facilitated diffusion of CO_2 but on the other hand it is no longer necessary to invoke such mechanisms as an essential part of the photosynthetic apparatus if the maximal velocity of the carboxylase is so much higher and its K_m (CO_2) so much lower than had been previously supposed.

References

Bahr, J. T. and Jensen, R. G. (1974) Ribulose diphosphate carboxylase from freshly ruptured spinach chloroplasts having an *in vivo* K_m [CO_2]. *Pl. Physiol. Lancaster* **53,** 39–44.

Bjorkman, O. and Gauhl, E. (1969) Carboxydismutase activity in plants with and without β-carboxylation photosynthesis. *Planta* **88,** 197–203.

Cockburn, W., Baldry, C. W. and Walker, D. A. (1967) Some effects of inorganic phosphate on O_2 evolution by isolated chloroplasts. *Biochim. biophys. Acta* **143,** 614–624.

Cooper, T. G., Filmer, D., Wishnick, M. and Lane, M. D. (1969) The active species of 'CO_2' utilized by ribulose diphosphate carboxylase. *J. biol. Chem.* **244,** 1081–1083.

Delieu, T. and Walker, D. A. (1972) An improved cathode for the measurement of photosynthesis by isolated chloroplasts. *New Phytol.* **71,** 201–225.

Gibbs, M., Latzko, E., Everson, R. G. and Cockburn, W. (1967) In *Harvesting the Sun,* ed. San Pietro, A., Greer, F. A. and Army, T. J. pp. 111–130, Academic Press, London.

Goldsworthy, A. (1968) Comparison of the kinetics of photosynthetic carbon dioxide fixation in maize, sugar cane and its relation to photorespiration. *Nature Lond.* **217,** 6.

Hatch, M. D. (1970) Mechanism and function of the C4 pathway of photosynthesis. In *Photosynthesis and Photophosphorylation,* ed. Hatch, M. D., Osmond, C. B. and Slatyer, R. O. p. 139. Wiley, New York.

Heath, O. V. S. (1969) *The Physiological Aspects of Photosynthesis.* Heinemann, London.

Heber, U. and Santarius, K. A. (1970) Direct and indirect transfer of ATP and ADP across the chloroplast envelope. *Z. Naturf.* **25b,** 718–728.

Hesketh, J. D. (1963) Limitations to photosynthesis responsible for differences among species. *Crop Sci.* **3,** 493–496.

Hind, G., Nakatani, H. Y. and Izawa, S. (1974) Light-independent redistribution of ions in suspensions of chloroplast. *Proc. natn. Acad. Sci. U.S.A.* **71,** 1484–1488.

Jensen, R. G. (1971) Activation of CO_2 fixation in isolated spinach chloroplasts. *Biochim. biophys. Acta* **234,** 360–370.

Jensen, R. G. and Bahr, J. T. (1972) In *Progress in Photosynthesis,* Proceedings of the 2nd International Congress of Photosynthesis, Stresa, 1971. ed. Forti, G., Avron, M. and Melandri, A. Vol. 3, pp. 1787–1794.W. Junk, The Hague.

Jensen, R. G. and Bassham, J. A. (1968) Photosynthesis by isolated chloroplasts. III. Light activation of the carboxylation reaction. *Biochim. biophys. Acta* **153,** 227–234.

Johnson, E. J. and Bruff, B. S. (1967) Chloroplast integrity and ATP-dependent CO_2-fixation in *Spinacia oleracea. Pl. Physiol. Lancaster* **42,** 1321–1328.

Kawashima, N. and Wildman, S. G. (1970) Studies on Fraction I protein. I. Effect of crystallization of Fraction I protein from tobacco leaves on ribulose diphosphate carboxylase activity. *A. Rev. Pl. Physiol.* **21,** 325–358.

Lin, D. C. and Nobel, P. S. (1971) Control of photosynthesis by Mg^{2+}. *Archs Biochem. Biophys.* **145,** 622–632.

LILLEY, R. McC. and WALKER, D. A. (1974) An improved spectrophotometric assay for ribulose diphosphate carboxylase. *Biochim. biophys. Acta* **358,** 226–229.
LILLEY, R. McC., HOLBOROW, K. and WALKER, D. A. (1974) Magnesium activation of photosynthetic CO_2 fixation in a reconstituted chloroplast system. *New Phytol.* **73,** 657–662.
LILLEY, R. McC. and WALKER, D. A. (1975) Carbon dioxide assimilation by leaves, isolated chloroplasts and ribulose bisphosphate carboxylase from spinach. *Pl. Physiol. Lancaster*. In press.
MILTHORPE, F. L. and MOORBY, J. (1974) *An Introduction to Crop Physiology*. Cambridge University Press, London.
PETERKOFSKY, A. and RACKER, E. (1961) The reductive pentose phosphate cycle. III. Enzyme activity in cell-free extracts of photosynthetic organisms. *Pl. Physiol. Lancaster* **36,** 409–414.
PON, N. G. (1959) Studies on the carboxydismutase system and related materials. Ph.D. thesis, University of California, Berkeley.
PON, N. G., RABIN, B. R. and CALVIN, M. (1963) Mechanism of the carboxydismutase reaction. I. The effect of preliminary incubation of substrate, metal ion and enzyme on activity. *Biochem. Z.* **338,** 7–19.
RABINOWITCH, E. I. (1956) *Photosynthesis and Related Processes*. Vol. 2, Part 2, Interscience, New York.
RAVEN, J. A. (1970) Exogenous inorganic carbon sources in plant photosynthesis. *Biol. Rev.* **45,** 167–221.
REEVES, S. G. and HALL, D. O. (1973) The stoichiometry ($ATP/2e^-$ ratio) of non-cyclic photophosphorylation in isolated spinach chloroplasts. *Biochim. biophys. Acta* **314,** 66–78.
STILLER, M. (1962) The path of carbon in photosynthesis. *A. Rev. Pl. Physiol.* **13,** 151–170.
STOKES, D. M., WALKER, D. A. and McCORMICK, A. V. (1972) Photosynthetic oxygen evolution in a reconstituted chloroplast system. In *Progress in Photosynthesis*, Proceedings of the 2nd International Congress of Photosynthesis, Stresa, 1971. ed. FORTI, G., AYRON, M. and MELANDRI, A., Vol. III, p. 1779. W. Junk, The Hague.
SUGIYAMA, T., NAKAYAMA, N. and AKAZAWA, T. (1968) Structure and function of chloroplast proteins. V. Homotropic effect of bicarbonate in RuDP carboxylase reaction and the mechanism of activation by magnesium ions. *Archs Biochem. Biophys.* **126,** 737–745.
WALKER, D. A. (1972) The affinity of ribulose diphosphate carboxylase for CO_2/bicarbonate. In *Progress in Photosynthesis*, Proceedings of the 2nd International Congress of Photosynthesis, Stresa, 1971. ed. FORTI, G., AVRON, M. and MELANDRI, A. Vol. III, p. 1773. W. Junk, The Hague.
WALKER, D. A. (1973) Photosynthetic induction phenomena and the light activation of ribulose diphosphate carboxylase. *New Phytol.* **72,** 209–235.
WALKER, D. A. (1974) Some characteristics of a primary carboxylating mechanism. In *Plant Carbohydrate Biochemistry*, Proceedings of the Tenth Symposium of the Phytochemical Society, Edinburgh, 1973, ed. PRIDHAM, J. B. pp. 7–26. Academic Press, London and New York.
WALKER, D. A. and CROFTS, A. R. (1970) Photosynthesis. *A. Rev. Biochem.* **39,** 389–428.
WEISSBACH, A., HORECKER, B. L. and HURWITZ, J. (1956) The enzymatic formation of phosphoglyceric acid from ribulose diphosphate and carbon dioxide. *J. biol. Chem.* **218,** 795–810.
WEST, K. R. and WISKICH, J. T. (1973) Evidence for two phosphorylation sites associated with the electron transport chain of chloroplasts. *Biochim. biophys. Acta* **292,** 197–205.
ZELITCH, I. (1971) *Photosynthesis, Photorespiration and Plant Productivity*. Academic Press, London.

AUTOTROPHIC ENDOSYMBIONTS OF INVERTEBRATES

D. C. Smith

Department of Botany, University of Bristol, Bristol, U.K.

Introduction

During the past two decades, there has been a substantial resurgence of interest in the experimental study of the symbiosis between certain lower invertebrates and their algal endosymbionts. These investigations have had the added impetus that the symbiotic theory of the origin of eukaryotic cell structure is now gaining wide acceptance. Studies of the interactions between modern symbionts and their host cells may have potential relevance to understanding cell function and organization since knowledge of how a cell regulates the activity of a symbiont might shed useful light upon how mechanisms for the control of organelles may have evolved. The study of autotrophic endosymbionts presents the particular experimental advantage that photosynthetic carbon fixation provides an excellent marker for investigation of the activity of the symbiont within the intact association.

The purpose of this paper is to discuss how the biochemical activities of autotrophic endosymbionts become regulated by host cells, with especial reference to photosynthesis and the release of photosynthate to the host. As with any new subject, a substantial amount of background experimental knowledge is first required before these problems can be satisfactorily investigated. For example, experience of different associations is needed to discover which is the best material for laboratory experiments. It is also necessary to show that symbionts do in fact release photosynthate to the host and to identify the compounds involved. Indeed, much of the work in the last two decades has been devoted to obtaining this kind of basic information, and it is only recently that investigators have begun to study the deeper problems of how symbionts become integrated into the activities of host cells. Three associations which have provided useful information are: *Hydra viridis*, *Convoluta roscoffensis* and *Elysia viridis*.

Hydra viridis

Green hydra offers many advantages for the laboratory study of symbiosis. It is simple to grow large quantities in culture. The algae, which occur packed each inside a vacuole in the gastrodermal cells, can be separated completely and without damage from animal tissue by centrifuging homogenates of the hydra, so facilitating studies of the transfer of substances between the symbionts. The algal symbiont is *Chlorella*, a genus about whose biochemistry much is already known. Some strains of *Hydra viridis* can be freed of algae (aposymbiotic strains) so that the effects of presence and absence of symbionts can be investigated.

Muscatine and Lenhoff (1965) illustrated the selective advantage of the symbionts to the host by showing that green hydra survived starvation in the light better than aposymbiotic strains, although with excess food supply, there was no difference between the growth of the two types. Muscatine (1965) showed that at least half of all the photosynthetically fixed carbon was released by the alga to the hydra, and that it was released almost entirely as maltose. Cernichiari *et al.* (1969) investigated the mechanism of maltose synthesis and release by the alga after isolation from the hydra. They concluded that maltose was synthesized at or near the algal cell surface by a modification of some transglycosylation mechanism. An outstanding feature of the system was that maltose synthesis and excretion was very sensitive to the pH of the medium (Fig. 1), and this led them to suggest that the animal might control the flow of sugar by regulating the pH in the vacuole surrounding each algal cell.

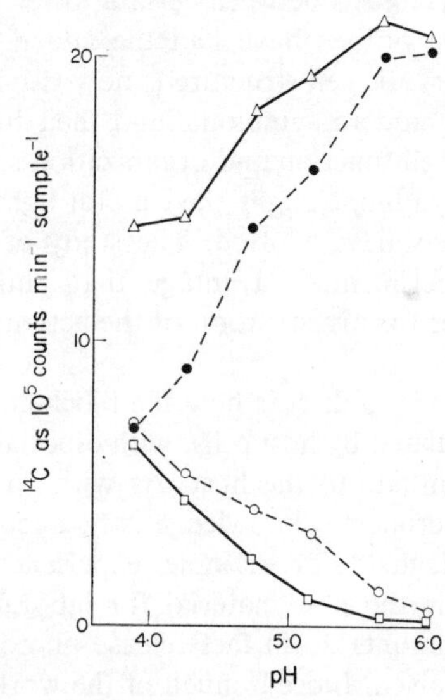

FIG. 1. Effect of pH upon fixation and release of ^{14}C by *Chlorella* isolated from *Hydra viridis* (data of Cernichiari *et al.*, 1969). —△—, total net ^{14}C fixed; – – ● – –, ^{14}C in cells; – – ○ – –, ^{14}C released to medium; —□—, ^{14}C maltose in medium.

This hypothesis has been recently tested by Dahlin and Smith (unpublished), who compared the flow of fixed carbon from alga to host in situations where the host might be expected to have either a high or low demand for carbon. These situations were created by pre-treating the hydra for periods of up to 7 days without food in either continuous light (i.e. low demand) or continuous darkness (high demand) before exposure to $NaH^{14}CO_3$ for a test period in the light. The results (Table 1) show that the proportion of fixed carbon moving from alga to hydra remains constant but ^{14}C fixation differs markedly in the two treatments. This suggests that carbon flow is not regulated by pH, despite the sensitivity of maltose release to small changes of external pH. It is not known if there was any change in number of algae per hydra during the two treatments, but any such change would be likely to be a reduction of algal number in the dark (cf. Pardy, 1974). Therefore, it appears as if there is some control over rate of carbon fixation, although the differences in ^{14}C fixation between light and dark pre-treatment might well result from differing dilutions of the isotope by $^{12}CO_2$ from animal respiration. At all events, the earlier suggestion of Cernichiari *et al.* (1969) must be discarded.

TABLE 1. *Hydra viridis*: EFFECT OF PRE-TREATMENT IN CONTINUOUS LIGHT OR CONTINUOUS DARKNESS ON THE SUBSEQUENT LEVEL OF NET ^{14}C FIXATION AND RELEASE OF FIXED ^{14}C FROM ALGA TO ANIMAL (data of Dahlin and Smith, unpublished)

Parameter	Net ^{14}C fixed (as counts min^{-1} $hydra^{-1}$)			% fixed ^{14}C released from alga to animal tissues	
Days	0	4	7	0	7
Continuous light		109	91		44
	112			40	
Continuous dark		123	154		44

Samples of 100 hydra incubated in 2 ml culture medium (cf Muscatine and Lenhoff, 1965) in 2.54 cm diam specimen tubes in continuous light (2.15 klx) or darkness at 18°C. For measurement of net ^{14}C fixation, 10 μCi $NaH^{14}CO_3$ added to each sample with illumination of 21.5 klx for 1 hr; animals then extracted in 80 per cent ethanol as described by Cernichiari *et al.* (1969). Release of fixed ^{14}C from alga to animal measured by estimating ^{14}C in algae and animal tissues after separation by homogenization and centrifugation (see Cernichiari *et al.* for details).

Despite its obvious suitability for laboratory experiments, green hydra may not be typical of other algal/invertebrate symbioses. For example, in no symbiosis involving algae other than *Chlorella* does release of photosynthate occur by modification of a surface enzyme system (Smith, 1974); certainly, there are no other symbiotic algae whose photosynthate release shows such marked pH sensitivity. Also, in all other algal/invertebrate associations, carbohydrate release ceases immediately or very soon after the algae are isolated, but in symbiotic *Chlorella*, it continues without decline for at least 24 hr. Smith (1974) suggested that this inability to revert to non-symbiotic behaviour may explain why the alga has so far proved unculturable.

Convoluta roscoffensis

This is a small acoel platyhelminth worm which occurs on certain sandy beaches in the Channel Islands and North Brittany coast. The symbiotic alga is *Platymonas convolutae*. The worm has a simple life cycle in which the eggs hatch to produce larvae; the larvae feed voraciously but only develop to maturity if they become infected with the alga (which is a component of the local phytoplankton). Once the worms reach maturity, they cease holozoic feeding. Since there is no apparent appreciable digestion of algae until the worms become senescent, it is assumed that mature worms represent a system in which there is a closed circulation of nutrients between symbiont and host. This is of considerable advantage compared to associations where the host continues holozoic feeding, and in which biochemical investigation of nutrient flow between symbionts is consequently complicated.

With *Convoluta roscoffensis*, laboratory studies of metabolite movement between the symbionts were initially difficult because separation of alga from host tissue was very unsatisfactory owing to the fragile nature of the algal cells; they lack a theca and mostly rupture during even gentle homogenization and centrifugation. Nevertheless, Muscatine *et al.* (1974) showed that the alga could fix carbon by photosynthesis very efficiently in the animal, and the fixed ^{14}C could be detected in animal products such as eggs and mucus, often appearing in the amino-acid moieties of hydrolysates of these products. Boyle and Smith (1975) found that photosynthesis in the worm could be inhibited by 3-(3,4-dichlorophenyl)-1,1-dimethylurea (DCMU) without immediate harm to the worm itself. If photosynthesis was so inhibited, then uric acid accumulation was easily detectable after about 2 weeks. They also showed that the association contains detectable amounts of uricase, and also glutamine synthetase. It is of interest that Gooday (1970) showed that algae cultured after isolation from the worm could live on uric acid as a sole nitrogen source, and he also noted the absence of uricase as reported for other platyhelminths. Thus, it seems that the algal symbiont may play a key role in recycling nitrogenous waste.

A substantial technical advance was made when Boyle and Smith (1975) discovered that *Convoluta* could be induced to eject most or even all of its algae by a brief exposure to carbon dioxide. This enabled comparisons to be made of freshly isolated algae with intact worms, and with those which had lost their algae. It could be shown that the amount of carbon moving from alga to worm was large and comparable to that in other algal/invertebrate associations. Substantial indirect evidence showed that most of the carbon moved as amino acids and amides—especially glutamine—rather than as carbohydrates. The principal product of algal photosynthesis was mannitol, but since the worms could not utilize mannitol, it formed a useful carbon store 'protected' from the

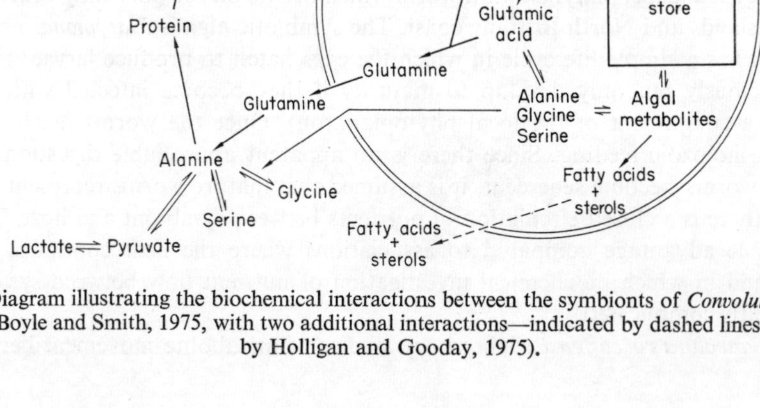

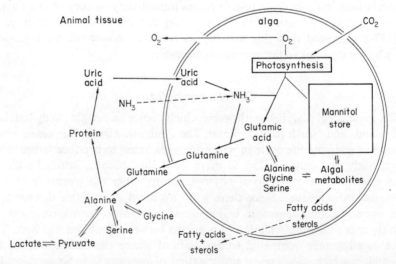

FIG. 2. Diagram illustrating the biochemical interactions between the symbionts of *Convoluta roscoffensis* (after Boyle and Smith, 1975, with two additional interactions—indicated by dashed lines—suggested by Holligan and Gooday, 1975).

animal. Within the worms, mannitol turns over slowly, but as soon as the algae are ejected, it turns over rapidly. The main conclusions of Boyle and Smith about nutrient cycling in *Convoluta*, as modified by the views of Holligan and Gooday (1975), are summarized in Fig. 2.

A key problem posed by *Convoluta* is the observation that while about 40 per cent of algal photosynthate passes to the animal, the alga releases very little photosynthate (less than 5 per cent) immediately after ejection. Why is the alga induced to release so much photosynthate in symbiosis? This is an important question raised by many biotrophic associations. In the case of most alga/invertebrate symbioses, there is evidence that animal tissues contain a 'factor' which induces photosynthate-release from the alga. The main characteristics of such 'photosynthate-releasing factors' as shown by the investigations of Muscatine (1967), Muscatine *et al.* (1972), and Trench (1971) on various symbiotic coelenterates and giant clams, are that (i) they stimulate increased release of specific compounds from freshly isolated algae, (ii) they are thermolabile, and (iii) they are not present in animals lacking symbionts. There is some evidence for the existence of such 'factors' in *Convoluta* since Muscatine *et al.* (1974) found that release of fixed ^{14}C from the alga was greater in homogenates of the worm than in isolation. However, both the fragility of the alga in *Convoluta* and its small size make it unsuitable for detailed investigations of the nature of the 'factors'. In recent work 'photosynthate-releasing factors' have been studied mainly in *Elysia viridis*.

Elysia viridis

Elysia viridis is a sacoglossan mollusc which feeds on the siphonaceous seaweed, *Codium fragile*. It acquires chloroplasts from the seaweed, which enter cells lining the digestive diverticulae. Once inside the animal cells, the chloroplasts behave essentially like the algal symbionts in *Hydra* and *Convoluta*, and can photosynthesize for up to 2 months (though always at a declining rate). Freshly collected molluscs have approximately the same rates of photosynthetic carbon fixation (per mg chlorophyll) as the seaweed itself, and as in *Convoluta*, 40 per cent or more of the carbon fixed moves from chloroplast to host (Trench *et al.*, 1973). The presence of chloroplasts is clearly of selective advantage in that molluscs starved in the light lose weight less rapidly than those starved in the dark, the difference in weight loss being of the right order to be accounted for by measured rates of photosynthesis (Hinde and Smith, 1975).

It is essential to stress that *Codium* chloroplasts are quite different in structure from higher plant chloroplasts, and are probably the most robust organelles yet known, being able to remain functional in a simple mineral medium for at least 7 days after isolation from the seaweed. After isolation, the chloroplasts release very little fixed carbon compared with the 40 per cent or more released when within the intact seaweed or mollusc. Moreover, the products released are different. In homogenates of the mollusc, which can carry out photosynthesis at a high rate, the 40 per cent of fixed carbon released appears mainly as glucose (Trench *et al.*, 1973), whereas, in isolated chloroplasts, the product released is mainly glycollic acid. Recently, Gallop (1974) has shown that the level of photosynthate released by isolated *Codium* chloroplasts can be substantially raised if the chloroplast suspension is combined with the clear supernatant of a centrifuged homogenate of *Elysia*; the products released are then glucose, alanine and also glycollic acid.

The clear supernatant of the *Elysia* homogenate contains a 'photosynthate-releasing factor' similar to those proposed for other symbiotic marine invertebrates. The supernatant is inactivated by boiling, it stimulates release of a few specific compounds rather than a single one, and is restricted to the chloroplast-containing tissues of the mollusc; supernatants obtained from homogenates of related non-symbiotic molluscs have no enhancing properties (Table 2).

TABLE 2. EFFECTS OF *Elysia* HOMOGENATES AND OTHER FACTORS UPON RELEASE OF PREVIOUSLY FIXED ^{14}C FROM ISOLATED *Codium* CHLOROPLASTS AND FROM *Anemonia* ZOOXANTHELLAE (based upon results of Gallop, 1974, 1975)

		Percentage of previously fixed ^{14}C released	
Experiment	Treatment	*Codium* chloroplasts	*Anemonia* zooxanthellae
1	*Elysia* homogenate	26.4	21.0
	Chloroplast suspension medium	9.5	14.7
	Anemonia homogenate	30.9	2.6
	Sea-water	6.0	1.8
2	*Elysia* homogenate of chloroplast-containing tissues	21.8	—
	Elysia homogenate of non-chloroplast containing tissues	7.5	—
3	*Limapontia* homogenate	5.7	—
	Codium homogenate	3.4	—
	Codium + *Elysia* homogenate	16.7	—
	Elysia hpmogenate	25.0	—

Chloroplast suspensions containing approx 45 μg chlorophyll/ml incubated with 20 μCi $NaH^{14}CO_3$/ml for 40–60 min at 15°C and 24.2 klx. Suspension then washed before incubation in desired treatment for 40–60 min at 17°C with 9.7 klx illumination. Other experimental details in Gallop (1974, 1975). *Limapontia* homogenate prepared from *L. capitata*, shown by Hinde and Smith (1974) to be non-symbiotic (whereas some forms of the related *L. depressa* are symbiotic). *Codium* + *Elysia* homogenate mixture was a mixture of the two in equal proportions (so that amount of *Elysia* homogenate in the mixture was half that in treatments with *Elysia* homogenate alone).

The presence of an enhancing 'factor' in the *Elysia* supernatant is supported by other observations made by Gallop (1974). She found that the *Elysia* supernatant enhanced the release of fixed carbon when combined with symbiotic algae (zooxanthellae) isolated from sea anemone (Table 2), and conversely, a supernatant prepared from homogenates of the sea anemone were active in combination with a suspension of isolated *Codium* chloroplasts. The enhancing effect of the *Elysia* and *Anemone* supernatants was not due to pH since release of fixed carbon from isolated chloroplasts showed no increase when they were incubated in media over a pH range of 3.0–10.0. Furthermore, there was no evidence for the presence of a 'factor' in *Codium* since homogenates of the seaweed were quite ineffective in stimulating release from isolated chloroplasts, and *Codium* homogenates did not inactivate the effect of *Elysia* homogenates (Table 2).

CONCLUSIONS

When in symbiosis, almost all autotrophs release substantial amounts of photosynthate to their hosts, but this release usually ceases immediately or very soon after isolation of the autotroph from the association. There is, therefore, some aspect of existence in the symbiosis which causes massive release of a few specific compounds from the autotroph. In so far as endosymbionts of invertebrates are concerned, the animal tissues of all symbiotic marine forms so far investigated appear to produce a 'factor' stimulating release of photosynthate. In the only freshwater association so far investigated, *Hydra viridis*, Cernichiari and Smith (unpublished) failed to detect such a 'factor'. For other reasons discussed above, it seems that despite its great convenience for laboratory study, *Hydra viridis* presents an anomalous case for the study of the control of photosynthate release, and that other associations may be more appropriate.

For so long as a 'photosynthate-releasing factor' remains unidentified and not even approximately characterized chemically, there must always be doubts as to whether it is some kind of artifact produced as a result of homogenization of tissues. The likelihood that such 'factors' in symbiotic invertebrates are artifacts has been reduced by the demonstration by Gallop (1974) that in *Elysia* they appear to be produced in response to the presence of symbiotic chloroplasts, and are absent from either non-symbiotic animals or from *Codium* itself. If 'photosynthate-releasing factors' are natural products of symbiont-containing cells, then the question of their nature arises; their thermolability is suggestive of proteins, but does not exclude other types of compound. Since they occur in a variety of different types of animals and are non-specific in action, it is difficult to believe that they are produced *de novo*. The simplest explanation would be that they are compounds naturally present in cells, but normally below the level of detection by the crude methods used. It could be visualized that their synthesis becomes markedly stimulated by the presence of symbionts. As Gallop (1974) has suggested, they might indeed have a natural function of regulating transport of materials across cell membranes.

In higher plant chloroplasts, it is believed that the outer envelope is very permeable to triose phosphates, and a transport system involving the shuttling of C_3 compounds in and out of the chloroplast has been suggested (e.g. Heber, 1974). So far, however, there is no evidence that such a system operates in *Codium* chloroplasts. Whatever system does operate, it appears to require some special feature of the host cell environment (the 'factor') which so far has not been achieved in isolation media, nor detected in the seaweed. It is likely, therefore, that the animal cell employs a different stimulus from the plant to induce transport of photosynthate out of the chloroplast.

In general, compounds released in response to 'factors' tend to be sugars, glycerol or neutral amino acids (Smith, 1974). There is an obvious advantage to an animal host in using not only photosynthetic carbon fixation but also the ability of most autotrophs of converting inorganic nitrogen or nitrogenous waste into usable amino acids. This is particularly well illustrated by *Convoluta*.

At first sight it might appear that study of 'photosynthate-releasing factors' has little relevance to modern problems of the regulation of photosynthate export from chloroplasts in plant cells. On the other hand, if such 'factors' turn out to be natural compounds which can bind to membranes, so modifying characteristics of transport, identification of the 'factor' might be a first step in interpreting the original evolution of a symbiotic alga into a chloroplast.

Finally, while photosynthate transport in symbiosis is an obvious subject for study because of the experimental ease with which it can be investigated, it is only one of a variety of aspects of host–symbiont interaction which may have wider significance in the understanding of general problems of cell biology. For example, recent investigation of how hosts 'recognize' their symbiotic alga upon infection may shed interesting light on the biology of cell surfaces (cf. Muscatine, 1975).

References

BOYLE, J. E. and SMITH, D. C. (1975) Biochemical interactions between the symbionts of *Convoluta roscoffensis*. *Proc. R. Soc.* B. **189,** 121–135.

CERNICHIARI, E., MUSCATINE, L. and SMITH, D. C. (1969) Maltose excretion by the symbiotic algae of *Hydra viridis*. *Proc. R. Soc.* B. **173,** 557–576.

GALLOP, A. (1974) Evidence for the presence of a 'factor' in *Elysia viridis* which stimulates photosynthate release from its symbiotic chloroplasts. *New Phytol.* **73,** 1111–1117.

GALLOP, A. (1975) *Chloroplast symbiosis*. D.Phil. thesis, Oxford University.

GOODAY, G. W. (1970) A physiological comparison of the symbiotic alga *Platymonas convolutae* and its free-living relatives. *J. mar. biol. Ass. U.K.* **50,** 199–208.

HEBER, U. (1974) Metabolite exchange between chloroplasts and cytoplasm. *A. Rev. Pl. Physiol.* **25,** 393–421.

HINDE, R. and SMITH, D. C. (1974) Chloroplast symbiosis and the extent to which it occurs in Sacoglossa. *Biol. J. Linn. Soc.* **6,** 349–356.

HOLLIGAN, P. M. and GOODAY, G. W. (1975) Symbiosis in *Convoluta roscoffensis*. In *Symbiosis* ed. JENNINGS, D. H. Proceedings of the 29th Symposium of the Society for Experimental Biology held in Bristol 1974. Cambridge Univ. Press in press.

MUSCATINE, L. (1965) Symbiosis of hydra and algae. III. Extracellular products of the algae. *Comp. Biochem. Physiol.* **16,** 77–92.

MUSCATINE, L. (1967) Glycerol excretion by symbiotic algae from corals and *Tridacna* and its control by the host. *Science N.Y.* **156,** 519.

MUSCATINE, L. (1975) Symbiosis in green hydra. In *Symbiosis* ed. JENNINGS, D. H. Proceedings of the 29th Symposium of the Society for Experimental Biology held in Bristol 1974. Cambridge Univ. Press. in press.

MUSCATINE, L. and LENHOFF, H. M. (1965) Symbiosis of hydra and algae. II. Effects of limited food and starvation on symbiotic and aposymbiotic hydra. *Biol. Bull. mar. biol. Lab. Woods Hole,* **129,** 316–328.

MUSCATINE, L., POOL, R. R. and CERNICHIARI, E. (1972) Some factors influencing selective release of soluble organic material by zooxanthellae from reef corals. *Mar. Biol.* **13,** 298.

MUSCATINE, L., BOYLE, J. E. and SMITH, D. C. (1974) Symbiosis of the acoel flatworm *Convoluta roscoffensis* with the alga *Platymonas convolutae*. *Proc. R. Soc.* B. **187,** 221–234.

PARDY, R. L. (1974) Some factors affecting the growth and distribution of the algal endosymbionts of *Hydra viridis*. *Biol. Bull. mar. biol. Lab. Woods Hole,* **147,** 105–118.

SMITH, D. C. (1974) Transport from symbiotic algae and symbiotic chloroplasts to host cells. In *Transport at the Cellular Level* ed. SLEIGH, M. A. and JENNINGS, D. H. Proceedings of the 28th Symposium of the Society for Experimental Biology held in London 1973. Cambridge University Press.

TRENCH, R. K. (1971) The physiology and biochemistry of zooxanthellae symbiotic with marine coelenterates. III. The effects of homogenates of host tissue on the excretion of photosynthetic products *in vitro* by zooxanthellae from two marine coelenterates. *Proc. R. Soc.* B **177,** 251–264.

TRENCH, R. K. T., BOYLE, J. E. and SMITH, D. C. (1973) The association between chloroplasts of *Codium fragile* and the mollusc *Elysia viridis*. II. Chloroplast ultrastructure and photosynthetic carbon fixation in *Elysia viridis*. *Proc. R. Soc.* B. **184,** 63.

INTERCHANGE OF METABOLITES IN BIOTROPHIC SYMBIOSES BETWEEN ANGIOSPERMS AND FUNGI

D. H. LEWIS

Department of Botany, University of Sheffield, Sheffield, U.K.

INTRODUCTION

The nutrition of fungi which derive their nutrients from higher plants may be either *biotrophic* or *necrotrophic*—that is, symbiotic fungi absorb soluble nutrients either directly from living cells of the host or from degradative products of host cells killed in advance of colonization (Thrower, 1966; Lewis, 1973a, 1974). Fungi, necrotrophic on higher plants, therefore essentially live off the *past* photosynthetic products of the plants they exploit whereas biotrophs utilize their *current* assimilates. The degradative behaviour of necrotrophs, which involves the massive production of cytolytic enzymes, has attracted a great deal of attention (see Wood, 1967; Albersheim *et al.*, 1969) and will not be considered further here.

As far as the infected angiosperms are concerned, the outcome of symbioses with biotrophic fungi may be antagonistic (i.e. the fungi are parasites) or mutualistic. The latter outcome only occurs when the fungal partner is root-infecting and possesses a mycelium which permeates the soil from the root. Large volumes of soil are thus exploited for mineral nutrients, many of which pass to the host. The host in turn acts as carbohydrate source for the fungus. This exchange of carbon compounds for mineral nutrients is the basis of the mutualism exhibited by ecto- and arbuscular-mycorrhizal associations (Lewis, 1973a, b; 1974) and probably also by ericoid mycorrhizas (Read and Stribley, 1973; Stribley and Read, 1974a, b).

Since, in a biotrophic symbiosis between an angiosperm and a fungus, any potential the latter possesses to degrade complex polymers of the former such as cellulose, matrical cell wall polysaccharides, proteins and lignin, is not expressed, it must mean that considerable quantities of simple soluble substances pass to the fungus. This drain on metabolites from the host may be considerable not only because the fungus has a high rate of respiration but also because massive synthesis of fungal tissue may occur, e.g. during spore production in the case of pathogenic fungi or during fruit body production by mycorrhizal fungi (see Harley, 1971 for calculations concerning the latter). Since the percentage of host cells that are infected may be small, it follows that the demand for assimilates by the fungus can only be met by dramatic alterations to the normal patterns of translocation of photosynthetic products in the host (see Scott, 1972; Lewis, 1973a, 1974 for references to effect of biotrophs on patterns of translocation).

Lewis (1974) used carbon source as a basis for the construction of a chequer-board of donor and recipient organisms involved in symbioses and pointed out that, when more information was available, it would be instructive to construct similar chequer-boards based upon the movement of specific compounds. Scott (1972), however, had earlier noted that surprisingly little attention had been paid to the chemical nature of compounds which move from higher plants to symbiotic fungi and the rate at which these compounds are transported.

This paper discusses the evidence relating to the identity, firstly, of the carbohydrate(s) which pass from autotroph to heterotroph in an antagonistic symbiosis, and secondly, of the mobile amino-compound(s) by which nitrogen is transferred from heterotroph to autotroph in a mutualistic symbiosis.

THE SYMBIOTIC ASSOCIATIONS CHOSEN AS MODEL SYSTEMS

In a discussion of the movement of carbohydrates from autotrophs to heterotrophs, Smith *et al.* (1969) pointed out that very similar mechanisms are involved in antagonistic and mutualistic symbioses as diverse as rust fungi on leaves and ectomycorrhizal fungi on roots and lichens. In view of this, a rust, *Puccinia poarum* Niels., was chosen as the model system to study carbon transfer between symbionts, since it is experimentally simple to label host metabolites selectively with ^{14}C by supplying $^{14}CO_2$ to infected, illuminated leaves. To study the movement of organic nitrogen from heterotroph to autotroph excised ectomycorrhizal roots of beech (*Fagus sylvatica* L.) were chosen since this symbiosis has been intensively studied and much background information is available (Harley, 1969). Furthermore, abundant uniform field material can be readily collected and sampled.

Puccinia poarum has a number of suitable attributes as a model system. It has two hosts, one a grass, *Poa pratensis* L. and the other a composite, *Tussilago farfara* L.; both are perennial and can be easily clonally propagated, in contrast to the cereal and woody hosts of rusts such as *Puccinia graminis* and *Puccinia coronata*. *Puccinia poarum* is a long-cycled, heteroecious rust which, from single basidiospores, produces large (up to 10 mm diameter), circular, discrete, pycnial/aecial pustules on *Tussilago*. These pustules, initially mono- and, later, di-karyotic, are amenable to individual study and sufficiently large to permit the assay of metabolites and enzyme activities in transects across them. Furthermore, healthy and infected tissue of the same leaf can be directly compared. The rust is unusual in undergoing two complete life cycles in the year so that infected material is available from the field for approximately 8 months of the year. This was important in initial stages of the work which relied on field material but, following the development of a simple procedure for inducing germination of teliospores, the rust can now be maintained on both hosts throughout the year under controlled conditions (McGee *et al.*, 1973).

MOVEMENT OF CARBOHYDRATE TO *Puccinia poarum* FROM ITS HOSTS

The pustules of rust fungi act as foci for the accumulation of many metabolites. Not only does the fungus derive nutrients by alteration of direction of transport in the phloem but also, since spore production results in rupture of host epidermis which increases water loss from the pustules, more nutrients are made available by the enhanced transpiration. Analysis of the major chemical components of healthy and infected tissue has made possible: (i) identification of fungal metabolites and of their precursors in the host and (ii) investigation of the metabolism and movement of these precursors from host to fungus.

Major carbohydrate components of healthy and infected tissue

Although some workers have recorded decreases in the carbohydrate content at the sites of biotrophic infection, most studies show a marked initial accumulation which may subsequently decline (see references in Long *et al.*, 1975). At infection sites in either host of *Puccinia poarum*, there is an accumulation of host sucrose, glucose, free fructose and fructose polymers (Table 1). In the *Poa* host, this accumulation of soluble carbohydrates

TABLE 1. INCREASE IN CONCENTRATION OF SOLUBLE CARBOHYDRATE AT INFECTION SITES IN LEAVES OF *Tussilago farfara* AND *Poa pratensis* INFECTED BY *Puccinia poarum*
(Values as the ratio, diseased:healthy)

	Diseased Stage	Fructose	Glucose	Sucrose	Fructans
(A)	on *Tussilago farfara**				
	Pycnial	2.1	7.0	0.6	1.2
	Aecial with aecia visible but not dehisced	2.0	9.5	1.4	2.7
	Aecial with < 50% aecia dehisced	3.4	9.5	1.1	2.8
	Aecial with > 50% aecia dehisced	3.7	14.0	1.6	4.0
(B)	on *Poa pratensis*†				
	Uredial	7.4	7.6	2.1	15.0

* Data from Holligan *et al.* (1973).
† A.K. Fung (unpublished)

is accompanied by increased starch deposition in host cells *around* the pustules as has been noted for other biotrophic associations. In the *Tussilago* host, on the other hand, no such accumulation of starch occurs but determinations of fructan content across the pustules reveal levels rising from 350 $\mu g/cm^2$ in tissue immediately surrounding the pustules to 2300 $\mu g/cm^2$ at the pustule centres (Holligan *et al.*, 1973). In *Tussilago*, the low levels of starch present show the typical diurnal variation of assimilatory starch (Holligan *et al.* 1974a) whereas in *Poa* (as in bean, clover and various other grasses) the starch induced around pustules shows characteristics of storage starch which is not subject to diurnal variations (Wang, 1961; Thrower, 1964). Concomitant with these alterations to the carbohydrate composition of host cells, the fungus accumulates mannitol, arabitol and glycogen (Holligan *et al.*, 1973, 1974a). On *Poa*, small quantities of sorbitol are also accumulated (A. K. Fung, unpublished).

Metabolism and movement of carbohydrate from hosts to fungus

When $^{14}CO_2$ is fed to healthy leaves or leaf discs of either of the *Puccinia* hosts, ^{14}C is

TABLE 2. DISTRIBUTION OF RADIOACTIVITY IN RUST INFECTED LEAF TISSUE OF *Tussilago farfara* FOLLOWING EXPOSURE TO $^{14}CO_2$ (data from Holligan *et al.*, 1974b)

Treatment	Tissue	Total ^{14}C in extracts (ct/min/cm^2)	^{14}C as % total in neutral soluble fraction* Sucrose	Hexose	Arabitol + Mannitol
15 min exposure to $^{14}CO_2$ in light	A. Discs (4 mm diam.) from host tissue away from pustules	48	89	8	1†
	B. Annuli (7 mm external diam.) from around sample C	46	42	55	0
	C. Discs (4 mm diam.) from pustule centres	24	57	13	17
15 min exposure to $^{14}CO_2$ in light followed by 24 hr in dark	A. Discs (4 mm diam.) from host tissue away from pustules	18	51	12	15†
	B. Annuli (7 mm external diam.) from around sample C	18	45	13	26
	C. Discs (4 mm. diam) from pustule centres	194	4	6	72

* Unidentified compounds adhering to base-line of chromatograms, sucrosylfructose and trehalose comprise remainder to 100%.

† The presence of [^{14}C]polyols in region A away from pustules may be due either to leakage from spores and hyphae or to the presence of very young pustules not visible to the naked eye.

TABLE 3. DISTRIBUTION OF RADIOACTIVITY, PHOTOSYNTHETICALLY FIXED FROM $^{14}CO_2$, WHICH IS LOST FROM HEALTHY LEAF DISCS OF *Tussilago farfara* INTO 'INHIBITION MEDIA' AT 20°C IN 20 hr. (Unpublished data of P. M. Holligan)

Medium	Total soluble ^{14}C	% ^{14}C lost into medium	Distribution of ^{14}C in medium (%) Fructose	Glucose	Sucrose	Sucrosylfructose
Water	984	9.7	47.0	13.6	32.5	6.9
0.5% Glucose	1093	23.7	45.3	47.9	6.5	0.3
0.5% Sucrose	1072	22.0	28.5	20.2	48.8	2.5

rapidly incorporated into sucrose with little radioactivity entering free hexoses (Yuen, 1969; Holligan *et al.*, 1974b). In diseased leaves or discs bearing single pustules, more $^{14}CO_2$ is incorporated into host hexoses and there is rapid transfer of photosynthetic products to the fungus. This can be shown autoradiographically and by analysis of the transfer of ^{14}C to fungal metabolites (Table 2).

Alterations in the pattern of translocation (i.e. direction of sucrose movement), the accumulation of sucrose at infection sites and the photosynthetic production of sucrose all suggest that the disaccharide is the carbon source utilized by the fungus. The particular accumulation of hexoses and enhanced incorporation of ^{14}C into them at infection sites further suggest that host sucrose is hydrolysed before or during transfer to the fungus. The mechanism of transfer has been investigated in three ways.

(1) If healthy illuminated leaf discs are infiltrated with water at 0°C after exposing them to $^{14}CO_2$ and then washed for 1 hr, [^{14}C]sucrose is the predominant carbohydrate in the washings (P. M. Holligan, unpublished). This result is interpreted to mean that [^{14}C]sucrose from photosynthesis is released from cells of the leaf disc into the 'free space' or apoplasm (cf. Hawker, 1965; Kursanov and Brovchenko, 1970), where it is free to diffuse into the bathing solution.

(2) The second approach has been to use the 'inhibition technique', originally devised by Drew and Smith (1967) to identify carbohydrates mobile between the symbionts of lichens. The technique involves labelling the photosynthetic products from $^{14}CO_2$ and either concurrently or subsequently supplying the unlabelled form of the putative mobile carbohydrate exogenously. The mobile [^{14}C]carbohydrate then accumulates in the bathing medium. Table 3 records the effects of exogenous glucose and sucrose on appearance of [^{14}C]carbohydrates in the medium when healthy leaf discs of *Tussilago* were exposed to $^{14}CO_2$. The presence of [^{14}C]hexoses in the bathing solution lacking added sugars in this experiment in contrast to the predominance of [^{14}C]sucrose when discs were washed at 0°C for 1 hr (see above) may be explained by the higher temperature (20°C) used and the longer duration of the treatment (20 hr). Under these conditions, much of the sucrose released is hydrolysed by the endogenous (as well as wound-response) invertase present. The alteration in distribution of ^{14}C between sucrose and hexoses caused by exogenous sucrose and glucose may be explained by release of intact sucrose and its absorption following hydrolysis. Exogenous sucrose inhibits hydrolysis of released [^{14}C]sucrose so that labelled-sucrose predominates in the bathing solution. Added glucose inhibits uptake of both [^{14}C]glucose and [^{14}C]fructose derived from [^{14}C]sucrose so that both labelled-hexoses appear in the medium. This pattern is consistent with that previously reported for infected discs of *Tussilago* by Yuen (1969) and Lewis (1970). They also showed that, when fructose is added, [^{14}C]fructose becomes the predominant sugar in the bathing solution. This is so since unlabelled-fructose inhibits uptake of [^{14}C]fructose only. Other experiments by P. M. Holligan revealed that leakage of sucrose was greater from infected tissues. These experiments with healthy and infected tissues therefore show that sucrose is released intact from host cells under both circumstances and that its utilization following hydrolysis by the rust is only an exaggerated version of the normal transfer process.

(3) That hydrolysis of sucrose is especially important in infected tissues is confirmed by their increased invertase activity (Table 4). In *Tussilago*, the enhancement is more than

tenfold at the centre of pustules but activity declines rapidly with increasing distance from the centre.

TABLE 4. ACTIVITY OF ACID INVERTASE, AT NEAR OPTIMAL pH VALUES, IN HEALTHY AND RUST-INFECTED LEAF TISSUE (data from Long *et al.*, 1975)

Tissue	Invertase activity μg reducing sugar/mg dry wt/hr
(A) *Poa pratensis*	
(a) Healthy segments (1 cm)	133
(b) Segments with uredial pustules (1 cm)	329
(B) *Tussilago farfara*	
(a) Healthy discs (5 mm diam.)	44
(b) Centre of aecial pustule (5 mm diam.)	495
(c) Annulus around (b) (5–9 mm diam.)	138
(d) Annulus around (c) (9–14 mm diam.)	59

In addition to providing hexoses which are metabolized by the rust, the increased activity of invertase also causes a change in the metabolism of host cells. In a mature healthy leaf, the mobile carbohydrate normally encountered in quantity by the cells is sucrose which is metabolized slowly if at all, since most is exported. In biotrophically-infected leaves, such cells are now presented with hexoses; these readily cross membranes and their carbon skeletons, after glycolytic degradation to triose phosphates, can enter chloroplasts and there be utilized as substrates for biosynthesis of starch (Walker, 1974). Since chloroplasts tend to be destroyed within pustules, such starch synthesis occurs in the first available chloroplasts i.e. in cells *around* pustules. Thus, biotrophic fungi induce a 'sugar-feeding' of leaf cells of the kind that has been conducted experimentally by plant physiologists for more than a century (see Pfeffer, 1900).

Starch synthesis in host cells possibly may also be enhanced by fungal sequestration of orthophosphate, e.g. as polyphosphate (MacDonald and Strobel, 1970; Bennett and Scott, 1971). Low levels of orthophosphate in host cells not only would relieve any allosteric inhibition of ADPG pyrophosphorylase (Preiss *et al.*, 1967) but would also encourage retention of triose phosphates in the chloroplast and discourage phosphorolysis of starch (Chen-She *et al.*, 1975). Within the chloroplast, phosphoglycerate, which stimulates ADPG-pyrophosphorylase, would then act as both effector and substrate for starch synthesis. The events leading to the formation of starch in biotrophically infected leaves are summarized in Fig. 1.

Since one of the enzymes necessary for fructan biosynthesis, fructan-fructan 1-fructosyltransferase, is possibly situated close to or within the tonoplast (Edelman and Jefford, 1968), accumulation of fructans in the chloroplast-free, but still viable, cells *within* pustules could occur. The synthesis stimulated by infection is again the natural equivalent of that elicited by 'sugar feeding' to leaves of certain Compositae (Chandorkar and Collins, 1972) including *Tussilago* (McGee and Lewis, unpublished). Figure 2 summarizes the events leading to the synthesis of fructans within the pustules on *Tussilago*. The

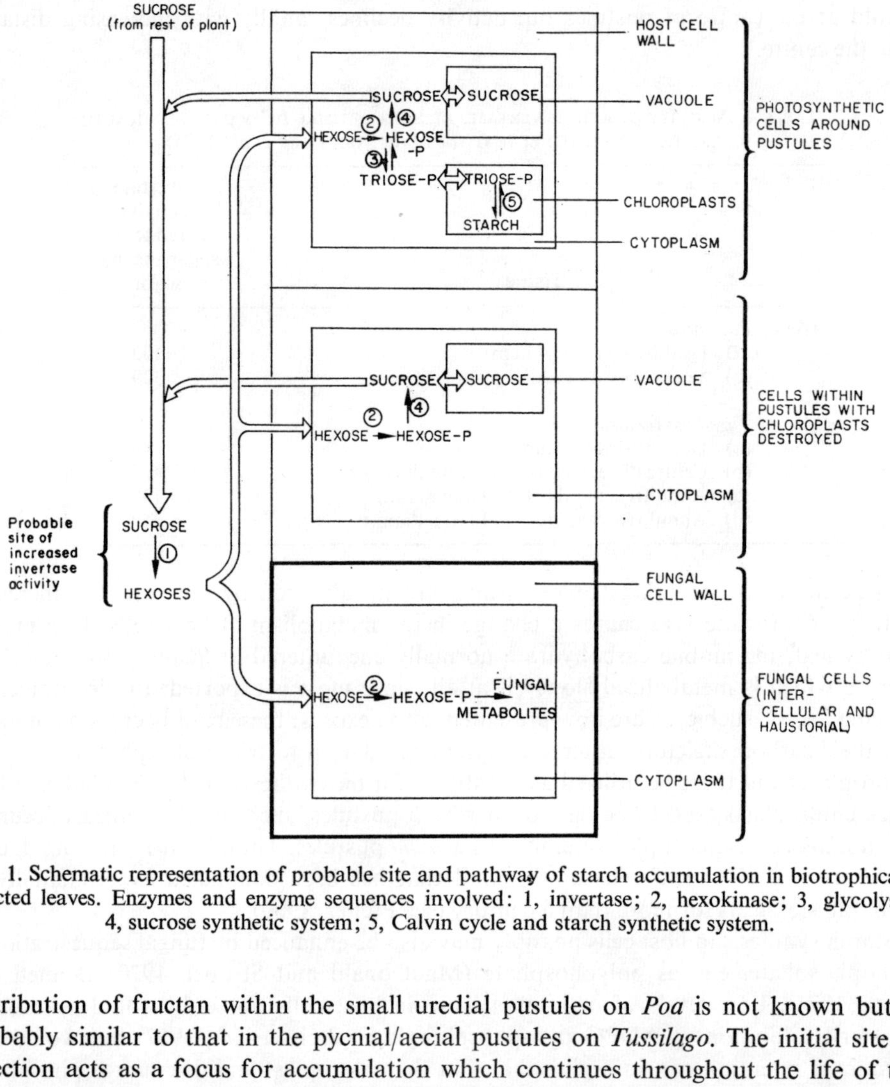

FIG. 1. Schematic representation of probable site and pathway of starch accumulation in biotrophically infected leaves. Enzymes and enzyme sequences involved: 1, invertase; 2, hexokinase; 3, glycolysis; 4, sucrose synthetic system; 5, Calvin cycle and starch synthetic system.

distribution of fructan within the small uredial pustules on *Poa* is not known but is probably similar to that in the pycnial/aecial pustules on *Tussilago*. The initial site of infection acts as a focus for accumulation which continues throughout the life of the pustule so that, in mature pustules, the centres contain the highest concentration.

It was emphasized above that *Puccinia poarum* had been chosen as a model system to investigate the nutrition of biotrophic fungi of higher plants. It seems probable that the schemes outlined in Figs. 1 and 2 are of broad applicability to biotrophic leaf pathogens (Long *et al.*, 1975) and it certainly appears as if hydrolysis of sucrose precedes uptake by the fungus in several kinds of mycorrhizal association (Lewis, 1975). Similarly, the conversion of hexoses to the fungal metabolites, acyclic polyols and/or trehalose, is widespread among pathogenic biotrophs of the Plasmodiophorales (Keen and Williams, 1969; Williams *et al.*, 1968), Peronosporales (Long and Cooke, 1974; Maclean, personal communication), Erysiphales (Edwards and Allen, 1966; Smith *et al.*, 1969), Clavicipitales (Thrower and Lewis, 1973) and other Uredinales (Daly *et al.*, 1962; Livne, 1964;

Smith *et al.*, 1969) as well as mutualistic mycorrhizal fungi (Lewis and Harley, 1965; Stribley and Read, 1974a).

Much remains to be learnt about enzymic control of the turnover of storage polysaccharides in biotrophically-infected leaves. Also the nature of the enhanced invertase requires investigation. The acid invertase is probably bound to cell walls and is likely to be of both fungal and host origin. As the hyphae of rusts in culture hydrolyse sucrose before uptake (Maclean, 1966), they must produce their own invertase. That of hosts could be stimulated both by the physical disruption of cells as intercellular hyphae grow

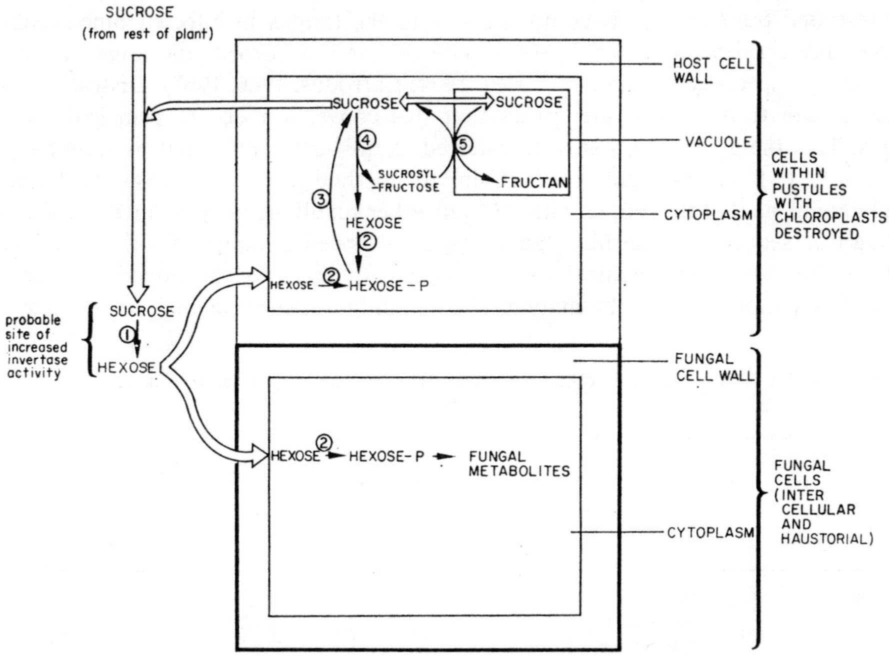

FIG. 2. Schematic representation of probable site and pathway of fructan accumulation in biotrophically infected leaves. Enzyme and enzyme sequences involved: 1, invertase; 2, hexokinase, 3, sucrose synthetic system; 4, sucrose–sucrose 1-fructosyl-transferase; 5, fructan–fructan 1-fructosyl-transferase.

and by the eruption of pustules through the epidermis since enhanced synthesis consequent upon injury to tissue is well documented (e.g. Edelman and Hall, 1964; Bacon *et al.*, 1965; Matsushita and Uritani, 1974). Also, increased synthesis in response to hormones produced by fungi is possible (cf. Kaufman *et al.*, 1973; Gaylor and Glasziou, 1969). It is important to gather further information on the invertases of both hosts and fungi since it is not inconceivable that differential inhibition of the enzymes could form the basis of a rational approach to the development of fungicides or fungistatic compounds—inhibition of utilization of sucrose would starve the fungus of its carbon source.

MOVEMENT OF ORGANIC NITROGEN FROM FUNGUS TO HOST IN ECTOMYCORRHIZAL ROOTS OF BEECH

As already indicated, bi-lateral exchange of major nutrients is the essence of mutualism. In ectomycorrhizal associations, the host supplies its symbiotic fungus with carbohydrate by mechanisms essentially similar to that discussed above for rusts (Smith *et al.*, 1969); evidence that other mutualistic mycorrhizas behave comparably has been discussed by Lewis (1975). Harley (1969) has reviewed the experimental evidence for the reverse movement of phosphorus where the mobile form of the element is orthophosphate (Harley and Loughman, 1963). Less work has been devoted to the movement of nitrogen compounds between the mycorrhizal symbionts although Melin and Nilsson (1952, 1953) demonstrated transfer of ^{15}N-compounds from the fungus in infected pine seedlings. Excised mycorrhizal roots of beech rapidly convert absorbed ammonia to amino-compounds, especially glutamine (Harley, 1964; Carrodus, 1966, 1967). Dissection of the fungal sheath from host tissue shows that this conversion occurs principally in the fungus. That the glutamine is then transferred to the host is indicated by recent experiments of C. P. P. Reid (unpublished). Using the method of Harley (1964), Reid labelled the glutamine pool of the fungus with ^{14}C and either simultaneously (termed 'continuous-feeding') or sequentially (termed 'pulse-chase') supplied a range of unlabelled amino acids to the excised mycorrhizal roots of beech. In this way, he hoped to 'trap' the mobile [^{14}C] amino acid in the manner of the 'inhibition experiments' described above.

TABLE 5. RELEASE OF [^{14}C]GLUTAMINE INTO 'INHIBITION MEDIA' BY MYCORRHIZAL ROOTS OF BEECH

Type of Experiment	Medium	Time hr	^{14}C in medium as % total soluble ^{14}C — 1st Series	2nd Series
'Pulse-chase'	Water	4	0.33	0.13
	Aspartate	4	1.72	—
	Glutamate	4	2.33	1.92
	Glutamine	4	1.36	1.34
'Continuous feeding'	Water	4	0.08	0.04
	Aspartate	4	0.62	—
	Alanine	4	0.87	—
	Glutamine	4	1.07	0.75
	Water	5	0.14	
	Glutamate	5	1.07	
	Glutamine	5	1.18	

(Unpublished data of C. P. P. Reid—for details see text).
(— = not tested)

Table 5 summarizes data from two series of experiments in which the amino-compounds were supplied at 1 per cent, a concentration close to the optimum for eliciting maximum release of ^{14}C to the medium in a range of symbiotic associations (Smith, 1974). There is negligible release of fixed ^{14}C into water but addition of an amino acid brings about a marked (up to twenty-fold) increase of ^{14}C in the medium. Also, chromatographic and electrophoretic analysis of the medium reveals that the principal ^{14}C-compound released with every amino acid tested is glutamine. These data are therefore consistent with the notion that glutamine is a principal mobile nitrogen compound from fungus to host in beech mycorrhizas.

The stimulated release of glutamine by alanine suggests that the amino acid transfer system A of Oxender (1972) is operating but, in view of the similar effects elicited by dicarboxylic amino acids, much more work is needed to characterize the transport system. Furthermore, the fact that one amino acid effects the release of another offers the opportunity of studying rates of release of specific amino acids, a situation impossible where the C_{12}-version of a compound has to be used to bring about release of its ^{14}C-counterpart (cf. the use of deoxyglucose to study glucose release from lichen algae (Hill and Smith, 1972)).

This general approach is also being applied to amino acid movement from host to fungus in rust infections of leaves (M. M. Burrell, personal communication). After characterization of mobile amino acids, essential to the fungus, the development of techniques to minimize this transfer again offers an alternative approach to the design of fungicides.

Acknowledgements: The work discussed above has been a co-operative effort involving Drs. P. M. Holligan, C. Chen, C. P. P. Reid and D. E. Long, Mr. A. K. Fung and Mrs. E. E. M. McGee. I am grateful to them and Drs. D. J. Maclean and M. M. Burrell for allowing me to quote their unpublished results. The work has been supported by grants from both Science and Agricultural Research Councils.

References

Albersheim, P., Jones, T. M. and English, P. D. (1969) Biochemistry of the cell wall in relation to infective processes. *A. Rev. Phytopath.* **7,** 171–194.

Bacon, J. S. D., MacDonald, I. R. and Knight, A. H. (1965) The development of invertase activity in slices of the root of *Beta vulgaris* L. washed under aseptic conditions. *Biochem. J.* **94,** 175–182.

Bennett, J. and Scott, K. J. (1971) Inorganic polyphosphates in the wheat stem rust fungus and in rust-infected wheat leaves. *Physiol. Pl. Path.* **1,** 185–198.

Carrodus, B. B. (1966) Absorption of nitrogen by mycorrhizal roots of beech. I. Factors affecting the assimilation of nitrogen. *New Phytol.* **65,** 358–371.

Carrodus, B. B. (1967) Absorption of nitrogen by mycorrhizal roots of beech. II. Ammonium and nitrate as sources of nitrogen. *New Phytol.* **66,** 1–4.

Chandorkar, K. R. and Collins, F. W. (1972) *De novo* synthesis of fructo-oligosaccharides in leaf discs of certain Asteraceae. *Can. J. Bot.* **50,** 295–304.

Chen-She, S-H., Lewis, D. H. and Walker, D. A. (1975) Stimulation of photosynthetic starch formation by sequestration of cytoplasmic orthophosphate. *New Phytol.* **74,** 383–392.

Daly, J. M., Inman, R. E. and Livne, A. (1962) Carbohydrate metabolism in higher plant tissues infected with obligate parasites. *Pl. Physiol. Lancaster* **37,** 531–539.

Drew, E. A. and Smith, D. C. (1967) Studies in the physiology of lichens. VIII. Movement of glucose from alga to fungus during photosynthesis in the thallus of *Peltigera polydactyla. New Phytol.* **66,** 389–400.

Edelman, J. and Hall, M. A. (1964) Effect of growth hormones on the development of invertase associated with cell walls. *Nature Lond.* **201,** 296–297.

Edelman, J. and Jefford, T. G. (1968) The mechanism of fructosan metabolism in higher plants as exemplified in *Helianthus tuberosus. New Phytol.* **67,** 517–531.

Edwards, H. H. and Allen, P. J. (1966) Distribution of products of photosynthesis between powdery mildew and barley. *Pl. Physiol. Lancaster* **41,** 683–688.

GAYLOR, K. R. and GLASZIOU, K. T. (1969) Plant enzyme synthesis: hormonal regulation of invertase and peroxidase synthesis in sugar cane. *Planta* **84,** 185–194.

HARLEY, J. L. (1964) Incorporation of carbon dioxide into excised beech mycorrhizas in the presence and absence of ammonia. *New Phytol.* **63,** 203–208.

HARLEY, J. L. (1969) *The Biology of Mycorrhiza* (2nd Edn.), Leonard Hill, London.

HARLEY, J. L. (1971) Fungi in ecosystems. *J. Ecol.* **59,** 653–668.

HARLEY, J. L. and LOUGHMAN, B. C. (1963) The uptake of phosphate by excised mycorrhizal roots of the beech. IX. The nature of the phosphate compounds passing into the host. *New Phytol.* **62,** 350–359.

HAWKER, J. S. (1965) The sugar content of cell walls and intercellular spaces in sugar-cane stems and its relation to sugar transport. *Aust. J. biol. Sci.* **18,** 959–969.

HILL, D. J. and SMITH, D. C. (1972) Lichen physiology. XII. The 'inhibition technique'. *New Phytol.* **71,** 15–30.

HOLLIGAN, P. M., CHEN, C. and LEWIS, D. H. (1973) Changes in the carbohydrate composition of leaves of *Tussilago farfara* during infection by *Puccinia poarum*. *New Phytol.* **72,** 947–955.

HOLLIGAN, P. M., CHEN, C., MCGEE, E. M. M. and LEWIS, D. H. (1974a) Carbohydrate metabolism in healthy and rusted leaves of coltsfoot. *New Phytol.* **73,** 881–888.

HOLLIGAN, P. M., MCGEE, E. M. M. and LEWIS, D. H. (1974b) Quantitative determination of starch and glycogen and their metabolism in leaves of *Tussilago farfara* during infection by *Puccinia poarum*. *New Phytol.* **73,** 873–879.

KAUFMAN, P. B., GHOSHEH, N. S., LACROIX, D., SONI, S. L. and IKUMA, H. (1973) Regulation of invertase levels in *Avena* stem segments by gibberellic acid, sucrose, glucose and fructose. *Pl. Physiol. Lancaster* **52,** 221–228.

KEEN, N. T. and WILLIAMS, P. H. (1969) Translocation of sugars into infected cabbage tissues during clubroot development. *Pl. Physiol. Lancaster* **44,** 748–754.

KURSANOV, A. L. and BROVCHENKO, M. I. (1970) Sugars in the free space of leaf plates: their origin and possible involvement in transport. *Can. J. Bot.* **48,** 1243–1250.

LEWIS, D. H. (1970) Physiological aspects of symbiosis between green plants and fungi. *Lichenologist* **4,** 326–336.

LEWIS, D. H. (1973a) Concepts in fungal nutrition and the origin of biotrophy. *Biol. Rev.* **48,** 261–278.

LEWIS, D. H. (1973b) The relevance of symbiosis to taxonomy and ecology, with particular reference to mutualistic symbioses and the exploitation of marginal habitats. In *The Interrelations of Taxonomy and Ecology* ed. HEYWOOD, V. H. pp. 151–172, Academic Press, London.

LEWIS, D. H. (1974) Micro-organisms and plants: the evolution of parasitism and mutualism. *Symp. Soc. gen. Microbiol.* **24,** 367–392.

LEWIS, D. H. (1975) Aspects of the carbon nutrition of mycorrhizas. In *Endomycorrhizas* ed. MOSSE, B., SANDERS, F. E. T. and TINKER, P. B. Proceedings of the International Symposium held at the University of Leeds, Academic Press, London.

LEWIS, D. H. and HARLEY, J. L. (1965) Carbohydrate physiology of mycorrhizal roots of beech. III. Movement of sugars between host and fungus. *New Phytol.* **64,** 256–269.

LIVNE, A. (1964) Photosynthesis in healthy and rust-affected plants. *Pl. Physiol. Lancaster* **38,** 614–621.

LONG, D. E. and COOKE, R. C. (1974) Carbohydrate composition and metabolism of *Senecio squalidus* L. leaves infected with *Albugo tragopogonis* (Pers.) S. F. Gray. *New Phytol.* **73,** 889–899.

LONG, D. E., FUNG, A. K., MCGEE, E. E. M., COOKE, R. C. and LEWIS, D. H. (1975) The activity of invertase and its relevance to the accumulation of storage polysaccharides in leaves infected by biotrophic fungi. *New Phytol.* **74,** 173–182.

MCGEE, E. E. M., HOLLIGAN, P. M., FUNG, A. K. and LEWIS, D. H. (1973) Maintenance of the rust, *Puccinia poarum*, on its alternate hosts under controlled conditions. *New Phytol.* **72,** 937–945.

MACDONALD, P. W. and STROBEL, G. A. (1970) Adenosine diphosphate-glucose pyrophosphorylase control of starch accumulation in rust infected wheat leaves. *Pl. Physiol. Lancaster* **46,** 126–135.

MACLEAN, D. J. (1966) Studies on the nutrition of the wheat stem rust fungus. B.Sc. Thesis, University of Sydney, Australia.

MATSUSHITA, K. and URITANI, I. (1974) Changes in invertase activity of sweet potato in response to wounding and purification and properties of its invertases. *Pl. Physiol. Lancaster* **54,** 60–66.

MELIN, E. and NILSSON, H. (1952) Transfer of labelled nitrogen from an ammonium source to pine seedlings through mycorrhizal seedlings. *Svensk botanisk tidskrift* **46,** 281–285.

MELIN, E. and NILSSON, H. (1953) Transfer of labelled nitrogen from glutamic acid to pine seedlings through the mycelium of *Boletus variegatus* (Sw.) Fr. *Nature Lond.* **171,** 134.

OXENDER, D. L. (1972) Membrane transport. *A. Rev. Biochem.* **41,** 777–814.

PFEFFER, W. (1900) *The Physiology of Plants*, Vol. 1, Clarendon Press, Oxford.

PREISS, J., GHOSH, H. P. and WITTKOP, J. (1967) Regulation of the biosynthesis of starch in spinach leaf chloroplasts. In *Biochemistry of the Chloroplasts* ed. GOODWIN, T. W., Vol. 2, p. 131, Academic Press, London.

READ, D. J. and STRIBLEY, D. P. (1973) Effect of mycorrhizal infection on nitrogen and phosphorus nutrition of ericaceous plants. *Nature Lond.* **244,** 81–82.

SCOTT, K. J. (1972) Obligate parasitism by phytopathogenic fungi. *Biol. Rev.* **47,** 537–572.

SMITH, D. C. (1974) Transport from symbiotic algae and symbiotic chloroplasts to host cells. In *Transport at the Cellular Level* ed. SLEIGH, M. A. and JENNINGS, D. H. Symposium of the Society for Experimental Biology, Vol. 28, pp. 485–520. Cambridge University Press.

SMITH, D., MUSCATINE, L. and LEWIS, D. (1969) Carbohydrate movement from autotroph to heterotroph in parasitic and mutualistic symbiosis. *Biol. Rev.* **44,** 17–90.

STRIBLEY, D. P. and READ, D. J. (1974a) The biology of mycorrhiza in the Ericaceae. III. Movement of carbon-14 from host to fungus. *New Phytol.* **73,** 733–741.

STRIBLEY, D. P. and READ, D. J. (1974b) The biology of mycorrhiza in the Ericaceae. IV. The effect of mycorrhizal infection on uptake of ^{15}N from labelled soil by *Vaccinium macrocarpon* Ait. *New Phytol.* **73,** 1149–1155.

THROWER, L. B. (1964) Photophosphorylation and starch metabolism in association with infections by obligate parasites. *Phytopath. Z.* **51,** 425–436.

THROWER, L. B. (1966) Terminology for plant parasites. *Phytopath. Z.* **56,** 258–259.

THROWER, L. B. and LEWIS, D. H. (1973) Uptake of sugars by *Epichloe typhina* (Pers. ex Fr.) Tul. in culture and from its host, *Agrostis stolonifera* L. *New Phytol.* **72,** 501–508.

WALKER, D. A. (1974) Chloroplast and cell—the movement of key substances across the chloroplast envelope. *International Review of Science, Biochemical Series I,* ed. NORTHCOTE, D. H., Vol. 11, pp. 1–49, Butterworths, London.

WANG, D. (1961) The nature of starch accumulation at the rust infection site in leaves of pinto bean plants. *Can. J. Bot.* **39,** 1595–1604.

WILLIAMS, P. H., KEEN, N. T., STRANDBERG, J. O. and MCNABOLA, S. S. (1968) Metabolite synthesis and degradation during clubroot development in cabbage hypocotyls. *Phytopathology* **58,** 921–928.

WOOD, R. K. S. (1967) *Physiological Plant Pathology*, Blackwell Scientific Publications, Oxford.

YUEN, C. (1969) The movement of carbohydrates from green plants to fungal symbionts. Ph.D. thesis, University of Sheffield.

NITRATE AND UREA ASSIMILATION BY ALGAE

P. J. SYRETT and J. W. LEFTLEY

Department of Botany and Microbiology, University College, Swansea, U.K.

INTRODUCTION

Sources of combined nitrogen for algal growth include the inorganic ions, ammonium nitrite, nitrate, and organic compounds such as urea, amino acids and purines. Of these, ammonium ions and nitrate ions are the most usual sources of nitrogen in laboratory cultures and are assumed to be the most important in nature. However, urea is an excellent source of nitrogen for the growth of many algae and the rapidity of its utilization suggests that it could be a significant natural source of nitrogen. Information about the ability of various algal species to utilize nitrogenous compounds and the mechanisms involved is given by Cain (1965), Thomas (1968), Naylor (1970), Healey (1973) and Morris (1974).

ASSIMILATION OF AMMONIUM

Ammonium is probably assimilated by algae, as by most other plants, chiefly through glutamate dehydrogenase which catalyses the reductive amination of α-oxoglutarate to glutamate. In the higher fungi, ammonium is assimilated by a NADP-linked glutamate dehydrogenase while a NAD-linked enzyme is responsible for the degradation, and hence the utilization, of glutamate (Lé John, 1971; Kinghorn and Pateman, 1973). The lower fungi have only a NAD-linked enzyme. In algae such as *Chlorella* and *Ankistrodesmus*, a NADP-linked glutamate dehydrogenase is readily detectable and probably functions in ammonium assimilation. In these algae NADH-linked glutamate dehydrogenase activity is generally low or absent. However, it has been reported that nitrate-grown cells of *Chlorella sorokiniana* contain only a NADH-specific glutamate dehydrogenase and that the NADPH-linked activity appears when ammonium is added (Israel *et al.*, 1974). The enzymes of algae grown with glutamate as nitrogen source have not been studied. Little is known, too, of the significance of alanine dehydrogenase in algal ammonium assimilation although a NADH-specific activity has been measured, e.g. in *Anabaena* (Haystead *et al.*, 1973).

Another important pathway of ammonium assimilation is the glutamate synthase reaction which combines ammonium with glutamate to form glutamine. The $-NH_2$ so incorporated can then be transferred to oxoglutarate so that two molecules of glutamate result; this reaction is catalysed by glutamine 2-oxoglutarate amino transferase (GOGAT). These are important reactions in some marine bacteria when grown with nitrate or low concentrations of ammonium as nitrogen source (Brown *et al.*, 1972). In these bacteria, with higher concentrations of ammonium, these enzymic activities are repressed and assimilation proceeds via glutamic dehydrogenase. Recent evidence suggests that this assimilatory system also operates in algae (Lea and Miflin, 1974, and personal communication).

ASSIMILATION OF NITRATE AND NITRITE

Assimilation of nitrate takes place by reduction of nitrate to nitrite, followed by reduction of nitrite to ammonium. The first step, catalysed by nitrate reductase, requires two electrons per nitrogen atom and the second, catalysed by nitrite reductase, requires six:

$$NO_3^- \xrightarrow{2e} NO_2^- \xrightarrow{6e} NH_4^+$$

These enzymes have been partially characterized in some algae, more especially *Chlorella*, and resemble the corresponding enzymes of higher plants. *Chlorella* nitrate reductase is a large molecule with a molecular weight of about 500,000 (Zumft *et al.*, 1969). Overall the enzyme catalyses the reduction of nitrate to nitrite with NADH as electron donor but two separate enzymic activities can be distinguished although the enzyme cannot be subdivided to give each separately. The first activity is a NADH-diaphorase activity in which NADH probably reduces a flavin in the enzyme molecule; the reduced flavin can then transfer electrons to an oxidant such as cytochrome c or it can transfer them, perhaps via a cytochrome, to the second component in the nitrate reductase complex. This component contains molybdenum and it is probably through molybdenum being alternately oxidized and reduced that electrons are transmitted to nitrate (Vega *et al.*, 1971; Stiefel, 1973). The second component can also be reduced directly by an exogenous electron donor such as methyl viologen.

Chlorella nitrate reductase is clearly a complex enzyme about which much remains unknown. There is evidence of associated cytochrome b type component (Vennesland and Jetschmann, 1971) which may function between the diaphorase and the molybdenum containing components. One interesting property of the enzyme is the reversible inactivation/activation which it undergoes. The enzyme is inactivated by incubation with NADH in the presence of cyanide and the mechanism of inactivation is probably that cyanide binds to and inactivates the reduced form of the enzyme (Solomonson, 1974; Solomonson *et al.*, 1974). The part of the enzyme that becomes reduced and inactive is the molybdenum component because the diaphorase activity of the enzyme is unaffected. Activity can be restored by oxidizing the enzyme with potassium ferricyanide. Sometimes, depending on the strain of *Chlorella* and on the prior treatment of the cells, the enzyme when isolated is largely in the inactive form and it is suggested that cyanide is a normal constituent in the organisms and is reponsible for this inactivation. There is also a suggestion that ADP is necessary for the inactivation reaction (Maldonado *et al.*, 1973) but there is some doubt about this (Solomonson, 1974). The reversible inactivation phenomenon, or something similar to it, has been reported of the nitrate reductases from two strains of *Chlorella* (Maldonado *et al.*, 1973; Solomonson *et al.*, 1973), a strain of *Chlamydomonas reinhardii* (Herrera *et al.*, 1972) and the thermophilic alga, *Cyanidium caldarium* (Rigano and Violante, 1973). Similar reversible inactivation of nitrate reductase has also been shown in cell-free extracts of the yeast, *Torulopsis nitratophila* and the bacterium, *Azotobacter chroococcum*, and it is also reported that spinach chloroplast nitrate reductase shows similar behaviour (Losada, 1975). These reports suggest that the phenomenon is a general one.

In contrast to nitrate reductase, nitrite reductase has a small molecule with a molecular weight of about 63,000 (Zumft *et al.*, 1969). The enzyme contains iron and, in cell-free extracts, reduces nitrite to ammonium with reduced ferredoxin as electron donor. NADPH acts as a reductant only if NADP-reductase and ferredoxin are present; therefore electrons from NADPH do not pass directly to the enzyme. No intermediates between nitrite and ammonium have been detected with any certainty. There are no reports of a reversible inactivation of nitrite reductase from algae but the enzyme from *Azotobacter* is inactivated by incubation with NADH and the inactivation is reversed by nitrite (Losada, 1974).

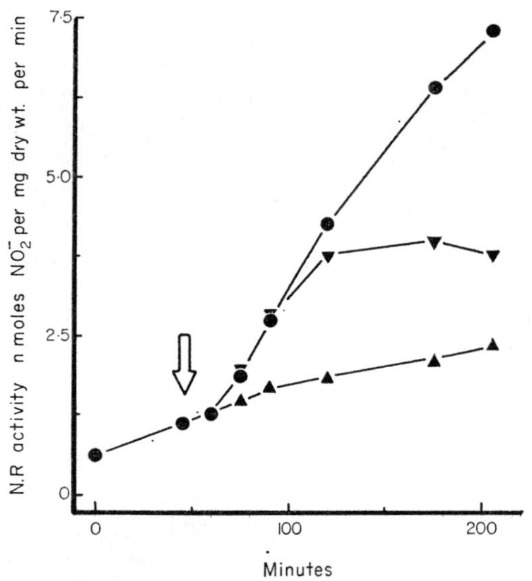

FIG. 1. Effect of NH_4^+ on induction of nitrate reductase (N.R.) in *Chlorella fusca*, strain 211/8p. Ammonium-grown algae (1.5 mg dry wt/ml) in 67 mM phosphate, pH 6.1, and 28 mM glucose were illuminated at 25°C and aerated with air/0.5 per cent CO_2. All cultures received KNO_3 (final conc. 1 mM) after 45 min (arrow). Samples were removed and assayed for nitrate reductase activity by the frozen cell method (Syrett, 1973). —▲—▲— NH_4^+ (final conc. 10 mM) added at zero time, —▼—▼— NH_4^+ (final conc. 10 mM) added at 45 min. i.e. with KNO_3, —●—●— no addition of NH_4^+.

REGULATION OF NITRATE ASSIMILATION IN ALGAE

In algae both nitrate reductase and nitrite reductase behave as inducible/repressible enzymes. Activities are high in nitrate-grown cells and very low in cells grown with ammonium as nitrogen source. The addition of ammonium to *Chlorella* in which nitrate reductase activity is being induced leads to cessation of nitrate reductase induction after a lag of about 1 hr (Fig. 1). This lag suggests that it is a product of ammonium assimilation, rather than ammonium itself which is the repressor of nitrate reductase induction;

alternatively, ammonium may inhibit the transcription of a messenger RNA for nitrate reductase synthesis. Repression by ammonium is important because appreciable levels of nitrate reductase are found in cells grown in the absence of nitrate, for example, with urea as nitrogen source, or in ammonium-grown cells that have been starved of nitrogen (Morris and Syrett, 1963; Rigano and Violante, 1973; Syrett and Hipkin, 1973). A similar induction of nitrate reductase activity on nitrogen-starvation is found in fungi (Morton and Macmillan, 1954).

In addition to preventing the formation of nitrate reductase activity, addition of ammonium immediately (i.e. within 2–3 min of addition) inhibits nitrate assimilation (Syrett and Morris, 1963; Thacker and Syrett, 1972a). Again, similar effects are observed with fungi (Morton, 1956). This effect also appears to be due to a product of ammonium assimilation, rather than to ammonium itself, because with carbon-starved cells which do not assimilate ammonium, addition of ammonium has no effect on nitrate reduction to nitrite.

The mechanism of this inhibitory effect of ammonium, or a product of ammonium assimilation, on nitrate assimilation is uncertain. It could be due to the combined inhibitory effects of amino acids and nucleotides on nitrate-reductase activity (Sims *et al.*, 1968). Or it could be due to inhibition of the uptake of nitrate (cf. Minotti *et al.*, 1969). Or the reversible inactivation of nitrate reductase may play a role. In some algal strains, the rapid inactivation of nitrate-reductase activity following addition of ammonium to *illuminated* cultures has been described (Losada, 1974). The inactivation is reversible *in vivo* by removal of ammonium. The enzyme is also reversibly inactivated when arsenate is added or cultures are deprived of oxygen. It is suggested that ammonium and arsenate here act by uncoupling photophosphorylation and that inactivation of nitrate reductase results from an accumulation of reductants and ADP within the cells (Losada, 1974). These phenomena are related to the reversible inactivation of nitrate reductase that can be demonstrated with cell-free extracts and purified enzyme preparations (see above) but their role in *in vivo* regulation of nitrate metabolism is not clear. With *Chlamydomonas reinhardii*, Sager strain 2192, inhibition of nitrate assimilation following addition of a small quantity of ammonium is immediate and complete, but inactivation of nitrate reductase in cell-free extracts is slow and not more than one-sixth of enzyme activity is lost (Fig. 2). Here, inactivation of nitrate reductase does not account for inhibition of nitrate assimilation.

Effect of light on nitrate assimilation by algae

Ever since Warburg and Negelein showed in 1920 that light stimulates nitrate reduction by *Chlorella*, the role of light in nitrate assimilation by chlorophyll-containing cells has been much discussed. It is not easy to ascertain the nature of the *in vivo* electron donor for a particular reductive reaction but there is, at present, no reason to doubt that NADH, or perhaps NADPH in some algae such as *Ankistrodesmus braunii* (Zumft *et al.*, 1972), is the natural electron donor for the reduction of nitrate to nitrite in algae and that reduced ferredoxin is the donor for the further reduction of nitrite. The effects of light on nitrate reduction by *Chlamydomonas* are consistent with this view. Nitrate is assimilated in darkness provided that cells contain a source of carbon such as stored carbohydrate,

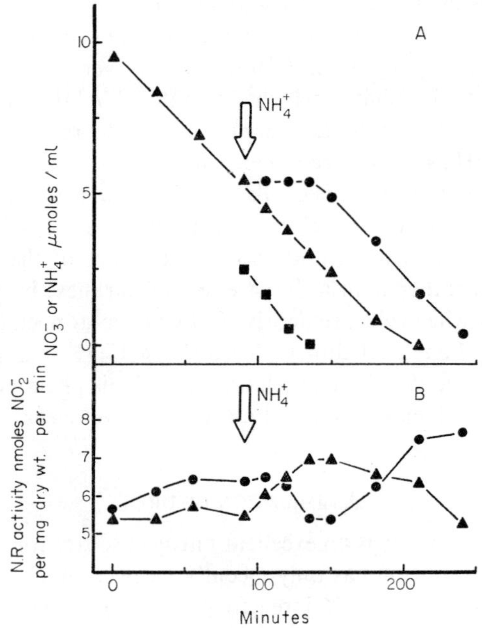

FIG. 2. Inhibition of nitrate assimilation by ammonium in *Chlamydomonas reinhardii*, strain 2192. Nitrate-grown organisms (0.4 mg dry wt/ml) were illuminated at 25°C and aerated with air/0.5 per cent CO_2. Two identical cultures were set up and samples removed for measurement of (A) nitrate in medium; (B) nitrate-reductase activity in organisms. NH_4^+ (final conc. 0.2 mM) was added to one culture (—●—●—) after 90 min. —▲—▲—, no ammonium added. (A) Disappearance of nitrate from medium. —■—■— shows disappearance of added NH_4^+. Note resumption of NO_3^- assimilation after NH_4^+ has gone. (B) NADH-nitrate reductase (N.R.) activity in organisms. Nitrate was measured by absorption at 210 nm, ammonium by the phenol-nitroprusside method and nitrate-reductase activity by the quantity of nitrite formed in a 10 min incubation of cell extract (made by freezing/thawing cells) with KNO_3 and NADH.

while illuminated cells that are depleted in endogenous carbon only assimilate nitrate or ammonium when they are allowed to fix CO_2 (Thacker and Syrett, 1972a). However, illuminated carbon-depleted cells, in the absence of CO_2, reduce nitrate to ammonium which accumulates. Such behaviour suggests that the cells can generate the reduced pyridine nucleotide required for nitrate reduction either by respiratory reactions in which carbon compounds are oxidized or by photochemical reduction. In contrast, the effect of light on reduction of nitrite is much greater. Carbon-depleted cells, deprived of CO_2, reduce nitrite rapidly to ammonium upon illumination and indeed such cells reduce nitrite much more quickly than cells supplied with CO_2. These results are consistent with the photochemical production of a reductant such as ferredoxin, which is readily available either for nitrite reduction or CO_2 fixation. The marked decrease in chlorophyll fluorescence in *Chlorella* and *Ankistrodesmus* when nitrite is added also shows that nitrite reduction can be very closely coupled to the electron transport system in the chloroplast (Kessler and Zumft, 1973). In contrast, addition of nitrate has no effect on chlorophyll fluorescence. There remains the question of how nitrite reduction takes place in darkness. It is assumed that ferredoxin is reduced with electrons derived from respiratory reactions via NADPH and NADP-reductase (Guerrero *et al.*, 1971). Recent work with the fungus, *Aspergillus nidulans*, suggests very close linkage between the oxidation of glucose-6-phosphate, which generates NADPH and nitrate reduction (Hankinson and Cove, 1974); in this organism nitrate reductase, as well as nitrite reductase, uses NADPH as the source of electrons.

Another phenomenon which connects light and nitrate assimilation is the disappearance of nitrate reductase activity which takes place when leaves of higher plants or algal cells are darkened. This loss of activity is a consequence of the prevention of photosynthetic CO_2 fixation rather than a direct effect of darkness because a similar loss of activity occurs in light when cells are deprived of CO_2 or stopped from fixing it by addition of 3-(3,4-dichlorophenyl)-1,1-dimethylurea (DCMU) (Kannangara and Woolhouse, 1967; Thacker and Syrett, 1972b). The relationship of this phenomenon to the reversible inactivation of nitrate reductase discussed above is not yet clear.

ASSIMILATION OF UREA

Although urea is well-known as an excellent nitrogen source for growth of many algae, the mechanism of its utilization has only recently received much attention. The work of Allison *et al.* (1954) on feeding [^{14}C]urea to *Nostoc* suggested that carbon from urea was assimilated only after its conversion to CO_2 although Ellner and Steers (1955) found some evidence for a direct incorporation of urea carbon into guanine by *Chlorella* and *Scenedesmus*. There are suggestions, too, that the incorporation of urea-N may proceed other than via prior conversion of urea-N to ammonium-N (Hattori, 1960; Baker and Thompson, 1962) but as yet no enzymic activity that results in the direct incorporation of urea carbon or nitrogen into cell constituents has been unequivocally demonstrated in algae.

A major difficulty in understanding algal urea metabolism was that several of the laboratory algae, e.g. *Chlorella*, which grew most prolifically with urea as nitrogen source, lacked the enzyme urease. Urease catalyses the reaction

$$CO(NH_2)_2 + H_2O = CO_2 + 2NH_3$$

and was generally thought to be responsible for the initial step in urea metabolism.

This situation was clarified when Roon and Levenberg in 1968 described, in yeasts and in *Chlorella*, the presence of another urea-degrading enzyme that requires ATP for activity. The enzyme was called ATP:urea amidolyase (UAL-ase) and the overall reaction catalysed is:

$$CO(NH_2)_2 + ATP + H_2O = CO_2 + 2NH_3 + ADP + P_i$$

Later work (Thompson and Muenster, 1971; Whitney and Cooper, 1972) showed that the overall reaction resulted from two enzymic activities which, in extracts from *Chlorella*, but not from yeasts, could be separated into two components. The first is a biotin-dependent carboxylation of urea to allophanate catalysed by urea:CO_2 ligase (or urea carboxylase) and the second, the breakdown of allophanate to give two molecules of NH_3 and two of CO_2, catalysed by allophanate hydrolase.

$$\text{Urea} + \text{ATP} + \text{HCO}_3^- \underset{\text{urea carboxylase}}{\overset{\text{Mg}^{2+}\ \text{K}^+}{\rightleftharpoons}} \text{allophanate} + \text{ADP} + \text{P}_i$$

$$\text{allophanate} \xrightarrow[\text{allophanate hydrolyase}]{} 2\text{NH}_3 + 2\text{CO}_2$$

In cell-free extracts the urea amidolyase system can be clearly distinguished from urease because

(a) urease activity is inhibited by hydroxyurea; that of UAL-ase is not,
(b) UAL-ase activity requires addition of ATP; urease activity is unaffected by addition of ATP,
(c) UAL-ase activity is inhibited by avidin which combines with biotin and this inhibition is overcome by excess biotin; avidin is without effect on urease activity.

Using these methods, Leftley and Syrett (1973) studied the distribution of urease and UAL-ase in a number of algae which could be grown in axenic cultures with urea as sole nitrogen source. Table 1 summarizes their conclusions and also includes other reports

TABLE 1. DISTRIBUTION OF UREASE AND UAL-ASE ACTIVITY IN CELL-FREE EXTRACTS FROM SOME UNICELLULAR ALGAE

	Urease	UAL-ase
Chlorophyceae		
Ankistrodesmus braunii		+
Asterococcus superbus		+
**Chlorella* (*fusca, ovalis* and *vulgaris*)		+
**Chlamydomonas reinhardii*		+
Dunaliella primolecta		+
**Golenkinia* sp.	?+	
Nannochloris coccoides		+
Nannochloris oculata	+	
Scenedesmus (*basilensis* and *obliquus*)		+
Stichococcus bacillaris		+
Prasinophyceae		
Tetraselmis subcordiformis	+	
Tetraselmis sp.	+	
Xanthophyceae		
**Monodus subterraneus*	+	
Chrysophyceae		
Monochrysis lutheri	+	
**Ochromonas malhamensis*	+	
Bacillariophyceae		
Phaeodactylum tricornutum	+	
Cryptophyceae		
Chroomonas salina	+	
Hemiselmis virescens	+	
Euglenophyceae		
**Euglena gracilis*	?+	
Myxophyceae		
**Nostoc muscorum*	+	
**Phormidium luridum*	+	
**Plectonema calothricoides*	+	

* Organisms not studied in our laboratory, i.e. report from literature.

from the literature. Two main points emerge. Firstly, no organism studied contains both enzymic activities; an organism has either one or the other. Secondly, with two exceptions, all the members of the Chlorophyceae which were examined contained UAL-ase; all other algae contained urease. Moreover, the two exceptions may be insignificant in that *Nannochloris oculata* should probably not be classed as a member of Chlorophyceae because it does not contain chlorophyll *b* (R. Guillard, personal communication) and the attribution of urease to *Golenkinea* is not well substantiated.

This distribution of enzyme activities has puzzling features. At present the only organisms known to possess UAL-ase are some yeasts and Chlorophyceae algae. Yet these two groups differ markedly on other metabolic patterns, e.g. in lysine biosynthesis (Vogel *et al.*, 1970). Moreover, the Chlorophyceae are, in many features of their biochemistry such as pigments and carbon storage products, much closer to higher plants than are the other algal groups. Yet no higher plant is known to possess UAL-ase while many have urease (Thompson and Muenster, 1974). It would be of interest to know if these enzymes are present in the larger members of the Chlorophyta which may be closer to the ancestors of the higher plants and, if so, how these enzymic activities are distributed.

A further puzzling feature concerns the function of UAL-ase and its requirement for ATP. There is no thermodynamic requirement for ATP since the breakdown of urea to NH_3 and CO_2 catalysed by urease proceeds to completion. One possibility is that UAL-ase is more susceptible to metabolic control than is urease. In bacteria urease appears to be inducible in some species but constitutive in others (Kaltwasser *et al.*, 1972). In flowering plants, high levels of urease are sometimes present, e.g. in seeds of *Canavalia* but little is known of the mechanism of control; urease activity is, however, inducible in the duckweed *Spirodela* (Bollard and Cook, 1968). Little work has been done in algae, but with *Phaeodactylum tricornutum* our own observations suggest that urease is constitutive in that it is present in ammonium and nitrate-grown cells. In contrast, in all the algae studied so far, at least one component of the UAL-ase complex is inducible in that it is not present unless the cells are grown with urea as nitrogen source. In *Chlamydomonas reinhardii* (strain y-1) both urea carboxylase activity and allophanate lyase activity are induced when urea is present but repressed in the presence of ammonium (Hodson *et al.*, 1975). In two *Chlorella* strains, urea carboxylase behaves as an enzyme induced by urea and repressed by ammonium whereas the allophanate lyase appears to be constitutive (Thompson and Muenster, 1974; Hodson *et al.*, 1975). Nevertheless, the result is that in both *Chlamydomonas* and *Chlorella* the breakdown of urea is under metabolic control.

Another difference between urease and UAL-ase which seemed to us of possible significance was the difference in the published values of the Michaelis constants (K_m) for the two enzymic activities. Published values of the K_m values for urease for a variety of organisms range from 3000 to 100,000 μM (Reithel, 1971). The few published values for UAL-ase are appreciably lower, e.g. 100 μM for *Candida* (Roon and Levenberg, 1968) and 200 μM for *Chlorella* (Adams, 1971). These observations suggested that UAL-ase might have a significantly greater affinity for the substrate urea, than has urease. Such a higher affinity might confer an advantage on the organism possessing it, in an environment where the urea concentration is low as indeed it is in sea-water. Little was

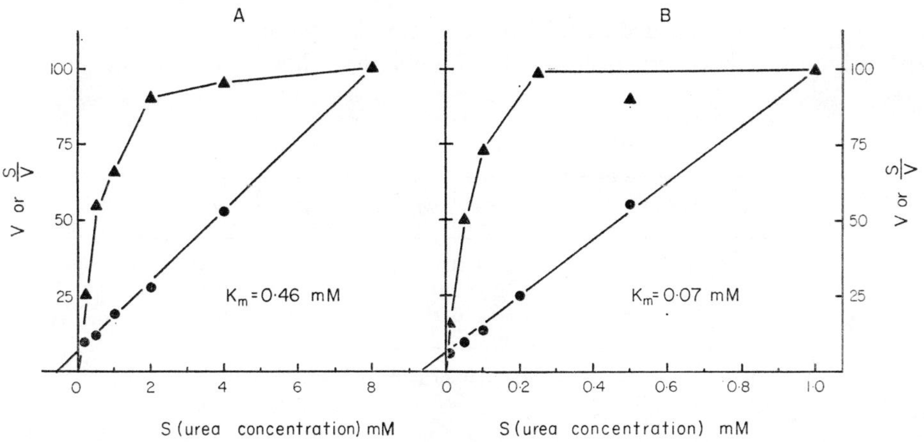

FIG. 3. Rate of reaction (V) versus urea concentration (S) for: (A) urease from *Phaeodactylum tricornutum* (B) UAL-ase from *Dunaliella primolecta*. V, —▲—▲—; S/V, —●—●—. For each enzyme activity, V and S/V values are given as a percentage of the maximum value. The value of the half-saturation constant, K_m, is given by the negative value of S when S/V is zero.

known of the value for the Michaelis constants of the algal enzymes. Accordingly we selected two marine algae for comparison. The first, *Dunaliella primolecta* (Cambridge Culture Collection, strain 11/34) contains UAL-ase, and the second *Phaeodactylum tricornutum* (strain 1052/6), urease.

Both organisms were grown at 15°C in ASP-2 medium (Provasoli *et al.*, 1957) with urea as nitrogen source. Cultures were illuminated by warm-white fluorescent tubes and aerated with air containing 0.5 per cent CO_2. Cell-free extracts from both organisms were partially purified by ammonium sulphate precipitation followed by desalting on Sephadex-G25. In both instances, about a five-fold purification of enzyme activity resulted. The enzyme activities were measured in 0.1 M tris-HCl buffer at pH 8.0 by measuring $^{14}CO_2$ released from [^{14}C]urea (Leftley and Syrett, 1973). Representative results are shown in Fig. 3; from them it can be seen that the K_m value for *Phaeodactylum* urease is about 460 μM and that of *Dunaliella* UAL-ase is about 70 μM. The value for UAL-ase is probably an overestimate since to convert [^{14}C]urea to the $^{14}CO_2$ which is measured requires two enzymic activities. Measurements of the K_m of the first enzyme alone, urea carboxylase, have not been made in algae but R. C. Hodson (personal communication) reports that the second enzyme, allophanate lyase, has a K_m of 60 μM for allophanate at pH 8.0 and 30°C.

The results show, however, that although *Phaeodactylum* urease has a K_m which is considerably lower than most of the published values for this enzyme, UAL-ase with a K_m of 70 μM, has an appreciably higher affinity for urea. In order to see, therefore, whether this difference in affinity between the enzymes in cell-free extracts was reflected in the behaviour of the two organisms, we compared their ability to metabolize urea at different concentrations and hence obtained half-saturation constants for urea decomposition by whole organisms (cf. Carpenter *et al.*, 1972). Hyperbola were fitted to the data of rate versus urea concentration by the iterative procedure of Bliss and James (1966) using an ICL 1904S computer and from the fitted curves values for the half-saturation constant were deduced. They were remarkably consistent from a number of experiments with both organisms (Table 2) and they show that, for both, the half-

TABLE 2. HALF-SATURATION CONSTANTS* (μM) FOR UREA DECOMPOSITION BY *Dunaliella primolecta* AND *Phaeodactylum tricornutum*

Experiment	*D. primolecta*	*P. tricornutum*
1	1.25 ± 0.06	1.37 ± 0.43
2	1.50 ± 0.16	1.46 ± 0.37
3	1.30 ± 0.06	1.22 ± 0.22

* The urea concentration (μM) that sustains half the maximum rate of decomposition at 20°C.

saturation constant for urea decomposition is about 1.4 μM. Thus whole organisms of the two species behave similarly and their affinity for urea is 50 to 300 times the affinity of the enzymes responsible for urea decomposition as measured in cell-free extracts. Consequently, we must conclude that our idea that the possession of UAL-ase instead of urease would enable cells to use more effectively urea at low concentrations is unfounded. Both organisms can metabolize urea rapidly at the same, rather low, concentration irrespective of their difference in enzymes.

ROLE OF UPTAKE MECHANISMS IN NITROGEN ASSIMILATION

The high affinity of whole cells of *Dunaliella* and *Phaeodactylum* for urea is relevant to conditions in nature; for example, in sea-water, urea concentrations range from zero to about 1 μM (Remsen, 1971). The simplest interpretation is that the organisms have an efficient uptake mechanism for urea which is saturated at quite low concentrations. McCarthy (1972) studied the uptake of urea by different species of marine phytoplankton and found half-saturation constants of 0.2–0.8 μM. Of particular interest was his finding that some clones, e.g. of the diatom *Ditylum brightwellii*, cannot metabolize urea but nevertheless took it up with the same efficiency as clones that do utilize it. This suggests that the uptake mechanism is distinct from the mechanism of urea metabolism.

Similar considerations show that uptake mechanisms are important in the assimilation of nitrate (see e.g., Eppley *et al.*, 1969; Eppley and Rogers, 1970). K_m values, for nitrate, for nitrate reductase are remarkably consistent at around 100–250 μM irrespective of the source of the enzyme. In contrast, for algae the half-saturation constants for nitrate assimilation by whole cells, or for growth with nitrate as nitrogen source, are much lower and range from 0.1–10 μM (Eppley *et al.*, 1969).

Nitrate uptake has received little attention. Heimer and Filner (1971) showed that cultured tobacco cells have a nitrate-inducible uptake system which is energy dependent and can concentrate nitrate eighty-fold. Schloemer and Garrett (1974a) have described a mechanism in the fungus *Neurospora* which is induced by nitrate or nitrite but is absent from fungus grown with ammonium or amino acids as nitrogen source. Ammonium inhibits the activity of this mechanism. These uptake mechanisms have half-saturation constants of 300–400 μM nitrate. Some algae must have nitrate-uptake mechanisms with much higher affinities for nitrate (Eppley *et al.*, 1969). *Ditylum brightwellii* has been studied by Eppley and Coatsworth (1968). This organism accumulates nitrate, probably in its vacuole which occupies about 85 per cent of the cell volume. Nitrate uptake has a half-saturation value of 2 μM and is not inhibited by ammonium. In *Ditylum*, the internal concentration of nitrate can be 40 μM. In smaller, less vacuolate, organisms it is not easy to demonstrate the presence of nitrate although we have been able to show its presence in nitrate-reductase deficient mutants of *Chlamydomonas*.

There is evidence too, for uptake systems for nitrite and ammonium (Eppley and Rogers, 1970; Schloemer and Garrett, 1974b). These as yet are even less well-characterized than the nitrate uptake system.

Effect of environment on mechanisms of uptake by algae

In fungi and higher plants the nitrate-uptake system appears to be inducible and is absent when nitrate is lacking. In *Neurospora* its appearance is inhibited by cycloheximide so, possibly, protein synthesis is required. A question of some significance is to what extent the characteristics of the nitrate-uptake system are affected by the external concentration of nitrate. Several reports show that phyto planktonic algae from coastal waters, where nitrate concentrations may be 50 μM or higher, have half-saturation constants for nitrate uptake of over 2 μM whilst those from open seas where nitrate concentrations are lower (<0.1 μM) have half-saturation constants of 0.5 μM or below (Eppley and Thomas, 1969; MacIsaac and Dugdale, 1969). Carpenter and Guillard (1971) showed that this difference extends to clones of the same species. Thus a strain of the diatom *Cyclotella nana* isolated from an estuary had a half-saturation constant for nitrate uptake of 1.9 μM while a strain of the same species isolated from the Sargasso Sea had a value of 0.38 μM. Such a difference could be of adaptive significance. We do not know whether the affinity of the nitrate-uptake system can be altered readily by change of environment. The clones studied by Carpenter and Guillard had been in laboratory culture in high salt media for several years and yet had retained the differences in half-saturation constants. This suggests that the differences are genetic and stable. In our laboratory we have compared normal and nitrogen-starved *Chlorella* but have been unable to detect differences in the half-saturation constant for nitrate uptake (R. J. Thomas, unpublished). In contrast, Thomas and Dodson (1974) found that the value of the half-saturation constant for growth of *Dunaliella* and *Gymnodinum splendens* on nitrate was dependent on the temperature at which the algae were grown. Whether the characteristics of the uptake system can alter relatively rapidly as the environment changes is relevant to the problem of the adaptation of algae to their natural environments. It is also relevant to the interpretation of results from laboratory studies. Several investigations of the growth of algae in nitrogen-limited chemostat cultures have now shown that they do not behave as classical chemostat theory would suggest (Caperon and Meyer, 1972; Eppley and Renger, 1974). In interpreting such studies it is necessary to know whether the half-saturation value for growth on the limiting nutrient is constant at all levels of nutrient, as classical theory assumes, or whether it is a variable. As yet the answer is in doubt.

CONCLUSION

More than 50 years have passed since Warburg first introduced *Chlorella* as a model organism for the study of photosynthesis and Warburg and Negelein used it to study the effect of light on nitrate assimilation. Since then, and more especially in the last 10 years, much has been learnt of the enzymology of nitrogen assimilation and of its control. Laboratory studies have been extended to the study of the assimilation of nitrogen by phytoplankton in nature and these studies have revealed new problems. The importance of uptake mechanisms for the assimilation of limiting algal nutrients such as nitrogen, phosphorus and silicon is now recognized. During the next decade attempts to understand the assimilation of nitrogen compounds must consider these mechanisms as well as the enzymic ones.

Acknowledgment: We wish to acknowledge the receipt of a grant from the Natural Environment Research Council which supported some of the original research described here.

REFERENCES

ADAMS, A. A. (1971) The regulation of nitrogen assimilation in *Chlorella*. Ph.D. thesis, University of Wales.

ALLISON, R. K., SKIPPER, H. E., REID, M. R., SHORT, W. A. and HOGAN, G. L. (1954) Studies on the photosynthetic reaction. II. Sodium formate and urea feeding experiments with *Nostoc muscorum*. *Pl. Physiol. Lancaster* **29,** 164–168.

BAKER, J. E. and THOMPSON, J. F. (1962) Metabolism of urea and ornithine cycle intermediates by nitrogen-starved cells of *Chlorella vulgaris*. *Pl. Physiol. Lancaster* **37,** 618–624.

BLISS, G. I. and JAMES, A. T. (1966) Fitting the rectangular hyperbola. *Biometrics* **22,** 573–602.

BOLLARD, E. G. and COOK, A. R. (1968) Regulation of urease in a higher plant. *Life Sci.* **7,** 1091–1094.

BROWN, C. M., MACDONALD-BROWN, D. S. and STANLEY, S. O. (1972) Inorganic nitrogen metabolism in marine bacteria; nitrogen assimilation in some marine Pseudomonads. *J. mar. biol. Ass. U.K.* **52,** 793–804.

BROWN, C. M., BURN, V. J. and JOHNSON, B. (1973) Presence of glutamate synthase in fission yeasts and its possible role in ammonia assimilation. *Nature New Biol.* **246,** 115–116.

CAIN, J. (1965) Nitrogen utilization in 38 freshwater Chlamydomonad algae. *Can. J. Bot.* **43,** 1367–1377.

CAPERON, J. and MEYER, J. (1972) Nitrogen-limited growth of marine phytoplankton. II. Uptake kinetics and their role in nutrient limited growth of phytoplankton. *Deep Sea Res.* **19,** 619–632.

CARPENTER, E. J. and GUILLARD, R. R. R. L. (1971) Intraspecific differences in nitrate half-saturation constants for three species of marine phytoplankton. *Ecology* **52,** 183–185.

CARPENTER, E. J., REMSEN, C. C. and WATSON, S. W. (1972) Utilization of urea by some marine phytoplankters. *Limnol. Oceanogr.* **17,** 265–269.

ELLNER, P. D. and STEERS, E. (1955) Urea as a carbon source for *Chlorella* and *Scenedesmus*. *Archs Biochem. Biophys.* **59,** 534–535.

EPPLEY, R. W. and COATSWORTH, J. L. (1968) Uptake of nitrate and nitrite by *Ditylum brightwellii*—kinetics and mechanisms. *J. Phycol.* **4,** 151–156.

EPPLEY, R. W. and RENGER, E. H. (1974) Nitrogen assimilation of an oceanic diatom in nitrogen-limited continuous culture. *J. Phycol.* **10,** 15–23.

EPPLEY, R. W. and ROGERS, J. N. (1970) Inorganic nitrogen assimilation of *Ditylum brightwellii*, a marine plankton diatom. *J. Phycol.* **6,** 4, 344–351.

EPPLEY, R. W. and THOMAS, W. H. (1969) Comparison of half-saturation 'constants' for growth and nitrate uptake by marine phytoplankton. *J. Phycol.* **5,** 365–369.

EPPLEY, R. W., ROGERS, J. N. and MCCARTHY, J. J. (1969) Half-saturation constants for uptake of nitrate and ammonium by marine phytoplankton. *Limnol. Oceanogr.* **14,** 6, 912–920.

GUERRERO, M. G., RIVAS, J., PANEQUE, A. and LOSADA, M. (1971) Mechanism of nitrate and nitrite reduction in *Chlorella* cells grown in the dark. *Biochem. biophys. Res. Commun.* **45,** 1, 82–89.

HANKINSON, O. and COVE, D. J. (1974) Regulation of the pentose phosphate pathway in the fungus *Aspergillus nidulans*—the effect of growth with nitrate. *J. biol. Chem.* **249,** 8, 2344–2353.

HATTORI, A. (1960) Studies on the metabolism of urea and other nitrogenous compounds in *Chlorella ellipsoidea*. III. Assimilation of urea. *Pl. Cell. Physiol. Tokyo* **1,** 107–115.

HAYSTEAD, A., DHARMAWARDENE, M. W. N. and STEWART, W. D. P. (1973) Ammonia assimilation in a nitrogen-fixing blue-green alga. *Pl. Sci. Lett.* **1,** 439–445.

HEALEY, F. P. (1973) Inorganic nutrient uptake and deficiency in algae. *CRC Crit. Rev. Microbiol.* **3,** 69–113.

HEIMER, Y. M. and FILNER, P. (1971) Regulation of the nitrate assimilation pathway in cultured tobacco cells. III. The nitrate uptake system. *Biochim. biophys. Acta* **230,** 362–372.

HERRERA, J., PANEQUE, A., MALDONADO, J. M., BAREA, J. L. and LOSADA, M. (1972) Regulation by ammonia of nitrate reductase synthesis and activity in *Chlamydomonas reinhardii*. *Biochem. biophys. Res. Commun.* **48,** 4, 994–1003.

HODSON, R. C., WILLIAMS, S. K. and DAVIDSON, W. R. (1975) Metabolic control of urea catabolism in *Chlamydomonas reinhardii* and *Chlorella pyrenoidosa*. *J. Bact.* **121,** 1022–1035.

ISRAEL, D. W., GRONOSTAJSKI, R. M. and SCHMIDT, R. R. (1974) Regulation of NADH- and NADPH-specific isozymes of glutamate dehydrogenase during the synchronous cell cycle of *Chlorella sorokiniana*. *Pl. Physiol. Lancaster* **53,** supplement, abstract 373.

KALTWASSER, H., KRÄMER, J. and CONGER, W. R. (1972) Control of urease formation in certain aerobic bacteria. *Arch. Mikrobiol.* **81,** 178–196.

KANNANGARA, C. G. and WOOLHOUSE, H. W. (1967) The role of carbon dioxide, light and nitrate in the synthesis and degradation of nitrate reductase in leaves of *Perilla frutescens*. *New Phytol.* **66,** 553–561.

KESSLER, E. and ZUMFT, W. G. (1973) Effect of nitrite and nitrate on chlorophyll fluorescence in green algae. *Planta* **111,** 41–46.

KINGHORN, J. R. and PATEMAN, J. A. (1973) NAD and NADP L-Glutamate dehydrogenase activity and ammonium regulation in *Aspergillus nidulans*. *J. gen. Microbiol.* **78,** 39–46.

LEA, P. J. and MIFLIN, B. J. (1974) Alternative route for nitrogen assimilation in higher plants. *Nature Lond.* **251,** 614–616.

LÉ JOHN, H. B. (1971) Enzyme regulation, lysine pathways and cell wall structures as indicators of major lines of evolution in fungi. *Nature Lond.* **231,** 164–168.

LEFTLEY, J. W. and SYRETT, P. J. (1973) Urease and ATP:Urea amidolyase activity in unicellular algae. *J. gen. Microbiol.* **77,** 109–115.

LOSADA, M. (1975) Interconversion of nitrate and nitrite reductase of the assimilatory type. In *Third International Symposium on the Metabolic Interconversion of Enzymes*. Springer-Verlag, Vienna. In press.

MCCARTHY, J. J. (1972) The uptake of urea by marine phytoplankton. *J. Phycol.* **8,** 216–222.

MACISAAC, J. J. and DUGDALE, R. C. (1969) The kinetics of nitrate and ammonia uptake by natural populations of marine phytoplankton. *Deep Sea Res.* **16,** 415–422.

MALDONADO, J. M., HERRERA, J., PANEQUE, A. and LOSADA, M. (1973) Reversible inactivation by NADH and ADP of *Chlorella fusca* nitrate reductase. *Biochem. biophys. Res. Commun.* **51,** 1, 27–33.

MINOTTI, P. L., WILLIAMS, D. C. and JACKSON, W. A. (1969) The influence of ammonium on nitrate reduction in wheat seedlings. *Planta* **86,** 267–271.

MORRIS, I. (1974) Nitrogen assimilation and protein synthesis. In *Algal Physiology and Biochemistry*. ed. Stewart, W. D. P. pp. 583–609. Blackwell Scientific Publications, Oxford.

MORRIS, I. and SYRETT, P. J. (1963) The development of nitrate reductase in *Chlorella* and its repression by ammonium. *Arch. Mikrobiol.* **47,** 32–41.

MORTON, A. G. (1956). A study of nitrate reduction in mould fungi. *J. exp. Bot.* **7,** 97–112.

MORTON, A. G. and MACMILLAN, A. (1954) The assimilation of nitrogen from ammonium salts and nitrate by fungi. *J. exp. Bot.* **5,** 232–252.

NAYLOR, A. W. (1970) Phylogenetic aspects of nitrogen metabolism in the algae. *Ann. N.Y. Acad. Sci.* **175,** 511–523.

PROVASOLI, L., MCLAUGHLIN, J. J. A. and DROOP, M. R. (1957) The development of artificial media for marine algae. *Arch. Mikrobiol.* **25,** 392–428.

REITHEL, F. J. (1971) Urease. In *The Enzymes*, ed. BOYER, P. D. Vol. IV, pp. 1–21, Academic Press, New York and London.

REMSEN, C. C. (1971) The distribution of urea in coastal and oceanic waters. *Limnol. Oceanog.* **16,** 732–739.

RIGANO, C. and VIOLANTE, U. (1973) Effect of nitrate, ammonia and nitrogen starvation on the regulation of nitrate reductase in *Cyanidium caldarium*. *Arch. Mikrobiol.* **90,** 27–33.

ROON, R. J. and LEVENBERG, B. (1968) A adenosine triphosphate-dependent avidin-sensitive enzymatic cleavage of urea. *J. biol. Chem.* **245,** 4593–4595.

SCHLOEMER, R. H. and GARRETT, R. H. (1974a) Nitrate transport system in *Neurospora crassa*. *J. Bact.* **118,** 1, 259–269.

SCHLOEMER, R. H. and GARRETT, R. H. (1974b) Uptake of nitrite by *Neurospora crassa*. *J. Bact.* **118,** 1, 270–274.

SIMS, A. P., FOLKES, B. F. and BUSSEY, A. H. (1968) Mechanisms involved in the regulation of nitrogen assimilation in microorganisms and plants. In *Recent Aspects of Nitrogen Metabolism in Plants*. ed. HEWITT, E. J. and CUTTING, C. V. pp. 91–114. Academic Press, London and New York.

SOLOMONSON, L. P. (1974) Regulation of nitrate reductase activity by NADH and cyanide. *Biochim. biophys. Acta* **334,** 297–308.

SOLOMONSON, L. P., JETSCHMANN, K. and VENNESLAND, B. (1973) Reversible inactivation of the nitrate reductase of *Chlorella vulgaris* Betjerinck. *Biochim. biophys. Acta* **308,** 32–43.

SOLOMONSON, L. P., LORIMER, G. H., GEWITZ, H. S. and VENNESLAND, B. (1974) A possible role of HCN in controlling nitrate reductase. *Pl. Physiol. Lancaster* **53** supplement, abstract 368.

STIEFEL, E. I. (1973) Proposed molecular mechanism for the action of molybdenum in enzymes; coupled proton and electron transfer. *Proc. natn. Acad. Sci., U.S.A.* **70,** 988–992.

SYRETT, P. J. (1973) Measurement of nitrate and nitrite-reductase activities in whole cells of *Chlorella*. *New Phytol.* **72,** 37–46.

SYRETT, P. J. and HIPKIN, C. R. (1973) The appearance of nitrate reductase activity in nitrogen-starved cells of *Ankistrodesmus braunii*. *Planta* **111,** 57–64.

SYRETT, P. J. and MORRIS, I. (1963) The inhibition of nitrate assimilation by ammonium in *Chlorella*. *Biochim. biophys. Acta* **67,** 566-575.

THACKER, A. and SYRETT, P. J. (1972a) The assimilation of nitrate and ammonium by *Chlamydomonas reinhardii*. *New Phytol.* **71,** 423–433.

THACKER, A. and SYRETT, P. J. (1972b) Disappearance of nitrate reductase activity from *Chlamydomonas reinhardii*. *New Phytol.* **71,** 435–441.

THOMAS, W. H. (1968) Nutrient requirements and utilization: algae. In *Metabolism, Biological Handbook*, ed. ALTMAN, P. L. and DITTMER, D. S. pp. 210–228. Federation of American Societies for Experimental Biology, Bethesda, Maryland.

THOMAS, W. H. and DODSON, A. N. (1974) Effect of interactions between temperature and nitrate supply on the cell-division rates of two marine phytoflagellates. *Mar. Biol.* **24,** 213–217.

THOMPSON, J. F. and MUENSTER, A-M. E. (1971) Separation of the *Chlorella* ATP:urea amidolyase into two components. *Biochem. biophys. Res. Commun.* **43,** 1049–1055.

THOMPSON, J. F. and MUENSTER, A-M. E. (1974) ATP-dependent urease: Characterization of and control in *Chlorella*; the search for it in higher plants. *Bull. R. Soc. N.Z.* **12,** 91–97.

VEGA, J. M., HERRERA, J., APARICIO, P. J., PANEQUE, A. and LOSADA, M. (1971) Role of molybdenum in nitrate reduction by *Chlorella*. *Pl. Physiol. Lancaster* **48,** 294–299.

VENNESLAND, B. and JETSCHMANN, C. (1971) The nitrate reductase of *Chlorella pyrenoidosa*. *Biochim. biophys. Acta* **229,** 554–564.

VOGEL, H. J., THOMPSON, J. S. and SHOCKMAN, G. D. (1970) Characteristic metabolic patterns of prokaryotes and eukaryotes. *Symp. Soc. gen. Microbiol.* **20,** 107–120.

WARBURG, O. and NEGELEIN, E. (1920) Über die Reduktion der Salpetersäure in grünen Zellen. *Biochem. Z.* **110,** 66–115.

WHITNEY, P. A. and COOPER, T. G. (1972) Urea carboxylase and allophanate hydrolyase. Two components of adenosine triphosphate: urea amidolyase in *Saccharomyces cerevisiae*. *J. biol. Chem.* **247,** 1349–1353.

ZUMFT, W. G., PANEQUE, A., APARICIO, P. J. and LOSADA, M. (1969) Mechanisms of nitrate reduction in *Chlorella*. *Biochem. biophys. Res. Commun.* **36,** 6, 980–986.

ZUMFT, W. G., SPILLER, H. and YEBOAH-SMITH, I. (1972) Eisengehalt und Electronen-donator-Spezifität der Nitratereductase von *Ankistrodesmus*. *Planta* **102,** 228–236.

ANABAENA CYLINDRICA, A MODEL NITROGEN-FIXING ALGA

W. D. P. STEWART

Department of Biological Sciences, University of Dundee, Dundee, U.K.

INTRODUCTION

The blue-green algae (Cyanophyceae) are oxygen-evolving, prokaryotic, photosynthetic, micro-organisms, some species of which fix nitrogen (see Fogg *et al.*, 1973). In 1923, when the Society for Experimental Biology was founded, the blue-green algae were regarded as microscopical curiosities, of little physiological significance, and as far as N_2 fixation was concerned, the outlook seemed bleak. In 1927, Waksman wrote 'It seems to be definitely established that blue-green algae do not fix nitrogen'; however, the following year, Drewes (1928) published a paper which showed that in axenic culture, two strains of *Anabaena* and one of *Nostoc* did indeed fix N_2. Despite this observation, study of N_2 fixation by blue-green algae remained very much a backwater and although Allison's group (see Allison *et al.*, 1937) carried out some excellent work on *Nostoc muscorum*, the pre-1940 period was dominated by work on N_2-fixing legumes (see Wilson, 1940) and it was not until 1942 when *Anabaena cylindrica* came on the scene that interest in N_2 fixation by blue-green algae was firmly re-awakened. This alga was isolated from a garden pond in Surrey by S. P. Chu, who was then a research student of F. E. Fritsch at East End College (now Queen Mary College) in London. The first physiological experiments on *Anabaena cylindrica* were reported by G. E. Fogg (1942) in the *Journal of Experimental Biology*. Since then N_2 fixation by blue-green algae, particularly the physiology, biochemistry, and the inter-relations of structure and function, has become a fascinating subject for research. In this paper, I would like to consider briefly current knowledge on N_2 fixation by *Anabaena cylindrica*, emphasizing, in particular, some recent data from my own laboratory. In so doing, I would not wish to minimize the contributions of other groups, who have all contributed substantially to knowledge in this field (see Bothe, 1970; Singh, 1972; Carr and Whitton, 1973; Fleming and Haselkorn, 1973, 1974; Fogg *et al.*, 1973; Stewart, 1973; Wolk, 1973).

CELL TYPES IN *Anabaena cylindrica*

Anabaena cylindrica is a photoautotrophic, unbranched, filamentous, heterocystous blue-green alga belonging to the family Nostocaceae. Early physiological studies were concerned primarily with the relationship between N_2 fixation and photosynthesis at the whole-organism level, and while important information was obtained (see Fogg, 1962), it was not until the organism was studied at the cellular level and the physiology of its constituent cells compared, that the biochemical complexity of this morphologically simple alga became obvious. The organism shows a maximum of three cell types, vegetative cells, heterocysts and akinetes (Fig. 1a). Each cell type is specialized in a different way. The vegetative cells fix CO_2 and evolve O_2 whereas the heterocysts are unable to fix CO_2 or evolve O_2, but can fix N_2 under aerobic conditions. The akinetes or perennating spores develop between the vegetative cells and heterocysts and obtain both fixed nitrogen and fixed carbon from them.

VEGETATIVE CELLS

The vegetative cells of *Anabaena cylindrica*, like those of other blue-green algae, have a higher-plant type of photosynthesis, in which water acts as the ultimate source of reductant, O_2 is evolved and CO_2 fixed according to the Calvin cycle pathway (see Coombs *et al.*, this volume). The photosynthetic pigments are located on the thylakoids which occur in the outer region of the cell (the chromatoplasm). These are chlorophyll *a*, carotenes, xanthophylls, and the characteristic accessory phycobiliprotein pigments, *c*-phycocyanin, *c*-phycoerythrin and *c*-allophycocyanin, which are located on the outer surface of the thylakoid membranes as macromolecular aggregations (see Fogg *et al.*, 1973). The exact location of the Calvin cycle enzymes is uncertain, although ribulose-bisphosphate carboxylase (E.C. 4.1.1.39) is thought to be present in the polyhedral bodies (carboxysomes) (Shively *et al.*, 1973) which occur in the centroplasm of the vegetative cells (Fig. 1c). Under conditions of carbon sufficiency, fixed carbon is stored in the form of polyglucoside bodies which are deposited in the interthylakoidal spaces. There is no unequivocal evidence, to date, that the vegetative cells of aerobic cultures of *Anabaena cylindrica* fix N_2 (see however, Wolk *et al.*, 1974), although the possibility exists that they may do so under microaerobic conditions, as in *Plectonema boryanum* (Stewart and Lex, 1970).

HETEROCYSTS

The heterocysts of blue-green algae do not fix CO_2 photosynthetically (see later) but fix N_2 (see Stewart, 1973). Despite the fact that Fritsch (1951) called such cells 'a botanical enigma', there is probably more known now about their structure and metabolism than about the metabolism of any other differentiated cell type in the Prokaryota. The evidence that heterocysts fix N_2 is of two sorts. First, there is the finding that isolated heterocysts fix N_2, and although the rates obtained in early studies were low, E. Tel-Or working in my laboratory on *Anabaena* has shown that nitrogenase activity equivalent to 46 per cent of the total filament activity can be detected in isolated heterocysts which contain only 10 per cent of the total filament protein, whereas only 5 per cent of the activity is associated with the remaining 90 per cent of the total filament protein. Wolk and Wojciuch (1971) using a different technique have previously concluded that heterocysts

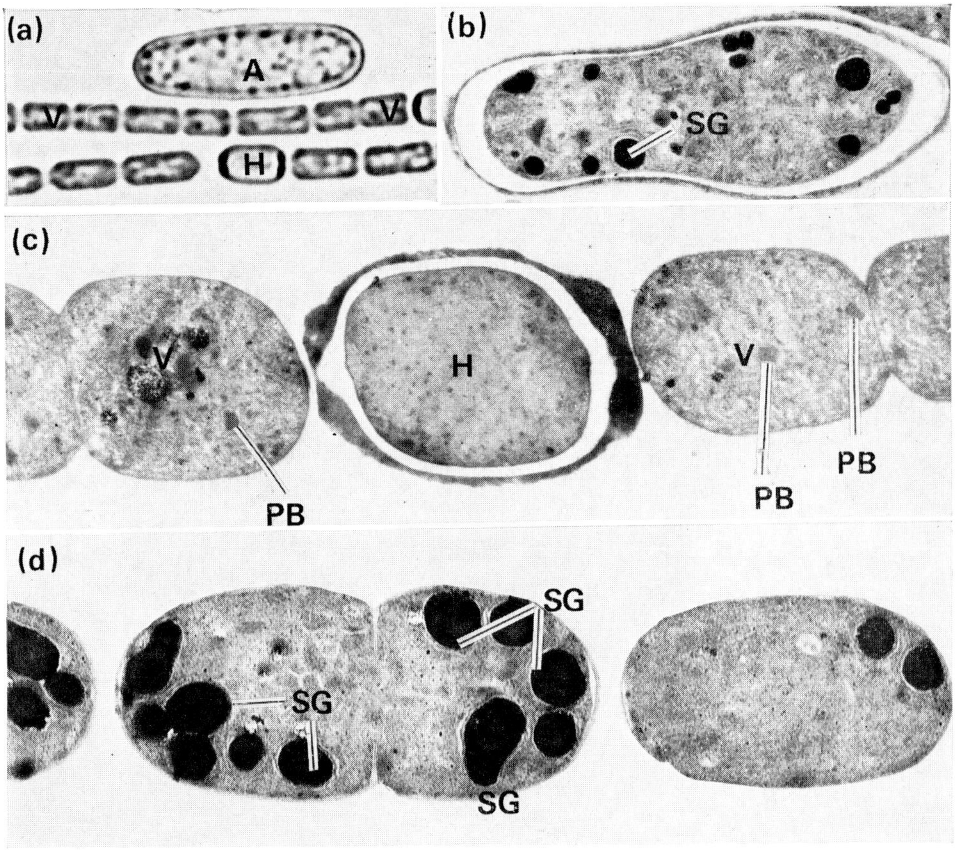

FIG. 1. (a) Light micrograph of *Anabaena cylindrica* showing vegetative cells (V), akinete (A), and heterocyst (H) (× 1800). (b) Electron micrograph of akinete of N_2-grown *Anabaena cylindrica* showing the presence of abundant structured granules (SG) (× 8000). (c) Electron micrograph of log phase *Anabaena cylindrica* showing the presence of polyhedral bodies (PB) in the vegetative cells (V) but not in the heterocysts (H) (× 10,000). (d) Electron micrograph of vegetative cells of ammonium-grown (50 mg l^{-1} of NH_4-N) *Anabaena cylindrica* showing the presence of abundant structured granules (SG) (× 10,000).

Facing page 236

may contribute up to 30 per cent of the total nitrogenase activity of intact filaments. Second, Fleming and Haselkorn (1973, 1974) have shown, by disc gel electrophoretic techniques, that proteins which are apparently similar to the subunits of the iron-molybdenum component of *Azotobacter* nitrogenase are present in the heterocysts, but not in the vegetative cells, of aerobic cultures of *Nostoc muscorum*.

The *Anabaena cylindrica* nitrogenase, like that of other organisms, requires a source of electrons and chemical energy as ATP, for N_2 reduction to occur. There also must be an O_2 protective mechanism, as the enzyme is extremely O_2 labile, and there must be a mechanism whereby NH_3, the end-product of N_2 fixation, or its product, is not allowed to accumulate and thus repress nitrogenase synthesis (see Stewart, 1973). There is now considerable evidence that electrons generated on the photolysis of water do not supply the reductant directly to nitrogenase. For example, there is no immediate inhibition of nitrogenase activity when 3-(3,4-dichlorophenyl)-1,1-dimethylurea (DCMU) (3×10^{-5} M), an inhibitor of photosystem II activity, is added to carbon-rich cultures, although there is a marked inhibition of photosynthetic CO_2 fixation (Table 1). In these experiments the carbon-rich cells were achieved by pre-incubating the alga under $Ar/O_2/CO_2$ (79.96/20.00/0.04, v/v) or Ar/CO_2 (99.96/0.04, v/v) (Lex and Stewart, 1973) (see also Donze *et al.*, 1974). Indeed it would be impossible for electrons from the photolysis of water to be donated directly to nitrogenase in heterocysts because the latter lack the capacity to split water photosynthetically (see Carr and Whitton, 1973).

It appears that the main function of photosystem II is to fix CO_2 and that the reduced carbon compounds provide electrons for nitrogenase. In the dark, $NADPH_2$ generated via the oxidative pentose phosphate pathway (see e.g. Pelroy *et al.*, 1972; Winkenbach and Wolk, 1973) or reduced ferredoxin generated from pyruvate via pyruvate:ferredoxin oxidoreductase (E.C.1.2.7.1) (Bothe and Falkenberg, 1973; Codd and Stewart, 1973) could supply electrons to nitrogenase. The latter enzyme, which is four to five times more active in N_2-grown cultures than in NH_4-grown cultures, has been investigated in detail by Bothe *et al.* (1974), who could find no evidence of a pyruvate dehydrogenase (E.C.

TABLE 1. EFFECT OF DCMU ON THE FIXATION OF $^{14}CO_2$ AND ON C_2H_2 REDUCTION BY *Anabaena cylindrica* FILAMENTS AFTER VARIOUS PRE-TREATMENTS

Pre-treatment	DCMU (3×10^{-5} M)	^{14}C fixation (cps)	% inhibition	nmol C_2H_4 sample^{-1} hr^{-1}	% inhibition
Ar/CO_2	−	1740		285	
(99.96/0.04)	+	350	80	278	2
$Ar/O_2/CO_2$	−	1650		184	
(79.96/20.00/0.04)	+	120	93	114	38

The suspensions were bubbled with the appropriate gas phase for 12 hr prior to the addition of $NaH^{14}CO_3$ (0.5 μCi/ml), or 10% C_2H_2, for 60 min, in the presence or absence of DCMU. The light intensity was 2000 lux, the temperature 25°C and each value is the mean of triplicate determinations. Pre-treatment in the presence of Ar/CO_2 would lead to the build up of a greater pool of carbon than pre-treatment under $Ar/O_2/CO_2$, because oxidative phosphorylation would be reduced. In air where N_2 fixation and assimilation as well as oxidative phosphorylation would occur during the pre-treatment, the pool of fixed carbon would be even lower, and in such cultures there was an 84 per cent inhibition of C_2H_2 reduction by DCMU within 60 min.

1.2.4.1) in *Anabaena cylindrica* (Bothe and Nolteernsting, 1975). We have recent evidence that pyruvate can supply electrons to nitrogenase *in vitro* (Table 2) in preparations where there is no evidence of pyruvate dehydrogenase activity (Table 3) and where pyruvate: ferredoxin oxidoreductase is present at a concentration high enough to account for the observed nitrogenase activity (Codd *et al.*, 1974). It must be pointed out, however, that pyruvate:ferredoxin oxidoreductase has not yet been reported in heterocysts. There is also evidence that in the light ferredoxin may provide reductant via electrons fed into photosystem I from a pool of organic compounds. Evidence supporting this view comes from the *in vitro* studies of Smith *et al.* (1971) and the *in vivo* studies of Lex and Stewart (1973). Further studies in this area are required.

There is early evidence that the main source of ATP for nitrogenase activity is cyclic photophosphorylation (see Fogg *et al.*, 1973) and we have obtained evidence supplementary to the data reported there, based on measurements of ATP pools by the luciferin-luciferase technique and also on ^{32}P-uptake studies. The ATP pool data show that:

(i) sufficient ATP for maximum nitrogenase activity can be generated in the light without any direct involvement of photosystem II;
(ii) photosystem I activity alone is sufficient to generate all the ATP necessary for maximum nitrogenase activity;
(iii) oxidative phosphorylation can supply only a small proportion of the necessary ATP in this photoautotrophic alga (it can supply much more in photoheterotrophs); and
(iv) substrate phosphorylation is unimportant as a source of ATP (Stewart and Bottomley, 1975; Bottomley and Stewart, unpublished).

TABLE 2. ACETYLENE REDUCTION BY EXTRACTS OF *Anabaena cylindrica* SUPPLIED WITH PYRUVATE AND VARIOUS COFACTORS

Reaction mixture	nmol C_2H_4 mg protein^{-1}min^{-1}
Complete: extract prepared in the presence of dithionite[a]	46.9
Complete: extract prepared in the absence of dithionite[b]	22.4
Dithionite omitted	2.4
Dithionite omitted, pyruvate[c] added	3.8
Dithionite omitted, ferredoxin[d] added	3.0
Dithionite omitted, pyruvate + ferredoxin added	14.2
Dithionite omitted, pyruvate + coenzyme A[e] added	3.6
Dithionite omitted, pyruvate + ferredoxin + coenzyme A added	20.0
Dithionite omitted, pyruvate + ferredoxin + coenzyme A + NADP[f] + ferredoxin–NADP reductase[g] added	18.8

[a] Prepared as in Haystead and Stewart (1972); [b] the preparations used in this and subsequent reactions were prepared anaerobically, but in the absence of added dithionite; [c] 15 μmol pyruvate per assay; [d] 10 μmol ferredoxin per assay; [e] 0.1 μmol coenzyme A per assay; [f] 2.5 μmol NADP per assay; [g] saturation amounts of ferredoxin–NADP reductase per assay. Each value is the mean of triplicate determinations.

We have recently investigated photophosphorylation by isolated heterocysts utilizing ^{32}P (see Scott and Fay, 1972) and obtained activities which are much higher than any hitherto reported (50 μmol ATP formed (mg chlorophyll $a)^{-1}hr^{-1}$). A full report of this work will be published elsewhere (see also Stewart *et al.*, 1975b).

The ability of heterocysts of *Anabaena cylindrica* to fix N_2, but not CO_2 (they receive fixed carbon from adjacent vegetative cells, see Fogg *et al.*, 1973; Wolk, 1973) seemed anomalous until it was realized (Winkenbach and Wolk, 1973) that the key Calvin cycle enzyme ribulose bisphosphate carboxylase was virtually inactive in heterocysts. We have confirmed this finding and shown further that polyhedral bodies (carboxysomes) which appear to be packets containing ribulose bisphosphate carboxylase in autotrophic prokaryotes (Shively *et al.*, 1973) are present in the vegetative cells and akinetes of *Anabaena cylindrica*, but are absent from heterocysts (Fig. 1c). They are also present in the vegetative cells, but not in the heterocysts of a variety of other blue-green algae which we have grown photoautotrophically, photoheterotrophically and dark heterotrophically (Stewart and Codd, 1975).

In *Anabaena cylindrica*, as in other N_2-fixing organisms, the end-product of N_2 fixation is ammonia, and under conditions where incorporation of ammonia into amino acids is blocked, ammonia is excreted into the medium. Table 4 shows typical results obtained when the analogue L-methionine-DL-sulphoximine (1 μM), is added to N_2-fixing and nitrogen-starved (incubation under $Ar/O_2/CO_2$:79.96/22.00/0.04, v/v) cultures of *Anabaena cylindrica.* These show that the addition of analogue results in the release of ammonia from N_2-fixing cells, but not from N_2-starved cells, which indicates that its release is not due to cell autolysis. Such released ammonia accounts for over half of the ammonia fixed, if it is assumed that the *in vivo* rate of C_2H_4 produced is one and a half times that of ammonia produced (Stewart *et al.*, 1968; Hardy *et al.*, 1971). The fact that there was no release of ammonia from the N_2-starved cells indicates that the ammonia released from N_2-fixing cells was not due to cell autolysis, and provides evidence in

TABLE 3. *In vitro* EVIDENCE FOR THE PRESENCE OF PYRUVATE FERREDOXIN OXIDOREDUCTASE, AND THE LACK OF PYRUVATE DEHYDROGENASE IN C_2H_2 REDUCING EXTRACTS OF *A. cylindrica*

Reaction mixture	μmol acetohydroxamate formed (mg protein)$^{-1}$ hr^{-1}
Complete for pyruvate ferredoxin oxidoreductase assay[a]	0.64
ATP omitted	0.39
Coenzyme A omitted	0.56
ferredoxin omitted	0.52
pyruvate omitted	0.01
NADP[b] + ferredoxin-NADP reductase[c] added	0.65
Ferredoxin omitted, NADP[b] + ferredoxin-NADP reductase[c] added	0.52

[a] Assay of Bothe and Falkenberg (1973); [b] 2.5 μmol per assay; [c] saturating amount. In such preparations NADP plus ferredoxin-NADP reductase was known to be limiting, so that the lack of response on their addition was not due to saturating concentrations already being present. For further details see Codd *et al.* (1974).

TABLE 4. PRODUCTION OF EXTRACELLULAR AMMONIA BY N_2-FIXING CULTURES OF *A. cylindrica* IN THE PRESENCE AND ABSENCE OF L-METHIONINE-DL-SULPHOXIMINE

Growth conditions	Analogue	Sample number	nmol NH_3 fixed (μg chlorophyll a)$^{-1}$hr^{-1}	nmol NH_3 excreted (μg chlorophyll a)$^{-1}$hr^{-1}
N_2-fixing	+	1	6.8	3.4
(incubated	+	2	8.2	4.1
in air)	+	3	7.8	3.5
	−	1	6.8	<0.02
	−	2	6.2	<0.02
	−	3	6.0	<0.02
Nitrogen-starved	+	1	0.0	<0.02
(incubated under	+	2	0.0	<0.02
$A/O_2/CO_2$:	+	3	0.0	<0.02
79.96/20.00/0.04,	−	1	0.0	<0.02
v/v)	−	2	0.0	<0.02
	−	3	0.0	<0.02

The alga was grown in aerated illuminated (3000 lux) continuous cultures at 28°C. Samples were then transferred to 25 ml batch cultures and pretreated for 24 hr with the appropriate gas phase. The analogue was then added for 24 hr and the material assayed for extracellular NH_3, chlorophyll a and acetylene reduction.

support of the view that ammonia is the end-product of N_2 fixation and that a pathway of its incorporation into cell metabolism is via glutamine synthetase-glutamate synthase.

The pathways of ammonia incorporation into amino acids have been investigated by *in vitro* enzyme assays, and kinetic studies involving ^{15}N as tracer. The results, which have been considered elsewhere (see Stewart *et al.*, 1975a), show (a) that glutamic dehydrogenase (E.C.1.4.1.3) activity is very low in N_2-fixing cells, suggesting perhaps that it plays only a minor role in ammonia incorporation, and (b) that glutamine synthetase (E.C.6.3.1.2) activity is high (48 nmol P_1 released (mg protein)$^{-1}$ min^{-1}), but glutamate-synthase activity (E.C.2.6.1.53) is low. There is good alanine-dehydrogenase activity (E.C.1.4.1.1). We have no evidence that glutamine synthetase regulates nitrogenase synthesis via an adenylylation-deadenylylation mechanism, as appears to happen in *Klebsiella* (Streicher *et al.*, 1975). Possible metabolic inter-relations in the heterocyst are summarized in Fig. 2.

VEGETATIVE CELL—HETEROCYST INTERACTIONS

The removal of ammonia from around the site of nitrogenase synthesis prevents repression of nitrogenase synthesis, and ^{15}N tracer studies have shown that just as there is a rapid transfer of carbon compounds from the vegetative cells to the heterocysts there also appears to be a rapid transfer of nitrogen from heterocysts to vegetative cells (Stewart *et al.*, 1969), but whether this is transferred as ammonia, or as amino acids, is uncertain (see Fig. 2, and Stewart *et al.*, 1975a). In the vegetative cells, the fixed nitrogen is used in general cell metabolism, excreted, or stored as the phycobiliprotein, phycocyanin, or as structured granules which are densely osmiophilic copolymers of

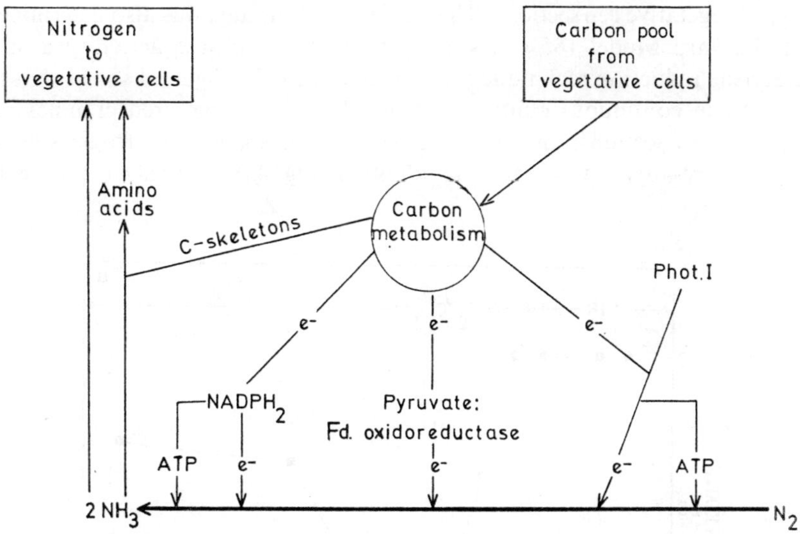

FIG. 2. Possible inter-relations between carbon metabolism and nitrogen metabolism in heterocysts of *Anabaena cylindrica.*

aspartic acid and arginine (see Fig. 3, Fig. 1d, and Fogg *et al.*, 1973). The formation of phycocyanin in the vegetative cells after provision of N_2 has been demonstrated convincingly in whole organisms by van Gorkom and Donze (1971) and the biochemistry of structured granules has been elucidated by Simon (1971, 1973a, b).

The relationship between the abundance of structured granules in the vegetative cells and the availability of nitrogen is shown by the data of Figs. 4 and 5 for *Anabaena flos-aquae*, an organism closely related to *Anabaena cylindrica.* In this study we measured the abundance of structured granules by counting in the electron microscope (Associated Electrical Industries EM801 at 60 kV), the mean number of structured granules present

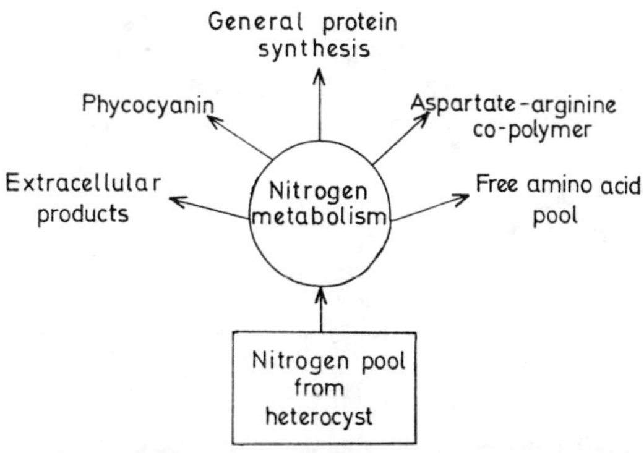

FIG. 3. Fate of nitrogen previously fixed in the heterocysts on transfer to the vegetative cells of *Anabaena cylindrica.*

in a thousand vegetative cell sections. The C:N ratio of the alga was also determined using a Hewlett–Packard Model 185 CHN analyser, and nitrogenase activity was measured by the acetylene reduction technique (Stewart *et al.*, 1967). Figure 4 shows that the alga growing on N_2 in continuous culture contains about fifty structured granules per thousand vegetative cell sections, and that these virtually disappear on nitrogen starvation of the culture, and re-appear when air is re-admitted (Fig. 4a). Figures 4a, b show that the

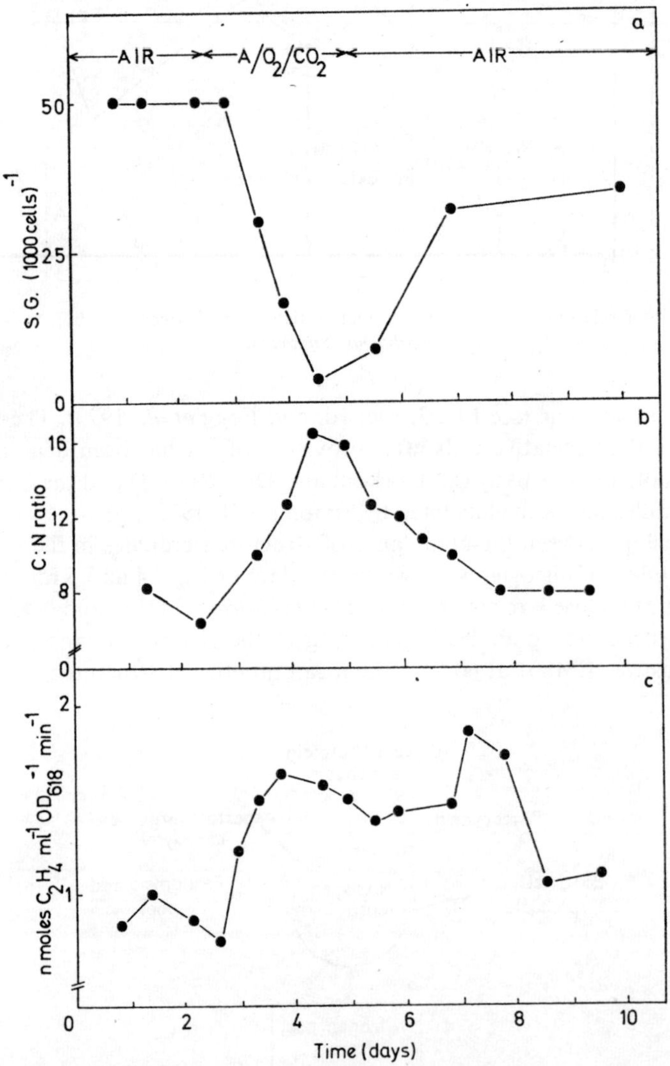

FIG. 4. Effect of nitrogen availability on (a) the abundance of structured granules, (b) C/N ratio and (c) acetylene reduction by *Anabaena flos-aquae* grown in continuous culture. The gas phase was either air, or $Ar/O_2/CO_2$ (79.96/20.00/0.04, v/v) as shown. The experiment was carried out at 28°C and 3000 lx and each value is the mean of triplicate determinations.

abundance of structured granules is inversely proportional to the C:N ratio of the cells, while Fig. 4c shows that nitrogenase activity is usually highest under conditions of nitrogen deficiency, probably because of depletion of nitrogenous compounds which partially inhibit nitrogenase synthesis.

AKINETES

Akinetes (see Figs. 1a, b) develop in *Anabaena cylindrica* particularly under adverse conditions. These perennating non-N_2-fixing structures develop adjacent to the heterocysts from which they receive fixed nitrogen, and where they may also receive fixed carbon from the vegetative cells. Figure 1b shows the presence of abundant structured granules in akinetes. Figure 5 shows the general relationship between the abundance of structured granules in filaments of *Anabaena flos-aquae* and the development of akinetes. It is seen that the numbers of structured granules decrease as log-phase algae move into the exponential phase of growth and remain fairly low during log phase, before increasing

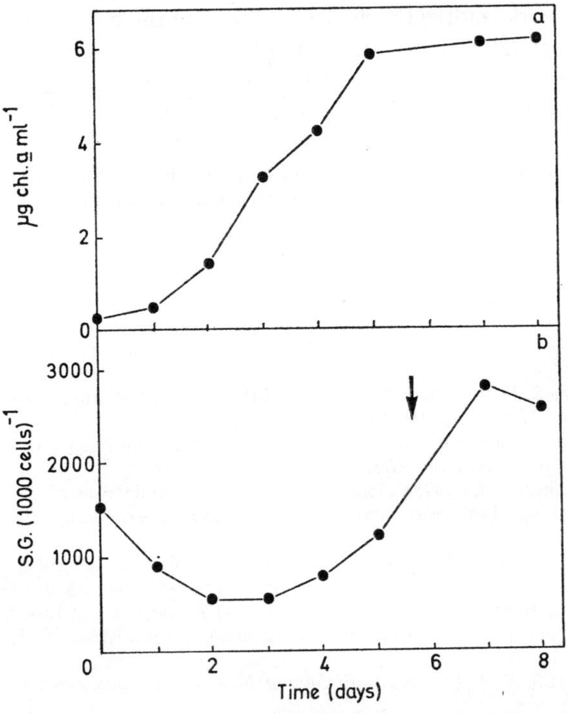

FIG. 5. Abundance of structured granules in *Anabaena flos-aquae* during growth of the organism in batch culture. Akinetes (or spores) were first noted in the culture at the time denoted by the arrow, and on days 7 and 8 structured granules in akinetes and vegetative cells were counted together. Growth of the organism (a) was measured in terms of chlorophyll *a* as previously described (Lyne and Stewart, 1973). The abundance of structured granules (b) was determined as in the text, and the experiment was carried out in 750 ml aliquots of medium in batch culture at 28°C and 3000 lx.

markedly in post-log-phase cultures. This latter increase is due to some increase in structured granules in the vegetative cells, but more especially to the development of spores which contain large numbers of structured granules. Under suitable environmental conditions, the spores germinate to produce heterocystous sporelings which develop into mature filaments. It may also be noted from a comparison of the data on structured granules in Figs. 4 and 5 that numbers are considerably higher in log-phase batch culture than in the more actively-growing continuous culture.

PERSPECTIVE

In the available space it has not been possible to deal in any detail with most of the points which I have mentioned. Nevertheless, I hope that this summary does indicate the ways in which N_2-fixing blue-green algae, such as *Anabaena cylindrica*, are organized structurally, physiologically and biochemically to provide an overall metabolism which enables such morphologically simple forms to compete successfully in many present-day habitats such as eutrophic waters and rice paddy soils. The stimulus to such studies was provided in no small way by members of the Society for Experimental Biology; much of the experimental work leading to current thinking on the blue-green algae was carried out in plant physiology laboratories, and if the bacteriologist now wishes to lay claim to the group (see e.g. Stanier *et al.*, 1971), this is perhaps a reflection that the work of the algal physiologist and biochemist has not gone unnoticed.

Acknowledgements: My own work reported here was supported by the Natural Environment Research Council and the Science Research Council. I thank Miss Gail Alexander and Miss L. Al-Ugaily for technical assistance.

REFERENCES

ALLISON, F. E., HOOVER, S. R. and MORRIS, H. M. (1937) Physiological studies with the nitrogen-fixing alga, *Nostoc muscorum*. *Bot. Gaz.* **98,** 433–463.

BOTHE, H. (1970) Photosynthetische Stickstoffixierung mit einem zellfreien Extract aus der Blaualge *Anabaena cylindrica*. *Ber. dtsch. bot. Ges.* **83,** 421–432.

BOTHE, H. and FALKENBERG, B. (1973) Demonstration and possible role of a ferredoxin-dependent pyruvate decarboxylation in the nitrogen-fixing blue-green alga *Anabaena cylindrica*. *Pl. Sci. Lett.* **1,** 151–156.

BOTHE, H. and NOLTEERNSTING, U. (1975) Pyruvate dehydrogenase complex, pyruvate:ferredoxin oxidoreductase and lipoic acid content in microorganisms. *Arch. Mikrobiol.* **102,** 53–58.

BOTHE, H., FALKENBERG, B. and NOLTEERNSTING, U. (1974) Properties and function of the pyruvate: ferredoxin oxidoreductase from the blue-green alga *Anabaena cylindrica*. *Arch. Mikrobiol.* **96,** 291–304.

CARR, N. G. and WHITTON, B. A. (1973) *The Biology of Blue-Green Algae*, Blackwell Scientific Publications, Oxford.

CODD, G. A. and STEWART, W. D. P. (1973) Pathways of glycollate metabolism in the blue-green alga *Anabaena cylindrica*. *Arch. Mikrobiol.* **94,** 11–28.

CODD, G. A., ROWELL, P. and STEWART, W. D. P. (1974) Pyruvate and nitrogenase activity in cell-free extracts of the blue-green alga *Anabaena cylindrica*. *Biochem. biophys. Res. Commun.* **61,** 424–431.

DONZE, M., RAAT, A. J. P. and VAN GORKOM, H. J. (1974) Supply of ATP and reductant to nitrogenase in the blue-green alga *Anabaena cylindrica*. *Pl. Sci. Lett.* **3,** 35–41.

DREWES, K. (1928) Über die Assimilation des Luftstickstoffe durch Blaualgen. *Zentbl. Bakt. ParasitKde Abt. II* **76,** 88–101.

FLEMING, H. and HASELKORN, R. (1973) Differentiation in *Nostoc muscorum.* Nitrogenase is synthesized in heterocysts. *Proc. natn. Acad. Sci. U.S.A.* **70,** 2727–2731.

FLEMING, H. and HASELKORN, R. (1974) The program of protein synthesis during heterocyst differentiation in nitrogen fixing blue-green algae. *Cell.* **3,** 159–170.

FOGG, G. E. (1942) Studies on nitrogen fixation by blue-green algae. 1. Nitrogen fixation by *Anabaena cylindrica* Lemm. *J. exp. Biol.* **19,** 78–87.

FOGG, G. E. (1962) Nitrogen fixation. In *Physiology and Biochemistry of Algae*, ed. LEWIN, R. A., Academic Press, London and New York.

FOGG, G. E., STEWART, W. D. P., FAY, P. and WALSBY, A. E. (1973) *The Blue-Green Algae*, Academic Press, London.

FRITSCH, F. E. (1951) The heterocyst: a botanical enigma. *Proc. Linn. Soc. Lond.* **162,** 194–211.

HARDY, R. W. F., BURNS, R. C., HEBERT, R. R., HOLSTEN, R. D. and JACKSON, E. K. (1971) Biological nitrogen fixation: a key to world protein. *Pl. Soil, Special Volume*, 561–590.

HAYSTEAD, A. and STEWART, W. D. P. (1972) Characteristics of the nitrogenase system of the blue-green alga *Anabaena cylindrica. Arch. Mikrobiol.* **82,** 325–336.

LEX, M. and STEWART, W. D. P. (1973) Algal nitrogenase, reductant pools and photosystem I activity. *Biochim. biophys. Acta* **292,** 436–443.

LYNE, R. L. and STEWART, W. D. P. (1973) Emerson enhancement of carbon fixation but not of acetylene reduction (nitrogenase activity) in *Anabaena cylindrica. Planta* **109,** 27–38.

PELROY, R. A., RIPPKA, R. and STANIER, R. Y. (1972) The metabolism of glucose by unicellular blue-green algae. *Arch. Mikrobiol.* **87,** 303–322.

SCOTT, W. E. and FAY, P. (1972) Phosphorylation and amination in heterocysts of *Anabaena cylindrica. Br. phycol. J.* **7,** 283–284.

SHIVELY, J. M., BALL, F. L. and KLINE, B. W. (1973) Electron microscopy of the carboxysomes (polyhedral bodies) of *Thiobacillus neapolitanus. J. Bact.* **116,** 1405–1411.

SIMON, R. D. (1971) Cyanophycin granules from the blue-green alga *Anabaena cylindrica*: A reserve material consisting of copolymers of aspartic acid and arginine. *Proc. natn. Acad. Sci. U.S.A.* **68,** 265–267.

SIMON, R. D. (1973a) The effect of chloramphenicol on the production of cyanophycin granule polypeptide in the blue-green alga *Anabaena cylindrica. Arch. Mikrobiol.* **92,** 115–122.

SIMON, R. D. (1973b) Measurement of the cyanophycin granule polypeptide contained in the blue-green alga *Anabaena cylindrica. J. Bact.* **114,** 1213–1216.

SINGH, R. N. (1972) *Physiology and Biochemistry of Nitrogen Fixation by Blue-Green Algae* (Final Technical Report) (1967–1972) Dept. Bot. Banaras Hindu University, Varanasi 5, India 66 pp.

SMITH, R. V., NOY, R. J. and EVANS, M. C. W. (1971) Physiological electron donor systems to the nitrogenase of the blue-green alga *Anabaena cylindrica. Biochim. biophys. Acta* **253,** 104–109.

STANIER, R. Y., KUNISAWA, R., MANDEL, M. and COHEN-BAZIRE, G. (1971) Purification and properties of unicellular blue-green algae (Order Chroococcales). *Bact. Rev.* **35,** 171–205.

STEWART, W. D. P. (1973) Nitrogen fixation by photosynthetic micro-organisms. *A. Rev. Microbiol.* **27,** 283–318.

STEWART, W. D. P. and BOTTOMLEY, P. (1975) The nitrogenase of blue-green algae. *Proceedings of the International Symposium on Nitrogen Fixation, Pullman, Washington, June* 1974 (in press).

STEWART, W. D. P. and CODD, G. A. (1975) Polyhedral bodies (carboxysomes) of nitrogen-fixing blue-green algae. *Br. phycol. J.* (in press).

STEWART, W. D. P. and LEX, M. (1970) Nitrogenase activity in the blue-green alga *Plectonema boryanum* strain 594. *Arch. Mikrobiol.* **73,** 250–260.

STEWART, W. D. P., FITZGERALD, G. P. and BURRIS, R. H. (1967) *In situ* studies on N_2 fixation using the acetylene reduction technique. *Proc. natn. Acad. Sci. U.S.A.* **58,** 2071–2078.

STEWART, W. D. P., FITZGERALD, G. P. and BURRIS, R. H. (1968) Acetylene reduction by nitrogen-fixing blue-green algae. *Arch. Mikrobiol.* **62,** 336–348.

STEWART, W. D. P., HAYSTEAD, A. and PEARSON, H. W. (1969) Nitrogenase activity in heterocysts of filamentous blue-green algae. *Nature Lond.* **224,** 226–228.

STEWART, W. D. P., HAYSTEAD, A. and DHARMAWARDENE, M. W. N. (1975a) Nitrogen assimilation and metabolism in blue-green algae. In *Nitrogen Fixation by Free-Living Micro-Organisms*, ed. STEWART, W. D. P., pp. 129–158, Cambridge University Press, Cambridge.

STEWART, W. D. P., ROWELL, P. and TEL-OR, E. (1975b) Nitrogen fixation and the heterocyst. *Biochem. Soc. Trans.* **3,** 357–361.

STREICHER, S. L., SHANMUGAN, K. T., AUSUBEL, F., MORANDI, C. and GOLDBERG, B. (1975) Regulation of nitrogenase synthesis by glutamine synthetase. *Proceedings of the International Symposium on Nitrogen Fixation, Pullman, Washington, June* 1974 (in press).

VAN GORKOM, H. J. and DONZE, M. (1971) Localization of nitrogen fixation in *Anabaena. Nature Lond.* **234,** 231–232.

WAKSMAN, S. A. (1927) *Principles of Soil Microbiology*, Bailliere, Tindall & Cox, London.

WILSON, P. W. (1940) *The Biochemistry of Symbiotic Nitrogen Fixation.* University of Wisconsin Press, Madison.

WINKENBACH, F. and WOLK, C. P. (1973) Activities of enzymes of the oxidative and the reductive pentose phosphate pathways in heterocysts of a blue-green alga. *Pl. Physiol. Lancaster* **52,** 480–483.

WOLK, C. P. (1973) Physiology and cytological chemistry of blue-green algae. *Bact. Rev.* **37,** 32–101.

WOLK, C. P. and WOJCIUCH, E. (1971) Photoreduction of acetylene by heterocysts. *Planta* **97,** 126–134.

WOLK, C. P., AUSTIN, S. M., BORTINS, J. and GALONSKY, A. (1974) Autoradiographic localization of ^{13}N after fixation of ^{13}N-labeled nitrogen gas by a heterocyst-forming blue-green alga. *J. Cell Biol.* **61,** 440–453.

REGULATION OF GLUTAMINE METABOLISM IN FUNGI WITH PARTICULAR REFERENCE TO THE FOOD YEAST *CANDIDA UTILIS*

A. P. SIMS

School of Biological Sciences, University of East Anglia, Norwich, U.K.

INTRODUCTION

The amide group of L-glutamine serves as the source of nitrogen for the biosynthesis of a large number of other compounds including histidine, tryptophan, carbamyl phosphate, glucosamine 6-phosphate, AMP, CTP and NAD. The enzyme glutamine synthetase (L-glutamate: ammonia ligase (ADP), E.C. 6.3.1.2) catalyses the synthesis of L-glutamine as follows.

$$\text{L-glutamate} + \text{ATP} + \text{NH}_4^+ \xrightarrow{\text{Mg}^{2+}} \text{L-glutamine} + \text{ADP} + \text{P}_i$$

This reaction favours the biosynthesis of L-glutamine with a free energy change of about -5.2 Kcal mole^{-1} at pH 7 (Alberty, 1968).

Glutamine synthetase is a particularly important enzyme since it catalyses the first step in a highly branched metabolic network that ultimately produces many of the important macromolecules of the cell. For this reason considerable research effort has been expended to investigate the control of glutamine biosynthesis and the mechanisms and control of the glutamine dependent reactions: many of these studies have been summarized in a volume edited by Prusiner and Stadtman (1973). Important reviews on the glutamine synthetase of mammals (Meister, 1974) microorganisms (Shapiro and Stadtman, 1970) and the bacterium *Escherichia coli* (Stadtman and Ginsburg, 1974) have also recently appeared and consequently I intend to limit this article principally to a discussion of the control of glutamine biosynthesis in fungi. For a useful general review of inorganic nitrogen metabolism, including the fungi, see Brown *et al.* (1974).

REGULATION OF GLUTAMINE METABOLISM IN FUNGI

Early thinking on the control of glutamine biosynthesis in fungi was largely influenced by the results of studies on the enzyme from *Escherichia coli*. Pioneer work in E. R. Stadtman's laboratory (see Woolfolk *et al.*, 1964, 1966) had led to the discovery of a mechanism of control (cumulative feedback inhibition) in which each metabolite derived from glutamine could partly inhibit the activity of the enzyme and which in combination could bring about its almost complete inhibition. The inhibitors behaved kinetically as if they were acting independently at separate sites on the enzyme molecule and the cumulative effects were explained by the existence of a large number of specific non-interacting binding sites on individual enzyme sub-units (for a recent summary see Stadtman *et al.*, 1972). Such a mechanism provided an effective means of controlling the supply of glutamine to a multibranched pathway since an excess of a single end product could inhibit only part of the enzyme activity; only when all end products were present in excess would complete inhibition of activity occur. Subsequent studies with glutamine synthetase from fungi (including *Neurospora crassa* and *Candida utilis*) and other sources (Hubbard and Stadtman, 1967) led to the suggestion that this mechanism of control was widespread and because of its importance it had been carefully conserved through evolution. Fairly detailed studies with the enzyme from *Neurospora crassa* (Kapoor and Bray, 1968; Kapoor *et al.*, 1968, 1969) further substantiated this suggestion: the slight differences in the pattern of inhibition noted between *Neurospora crassa* and *Escherichia coli* were reconciled to differences in the metabolism of the two organisms.

In the meantime continued investigation of glutamine synthetase from *Escherichia coli* had led to the discovery of another sophisticated mechanism of control involving enzyme dependent deactivation of the synthetase (Mecke *et al.*, 1966). It was subsequently shown that covalent modification of the enzyme by adenylation and de-adenylation of specific tyrosyl residues produced marked changes in its catalytic potential, susceptibility to feedback inhibition and divalent metal specificity (Shapiro *et al.*, 1967; Wulff *et al.*, 1967). The possibility that a similar mechanism of control operated in fungi was excluded by the observations of Gancedo and Holzer (1968); rapid inactivation of synthetase did not

TABLE 1. COMPARISON BETWEEN THE *in vivo* RATE OF GLUTAMINE SYNTHESIS IN *Candida utilis* MEASURED WITH ^{15}N AND THE RATE CALCULATED FROM THE ENZYME CONTENT OF THE YEAST

(a) Yeast growing on NH_3 (25°C)		
Measured rate of glutamine synthesis†		0.064 mg N/g dry wt. yeast
Rate of synthesis calculated from enzyme content assuming all substrates are present at saturating concentrations†		0.140 mg N/g dry wt. yeast
An estimate of the possible extent of enzyme substrate saturation in the cell was obtained from the mean cellular concentrations of substrates		
substrate	mean cell. conc. (mM)	% saturation of enzyme
NH_3*	1.45	91
glutamate†	74.00	95
Mg^{2+}‡	6.60	97
ATP‡	4.10	93
Corrected rate of glutamine synthesis		0.112 mg N/g dry wt. yeast
(b) Yeast growing on glutamate (25°C)		
In vivo rate of glutamine biosynthesis†		0.060 mg N/g dry wt. yeast
Rate of synthesis calculated from enzyme content with full substrate saturation		0.634 mg N/g dry wt. yeast
substrate	mean cell. conc. (mM)	% saturation of enzyme
NH_3*	0.12	14
glutamate†	100.00	100
Mg^{2+}‡	6.60	97
ATP‡	4.10	93
Corrected rate of glutamine biosynthesis		0.082 mg N/g dry wt. yeast

* data from Sims *et al.* (1968)
† data from Sims and Ferguson (1974)
‡ unpublished data (Box and Sims).

occur when ammonia was added to yeast(s) having de-repressed levels of enzyme (see however, Ferguson and Sims, 1971). It should be recalled that at this time most of the ideas concerning the mechanisms of control of glutamine synthesis in fungi were based on the properties of the isolated enzyme.

My interest in the control of glutamine metabolism stemmed from a realization that it was possible to apply ^{15}N kinetic methods and directly measure the rate of synthesis of the amino acid in living cells (for details of the kinetic methods see Sims and Folkes, 1964; Folkes and Sims, 1974). It was possible therefore directly to test the system of cumulative feedback inhibition proposed for *Candida utilis*.

FEEDBACK CONTROL AS A MECHANISM OF REGULATING GLUTAMINE BIOSYNTHESIS IN FOOD YEAST

Metabolic pathways subject to direct feedback inhibition are likely to possess the following characteristics (see Sims *et al.*, 1968).

(a) The steady-state rate of synthesis realized from the metabolic pathway is unlikely to represent more than a fraction of the total potential activity of the enzyme within the cell.

(b) A reduction in the cellular concentration of feedback inhibitors should increase the rate of synthesis above the steady-state value and be accompanied by an increase in activity per unit of enzyme.

Isotope kinetic methods were used to ascertain whether these properties were features of the glutamine pathway in food yeast (Sims and Ferguson, 1974). It can be seen (Table 1) that in yeast growing under steady-state conditions on ammonia there is not a large excess of catalytic activity, instead there is a close match between the *in vivo* rate of glutamine synthesis and the amount of enzyme. Similarly although the level of enzyme is much higher in yeast growing on glutamate, when an allowance is made for the reduced concentration of NH_3 in the cell, the rate calculated for amide synthesis agrees closely with the determined value. Evidently then in steady-state cells little increase in amide synthesis is possible without a corresponding increase in the level of the enzyme or its substrates.

The effect of a reduction of end product(s) on the rate of the process was examined by exposing yeast to a short period of nitrogen-depletion. Changes of *in vivo* synthesis could then be measured after re-supply of $^{15}NH_3$ to the culture (Fig. 1).

Several important conclusions were drawn from these data. The maximum rate of glutamine synthesis attained was only 30 per cent above the steady-state value even though the mean cellular concentration of glutamine and many of its end products had been reduced by up to 70 per cent of their steady-state values. In fact, when allowance was made for the slight increase in enzyme amount that had occurred during nitrogen-starvation, the actual rate of amide synthesis per unit of enzyme was only 60 per cent of the steady-state value. The kinetics of these changes were not characteristic of a system under feedback control since the maximum response was delayed until about 8 min after the addition of NH_3 to the culture during which time the level of enzyme actually fell.

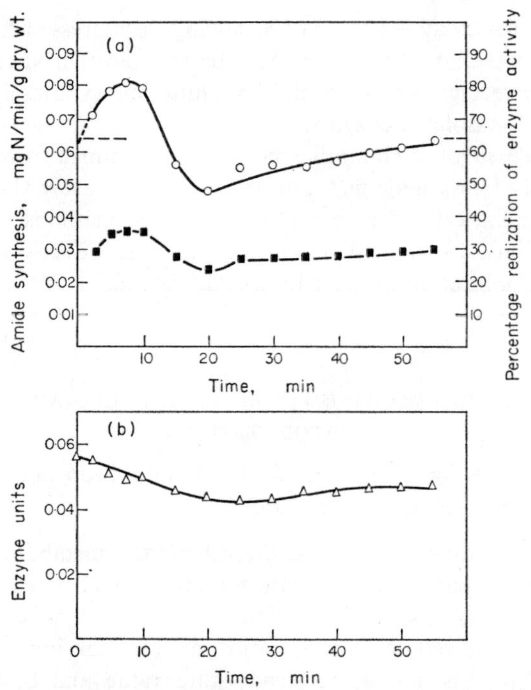

FIG. 1. Changes in the *in vivo* rate of amide synthesis (a) and in the level of glutamine synthetase (b) in *Candida utilis* after a short period of ammonia deprivation. The enzyme activity data shown in (b) were used to establish the synthetic potential of the yeast at different times, assuming the enzyme operated in the absence of any feedback control or substrate limitation. The observed rate at a particular time was then expressed as a percentage of potential rate at that time to determine the extent to which its activity was realized. The dashed line in (a) shows the steady-state rate of glutamine amide synthesis on ammonia. ○, Rate of glutamine synthesis; ■, percentage realization of synthetic potential; and △, activity of glutamine synthetase. The percentage realization of synthetic potential during steady-state growth on ammonia was 50. (Data from Sims and Ferguson, 1974.)

Subsequent investigation established that the concentration of free NH_3 in the cell was at its highest at this time.

All these recent observations fully supported an earlier suggestion that the glutamine pathway was not subject to feedback control (Sims *et al.*, 1968) but instead that it was regulated by the availability of enzyme substrates particularly NH_3 and glutamate. When, in other experiments, the pool of free glutamate was reduced further, the rate of glutamine synthesis could be positively correlated with the level of free glutamate in the cell.

In view of our failure to detect any significant *in vivo* end product inhibition on the glutamine pathway we decided to re-examine the effects of glutamine and possible end products on the activity of partially purified yeast enzyme. Two different assay procedures were used for this investigation.

(1) *Synthetase assay*. This assays the reaction in the physiological direction and is Mg^{2+} dependent in yeast.

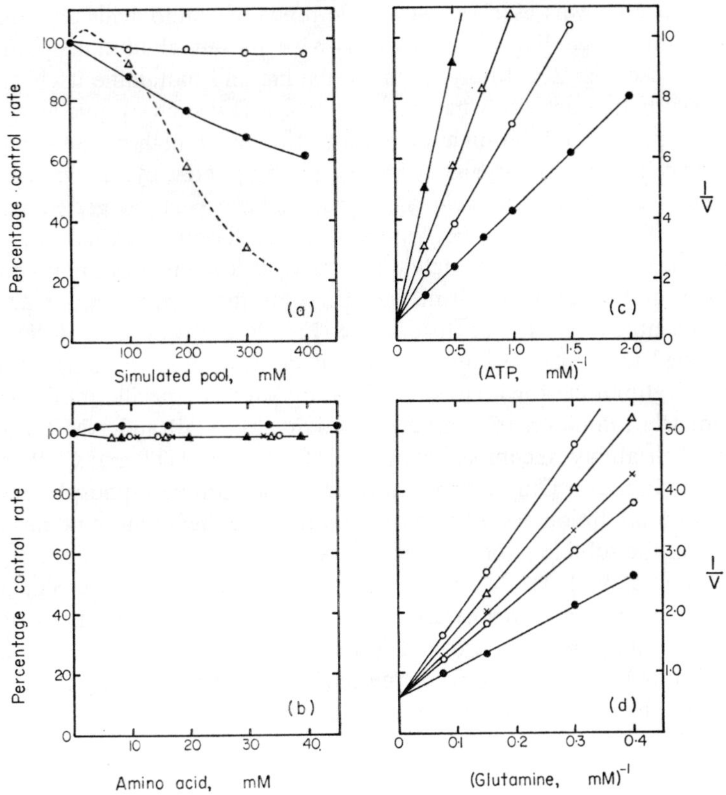

FIG. 2. Effect of amino acids and nucleotides on purified glutamine synthetase of *Candida utilis*. (For details of synthetase and transferase assay see Ferguson and Sims, 1974a.) (a) Effect of an amino acid mixture simulating the soluble pool of *C. utilis* (details Sims *et al.*, 1968) on enzyme activity. ○ Synthetase; ● transferase. Their effect on the NADPH glutamate dehydrogenase (. . . △ . . .) is given for purposes of comparison. (b) Effect of individual L-amino acids on synthetase activity. ● glutamine; ○ alanine; △ glycine; ▲ histidine; × serine. (c) Inhibition of synthetase by nucleotides. ● Control K_m ATP = 3.6 mM; ○ 1 mM AMP; △ 2 mM AMP; ▲ 4 mM AMP. K_i AMP = 1.6 mM. Competitive effects were also observed with ADP. GTP, ITP, UTP, CTP (2 mM) were inactive in the assay and did not inhibit the enzyme. (d) Inhibition of transferase activity by L-amino acids. ● Control, K_m glutamine = 8 mM; ○ alanine at 15 and 45 mM, K_i = 26 mM; △ glycine 40 mM; × serine 40 mM. Aspartate, lysine, histidine, arginine, arginine acid proline were also shown competitively to inhibit the enzyme. 40 mM Asparagine, threonine, leucine, glutamate and valine did not. Unpublished data of Sims and Ferguson.

$$\text{ATP} + \text{L-glutamate} \begin{array}{c} + NH_2OH \\ \\ + NH_3 \end{array} \longrightarrow \begin{array}{c} \text{L-}\gamma\text{ glutamyl hydroxamate} \\ \\ \text{L-glutamine} \end{array} + \text{ADP}$$

Identical results were obtained when NH_3 or NH_2OH were used as substrates.

(2) *Transferase assay*. This is 6 times more sensitive than the synthetase assay and is frequently used. Mg^{2+} is inactive in this assay which appears to be Mn^{2+} specific.

$$\text{L-glutamine} + NH_2OH \xrightarrow[\text{ADP AsO}_4^{2+}]{Mn^{2+}} \text{L-}\gamma\text{glutamyl hydroxamate} + NH_3$$

The synthetase assay was used to test a wide range of amino acids and amides singly and in combination including all those known to inhibit glutamine synthetase from *Escherichia coli* (see Fig. 2a). Many, such as alanine and glutamine itself produced no inhibition of synthetase activity whatsoever, whilst a few such as glycine and histidine produced only 5 per cent inhibition at a concentration much higher than would normally be found in the cell (Fig. 2b). Organic acids, sugar phosphates, pyridoxamine phosphate, DPN, shikimic acid, anthranilic acid all of which have been suggested as regulatory ligands had no effect on the enzyme from yeast. Results obtained with the transferase assay on the other hand, were strikingly different since a number of amino acids (not all end products of glutamine metabolism) including glycine, serine, alanine and aspartate produced appreciable inhibition of the enzyme (Fig. 2d). However, in all these instances it was established that enzyme inhibition could be attributed to competition between amino acid and glutamine for a substrate binding site: no specific inhibitor sites were detected. Similarly inhibition of synthetase activity by ADP and AMP also probably resulted from their ability to compete with ATP. GTP, ITP, UTP and CTP had no effect on the enzyme (Fig. 2c). Only glucosamine and glucosamine 6-phosphate showed the type of saturation inhibition described as cumulative. Neither compound has been found to accumulate in the soluble pool (unpublished data).

Results of studies with inhibitors on the yeast enzyme are in agreement with the physiological data so long as synthetase activity is considered. Since many of the effects of inhibitors described by Hubbard and Stadtman (1967) and by Kapoor and his co-workers (1968, 1969) were revealed when measuring transferase activity there are grounds for doubting the physiological significance of their results.

GLUTAMINE SYNTHESIS IN YEAST WITH ELEVATED LEVELS OF ENZYME AND SUBSTRATES

If the glutamine pathway is not subject to direct feedback control, and amide synthesis is determined largely by the level of enzyme and its substrates, then the addition of NH_3 to yeast growing on glutamate should produce *in vivo* rates that directly reflect the higher enzyme content of the yeast. This appears to be the case (Fig. 3). Addition of $^{15}NH_3$ to yeast grown on glutamate did initially yield *in vivo* rates six-fold higher than the steady-state rate on NH_3. Clearly if these high rates were maintained indefinitely extensive diversion of glutamate, 2 oxo-glutarate, NADPH and ATP into glutamine would ultimately affect a large number of other metabolic and physiological processes.

In fact there was a rapid fall in the levels of glutamate and NH_3 and this was accompanied by a fall in enzyme level and a subsequent decrease in the rate of amide synthesis to approach the steady-state value on NH_3 (Fig. 4). The relative contributions of enzyme and of substrate to this change in rate were assessed by reference to data on enzyme affinities. These data allowed estimates of the effect of substrate concentration on enzyme activity to be made throughout the experiment, from which it was possible, in conjunction with measurements of enzyme level, to predict the overall rate of glutamine synthesis. These values could then be compared with experimentally determined rates.

For this nitrogen-transition experiment, during which there were phases of increased saturation of enzyme by NH_3, enzyme inactivation, and substrate limitation (glutamate

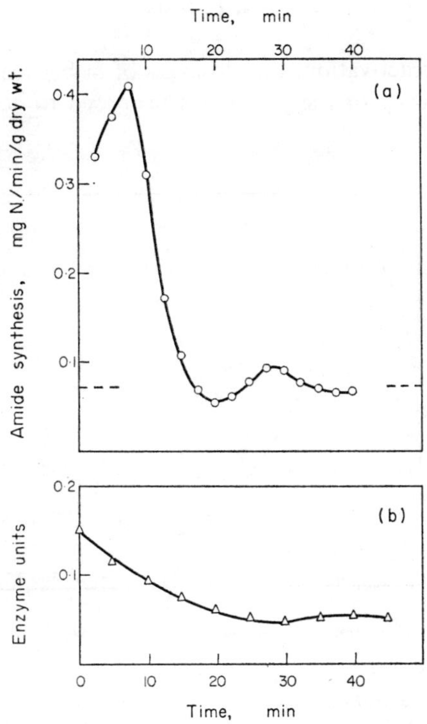

FIG. 3. Changes in the *in vivo* rate of amide synthesis and in the level of glutamine synthetase after the addition of ammonia to yeast (*Candida utilis*) growing on glutamate. (a) Changes in the *in vivo* rate of glutamine-amide synthesis. The steady-state rate on ammonia is indicated by the dashed line. (b) Changes in the level of enzyme. (Data from Sims and Ferguson, 1974.)

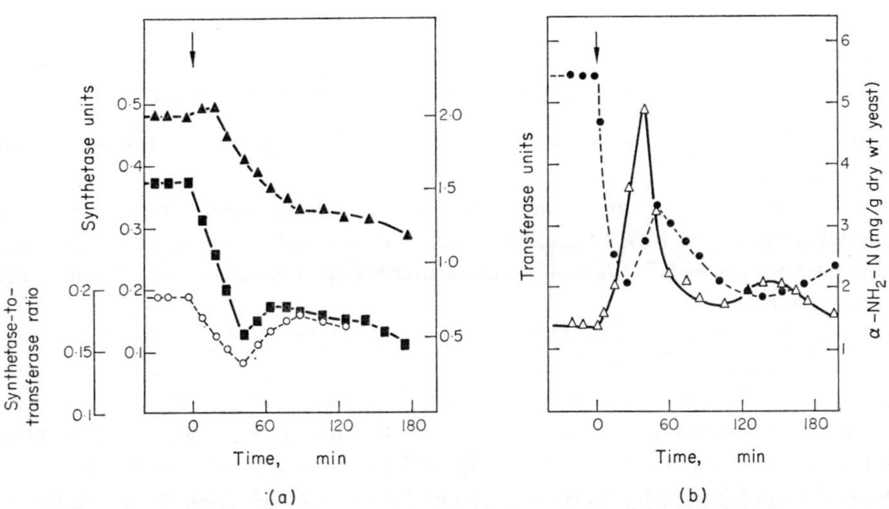

FIG. 4. Changes in (a) specific activity of glutamine synthetase and (b) pools of glutamate and glutamine in *Candida utilis* after a change in nitrogen source from glutamate to ammonia. ■, Synthetase activity; ▲, transferase activity; ●, synthetase transferase ratio; ○, glutamate; △, glutamine. Arrows denote line of transfer from glutamate to ammonia. Data from Sims *et al.* (1974b).

and NH_3) followed by reactivation, the closeness of fit between predicted and observed rates revealed the adequacy of the postulates proposed to account for the changes in amide synthesis (Fig. 5).

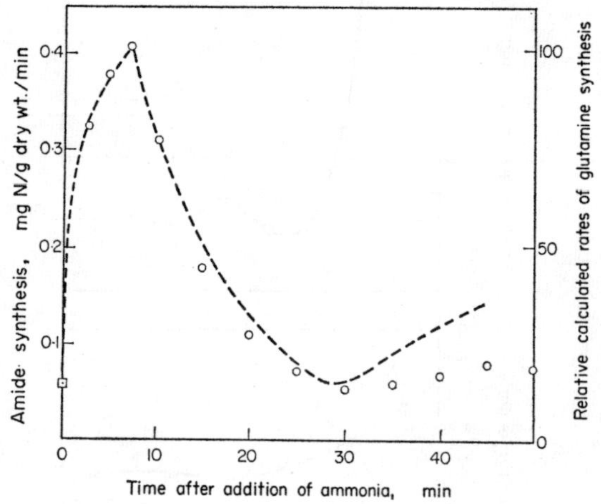

FIG. 5. A comparison of the *in vivo* rates of amide synthesis in *Candida utilis* with values calculated from measurements of enzyme concentrations after allowing for changes in substrate availability. The value calculated from the enzyme data at 8 min was matched to the *in vivo* rate determined at this time and all the other values are shown relative to this point. – – –, Steady-state rate of glutamine synthesis in yeast growing in NH_3. (Data reproduced by permission from Sims and Ferguson (1974) *J. gen. Microbiol.* **80,** 143–158.)

SUBUNIT STRUCTURE OF YEAST GLUTAMINE SYNTHETASE

The possible influence of the subunit structure of the enzyme on the mechanism of enzyme modulation was examined as described in the legend of Table 2. The yeast enzyme was shown to be an octameric globular protein (sediment coefficient $s = 15.4$, frictional ratio $f/f_0 = 1.29$, molecular weight 390,000 daltons) comprising two weakly bound tetramers ($s = 8.7, f/f_0 = 1.35$, molecular weight 180,500 daltons) having reduced synthetase activity.

Analyses of protein extracts made during a glutamate to NH_3 transition experiment revealed that loss of synthetase activity was associated with a reduction in the amount of the octamer and with formation of appreciable amounts of enzyme tetramer. Studies with isolated tetramers revealed that they could reversibly re-associate *in vitro* to produce enzyme with normal biosynthetic activity and this suggested a possible basis for the modulation of enzyme activity. In contrast, however, an earlier study involving inhibitors of protein synthesis (Ferguson and Sims, 1974a) had suggested that enzyme deactivation was probably an irreversible process and that any restoration of activity required the *de novo* synthesis of enzyme. Clearly it was of great importance to ascertain whether freely-reversible enzyme interconversion could occur within the milieu of the cell.

TABLE 2. MOLECULAR PARAMETERS OF YEAST GLUTAMINE SYNTHETASE

Source and treatment of enzyme	Sedimentation coefficient (× 10¹³ s)	Stokes's radius (nm)	Apparent molecular weight	Frictional ratio	Synthetase to transferase activity ratio	pH optimum of enzyme
Crude extracts*	15·4	6·23	394000	1·29	0·17	6·5
Purified enzyme	15·4	6·23	394000	1·29	0·17	6·5
Urea dissociation	14·2	6·23	363000§	1·32	0·17	6·5
of purified enzyme†	8·7	5·04	180400	1·35	0·15	6·5
	5·8†	—	90700	—	—	—
	3·8†	—	45500	—	—	—
'Deactivated'	14·2	6·23	363000	1·32	0·17	6·5
enzyme	8·7	5·04	180400	1·35	0·08	5·8
preparations‡	5·8	—	90700	—	—	—
	3·8	—	45500	—	—	—

The figures are average values taken from at least three separate experiments.

* Crude enzyme extracts were obtained from yeast grown on aspartate, glutamate, alanine, ammonia and glutamine (all 10 mM) as sole sources of nitrogen. In all instances there were traces of activity (about 5 %) in a position corresponding to a 9 *s* component.

† The enzyme was pre-dissociated with 0·05 M-tris-urea (pH 9·6). Similar results were obtained when 0·03 M-sodium dodecyl sulphate was used.

‡ Deactivated enzyme was obtained by transferring glutamate grown yeast to glutamine (10 mM) for 1 h. Identical results were obtained when the enzyme was deactivated by transferring a culture to NH_4^+ (5 mM) or to a medium without a source of carbon.

§ The observed decrease in apparent molecular weight could be accounted for by an increase in $\bar{V}$ of the protein from 0·721 to 0·743 cm³/g.

Data from Sims *et al.* (1974a).

REVERSIBLE REASSOCIATION OF ENZYME TETRAMERS WITH DIMINISHED CATALYTIC ACTIVITY IN THE MODULATION OF GLUTAMINE SYNTHETASE ACTIVITY

In view of the uncertainties inherent in the interpretation of results from experiments with metabolic inhibitors, the problem of enzyme interconversion was tackled in another way. The yeast was pulse-labelled with a radioactive amino acid and changes assessed in the ^{14}C content of the enzyme during its deactivation and reactivation (for details see legend of Fig. 6).

Some important conclusions can be drawn from this experiment:

(i) a net movement of ^{14}C isotope back into active enzyme accompanies the *in vivo* reactivation of the enzyme,

(ii) the amount of isotope returning to the octamer corresponds almost exactly to that predicted if tetramers present in the cell were to fully re-associate,

(iii) only about 50 per cent of the ^{14}C disintegrations originally present in active enzyme returned after reactivation; appreciable irreversible destruction must also accompany deactivation of the enzyme

(iv) a reduction in the ^{14}C specific activity of the enzyme occurs during enzyme deactivation. Since this reduction is appreciably greater than can be accounted for by continued enzyme synthesis it seems probable that it arose as the consequence of rapid equilibration between unequally-labelled octamers and tetramers during deactivation.

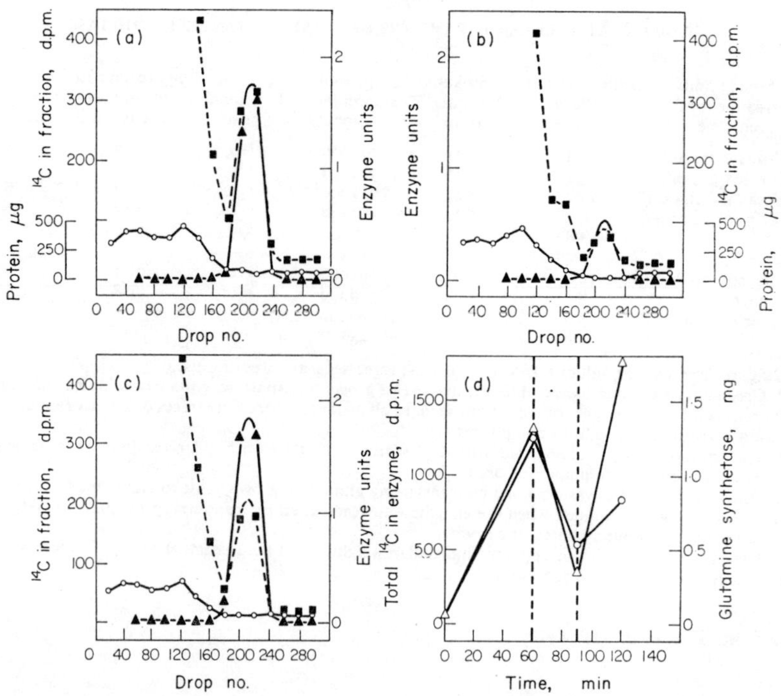

FIG. 6. Changes in the ^{14}C content of glutamine synthetase in *Candida utilis* during enzyme deactivation and reactivation. Yeast was pulse-labelled with [U-^{14}C]isoleucine and the enzyme deactivated by transferring the culture to an unlabelled medium containing glutamine. The culture was subsequently transferred to glutamate to reactivate the glutamine synthetase. Enzyme was isolated after the different treatments and fractionated so that changes in its ^{14}C content could be measured. (a) Sucrose density centrifugation of partly purified enzyme octamer (15.4S) isolated from yeast after 60 min exposure to isotope. Fractions were assayed for transferase activity (▲), ^{14}C content (■) and protein (○). (b) Changes in transferase activity and ^{14}C content of the octamer after 30 min deactivation; symbols as in (a). (c) Return of ^{14}C into the enzyme following reactivation of the enzyme; symbols as in (a). (d) Kinetics of changes in enzyme and its ^{14}C content after 30 min deactivation and reactivation. △, Total enzyme; ○, total ^{14}C. An enzyme unit is defined throughout as μmol hydroxamate formed/min/mg protein (30°C). The dashed lines indicate the times of deactivation and reactivation. Data from Sims *et al.* (1974b).

These observations confirmed that reversible interconversion could occur between the two enzyme forms during changed physiological conditions and a model incorporating the observations is shown in Fig. 7.

DIRECT PARTICIPATION OF METABOLITES IN ENZYME INTERCONVERSION

Recent studies have established that the fine control of glutamine synthetase involves a mechanism in which enzyme substrates and metabolites not necessarily related to the biosynthetic pathway determine the extent to which the active enzyme (15.4s component) is converted to its 'resting' form (8.4s) (Sims *et al.*, 1974b). Results from a series of *in vitro* experiments in which active enzyme octamers and 'resting' tetramers were exposed to a number of cell metabolites are shown in Table 3.

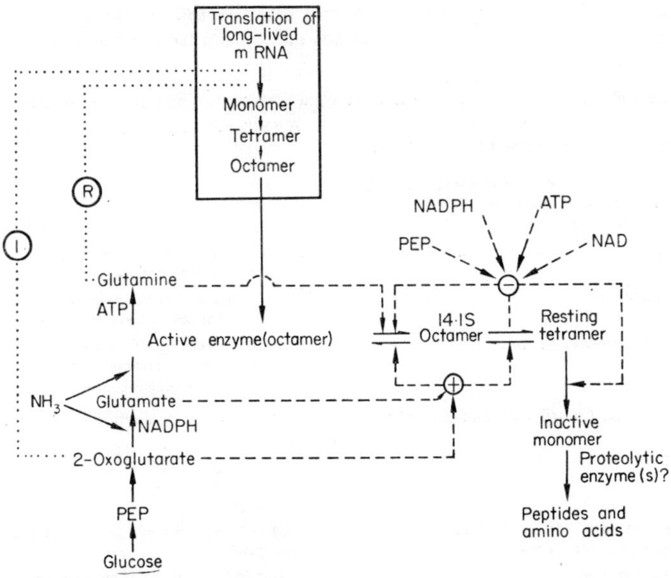

FIG. 7. A possible system of controls concerned with the regulation of glutamine biosynthesis in *Candida utilis*. This scheme interprets the enzyme-labelling data and summarizes present knowledge of the action of a number of regulatory ligands on the activity and synthesis of glutamine synthetase. It is suggested that the enzyme is assembled from component monomers into an octameric protein consisting of two loosely-bound half molecules and is subsequently transferred to a site where it functions. The biosynthetic activity of the enzyme at this second site is dependent upon the concentration of effectors in the cytosol and under normal conditions the enzyme is partly dissociated to provide a reserve of 'resting' enzyme. Enzyme inactivation involves the irreversible dissociation of these 'tight' enzyme tetramers. ···Ⓘ···, Co-inducer of enzyme synthesis; ···Ⓡ··· co-repressor of enzyme synthesis;---⊕---, regulatory ligands that can affect the re-association of 'resting' tetramers into active enzyme; ---⊖---, regulatory ligands that promote the dissociation of the active enzyme (after Sims *et al.*, 1974b).

Low concentrations of NAD, NADPH and ATP caused significant dissociation of the active enzyme as indicated by the formation of enzyme monomers. Glutamine promoted the conversion of active enzyme into a 'relaxed' but active form (14.1s) (see Sims *et al.*, 1974a). 2-Phosphoenol pyruvate, in addition to facilitating the formation of relaxed enzyme promoted the formation of enzyme tetramers. A physiological role for glutamate and Mg^{2+} in enzyme modulation was indicated since they prevented the dissociation of the enzyme into inactive octamers even in the presence of 1 M urea; they also promoted the reassociation of enzyme tetramers. 2-Oxoglutarate protected the enzyme from dissociation but could not promote the reassociation of 'resting' tetramers.

Mechanisms of control of yeast glutamine synthetase

It is now possible to recognize three independent mechanisms concerned with the regulation of glutamine biosynthesis in food yeast:

(i) a sensitive mechanism for controlling the *de novo* rate of enzyme synthesis,
(ii) fine control of enzyme activity by the interconversion of active enzyme into tetramers with diminished catalytic activity, and

TABLE 3. EFFECTS OF METABOLIC INTERMEDIATES ON THE DISSOCIATION AND RE-ASSOCIATION OF GLUTAMINE SYNTHETASE IN THE FOOD YEAST, *Candida utilis*. Data from Sims *et al.* (1974b).

I. Centrifugation of purified enzyme in sucrose gradients containing various possible effectors	
1 mM-Glucose 1-phosphate, 1 mM-glucose 6-phosphate, 1 mM-fructose 1,6-diphosphate, 5 mM-pyruvate, 2 mM ADP, 2 mM-AMP, 50 mM-D-glucosamine, 2 mM-3′5′-cyclic AMP, 1 mM-acetyl CoA, 2 mM-citrate, or 2 mM-DL-isocitrate	A 100 % recovery of enzyme as 15·4 *s* component
2 mM-NAD, 2 mM-NADPH or 2 mM-ATP	About 50–60 % of enzyme recovered as 15·4 *s* component. Enzyme monomers were also detected
50 mM-Glutamine	An 85 % recovery of activity as the 14·1 *s* component. Monomers were also detected
2 mM-2-Phosphoenolpyruvate	Only 40 % recovery of enzyme as 14·1 *s* component. Enzyme tetramers (8·4 *s*) and monomers detected in appreciable amounts
II. Centrifugation of purified enzyme in sucrose gradients containing 1·0 M urea plus various effectors	
Urea alone	About 50 % recovery of activity; 14·1 *s* and 8·4 *s* components recovered in approximately equal amounts
100 mM-Glutamate + 1 mM-Mg^{2+} or 6 mM-oxoglutarate + 1 mM-Mg^{2+}	A 100 % recovery of activity as a single 15·4 *s* component
6 mM-Oxoglutarate	About 60 % recovery of enzyme activity peaking at about 11 *s*. This indicated only slight protection against dissociation was afforded by this compound in the absence of Mg^{2+}
III. Centrifugation of an extract containing enzyme tetramers in sucrose gradients containing various effectors	
Control (no addition)	Approximately equal amounts of 14·1 and 8·4 *s* present
1 mM-Mg^{2+}	About 200 % increase in the 15·4 *s* found. A slight increase in the recovery of the 8·4 *s* component also evident.
100 mM-Glutamate + 1 mM-Mg^{2+}	A 400 % increase in 15·4 *s* component observed together with a decrease in the 8·4 *s* component
6 mM-Oxoglutarate + 1 mM-Mg^{2+}	An increase in the amount of 8·4 *s* component was observed. The amount of octamer was less than in the control experiment, indicating that oxoglutarate did not enhance re-association of the enzyme

(iii) enzyme destruction achieved by the irreversible dissociation of tetramers into inactive subunits.

'Coarse' control has been shown to involve a mechanism of enzyme synthesis sensitive to changes in the amount of the end product glutamine (Ferguson and Sims, 1974b, Sims *et al.*, 1974b). Isotope incorporation studies are consistent with control of enzyme synthesis being effected at the level of enzyme translation. No delay of *de novo* enzyme synthesis was observed when yeast was transferred from glutamine to glutamate; had mRNA synthesis to be increased nearly forty-fold to accommodate the observed increase in rate, a significant delay in the incorporation of isotope into protein would have been expected. Again, after 'step down' from glutamate to glutamine the *de novo* rate of enzyme synthesis was decreased immediately to a lower linear rate. There was no suggestion that the rate decayed exponentially as would be expected if the mRNA synthesized had a short but finite half-life.

'Fine' control of glutamine synthetase appears to be based on a mechanism in which interplay between enzyme substrates and other metabolites not necessarily related to the pathway determines the extent to which the enzyme is converted into a 'resting' form. This mechanism can increase or decrease the activity of the biosynthetic pathway and its operation provides an explanation for the oscillations in enzyme level observed when the nitrogen supply to the organisms is withheld or suddenly increased (see Fig. 4). There is good evidence to suggest that this mechanism can operate under widely differing conditions of growth. Analysis of crude enzyme extracts prepared from exponentially dividing cultures of yeast grown on a number of nitrogen sources revealed the presence of significant amounts of enzyme tetramers (Sims *et al.*, 1974a).

Sufficient clarification of the identity of certain regulatory ligands and of the nature of their interaction with the enzyme has emerged to propose a plausible mechanism for control of enzyme activity (Fig. 7). Good agreement exists between *in vivo* and *in vitro* observations to suggest that glutamate can actively promote enzyme reassociation. There is also evidence to indicate that 2-oxoglutarate can stabilize the active form of the enzyme (there is now some evidence that it plays a similar role in the living cell and is probably a co-inducer of enzyme synthesis (Box and Sims, unpublished)), and this could account for the earlier observation that continued catabolism of carbon intermediates was necessary to prevent enzyme deactivation (Ferguson and Sims, 1974b).

Negative effectors are also implicated in the system of control. Thus 2-phosphoenol pyruvate is very effective in enhancing the dissociation of the enzyme, whilst structurally dissimilar compounds such as NAD, NADPH and ATP are also active. The observation that if either 2-oxoglutarate or glutamate is present the enzyme can maintain its active configuration in conditions that normally lead to its complete dissociation, suggests a possible basis for enzyme modulation. The cumulative effect of negative effectors would ensure that the enzyme is normally dissociated but for the presence of 2-oxoglutarate and glutamate. Failure adequately to maintain the synthesis of these two compounds would automatically produce enzyme deactivation and not require an elevation in the level of negative effectors.

Finally a comment concerning the destruction of the enzyme. Isotope and inhibitor studies have revealed that when conditions of culture of the yeast are so changed that a large adjustment in enzyme level becomes necessary then an irreversible dissociation of 'resting' tetramers is observed. The observation that enzyme monomers do accumulate within the cell and yet cannot be used to produce active enzyme suggests that conditions within the cytosol are not suitable for the assembly of enzyme monomers: presumably instead conditions favour the proteolytic attack of enzyme monomers.

Regulation of glutamine biosynthesis in other fungi

Much remains to be done before the systems of control operating in other fungi are fully understood. Certainly it would be very misleading to generalize from the results of a study on a single fungus especially since it is known that the mechanisms controlling glutamine biosynthesis vary within the bacteria (for *Escherichia coli* see Ginsburg and Stadtman, 1973; for *Bacillus subtilis* see Deuel and Prusiner, 1974).

However, it is evident that certain features of the food yeast system are observed in a number of other fungi. Thus rapid deactivation of synthetase has already been observed in several budding yeasts (Ferguson and Sims, 1971) and also in *Neurospora crassa* and *Aspergillus nidulans* (unpublished data of J. Toone and V. Box). Interestingly, preliminary studies on glutamine synthetase from *N. crassa* (Kapoor *et al.*, 1969) indicate that the enzyme has a slightly lower molecular weight (350,000 daltons) than the yeast enzyme but that it too is probably comprised of two weakly bound half-mers. It is clear, however that control of glutaminc biosynthesis in the fission yeast *Schizosaccharomyces pombe* does not appear to conform to the same pattern: thus, in common with a number of other nitrogen-assimilatory enzymes, derepression of the synthesis of glutamine synthetase does not accompany the reduction in the glutamine pool (Hemmings, Box and Sims, unpublished observations). It may be relevant that an alternative route of glutamate synthesis involving the glutamine pathway has been suggested for this yeast (Brown *et al.*, 1973).

Acknowledgements: I am grateful to the Science Research Council for financial support. I would like to thank Professor B. Folkes, Dr. A. R. Ferguson and B. Hemmings for stimulating discussion and collaboration in the work described: without Veronica Box and Jennifer Toone and their skilled technical assistance few of the ideas would have been realized. D. Walls has given excellent technical help and has prepared the figures.

REFERENCES

ALBERTY, R. A. (1968) Effect of pH and metal ion concentration on the equilibrium hydrolysis of adenosine triphosphate to adenosine diphosphate. *J. biol. Chem.* **243,** 1337–1343.

BROWN, C. M., BURN, V. J. and JOHNSON, B. (1973) Presence of glutamate synthase in fission yeasts and its possible role in ammonia assimilation. *Nature New Biol.* **246,** 115–116.

BROWN, C. M., MACDONALD-BROWN, D. S. and MEERS, J. L. (1974) Physiological aspects of microbial inorganic nitrogen metabolism. *Adv. microb. Physiol.* **11,** 1–52.

DEUEL, T. F. and PRUSINER, S. (1974) Regulation of glutamine synthetase from *Bacillus subtilis* by divalent cations, feedback inhibitors and L-glutamine. *J. biol. Chem.* **249,** 265–274.

FERGUSON, A. R. and SIMS, A. P. (1971) Inactivation *in vivo* of glutamine synthetase and NAD-specific glutamate dehydrogenase: its role in the regulation of glutamine synthesis in yeasts. *J. gen. Microbiol.* **69,** 423–427.

FERGUSON, A. R. and SIMS, A. P. (1974a) The regulation of glutamine metabolism in *Candida utilis*: the inactivation of glutamine synthetase. *J. gen. Microbiol.* **80,** 173–185.

FERGUSON, A. R. and SIMS, A. P. (1974b) The regulation of glutamine metabolism in *Candida utilis*: the role of glutamine in the control of glutamine synthetase. *J. gen. Microbiol.* **80,** 159–171.

FOLKES, B. F. and SIMS, A. P. (1974) The significance of amino acid inhibition of NADP-linked glutamate dehydrogenase in the physiological control of glutamate synthesis in *Candida utilis. J. gen. Microbiol.* **82,** 77–95.

GANCEDO, C. and HOLZER, H. (1968) Enzyme inactivation of glutamine synthetase in *Enterobacteriaceae. Eur. J. Biochem.* **40,** 190–192.

GINSBURG, A. and STADTMAN, E. R. (1973) Regulation of glutamine synthetase in *Escherichia coli.* In *The Enzymes of Glutamine Metabolism.* ed. PRUSINER, S. and STADTMAN, E. R. pp. 9–43. Academic Press, New York and London.

HUBBARD, J. S. and STADTMAN, E. R. (1967) Regulation of glutamine synthetase. II. Patterns of feedback inhibition in micro-organisms. *J. Bact.* **93,** 1045–1055.

KAPOOR, M. and BRAY, D. F. (1968) Feedback inhibition of glutamine synthetase of *Neurospora crassa* by nicotinamide-adenine dinucleotide. *Biochemistry* **7,** 3583–3589.

KAPOOR, M., BRAY, D. F. and WARD, G. W. (1968) Anthranilic acid as a feedback inhibitor of glutamine synthetase of *Neurospora crassa. Archs. Biochem. Biophys.* **128,** 810–814.

KAPOOR, M., BRAY, D. F. and WARD, G. W. (1969) Glutamine synthetase of *Neurospora crassa*. Inactivation by urea and protection by some substrates and allosteric effectors. *Archs. Biochem. Biophys.* **134,** 423–433.

MECKE, D., WULFF, K., LIESS, K. and HOLZER, H. (1966) Characterization of a glutamine synthetase inactivating enzyme from *Escherichia coli. Biochem. biophys. Res. Commun.* **20,** 452–458.

MEISTER, A. (1974) Glutamine synthetase of mammals. In *The Enzymes*, ed. BOYER, P. D. 3rd ed. Vol. 10. pp. 699–754. Academic Press, New York.

PRUSINER, S. and STADTMAN, E. R. (1973) *The Enzymes of Glutamine Metabolism*. Academic Press, New York and London.

SHAPIRO, B. M. and STADTMAN, E. R. (1970) The regulation of glutamine synthesis in microorganisms. *A. Rev. Microbiol.* **24,** 501–524.

SHAPIRO, B. M., KINGDON, H. S. and STADTMAN, E. R. (1967) Regulation of glutamine synthetase. VII. Adenyl glutamine synthetase: a new form of the enzyme with altered regulatory and kinetic properties. *Proc. natn. Acad. Sci.* **58,** 642–649.

SIMS, A. P. and FERGUSON, A. R. (1974) The regulation of glutamine metabolism in *Candida utilis*: studies with $^{15}NH_3$ to measure *in vivo* rates of glutamine synthesis. *J. gen. Microbiol.* **80,** 143–158.

SIMS, A. P. and FOLKES, B. F. (1964) A kinetic study of the assimilation of ^{15}N-ammonia and the synthesis of amino acids in an exponentially growing culture of *Candida utilis. Proc. R. Soc.* B **159,** 479–502.

SIMS, A. P., FOLKES, B. F and BUSSEY, A. H. (1968) Mechanisms involved in the regulation of nitrogen assimilation in micro-organisms and plants. In *Recent Aspects of Nitrogen Metabolism in Plants*, First Long Ashton Symposium, 1967, ed. HEWITT, E. J. and CUTTING, C. V. pp. 91–114. Academic Press, London and New York.

SIMS, A. P., TOONE, J. and BOX, V. (1974a) The regulation of glutamine synthesis in the food yeast *Candida utilis*: the purification and subunit structure of glutamine synthetase and aspects of enzyme deactivation. *J. gen. Microbiol.* **80,** 485–499.

SIMS, A. P., TOONE, J. and BOX, V. (1974b) The regulation of glutamine metabolism in *Candida utilis*: mechanisms of control of glutamine synthetase. *J. gen. Microbiol.* **84,** 149–162.

STADTMAN, E. H. and GINSBURG, A. (1974) The glutamine synthetase of *Escherichia coli*: structure and control. In *The Enzymes*, ed. BOYER, P. D. 3rd ed., Vol. 10. pp. 754–807. Academic Press, New York.

STADTMAN, E. R., GINSBURG, A., ANDERSON, W. B., SEGAL, A., BROWN, M. S. and CIARDI, J. E. (1972) *The Molecular Basis of Biological Activity*, ed. GAEDE, K., HORECKER, B. L. and WHELAN, W. J. pp. 127–180. Academic Press, New York.

WOOLFOLK, C. A. and STADTMAN, E. R. (1964) Cumulative feedback inhibition in the multiple end product regulation of glutamine synthetase activity in *Escherichia coli. Biochem. biophys. Res. Commun.* **17,** 313–319.

WOOLFOLK, C. A., SHAPIRO, B. and STADTMAN, E. R. (1966) Regulation of glutamine synthetase. I. Purification and properties of glutamine synthetase from *Escherichia coli. Archs. Biochem. Biophys.* **116,** 177–192.

WULFF, K., MECKE, D. and HOLZER, H. (1967) Mechanism of the enzymatic inactivation of glutamine synthetase from E-coli. *Biochem. biophys. Res. Commun.* **28,** 740–745.

AMINO ACIDS: OCCURRENCE, BIOSYNTHESIS AND ANALOGUE BEHAVIOUR IN PLANTS

L. FOWDEN

Rothamsted Experimental Station, Harpenden, Herts, U.K.

INTRODUCTION

Experimental biology has been strongly influenced during the past 25 years by the development of many new analytical techniques, including the various forms of chromatography, mass spectrometry and nuclear magnetic resonance spectroscopy. These have permitted more precise and sensitive detection, identification and assay of natural products from both plant and animal materials. Identifications of new compounds from plants have been more prolific than from animals, possibly because a far larger number of plant species have been examined. Every important group of natural products has increased in numerical and structural complexity during this period. The amino acids form an excellent example of this situation. In 1950, before the widespread adoption of chromatography, the twenty normal constituents of protein had been recognized but less than ten other natural amino acids were known. By April 1974, E. A. Bell (personal communication) had counted 374 naturally-occurring amino acids: over 200 of these had their origin in higher plants.

STRUCTURAL RELATIONSHIPS OF THE NEWER AMINO ACIDS

This much enlarged group can still be divided in ways frequently adopted for the protein amino acids; large numbers of dicarboxylic, basic, sulphur-containing, and aromatic amino acids are now recognized. The aromatic group includes compounds containing a variety of heterocyclic ring systems, unknown in the protein amino acids, and usually attached as β-substituents to an alanine moiety. Similarly proline, the original imino acid, has become the parent compound of a group having more than forty members; the majority possess the same 5-atom pyrrolidine ring, but imino acids based on azetidine and piperidine rings also occur in plants. It also has been convenient to create new structural groups including one of unsaturated aliphatic amino acids (where the carbon skeleton contains ethylenic or acetylenic linkages) and another based on compounds having a cyclopropane ring as a characteristic feature. A more detailed account of the chemistry of the plant amino acids appeared in a recent review (Fowden, 1970).

In addition to these general relationships, more specific structural links exist between many of the newer amino acids and their protein counterparts: these may take the form of (i) homology, (ii) simple substitution of protons by larger groups, and (iii) isosterism.

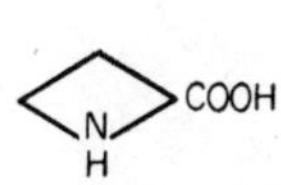

(I)

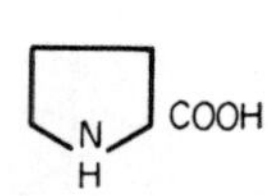

(II)

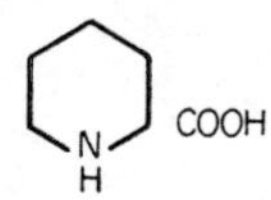

(III)

Homology is commonly encountered, and is exhibited even among the protein amino acids, e.g. aspartic and glutamic acids. The series has been extended by the isolation of α-aminoadipic and α-aminopimelic acids from plants. Similarly, four α,ω-diamino acids having linear C skeletons with from three to six atoms are known. Another interesting homologous group consists of the imino acids, azetidine-2-carboxylic acid (I), proline (II), and pipecolic acid (III).

Substitution is a pragmatic concept, useful for chemical classification but of lesser importance in indicating biosynthetic relationships. A heterogeneous collection of compounds may be regarded as derived from the protein amino acids by substitution at the aliphatic or aromatic carbon skeleton. For example glutamic acid features as the parent compound of a large group of dicarboxylic amino acids that include its γ-hydroxy, γ-methyl, γ-methylene, γ-ethyl, γ-ethylidene, and γ-hydroxy-γ-methyl derivatives. Ring substitution in phenylalanine or tyrosine by methyl, hydroxymethyl, carboxyl, amino, aminomethyl or additional hydroxyl groups generates a further large group of plant amino acids.

Compounds are isosteric when their molecules have (almost) the same size and spatial configuration. Adjacent members of a homologous series may behave as isosteres, e.g. in some metabolic systems little discrimination may occur between two homologues; similarly, simple substitution may cause little change in a molecule's overall shape. Several interesting examples of isosterism, involving replacement in a protein amino acid of one atom by another (or by a group of atoms), occur among plant constituents. Replacement of C by N, O or S produces only small changes of molecular configuration,

$H_2NOCCH_2CH_2CH(NH_2)COOH$ (IV)

$H_2NOCNHCH_2CH(NH_2)COOH$ (V)

(VI)

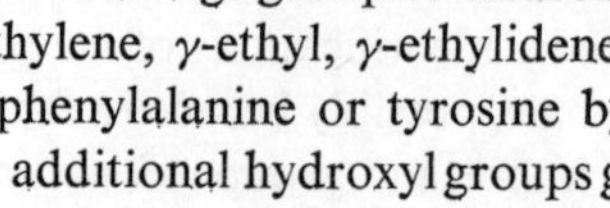

(VIII)

$H_2N(CH_2)_2CH_2CH_2CH(NH_2)COOH$

(VII)

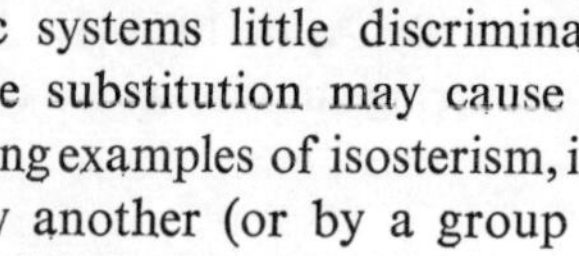

(IX)

$H_2N(CH_2)_2SCH_2CH(NH_2)COOH$

and so glutamine (IV) and albizziine (V), arginine (VI) and canavanine (VII), and lysine (VIII) and *S*-aminoethylcysteine (IX) represent pairs of isosteres. Canavanine occurs widely in the Papilionoideae, albizziine is restricted to fewer legume genera but is important in many *Albizzia* and *Acacia* spp. (see Fowden, 1970), whilst *S*-aminoethylcysteine is known only in the fungus, *Rozites caperata* (Warin *et al.*, 1969).

A different type of replacement, S by Se, is encountered in amino acids present in certain selenium-accumulator plants, typified by selected *Astragalus* or *Oxytropis* spp. Whilst selenocysteine and selenomethionine probably represent minor constituents of all such accumulator species, *Se*-methylselenocysteine and its γ-glutamylpeptide form the major Se-containing amino compounds.

Many of the newer compounds, related structurally to the protein amino acids by homology, substitution or isosterism, show affinity for enzyme systems involved in the basic processes of amino acid and protein biosynthesis; they frequently act as competitive substrates or as end-product regulators, and so behave as metabolic antagonists. Such compounds are commonly called analogues of the normal metabolites, and they often produce growth-inhibitory or toxic effects in organisms other than the plants synthesizing them (see later; also Fowden *et al.*, 1967).

Occurrence and biosynthesis

The identification of a new compound in a plant frequently follows from a survey of a group of related species: although the screening technique now invariably involves a sensitive chromatographic or electrophoretic procedure, compounds can only be detected in crude plant tissue extracts if their concentration exceeds a critical threshold value. Substances present in lower amount will not be detected unless some fractionation of the extract is achieved. Statements about the distribution of amino acids then must have an element of uncertainty, and many may occur more widely in the plant kingdom than we suspect at present. This possibility may be illustrated by the example of azetidine-2-carboxylic acid. The imino acid was isolated initially in 1955 from *Convallaria majalis* and subsequently shown to be an important constituent of many members of the Liliaceae. Much later it was recognized in significant amounts in a few legumes, i.e. in members of a family quite unrelated to the Liliaceae. Now there is a further record of its occurrence in extracts of sugar beet, but in concentrations so minute that it was identified only after fractionation on an industrial scale. Commercial batch cation-exchange processing of the waste nitrogenous fraction resulting during refining of 10^5 tonnes of sugar beet gave about 30 kg azetidine-2-carboxylic acid, i.e. a yield of 0.3 mg/kg beet (Fowden, 1972). The imino acid may occur in similar concentration in many other species, but its presence would pass unnoticed because routine screening methods are too insensitive to detect the compound in such low amounts.

Indirect support for this idea comes from recent work with beech-nuts (*Fagus sylvatica*) and tobacco (*Nicotiana tabacum*). Larsen (1974) described the isolation of *N*-(3-amino-3-carboxypropyl)azetidine-2-carboxylic acid (X) and *N*-(*N*-(3-amino-3-carboxypropyl)-3-amino-3-carboxypropyl)azetidine-2-carboxylic acid (XI) from *Fagus sylvatica* fruits. Although he could not detect any free azetidine-2-carboxylic acid, he showed that compounds (X) and (XI) were formed readily when a solution of imino acid was heated.

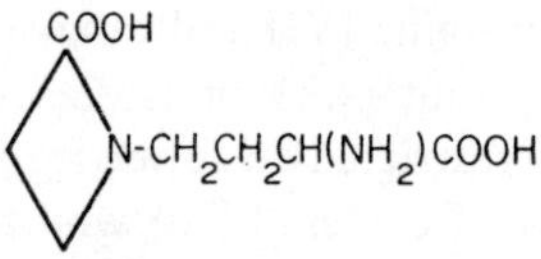

(X)

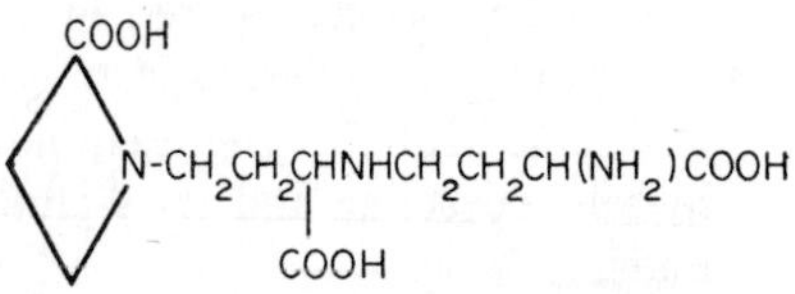

(XI)

Slightly alkaline conditions favoured the polymerizations, although they proceeded in neutral solution. These observations raise the possibility that the *in vivo* syntheses of (X) and (XI) occur by a similar process, which would be dependent upon the presence of free azetidine-2-carboxylic acid in *Fagus sylvatica*. Larsen established that nicotianamine, previously isolated from tobacco by Japanese workers, is identical to (XI).

Conclusions about the distribution of other amino acids are subject to similar uncertainty. Some compounds, such as γ-aminobutyric acid, seem to be present in all plants, whilst others are known only in the single species from which their isolation was reported, e.g. *S*-aminoethylcysteine. However, a majority of the newer amino acids appear to be characteristic of small groups of related species, and are sometimes confined to particular tribes of a family, or sections of a genus. The first type of distribution pattern is exhibited by β-pyrazol-1-ylalanine in the Cucurbitaceae (Dunnill and Fowden, 1965), whilst the ability to synthesize lathyrine resides within sub-generic groups of *Lathyrus* (Bell, 1962).

Our present knowledge of processes involved in biosynthesis of the large group of non-protein amino acids is very inadequate. Conceivably many parallels exist with reactions of intermediary nitrogen metabolism leading to the protein amino acids. Transamination, which represents the last step in the synthesis of almost half the protein amino acids, certainly is implicated in the metabolism of many of the newer amino acids. The pathways of assimilation of sulphur and selenium into amino acids also show analogies, whilst evidence from animal systems suggests that the same enzymes are responsible for the synthesis of both proline and pipecolic acid from their respective C_5 and C_6 precursors. Perhaps non-specific enzymic catalysis may be implicated more widely in non-protein amino acid biosynthesis, especially in situations where compounds occur in very low concentration. Thus the production of azetidine-2-carboxylic acid by sugar beet could result from the aberrant action of proline biosynthetic enzymes; alternatively the extremely low concentration of the imino acid in beet could result from almost complete repression or inhibition of a specifically-catalysed C_4 pathway.

Although many of the aromatic and heterocyclic amino acids are β-substituted alanines, several distinct pathways are involved in biosynthesis. Whilst the normal shikimate pathway leads to phenylalanine, tyrosine and tryptophan, several *m*-carboxy derivatives are formed via a modified shikimate pathway in which the carboxy group of the original shikimate molecule is retained. 2,4-Dihydroxy-6-methylphenylalanine (orcylalanine) arises in a completely different way: the substituted phenyl ring follows from acetate (malonyl-CoA) condensation, whilst serine (or perhaps *O*-acetylserine) provides the C_3 alanine moiety. Recently, Murakoshi and his collaborators in Japan have established the more general importance of *O*-acetylserine as the donor of the C_3 side chain for several other β-substituted alanines. The activated serine derivative condenses with a variety of heterocyclic ring compounds, catalysed by an apparently non-specific enzyme resembling tryptophan synthetase (see also Smith and Chang, 1973).

Analogue function

Amino acids that do not form constituents of protein are frequently classed as a further group of plant secondary products. This terminology generally implies that the compounds do not participate in metabolic processes vital to normal cell growth, i.e. that the role of a compound in the plant producing it is obscure. Whilst this is true for most non-protein amino acids, a few nevertheless are strikingly active as growth inhibitors and/or toxic compounds. In some cases, the amino acid was isolated after a protracted search for the toxic factor causing an endemic dietary disease of man or animals. β-(Methylenecyclopropyl)alanine (hypoglycin A) produces the hypoglycaemic condition characterizing the 'vomiting sickness' of Jamaica following ingestion of unripe akee (*Blighia sapida*) fruit. A similar approach established β-oxalyl-α,β-diaminopropionic acid as the principal toxic factor in seeds of several *Lathyrus* species causing 'lathyrism' in man in India, whilst β-cyanoalanine was confirmed as the neurolathyrogen present in *Vicia sativa* seed. Other amino acids identified as toxins of food grains or forages include mimosine (*Leucaena* spp.), indospicine (an isostere of arginine from *Indigofera spicata*), and *Se*-methylselenocysteine (*Astragalus* spp.). Each compound disrupts some facet of normal metabolism but the nature of the antagonism is often complex, and sometimes indirect (see the review of hypoglycin toxicity by Bressler *et al.* (1969). Direct analogue effects can be exerted on the processes of amino acid and protein biosynthesis, or upon systems responsible for uptake of amino acids into cells. Often the analogue acts in direct competition with the normal metabolite by binding at the active sites of permeases or biosynthetic enzymes: in many cases analogues behave as alternative substrates, but some exert only inhibitory activity. There is also considerable evidence indicating that analogue molecules may mimic the action of normal end-products of reaction pathways: they may regulate rates of synthesis either by inhibiting key enzymes catalysing early steps in the pathway or by repressing the production of new molecules of these enzymes. Examples of these types of analogue behaviour follow.

Analogue competition with permease enzymes

The uptake of amino acids into bacterial cells is catalysed by a series of permeases showing high degrees of substrate specificity. Active uptake into cells of higher plant tissues probably involves similar but less specific permeases. Studies with bacterial systems have provided numerous instances of analogue competition at the level of permease action; the analogue may be accumulated within the bacterial cell at the expense of the normal metabolite, and growth inhibition may follow in heterotrophs from such a limitation of an essential cell component. In some instances the analogue also may behave as an intracellular antimetabolite so enhancing the degree of growth inhibition. Azetidine-2-carboxylic acid, and other proline analogues such as 3,4-dehydroproline and 3,4-methanoproline, show a high affinity for the proline permease of *Escherichia coli* or *Salmonella typhimurium*, and each imino acid produces marked growth inhibition of 'wild-type' bacterial cultures. However, mutant bacterial strains, selected for their ability to overcome the growth inhibitory effects of analogues, possess modified forms of the proline permease which exhibit little or no affinity for the analogue molecules (Rowland and Tristram, 1972). Canavanine displays a similar behaviour as an antagonist of arginine uptake into bacterial cells, whilst specific mutation within the phenylalanine permease affords protection to strains of *Neurospora crassa* resistant to inhibition by *p*-fluorophenylalanine.

Analogue effects upon amino acid biosynthesis

Analogue molecules may affect biosynthetic reactions either by acting as direct competitive inhibitors of particular enzymes or by behaving as 'false' end-product regulators mediating the activity or synthesis of early key enzymes of the pathway. Examples of each type are considered below:

Recently my collaborator (P. J. Lea) has achieved a considerable purification of an asparagine synthetase from seedlings of field lupin and other legumes. The enzyme catalyses a glutamine-dependent synthesis of asparagine represented as:

$$\text{Aspartate} + \text{glutamine} \xrightarrow[\text{Mg}^{2+}]{\text{ATP}} \text{asparagine} + \text{glutamate.}$$

Albizziine (at 6 mM conc.), an isostere of glutamine, acts as a competitive inhibitor of the reaction, especially when the asparagine synthetase originates from lupin. The reaction is still strongly inhibited when an equivalent concentration of glutamine is present as substrate. However, albizziine-producing species such as *Albizzia lophantha* and *Acacia farnesiana* possess asparagine synthetases less affected by albizziine. The analogue may accumulate in these plants in amounts equivalent to 6 mM, i.e. the concentration used in the inhibition studies reported in Table 1. Albizziine produces no significant inhibition

TABLE 1. INHIBITION OF ASPARAGINE SYNTHETASE BY 6 mM ALBIZZIINE

Glutamine conc. mM	Source and activity of enzyme: *Lupinus*	*Acacia*	*Albizzia*
0.012	2.7	11.3	0
0.012*	6.8	24.5	25
1.2†	20.7	66.4	97.2
6.0	28.7	96.4	109.9

Activities are expressed as percentages of uninhibited controls.
* Conc. saturating enzyme from *Lupinus*.
† Conc. saturating enzymes from *Acacia* and *Albizzia*.

of the *Albizzia* enzyme when the glutamine concentration is just sufficient to saturate the enzyme. These data provide an example of a more general situation in which enzymes from species producing analogue molecules often exhibit lowered affinities for the analogues, either as competitive substrates or inhibitors, when compared with the similar enzymes from species unable to synthesize the analogues.

Proline acts as an end-product inhibitor of the formation of Δ^1-pyrroline-5-carboxylate from glutamate in several systems including the bacteria, *Escherichia coli* and *Salmonella typhimurium*. A number of analogues including azetidine-2-carboxylic acid, 3,4-dehydroproline and 3,4-methanoproline mimic this feedback inhibitory activity of proline. The *cis*-L isomer of 3,4-methanoproline is a particularly potent inhibitor. When present at 1.25×10^{-4} M concentration it completely suppressed Δ^1-pyrroline-5-carboxylate synthesis in *Escherichia coli* cultures for at least 3 hr; a higher concentration of L-proline (10^{-3} M) was required to produce total inhibition, and the culture escaped from inhibition by proline after 2.5 hr (Rowland and Tristram, 1972). The *trans*-L isomer of 3,4-methanoproline was a much less effective inhibitor of growth and Δ^1-pyrroline-5-carboxylate synthesis in *Escherichia coli*.

An example of false feed-back inhibition of a higher plant enzyme is seen in preliminary studies of B. J. Miflin and M. Wade (personal communication). They have demonstrated that maize contains an aspartyl kinase (the first enzyme of the lysine biosynthetic pathway) that is inhibited completely by 1 mM lysine. A lysine analogue, *S*-aminoethyl-L-cysteine (IX), when used at the same concentration caused 80 per cent inhibition of the enzyme.

Analogue affinity for aminoacyl-tRNA synthetases

Growth inhibitory or toxic effects of analogues may be due in part to an incorporation of analogue residues into protein molecules in place of normal amino acid constituents. Inhibition of radicle development in *Phaseolus aureus* by azetidine-2-carboxylic acid was determined by the degree of replacement of proline residues by those of the analogue in newly synthesized protein molecules. Canavanine and *p*-fluorophenylalanine cause similar inhibitions of bacterial growth by their respective replacement of arginine and phenylalanine residues in the microbial protein. Replacements of this type require that the appropriate aminoacyl-tRNA synthetases accept the analogues as competitive substrate molecules. However, there is no evidence establishing that analogues are incorporated into the protein molecules of plant species synthesizing them: indeed, nature seems often to have devised specific methods, based on discriminatory synthetase activities, for excluding them from the tissue proteins of 'producer' plants.

Good examples of species differences in the substrate specificity of aminoacyl-tRNA synthetases occur with the glutamyl- and prolyl- enzymes. The glutamyl-tRNA synthetase from *Phaseolus aureus* accepts the L-isomers of both *threo* and *erythro* forms of γ-hydroxy- and γ-methyl-glutamic acids as substrates: the substituted glutamic acids have not been detected as constituents of this species. *Threo*-γ-hydroxy-L-glutamic acid is produced by *Hemerocallis fulva* and *erythro*-γ-methyl-L-glutamic acid forms an important component of *Caesalpinia bonduc*: the alternate diastereoisomers are not recognized as higher plant products. Table 2 shows that glutamyl-tRNA synthetase preparations from these species fail completely to activate their own products, although in each case the non-natural diastereoisomer shows some affinity for the enzyme (Lea and Fowden, 1972).

TABLE 2. ACTIVATION OF GLUTAMIC ACID AND ITS ANALOGUES BY GLUTAMYL-tRNA SYNTHETASES FROM VARIOUS PLANT SOURCES

Substrate		Sources and kinetic parameters of enzymes		
		Phaseolus aureus seed	*Hemerocallis fulva* leaf	*Caesalpinia bonduc* seed
L-Glutamic acid	K_m	7.2	5.2	9.3
	V_{max}	100	100	100
erythro-γ-Methyl-L-glutamic acid	K_m	15	28	∞
	V_{max}	68	40	0
threo-γ-Methyl-L-glutamic acid	K_m	—	∞	—
	V_{max}	55	0	20
threo-γ-Hydroxy-L-glutamic acid	K_m	21	∞	52
	V_{max}	54	0	24
erythro-γ-Hydroxy-L-glutamic acid	K_m	—	—	∞
	V_{max}	58	34	0

K_m values are expressed as mM conc.
V_{max} values are expressed as percentages of values determined for glutamic acid.

A more detailed study of prolyl-tRNA synthetases from a variety of plants has led to a better understanding of the ways in which an altered imino acid substrate specificity may be related to the molecular configuration of the enzymes (Norris and Fowden, 1972). This synthetase has been purified from a variety of plants; some like *Convallaria majalis* or *Delonix regia* produce large amounts of the proline analogue azetidine-2-carboxylic acid, whilst others were selected as non-producers of the imino acid (*Phaseolus aureus* and *Ranunculus bulbosus*). Enzyme was prepared also from sugar beet (*Beta vulgaris*) seedlings, a species producing minute amounts of azetidine-2-carboxylic acid (see earlier). The affinity of each synthetase preparation for a range of imino acids was tested, and V_{max} and K_m values were determined. In general, prolyl-tRNA synthetases from plants producing azetidine-2-carboxylate showed no substrate affinity for the imino acid, but they invariably activated *cis*-3,4-methano-L-proline, a slightly larger molecule than the normal substrate proline. Synthetases from non-producer species behaved in a contrasting manner, activating azetidine-2-carboxylate efficiently but showing no affinity for methanoproline (Table 3). Such results suggest that the geometry of the active site of the synthetase from plants discriminating against azetidine-2-carboxylic acid is such that molecules larger than proline can still bind to the enzyme forming the correct configuration alignment between the carboxyl group of the imino acid and the α-PO_4^{3-} group of ATP. By displaying this flexibility, the fit of analogue molecules smaller than proline is presumably too loose for effective ligand bond formation. Conversely, the active sites of prolyl-tRNA synthestases from species not producing azetidine-2-carboxylic acid are presumed to be smaller, thereby facilitating binding of analogues less bulky than proline but imposing constraints upon the correct positioning of large molecules at the active centre. Such conformational differences at the active site presumably have their origin in evolutionary mutations; a single mutation leading to the alteration of only one amino acid residue within the active site might be sufficient to explain the altered properties of prolyl-tRNA synthetases from azetidine-2-carboxylate-producing plants.

TABLE 3. ACTIVATION OF IMINO ACIDS BY PROLYL-tRNA SYNTHETASES FROM VARIOUS PLANT SPECIES

Plant source		L-Proline	L-Azetidine-2-carboxylic acid	*cis*-3,4-Methano-L-proline
Delonix regia	K_m	0.18	—	4.6
	V_{max}	100	~0	22
Convallaria majalis	K_m	0.45	—	~2.5
	V_{max}	100	~0	36
Beta vulgaris	K_m	0.45	2.2	—
	V_{max}	100	73	<3
Phaseolus aureus	K_m	0.14	1.4	—
	V_{max}	100	55	<2
Ranunculus bulbosus	K_m	0.29	2.0	∞
	V_{max}	100	66	0

K_m values are expressed as mM conc.
V_{max} values are expressed as percentages of values determined for proline.

A further striking difference in the properties of the two types of prolyl-tRNA synthetase is apparent when the thermal stabilities of the enzymes from *Delonix regia* and *Phaseolus aureus* are compared. Figure 1 shows the activity remaining after heating enzyme preparations for 7 min at a range of temperatures. In the absence of substrates, activators and thiol group protectors, the *Delonix* enzyme is much more thermolabile than that from *Phaseolus*. Addition of proline (20 mM) effects a marked stabilization of *Delonix* enzyme, whilst conferring a degree of protection upon the *Phaseolus* enzyme. Azetidine-2-carboxylic acid also affords considerable protection against heat denaturation of the enzymes (Norris and Fowden, 1973). Both types of synthetase were unstable

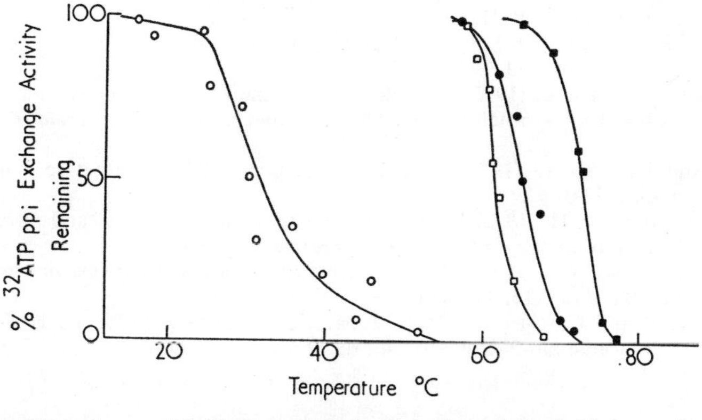

FIG. 1. Loss of activity when prolyl-tRNA preparations are heated for 7 min at various temperatures. Enzyme from *Delonix regia*: ○—○, in buffer alone; □—□, in presence of 20 mM proline. Enzyme from *Phaseolus aureus*: ●—●, in buffer; ■—■, in presence of 20 mM proline.

during fractionation at 2–4°C, and detailed studies of cold-lability have revealed that the *Delonix* enzyme lost over 90 per cent of its activity during storage for 20 min at 2°C. Polyols, especially glycerol, provide considerable protection against inactivation during cold storage. The changes in the activity of the *Phaseolus* enzyme on storage at 2°C were less extreme. Both enzymes could be taken through several cycles of inactivation–reactivation, if cooling and subsequent warming to 40°C were performed in the presence of sulphydryl-reducing reagents (Norris and Fowden, 1974).

CONCLUSION

In a period of about 25 years, a large new group of plant products has been recognized. Although the new amino acids initially tended to be regarded as metabolic curiosities, later investigations have revealed how many of the compounds may serve as storage or transport forms of nitrogen within plants, as products conferring advantages for the survival and ecological adaptation of the producer species, and as substances possessing analogue properties that can be usefully applied in studies concerned with the regulatory control of metabolic processes.

REFERENCES

BELL, E. A. (1962) Associations of ninhydrin-reacting compounds in the seeds of 49 species of *Lathyrus*. *Biochem. J.* **83,** 225–229.

BRESSLER, R., CORREDOR, C. and BRENDEL, K. (1969) Hypoglycin and hypoglycin-like compounds. *Pharmac. Revs.* **21,** 105–130.

DUNNILL, P. M. and FOWDEN, L. (1965) The amino acids of seeds of the Cucurbitaceae. *Phytochemistry* **4,** 933–944.

FOWDEN, L. (1970) The non-protein amino acids of plants. In *Progress in Phytochemistry*, ed. REINHOLD, L. and LIWSCHITZ, Y. vol. 2, pp. 203–266, Wiley, London.

FOWDEN, L. (1972) Amino acid complement of plants. *Phytochemistry* **11,** 2271–2276.

FOWDEN, L., LEWIS, D. and TRISTRAM, H. (1967) Toxic amino acids: their action as antimetabolites. *Adv. Enzymol.* **29,** 89–163.

LARSEN, P. O. (1975) Non-protein amino acids derived from primary and secondary biosynthetic pathways. *Phytochemistry* in press.

LEA, P. J. and FOWDEN, L. (1972) Stereospecificity of glutamyl-tRNA synthetase isolated from higher plants. *Phytochemistry* **11,** 2129–2138.

NORRIS, R. D. and FOWDEN, L. (1972) Substrate discrimination by prolyl-tRNA synthetase from various higher plants. *Phytochemistry* **11,** 2921–2935.

NORRIS, R. D. and FOWDEN, L. (1973) A comparison of the thermal stability and substrate binding constants of prolyl-tRNA synthetase from *Phaseolus aureus* and *Delonix regia*. *Phytochemistry* **12,** 2109–2121.

NORRIS, R. D. and FOWDEN, L. (1974) Cold-lability of prolyl-tRNA synthetase from higher plants. *Phytochemistry* **13,** 1677–1687.

ROWLAND, I. and TRISTRAM, H. (1972) Inhibition of bacterial growth by *cis*- and *trans*-3,4-methano-L-prolines: mechanism of toxicity. *Chem.-Biol. Interac.* **4,** 377–388.

SMITH, L. W. and CHANG, F-Y. (1973) Aminotriazole metabolism in *Cirsium arvense* (L) Scop. and *Pisum sativum* L. *Weed Res.* **13,** 339–350.

WARIN, P., JADOT, J. and CASIMER, F. (1969) Séparation et caractérisation de la *S*-(amino-2′-éthyl) cystéine et de la *S*-(acetamido-2′-éthyl)cystéine de *Rozites caperata*. *Extrait Bull Soc. r. Sci. Liége* **38,** 280–287.

QUANTITATIVE REGULATION OF GENE ACTIVITY

J. Ingle, J. N. Timmis and J. R. Gore

Department of Botany, University of Edinburgh, Mayfield Road, Edinburgh, U.K.

Introduction

Certain genes in eukaryotic cells are present as multiple copies. The DNA sequences coding for the two high molecular weight ribosomal-RNAs (rRNA), the 5S rRNA, the transfer RNAs and histones have been studied, and their redundancy within the genome investigated. The presence within the genome of other DNA sequences having similar degrees of repetition and complexity suggests that other genes may also exist as multiple copies. Although regulation of single copy genes may readily be envisaged in terms of an on/off mechanism, the regulation of activity of multiple gene sequences must be under a more complex quantitative control. The DNA sequences coding for the high molecular weight rRNAs are a suitable system for studying gene utilization, since the redundancy of the gene may be readily and accurately established, and the amount of gene product (the stable rRNAs) easily determined. The experiments described in this paper were undertaken to establish whether the utilization of rRNA genes is directly controlled in plant cells, and to investigate the mechanism of such control.

Positive control of cellular rRNA content

The redundancy of the rRNA gene varies from one to thirty thousand in the range of plant species studied (Ingle *et al.*, 1975). The total growth of a species does not appear to be limited by rRNA gene redundancy, as attested by a plant of Jerusalem artichoke (9 ft in height) containing 1600 copies of the gene, and a hyacinth (only 9 in. high) with 17,000 genes. Even within closely related species total protein synthesis, and hence plant growth, is not necessarily correlated with rRNA gene redundancy. *Nicotiana tabacum*, with 2200 genes, outgrows by several fold *Nicotiana glutinosa*, which has 3200 genes. Such observations themselves suggest that the efficiency of rRNA gene utilization varies considerably between different species. A direct analysis of the amount of rRNA gene product in species containing a wide variation of gene redundancy confirms this indirect conclusion. Three species, Jerusalem artichoke, melon and swisschard, containing few genes, two species, sunflower and cucumber, with an intermediate number, and three, wheat, onion and hyacinth, containing a high number of rRNA genes were selected (Table 1). Measurements were made on the meristematic region of the root tip, varying in size from 1 to 3 mm dependent on the extent of the meristem. The ratio of total RNA/DNA was calculated from the amounts of DNA and RNA present in the root tip, and the amount of RNA per telophase cell estimated from this ratio and the amount of DNA per

TABLE 1. RELATIONSHIP BETWEEN rRNA GENE REDUNDANCY AND rRNA GENE PRODUCT IN A RANGE OF HIGHER PLANTS

	rRNA genes/telophase nucleus	DNA/tip (10^{-6} g)	RNA/tip (10^{-6} g)	RNA/DNA	DNA/telophase cell (10^{-12} g)	RNA/telophase cell (10^{-12} g)	RNA/gene (10^{-15} g)
Artichoke	1580	2.44	9.90	4.0	24.0	97	62.0
Melon	2000	0.17	4.74	27.9	1.9	53	26.5
Swisschard	2300	0.10	1.48	14.4	2.5	36	15.6
Sunflower	7600	0.46	6.27	13.6	10.0	136	17.9
Cucumber	8800	0.07	4.20	58.2	2.0	116	13.3
Wheat	12700	1.05	5.02	4.8	30.0	144	11.3
Onion	13300	4.94	9.30	1.9	32.0	61	4.6
Hyacinth	16700	6.50	13.40	2.1	49.0	101	6.1

Tips (1–3 mm long) were removed from rapidly growing roots of melon, swisschard, sunflower, cucumber and wheat seedlings, Jerusalem artichoke tubers, and onion and hyacinth bulbs. Total DNA, and total RNA, per root tip was determined as described in Timmis and Ingle (1975). DNA per telophase nucleus was determined by comparative microdensitometry of feulgen stained preparations (see Ingle *et al.*, 1975; Timmis *et al.*, 1972).

telophase cell, determined from quantitative measurements of feulgen stained nuclei (McLeish and Sunderland, 1961). It is immediately apparent (Table 1) that the variation in RNA per cell within this selection of plant species is only four-fold, compared with a twenty-five-fold variation in genome size and a ten-fold variation in rRNA gene redundancy. Total RNA accumulated (which is proportional to rRNA accumulation since approximately 80 per cent of the total RNA is rRNA) when expressed per rRNA gene illustrates that the genes in artichoke, melon and swisschard are used more efficiently than those in wheat, onion and hyacinth. Artichoke, with only one tenth the rRNA gene redundancy of hyacinth, accumulates a similar amount of rRNA per cell, suggesting that its genes are utilized ten times as efficiently.

Such an analysis indicates that the rRNA genes are used with varying efficiencies in different plant species, but inherent differences between the various species prevent a more detailed interpretation. However, a more complete study is possible on a single species which contains a range of rRNA gene redundancies. Such a species is *Hyacinthus orientalis,* which is available as various euploid and aneuploid varieties (Darlington *et al.*, 1951). The rRNA gene redundancy is proportional to the number of nucleolar

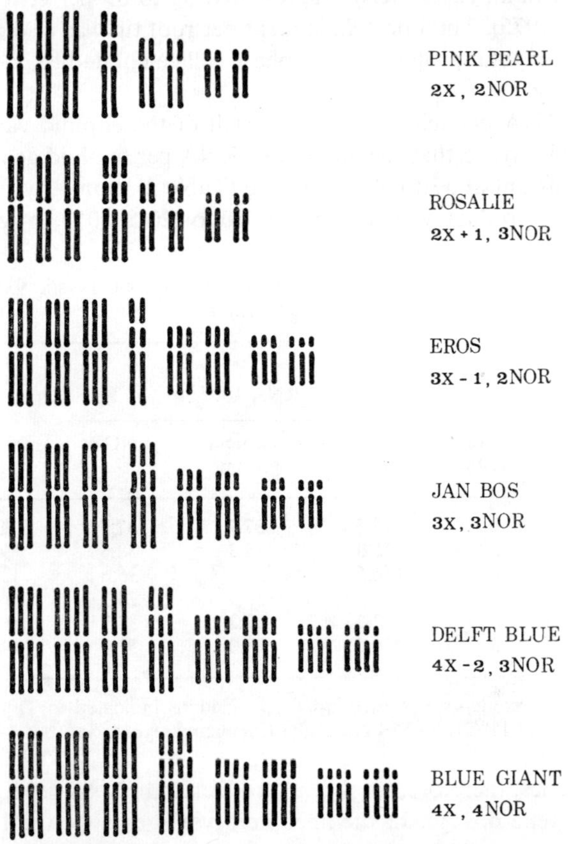

FIG. 1. Karyotypes of three euploid and three aneuploid varieties of *Hyacinthus orientalis*. The NOR is shown as the secondary constriction on a long chromosome.

organizing regions (NOR), which are the location of the rRNA genes within the nucleus, although the number of copies of the gene per NOR is slightly less (20 per cent) in the aneuploids than in the euploids (Timmis *et al.*, 1972). The varieties used were a diploid, triploid and tetraploid, containing 2, 3 and 4 NORs respectively, and three aneuploids either lacking one nucleolus organizing chromosome or containing an additional one (Fig. 1). For example the variety Rosalie ($2x + 1$) has essentially a diploid genome carrying 3 NORs, whereas in the variety Eros ($3x - 1$), which has close to a triploid genome, there are only 2 NORs. It was therefore possible to determine whether the tetraploid, with twice the gene dosage of the diploid, made twice as much gene product, and whether the amount of rRNA product in the aneuploids was directly proportional to their rRNA gene dosage.

The hyacinth bulbs were cultured in aerated water at 11°C and 2 to 3 mm tips were removed from actively-growing roots for analysis. Both the DNA and the total RNA content per root tip were determined as discussed previously to obtain the amount of rRNA gene product per cell, and hence the amount per haploid genome or NOR. Total nucleic acid preparations from each of the six hyacinth varieties were fractionated by gel electrophoresis and in all cases rRNA represented 86 to 89 per cent of the total RNA (Timmis and Ingle, 1975). Total protein content per root tip was also determined in each variety and values for total protein per telophase cell calculated to give a measure of the total genome activity.

The amount of RNA per telophase cell in each of the euploid varieties increased in proportion to the ploidy, so that the amount of RNA per haploid genome, or per NOR, was remarkably constant at 37 to 38 $\times 10^{-12}$ g (Table 2). Similarly the total amount of protein, or total gene product, was constant at 310 to 320 $\times 10^{-12}$ g per haploid genome.

TABLE 2. RELATIONSHIP BETWEEN GENE PRODUCT AND GENE DOSAGE IN SIX VARIETIES OF HYACINTH (*Hyacinthus orientalis*)

			RNA 10^{-12} g			Protein 10^{-12} g	
Variety	Genome	No. of NORs	/telophase cell	/haploid genome	/NOR	/telophase cell	/haploid genome
Pink Pearl	$2x$	2	75.5	37.7	37.7	612	306
Jan Bos	$3x$	3	115.0	38.3	38.3	947	316
Blue Giant	$4x$	4	146.5	36.6	36.6	1285	321
Rosalie	$2x+1$	3	72.7	36.3	24.2	660	330
Eros	$3x-1$	2	117.5	39.2	58.7	950	317
Delft Blue	$4x-2$	3	156.0	39.0	52.0	1430	357

Total protein and RNA contents per root tip determined as indicated in Table 1. Bulbs grown in aerated water for 17 days at 11°C. NOR = Nucleolar organizing region.

In the euploid varieties, therefore, total gene product and rRNA gene product were both directly related to gene dosage. In the aneuploid varieties also, total protein and total RNA values were remarkably constant when expressed on a nominal haploid genome basis, and not significantly different from the euploid values. However, the amount of

RNA per NOR varied between the aneuploids and differed from that obtained in the euploids. Eros cells made the triploid amount of rRNA although they possess only 2 NORs, while those of Rosalie, which possess three NORs, made only the diploid amount of rRNA. The rRNA genes in Eros are therefore used more efficiently than those in Rosalie.

The absolute amount of rRNA and of protein per haploid genome varied with the age of the root (Table 3), but the over- and under-production of rRNA per NOR in cells of

TABLE 3. REPRODUCIBILITY OF THE GENE/GENE PRODUCT RELATIONSHIP IN HYACINTH

Variety	Age of roots days	Protein 10^{-12} g/ haploid genome	RNA 10^{-12} g/ haploid genome	RNA 10^{-12} g/ NOR
Pink Pearl	10	440	50.0	50.0
Rosalie		390	48.5	32.3
Eros		375	50.6	75.1
Pink Pearl	11	357	48.5	48.5
Rosalie		390	46.0	30.7
Eros		440	48.1	72.5
Pink Pearl	17	306	37.7	37.7
Rosalie		330	36.3	24.2
Eros		317	39.1	58.7

Determinations made on root tips as indicated in Table 1. Bulbs grown in aerated water for 10, 11, 17 days at 11°C.

the two aneuploids Eros and Rosalie, compared with the euploid Pink Pearl, was remarkably reproducible. The constancy of the amount of rRNA per haploid genome in the aneuploids containing varying dosages of rRNA genes indicates that the cellular rRNA content is positively regulated, and that it is controlled by the total genome rather than by the specific gene dosage. The results also show that in the euploids the rRNA genes are not operating at maximum capacity, since cells of the aneuploid Eros operate at a greater efficiency.

MECHANISM OF rRNA REGULATION

Having established that rRNA genes are used 2.4 times more efficiently in Eros compared with Rosalie cells, it is pertinent to consider the level at which this regulation occurs. The regulation may be at the level of transcription, Eros transcribing more copies of the rRNA gene than Rosalie. It may be that similar transcription occurs in both varieties, but that in Eros the transcription product is processed and conserved as stable rRNA, whereas in Rosalie the transcription product is not conserved but degraded. Alternatively, transcription and conservation may be similar in both varieties, but the ribosomes may be less stable and turned over in Rosalie.

As a preliminary investigation aimed at distinguishing between these three possibilities, a system showing a greater difference in rRNA accumulation was studied. DNA synthesis and cell division in explants of Jerusalem artichoke tuber are both dependent on

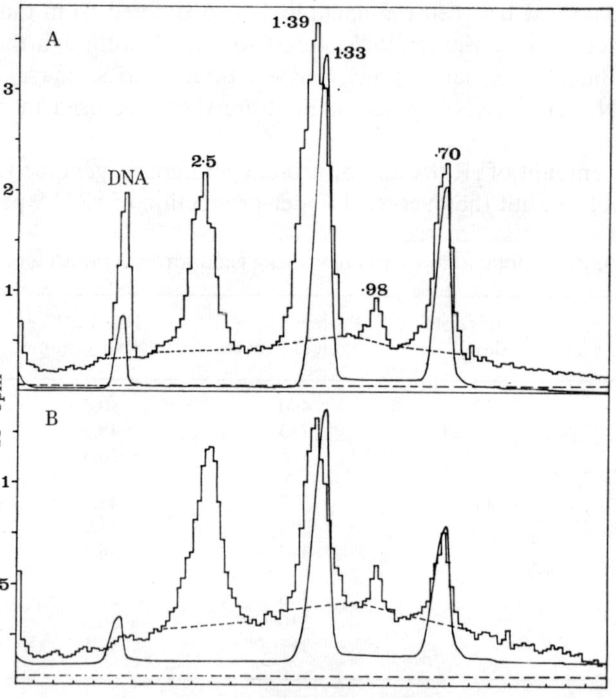

FIG. 2. Polyacrylamide gel fractionation of RNA synthesized by Jersusalem artichoke tuber explants cultured in the presence and absence of 2,4-D. Total nucleic acid was prepared from explants cultured in the presence (A) and absence (B) of 2,4-D for 68 hr at 25°C in darkness. Explants were labelled with [^{32}P]orthophosphate for the final 60 min. The nucleic acid was fractionated on 2.4 per cent polyacrylamide gels for 3.5 hr at 50 V. The values refer to the molecular weight $\times$ 10^6 of the artichoke rRNA components determined relative to *Escherichia coli* rRNA. The continuous scan is the E_{265} and the histogram the cpm per 0.5 mm slice.

the presence of 2,4-dichlorophenoxyacetic acid (2,4-D) in the culture medium (Yeoman and Mitchell, 1970). Continued rRNA accumulation is also dependent on exogenous 2,4-D although in the early stages of culture accumulation is independent of the auxin (Gore and Ingle, 1974). Thus, after a suitable period of culture, e.g. 60 hr, explants growing in the presence of 2,4-D are accumulating rRNA whereas those cultured without 2,4-D accumulate no rRNA. However, the number of rRNA genes is unaffected by the conditions of culture (Ingle and Sinclair, 1972) so the difference in rRNA accumulation is the result of regulation of gene activity.

Analysis of RNA synthesized in explants cultured in the presence and absence of 2,4-D for 68 hr showed that in both cases the rRNA genes were being transcribed into the large 2.5 $\times$ 10^6 molecular weight precursor rRNA (Fig. 2). The lack of rRNA accumulation was not due therefore simply to lack of transcription. Uptake of [^{32}P]orthophosphate was similar under both conditions of culture, but total incorporation of the labelled phosphate into nucleic acid was about halved in the absence of 2,4-D (Table 4). The similar percentage of radioactivity present in the 2.5 $\times$ 10^6 precursor rRNA under both cultural conditions indicates therefore that transcription of rRNA genes is halved when explants are cultured in the absence of 2,4-D. This assumes that the amount of incorporation is proportional to synthesis. Less of the 2.5×10^6 molecular weight precursor was processed and conserved as stable rRNA in explants cultured in the absence of 2,4-D (Table 4). This decreased conversion of precursor to stable rRNA is

TABLE 4. GROWTH, PHOSPHATE UPTAKE AND RNA SYNTHESIS IN JERUSALEM ARTICHOKE TUBER EXPLANTS CULTURED IN THE PRESENCE AND ABSENCE OF 2,4-DICHLOROPHENOXYACETIC ACID (2,4-D)

	Plus 2,4-D	Minus 2,4-D
No. of cells/explant	68,300	24,800
[^{32}P]orthophosphate uptake/60 min (cpm/cell)	58	71
Total incorporation into nucleic acid/60 min (10^{-3} cpm/cell)	234	134
Percentage distribution of radioactivity		
DNA	5.5	0
2.5×10^6 precursor rRNA	13.7	11.6
1.39×10^6 precursor rRNA + 1.33×10^6 rRNA	23.0	10.2
0.98×10^6 precursor rRNA	1.9	1.5
0.7×10^6 rRNA	9.5	3.9
polydisperse RNA	47.0	72.8

Explants were cultured in the presence and absence of 2,4-D at 25°C in darkness for 68 hr. Uptake into the final aqueous phase of the nucleic acid preparation, and incorporation of [^{32}P]orthophosphate (50 μCi/ml) into nucleic acid, was determined during the terminal 60 min. Nucleic acid was fractionated on 2.4 per cent polyacrylamide gels for 3.5 hr at 50 V, and the distribution of RNA components was calculated excluding the 5S rRNA, tRNA and polydisperse RNA of less than 0.3×10^6 daltons (see Fig. 2).

reflected in the increased percentage of polydisperse RNA. The lack of accumulation of rRNA in the artichoke explants cultured in the absence of 2,4-D is therefore due to reduced transcription, reduced conversion of the transcription product and possibly increased turnover of the final rRNA product. Such post-transcriptional regulation of rRNA accumulation has been demonstrated in resting lymphocytes (Cooper and Gibson, 1971), in unfertilized *Urechis* eggs (Das *et al.*, 1970) and in differentiating yolk sac erythroid cells (Fantoni *et al.*, 1972).

Transcription and processing of rRNA was similarly studied in the $2x$, $2x + 1$ and $3x - 1$ hyacinth varieties. Two measures of transcription were obtained, incorporation of [^{3}H]uridine into the 2.5×10^6 molecular weight precursor rRNA during a 3 or 4 hr pulse, and incorporation into the stable 1.3×10^6 and 0.7×10^6 molecular weight rRNA during a 16 hr chase subsequent to the pulse, both expressed on a genome basis. In three experiments transcription was least in Rosalie ($2x + 1$), in which only the diploid amount of rRNA is produced despite the three NORs present, and highest in Eros ($3x - 1$), in which only two NORs are present but the triploid amount is produced (Table 5). The similarity in relative transcription rates, whether on the basis of incorporation into precursor or stable rRNA, indicates that similar processing and conservation of the precursor occurs in all three varieties. This is consistent with the constancy of the conservation ratio (i.e. incorporation into stable 1.3×10^6 mol. wt. rRNA during the

TABLE 5. TRANSCRIPTION AND PROCESSING OF rRNA IN THE THREE HYACINTH VARIETIES, PINK PEARL, ROSALIE AND EROS

	Experiment I (3 hr)	Experiment II (4 hr)	Experiment III (4 hr)
Transcription/NOR			
2.5×10^6 precursor rRNA			
Pink Pearl	—	2.7	1.5
Rosalie	0.47	1.0	0.9
Eros	0.64	—	3.4
$1.3 + 0.7 \times 10^6$ stable rRNA			
Pink Pearl	—	59	81
Rosalie	15	44	40
Eros	23	—	197
Conservation ratio ($1.3 \times 10^6/2.5 \times 10^6$)			
Pink Pearl	—	—	28.0
Rosalie	19.0	—	22.2
Eros	22.9	—	27.0

Hyacinth bulbs were cultured in aerated water for 15 days at 11°C. [^{3}H]uridine (25 μCi/ml) and 50 μg/ml chloramphenicol were added for the terminal 3 or 4 hr. Half the roots were removed for analysis and the remainder rinsed and transferred to unlabelled uridine for a 16 hr chase. Total nucleic acid was prepared from the root tips and fractionated by polyacrylamide gel electrophoresis. The amount of radioactivity in the 2.5×10^6 precursor rRNA (pulse) or the $1.3 + 0.7 \times 10^6$ rRNAs (chase) was determined and expressed on a genome, and hence NOR basis.

chase divided by incorporation into the 2.5×10^6 mol. wt. precursor rRNA during the pulse) in the three varieties (Table 5). The comparison of absolute incorporation of [^{3}H]uridine into root tips of different bulbs is fraught with technical problems, but the qualitative consistency of the incorporation data in the three experiments, and the lack of evidence for any post-transcriptional control, suggest that the greater efficiency of utilization of rRNA genes in Eros compared with Rosalie is regulated at the level of transcription. The differences in regulation of rRNA accumulation observed between the hyacinth varieties and the Jerusalem artichoke system suggest that different levels of control may function under different conditions, different genomes being regulated at the transcriptional level with more temporal internal regulation residing at post-transcriptional levels.

REFERENCES

COOPER, H. L. and GIBSON, E. M. (1971) Control of synthesis and wastage of ribosomal ribonucleic acid in lymphocytes. *J. biol. Chem.* **246,** 5059–5066.

DARLINGTON, C. D., HAIR, J. B. and HURCHOMBE, R. (1951) The history of the garden hyacinths. *Heredity* **5,** 233–252.

DAS, N. K., MICON-EASTWOOD, J., RAMAMURTHY, G. and ALFERT, M. (1970) Sites of synthesis and processing of ribosomal RNA precursors within the nucleolus of *Urechis caupo* eggs. *Proc. natn. Acad. Sci. U.S.* **67,** 968–975.

FANTONI, A., BORDIN, S. and LUNADEI, H. (1972) Control of ribosomal RNA maturation in differentiating yolk sac erythroid cells. *Cell Differ.* **1,** 219–228.

GORE, J. R. and INGLE, J. (1974) RNA polymerase activities in Jerusalem artichoke tissue. *Biochem. J.* **143,** 107–113.

INGLE, J. and SINCLAIR, J. (1972) r-RNA genes in plant development. *Nature Lond.* **235,** 30–32.
INGLE, J., TIMMIS, J. N. and SINCLAIR, J. (1975) The relationship between satellite DNA, r-RNA gene redundancy and genome size in plants. *Pl. Physiol. Lancaster* **55,** 496–501.
MCLEISH, J. and SUNDERLAND, N. (1961) Measurements of DNA in higher plants by feulgen photometry and chemical methods. *Expl Cell Res.* **24,** 527–540.
TIMMIS, J. N. and INGLE, J. (1975) The quantitative regulation of gene activity. *Pl. Physiol. Lancaster.* In press.
TIMMIS, J. N., SINCLAIR, J. and INGLE, J. (1972) rRNA genes in euploids and ancuploids of hyacinth. *Cell Differ.* **1,** 335–339.
YEOMAN, M. M. and MITCHELL, J. P. (1970) Changes accompanying the addition of 2,4-D to excised Jerusalem artichoke tuber tissue. *Ann. Bot.* **34,** 799–810.

THE SEARCH FOR PLANT MESSENGER RNA

R. J. Ellis

Department of Biological Sciences, University of Warwick, Coventry, U.K.

The nature of the problem

Messenger RNA was defined first by Monod *et al.* (1962) as a cytoplasmic RNA which determines the sequence of amino acids in a polypeptide chain. This concept was applied initially to prokaryotic cells and viruses, but its ubiquity has since been extended to animals. In 1971 the first convincing report of the globin messenger was published by Lockard and Lingrel, and since then there has been an explosion of publications in the area of animal messengers, their transcription, processing, and translation (Matthews, 1973; Darnell *et al.*, 1973). At least a dozen different messengers have been isolated from animal tissues, and their availability has opened up a whole new fascinating area of research between the disciplines of cell biology and biochemistry (Gurdon, 1974). This advance has not been matched by any comparable number of studies with plant tissues, and only in the past few months has there been any success in the detection of specific messenger RNA molecules from plant sources, other than viral messengers.

Why be interested in messenger RNA?

Molecular biologists argue that the most important problem to study today is the process of development (e.g. Watson, 1970), not only because the ability to develop is the key to the success of multicellular organisms, but also because developmental biology is a field where new concepts are needed, and thus it presents a challenge which some other fields of biology do not. We have no *theory* of development in the same way that we have, for example, a Mendelian theory of genetics, but we do know that the immediate cause of development is the spectrum of proteins in a given cell. One cell develops differently from another by virtue of a different protein complement. There are at least six ways in which the protein spectrum of a cell can be altered:

(1) selective *transcription* of the genome resulting in the production of different sets of messengers,
(2) selective *degradation* of messengers,
(3) selective *transport* of messengers from nucleus to cytoplasm,
(4) selective *translation* of messengers, some remaining masked at certain stages of development,
(5) selective *assembly* or *activation* of protein subunits into the functional quaternary structure,
(6) selective *breakdown* of proteins.

One, all, or any combination of these processes may be important in a given situation. Most of these processes involve messenger RNA, and therefore to dissect and evaluate their role in development, we must have the technology to measure specific messengers.

WHAT ARE THE PROBLEMS IN MEASURING SPECIFIC MESSENGERS?

It is noticeable that success in the isolation of animal messengers has depended upon the use of cells which make only one or a small number of proteins at a given time, so that the messenger RNA fraction is enriched in one or a few species (Mathews, 1973). It is also clear that the only reliable way of measuring a given messenger is to determine the production of the protein it encodes in a heterologous cell-free system. The system must be heterologous, otherwise endogenous messengers may be measured; two suitable systems are those derived from *Escherichia coli* (Capecchi, 1966; Modelell, 1971) and from wheat germ (Marcus, 1972; Allende, 1969; Roberts and Paterson, 1973). The product must be assayed by a rigorously specific method. The chequered history of the search for the globin messenger shows that mere stimulation of the incorporation of labelled amino acids into protein is neither a sufficient nor, surprisingly, a necessary criterion for a messenger. Experience suggests that the best analytical method for the labelled product is electrophoresis on sodium dodecylsulphate (SDS) polyacrylamide gels (e.g. Blair and Ellis, 1973; Eaglesham and Ellis, 1974). Since SDS gels fractionate on the basis of molecular weight only, a more specific method must subsequently be employed for convincing identification. Tryptic and chymotryptic peptide mapping is probably the best method available.

WHAT SITUATIONS IN PLANT TISSUES ARE FAVOURABLE FOR DETECTING MESSENGERS?

Plant tissues do not show the same degree of differentiation in terms of protein as animal tissues; plants show their major differences much more in terms of carbohydrates. There are, however, two situations in which plants produce large amounts of one protein sufficient to permit detection of messengers. The first is in the nitrogen-fixing nodule of legumes. These nodules can accumulate enough leghaemoglobin to appear pink to the naked eye, and the messenger for this protein has been isolated from soybean nodule polysomes (Verma *et al.*, 1974). The second example is the chloroplast protein known as Fraction I, and in the rest of the paper I will discuss our studies on the synthesis of this protein in the general context of the work of the Chloroplast Group at the University of Warwick.

PROPERTIES OF FRACTION I PROTEIN

Fraction I is a protein of more than usual interest (see Kawashima and Wildman, 1970; Ellis, 1973a for reviews). The purified protein has two related enzymatic activities, ribulosediphosphate carboxylase and ribulosediphosphate oxygenase; it is thus implicated in both photosynthesis and photorespiration. The protein is found in all organisms which use carbon dioxide as a source of carbon (McFadden, 1973). It is a highly soluble protein, located in the stroma of chloroplasts, and in centrifuged extracts of prokaryotic cells it appears in the supernatant fraction. The most striking feature of Fraction I protein is its abundance; it accounts for up to 50 per cent of the total soluble protein of leaf extracts, and could well be the most abundant protein in nature.

The structure of Fraction I protein from eukaryotes is complex. The molecular weight of the undenatured protein is usually just over 5×10^5 daltons. Treatment with urea, SDS, or alkaline pH dissociates the molecule into two types of subunit. The large subunit has a molecular weight in the range $5.2–6.0 \times 10^4$ daltons, while the small subunit has a molecular weight in the range $1.2–2.4 \times 10^4$ daltons. This large difference in molecular weight between subunits allows easy separation on SDS polyacrylamide gels. There is now good evidence that the catalytic site resides on the large subunit (Nishimura and Akazawa, 1973). It has recently been found that each subunit from tobacco Fraction I protein, although homogenous on SDS polyacrylamide gels, can be resolved into several bands on electrofocusing gels (Sakano *et al.*, 1974). We have confirmed this result with Fraction I protein from *Pisum sativum*, but the detailed nature of the differences between the bands is not known.

A MODEL FOR THE SYNTHESIS OF FRACTION I PROTEIN

Studies on the synthesis of Fraction I protein have revealed a fascinating interplay between different cellular compartments. Our current model for the synthesis of Fraction I protein is shown in Fig. 1. The large subunit is envisaged as being both encoded and synthesized within the chloroplasts; it is assembled on free ribosomes and moreover is the only readily detectable soluble product of protein synthesis. The small subunit, on the other hand, is encoded in nuclear DNA, and synthesized by cytoplasmic ribosomes.

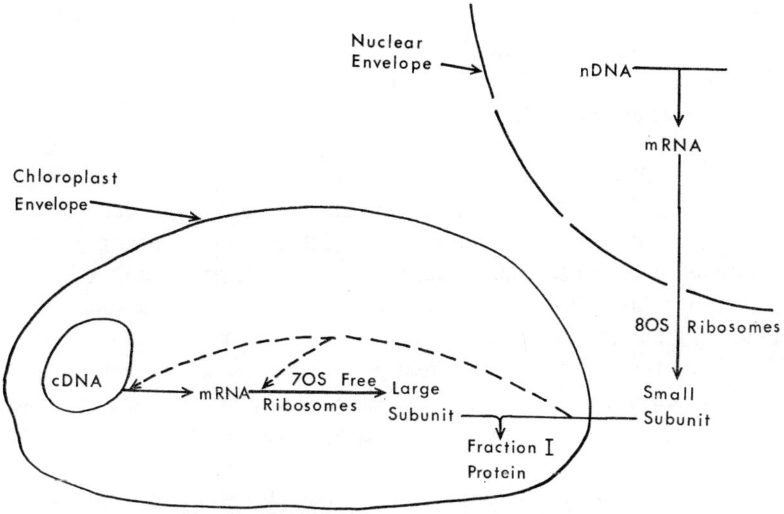

FIG. 1. Current model for the co-operation of nuclear and chloroplast genomes in the synthesis of Fraction I protein. nDNA and cDNA stand for nuclear and chloroplast DNA respectively, and the dashed lines indicate possible control points at which the small subunit may control the synthesis of the large subunit. Modified from Ellis (1973a) and Kawashima and Wildman (1972).

The small subunit then passes through the envelope of the chloroplast and combines with the large subunit to form Fraction I protein. The small subunit must be present for initiation of either transcription or translation of the large subunit messenger.

Evidence for the model

Evidence about the mode of synthesis of Fraction I protein has been obtained by three complementary methods. In the first, cells making chloroplasts are treated with compounds which one hopes will selectively inhibit either transcription of chloroplast DNA or translation of messages by chloroplast ribosomes. Inhibitors of transcription such as rifampicin have not proved useful in higher plants, and while inhibitors of translation such as chloramphenicol and cycloheximide have been more effective, conflicting conclusions have been drawn by different workers (Ellis *et al.*, 1973). This conflict has probably arisen because chloramphenicol and cycloheximide are not totally specific for protein synthesis in some plant tissues (Ellis, 1963; Ellis and MacDonald, 1970; Wilson and Moore, 1973). Nevertheless the bulk of the inhibitor evidence suggests that of all the *soluble* chloroplast proteins, Fraction I is the *only* one that requires chloroplast ribosomes for its synthesis (Boulter *et al.*, 1972), and since this synthesis is inhibited by cycloheximide (Ellis, 1973a) or 2-(4-methyl-2,6-dinitroanilino)-*N*-methyl-propionamide (Ellis, 1975) both of which arc cytoplasmic ribosome inhibitors, the simplest interpretation is that both cytoplasmic and chloroplast ribosomes are required. I must emphasize, however, that inhibitor experiments are never more than suggestive. They never really tell us more than that the activity of chloroplast ribosomes is required for a certain protein to accumulate in the chloroplast; they do not tell us that the chloroplast ribosomes actually synthesize that protein.

The second method that has been used for looking at the synthesis of Fraction I protein is the genetic approach. Many mutations which affect chloroplast components are inherited in a Mendelian fashion, and are presumed to reside in nuclear genes (Levine and Goodenough, 1970; Kirk, 1972), but certain chloroplast defects in higher plants are inherited via the maternal line only; this is correlated with the absence of plastids from the pollen tube, and the genes concerned have long been postulated to reside in chloroplast DNA (Kirk, 1972). Wildman and colleagues have studied amino acid sequences in several chloroplast proteins isolated from tobacco mutants and have provided conclusive evidence that a mutation in the large subunit of Fraction I is inherited via chloroplast DNA, whereas a mutation in the small subunit is inherited via nuclear DNA (Chan and Wildman, 1972; Kawashima and Wildman, 1972; Sakano *et al.*, 1974). This type of study has great potential especially in relation to cultivated plants among which many varieties are available. A limitation of the approach is that while it can tell us where a protein is encoded, it cannot tell us where the protein is synthesized.

This brings me to the third and most direct approach for studying synthesis of Fraction I protein—the identification of protein and RNA molecules made by isolated chloroplasts and cell-free cytoplasmic systems. When we started work in 1970 all previous attempts to persuade isolated chloroplasts to produce discrete identifiable products had been unsuccessful. In my laboratory, we have developed methods in which isolated chloroplasts and chloroplast ribosomes from pea (*Pisum sativum*) and spinach (*Spinacia oleracea*) make discrete protein and RNA molecules (Blair and Ellis, 1973; Eaglesham and Ellis, 1974; Hartley and Ellis, 1973; Ellis, 1973b).

Protein synthesis by isolated chloroplasts

The rationale of our approach is that in order to produce identifiable proteins, we must use conditions in which correct termination and release of the polypeptide chains occur; initiation is not a requirement for purposes of identification. Such conditions are

Table 1. Characteristics of light-driven protein synthesis by isolated pea chloroplasts

Energy Source	Treatment	Incorporation
Light	Complete	100
None	Zero time	0.5
None	Complete	3
ATP + ATP-generating system	Complete	50
Light + ATP + ATP generating system	Complete	125
Light	Lysed	3
Light	+ Ribonuclease (30 μg/ml)	95
Light	+ CCCP (5 μM)	6
Light	+ DCMU (1 μM)	38
Light	+ D-threo chloramphenicol (50 μg/ml)	5
Light	+ Lincomycin (5 μM)	25
Light	+ Cycloheximide (100 μg/ml)	100
Light	+ Actinomycin D (10 μg/ml)	100
Light	K^+ replaced by Na^+	5
Light	K^+ reduced to 25 mM	25

Pea seeds (*Pisum sativum* var. Feltham First) were grown in compost for 7–10 days under a 12 hr photoperiod of 2000 lux white light. Apical leaves (15 g) were homogenized for 4 sec (Polytron homogenizer) in 100 ml of a sterile ice slurry containing 0.35 M sucrose, 25 mM HEPES-NaOH, 2 mM EDTA and 2 mM sodium isoascorbate (pH 7.6). The homogenate was immediately strained through 8 layers of muslin and centrifuged at 2500 *g* for 1 min at 0°C. The pellet was resuspended by cotton wool in 1–5 ml of sterile 0.2 M KCl, 66 mM tricine-KOH, 6.6 mM $MgCl_2$ (pH 8.3). The suspension contains 40–50 per cent intact chloroplasts as judged by phase-contrast microscopy. Chloroplasts (300 μl) were incubated in a final volume of 500 μl with 0.5 μCi of either [^{14}C]-L-leucine (3 μM) or [^{35}S]-L-methionine (36 nM). Tubes were illuminated at 20°C with filtered red light (4000 lux) for 40 min. Protein was extracted, and counted by liquid scintillation spectrometry. Incorporation by the complete light-driven system is called 100, and represents a rate of incorporation of between 0.5 and 1.0 nmole [^{14}C]leucine per mg chlorophyll per hr. The ATP and ATP-generating system consists of 2 mM ATP, 5 mM creatine phosphate and 100 μg/ml creatine phosphokinase. Chloroplasts were lysed by resuspending in 25 mM tricine-KOH, 10 mM $MgSO_4$, 5 mM 2-mercaptoethanol (pH 8.0), followed by addition of KCl to 0.2 M.

more likely to be met in intact chloroplasts rather than in lysed systems that are commonly used. We therefore used methods that were originally developed to isolate intact chloroplasts capable of high rates of photosynthetic CO_2 fixation. It is characteristic of such preparations that they use light as the source of energy in the absence of added cofactors, such as ADP or ferredoxin. The preparation of intact chloroplast suspensions is described in Table 1, which also shows characteristics of the ability of the preparations to incorporate labelled amino acids into protein. In the pea system, light can only be partially replaced as an energy source by added ATP and an ATP-generating system, while the use of both ATP and light produces only a slight stimulation of incorporation. Incorporation in the light is destroyed by lysis of the chloroplasts or by addition of inhibitors of photophosphorylation to the intact system. Antibiotics known to inhibit protein synthesis by prokaryote ribosomes also inhibit incorporation by the intact system. In contrast, incorporation is not inhibited by cycloheximide nor by ribonuclease. Actinomycin D also fails to inhibit incorporation when added at a concentration of 10 μg/ml, but does inhibit light-dependent incorporation of uridine into RNA in the intact system by about 85 per cent. The system shows an absolute requirement for K^+ ions which cannot be replaced by Na^+ ions.

We conclude from all these characteristics that protein synthesis proceeds in intact chloroplasts only, that it is being powered by photophosphorylation, and probably involves messenger RNA synthesized before the chloroplasts were isolated. The question now is—does this system make discrete proteins?

Analysis of products of protein synthesis by isolated chloroplasts

After incubation with [^{35}S]-L-methionine or [^{14}C]-L-leucine, chloroplasts are dissolved in SDS and analysed by SDS polyacrylamide gel electrophoresis. Six discrete labelled peaks are found, all of which are removed by treatment with pronase. If the chloroplasts are osmotically lysed after incubation, and the green membrane fraction centrifuged down, five of these labelled proteins remain with the membranes (Eaglesham and Ellis,

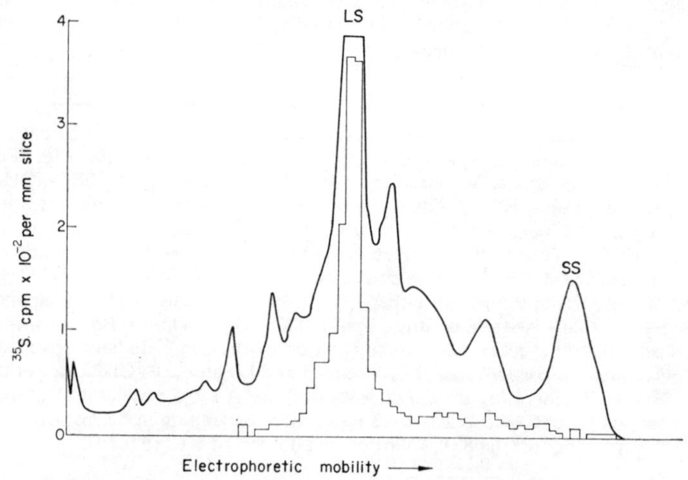

FIG. 2. SDS gel electrophoresis of the soluble fraction from intact isolated chloroplasts incubated with [^{35}S]methionine. Pea chloroplasts were isolated and incubated with [^{35}S]methionine in the light as described in Table 1. Chloroplasts were then lysed, centrifuged at 150,000 *g* for 1 hr, and the supernatant fraction run on 10 per cent SDS polyacrylamide gels as described by Blair and Ellis (1973). Gels were stained in Amido Black and sliced into 1 mm portions for counting by scintillation spectrometry. The smooth line represents absorbance at 620 nm, and the histogram indicates radioactivity in each 1 mm slice. LS and SS denote the large and small subunits of Fraction I protein respectively.

1974). The remaining peak, which is the most highly labelled, is not sedimented even when free ribosomes are removed by centrifugation (Blair and Ellis, 1973). Figure 2 shows the pattern obtained when the soluble phase of incubated chloroplasts is analysed on an SDS gel. One labelled peak is seen which moves coincident with the large subunit of Fraction I protein; the small subunit is not labelled, although it can be labelled with both methionine and leucine if these are fed to intact pea shoots.

This very clear result is highly reproducible; it has been obtained in over fifty experiments with isolated pea chloroplasts, and has also been found with spinach and barley chloroplasts. The same clear picture is seen if as much as 500 μCi of the highest specific activity [^{35}S]methionine available is used in each incubation tube. Very slight traces of labelled material with mobilities greater than the major peak are sometimes seen but are too poorly labelled to permit analysis. To confirm that the major peak is due to the large subunit, tryptic peptides of authentic large subunit produced by feeding [^{35}S]methionine to pea shoots were compared with those prepared from the soluble labelled products of isolated pea chloroplasts. The excellent correspondance of peptide maps we regard as convincing evidence that chloroplast ribosomes make the large subunit of Fraction I protein as the sole detectable soluble product (Blair and Ellis, 1973). Confirmatory evidence has since been reported for wheat chloroplasts, although the immunological methods used in this case did not give such clear-cut results (Gooding *et al.*, 1973).

The large subunit is made by free chloroplast ribosomes

In isolated pea chloroplasts the ratio of membrane-bound ribosomes to free ribosomes is about 1:2, but *in vivo* the ratio is probably nearer 1:4 (Tao and Jagendorf, 1973). It is possible that the two classes of ribosomes have distinct roles; free ribosomes might synthesize the large subunit of Fraction I protein as their sole product, while bound ribosomes might synthesize only membrane-bound proteins. We have tested this possibility by persuading lysed and fractionated chloroplasts to synthesize discrete proteins.

The method devised consists of isolating chloroplasts as described in Table 1, but then resuspending them in a hypotonic buffer containing 25 mM tricine-KOH, 10 mM $MgSO_4$ and 5 mM 2-mercaptoethanol (pH 8.0). This causes immediate lysis as judged by phase contrast microscopy. The lysed preparation will incorporate [^{35}S]methionine into protein if given ATP and GTP, but in contrast to intact chloroplasts does not use light energy (Table 2). Incorporation is sensitive to ribonuclease and D-threochloramphenicol, and requires 100 mM KCl for optimal rates (Table 2). The rate of incorporation by lysed chloroplasts is similar to that by intact chloroplasts. Analysis of the products of protein synthesis by such lysed preparations shows that the electrophoretic pattern is very similar to that given by intact chloroplasts using light energy (Fig. 3). The most highly labelled peak runs coincident with the large subunit of Fraction I protein.

When lysed chloroplasts are fractionated by centrifugation at 38,000 *g* for 5 min, both the colourless supernatant fraction and the green membrane fraction incorporate [^{35}S]methionine into protein when supplied with ATP and GTP (Table 3). If the membrane fraction is washed by resuspension in lysis medium and recentrifuged, its activity is much reduced. This is expected since soluble factors such as tRNA and amino acid activating enzymes will be removed by this procedure. However, the activity of the

TABLE 2. CHARACTERISTICS OF PROTEIN SYNTHESIS BY LYSED PEA CHLOROPLASTS

Energy Source	Treatment	Incorporation
ATP	Complete	100
Light	Complete	3
Light + ATP	Complete	80
ATP	+ Ribonuclease (30 μg/ml)	6
ATP	+ D-threochloramphenicol (150 μg/ml)	18
ATP	+ Cycloheximide (300 μg/ml)	90
ATP	10 mM K^+	36
ATP	200 mM K^+	50
ATP	100 mM Na^+ in place of K^+	15

Chloroplasts were lysed as described in the text and incubated with [^{35}S]methionine, 2 mM ATP, 0.2mM GTP and 100 mM KCl for 40 min at 25°C. The amount of labelled protein formed in the complete ATP-driven system is called 100.

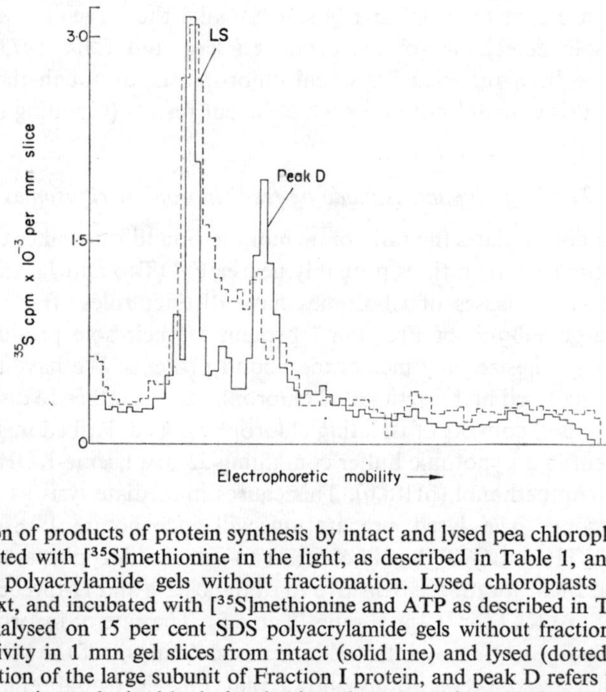

FIG. 3. Comparison of products of protein synthesis by intact and lysed pea chloroplasts. Intact chloroplasts were incubated with [^{35}S]methionine in the light, as described in Table 1, and then analysed in 15 per cent SDS polyacrylamide gels without fractionation. Lysed chloroplasts were prepared, as described in the text, and incubated with [^{35}S]methionine and ATP as described in Table 2; the preparation was then analysed on 15 per cent SDS polyacrylamide gels without fractionation. Histograms represent radioactivity in 1 mm gel slices from intact (solid line) and lysed (dotted line) chloroplasts. LS marks the position of the large subunit of Fraction I protein, and peak D refers to the major membrane-bound protein synthesized by isolated chloroplasts. Data from Eaglesham and Ellis (1974).

washed membrane fraction can be partially restored by supplementing with a supernatant prepared by centrifugation of lysed chloroplasts at 200,000 *g* for 3 hr. This high-speed supernatant has no protein synthetic activity of its own, since it lacks ribosomes (Table 3).

The products of protein synthesis by free chloroplast ribosomes, and by membrane-bound ribosomes supplemented by the high-speed supernatant fraction, have been

TABLE 3. PROTEIN SYNTHESIS BY FRACTIONS PREPARED FROM LYSED PEA CHLOROPLASTS

Fraction	Incorporation
Total lysed	100
38,000 *g* supernatant	31
38,000 *g* pellet	18
Washed 38,000 *g* pellet	2.6
200,000 *g* supernatant	0
Washed pellet + 200,000 *g* supernatant	10

Chloroplasts were lysed and fractionated as described in the text, and incubated as described in Table 2 for the ATP-driven system. Results are expressed relative to the incorporation given by lysed unfractionated chloroplasts.

analysed on SDS polyacrylamide gels. Free ribosomes make a product which has the same electrophoretic mobilities as the large subunit of Fraction I protein (Fig. 4); its identity has been confirmed by tryptic peptide analysis (Fig. 5). Free ribosomes also make some lower molecular weight products which have the same mobility as membrane-bound products made by intact chloroplasts, but in this case these products are largely soluble, and do not centrifuge down at 200,000 *g* for 3 hr. Identities of these lower molecular weight products are unknown; they are not proteolytic breakdown products of the

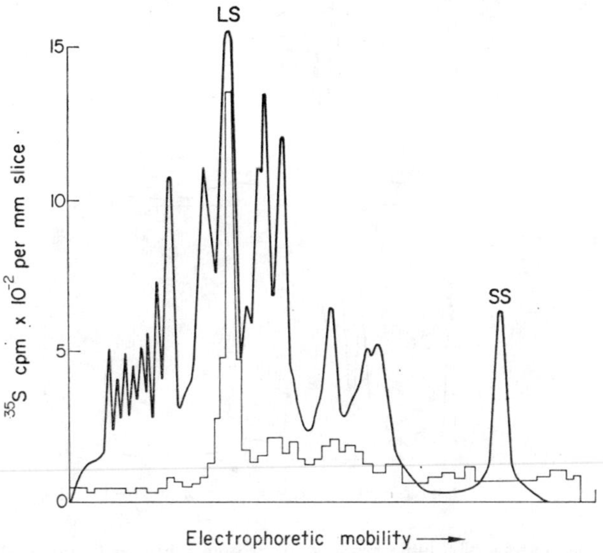

FIG. 4. Soluble products of protein synthesis by free chloroplast ribosomes. The supernatant fraction from lysed pea chloroplasts was incubated with [^{35}S]methionine and ATP, as described in Table 3. The preparation was then centrifuged at 200,000 *g* for 3 hr to remove ribosomes, and analysed on 15 per cent SDS polyacrylamide gels. Gels were stained in Coomassie Blue, and sliced into 1 mm fractions. The smooth line represents absorbance at 620 nm, and the histogram radioactivity in the gel slices. LS and SS mark the large and small subunits of Fraction I protein respectively.

large subunit, and experiments with cycloheximide and chloramphenicol suggest they are not synthesized by contaminating cytoplasmic ribosomes. One possibility is that they are soluble forms of membrane proteins which fail to attach to the membranes in lysed chloroplasts. The products of protein synthesis by the supplemented bound ribosomes are all membrane-bound, and do not include the large subunit of Fraction I protein.

The results of this series of experiments suggest that there is a division of labour in the chloroplast—the large subunit of Fraction I protein is made only by free ribosomes but both types of ribosomes may be implicated in the synthesis of the membrane-bound proteins.

Evidence about the small subunit

We have never observed any incorporation of labelled amino acids into the small subunit of Fraction I protein by isolated intact or lysed pea chloroplasts. It has been reported that incorporation of labelled amino acids into the small subunit by greening barley leaves is inhibited preferentially by cycloheximide (Criddle *et al.*, 1970). It seems reasonable to conclude that the small subunit is synthesized by cytoplasmic ribosomes, and recently positive *in vitro* evidence for this has been reported. Gray and Kekwick (1974)

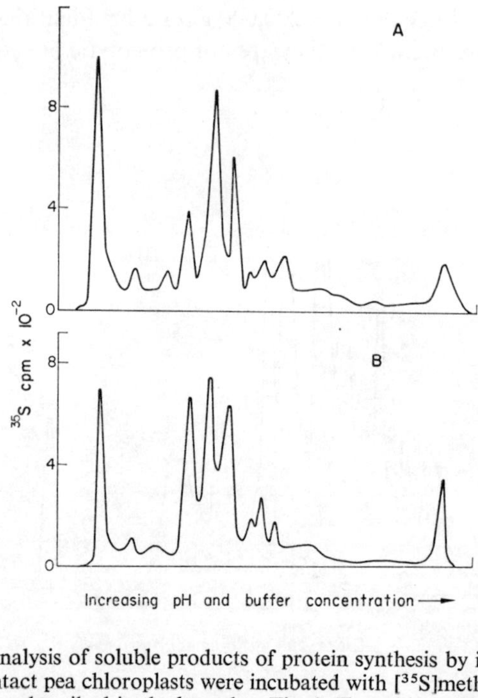

FIG. 5. Tryptic peptide analysis of soluble products of protein synthesis by intact chloroplasts and free chloroplast ribosomes. Intact pea chloroplasts were incubated with [^{35}S]methionine in the light, and the soluble fraction isolated as described in the legend to Fig. 2. Free chloroplast ribosomes were prepared as described in the text, and incubated with [^{35}S]methionine and ATP; the incubated mixture was then centrifuged at 200,000 *g* for 3 hr and the soluble fraction removed. Soluble fractions were digested with trypsin, and peptides separated on an Aminex A5 ion-exchange column by means of an increasing gradient of pH and buffer concentration. (A) product of intact chloroplasts; (B) product of free ribosomes.

found that cytoplasmic polysomes isolated from greening *Phaseolus* leaves synthesize protein which can be precipitated by an antiserum against the small subunit; tryptic peptide analysis was not reported, and would be desirable.

Two remaining features of our model are more speculative. The dashed lines in Fig. 1 represent one way in which the synthesis of the two subunits in the different cellular compartments may be integrated. The small subunit may be a positive factor required for initiation of either transcription or translation of the messenger RNA for the large subunit. The only evidence for this idea comes from the rapid inhibition of incorporation of labelled [^{35}S]methionine into the large subunit in pea shoots by cytoplasmic ribosomal inhibitors, and other explanations cannot be ruled out (Ellis, 1975). We have made attempts to stimulate *in vitro* synthesis of the large subunit by adding the small subunit to isolated chloroplasts. These attempts were unsuccessful but a negative result is not necessarily meaningful because preparation of the small subunit requires denaturing conditions. The other problem is the mechanism by which the small subunit, along with all other soluble proteins of the chloroplasts which are made on cytoplasmic ribosomes, passes through the chloroplast envelope. This mechanism must be specific, and since neither pores nor pinocytosis have been reported for chloroplasts, I am inclined to think in terms of protein carriers. The chloroplast envelope might contain a protein which recognizes a site common to all those chloroplast proteins made in the cytoplasm; this would ensure the required specificity, but how such a complex could move the adsorbed protein across the outer membrane is as obscure as in the case of inorganic ion uptake. We have shown that pea chloroplast envelopes contain a spectrum of proteins that is different from the spectrum of proteins from chloroplast lamellae; at least two of the envelope proteins are synthesized by chloroplast ribosomes (Joy and Ellis, 1975). Experiments to persuade isolated chloroplasts to take up labelled proteins are underway in my laboratory.

DETECTION OF THE MESSENGER RNA FOR THE LARGE SUBUNIT

Our discovery that the large subunit of Fraction I protein is a major product of protein synthesis by chloroplast ribosomes prompted us to see whether we could detect the messenger for this subunit by adding chloroplast RNA to a heterologous protein-synthesizing system. This work was carried out by M. R. Hartley and Annabel Wheeler on spinach chloroplasts, because spinach RNA is less degraded on isolation than pea RNA. Nucleic acid was extracted from washed chloroplasts by the phenol-detergent method of Parish and Kirby (1966) with modifications described by Hartley and Ellis (1973); DNA was removed by incubation with RNAase-free DNAase. For the cell-free protein-synthesizing system, the S-30 preparation from *Escherichia coli* was used (Capecchi, 1966), because of the similarities between chloroplast and prokaryote protein synthesis. To check that our S-30 extracts were correctly translating added messenger RNA, we first studied the translation of a well-characterized messenger RNA—that from bacteriophage MS2. Addition of MS2 RNA to the S-30 extract stimulated [^{35}S]methionine incorporation into protein ten-fold (Table 4), and analysis of the products on SDS polyacrylamide gels showed that a major discrete peak of molecular weight about 14,000 appeared; this peak coelectrophoresed exactly with MS2 coat protein. We conclude from this result that initiation, translation, and termination from exogenous phage messenger RNA is occurring correctly in our S-30 extracts. When spinach chloroplast RNA is added to the S-30 system, a ten-fold stimulation of protein synthesis is again observed (Table 4). The greater amount of chloroplast RNA, relative to MS2 RNA, required to achieve saturation of incorporation probably reflects the fact that the chloroplast RNA is largely composed of ribosomal and transfer RNA. We have used unfractionated chloroplast RNA in these experiments because of the difficulty of

TABLE 4. STIMULATION BY ADDED CHLOROPLAST RNA OF PROTEIN SYNTHESIS BY AN *Escherichia coli* CELL-FREE SYSTEM

Treatment	Incorporation into protein (cpm)	Incorporation as % of endogenous
Endogenous	13,590	100
Zero time	551	4
+ MS2 RNA (2.5 μg)	29,864	220
+ MS2 RNA (17.5 μg)	144,201	1004
+ MS2 RNA (25.0 μg)	139,725	1002
+ spinach chloroplast RNA (250 μg)	99,150	732
+ spinach chloroplast RNA (500 μg)	132,450	975
+ spinach chloroplast RNA (1000 μg)	112, 783	830

An S-30 extract was prepared from *Escherichia coli* by the method of Capecchi (1966) and 150 μg protein added to a final volume of 0.2 ml with 3 mM ATP, 2 mM GTP, 5 mM phosphoenolpyruvate, 1 μg pyruvate kinase, 10 mM magnesium acetate, 50 mM tris-HCl (pH 7.8), 10 mM glutathione, 20 μg citrovorum, and 1 μCi [^{35}S]-L-methionine (specific activity 120–250 Ci/mmole). Mixtures were incubated for 15 min at 35°C.

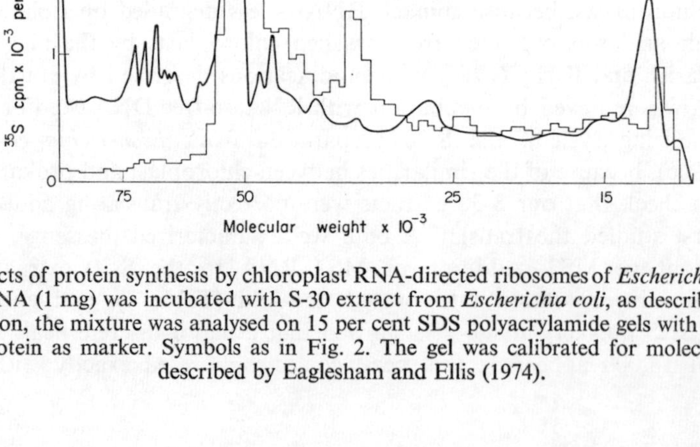

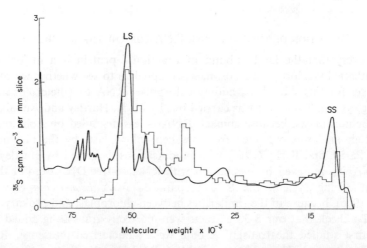

FIG. 6. Products of protein synthesis by chloroplast RNA-directed ribosomes of *Escherichia coli*. Spinach chloroplast RNA (1 mg) was incubated with S-30 extract from *Escherichia coli*, as described in Table 4. After incubation, the mixture was analysed on 15 per cent SDS polyacrylamide gels with added spinach Fraction I protein as marker. Symbols as in Fig. 2. The gel was calibrated for molecular weight as described by Eaglesham and Ellis (1974).

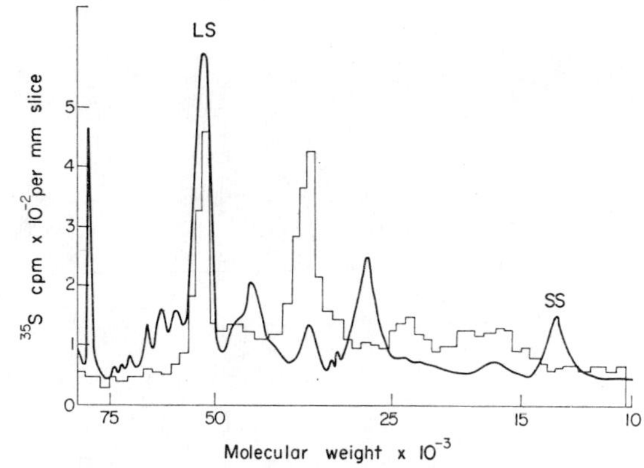

FIG. 7. Products of protein synthesis by isolated intact spinach chloroplasts. Chloroplasts were isolated from young spinach leaves by the methods described in Table 1, and incubated with [^{35}S]methionine in the light for 30 min at 20°C. An aliquot containing 30 μg chlorophyll was analysed on a 15 per cent SDS polyacrylamide gel. Symbols as in Fig. 2.

predicting the molecular weight of a messenger RNA from the molecular weight of the protein it encodes (Mathews, 1973). Analysis of the products of chloroplast RNA-stimulated protein synthesis on SDS gels reveals the pattern shown in Fig. 6. Two discrete products are seen, one with a molecular weight of about 52,000 and having the same mobility as the large subunit of Fraction I protein, and the other with a molecular weight of about 35,000. This pattern is very similar to the pattern of proteins synthesized by intact isolated spinach chloroplasts (Fig. 7) where the product of molecular weight about 35,000 is the most highly labelled membrane-bound protein. This similarity between the products of isolated chloroplasts and of the chloroplast RNA-directed heterologous system suggests that the chloroplast messengers are being correctly translated.

To be certain that the product of molecular weight 52,000 made by the heterologous system really is the large subunit of Fraction I protein, it was eluted from SDS gels on a preparative scale, and hydrolysed with chymotrypsin. The resulting peptides were separated by ion exchange chromatography, and their elution profile compared with that of peptides from authentic large subunit, prepared after feeding [^{35}S]methionine to intact spinach plants. The similarities between the two profiles leaves us in no doubt that the added chloroplast RNA is programming the bacterial ribosomes to synthesize the large subunit of Fraction I protein (Fig. 8).

PROSPECTS

Detection of messenger RNA for the large subunit of Fraction I protein opens up two immediate research prospects:

(1) establishment of a quantitative assay for the messenger RNA, so that variations in its amount in different developmental situations can be measured; light-stimulated synthesis of Fraction I protein in etiolated plants will be of especial interest in this regard,

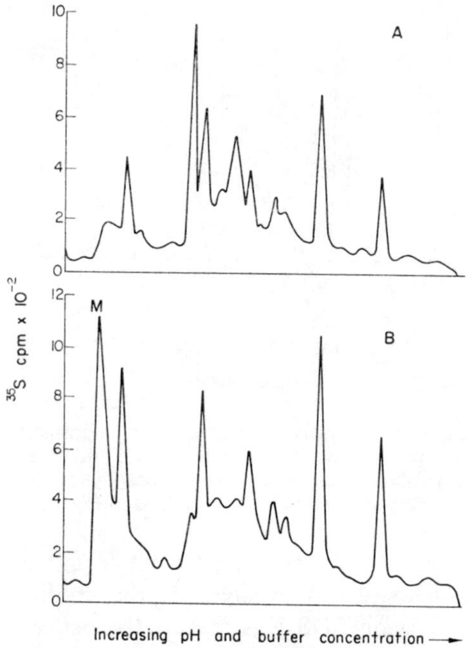

FIG. 8. Chymotrypic peptide analysis of products of chloroplast RNA-directed protein synthesis by *Escherichia coli* ribosomes. (A) elution pattern of chymotryptic peptides from [^{35}S]methionine labelled large subunit of Fraction I protein extracted from spinach plants; (B) elution pattern of chymotryptic peptides prepared from the large subunit area of the gel shown in Fig. 6. The peak labelled M is free [^{35}S]methionine which comes through the procedure in the *in vitro* case.

(2) purification of the messenger RNA, so that its properties can be studied. Availability of purified messenger RNA will allow us to study the translational control possibilities suggested in Fig. 1, and will enable us to check by DNA/RNA hybridization the presumed location of the gene for this messenger in chloroplast DNA. We are currently engaged in purifying this messenger, but are hampered by its failure to bind to oligo dT-cellulose under conditions where poly A binds. Since poly A tracts have been detected in animal messengers, but not in those from prokaryotes, this result is not unexpected in view of the other similarities between prokaryotes and chloroplasts. The lack of poly A does, however, preclude us from using the quickest method for purifying messengers.

It is my belief that detection of the messenger RNA for the large subunit of Fraction I protein will open up a new area of research on the molecular events underlying the development of chloroplasts. The abundance of the *small* subunit of Fraction I protein also offers the prospect of isolating a cytoplasmic messenger RNA from plants.

REFERENCES

ALLENDE, J. E. (1969) Protein biosynthesis in plant systems. In *Techniques in Protein Biosynthesis*, ed. CAMPBELL, J. N. and SARGENT, J. R., Vol. 2, pp. 55–100, Academic Press, London and New York.

BLAIR, G. E. and ELLIS, R. J. (1973) Protein synthesis in chloroplasts. I. Light-driven synthesis of the large subunit of Fraction I protein by isolated pea chloroplasts. *Biochim. biophys. Acta* **319,** 223–234.
BOULTER, D., ELLIS, R. J. and YARWOOD, A. (1972) Biochemistry of protein synthesis in plants. *Biol. Rev.* **47,** 113–175.
CAPECCHI, M. R. (1966) Cell-free protein synthesis programmed with R17 RNA: identification of two phage proteins. *J. molec. Biol.* **21,** 173–193.
CHAN, P. and WILDMAN, S. G. (1972) Chloroplast DNA codes for the primary structure of the large subunit of Fraction I protein. *Biochim. biophys. Acta* **277,** 677–680.
CRIDDLE, R. S., DAU, B., KLEINKOPF, G. E. and HUFFAKER, R. C. (1970) Differential synthesis of ribulosediphosphate carboxylase subunits. *Biochem. biophys. Res. Commun.* **41,** 621–627.
DARNELL, .J E., JELINEK, W. R. and MOLLOY, F. R. (1973) Biogenesis of mRNA; genetic regulation in mammalian cells. *Science N.Y.* **181,** 1215–1221.
EAGLESHAM, A. R. J. and ELLIS, R. J. (1974) Protein synthesis in chloroplasts. 2. Light-driven synthesis of membrane proteins by isolated pea chloroplasts. *Biochim. biophys. Acta* **335,** 396–407.
ELLIS, R. J. (1963) Chloramphenicol and the uptake of salt in plants. *Nature Lond.* **200,** 596.
ELLIS, R. J. (1973a) Fraction I protein. *Comment. Pl. Sci.* **4,** 29–38.
ELLIS, R. J. (1973b) The biogenesis of chloroplasts: protein synthesis by isolated chloroplasts. *Biochem. Soc. Trans.* **1,** 13–16.
ELLIS, R. J. (1975) Inhibition of chloroplast protein synthesis by lincomycin and 2-(4-methyl-2,6-dinitroanilino)-N-methyl-proprionamide. *Phytochemistry* **14,** 89–93.
ELLIS, R. J. and MACDONALD, I. R. (1970) Specificity of cycloheximide in higher plant systems. *Pl. Physiol. Lancaster* **46,** 227–232.
ELLIS, R. J., BLAIR, G. E. and HARTLEY, M. R. (1973) The nature and function of chloroplast protein synthesis. In *Nitrogen Metabolism in Plants,* ed. GOODWIN, T. W. and SMELLIE, R. M. S., *Biochem. Soc. Symp.* **38,** 137–162, The Biochemical Society, London.
GOODING, L. R., ROY, H. and JAGENDORF, A. T. (1973) Immunological identification of nascent subunits of wheat ribulose diphosphate carboxylase on ribosomes of both chloroplast and cytoplasmic origin. *Archs Biochem. Biophys.* **159,** 324–335.
GRAY, J. C. and KEKWICK, R. G. O. (1974) The synthesis of the small subunit of ribulose 1,5-biphosphate carboxylase in the French bean *Phaseolus vulgaris. Eur. J. Biochem.* **44,** 491–500.
GURDON, J. B. (1974) *The Control of Gene Expression in Animal Development,* Clarendon Press, Oxford.
HARTLEY, M. R. and ELLIS, R. J. (1973) Ribonucleic acid synthesis in chloroplasts. *Biochem. J.* **134,** 249–262.
JOY, K. W. and ELLIS, R. J. (1975) Protein synthesis in chloroplasts. 4. Polypeptides of the chloroplast envelope. *Biochim. biophys. Acta* **378,** 143–151.
KAWASHIMA, N. and WILDMAN, S. G. (1970) Fraction I protein. *A. Rev. Pl. Physiol.* **21,** 325–358.
KAWASHIMA, N. and WILDMAN, S. G. (1972) Studies on Fraction I protein. IV. Mode of inheritance of primary structure in relation to whether chloroplast or nuclear DNA contains the code for a chloroplast protein. *Biochim. biophys. Acta* **262,** 42–49.
KIRK, J. T. O. (1972) The genetic control of plastid formation; recent advances and strategies for the future. *Sub-cell. Biochem.* **1,** 333–361.
LEVINE, R. P. and GOODENOUGH, V. W. (1970) The genetics of photosynthesis and of the chloroplast in *Chlamydomonas reinhardii, A. Rev. Genet.* **4,** 397–408.
LOCKARD, R. E. and LINGREL, J. B. (1971) Identification of mouse haemoglobin messenger RNA, *Nature Lond.* **233,** 204–206.
MARCUS, A. (1972) Protein synthesis in extracts of wheat embryo. In *Protein Biosynthesis in Nonbacterial Systems,* ed. LAST, J. A. and LASKIN, A. I. pp. 128–145, Marcel Dekker, New York.
MATHEWS, M. B. (1973) Mammalian messenger RNA. In *Essays in Biochemistry,* ed. CAMPBELL, P. N. and DICKENS, F., Volume 9, pp. 59–102, Academic Press, London and New York.
MCFADDEN, B. A. (1973) Autotrophic CO_2 assimilation and the evolution of ribulose diphosphate carboxylase. *Bacteriol. Rev.* **37,** 3, 289–319.
MODELELL, J. (1971) The S-30 system from *Escherichia coli,* In *Protein Synthesis in Bacterial Systems,* ed. LAST, J. A. and LASKIN, A. I., Marcel Dekker, New York.
MONOD, J., JACOB, F. and GROS, F. (1962) Structural and rate-determining factors in the biosynthesis of adaptive enzymes, In *The Structure and Biosynthesis of Macromolecules,* ed. BELL, D. J. and GRANT, J. K. *Biochem. Soc. Symp.* **21,** 104–131, Cambridge University Press.
NISHIMURA, M. and AKAZAWA, T. (1973) Further proof for the catalytic role of the large subunit in the spinach leaf ribulose-1,5-diphosphate carboxylase. *Biochem. biophys. Res. Commun.* **54,** 842–848.
PARISH, J. H. and KIRBY, K. S. (1966) Reagents which reduce interactions between ribosomal RNA and rapidly labelled RNA from rat liver. *Biochim. biophys. Acta* **129,** 554–562.

ROBERTS, B. E. and PATERSON, B. M. (1973) Efficient translation of tobacco mosaic virus RNA and rabbit globin 9S RNA in a cell-free system from commercial wheat-germ. *Proc. natn. Acad. Sci. U.S.A.* **70,** 2330–2334.

SAKANO, K., KUNG, S. D. and WILDMAN, S. G. (1974) Identification of several chloroplast DNA genes which code for the large subunit of *Nicotiana* Fraction I proteins. *Molec. gen. Genet.* **130,** 91–97.

TAO, K. J. and JAGENDORF, A. T. (1973) The ratio of free to membrane-bound chloroplast ribosomes. *Biochim. biophys. Acta* **324,** 518–532.

VERMA, D. P. S., NASH, D. T. and SCHULMAN, H. M. (1974) Isolation and *in vitro* translation of leg-haemoglobin mRNA from soybean root nodules. *Pl. Physiol. Lancaster*. Suppl. June 1974, p. 9.

WATSON, J. D. (1970) Embryology at the molecular level. In *Molecular Biology of the Gene*, 2nd ed. pp. 506–560, W. A. Benjamin, New York.

WILSON, S. B. and MOORE, A. L. (1973) The effects of protein synthesis inhibitors on oxidative phosphorylation by plant mitochondria, *Biochim. biophys. Acta* **292,** 603–610.

MOLECULAR BASIS OF CELL DIFFERENTIATION IN *BLASTOCLADIELLA EMERSONNII*

C. J. LEAVER and J. S. LOVETT*

Department of Botany, University of Edinburgh, Edinburgh, U.K.

INTRODUCTION

Since its discovery in 1951 by Cantino, the non-filamentous aquatic fungus, *Blastocladiella emersonii* has proved to be a model system for the experimental study of cell differentiation. The fungus can be grown in large scale, synchronous cultures, and by subtle changes in the culture medium entire cell populations induced to initiate and complete differentiative events under highly controlled conditions. Cells can be obtained in a sequence of differentiative states in quantities sufficient to permit meaningful biochemical analysis, and changes occurring at the molecular level can thus be correlated with observable morphological events within a single cell.

The methodology of culture has been described by Lovett (1967), and effects of cellular and environmental variables on synchrony and rate of zoospore germination studied by Soll and Sonneborn (1972). Early attempts to relate structural changes to protein and RNA synthesis have been reported by Murphy and Lovett (1966), Lovett (1968) and Soll and Sonneborn (1971a, b). In this paper we concentrate on recent work designed to elucidate the mechanism by which synthesis of both RNA and protein is regulated in the fungus and in particular how regulation of these macromolecules participates in the control of specific biochemical and cytological events during differentiation of the cell.

THE LIFE CYCLE

Early work on the life cycle of the fungus was summarized by Cantino and Lovett in 1964. Since then observations made in the light microscope have been substantiated and extended in a series of electron microscopy studies (Fuller, 1966; Reichle and Fuller, 1967; Lessie and Lovett, 1968; Lovett, 1968; Soll *et al.*, 1969; Truesdell and Cantino, 1971; Barstow and Lovett, 1974a, b) and we now have a composite picture of the main intracellular events.

The life cycle consists essentially of four distinct phases which may be termed, the motile zoospore, germination, growth and sporulation phases. Germination results in the transformation of motile zoospores into sessile germlings (encystment), which after a period of growth and differentiation develop into zoosporangia each containing many zoospores.

*Present address: Department of Biological Sciences, Purdue University, West Lafayette, Indiana, U.S.A.

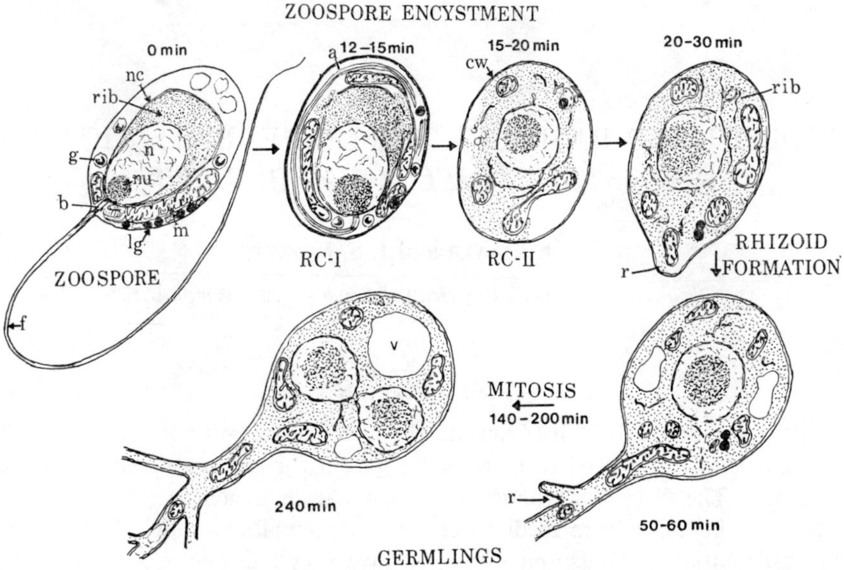

FIG. 1. Diagrammatic representation of zoospore germination and early development in *Blastocladiella emersonii*. Symbols: b, basal body; g, gamma particle; m, mitochondrion; rib, ribosomes; nc, nuclear cap; nu, nucleolus; n, nucleus; lg, lipid granules; a, flagellar axoneme; cw, cell wall; r, rhizoid; f, flagellum; v, vacuole; RC-I, round cell I; RC-II, round cell II.

Motile zoospore phase

The motile uninucleate zoospores are variably ovoid in shape (Figs. 1, 2A), measure 7–9 μm, and are propelled by a posterior flagellum of the whiplash type. The zoospore is characterized by the absence of a cell wall and contains a highly organized assembly of cell organelles. The principal features of this assembly are:

(i) a single large mitochondrion which surrounds the basal body and 'rootlets' of the flagellum,
(ii) lipid and other unidentified granules restricted to a narrow zone between the outer membrane and mitochondrion,
(iii) small cup-shaped granules enclosed in vesicles (gamma particles), and
(iv) a prominent cap which encloses the anterior two-thirds of the nucleus—the nuclear cap is bounded by a typical double-layered membrane and all the cytoplasmic ribosomes carried over from the previous growth-phase plants are confined within it.

Zoospores can swim about for relatively long periods in the absence of exogenous nutrients; they respire and metabolize endogenous reserves (largely glycogen, see Suberkropp and Cantino, 1973) but show neither detectable synthesis of DNA, RNA and protein, nor do they grow.

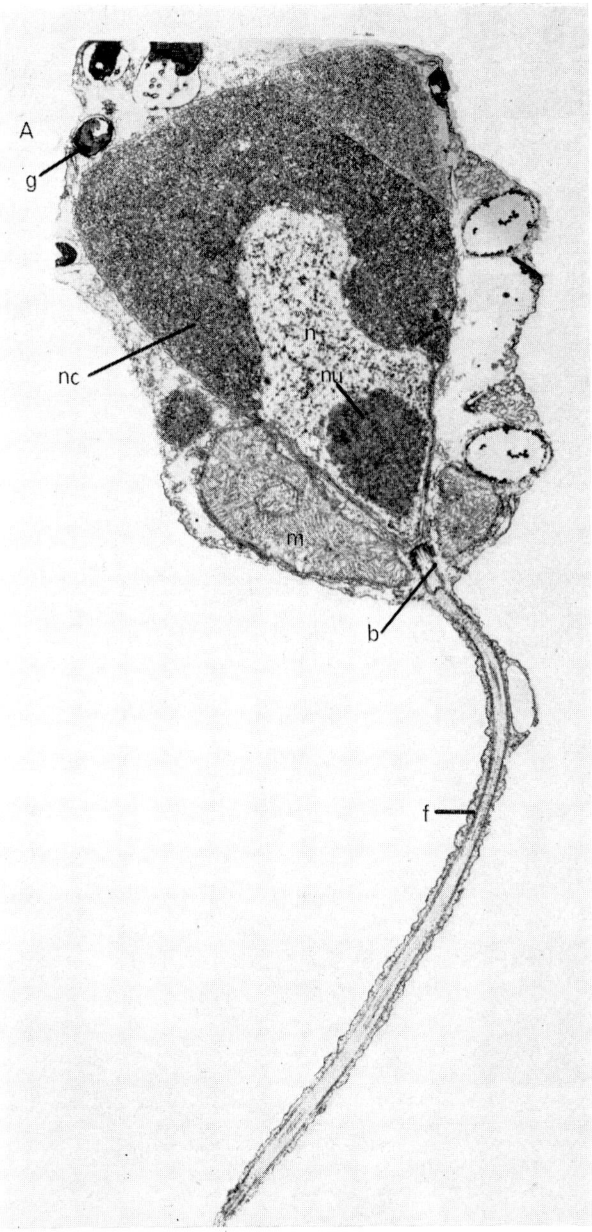

FIG. 2A. Electron micrograph to show intracellular organization of a zoospore of *Blastocladiella emersonii*. Symbols as in Fig. 1.

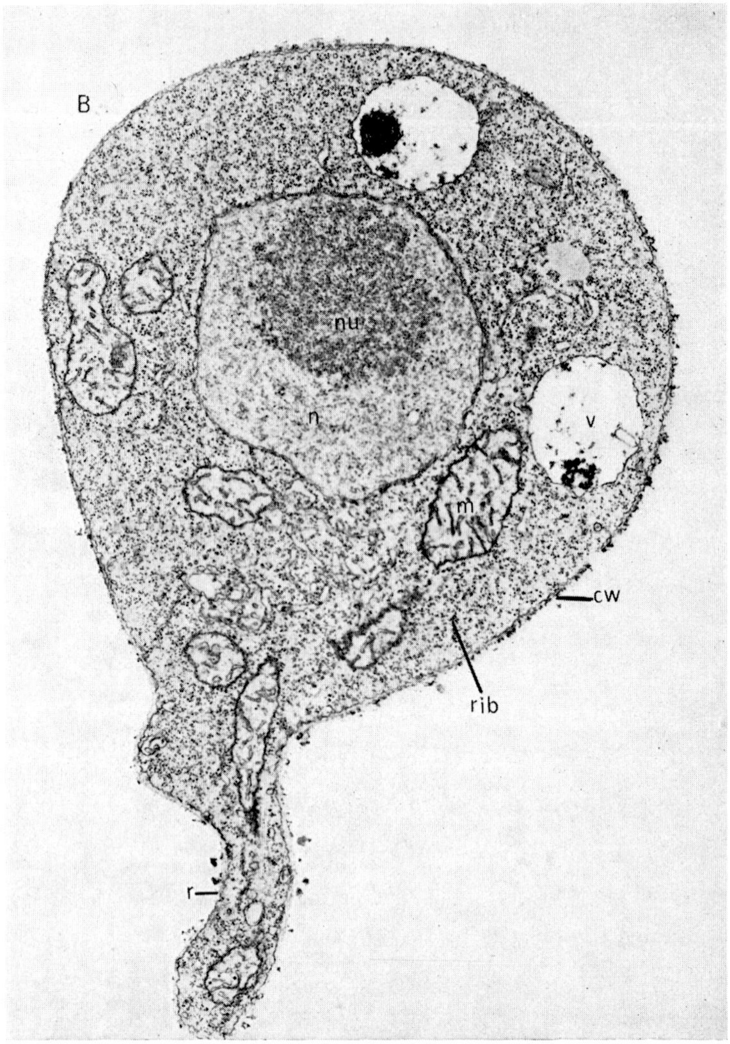

FIG. 2B. Electron micrograph to show intracellular organization of a germling of *Blastocladiella emersonii* at 60 min. Symbols as in Fig. 1.

Germination phase

Synchronous germination is induced by inoculating zoospores into an appropriate growth medium and aerating the resulting suspensions in a suitable culture vessel. Transformation of motile cells into sessile germlings is accompanied by a series of abrupt changes in cell organization which have been referred to by Soll *et al.* (1969) as the *round cell I* (RC-I), *round cell II* (RC-II) and *germling* stages (see Fig. 1).

RC-I

Germination commences by deposition of a thin, rigid chitinous wall around the zoospore. At the same time, the flagellum retracts and is incorporated within the wall (Lovett, 1968; Soll *et al.*, 1969; Truesdell and Cantino, 1971). The single mitochondrion begins to elongate, the gamma particles disappear, and lipid droplets are dispersed. Completion of RC-I is identified by commencement of disorganization of the nuclear-cap membrane.

RC-II

The nuclear cap is completely broken down and the enclosed ribosomes dispersed throughout the cell. The flagellar apparatus is also broken down, while the mitochondrion becomes highly elongate with branches permeating throughout the cell (Bromberg, 1974). Nucleolar activity is also indicated by a significant increase in size of the organelle.

Germling

After dispersal of the nuclear cap, formation of a germ tube (precursor of the rhizoid system) commences as a localized outgrowth. Further branching of the mitochondrion takes place, and also enlargement of the nucleolus. Altogether there is a three to four-fold increase in nucleolar size (cf. Fig. 2A, B).

Transformation from naked zoospore to walled germling possessing a primary rhizoid occupies less than 60 min. Changes associated with RC-I begin 5–7 min after inoculation (T_{50} = 13.5 min, Soll *et al.*, 1969) and dispersal of ribosomes (RC-II) is complete in most cells by 22–25 min (Soll *et al.*, 1969; Soll and Sonneborn, 1971b). Germ tube formation commences after 20–22 min and by 40 min nearly all cells bear a primary rhizoid (T_{50} = 31 min, Soll *et al.*, 1969).

Growth phase

Germination is succeeded by an exponential phase of growth during which both cell size and nuclear number per cell increase. The basal rhizoid system is further elaborated. Dry weight per cell increases linearly with time after the first 60 min of culture, and a net increase in protein per cell is detectable after 70 min (Lovett, 1968). Net increase in RNA per cell, on the other hand, is detectable after only 40 min.

Sporulation phase

Synchronous differentiation of zoosporangia is induced by transferring growth-phase cells into a dilute solution of inorganic salts (Murphy and Lovett, 1966). Induction commences by the formation of (i) a basal septum, which cuts off the anucleate rhizoid system from the rest of the cell, and (ii) an apical discharge papilla from which the zoospores ultimately escape from the sporangium. Over the first 2 hr after transfer to sporulating medium, the contents of the multinucleate part of the cell are transformed into about 256 zoospores having the characteristic assembly of organelles already described. Cytoplasmic cleavage results in the formation of a series of uninucleate prespore cells. Each prespore becomes associated with a developing flagellar complex, and aggregates of ribosomes, produced during the preceding growth phase around each nucleus, become localized within the membrane-bound nuclear cap. All the mitochondria cut off in each prespore fuse to yield a single large organelle, and gamma particles appear for the first time (Lessie and Lovett, 1968; Barstow and Lovett, 1974b). Sporangium formation is complete between 15 and 16 hr after transfer of growth-phase cells to sporulation medium, and discharge through the apical papilla is complete by about 19.5 hr.

CELLULAR REORGANIZATION IN THE ABSENCE OF PROTEIN SYNTHESIS

Many of the ultrastructural changes of the RC-I and RC-II stages take place in the absence of new RNA and protein synthesis (Lovett, 1968; Soll *et al.*, 1971a, b). Thus, inclusion of cycloheximide (inhibitor of translation) in the growth medium, while it blocks incorporation of leucine into proteins of encysting spores, fails to block disorganization of the nuclear cap, synthesis of the cell wall and several other structural changes (see Table 1). In the presence of actinomycin D (inhibitor of transcription), germination proceeds even further. Up to 40–50 min after inoculation of zoospores, the inhibitor blocks incorporation of uracil into RNA, but has no effect upon either incorporation of leucine into proteins or cellular reorganization. Later, after the commencement of germ tube formation, the inhibitor blocks leucine incorporation and causes a sharp arrest of development. Enlargement of the nucleolus appears to be the only morphological criterion of germination blocked by actinomycin (Table 1).

It is evident that most of the structural changes associated with encystment and germination result simply from internal rearrangements of existing membranous structures, and though cell wall deposition is a *de novo* event, this process too is probably independent of new protein synthesis. The terminal enzyme in the formation of cell wall

TABLE 1. EFFECT OF CYCLOHEXIMIDE AND OF ACTINOMYCIN D ON ULTRASTRUCTURAL CHANGES DURING ZOOSPORE GERMINATION IN *Blastocladiella emersonii*

	Control	Cycloheximide	Actinomycin D
Initial cell wall	+	+	+
Disintegrated nuclear cap and ribosome dispersal	+	+	+
Many mitochondrial profiles	+	+	+
Retraction of flagellum	+	+	+
Disappearance of flagellar axoneme	+	−	+
Nucleolar expansion (3–4 fold)	+	±	−
Development of Golgi apparatus	+	−	+
Presence of germ tube (primary rhizoid development)	+	−	+

polymers, chitin synthetase, is already present in zoospores before germination (Camargo *et al.*, 1967; Cantino and Myers, 1972). These observations suggest that the first essential protein synthesis takes place with stored messenger RNA (mRNA) and preformed ribosomes released from the zoospore nuclear cap, while continued rhizoid elongation after 45 min requires new RNA synthesis.

If such is the case this is yet another example of a situation in which stable mRNAs are synthesized at one stage of development but function at another, which further implies the presence of regulatory devices at the level of translation.

Activation of protein synthesis by assembly of pre-existing ribosomes on to stored mRNA has been reported in asexual spores of several fungi (Sussman and Douthit, 1973) and there is an interesting parallel in several animal, and possibly higher plant, embryonic systems. For instance, activation of stored mRNA at fertilization leads to a sharp increase in the rate of protein synthesis in echinoderms (Rinaldi and Monroy, 1969) though not in amphibians (Smith *et al.*, 1966). Recent evidence has also raised the possibility that stored mRNA may be chemically altered after fertilization by polyadenylation in the cytoplasm (Wilt, 1973; Slater *et al.*, 1973). The essential function dependent upon the stored mRNA remains uncertain, although evidence has been provided for a role in histone (Nemer and Lindsay, 1969) and microtubular protein (Raff *et al.*, 1972) synthesis.

Rates of protein synthesis during germination

Inhibitor studies unfortunately give little information about actual rates of synthesis once synthesis has been activated. However estimates of the time required for synthesis of an average-sized protein (average ribosome transit times) have been made on cells in the early germination (30 min) and growth (120 min) phases. These estimates show that despite the initial rapid rise in polysomes (see Fig. 3), cells at 30 min synthesize protein at only one-fifth of the rate observed after 120 min (after correction for transit rate and reduced polysome levels) (Leaver and Lovett, 1974).

Loss of a translation inhibitor during germination

The rates of protein synthesis during germination and the evidence for reduced translation rates are consistent with the gradual decrease in the level of ribosome-associated translation inhibitor during the first 2 hr (Adelman and Lovett, 1974a). This inhibitor was first described by Schmoyer and Lovett (1969) who found that the zoospore nuclear caps contained fully charged aminoacyl-tRNAs and aminoacyl-tRNA synthetases, but inactive 80 S ribosomes when assayed *in vitro*. The isolated ribosomes were functional only after washing with 0.5 M KCl-buffer, the inhibitor being removed by the washing procedure. The inhibitor (as yet unidentified) complexes with, and inactivates, ribosomes isolated from growth-phase cells (Adelman and Lovett, 1974b). It is of interest to note that when growth-phase cells are induced to differentiate zoospores in the late log phase, there is a sharp decrease in the level of polysomes, correlated with a rapid accumulation of the ribosome associated inhibitor (Adelman and Lovett, 1974a). The authors propose that the inhibitor is produced in response to the same environmental

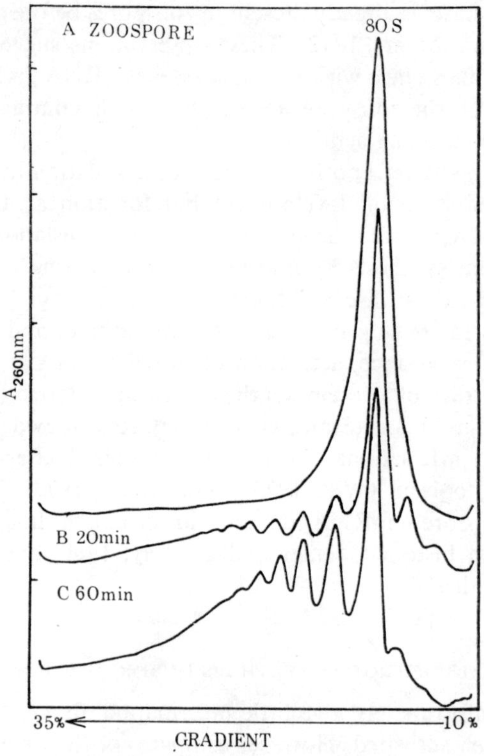

FIG. 3. Changes in ribosome distribution during encystment and early development of *Blastocladiell emersonii* cells. Representative gradient profiles of the ribosomes removed from cultures at: (A) zero time, zoospores (0 per cent polysomes), (B) 20 min (49 per cent polysomes), (C) 60 min (70 per cent polysomes).

trigger which induces the fungus to differentiate sporangia and to produce zoospores. Their working hypothesis is that the inhibitor modulates translation by binding to polysomes and then progressively inactivates 80 S ribosomes as they are released by run off. Such an inactivation might also be expected to restrict translation of newly synthesized mRNA transcripts which are stored in the zoospore and translated at germination, and/or to prevent turnover of inactive 80 S ribosomes until the inhibitor is released from the zoospores during germination.

POLYSOME FORMATION DURING GERMINATION

Protein synthesis appears to be initiated as a result of disorganization of the nuclear cap and release of the contents into the rest of the cell. Elsewhere we have shown that cells in RC-II synthesize protein which is required for completion of both structural and biosynthetic events of RC-II, and for subsequent development of the rhizoid system (Leaver and Lovett, 1974). Further information on the activation process has come from studies of polysome formation.

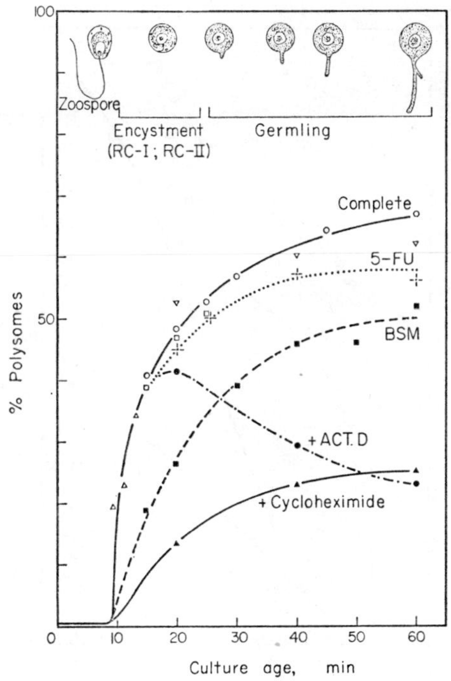

FIG. 4. Changes in the percentage of polysomes during encystment and early development of *Blastocladiella emersonii* cells cultured in an inorganic salt medium (BSM), complete growth medium, and complete growth medium plus various inhibitors. ▽—▽ △—△ ○—○ □—□, complete medium; ■---■, BSM; + · · · · +, complete medium plus 325 μg/ml 5-fluorouracil; ● - · - ●, complete medium plus 20 μg/ml actinomycin D; ▲—▲, complete medium plus 5 μg/ml cycloheximide. In complete medium cells were cultured at densities ranging from 1.69×10^9/l to 3×10^9/l; BSM and inhibitor-treated cultures were inoculated at densities ranging from 2.61×10^9 to 2.68×10^9 cells/l.

From the beginning of encystment (8 min after inoculation of zoospores into growth medium) to completion of RC-II (20 min) the percentage of ribosomes present as polysomes increases from zero to nearly 50 per cent (curves A,B, Fig. 3). After the initial rapid rise, the rate of polysome accumulation decreases, and over the ensuing 100 min of culture the percentage of polysomes increases to only 87 per cent (curve C, Fig. 3). If, on the other hand, zoospores are inoculated into a basal inorganic salts medium (BSM) polysome assembly is slower and reaches a maximum of about 50 per cent within 60 min (Fig. 4). Such spores undergo normal encystment and rhizoid formation but fail to enter the growth phase. The resulting germlings do not increase in dry weight, RNA or protein content and the nucleus fails to divide. Cycloheximide in the presence of complete growth medium reduces polysome assembly, and the final level of polysomes attained, still further, while actinomycin D permits normal assembly for about 20 min of culture (Fig. 4) after which the level of polysomes decreases exponentially ($T_{\frac{1}{2}}$ = 45 min). The analogue 5-fluorouracil, which selectively inhibits synthesis of ribosomal RNA (rRNA) but not of polydisperse RNA (thought to contain messenger RNA (mRNA)) in other fungi (De Kloet, 1968) has no effect on the initial rapid assembly of polysomes and only a slight depressing effect on the final level attained.

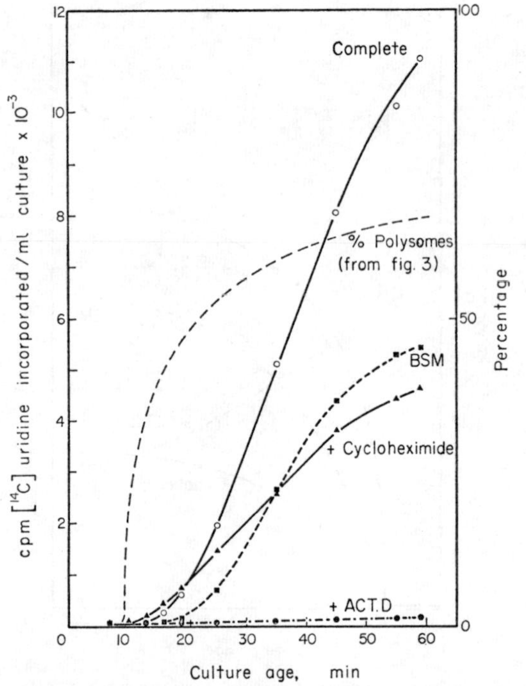

FIG. 5. Time course of incorporation of [^{14}C]uridine during encystment and early development of *Blastocladiella emersonii* cells cultured in an inorganic salt medium (BSM), complete growth medium and complete medium plus either cycloheximide or actinomycin D. Incorporation was measured in the same cultures used to obtain the results given in Fig. 3B. In each case 5 μCi of [^{14}C]uridine (42.3 mCi/mmole) were added 5 min after inoculation of zoospores. On each sampling occasion, 1 ml aliquots were harvested, mixed with 1 ml cold 10 per cent TCA, and left for 30 min. Each sample was then washed with cold 5 per cent TCA on glass fibre filters, dried, and counted. ○—○, complete medium; ■ — — —■, BSM; ▲—▲, complete medium with cycloheximide; ●—·—●, complete medium with actinomycin D.

The rapid formation of polysomes almost certainly occurs by attachment of newly released nuclear cap 80S monosomes (or their subunits) to mRNA molecules. Such a rapid rise would be expected to be accompanied by a concomitant increase in initiation events, and perhaps the sudden activation of initiation factors. Support for this assumption is provided by results obtained with the selective initiation inhibitor D-MDMP [2-(4-methyl-2,6-dinitroanilino)-*N*-methylpropionamide; Weeks and Baxter, 1972]. When added to the medium at 5×10^{-5} M it prevents development beyond the RC-II stage, delays the start of amino acid incorporation, and reduces the final rate of incorporation to only 12 per cent of that in cells treated with the inactive L-isomer.

RNA SYNTHESIS DURING GERMINATION AND EARLY DEVELOPMENT

Effects of the composition of the culture medium and of inhibitors on incorporation of [^{14}C]uridine into RNA are indicated by the data of Fig. 5. In the presence of complete growth medium, incorporation begins during the RC-I stage. Between 15 and 25 min after inoculation, the rate of incorporation increases exponentially, but linearly thereafter. Actinomycin D completely inhibits incorporation but has no effect upon either the initial rapid assembly of polysomes or the rate of amino acid incorporation (Lovett, 1968; Soll *et al.*, 1971a). The numerically equal rates of polysome increase in control cells and of polysome decrease in actinomycin D-treated cells, after 20 min, suggest that the slower, post-20 min rise in polysomes represents the appearance of newly transcribed mRNA. Inhibition of uridine incorporation by cycloheximide can probably be ascribed to an indirect inhibitor effect on synthesis of protein, shown to be necessary for synthesis and processing of rRNA (Lovett, unpublished observations).

If cells are pulse-labelled during germination and the resulting RNA analysed by gel electrophoresis, it is clear that despite the delay in rRNA accumulation until after 40 min (Lovett, 1968) rRNA synthesis begins at an early stage (Leaver and Lovett, 1974). However, processing of rRNA precursors is severely restricted initially when the rate of incorporation is rising. Before about 40 min there is an accumulation of radioactivity in the precursor-rRNA, while after 45–50 min a similar pulse does not give rise to such an accumulation.

Cell fractionation studies suggest that much of the newly synthesized rRNA might still be localized in the nucleus before completion of RC-II. This early accumulation of precursor-rRNA could result either from a feed-back control of processing owing to the restricting rate of protein synthesis or to a delay in the processing of ribosome subunits. The second possibility seems more likely in view of the extensive enlargement of the nucleolus during germination (Stenrum, 1972; Leaver and Lovett, 1974). This suggests that the reduced rate of precursor processing could be due to the need for re-organization and re-expansion of the contracted nucleolus (contraction occurs when rRNA synthesis ceases prior to zoospore differentiation). When the RNA of cytoplasmic ribosomes was extracted and analysed, it was found that the first newly synthesized RNA to enter the polysomes was polydisperse in size (and probably represented at least in part mRNA) and increased rapidly after 20 min. Entry of newly synthesized rRNA into cytoplasmic ribosomes was delayed until after 25–30 min.

These observations were confirmed by continuous-labelling studies (Fig. 6) which showed that increase in specific activity of cytoplasmic ribosomes lagged behind the increase in RNA synthesis and did not show any appreciable change until after 20 min when most of the initial rapid polysomes formation was complete.

Despite the activation of RNA synthesis at an early stage of germination, it is not until after the 45–50 min stage that an increase in total RNA per cell can be detected and actinomycin D causes developmental arrest (Lovett, 1968). This was originally thought to imply a stage specific requirement for rRNA and ribosome synthesis for subsequent growth and development. However, zoospores will germinate and develop normally up to around the 2 hr stage in an inorganic salts medium (BSM). These starved cells do not begin to increase in RNA or protein content at 45 min, as do cells in complete medium, even though they form a normal rhizoid system (the stage blocked by actinomycin D). Analysis of the pattern of RNA synthesis during germination of starved cells suggests that the new rRNA synthesized turns over and fails to enter new ribosome subunits. Thus a net increase in neither protein nor RNA is required for cells to proceed beyond the 45 min stage. Since rRNA synthesis is also now known to start well before this time,

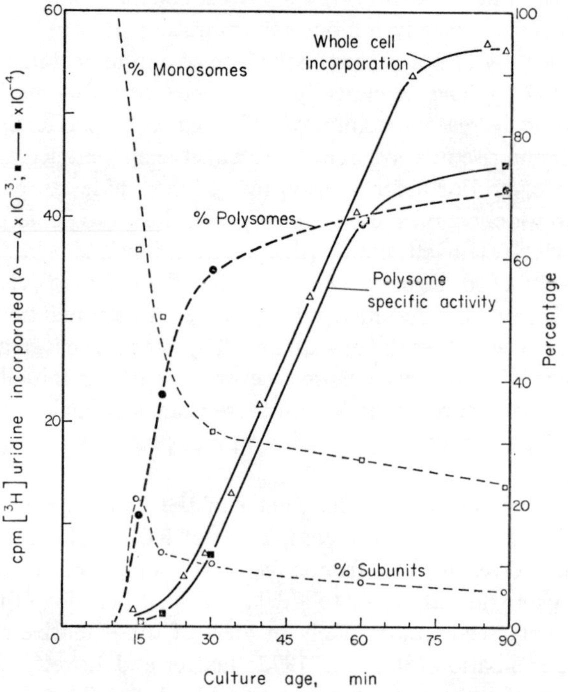

FIG. 6. Time course of incorporation of [^{3}H]uridine during encystment and development of *Blastocladiella emersonii* cells cultured in complete growth medium. One litre of medium containing 1 mCi of [^{3}H]uridine (29.3 mCi/mmole) was inoculated with zoospores (2.19×10^9/l) at zero time. One millilitre samples were removed at the times indicated to measure whole-cell incorporation, and large samples were removed at 15, 20, 30, 45, 60, 90 min for polysome extraction and analysis, ●—●, per cent polysomes; □—□, per cent 80S monosomes; ○— — — — ○, per cent ribosomal subunits; △—△, whole-cell uridine incorporation expressed as cpm $\times$ 10^3/2.19 $\times$ 10^4 cells; ■—■ polysome specific activity expressed as cpm/unit of A_{260} in the polysome regions of sucrose gradients.

the actinomycin D block suggests a requirement for either new mRNA synthesis at 45 min (perhaps to replace the zoospore mRNA which has turned over) or for some new, as yet undetected, event.

The preferential accumulation of RNA over protein between 40 and 80 min results in a rapid rise in the RNA/protein ratio per cell, from 0.32 in the zoospore (Murphy and Lovett, 1966) to a stable value of 0.41 characteristic of log-phase plants (Fig. 7). Thus, an early activity of the fungus is to re-adjust its ribosome content to a level that can support maximal rate of protein synthesis and rapid growth. This situation has parallels in bacteria where the RNA/protein ratios of cells are characteristic for any specific growth rate (Maaløe and Kjeldgaard, 1966).

Re-adjustment of the RNA/protein ratio of cells in complete medium provides a reasonable explanation for our failure to detect an increase in protein content prior to 80 min. On the other hand, the observed amino acid incorporation and presence of polysomes clearly indicates that some protein synthesis nevertheless occurs. It seems

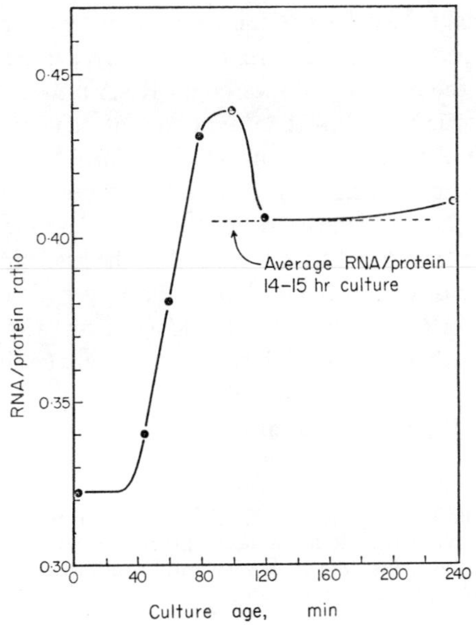

FIG. 7. Changes in the RNA/protein ratio during germination and early development of *Blastocladiella emersonii* cells cultured in complete growth medium. Data from Murphy and Lovett (1966), and Lovett (1968).

probable that a significant proportion of the new protein will be ribosomal structural protein, which agrees with our findings that these proteins represent about 20 per cent of the new proteins in ribosomes after a short 6 min pulse at 30 min culture age (Leaver and Lovett, 1974).

The mechanism by which eukaryotes regulate rates of major RNA and protein synthesis is not well understood; and this may be in part due to the fact that most of the research has been done on higher plants and an'mals which do not normally need to respond to rapid changes in environment. Unequivocal evidence in favour of translational control of protein synthesis would be a valuable contribution towards our understanding of the mechanism of control of synthesis in differentiating eukaryotic cells. Asexual spores of fungi are semi-dormant cells in which protein synthesis is either slight or completely absent (Sussman and Douthit, 1973). The presence of various components of the synthetic systems has been studied in only a handful of species, but without success in determining the cause of inactivity. In fact the data are not inconsistent with the view that all factors required for synthesis are present, though the entire spectrum has not been examined in a single organism (Lovett, 1974). It is evident that all factors for protein synthesis must be stored in at least minimal amounts, in order for the spore to reinitiate synthesis when it germinates. If, as seems to be the case, other factors such as charging levels of aminoacyl-tRNA or nucleotide triphosphates are not limiting, it is logical to assume that synthesis is controlled in these cells by (i) action of translation or elongation inhibitors, (ii) compartmentalization, or (iii) other mechanisms which prevent initiation in the presence of the required factors. Zoospores of *Blastocladiella* provide excellent material for analysis of factors regulating initiation of protein synthesis since the process seems too rapid to require much, if any, synthesis of new factors.

The observation that the majority of messenger RNA molecules isolated from eukaryotes contain a polyadenylic acid [Poly (A)] sequence at or near its 3′ terminus has led to the development of several methods for purification of mRNAs from bulk cellular RNA. Using such techniques, Lovett and Wilt (unpublished observations, 1974) have shown that about 3 per cent of the total zoospore RNA is associated with Poly (A) and furthermore that 80 per cent of this Poly (A) is localized in the nuclear cap. Further analysis of the distribution and fate of this Poly (A) RNA during germination, together with an *in vitro* assessment of its mRNA function, should confirm whether this is, in fact, the stored zoospore mRNA. The study of how this participates in the control of protein synthesis should be of more general significance to eukaryotic development as well, since stored messenger RNA is apparently a common feature in embryological and other simpler systems.

Acknowledgements: We would like to thank Dr. W. E. Barstow for providing the electron micrographs and W. Foster for excellent photographic assistance. This work was supported by U.S. Public Service grant A.104783 from the National Institute of Allergy and Infectious Diseases and a travel grant by the European Molecular Biology Organization to J.S.L.

REFERENCES

ADELMAN, T. G. and LOVETT, J. S. (1974a) Evidence for a ribosome-associated translation inhibitor during differentiation of *Blastocladiella emersonii*. *Biochim. biophys. Acta* **335,** 236–245.

ADELMAN, T. G. and LOVETT, J. S. (1974b) Ribosome function *in vitro* and *in vivo* during the life cycle of *Blastocladiella emersonii*. *Biochim. biophys. Acta* **349,** 240–249.

BARSTOW, W. E. and LOVETT, J. S. (1974a) Apical vesicles and microtubules in rhizoids of *Blastocladiella emersonii*: Effects of Actinomycin D and cycloheximide on development during germination. *Protoplasma* **82,** 103–117.

BARSTOW, W. E. and LOVETT, J. S. (1974b) Formation of gamma particles during zoosporogenesis in *Blastocladiella emersonii*. *Mycologia* In press.

BROMBERG, R. (1974) Mitochondrial fragmentation during germination in *Blastocladiella emersonii*. *Devl Biol.* **36,** 187–194.

CAMARGO, E. P., DIETRICH, C. P., SONNEBORN, D. R. and STROMINGER, J. L. (1967) Biosynthesis of chitin in spores and growing cells of *Blastocladiella emersonii*. *J. biol. Chem.* **242,** 3121–3128.

CANTINO, E. C. (1951) Metabolism and morphogenesis in a new *Blastocladiella*. *Antonie Von Leeuwenhoek. J. Microbiol. Serol.* **17,** 325–362.

CANTINO, E. C. and LOVETT, J. S. (1964) Non-filamentous aquatic fungi: model systems for biochemical studies of morphological differentiation. In *Advances in Morphogenesis* Vol. 3. ed. ABERCROMBIE, M. and BRACHET, J., pp. 33–93. Academic Press, New York.

CANTINO, E. C. and MYERS, R. B. (1972) Concurrent effect of visible light on γ-particles, chitin synthetase and encystment capacity in zoospores of *Blastocladiella emersonii*. *Arch. Mikrobiol.* **83,** 203–215.

DE KLOET, S. R. (1968) Effect of 5-fluorouracil and 6-azauracil on the synthesis of ribonucleic acid and protein in *Saccharomyces cerevisiae*. *Biochem. J.* **106,** 167–178.

FULLER, M. S. (1966) Structure of the uniflagellate zoospores of aquatic phycomycetes. In 18*th Symposium Colston Research Society*. p. 67–84. Butterworths, London.

LEAVER, C. J. and LOVETT, J. S. (1974) An analysis of protein and RNA synthesis during encystment and out-growth (germination) of *Blastocladiella* zoospores. *Cell Differ.* **3,** 165–192.

LESSIE, P. E. and LOVETT, J. S. (1968) Ultrastructural changes during sporangium formation and zoospore differentiation in *Blastocladiella emersonii*. *Am. J. Bot.* **55,** 220–236.

LOVETT, J. S. (1967) Aquatic Fungi, In *Methods in Developmental Biology* ed. WILT, F. W. and WESSELS, N. K. pp. 341–358. T. Y. Crowell Co., New York.

LOVETT, J. S. (1968) Reactivation of ribonucleic acid and protein synthesis during germination of *Blastocladiella* zoospores and the role of the ribosomal nuclear cap. *J. Bact.* **96,** 962–969.

LOVETT, J. S. (1974) Regulation of protein metabolism during spore germination. In *The Fungus Spore*. ed. HESS, W. M. and WEBER, D. J. Chapter 6. J. Wiley, New York. In press.

MAALØE, O. and KJELDGAARD, N. O. (1966) *Control of Macromolecular Synthesis*. p. 284. Benjamin. New York and Amsterdam.

MURPHY, M. N. and LOVETT, J. S. (1966) RNA and protein synthesis during zoospore differentiation in synchronized cultures of *Blastocladiella*. *Devl Biol*. **14,** 68–95.

NEMER, M. and LINDSAY, D. T. (1969) Evidence that the s-polysomes of early sea urchin embryos may be responsible for the synthesis of chromosomal histones. *Biochem. biophys. Res. Commun*. **35,** 156–160.

RAFF, R. A., COLOT, H. V., SELVIG, S. E. and GROSS, P. R. (1972) Oogenetic origin of messenger RNA for embryonic synthesis of microtubule proteins. *Nature Lond*. **235,** 211–214.

REICHLE, R. E. and FULLER, M. S. (1967) The fine structure of *Blastocladiella emersonii* zoospores. *Am. J. Bot*. **54,** 81–92.

RINALDI, A. M. and MONROY, A. (1969) Polyribosome formation and RNA synthesis in the early post-fertilization stages of the sea urchin egg. *Devl Biol*. **19,** 73–86.

SCHMOYER, I. R. and LOVETT, J. S. (1969) Regulation of protein synthesis in zoospores of *Blastocladiella*. *J. Bact*. **100,** 854–864.

SLATER, I., GILLESPIE, D. and SLATER, D. W. (1973) Cytoplasmic adenylation and processing of maternal RNA. *Proc. natn. Acad. Sci. U.S.A*. **70,** 406–411.

SMITH, L. D., ECKER, R. E. and SUBTELNY, S. (1966) The initiation of protein synthesis in eggs of *Rana pipiens*. *Proc. natn. Acad. Sci. U.S.A*. **56,** 1724–1728.

SOLL, D. R. and SONNEBORN, D. R. (1971a) Zoospore germination in *Blastocladiella emersonii*. III. Structural changes in relation to protein and RNA synthesis. *J. Cell Sci*. **9,** 679–699.

SOLL, D. R. and SONNEBORN, D. R. (1971b) Zoospore germination in *Blastocladiella emersonii*: Cell differentiation without protein synthesis. *Proc. natn. Acad. Sci. U.S.A*. **68,** 459–463.

SOLL, D. R. and SONNEBORN, D. R. (1972) Zoospore germination in *Blastocladiella emersonii*. IV. Ion control over cell differentiation. *J. Cell Sci*. **10,** 315–333.

SOLL, D. R., BROMBERG, R. and SONNEBORN, D. R. (1969) Zoospore germination in the water mold, *Blastocladiella emersonii*. I. Measurement of germination and sequence of subcellular morphological changes. *Devl Biol*. **20,** 183–217.

STENRUM, U. (1972) Relationship between nucleolar size and the synthesis and processing of pre-ribosomal RNA in the liver of the rat. *FEBS Symp*. **24,** 131–141.

SUBERKROPP, K. S. and CANTINO, E. C. (1973) Utilisation of endogenous reserves by swimming zoospores of *Blastocladiella emersonii*. *Arch. Mikrobiol*. **89,** 205–221.

SUSSMAN, A. S. and DOUTHIT, H. A. (1973) Dormancy in microbial spores. *A. Rev. Pl. Physiol*. **24,** 311–352.

TRUESDELL, L. C. and CANTINO, E. C. (1971) The induction and early events of germination in the zoospore of *Blastocladiella emersonii*. *Curr. Top. Devl Biol*. **6,** 1–44.

WEEKS, D. P. and BAXTER, R. (1972) Specific inhibition of peptide-chain initiation by 2-(4-methyl-2,6-dinitroanilino)-N-methyl-propionamide. *Biochemistry* **11,** 3060–3064.

WILT, F. H. (1973) Polyadenylation of maternal RNA of sea urchin eggs after fertilization. *Proc. natn. Acad. Sci. U.S.A*. **70,** 2345–2349.

PROTEIN TURNOVER IN PLANTS

D. D. Davies and T. J. Humphrey

School of Biological Sciences, University of East Anglia, Norwich, U.K.

Introduction

The study of protein turnover in plants was initiated by the work of Gregory and Sen (1937) who proposed that the rate of carbon dioxide production is controlled by the operation of a protein cycle. Their proposal involved two components: (1) that proteins undergo a cycle of synthesis and degradation, i.e. protein turnover; (2) that protein degradation is linked to the production of carbon dioxide. Direct evidence for protein turnover in plants was obtained by Vickery *et al.* (1940) with intact tobacco plants and by Steward *et al.* (1956, 1958) with carrot root explants.

Quantitative aspects of the link between protein degradation and carbon dioxide production have been examined by Bidwell *et al.* (1964) who suggest that as much as 27 per cent of the CO_2 produced by carrot root explants may be derived from protein; others argue that there is little evidence to suggest that significant amounts of carbon dioxide are derived from protein (James, 1957). However, the extent to which protein turnover contributes to respiration is not central to the concept of turnover itself. An excellent review of protein turnover in plants has recently appeared (Huffaker and Peterson, 1974); in this paper we concentrate discussion on the quantitative aspects of protein turnover.

Measurement of protein turnover

Protein turnover is usually assessed from changes that occur in specific activity of protein and amino acids after treatment of tissues with radioactive intermediates. The special difficulties associated with applying these methods to plants were noted by Steward *et al.* (1958) who drew attention to the important role of storage and metabolic pools in evaluating the recycling of protein amino acids and in determining the specific activity of amino acids at the site of protein synthesis. Some authors have attempted to minimize the effects of recycling by using labelled glutamate (Holmsen and Koch, 1964), whilst others have used labelled leucine to minimize the effect of storage pools on the specific activity of the amino acid (Kemp and Sutton, 1971). A different approach was developed by Hellebust and Bidwell (1963) who derived simple equations for measuring turnover and these equations were used to analyse data obtained in label-chase experiments (Bidwell *et al.*, 1964). Change is an essential part of these experiments. Proteins and amino acids change in both amount and specific activity. The analysis calls for calculus but the authors attempt to fit their data to simple algebraic equations and in our view, draw erroneous conclusions. A detailed criticism of the equations is not appropriate in this paper, but we must comment on the interpretation of the label-chase experiments given in the Bidwell *et al.* (1964) paper. As an example, we cite the experiment in which ^{14}C-sugar was supplied to carrot explants for 1 week. The explants were then transferred to a medium containing unlabelled sugar and the specific activities of the labelled sugars, protein amino acids and CO_2 were measured at various time intervals (Table 1).

TABLE 1. SPECIFIC ACTIVITIES OF CARBON IN SOLUBLE SUGARS, PROTEIN AMINO ACIDS AND CARBON DIOXIDE AT VARIOUS TIMES FOLLOWING THE REPLACEMENT OF ^{14}C-SUGAR BY UNLABELLED SUGAR (after Bidwell *et al.*, 1964)

	Specific activities at times (hr)				
	0–12	12–36	36–96	96	168
Sugars	16.1	10.7	7.8	6.2	3.2
Protein amino acids	15.2	10.7	7.4	6.9	5.2
Carbon dioxide	15.8	11.2	9.1	7.4	5.6

Bidwell *et al.* interpret these results as follows:

> The specific activity of CO_2 was initially below that of the soluble sugars in the tissue, but progressively it rose above this value. This could be brought about by the utilization of a source of carbon for respiration in addition to the sugars, which in the later time periods had a higher specific activity than sugars that is, the proteins. The average specific activity of the protein amino acids was probably slightly below the true value for those protein amino acids which turned over rapidly because of the presence of some more stable protein of lower specific activity. Another effect of protein turnover, that is the decrease in specific radioactivity of proteins with time, is clearly to be seen.

We draw exactly the opposite conclusion from this experiment! During the chase period of a label-chase experiment, the specific activity of the precursor will fall eventually to zero as unlabelled precursor replaces ^{14}C-labelled precursor. The specific activity of the product will initially continue to rise as ^{14}C from intermediates accumulates in the product, but when unlabelled carbon from the precursor reaches the product the specific activity will decline. This situation is shown diagrammatically in Fig. 1.

The data of Table 1 show a perfectly good precursor–product relationship between sugar and carbon dioxide but not between protein and carbon dioxide as claimed by Bidwell *et al.*

The analysis may be helped by considering the chase part of the experiment as a ^{12}C-*label* experiment. When the carrot explant is placed on ^{12}C-sugar the precursor–product relationship requires that the specific activity of the product (CO_2) must be less than the specific activity of the precursor (sugar or protein). If the precursor molecule has to have a *higher* specific activity of ^{12}C than does CO_2, the precursor must have a *lower* specific activity with respect to ^{14}C. Sugar, but not protein shows this labelling pattern expected for a precursor of CO_2.

The pioneering experiments of Steward and his collaborators clearly established the

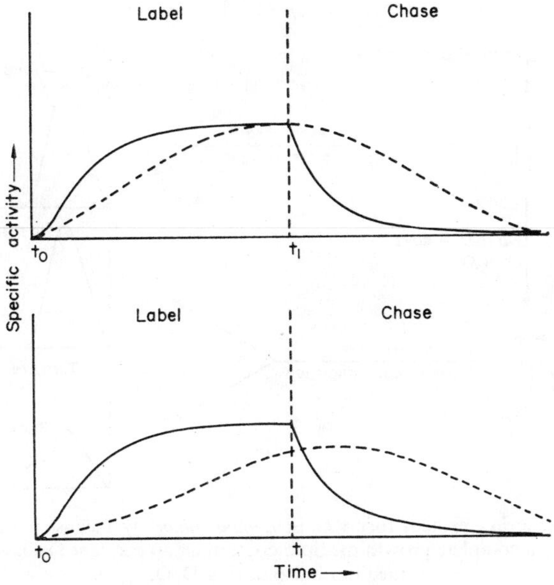

FIG. 1. Change of specific activity with time in a label-chase experiment. At time t_0 the tissue is incubated with a ^{14}C-precursor (e.g. ^{14}C-sugar). At time t_1 the tissue is transferred to the corresponding unlabelled compound. The curves show the expected changes in specific activity of the precursor (solid line) and the product CO_2 (dashed line). Upper graph, precursor and product in isotopic equilibrium before beginning of chase. Lower graph, specific activity of product less than that of precursor before beginning of chase.

reality of protein turnover in plants and identified the problems associated with measuring turnover. However, we question the validity of applying the algebraic equations of Hellebust and Bidwell (1963) to a label-chase experiment. A kinetic analysis of these experiments is exceedingly difficult and an alternative approach is necessary. In previous work from this laboratory, Trewavas (1972a, b) sought to overcome the difficulties by measuring the specific activity of amino-acyl t-RNA. However, this method is technically difficult and we have therefore explored further methods.

HEAVY WATER METHOD

One approach taken in collaboration with A. Boudet is a sort of Meselson and Stahl (1958) experiment as illustrated in Fig. 2. Plants of *Lemna minor* are first grown in a nutrient solution containing heavy water (50% D_2O) and then in the same solution lacking D_2O. The heavy and light proteins respectively produced can be extracted and then separated by isopycnic centrifugation. Caesium chloride solutions are usually used to obtain the necessary density gradients, but to achieve higher resolution we have used potassium bromide solutions. If a ^{14}C-amino acid is given as a pulse at the end of the period in D_2O, label appears in heavy protein and if protein degradation occurs when the plants are transferred to H_2O, label is lost from the heavy protein and hence from the

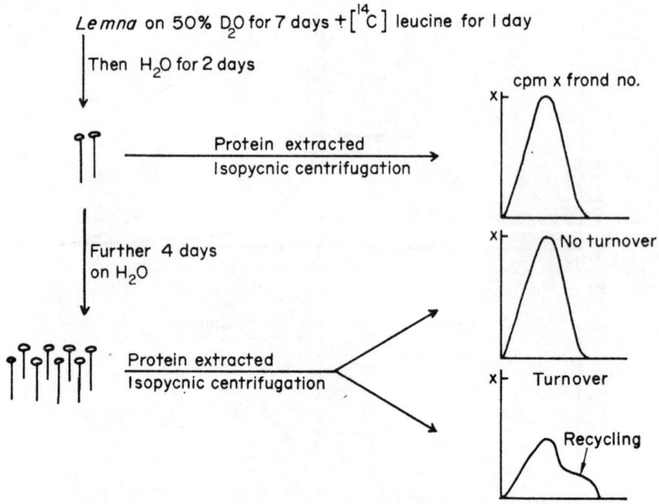

FIG. 2. Method of measuring protein turnover in *Lemna minor* by means of density and radioactive labelling. D_2O refers to a complete growth medium containing 50 per cent D_2O. H_2O refers to the same medium but omitting D_2O.

peak obtained on isopycnic centrifugation. If re-cycling of the amino acids occurs, label appears in the light-protein peak. If there is no protein turnover label will stay in the heavy-protein peak. Protein degradation can therefore be measured by following the decrease in radioactivity associated with the heavy-protein peak.

The method has a number of disadvantages—it is costly, time consuming and cannot be used to measure proteins with short half lives— in the case of *Lemna minor* the method cannot be used for proteins with a half life less than about 2 days. Furthermore the high concentration of D_2O (50%) produces some growth inhibition and we have therefore developed an alternative method.

TRITIATED WATER METHOD

When plants of *Lemna minor* are supplied with tritiated water ($T_2O \equiv {}^3H_2O$) tritium enters rapidly and exchanges with the hydrogen atoms of amino acids in a number of ways (Fig. 3). For present purposes, the most important reaction is the transaminase-catalysed exchange-reaction which involves the hydrogen on the α carbon atom of the amino acids.

$$R\cdot CHNH_2\cdot COOH \xrightarrow[\text{phosphate}]{\text{pyridoxal}} R\cdot CO\cdot COOH \xrightarrow[\text{phosphate} + T^+]{\text{pyridoxamine}} R\cdot CTNH_2\cdot COOH$$

Where R·CO·COOH represents keto acids which can undergo the keto-enol equilibrium, tritium will also exchange with hydrogen on the β carbon atom.

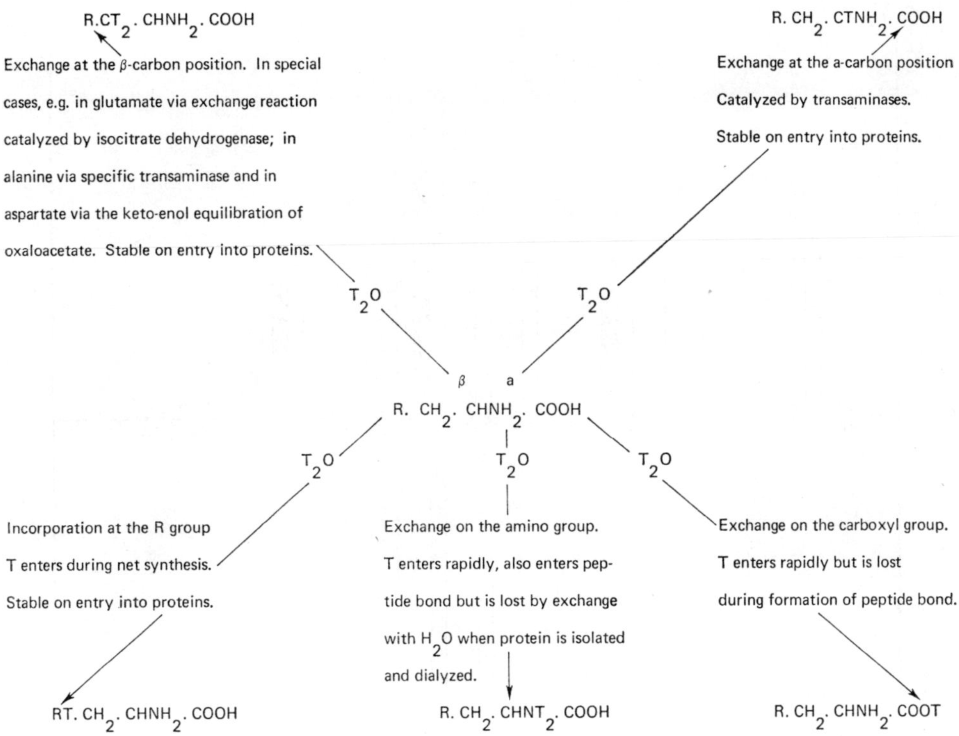

R.CT2 . CHNH2 . COOH
Exchange at the β-carbon position. In special cases, e.g. in glutamate via exchange reaction catalyzed by isocitrate dehydrogenase; in alanine via specific transaminase and in aspartate via the keto-enol equilibration of oxaloacetate. Stable on entry into proteins.
R. CH2. CTNH2. COOH
Exchange at the a-carbon position Catalyzed by transaminases. Stable on entry into proteins.
T2O
T2O
β
a
R. CH2. CHNH2. COOH
T2O
T2O
T2O
Incorporation at the R group T enters during net synthesis. Stable on entry into proteins.
Exchange on the amino group. T enters rapidly, also enters peptide bond but is lost by exchange with H2O when protein is isolated and dialyzed.
Exchange on the carboxyl group. T enters rapidly but is lost during formation of peptide bond.
RT. CH2. CHNH2. COOH
R. CH2. CHNT2. COOH
R. CH2. CHNH2. COOT

FIG. 3. Reactions which lead to rapid labelling of amino acids in cells exposed to tritiated water.

When the plants are transferred to a tritium-free medium, the transaminase-catalysed exchange-reaction ensures the rapid release of tritium from the free cytoplasmic amino acids (Fig. 4). However, if the tritiated amino acids become incorporated into protein the tritium is no longer exchangeable.

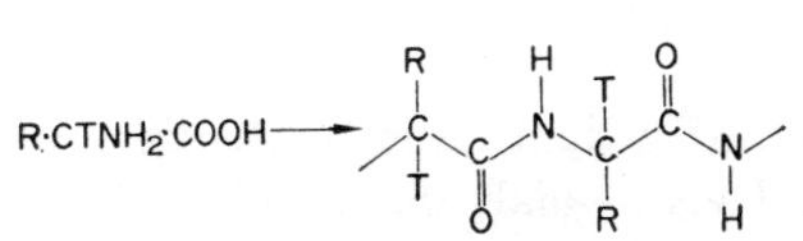

The incorporation of tritium into protein via these reactions represents the gross synthesis of protein.

Tritium will of course also exchange—non-enzymically—with hydrogen in the amino groups of amino acids and when these amino acids become incorporated into protein the tritium is associated with the peptide link

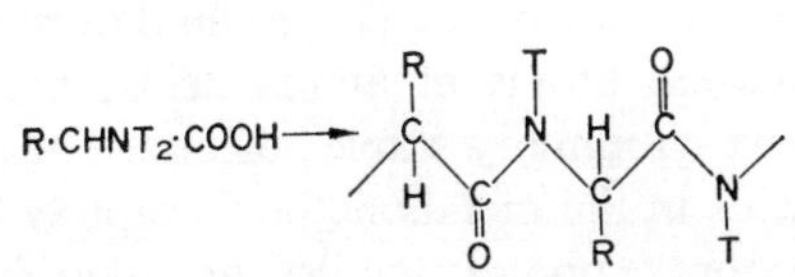

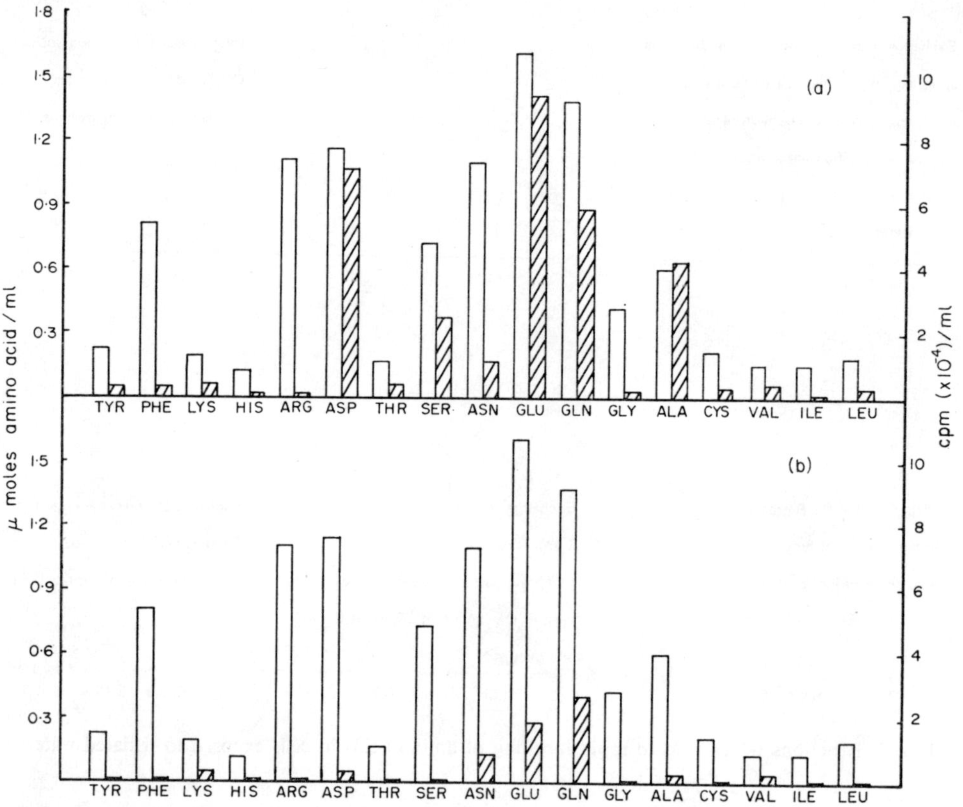

FIG. 4. Label chase of soluble amino acids in *Lemna minor*. Histograms representing the concentration of each amino acid (open columns) and the distribution of tritium in the acids (hatched columns) after (a) 20 min exposure of plants to T_2O and (b) 20 min in T_2O followed by 30 min in tritium-free medium. Amino acids were separated by means of a Technicon amino-acid analyzer.

However, these peptide-linked tritium atoms exchange rapidly with water, especially under alkaline conditions. Thus by dialysing proteins labelled in various positions, the tritium associated with amide groups can be eliminated whilst tritium associated with α-carbon atoms is retained (Fig. 5).

The broad specificity of the enzyme-catalysed exchange-reactions is verified by the observed incorporation of tritium into almost all of the protein amino acids—the most noticeable exception being arginine (Fig. 6).

These considerations suggest a new method of measuring protein turnover. *Lemna* plants are exposed to T_2O for a short period (15–20 min) during which tritium exchanges with hydrogen on the α-carbon atoms of amino acids; proteins formed from these tritiated amino acids will have tritium in a stable position. The plants are then transferred to H_2O; tritiated amino acids in the metabolic pools rapidly lose their tritium whereas tritiated amino acids in the storage pools (vacuoles) do not exchange their tritium because

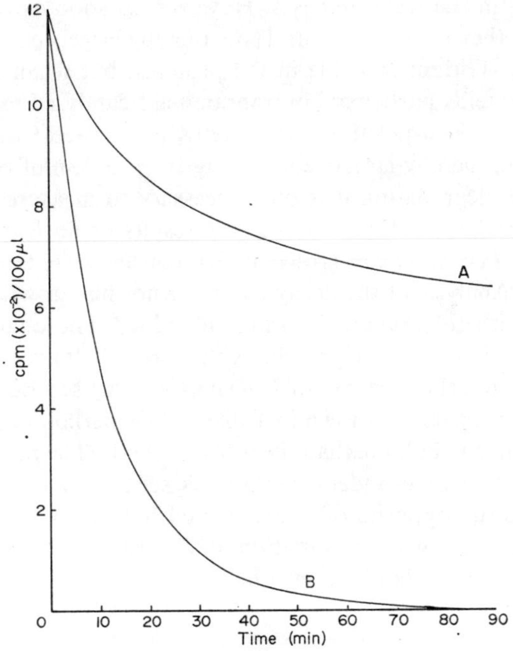

FIG. 5. Effect of dialysis on tritium-labelled protein in *Lemna minor*. Curve A represents protein synthesized in presence of T_2O (20 min) (α-labelling). Curve B represents protein equilibrated with T_2O for 120 min (amide-labelling).

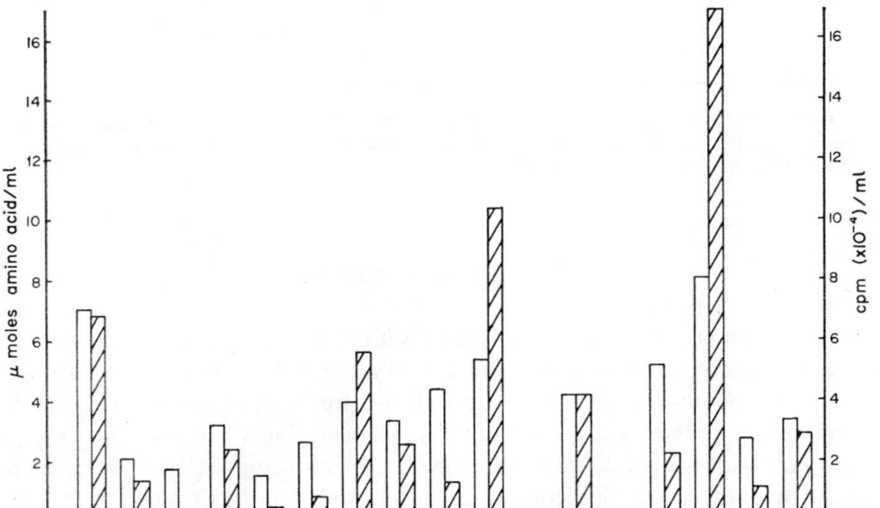

FIG. 6. Histograms representing the concentration of protein amino acids (open columns) in *Lemna minor* and the distribution of tritium in each acid (hatched columns) after exposing plants to T_2O for 20 min.

they are separated from the transaminases. However, as soon as tritiated amino acids leak from the vacuole they equilibrate with H_2O, thereby largely eliminating the problem of compartmentation. Tritium is stable in the proteins, but when protein degradation occurs the free amino acids participate in transaminase-catalysed exchange-reactions so that tritium is lost from the α-position and proteins synthesised from these amino acids do not contain tritium, thereby largely eliminating the problem of re-cycling.

To measure protein degradation it is only necessary to measure the radioactivity in protein as a function of time. Figure 7 presents results of an experiment designed to measure turnover in *Lemna minor* grown under constant light in the presence and absence of nutrients. Analysis of the decay curve for normal growth conditions permits fractions of proteins with different half lives to be identified. The logarithmic plot of Fig. 7 implies the presence of two groups of proteins with *average* half lives of 3 days and 7 days.

We have also measured the turnover of *Lemna* protein by several other methods. The results of these measurements are shown in Table 2. The method of Bidwell *et al.* (1964) clearly gives a much shorter half life than the other methods. The rate of protein degradation as measured by two independent methods is seen to be fastest when growth is slowest. This result is the opposite of that reported for carrot root explants (Bidwell *et al.*, 1964) but is consistent with the situation in bacteria where protein turnover is low in rapidly growing cultures (Mandelstam, 1960).

TABLE 2. COMPARISON OF HALF LIVES OF *Lemna minor* SOLUBLE PROTEIN OBTAINED BY VARIOUS METHODS

Condition	Bidwell *et al.* (1964)	Kemp and Sutton (1971)	Trewavas (1972a)	Tritiated water	Heavy water
Normal growth	0.88 days	12.25 days	7 days	Fraction A 3 days Fraction B 7 days	4 days
Starvation	—	—	2 days	2 days	

Plants were grown under constant conditions and protein turnover was measured by the methods listed above except that the data of Trewavas (1972a) were taken directly from his paper since the *Lemna* strain and conditions of growth were identical.

FURTHER DEVELOPMENTS

The tritiated-water method can only be used with relatively short exposures to tritiated water, because on prolonged exposure to tritiated water, tritium enters non-exchangeable positions in amino acids and re-introduces all the problems of compartmentation and re-cycling. These problems are avoided when *Lemna* plants are exposed to tritiated water for periods less than 20 minutes. However, by reducing the time of exposure we restrict the amount of tritium incorporated into protein and this makes it very difficult to extend the method for measurement of turnover of individual enzymes. We are therefore examining a modification of our tritiated-water method.

Tritium in positions other than the α-position of amino acids represents 'noise' or experimental error which increases with labelling time. We need a method of filtering out

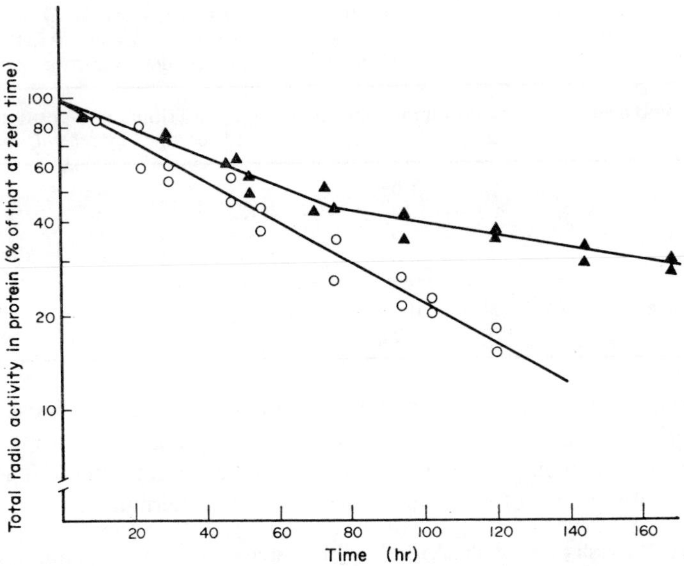

FIG. 7. Protein turnover in *Lemna minor* measured in the presence (▲) and absence (○) of nutrients. Plants were incubated in either complete growth medium plus T_2O (5 mCi/ml) or water plus T_2O for 20 min., then transferred to either complete medium minus T_2O (▲) or water minus T_2O (○).

this noise—a method which measures tritium incorporated into the α-position and ignores tritium in all other positions. There are a number of possible ways of doing this and we are currently examining some of them. Here we briefly report one of these methods.

When an amino acid is treated with acetic anhydride and acetic acid, it undergoes *N*-acetylation followed by ring closure to form an oxazolone ring. The opening and closing of this ring is the basic mechanism used by biochemists for the racemization of amino acids—we use it to measure the tritium present in the α-position of amino acids. The tritiated amino acid is heated with acetic anhydride and acetic acid, then neutralized with KOH; the mixture is distilled and the tritiated water counted. The results obtained by applying this method to amino acids with known patterns of labelling are shown in Table 3.

To use this method for measuring protein turnover, the protein must be hydrolysed, the amino acids racemized and the tritium counted. The method permits incorporation of large amounts of tritium into protein and preliminary studies indicate that it gives turnover values close to those reported in this paper.

FUTURE DEVELOPMENTS

We have been concerned to develop methods of measuring protein turnover which do not require complex mathematical analysis and which are sufficiently simple to be used as a routine method in physiological studies. The effects of hormones on protein syn-

TABLE 3. PERCENTAGE OF TRITIUM REPORTED IN THE α-CARBON POSITION OF SEVERAL TRITIATED AMINO ACIDS SUPPLIED BY THE RADIOCHEMICAL CENTRE, AMERSHAM, COMPARED WITH THE PERCENTAGE DETERMINED IN EACH BY THE RACEMIZATION METHOD

Amino acid	% Tritium in α-position (reported by Amersham)	% Tritium in α-position (determined by racemization)
Isoleucine	0	6
Lysine	0	7
Arginine	0	4
Phenylalanine	0	5
Glycine	100	100
Glutamate	100	96
Serine	0	17

thesis and enzyme levels have been widely reported and many authors have pointed out the possible critical importance of protein degradation. It is hoped that the present methods will prove useful in physiological studies and that by coupling to immunological methods, the turnover rates of individual proteins can be determined.

Acknowledgement: We wish to thank the Agricultural Research Council for supporting this work.

REFERENCES

BIDWELL, R. G. S., BARR, R. and STEWARD, F. C. (1964) Protein synthesis and turnover in cultured plant tissue: sources of carbon for synthesis and the fate of protein breakdown products. *Nature Lond.* **203** 367–373.

GREGORY, F. G. and SEN, P. K. (1937) Physiological studies in plant nutrition. VI. The relation of respiration rate to the carbohydrate and nitrogen metabolism of the barley leaf as determined by nitrogen and potassium deficiency. *Ann. Bot.* NS **1,** 521–561.

HELLEBUST, J. A. and BIDWELL, R. G. S. (1963) Protein turnover in wheat and snapdragon leaves. Preliminary investigations. *Can. J. Bot.* **41,** 969–983.

HOLMSEN, J. W. and KOCH, A. L. (1964) An estimate of protein turnover in growing tobacco plants. *Phytochemistry* **3,** 165–172.

HUFFAKER, R. C. and PETERSON, L. W. (1974) Protein turnover in plants and possible means of its regulation. *A. Rev. Pl. Physiol.* **25,** 363–392.

JAMES, W. O. (1957) Reaction paths in the respiration of the higher plants. *Adv. Enzymol.* **18,** 281–318.

KEMP, J. D. and SUTTON, D. W. (1971) Protein metabolism in cultured plant tissues. Calculation of an absolute rate of protein synthesis, accumulation and degradation in tobacco callus *in vivo*. *Biochemistry* **10,** 81–88.

MANDELSTAM, J. (1960) The intracellular turnover of protein and nucleic acids and its role in biochemical differentiation. *Bact. Rev.* **24,** 289–308.

MESELSON, M. and STAHL, F. W. (1958) The replication of DNA in *E. coli*. *Proc. natn. Acad. Sci. U.S.A.* **44,** 671–682.

STEWARD, F. C., BIDWELL, R. G. S. and YEMM, E. W. (1956) Protein metabolism, respiration and growth: a synthesis of results from the use of ^{14}C-labelled substrates and tissue cultures. *Nature Lond.* **178,** 734–738.

STEWARD, F. C., BIDWELL, R. G. S. and YEMM, E. W. (1958) Nitrogen metabolism, respiration and growth of cultured plant tissue. *J. exp. Bot.* **9,** 11–51.

TREWAVAS, A. (1972a) Determination of rates of protein synthesis and degradation in *Lemna minor*. *Pl. Physiol. Lancaster* **49,** 40–46.

TREWAVAS, A. (1972b) Control of the protein turnover rates in *Lemna minor*. *Pl. Physiol. Lancaster* **49,** 47–51.

VICKERY, H. B., PUCHER, G. W., SCHOENHEIMER, R., and RITTENBERG, D. (1940) The assimilation of ammonia nitrogen by the tobacco plant: a preliminary study with isotopic nitrogen. *J. biol. Chem.* **135,** 531–539.

COMMENT BY PROFESSOR D. D. DAVIES

A long discussion with Professor Bidwell failed to resolve our differences. We continue to believe that

(1) Bidwell *et al.* misinterpret the kinetics of label-chase experiments,
(2) we correctly follow their instructions in the application of their equations, and
(3) their equations are not applicable in a label-chase situation.

In view of these fundamental differences, it seemed fair to invite Professor Bidwell to comment.

This note represents the views of both signatories although its detail records the points which one of us (RGSB) had the opportunity to discuss at length with Professor Davies and Dr. Humphrey.

Professor Davies and Dr. Humphrey have disagreed with the interpretation of some of the data in an earlier paper (Bidwell *et al.*, 1964) on two grounds: our supposed misuse of the algebraic formulae developed by Hellebust and Bidwell (1963), and the criticized methodology of our approach to the chase portion of a pulse-chase experiment. In our view these disagreements derive from the misuse of our formulae which may be incorrectly applied so as to generate obviously meaningless data, and also from certain misinterpretations of our stated aims which were to provide limitations, as set by the data, upon the possible rates of protein breakdown and the extent of its contribution to respiration. We did not claim to establish accurate values. At the time the elegant techniques described by Davies and Humphrey were not available and it was only possible to set upper and lower limits. Nevertheless the data did establish, like those from earlier papers, that protein turnover does occur in carrot explants, that much of the carbon derived from protein breakdown is respired, and that the rate of turnover is much greater in fast- than in slow-growing explants.

The principle criticisms of Davies and Humphrey are (i) that we misinterpreted the data of table 7 of our own paper (Table 1 of their paper); (ii) that the use of the algebraic formulae leads to high results for turnover (column 1, Table 2 of their paper); and (iii) that we should not have used the formulae, derived for a labelling experiment, to evaluate data from a pulse-chase experiment. They suggest (cf. their Fig. 1) that a different kinetic approach is necessary for a chase experiment. However, regardless of whether ^{14}C is entering or leaving the pools, the specific activity (SA) of a product can never be very different from that of its immediate precursor. The fact that the SA of CO_2 was considerably higher than that of sugars and fell more slowly suggests that a substantial amount of ^{14}C must have been present in intermediates that were not derived immediately from sugars, but which yielded CO_2. The radioactivity of the conventional respiratory intermediates was so small that it could not be detected; the only other large pool of ^{14}C was protein. Davies and Humphrey pointed out that we did not correct for the size of the CO_2 pool that accumulated during the sampling period. Making this correction would have the effect of bringing the SA of CO_2 at any given time to a value such that its source could not easily be determined from the data in this Table. However, this does not reverse the conclusion, which still stands on other evidence.

In the calculation of turnover as now conducted by Davies and Humphrey in their Table 2, they supplied ^{14}C-amino acids and applied our formulae, but they used the SA of the free amino acid in the leaf as the SA of the source for synthesis (SA*s*). We would maintain that the total free amino acid rarely represents the immediate source of the protein amino acid. Instead, the SA*s* could be higher (by preferential use of the exogenously supplied amino acid) or lower (by dilution with carbon from sugar). The point is, however, that when applied appropriately in the right context the formulae can provide useful information, although care must be taken to select suitable situations to which they are applicable.

For the third point: although the formulae were derived for a labelling experiment, they apply equally in an algebraic sense to a chase experiment, if the changes in ΔA or ΔC are given the appropriate sign. Again, however, judgment should be exercised to determine which of the formulae should be used in any given situation.

R. G. S. BIDWELL, *Dep. of Biology, Queen's University, Kingston, Ontario.*

and

F. C. STEWARD, *Dept. of Biological Sciences, University of the State of New York at Stony Brook, Long Island, New York.*

PHOTOCONTROL OF ENZYME ACTIVITY

H. SMITH, T. H. ATTRIDGE and C. B. JOHNSON

Department of Physiology and Environmental Studies, University of Nottingham School of Agriculture, Sutton Bonington, Loughborough, Leics., U.K.

INTRODUCTION

When seedlings which have been grown in total darkness are transferred to light, a profound metabolic and developmental change occurs. Stem elongation is suppressed, leaf expansion stimulated, chlorophyll synthesis and chloroplast development initiated and the activities of a large number of enzymes are increased. This phenomenon of de-etiolation presents an unparalleled opportunity to investigate the control mechanisms which regulate development, and in particular allows us to focus onto the precise details of the developmental regulation of gene expression.

The two wavelength regions most effective in causing de-etiolation are the blue (400–470 nm) and the far-red (700–760 nm). The blue light responses are mediated by an, as yet, unidentified photoreceptor, which is probably a flavoprotein, whilst the far-red responses are mediated by phytochrome. Very little is known about the mechanism of action of the blue light-absorbing photoreceptor, but much more is known of phytochrome (see Briggs and Rice (1972) for review). Phytochrome is probably a component of certain cell membranes, and under the influence of light it is thought to regulate the properties of those membranes, possibly to control the intracellular distribution of critical metabolites or hormones. These changes are followed by a number of different metabolic changes, including the modulation of the extractable activity of a number of enzymes. Table 1 gives a list of some of the enzymes whose extractable activities are known to be controlled by phytochrome. Blue light appears to operate in a similar manner, since many of the enzymes in Table 1 are also increased in activity by blue light treatment.

Thus the primary action of phytochrome is probably at the membrane level, but the secondary effects on extractable enzyme activities must involve the control of gene expression. In 1964 Mohr suggested that phytochrome operated by controlling gene expression at the transcription step (Hock and Mohr, 1964; Mohr, 1966). The evidence for this view was based almost wholly on experiments with inhibitors of nucleic acid and protein synthesis. For example, the phytochrome-mediated increase in anthocyanin synthesis in mustard seedlings could be stopped at any point in the time course by puromycin, implying that continued protein synthesis was necessary. On the other hand, actinomycin D only exerted significant inhibition if it was given at or before the onset of light treatment (Rissland and Mohr, 1967). These results were taken as evidence that m-RNA synthesis was necessary at the time at which the environmental stimulus was

TABLE 1. ENZYMES WHOSE EXTRACTABLE ACTIVITY IS KNOWN TO BE UNDER PHYTOCHROME CONTROL

Enzyme	Plant
Intermediary metabolism	
NAD-kinase (*in vitro*)	Pea
Lipoxygenase	Mustard
Amylase	Mustard
Ascorbic acid oxidase	Mustard
Galactosyl transferase	Mustard
NAD^+-linked glyceraldehyde 3-phosphate dehydrogenase	Bean
Nitrate reductase	Pea
Nucleic acid and protein metabolism	
RNA polymerase (nuclear)	Pea
Ribonuclease	Lupin
Amino acid activating enzymes	Pea
Photosynthesis and chlorophyll synthesis	
Ribulose-1,5-diphosphate carboxylase	Bean
Transketolase	Rye
$NADP^+$-linked glyceraldehyde phosphate dehydrogenase	Bean
Alkaline fructose 1,6-diphosphatase	Pea
Inorganic pyrophosphatase	Maize
Adenylate kinase	Maize
Succinyl CoA synthetase	Bean
Peroxisome and glyoxisome enzymes	
Peroxidase	Mustard
Glycollate oxidase	Mustard
Glyoxylate reductase	Mustard
Isocitrate lyase	Mustard
Catalase	Mustard
Secondary product synthesis	
Phenylalanine ammonia-lyase	Peas (many others)
Cinnamate hydroxylase	Peas

first perceived. Other workers obtained similar results from experiments on other phytochrome-mediated phenomena (Carr and Reid, 1966). Inhibitors, however, have several drawbacks, and thus considerable effort has been expended in recent years in attempts to obtain *direct* evidence on the mechanism underlying the photocontrol of enzyme activity.

The possible control points in the expression of genetic information are enumerated in the following familiar scheme:

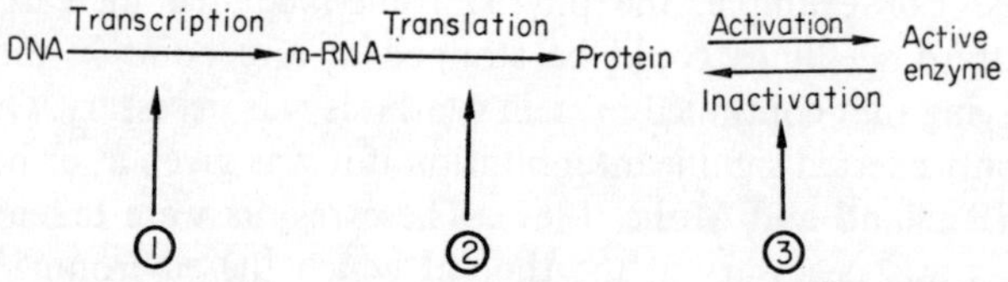

Control point 3 is often overlooked, but it is quite clear that a structural gene cannot be considered expressed until the enzyme it codes for is formed and active. It is likely that most enzymes are active in the form in which they are synthesized, but there is now commanding evidence that some enzymes are subject to chemical modification which renders them either active or inactive (Holzer *et al.*, 1972; Marcus, 1971). Thus the activation/inactivation of enzymes must be considered as a potential control point in the regulation of gene expression.

Testing for enzyme synthesis

Whether or not an observable increase in enzyme activity involves an increased rate of synthesis of that enzyme is not easy to determine. If the use of inhibitors is rejected, as it should be, there are only three available procedures: immunology, radioactive labelling, and density labelling. Of these, immunology and radioactive labelling necessitate the purification of the enzyme to a single homogeneous protein, and at least for the immunological methods, considerable amounts of pure enzyme must be prepared. Density labelling, on the other hand, does not require enzyme purification, and recent advances in the technique (Johnson *et al.*, 1973) have enabled the study of even relatively unstable enzymes by this method.

Theory of density labelling

Density labelling involves the use of stable heavy isotopes to label newly synthesized protein molecules; when a protein has incorporated a substantial amount of heavy isotope, its buoyant density is increased and it can be distinguished from non-labelled, but otherwise identical, molecules by centrifugation to equilibrium in gradients of caesium chloride (Hu *et al.*, 1962). The isotopes which may be used are ^{2}H (or deuterium), ^{13}C, ^{15}N and ^{18}O. Of these, ^{2}H, ^{13}C and ^{15}N can only be incorporated into proteins through *de novo* synthesis of the amino acid constituents. ^{18}O, on the other hand, can also be incorporated into amino acids by hydrolysis of existing protein. ^{2}H is also incorporated into amino acids upon hydrolysis of protein, but is exchanged when the protein is subsequently extracted in an aqueous (H_2O) medium. Only protons attached to C atoms (other than the α-C) are not exchangeable.

In the experiments described below, only ^{2}H has been used as the density label, although other workers have used ^{18}O (Filner and Varner, 1967) and ^{15}N (Zielke and Filner, 1971) with success.

The scheme in Fig. 1 is a simple representation of the different cellular pools which may be involved in determining the flow of ^{2}H into any particular enzyme. The only way in which heavy isotope can be incorporated into a protein is by synthesis of that protein from labelled amino acids. The extent of incorporation, on the other hand, is not merely a function of the rate of either protein or amino acid synthesis. Hydrolysis of unlabelled storage protein in large amounts will dilute the labelled amino acids in the amino acid pool and thus lower the rate of incorporation of label into the enzyme. Similarly, activation of pre-existing inactive enzyme will dilute newly-synthesized labelled enzyme, and

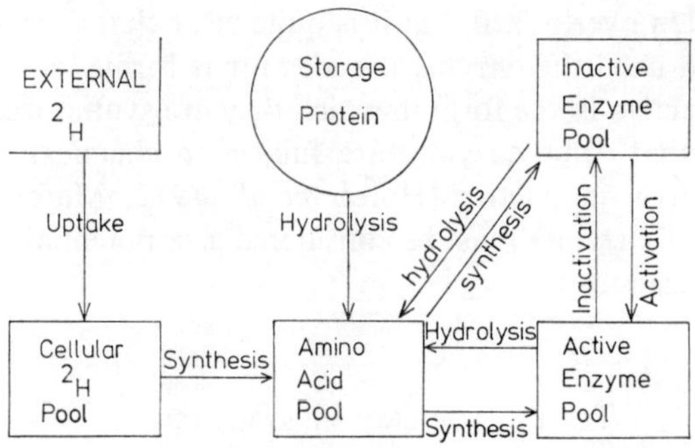

FIG. 1. A scheme showing the various processes which contribute to the determination of the buoyant density of an enzyme under non-steady-state density labelling conditions.

thus reduce the average amount of label incorporated. Furthermore, changes in the rate of degradation or inactivation of the enzyme will affect the mean buoyant density.

Attridge *et al.* (1974) have argued that an increased rate of enzyme synthesis can only be detected by density labelling if the enzyme has a low degradation rate (i.e. turns over slowly). In this case, the mean buoyant density of the enzyme increases slowly with time as more and more of the enzyme molecules become labelled. In comparative experiments treatments which cause an increased rate of synthesis will accelerate the rate of increase in buoyant density. When turnover is very rapid, however, the rate of labelling of the enzyme pool will be determined only by the rate of equilibration of the ^2H_2O with the amino acid pool. Thus, treatments which increase the rate of synthesis will not accelerate the rate of increase in buoyant density.

On the other hand, activation of a pre-existing pool of inactive enzyme molecules can most readily be detected when turnover rate is high. In this case the buoyant density of the enzyme in the untreated plants increases rapidly, whilst in the treated plants it increases more slowly owing to dilution with unlabelled enzyme from the inactive pool.

When an enzyme is centrifuged to equilibrium in a caesium chloride gradient, a Gaussian distribution is obtained. In situations where enzyme synthesis is occurring in the presence of a recently-added density label, the enzyme population will be heterogeneous with respect to buoyant density and a non-Gaussian distribution, possibly with a wider half-bandwidth, may be obtained. Thus, treatments which regulate the rate of synthesis or of activation may be expected to have effects on the bandwidths of the enzyme distributions. There are five factors which may significantly affect the bandwidth:

(1) the $t_{\frac{1}{2}}$ of the enzyme;
(2) the rate of incorporation of the density label into the enzyme molecule;
(3) the duration of exposure to heavy isotope;
(4) the protein loading on the gradient; and
(5) the conditions of assay of the enzyme.

If four of these variables can be kept constant, conclusions regarding the fifth can sometimes be made from measurements of bandwidths. These conditions usually can be met only in a steady state situation when bandwidth measurements can be used to estimate the $t_{\frac{1}{2}}$ of an enzyme (e.g. Quail *et al.*, 1973a, b; Zielke and Filner, 1971). However, where the enzyme activities in two treatments differ significantly it is, in practice, necessary to vary either the protein loading or the assay conditions between treatments. Furthermore, in investigations of the kind under discussion it is not unreasonable to assume that factors (1) and (2) may vary simultaneously. Changes in buoyant density values of the peaks obtained in density-labelling experiments are much better indicators of the mechanisms underlying changes in enzyme activity because

(1) the values are largely unaffected by variability in protein loading and enzyme assay conditions, and
(2) increasing incorporation of isotope always leads to an increase in buoyant density, whereas the effect on bandwidth is first an increase, then a decrease at a rate which is a function of the $t_{\frac{1}{2}}$ of the enzyme.

Thus, under the conditions obtaining in the experiments discussed here the significance of data obtained from bandwidth measurements is much less than that from changes in buoyant density. A greater increase in buoyant density in stimulated material than in the control is indicative of an increased rate of enzyme synthesis, whereas the reverse indicates that an increased rate of synthesis is not involved.

BLUE-LIGHT-MEDIATED ACTIVATION OF PHENYLALANINE AMMONIA-LYASE IN GHERKIN SEEDLINGS

When etiolated gherkin seedlings are irradiated with continuous blue light, a large but transient increase in the extractable activity of phenylalanine ammonia-lyase occurs. The increase begins after a lag of 1.5 hr, reaches a maximum at about 4 hr, after which the enzyme activity declines rapidly to the dark level. When seedlings were density labelled with 99 per cent ^2H_2O for the 4 hr treatment period, a rather remarkable result was obtained (Fig. 2) (Attridge and Smith, 1974). The buoyant density of the native lyase from gherkin is 1.295 kg/l. After 4 hr labelling in darkness the buoyant density had risen to 1.300 kg/l.; in the presence of blue light, on the other hand, there was no increase in buoyant density. At face value this means that in darkness, considerable synthesis of the lyase occurred without any increase in activity, whereas blue light caused a large increase in activity in the apparent absence of lyase synthesis. Control experiments had shown that no increase in the total amino acid pool occurred during the 4 hr blue light treatment, and thus it was concluded that the increase in activity must be due either to activation of pre-existing inactive enzyme or to a decrease in the turnover of the lyase. Previously it had been shown that treatment with high concentrations of a protein synthesis inhibitor, cycloheximide, also caused large increases in lyase activity in the absence of light treatment (Attridge and Smith, 1973). When plants treated in this way were incubated in ^2H_2O exactly analogous results were obtained: untreated controls showed a rise in buoyant density whilst the lyase from cycloheximide treated plants had the same buoyant density as the native enzyme (Attridge and Smith, 1974). Thus, in the absence of protein synthesis, a large increase in enzyme activity was obtained. These data seem to be best interpreted in terms of the blue light mediated activation of pre-existing inactive enzyme. Further evidence which supports this view comes from the

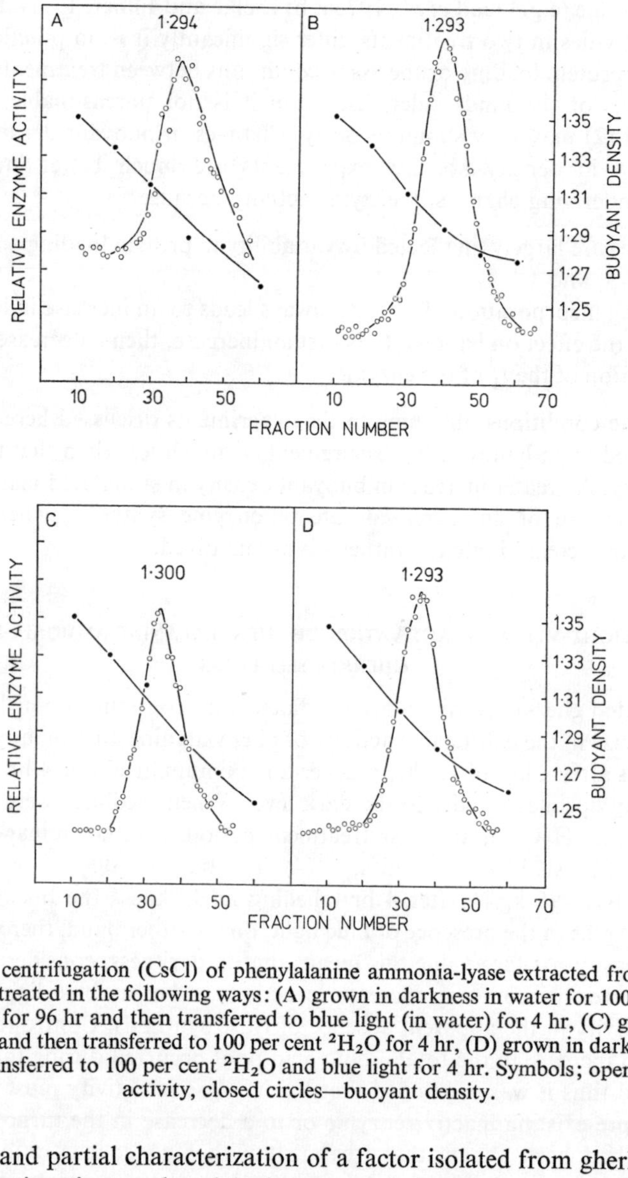

FIG. 2. Isopycnic centrifugation (CsCl) of phenylalanine ammonia-lyase extracted from hypocotyls of gherkin seedlings treated in the following ways: (A) grown in darkness in water for 100 hr, (B) grown in darkness in water for 96 hr and then transferred to blue light (in water) for 4 hr, (C) grown in darkness in water for 96 hr and then transferred to 100 per cent ^2H_2O for 4 hr, (D) grown in darkness in water for 96 hr and then transferred to 100 per cent ^2H_2O and blue light for 4 hr. Symbols; open circles—enzyme activity, closed circles—buoyant density.

demonstration and partial characterization of a factor isolated from gherkin hypocotyls which reversibly inactivates phenylalanine ammonia-lyase *in vitro* (French and Smith, 1975). There are indications that the quantity of inactivator present in the tissues is responsive in an appropriate way to the blue light treatment.

Recently, we have shown that the rate of synthesis of the lyase can be controlled by feedback repression by the reaction end products (Johnson *et al.*, 1975). Application of cinnamic acid to hypocotyls prevents increases in lyase activity and also markedly reduces the incorporation of ^2H into the enzyme. Engelsma and Meijer (1965) have shown that blue light treatment leads to an increase in the tissue concentration of the lyase reaction products. Consequently, the observed depression of buoyant density in Fig. 2 may be partially due to an actual inhibition of lyase synthesis by feedback repression.

PHYTOCHROME-MEDIATED ACTIVATION OF PHENYLALANINE AMMONIA-LYASE IN MUSTARD COTYLEDONS

Photomorphogenesis in mustard seedlings has been extensively investigated by Mohr and his colleagues, and the gene expression hypothesis of phytochrome action is largely based on the control of anthocyanin synthesis and of phenylalanine ammonia-lyase levels in these seedlings. When dark-grown mustard cotyledons are irradiated with continuous far-red light, a slow but substantial increase in extractable phenylalanine ammonia-lyase activity occurs. This increase also has a lag of about 1.5 hr normally, but it reaches a maximum after about 24 to 30 hr. In the presence of 100 per cent ^2H_2O, the time course is altered, with a much longer lag normally being seen. Figure 3 shows the distributions of the lyase in CsCl gradients after density labelling in darkness or far-red light for a period of 40 hr, at which point lyase activity is at its peak in the far-red treated seedlings. Clearly, a substantial increase in buoyant density had occurred in both treatments, demonstrating synthesis of the enzyme. However, the enzyme from the far-red

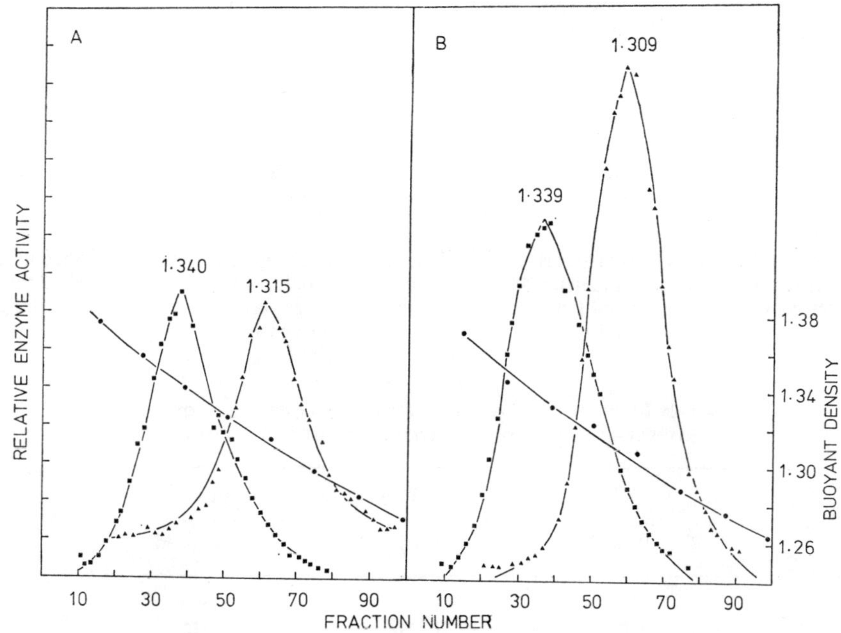

FIG. 3. Isopycnic centrifugation (CsCl) of phenylalanine ammonia-lyase and acid phosphatase extracted from cotyledons of mustard seedlings grown in darkness in water for 48 hr and then treated in the following ways: (A) transferred to 100 per cent ^2H_2O and harvested 40 hr later, or (B) transferred to 100 per cent ^2H_2O and irradiated with far-red light and harvested 40 hr later. Symbols; squares—acid phosphatase, triangles—phenylalanine ammonia-lyase, circles—buoyant density.

treated seedlings had a much lower buoyant density than that from the dark control. This was seen at all times of labelling whenever the activity of the lyase was significantly greater in far-red treated than in dark treated seedlings. These data cannot be interpreted in terms of a phytochrome-mediated increase in lyase synthesis.

PHYTOCHROME-MEDIATED INCREASED SYNTHESIS OF ASCORBIC ACID OXIDASE IN MUSTARD COTYLEDONS

Using exactly the same approach as described above, Attridge (1974) has recently shown that phytochrome does stimulate the synthesis of ascorbic acid oxidase in mustard cotyledons. Representative density labelling data are shown in Fig. 4, demonstrating that the buoyant density of the oxidase is higher in far-red treated seedlings than in dark controls. Table 2 shows the increase in buoyant density of the enzyme with increasing

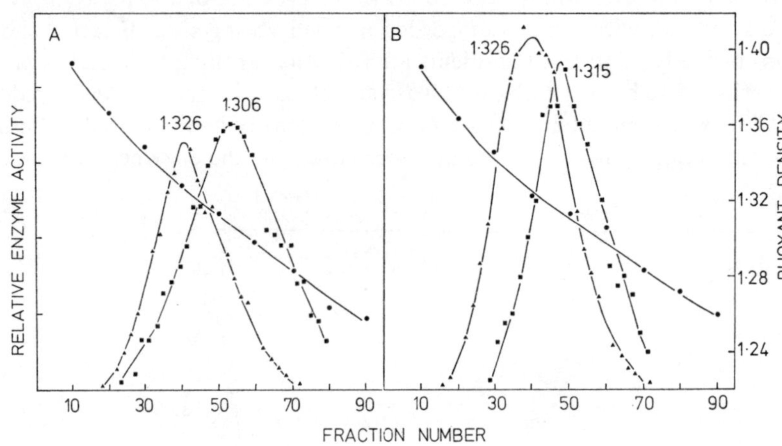

FIG. 4. Buoyant densities (CsCl) of ascorbic acid oxidase and acid phosphatase extracted from cotyledons of plants grown in darkness in water for 48 hr and then transferred to (A) 100 per cent ^2H_2O for 6 hr, (B) 100 per cent ^2H_2O and far-red light for 6 hr. Symbols: squares—ascorbic acid oxidase, triangles—acid phosphatase, circles—buoyant density.

TABLE 2. INCREASE IN BUOYANT DENSITY OF ASCORBIC ACID OXIDASE IN FAR-RED AND DARK TREATED MUSTARD COTYLEDONS

Time in ^2H_2O	Buoyant Density (Kg/l.)	
	Dark	Far-red
Native	1.302	1.303
6 hr	1.306	1.315
12 hr	1.310	1.315
24 hr	1.312	1.316

Seedlings were grown in H_2O for 48 hr in the dark, then transferred to ^2H_2O at the onset of light treatment.

time of far-red treatment. Obviously, ascorbic acid oxidase is turning over relatively slowly, allowing the greater rate of synthesis under far-red to be reflected in increased buoyant density. These results have recently been confirmed by Acton *et al.* (1974).

Photocontrol of gene expression

In arriving at the conclusions described above, we have placed most reliance on the differences in buoyant density between enzymes from treated and untreated seedlings, using as internal marker an enzyme, acid phosphatase, whose extractable activity is not affected by light and whose turnover is slow. Bandwidth measurements have not been used for the reasons stated above. Other workers (Acton and Schopfer, 1974, and in press; Acton *et al.*, 1974) have put forward the view that positive interpretations of density labelling data can only be made by use of bandwidth measurements. In their work, however, they have reduced the protein loading to such an extent, in order to measure bandwidths reliably, that they have been forced to assay phenylalanine ammonia-lyase over a 5-day reaction time, during which period the assay becomes decidedly non-linear. In our experiments, relatively high protein loadings have been used to keep the assay time short; we have shown, however, that the high loadings do not affect the buoyant densities (Attridge *et al.*, 1974).

Given these terms of reference, therefore, we can attempt to interpret the available data. In the case of phenylalanine ammonia-lyase in gherkin seedlings, a rather unexpected conclusion has been reached. Blue light appears to cause the activation of pre-existing lyase, possibly by affecting the formation, or the properties, of the specific lyase inactivator. The increased cellular activity of the enzyme leads to an increase in the concentration of the reaction products, which then feedback to repress lyase synthesis. Thus, an increased enzyme activity is accompanied by a decreased rate of synthesis. On this view, control of gene expression is exerted at control point 3 in Fig. 1. The mechanism of the feedback repression is not known, but it may be either at transcription, as in bacteria, or at translation. This model for the blue-light-mediated control of phenylalanine ammonia-lyase is shown in Fig. 5.

The photocontrol of gene expression in mustard cotyledons is also intriguing. In this case we have one enzyme, ascorbic acid oxidase, whose synthesis is increased by far-red light, and another, the lyase, whose synthesis appears not to be increased by far-red

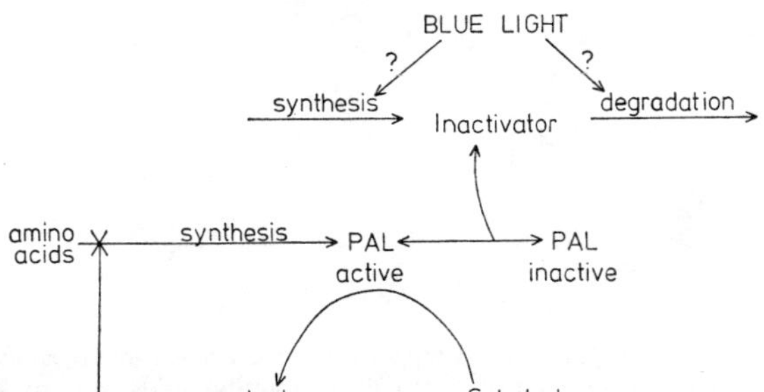

FIG. 5. A model to account for the observed data on the regulation of phenylalanine ammonia-lyase activity by blue light in gherkin seedlings.

light. It is not possible to rule out decreased turnover as being the cause of the increase in lyase activity; however, it seems most unlikely. The half-life of phenylalanine ammonia-lyase in mustard is probably between 2 and 4 hr; thus, since there is no change in activity in dark grown seedlings, the rate of synthesis must be such as to produce exactly half of the existing pool every 2 to 4 hr. Under far-red light, in the absence of ^2H_2O, there is a thirty-fold increase in activity within 20 hr. In the absence of an increased rate of synthesis, a total prevention of degradation would only yield increases of between three and a half- and five and a half-fold. The half-life would need to be as little as 20 min to achieve the observed increase. Thus, phytochrome-mediated decrease of turnover alone cannot account for the increased activity.

A further possibility is that treatment with far-red light causes a massive hydrolysis of unlabelled storage protein which dilutes the labelled amino acids and thus reduces the incorporation of heavy isotope into the lyase. This does not apparently happen in the case of ascorbate oxidase. However, we tested this possibility by determining the level of labelling of the amino acid pool. The total amino acids were extracted from seedlings which had been treated with ^2H_2O in the presence or absence of far-red light, and the purified amino acids fed as the sole nitrogen source to cultures of *Escherichia coli*. The bacterial cultures were grown for 24 hr and extracted for protein which was subjected to isopycnic ultracentrifugation in the normal way. Assays of β-galactosidase were made on the fractions and the results are presented in Fig. 6. These data show that the amino acid pools of both dark grown and far-red treated seedlings were identically labelled after 40 hr in ^2H_2O. On this basis, the large difference in buoyant density shown in Fig. 3 cannot be due to increased hydrolysis of storage protein under far-red light.

We are therefore left with two possible hypotheses, one simple and the other complicated. The complicated hypothesis states that phytochrome increases the rate of synthesis of the lyase and decreases its rate of degradation; thus the newly synthesized molecules are diluted by the existing molecules causing a depression in the buoyant density relative

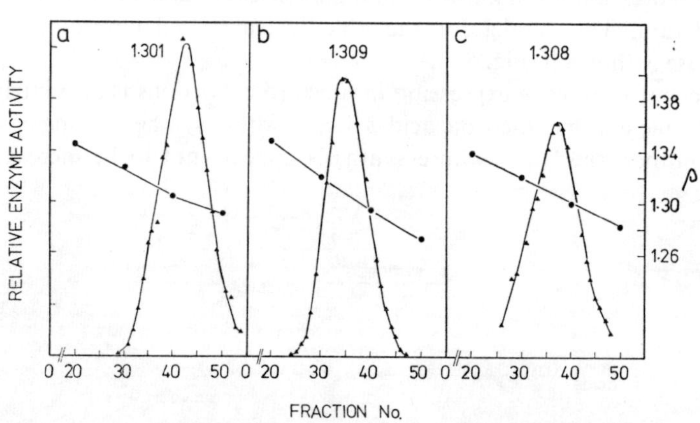

Fig. 6. Isopycnic centrifugation (CsCl) of β-galactosidase extracted from *Escherichia coli*. (a) native enzyme, (b) enzyme extracted from bacteria supplied with amino acids from mustard cotyledons which had been treated with 100 per cent ^2H_2O for 40 hr, (c) enzyme extracted from bacteria supplied with amino acids from mustard cotyledons which had been treated with 100 per cent ^2H_2O and exposed to far-red light for 40 hr. Symbols: ▲, enzyme activity, ●, buoyant density values.

to the dark control. Over the 40 hr labelling period, however, the enzyme activity increased by about twenty-fold and thus 95 per cent of the enzyme, on this hypothesis, would have been synthesized in the presence of ^2H_2O. Such a small dilution would not be expected to have a detectable effect on the buoyant density at this late stage of labelling, yet a marked depression is seen (Fig. 3). The simple hypothesis is that far-red light causes the activation of large amounts of pre-existing inactive enzyme which substantially dilute the newly-synthesized enzyme and results in the large drop in buoyant density observed. The simple hypothesis seems to fit the facts more successfully.

The intriguing feature of these conclusions, if they are valid, is that phytochrome, in the same tissue and under the same conditions, appears to control the synthesis of one enzyme, and the activation of another. On this basis, phytochrome cannot be considered to control gene expression in only one way. Indeed, there is much evidence accumulating to suggest that phytochrome controls enzyme levels in a multivalent manner. In work as yet unpublished, we have shown that phytochrome regulates the formation of polyribosomes even when RNA synthesis is almost completely prevented by 3′-deoxyadenosine (Smith, 1975). This evidence suggests that phytochrome controls the availability of pre-existing, stored messenger-RNA for assembly into polyribosomes. In addition, Travis *et al.* (1974) have recently shown that certain properties of ribosomes in relation to their capacity for *in vitro* synthesis of protein are regulated by phytochrome.

There is, therefore, evidence for phytochrome control of gene expression by regulation of (i) m-RNA availability and polyribosome assembly, (ii) ribosome properties and capacity for protein synthesis, (iii) enzyme synthesis, and (iv) enzyme activation. On the other hand, there is no direct evidence as yet that phytochrome regulates the synthesis of messenger-RNA. It is difficult to integrate all these facts into an acceptable view of a single mechanism of phytochrome action. An attractive possibility (Fig. 7) is that the primary function of phytochrome at the membrane is to release a critical metabolite which acts as a 'second messenger' to initiate the many diverse secondary processes which together constitute photomorphogenesis.

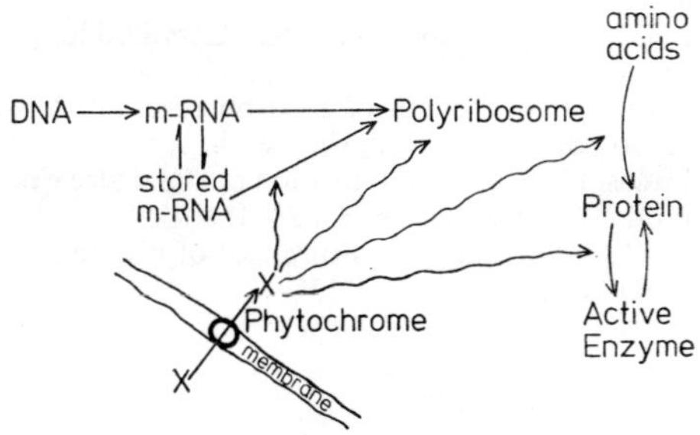

FIG. 7. Second-messenger model of phytochrome action. × = the postulated second-messenger molecule.

REFERENCES

ACTON, G. J. and SCHOPFER, P. (1974) Phytochrome-induced synthesis of ribonuclease *de novo* in lupin hypocotyl sections. *Biochem. J.* **142,** 449–455.

ACTON, G. J., DRUMM, H. and MOHR, H. (1974) Control of synthesis *de novo* of ascorbate oxidase in the mustard seedling (*Sinapis alba*) by phytochrome. *Planta* **121,** 39–50.

ATTRIDGE, T. H. (1974) Phytochrome-mediated synthesis of ascorbic acid oxidase in mustard cotyledons. *Biochim. biophys. Acta* **362,** 258–265.

ATTRIDGE, T. H. and SMITH, H. (1973) Evidence for a pool of inactive phenylalanine ammonia-lyase in *Cucumis sativus* seedlings. *Phytochemistry* **12,** 1569–1574.

ATTRIDGE, T. H. and SMITH, H. (1974) Density-labelling evidence for the blue-light-mediated activation of phenylalanine ammonia-lyase in *Cucumis sativus* seedlings. *Biochim. biophys. Acta* **343,** 452–464.

ATTRIDGE, T. H., JOHNSON, C. B. and SMITH, H. (1974) Density-labelling evidence for the phytochrome-mediated activation of phenylalanine ammonia-lyase in mustard cotyledons. *Biochim. biophys. Acta* **343,** 440–451.

BRIGGS, W. R. and RICE, H. V. (1972) Phytochrome: chemical and physical properties and mechanism of action. *A. Rev. Pl. Physiol.* **23,** 293–334.

CARR, D. J. and REID, D. M. (1966) Actinomycin-D inhibition of phytochrome-mediated responses. *Planta* **69,** 70–78.

ENGELSMA, G. and MEIJER, G. (1965) The influence of light of different spectral regions on the synthesis of phenolic compounds in gherkin seedlings in relation to photomorphogenesis. I Biosynthesis of phenolic compounds. *Acta bot. neerl.* **14,** 54–72.

FILNER, P. and VARNER, J. E. (1967) A test for *de novo* synthesis of enzymes: density labelling with H_2O^{18} of barley α-amylase induced by gibberellic acid. *Proc. natn. Acad. Sci. U.S.A.* **58,** 1520–1526.

FRENCH, C. J. and SMITH, H. (1975) An inactivator of phenylalanine ammonia-lyase from gherkin hypocotyls. *Phytochemistry* in press.

HOCK, B. and MOHR, H. (1964) Die Regulation der O_2-Aufnahme von Senfkeimlingen (*Sinapis alba*) durch Licht. *Planta* **61,** 209–228.

HOLZER, H., KATSUNUMA, T., SCHOTT, E. G., FERGUSON, A. R., HASILIK, A. and BETZ, H. (1972) Studies on a tryptophan synthase inactivating system from yeast. *Adv. Enzyme Reg.* **11,** 53–60.

HU, A. S. L., BOCK, R. M. and HALVORSON, H. O. (1962) Separation of labelled from unlabelled proteins by equilibrium density gradient centrifugation. *Analyt. Biochem.* **4,** 489–504.

JOHNSON, C. B., ATTRIDGE, T. H. and SMITH, H. (1973) Advantages of the fixed angle rotor for the separation of density labelled from unlabelled proteins by isopycnic centrifugation. *Biochim. biophys. Acta* **317,** 219–230.

JOHNSON, C. B., ATTRIDGE, T. H. and SMITH, H. (1975) Regulation of phenylalanine ammonia-lyase synthesis by cinnamic acid: its implication for the light-mediated regulation of the enzyme. *Biochim. biophys. Acta* **385,** 11–19.

MARCUS, A. (1971) Enzyme induction in plants. *A. Rev. Pl. Physiol.* **22,** 313–336.

MOHR, H. (1966) Differential gene activation as a mode of action of phytochrome-730. *Photochem. Photobiol.* **5,** 469–483

QUAIL, P. H., SCHÄFER, E. and MARMÉ, D. (1973a) *De novo* synthesis of phytochrome in pumpkin hooks. *Pl. Physiol. Lancaster* **52,** 124–127.

QUAIL, P. H., SCHÄFER, E. and MARMÉ, D. (1973b) Turnover of phytochrome in pumpkin cotyledons. *Pl. Physiol. Lancaster* **52,** 128–131.

RISSLAND, I. and MOHR, H. (1967) Phytochrom-induzierte Enzymbildung (Phenylalanine-desaminase), ein schnell ablaufender Prozess. *Planta* **77,** 239–249.

SMITH, H. (1975) Phytochrome-mediated assembly of polyribosomes in etiolated bean leaves: evidence for post transcriptional regulation of development. (in press).

TRAVIS, R. L., KEY, J. L. and ROSS, C. W. (1974) Activation of 80S maize ribosomes by red light treatment of dark-grown seedlings. *Pl. Physiol. Lancaster* **53,** 28–31.

ZIELKE, H. R. and FILNER, P. (1971) Synthesis and turnover of nitrate reductase induced by nitrate in cultured tobacco cells. *J. biol. Chem.* **246,** 1772–1779.

NITROGEN ASSIMILATION AND PROTEIN SYNTHESIS IN PLANT CELL CULTURES

H. E. STREET, A. R. GOULD and JOHN KING*

Botanical Laboratories, School of Biological Sciences, University of Leicester, Leicester, U.K.

INTRODUCTION

This paper is concerned with some recent technical advances in the manipulation of plant cell cultures and with experimental work illustrating that these advances make such culture systems particularly suited for studies on aspects of nitrogen assimilation and protein synthesis.

Plant cell cultures can now be grown in a highly reproducible manner in batch cultures. They consist of free cells and small aggregates of cells; almost all the cells (normally well in excess of 96 per cent) participate in culture growth. The cultures show, during each passage, a growth cycle (in terms of cell number and biomass increase) essentially similar to that of micro-organisms in batch culture (Street, 1973a). In such cultures, particularly when initiated from a relatively high cell density, the culture environment is constantly changing as growth proceeds. In a series of studies it has been shown that the

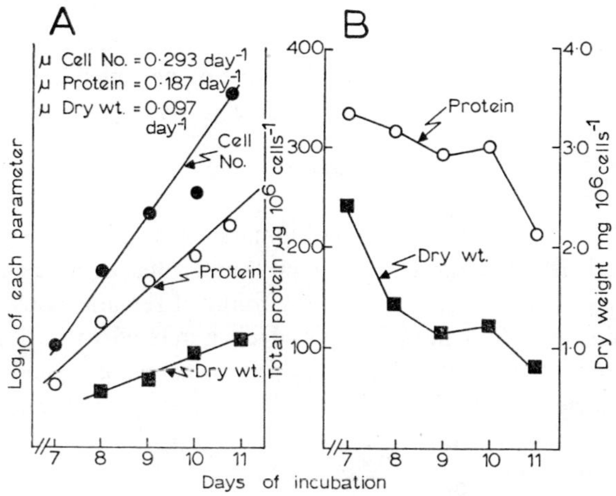

FIG. 1. Unbalanced growth of sycamore cells during the exponential phase of batch culture. (A) Semi-logarithmic plot showing rate of change of cell number, total protein, and cell dry weight per unit volume of culture. The slope of the line of best fit (calculated by linear regression analyses, $p<0.01$) was used to determine the specific growth rate (μ) of each parameter (plotting $\log_{10}x$ against t gives slope $= \mu/2.303$). (B) Changes in total protein and dry weight 10^6 cells^{-1} with time calculated from data in (A). Data from P. J. King and Street (1973).

* Present address: Department of Biology, University of Saskatchewan, Saskatoon, Canada S7N OWO.

rates of physiological processes, levels of metabolites and activities of individual enzymes show individually distinct patterns of change as the culture passes through the sequential stages of its growth cycle (King and Street, 1973). Peaks of activity of respiration, in levels of macromolecules (RNA, protein, starch) and soluble metabolites (sugars, amino acids, nucleotides) and in activity of a wide range of enzymes occur at predictable and different points in the growth cycle. The cells in such cultures pass through a succession of physiological states characterized by a changing balance between cell division and cell expansion and between the diverse biochemical pathways involved in cell growth. Even during the short period of exponential growth the cells never achieve a steady state (a state of balanced growth) as evidenced by the observation that cell division is uncoupled from cell dry weight and protein content (Fig. 1); the cells are changing in composition even during the period when mean generation time is constant. These observations raised the question of whether balanced growth could be achieved in plant cell cultures if a constant culture environment could be established by the development of an open continuous culture system. A system operational either as a chemostat or as a turbidostat, providing appropriate conditions of aeration and agitation for the growth of plant cells in suspension, and capable of long-term aseptic operation was described in 1971 (Wilson *et al.*). Subsequent work established, in conformity with expectation from the work of Monod (1950) and Novick and Szilard (1950) with micro-organisms, that steady states of growth could be achieved in plant cell suspensions by continuous dilution of a fixed volume of culture with fresh medium at a chosen rate and a balancing harvest of culture (King and Street, 1973; King *et al.*, 1973).

STEADY STATES IN CHEMOSTAT CULTURE

In a chemostat the relationship between change in cell density (cells per unit volume $= x$) and time (t) at a fixed dilution rate (volume of new medium added per unit time expressed as a fraction of the total culture volume $= D$) is given by the equation

$$\frac{dx}{dt} = \mu x - Dx \tag{1}$$

where μ = specific growth rate (increase in biomass per unit biomass per unit time). When equilibrium is reached, $dx/dt = 0$, $\mu = D$ and x has a value characteristic of the dilution rate. At equilibrium the nutritive environment remains constant, each nutrient achieving an equilibrium concentration (s) which is related to its input concentration (S_R) and the rate of its consumption (given by $\mu x/Y$ where Y is a yield coefficient = cells formed per unit nutrient used). The relationship is expressed by the equation

$$\frac{ds}{dt} = DS_R - Ds - \frac{\mu x}{Y} = 0 \tag{2}$$

The equilibrium state is reached because one particular nutrient (and which depends upon the composition of the culture medium) becomes the limiting nutrient and determines the value of μ. Here the relationship is given by

$$\mu = \mu_{max} \frac{s}{K_s + s} \tag{3}$$

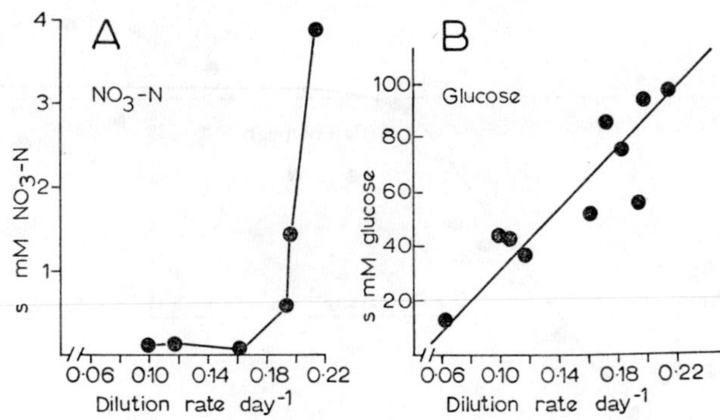

FIG. 2. Steady-state levels (s) of nitrate (A) and glucose (B) in chemostat cultures of sycamore at different dilution rates. Medium of Stuart and Street (1969). Data from P. J. King and Street (1973).

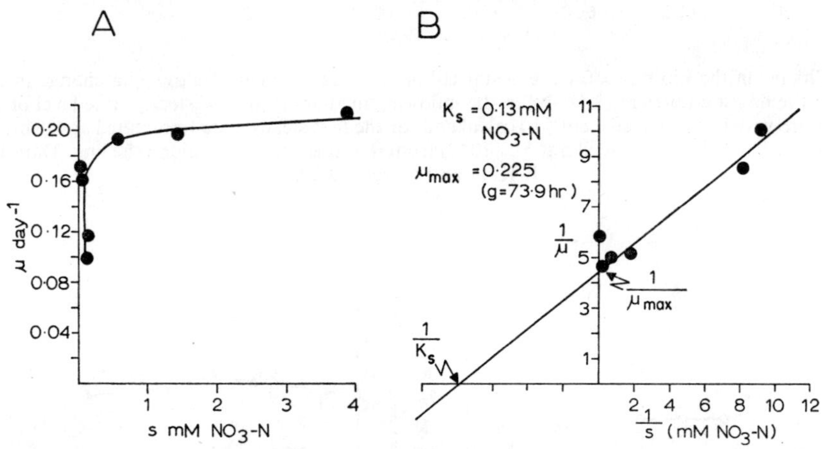

FIG. 3. (A) Relationship between specific growth rate (μ) and the steady state level (s) of nitrate-N in chemostat cultures of sycamore. (B) Double reciprocal plot of data of (A) showing calculation of K_s and μ_{max}. Data from P. J. King and Street (1973).

where μ_{max} = the maximum specific growth when no nutrient is limiting, at the temperature and aeration conditions adopted, and K_s = a 'saturation constant' numerical y equal to the concentration of the limiting nutrient at which the specific growth (μ) = 0.5 μ_{max}.

In the experiments now to be described cell cultures of sycamore (*Acer pseudoplatanus* L) were grown in the medium of Stuart and Street (1969) and in this medium it can be shown that nitrogen is the limiting nutrient. Figure 2 shows the influence of dilution rate on s for nitrate (A) and for glucose (B) (nitrate showing depletion and saturation over the dilution rate range tested). Figure 3 depicts the relationship between μ and s for nitrate, Fig. 3B being a Lineweaver–Burk plot giving a value for μ_{max} of 0.225 day^{-1} (a value in good agreement with μ_{max} as tested in the turbidostat system when x is set at a low value) and K_s (NO_3-N) = 0.13 × 10^{-3} M. By rendering phosphorus as the limiting nutrient, Wilson (1971) obtained a similar relationship with K_s (phosphorus) = 0.032 × 10^{-3} M.

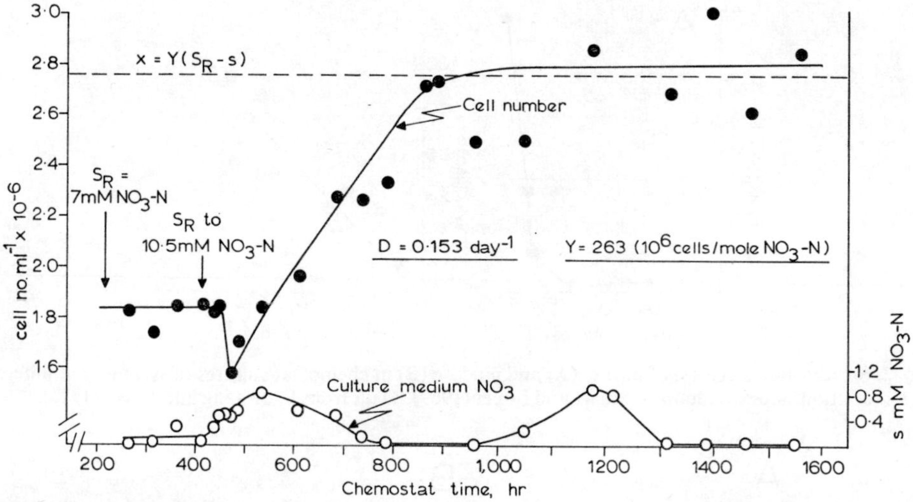

FIG. 4. Change in the biomass of a chemostat culture of sycamore cells following a change in concentration of the limiting nutrient (NO_3-N) in the inflowing medium (S_R). $s =$ steady state level of NO_3-N in the culture vessel. Yield coefficient (Y) calculated for the first steady state terminated at 430 hr. Dotted line shows predicted value of x (cell $ml^{-1} \times 10^{-6}$) from the equation shown along the line. Data from P. J. King and Street (1973).

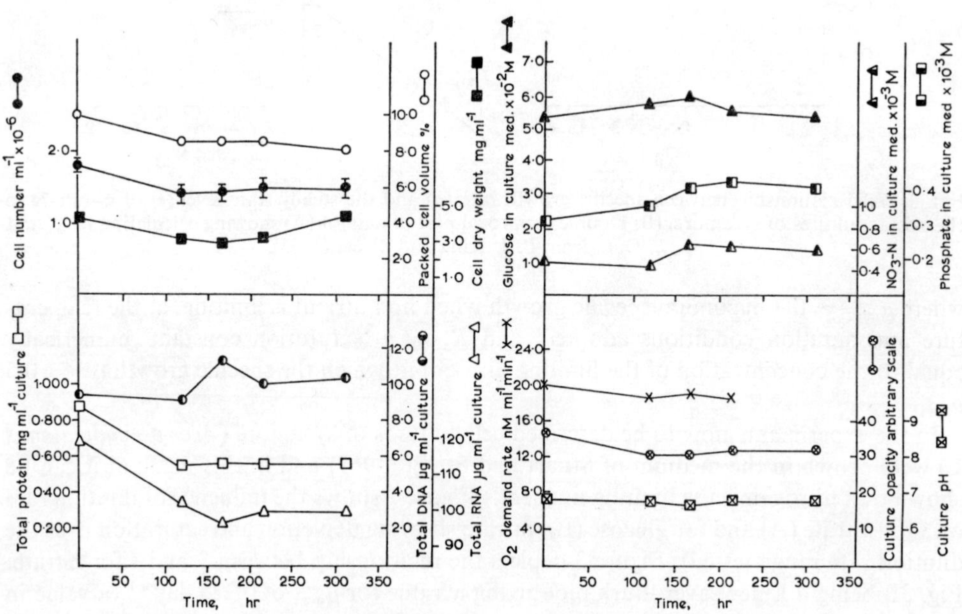

FIG. 5. A steady state established in a chemostat of sycamore cells. Steady state achieved 125 hr after commencing a dilution rate of 0.194 day^{-1} (doubling time 178 hr). Data from P. J. King and Street (1973).

This draws attention to the relatively low affinity of the sycamore cells for nitrate. The conformity of nitrate-limited cultures to chemostat theory can be further illustrated by obtaining a steady state at a given value of S_R nitrate and then changing this to a new value and measuring the new equilibrium value of x and seeing how far this corresponds to the theoretically expected value from the equation

$$x = Y(S_R - s) \tag{4}$$

which derives from eqn. (2) when, as at equilibrium $\mu = D$. The result of such an experiment is shown in Fig. 4.

The experiment just described illustrates the achievement of steady state cell densities (x) in such chemostat cultures. The demonstration that such steady state densities are of cells in a balanced state of growth and metabolism is illustrated in Fig. 5, which presents data for a steady state established at $D = 0.194$ day^{-1} and monitored over 300 hr. Attention is drawn to the steady state values for cell number, packed cell volume, cell dry weight, protein, DNA and RNA, oxygen demand, culture pH and opacity and the equilibrium levels in the culture medium of glucose, phosphate and nitrate. Such steady states are also characterized by steady state levels of activity of enzymes; this is illustrated in Fig. 6 for enzymes involved in nitrogen assimilation. Levels of activity of enzymes concerned in respiratory metabolism are shown for two steady states established at different dilution rates (and hence different cell doubling times) in Fig. 7. These data illustrate that with an increase in growth rate some enzyme activities may remain almost constant but others show a pronounced shift to an absolute (or relatively) higher or lower and again steady level.

Steady state levels of metabolites are also established and these can be shifted to new levels by alterations in dilution rate or by alterations of nutrient supply at a fixed dilution rate (fixed growth rate). This latter situation is illustrated in Tables 1 and 2 where, at very similar dilution rates, the cells received in the incoming medium either the standard nitrogen supply (7 μmol NO_3-N + 6.6 μmol urea-N) or the whole of their nitrogen as urea (19.8 μmol urea-N). Table 1 shows that the yields of cell fresh and dry weight per unit nitrogen utilized were similar for both cultures but that in other respects they were highly contrasted; the urea-grown culture containing a lower cell density (x value) of much larger cells (packed cell volume per 10^6 cells) and yielding almost twice the amount of protein (μg protein ml^{-1}). Table 2 shows that the amino acid pool of the urea-grown cells was 8 times that of the control cells; though this is in part a reflection of the contrast in mean cell size, even in terms of μmol amino acid per μl of cells, the values are for urea-grown, 3.1 and for control cells, 1.6. The detailed amino acid analyses also presented in Table 2 show the relatively high levels of asparagine and ammonia (per cent total amino acids columns) in the urea-grown cells, and of glutamic acid and alanine in control cells. The two analyses performed on the urea-grown cells (taken respectively after 129 days (Sample (i)) and after 147 days (Sample (ii)) from commencement of dilution) show reasonable general agreement but it can perhaps be inferred that levels of some amino acids (asparagine, aspartic acid, ammonia) may be subject to fluctuations during a generally steady state. This suggests that 'steady states' as at present achieved should be more critically investigated for short-term fluctuations in enzyme activities and metabolite levels.

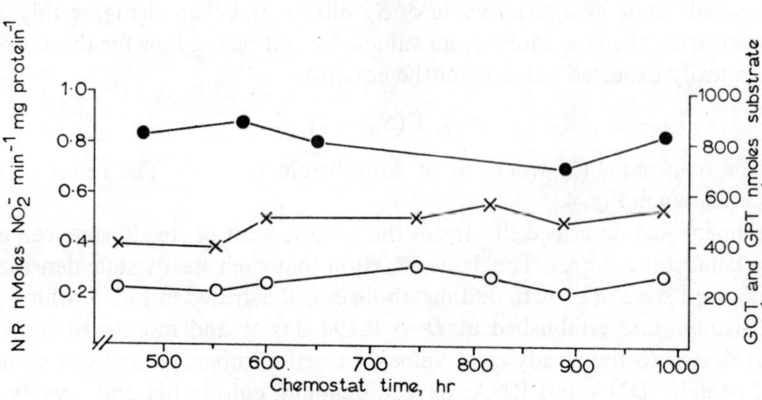

FIG. 6. Steady state activities of nitrate reductase (NR—●—) (E.C.1.6.6.1), glutamate-oxaloacetate transaminase (GOT—×—) (E.C.2.6.1.1) and, glutamate-pyruvate transaminase (GPT—○—) (E.C. 2.6.1.2) in a chemostat culture of sycamore cells. $D = 0.094$ day^{-1}, $x = 2.32 \times 10^6$ cells ml^{-1}, total protein = 705 μg ml^{-1}. Data from Young (1973).

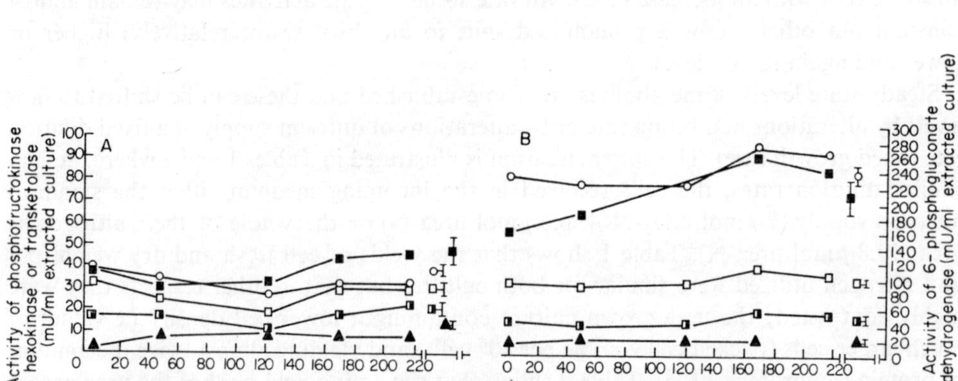

FIG. 7. Activities of respiratory enzymes measured during two steady states (A) and (B) maintained at different growth rates in a chemostat culture of sycamore cells. (A) $D = 0.084$ day^{-1}; cell number = 3.5×10^6 cells ml^{-1}; doubling time (*td*) = 380 hr. (B) $D = 0.145$ day^{-1}; cell number = 2.9×10^6 cells ml^{-1}; *td* = 175 hr. Enzymes: phosphofructokinase (E.C.2.7.1.11) (—■—■—); hexokinase, pH 8.5 (E.C.2.7.1.1) (—□—□—), 6-phosphogluconate dehydrogenase (E.C.1.1.1.44) (—○—○—)-; transketolase (E.C.2.2.1.1) (—▲—▲—). The length of the vertical lines = 2 × SE plotted with mean as mid point. Redrawn from Fowler and Clifton (1974).

TABLE 1. CHARACTERISTICS OF TWO STEADY STATE SYCAMORE CELL CULTURES SUPPLIED WITH NITRATE PLUS UREA (1) OR WITH UREA ONLY (2) AS SOURCE OF NITROGEN
(Previously unpublished data of John King for Culture 2, data for Culture 1 from Young, 1973)

Parameter	Culture 1	Culture 2
Nitrogen source (S_R in μmol ml^{-1})	7.0 NO_3 / 6.6 urea	19.8 urea
Dilution rate (D) (culture vol. day^{-1})	0.143	0.133
Doubling time (td) hr	116	125
Cell density (x) 10^6 cells ml^{-1}	1.89 ± 0.05(9)†	0.91 ± 0.02(25)
Cell fresh weight		
mg ml^{-1}	32.1 ± 1.6(8)	44.0 ± 0.3(20)
mg 10^6 cells^{-1}	17.0*	48.2 ± 3.6(20)
Cell dry weight		
mg ml^{-1}	3.05 ± 0.11(8)	4.70 ± 0.10(19)
mg 10^6 cells	1.6*	5.1 ± 0.1(19)
Packed cell volume %	8.6*	17.0 ± 0.5(21)
Total protein		
mg ml^{-1}	645 ± 16(9)	1125 ± 6(5)
μg 10^6 cells^{-1}	341*	1154 ± 43(5)
Equilibrium N concentration(s) in spent medium		
μmol ml^{-1}	0.17 ± 0.01(7)	0.90 ± 0.06(18)
Cell fresh weight (mg) per μmol N utilized	2.4	2.3
Cell dry weight (mg) per μmol N utilized	0.23	0.25

* Data calculated from Young (1973). † SE and, in brackets, number of samples analysed.

A second example of a change in amino acid levels is illustrated by an experiment in which cells in a steady state in the standard medium were perturbed by the addition of glutamate to the inflowing medium (5 μmol glutamate-N ml^{-1}) (Fig. 8). The data are selected for several amino acids to show that certain amino acids remained constant (e.g. cysteine and methionine), whereas others showed a rise in level. This rise applied particularly to alanine, glutamic acid and serine, but smaller increases in levels of valine, proline and histidine also occurred. In all cases the new steady state levels were reached in 24–36 hr (i.e. within a fraction of the doubling time). Simultaneously, changes to new steady state levels of activity were also shown to occur in certain enzymes involved in nitrogen assimilation and amino acid synthesis (Young, 1973). The three transferases, glutamate-oxaloacetate (E.C.2.6.1.1), glutamate-pyruvate (E.C.2.6.1.2) and γ-glutamyl (E.C.6.3.1.2) increased in activity, whereas there was some decline in urease (E.C.3.5.1.5) and nitrate reductase (E.C.1.6.6.1), and correlated with these changes a rise in the equilibrium levels of both urea and nitrate in the medium.

The above experimental work illustrates how cells containing very different protein and amino acid levels, and differing in the levels of activity of enzymes involved in nitrogen metabolism, can be obtained either by alteration of the dilution rate using a culture medium of fixed composition or by qualitative alteration of the nitrogen supply at a fixed dilution rate. However, equally striking changes in nitrogen metabolism can also be achieved by modifying the culture medium so that other major nutrients act as limiting factors. Table 3 shows how by making carbohydrate supply (glucose) the limiting factor the partition of cell carbon between protein and other cell constituents can be altered, and cells of exceptionally high protein content per unit cell dry weight obtained.

TABLE 2. FREE AMINO ACID CONTENTS OF ETHANOLIC EXTRACTS OF SYCAMORE CELLS RECOVERED FROM CULTURE 1 AND CULTURE 2 OF TABLE 1

Amino acid	Culture 1		Culture 2			
			(i)		(ii)	
	(a)	(b)	(a)	(b)	(a)	(b)
cysteic acid-urea	—	—	12.2	2.1	7.4	1.2
aspartic acid	4.0	5.5	44.9	7.9	29.3	4.9
threonine	4.0	5.5	22.0	3.9	44.7	7.4
serine	7.0	9.6	68.3	12.0	59.2	9.8
asparagine	—	—	227.0	39.8	185.2	30.8
glutamic acid	26.0	35.6	62.5	11.0	66.1	11.0
proline	1.0	1.4	—	—	—	—
glycine	1.0	1.4	8.3	1.5	11.1	1.8
alanine	17.0	23.3	40.6	7.1	66.4	11.1
valine	3.0	4.1	7.4	1.3	9.2	1.5
cysteine	0.2	0.4	1.1	0.2	1.5	0.3
methionine	—	—	0.8	0.1	1.1	0.2
isoleucine	1.0	1.4	2.3	0.4	2.7	0.5
leucine	1.0	1.4	3.5	0.6	4.7	0.8
tyrosine	0.3	0.4	2.6	0.5	3.3	0.6
phenylalanine	1.0	1.4	5.6	1.0	4.8	0.8
ammonia	2.0	2.7	24.8	5.0	65.9	11.0
γ-aminobutyric acid	—	—	4.0	0.7	—	—
ornithine	1.0	1.4	1.1	0.2	3.1	0.1
lysine	0.9	1.3	14.3	2.5	17.6	2.9
histidine	1.1	1.5	5.1	0.9	6.3	1.1
tryptophan	—	—	4.4	0.8	5.9	1.0
arginine	1.0	1.4	3.6	0.6	5.1	0.9
Total	72.5		569.2		601.2	

a = μmol amino acid 10^9 cells^{-1}; b = percentage total amino acid pool. Samples (i) and (ii) of culture 2 were taken 129 and 147 days from start of culture. Previously unpublished data of John King for Culture 2; data for Culture 1 from Young (1973).

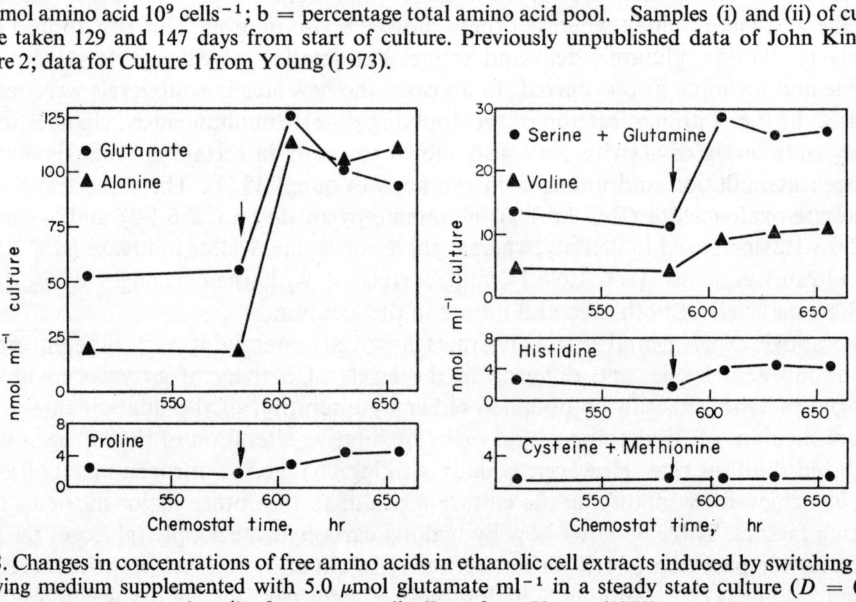

FIG. 8. Changes in concentrations of free amino acids in ethanolic cell extracts induced by switching to an inflowing medium supplemented with 5.0 μmol glutamate ml^{-1} in a steady state culture (D = 0.215 day^{-1}) of sycamore cells. Data from Young (1973).

TABLE 3. COMPARISON OF THE COMPOSITION OF SYCAMORE CELLS IN CHEMOSTAT STEADY STATES AT THE SAME GROWTH RATE BUT WITH DIFFERENT LIMITING NUTRIENTS
(previously unpublished work by P. J. King)

Parameter	State A Nitrate limited		State B* Glucose limited	
Nutrient data				
D, Dilution rate (day^{-1})	0.152		0.152	
x, cell number (ml^{-1})	1.86×10^6		0.74×10^6	
SR_1, glucose concentration (molar)				
input	11.0×10^{-2}		1.85×10^{-2}	
output	3.7×10^{-2}		8.7×10^{-4}	
SR_2, nitrate concentration (molar)				
input	8.0×10^{-3}		8.0×10^{-3}	
output	8.1×10^{-5}		3.3×10^{-3}	
Biomass data				
Mean dry weight of each fraction and standard errors of the means				
	10^{-9}g cell^{-1}	% total	10^{-9}g cell^{-1}	% total
Starch	0.228 ± 0.013	12	0.087 ± 0.010	16
Protein	0.427 ± 0.018	23	0.369 ± 0.003	67
Others†	1.218 ± 0.075	65	0.096 ± 0.029	17
Total	1.841 ± 0.228		0.552 ± 0.021	

* State B was achieved from State A by changing the input glucose concentration from 11.0×10^{-2} M to 1.85×10^{-2} M after 672 hr operation in State A.

† The weights of other unidentified fractions were calculated by subtraction of starch and protein from total dry weight.

The full range of states of balanced growth which can be achieved by the chemostat culture of plant cells has yet to be explored. This applies particularly to steady states in which carbon and nitrogen may be channelled into the many secondary products of plant cells. The range of such steady states will, however, be limited by the consideration that for any dilution rate the cells must be able to achieve a matching growth rate. If this is not achieved, cell density will continue to decline and the culture suffer wash-out. For the same reason chemostat cultures can only be operated at growth rates not too close to the maximum specific growth rate (μ_{max}). For studies at μ_{max}, for instance for studies on the hormonal regulation of aspects of nitrogen metabolism where primary nutrients are not limiting, recourse must be made to a turbidostat culture system (Wilson *et al.*, 1971).

The chemostat steady state cultures described above represent a new system for work in the field of cellular physiology and biochemistry. However, as of now, their value for studies in the regulation of aspects of nitrogen metabolism has only begun to be explored. They can obviously provide us with large populations (a 4 l. chemostat of sycamore cells can contain up to 1.2×10^{10} cells) of 'uniform' cells grown with a particular study in mind. Thus Filner (1966) and Young (1973) have used cells from tobacco and sycamore respectively to study the induction of nitrate reductase and to present evidence that glutamate can depress the induced synthesis. Their work also indicates that nitrate reductase is present at the lowest activity of the assimilating enzymes but has activity sufficient to effect the recorded level of assimilation, thereby supporting the view that nitrate reductase regulates the rate of nitrogen assimilation in nitrate-fed cells. More recently Heimer and Filner (1971) with tobacco, and one of us (J.K.) with sycamore, have used cells grown in the absence of nitrate to study the kinetics of nitrate uptake. The cells used in the latter study were harvested from the steady state culture on urea-N described in Tables 1 and 2. They were prevented from embarking upon nitrate reductase induction during the nitrate uptake experiments by sodium tungstate (10^{-4} M) which was shown by fluorescein diacetate staining (Widholm, 1972) to have no adverse effect on viability and regrowth of cells following washing and suspension in new medium. The effectiveness of the tungstate was demonstrated by the recovery of 90 per cent of the nitrate taken up after 25 hr, under conditions where nitrate was accumulated (internal nitrate 4.78×10^{-6} M, external nitrate 0.46×10^{-6} M at 24 hr). Nitrate uptake was studied over the concentration range 10^{-3} to 10^{-6} M. When the urea-grown cells (total amino acid content 570–600 μmol 10^9 cells^{-1}) were starved of nitrogen for 24 hr their total amino acid content fell (to about 320 μmol 10^9 cells^{-1}), mainly as a result of decreases in the levels of aspartic acid, asparagine, serine and alanine. Such 'starved' cells showed a higher rate of uptake of nitrate than 'unstarved' cells during the first 3 hr of uptake (Fig. 9), suggesting an interaction between soluble organic nitrogen and expression of the nitrate uptake capacity. The limits of sensitivity of the nitrate assay (Montgomery and Dymock, 1961) made it difficult to determine whether 'unstarved' cells showed this lower rate of uptake during the first 10 to 15 min after being challenged with nitrate or whether in both cases there was an interval between the nitrate challenge and the activation of the uptake system. Measurements of velocity (V) of nitrate uptake over the first 3 hr by 'starved' cells exposed to various concentrations (s) of nitrate (Fig. 10) gave an apparent $K_m = 4.3 \times 10^{-5}$ M and an apparent V_{max} of 22.2 nmol NO_3 hr^{-1}10^6 cells^{-1}. Sodium azide (10^{-3} M) and dinitrophenol (10^{-4} M) completely inhibited this nitrate uptake. Cycloheximide (10^{-4} M) inhibited nitrate uptake (by 35 per cent at 10^{-4} M NO_3) during the first 2 hr, but subsequently uptake proceeded at the normal rate. This is of interest because inhibition of protein synthesis (synthesis of phenylalanine ammonia lyase and incorporation of [^{14}C]leucine) by cycloheximide (10^{-4} M) has been shown to be transient in cultured rose cells (Davies and Exworth, 1973). The K_m value reported above is an order of magnitude less than that calculated by Heimer and Filner (1971) for cultured tobacco cells though similar to that reported by van den Honert and Hooymans (1955) from studies of nitrate uptake by maize seedlings. From our chemostat culture studies it emerges that the growth limiting nitrate level ($K_s = 0.13 \times 10^{-3}$ M) is much higher than the saturating level for the uptake system studied ($K_m = 4.3 \times 10^{-5}$ M) so that in such cultures nitrate assimilation rather than nitrate uptake would appear to operate as the growth limiting factor.

Although the K_m value for nitrate uptake quoted above can maintain nitrate uptake at an appropriate rate for the control chemostat culture (Culture 1) detailed in Table 1 (rate 42 nmol hr^{-1} ml^{-1}) we know that, at μ_{max} and in the presence of a nitrate level close to the input level (7×10^{-3} M), nitrate uptake by sycamore cells can reach 190 nmol hr^{-1} 10^6 cells^{-1} in contrast to the V_{max} reported here of 22.2 nmol hr^{-1} 10^6 cells^{-1}. These studies have therefore raised a number of interesting questions relating to nitrate

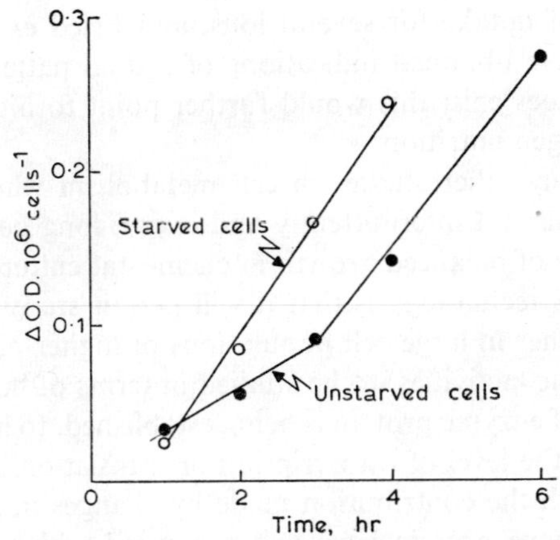

FIG. 9. Accumulation of nitrate (10^{-4} M) in presence of 10^{-4} M Na tungstate by urea-grown sycamore cells (see Table 1) (—●—) ('unstarved') and by these cells after 24 hr culture in nitrogen-omitted medium (—○—) ('starved'). Nitrate assay (Young, 1973) expressed as change in optical density at 550 nm 10^6 cells^{-1} (Montgomery and Dymock, 1961). Previously unpublished data of John King.

uptake accessible to and requiring further study with the sycamore system: does the nitrate uptake system decay in the absence of challenging nitrate, is protein synthesis essential for its activity, is it subject to control by particular amino acids and is there a second uptake system which only operates at a significant rate at levels of external nitrate higher than these tested here? In relation to this last question Epstein (1972) has

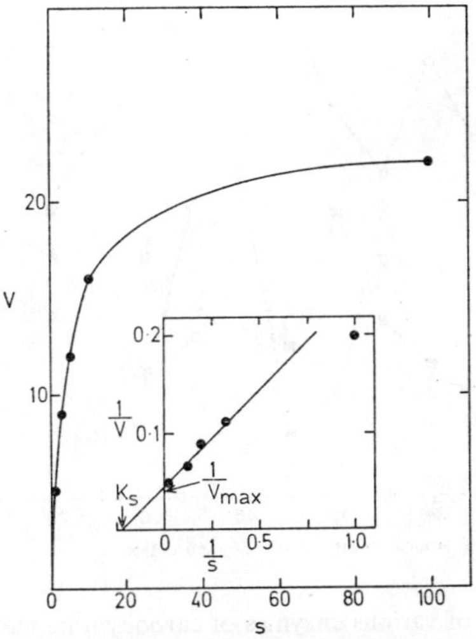

FIG. 10. Velocity of nitrate uptake (V as nmol NO_3 accumulated hr^{-1} 10^6 cells^{-1}) by a suspension culture of sycamore cells (averaged over a period of 1–4 hr from challenge of cells with nitrate) at various nitrate concentrations (s as multiples of 10^{-5} M NO_3^-) and in presence of 10^{-4} M Na tungstate. Insert is a double reciprocal plot ($1/V$ again $1/s$). Previously unpublished data of John King.

reported two systems of uptake for several ions, and Fried *et al.* (1965) in work with excised roots of rice have obtained indications of a dual pattern of nitrate uptake. If such a second system does exist this would further point to nitrate assimilation as the controlling step in nitrogen nutrition.

There are clearly many other studies in cell metabolism where it will be extremely valuable to be able to harvest intermittently and over a long period a large population of cells in a chosen state of balanced growth in chemostat culture. However perhaps the novel aspect of this new technology is that it will permit study of the transitions from one steady state to another in large cell populations of higher plant cells; it should also enable changes in enzyme 'activities' to be studied in terms of 'fine' and 'coarse' control and, where a new level of enzyme protein is being established, to locate whether enhanced synthesis is achieved at the level of transcription or translation. For individual enzymes we need to examine both the contribution made by changes in synthesis and turn-over rates to changes in enzyme protein level and the speed with which such changes can operate. It is already clear that such studies will from the beginning present us with unexpected and intriguing observations. Fowler and Clifton (1974, 1975) have studied over short time intervals the changes in enzyme activities which occur between steady states such as those depicted in Fig. 7. Their results show pronounced oscillations in activity in the transition from one steady state level to another and that the period of the oscillations is characteristic of the enzyme concerned (Fig. 11).

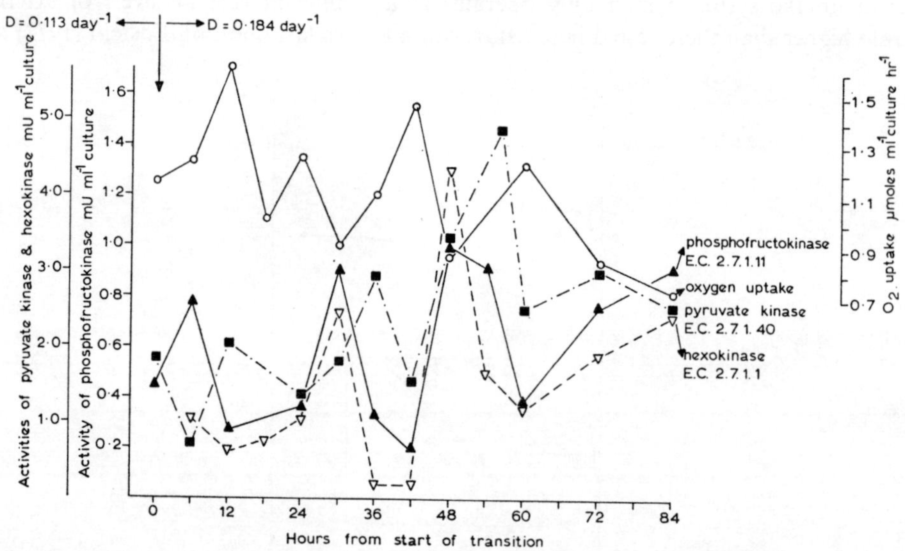

FIG. 11. Oscillations in activities of various enzymes of carbohydrate metabolism and in oxygen uptake during a transition from one steady state (D = 0.113 day^{-1}, td = 147 hr, cell number = 1.83 × 10^6 ml^{-1}) to a second steady state (D = 0.184 day^{-1}, td = 96 hr, cell number = 1.02 × 10^6 ml^{-1}) in a chemostat culture of sycamore cells. Methods of assay as in Fowler and Clifton (1974). Data from Fowler and Clifton (1975).

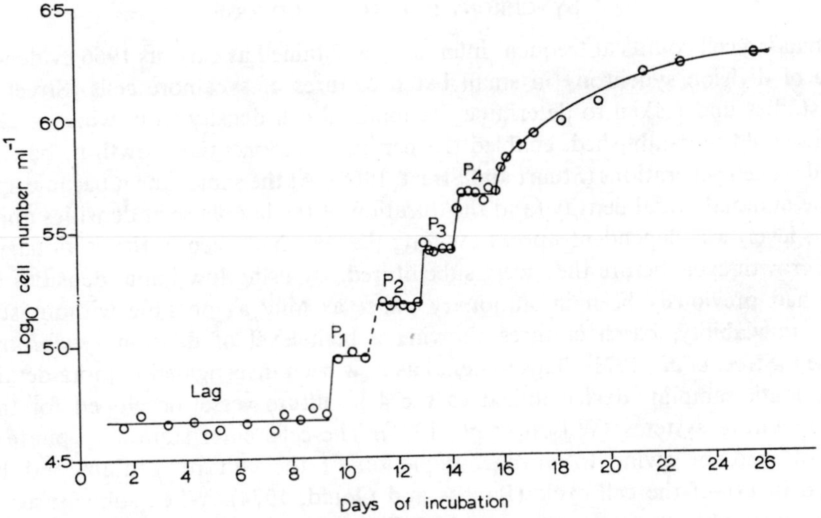

FIG. 12. Cell number growth curve for a synchronous culture of sycamore cells. Data from Gould and Street (1975).

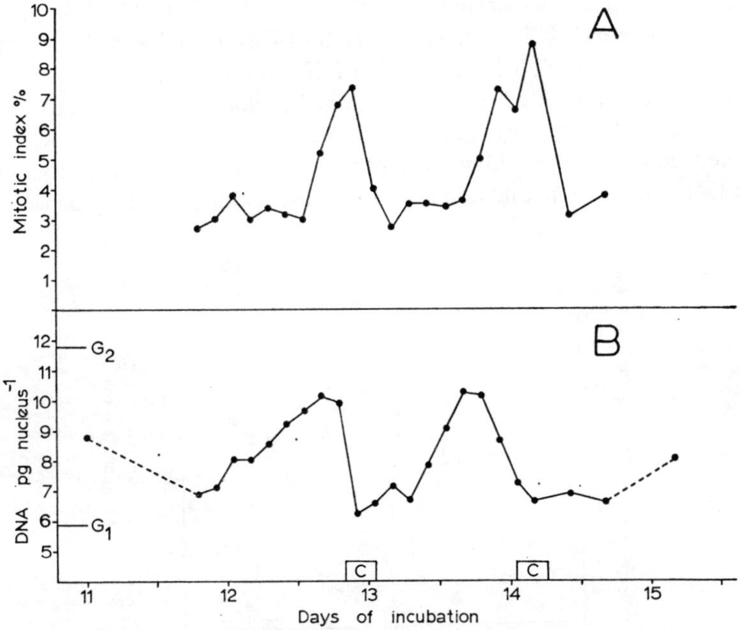

FIG. 13. (A) Fluctuations in mitotic index (per cent nuclei in mitosis) covering the second and third plateaux (P_2 and P_3 of the growth curve shown in Fig. 12). (B) Changes in average nuclear DNA contents (from microdensitometry of Feulgen stained nuclei) for the same period of the growth curve (Fig. 12) as in (A). G1 and G2 values for nuclear DNA of the sycamore cell line are shown for reference. C = periods over which the synchronous increases in cell number occur. Data from Gould and Street (1975).

SYNCHRONY IN BATCH CULTURES

By making cell counts at frequent intervals we obtained as early as 1966 evidence for a degree of division synchrony in small batch cultures of sycamore cells (Street, 1968). Then studies undertaken to determine the minimal cell density from which such batch cultures could be established, enabled the period of exponential growth to be extended beyond six cell generations (Stuart and Street, 1969). At the same time it became apparent that the minimal initial density (and the duration of the lag phase at densities approaching this level) was dependent upon how long the cells had been in the stationary phase of the growth cycle before they were subcultured. By using low initial densities of cells which had previously been in stationary phase as long as possible without suffering decline in viability, batch cultures showing a high level of division synchrony were obtained (Street *et al.*, 1971). This system has now been investigated in more detail using an automatic sampling device linked to the 4 l. culture vessel developed for the continuous culture systems (Wilson *et al.*, 1971). The cells enter stationary phase in our standard medium owing to nitrogen-depletion of the culture medium and become arrested in G1 of the cell cycle (Bayliss and Gould, 1974). When subcultured at low density, they embark upon a sequence of divisions which are synchronous for DNA replication (S-phase), mitosis and cytokinesis (Figs. 12, 13) (Gould and Street, 1975).

These cultures also show synchrony of metabolic events as evidenced by periodicity in respiration, [^{14}C]thymidine incorporation, and activity of the enzymes, thymidine kinase (TK) (E.C.2.7.1.21), asparate transcarbamylase (ATC) (E.C.2.1.3.2) (Fig. 14) and succinic dehydrogenase (E.C.1.3.99.1) (King *et al.*, 1974). High TK activity was restricted to a fraction of the cell cycle which included DNA replication. Yeoman and Aitchison (1973a) have also demonstrated an S-phase linked periodicity of TK activity during the transient mitotic synchrony they induce in explants of Jerusalem artichoke tuber. ATC, which is a key enzyme in pyrimidine biosynthesis, shows a peak of activity in G2, out of phase with DNA replication, and at a time when nucleotide pools may be low. It remains

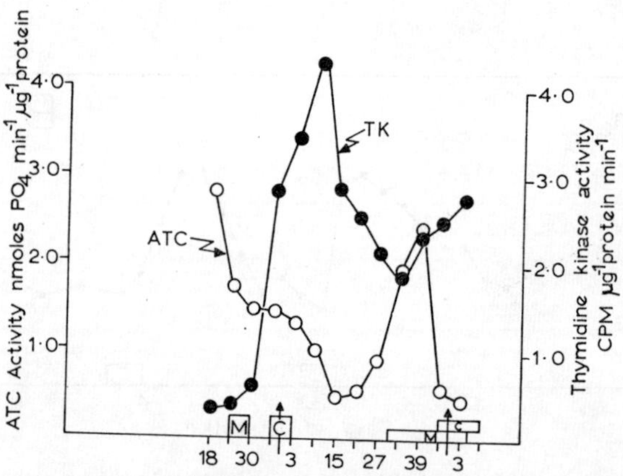

FIG. 14. Changes in activities of thymidine kinase (TK) and aspartate transcarbamylase (ATC) during a cell cycle of a synchronous sycamore culture. C = periods over which the synchronous increases in cell number occur. M = duration of mitosis derived from mitotic index data. Data from P. J. King *et al.* (1974).

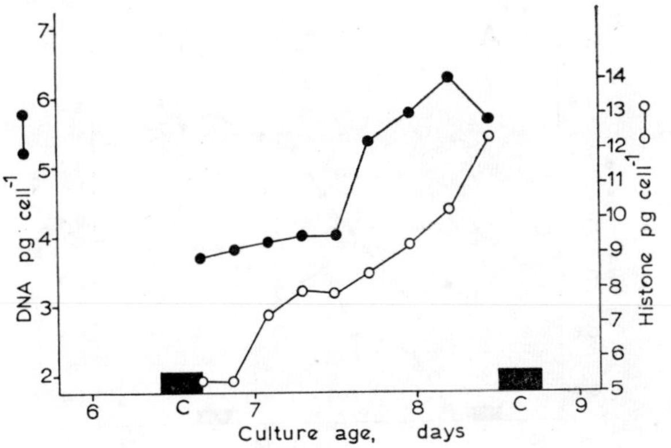

FIG. 15. Changes in contractable DNA and total histone in cultured sycamore cells during the interphase between two synchronous increases in cell number (*C*). Cells ground in sand in glass mortar and exhaustively washed in saline citrate buffer (Kovács, 1971). Histones extracted from cell pellet with 0.25 *N* HCl and estimated by method of Lowry *et al.* (see Layne, 1957) using calf thymus histone as standard. DNA extracted and estimated by method of Short *et al.* (1969) using calf thymus DNA as standard. Previously unpublished data of Kovács and Gould.

to be shown whether the changing levels of such enzymes during the cell cycle is the outcome of periodicity in their synthesis, or in the activity of endogenous inhibitors, activators or stabilizers, and how far gene-dosage effects determine rates of synthesis. However, it has been shown in the Jerusalem artichoke system (Yeoman and Aitchison, 1973b; Yeoman, 1974) that the stepped increase in glucose-6-phosphate dehydrogenase activity (E.C.1.1.1.49) involves *de novo* synthesis of the enzyme (deuterium-labelling technique of Hu *et al.*, 1962).

Recently one of us (A.R.G.) in collaboration with E. Kovacs (see Kovacs, 1971) has opened up the study of histone biosynthesis in synchronous sycamore cell cultures. These proteins are of particular interest because of their controversial role in chromosome structure and the regulation of gene activity. When the changes in histone and DNA content per cell were followed through a cell cycle there appeared to be a linear increase in histone and (as already indicated) a stepped rise in DNA (S-phase) (Fig. 15). This indicates that the histone-DNA ratio changes considerably through the cell cycle and gives no evidence of a strong coupling between histone and DNA synthesis. To examine this situation further a study was undertaken in which the total protein and histone fractions were pulse-labelled with [^{3}H]-L-lysine and [^{14}C]-L-arginine at intervals through a cell cycle. The data from this culture (Fig. 16) confirmed the linear increase in total histone and gave data consistent with previous evidence (King *et al.*, 1974) that there are no major fluctuations in the rising rate of total protein accumulation during interphase (Fig. 16B). However, the data for the incorporation of label into the histone fraction (Fig. 16C) revealed a peak of incorporation coincident with the period of rapid DNA synthesis (a finding which coincides with smaller changes in the rate of incorporation of label into total protein). This strongly suggests that there is a phase of enhanced histone turnover associated with the S-phase, a finding in line with evidence, from other systems, of periods of rapid histone turnover during the cell cycle (Prescott, 1964; Busch *et al.*, 1964; Das and Alfert, 1968). Clearly a better understanding of the phenomenon now

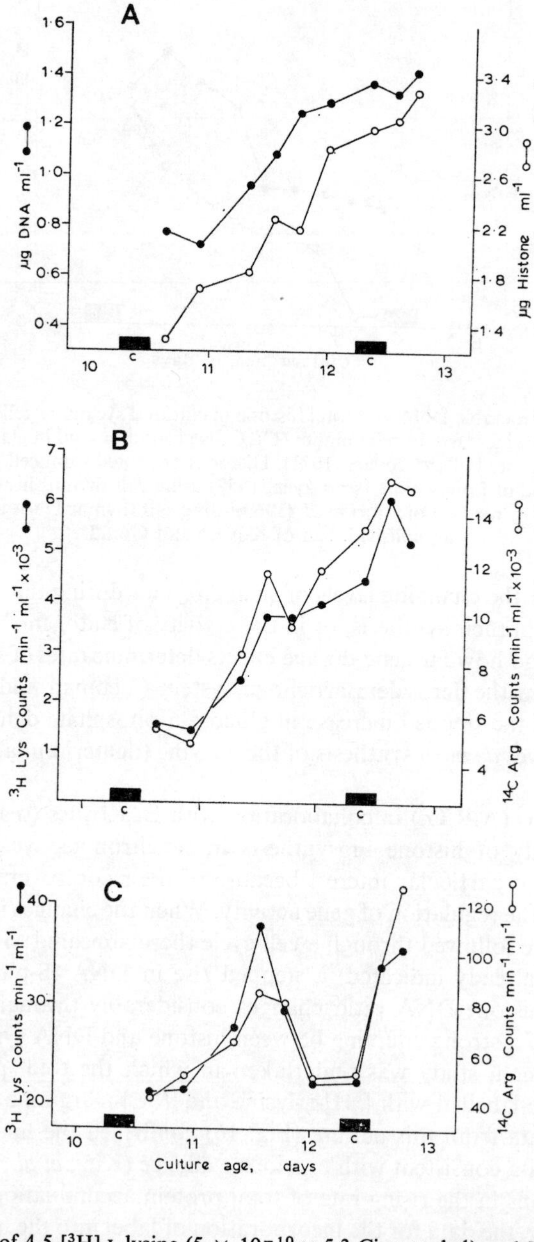

FIG. 16. Incorporation of 4-5-[^{3}H]-L-lysine (5×10^{-10} M 5.3 Ci m mol^{-1}) and [^{14}C]-L-arginine (3.5×10^{-9} M 318 Ci m mol^{-1}) into the total cold 5 per cent TCA-insoluble fraction and into histone fraction by pulse labelling (20 min) of culture samples withdrawn at intervals during a cell cycle of a synchronous sycamore cell culture. C = duration of cytokinesis. (A) Extractable DNA and total histone extracted as described in the legend of Fig. 15 and expressed as μg ml^{-1}. (B) Incorporation of label into total cold 5 per cent TCA-insoluble fraction. (C) Incorporation of label into total histone (extracted as per legend of Fig. 15). Previously unpublished data of Gould.

calls for fractionation of the histones of the sycamore cell to see whether rapid turnover of particular histones is coupled with DNA replication.

SELECTION OF MUTANT CELL LINES

The field of this paper was chosen, we hope in keeping with the predictive theme of the 50th Anniversary Meeting, as 'one full of promise but awaiting fulfilment.' Now we shall take this looking into the future a stage further. At least in the two preceding sections, we were able to describe technology ready and beginning to be exploited; here the emphasis is still at the level of trying to develop the essential technology.

A number of studies have shown that cell lines can be obtained with enhanced resistance to antimetabolites. In certain instances these isolations have followed treatment of the cell cultures with a mutagen (ethylmethane sulphonate, EMS (0.25 per cent), or N-methyl-N^1-nitro-nitrosoguanidine (400 μg ml^{-1}) or N-3-nitrophenyl-N^1-phenylurea (10^{-6} M)) but although the variants have occurred at frequencies such as might be expected from mutation (10^{-6}–10^{-7}) the effectiveness of the mutagens in producing the variants has not been critically investigated (Street, 1975). Amongst such 'mutant' cell lines have been lines resistant to amino acids and amino acid analogues e.g. threonine (Heimer and Filner, 1970), methionine sulphoxime (Carlson, 1973), DL-5-methyl-tryptophan (Widholm, 1972), *p*-fluorophenylalanine (Widholm, 1974), aminoethyl-cysteine and crotylglycine (Schaeffer, 1974). In the case of the threonine-resistant line it was shown that its nitrate uptake system was much less sensitive to inhibition by threonine and by several other amino acids than the parent cell line. In the case of the lines resistant to methyltryptophan and *p*-fluorophenylalanine it was shown that they accumulated much higher endogenous levels respectively of tryptophan and phenylalanine, and that in the case of the methyltryptophan resistant lines, their anthranilate synthetase was resistant to inhibition by tryptophan and tryptophan analogues. The cell line resistant to methionine sulphoxime was reported to produce five times the normal level of free methionine although the significance of this is not entirely clear since the sulphoxime is regarded as acting as a glutamine and not as a methionine analogue (Meins and Abrams, 1972). Studies of the steady state amino-acid pools and the protein composition of such cell lines could be of considerable interest because as Miflin (1973) has pointed out 'In many crop plants the levels of some of the nutritionally essential amino acids (from the animal standpoint) are low. Since these amino acids are also those whose synthesis is closely regulated, it is important to know, and perhaps to alter, the characteristics of this regulation.'

Attempts to obtain auxotrophic mutant cell lines are handicapped by lack of appropriate selection techniques. Although low density agar plating can now be carried out with high plating efficiencies (Street, 1973b), an effective replicate plating technique applicable to plant cell colonies on agar has yet to be devised. Carlson (1970) has reported the isolation of six leaky auxotrophs following EMS treatment of a haploid tobacco culture (tobacco is an amphidiploid hence perhaps the leaky nature of the 'haploid' mutants). These included lines requiring for growth equal to that of the parent culture the addition, in one case, of arginine and in another, of lysine or proline. After mutagenesis, Carlson treated the cells with 10^{-5} M 5-bromodeoxyuridine (BUdR) over 36 hr with the object of enriching the surviving population in auxotrophs. This use of BUdR was prompted by the evidence from work with mammalian cell cultures that growing cells (as opposed to non-growing cells) were killed by application of BUdR in the dark followed by exposure to light (Puck and Kao, 1967).

It would obviously greatly assist the isolation of auxotrophs from plant cell cultures if BUdR could be generally used as a selective lethal agent. Work by K. J. Mansfield and J.-P. Zryd in our laboratory and later by Zryd in Lausanne has shown that BUdR used as described by Carlson (1970) is not selectively lethal to actively growing as compared to non-growing sycamore cells. However, using the normal (2,4-d dependent) and an auxin-autotrophic strain (Lescure, 1970) and bringing the normal cells to a non-growing state by 2,4-d omission, the desired selectivity could be achieved (without any light exposure) by a 72 hr treatment with 5×10^{-5} M BUdR, 5×10^{-5} M uridine (inhibiting BUdR degradation?) plus 10^{-6} fluorodeoxyuridine (a potent inhibitor of the thymidine synthetase which converts deoxyuridine monophosphate to thymidine monophosphate). Further work is therefore justified on this approach to auxotroph enrichment.

Once a satisfactory procedure for selection has been developed it will be important to be able to obtain a haploid culture of the species, to be able to preserve haploid auxotrophs derived from it against cytological breakdown in culture, and finally if possible to regenerate plants from the auxotrophs in order to study inheritance of the mutant character. Anther culture (Sunderland, 1973) and the culture of isolated microspores (Nitsch, 1974) meet the first of these conditions. Freeze-preservation holds promise of meeting the second condition (Nag and Street, 1973) and already regeneration of plants from cultured cells is possible for a very wide range of species (Murashige, 1974).

The availability of amino-acid-requiring cell lines would not only extend our knowledge of synthetic pathways but enable us to study the effect on growth and nitrogen metabolism of pre-determined physiological levels of the amino acids in question by appropriate amino-acid feeding via chemostat culture. Phytohormone-minus lines would permit studies on the hormonal regulation of protein metabolism. Thus the availability of mutant lines in which normal regulatory mechanisms are disturbed or particular biochemical steps negated would add an additional dimension to the experimental approaches outlined in the two preceding sections of this paper.

Acknowledgement: This work was supported by a Science Research Council Grant to one of us (H.E.S.). One of us (J.K.) was able to participate by award of leave of absence by the Science Faculty of the University of Saskatchewan, one of us (A.R.G.) by a Science Research Council Research Studentship.

REFERENCES

BAYLISS, M. W. and GOULD, A. R. (1974) Studies on the growth in culture of plant cells. XVIII. Nuclear cytology of *Acer pseudoplatanus* suspension cultures. *J. exp. Bot.* **25,** 772–783.

BUSCH, H., STEELE, W. J., HNILICA, L. S. and TAYLOR, C. (1964) Metabolism of histones. In *The Nuclear Histones*. ed. BONNER, J. and TSAO, P. Holden-Day, San Francisco.

CARLSON, P. S. (1970) Induction and isolation of auxotrophic mutants in somatic cell cultures of *Nicotiana tabacum. Science N.Y.* **168,** 487–489.

CARLSON, P. S. (1973) Methionine sulfoximine resistant mutants of tobacco. *Science N.Y.* **180,** 1366–1368.

DAS, N.K. and ALFERT, M. (1968) Cytochemical studies on the concurrent synthesis of DNA and histone in primary spermatocytes of *Urechis caupo. Expl Cell Res.* **49,** 51–58.

DAVIES, M. E. and EXWORTH, C. P. (1973) Transient inhibition by cycloheximide of protein synthesis in cultured plant cell suspensions: a dose response paradox. *Biochem. biophys. Res. Commun.* **50,** 1075–1080.

EPSTEIN, E. (1972) The dual pattern of transport. In *Mineral Nutrition of Plants—Principles and Perspectives*, Wiley, New York.

FILNER, P. (1966) Regulation of nitrate reductance in cultured tobacco cells. *Biochim. biophys. Acta* **118,** 299–310.

FOWLER, M. W. and CLIFTON, A. (1974) Activities of enzymes of carbohydrate metabolism in cells of *Acer pseudoplatanus* L maintained in continuous (chemostat) culture. *Eur. J. Biochem.* **45,** 445–450.

FOWLER, M. W. and CLIFTON, A. (1975) Oscillations in carbohydrate metabolism in continuous cultures of plant cells. *Biochem. Trans.* in press.

FREID, M., ZSOLDOS, F., VOSE, P. B. and SHATOKHIN, I. L. (1965) Characterising the NO_3 and NH_4 uptake process of rice roots by use of ^{15}N labelled NH_4NO_3. *Physiologia Pl.* **18,** 313–320.

GOULD, A. and STREET, H. E. (1975) Kinetic aspects of synchrony in suspension cultures of *Acer pseudoplatanus*, L. *J. Cell Sci.* **17,** 337–348.

HEIMER, Y. M. and FILNER, P. (1970) Regulation of the nitrate assimilation pathway of cultured tobacco cells. II. Properties of a variant cell line. *Biochim. biophys. Acta* **215,** 152–165.

HEIMER, Y. M. and FILNER, P. (1971) Regulation of the nitrate assimilation pathway in cultured tobacco cells. III. The nitrate uptake system. *Biochim. biophys. Acta.* **230,** 362–372.

HU, A. S., BOCK, R. M. and HALVORSON, H. O. (1962) Separation of labelled from unlabelled proteins by equilibrium density gradient sedimentation. *Analyt. Biochem.* **4,** 489–504.

KING, P. J. and STREET, H. E. (1973) Growth patterns in cell cultures. In *Plant Tissue and Cell Culture.* ed. STREET, H. E. pp 269–337. Blackwell Scientific, Oxford.

KING, P. J., MANSFIELD, K. J. and STREET, H. E. (1973) Control of growth and cell division in plant cell suspension cultures. *Can. J. Bot.* **51,** 1807–1823.

KING, P. J., COX, B. J., FOWLER, M. W. and STREET, H. E. (1974) Metabolic events in synchronized cell cultures of *Acer pseudoplatanus*, L. *Planta* **117,** 109–122.

KOVACS, E. I. (1971) DNA, RNA, total protein and histone investigations in tobacco plants of genetically tumorous and normal condition. *Acta bot. hung.* **17,** 91–97.

LAYNE, E. (1957) Spectrophotometric and turbidometric methods for measuring proteins. Protein estimation with Folin-Ciocalteu reagent. In *Methods in Enzymology.* ed. KAPLAN, C. Vol. 3. Academic Press, New York.

LESCURE, A. M. (1970) Mutagénèse de cellules végétals cultivées *in vitro.* Methodes et resultats. *Soc. bot. Fr. Mem.* 353–365.

MEINS, F. JR. and ABRAMS, M. L. (1972) How methionine and glutamine prevent inhibition of growth by methionine sulfoxime. *Biochim. biophys. Acta* **266,** 307–311.

MIFLIN, B. J. (1973) Amino acid biosynthesis and its control in plants. In *Biosynthesis and its Control in Plants.* ed. MILBORROW, B. V. Academic Press, London.

MONOD, J. (1950) La technique de culture continuee. Theorie et application. *Annls Inst. Pasteur Paris,* **79,** 390–410.

MONTGOMERY, H. A. C. and DYMOCK, J. F. (1961) The determination of nitrate in water. *Analyst* **86,** 414–416.

MURASHIGE, T. (1974) Plant propagation through tissue culture. *A. Rev. Pl. Physiol.* **25,** 135–166.

NAG, K. K. and STREET, H. E. (1973) Carrot embryogenesis from frozen cultured cells. *Nature Lond.* **245,** 270–272.

NITSCH, C. (1974) La culture de pollen isole sur milieu synthetique. *C. r. hebd. Séanc. Acad. Sci. Paris.* **278,** 1031–1034.

NOVICK, A. and SZILARD, L. (1950) Description of the chemostat. *Science N.Y.* **112,** 715–716.

PRESCOTT, D. M. (1964) Turnover of chromosomal and nuclear proteins. In *The Nuclear Histones.* ed. BONNER, J. and TSO, P. Holden-Day, San Francisco.

PUCK, T. and KAO, F. (1967) Genetics of somatic mammalian cells. V. Treatment with 5-bromodeoxyuridine and visible light for isolation of nutritionally deficient mutants. *Proc. natn. Acad. Sci. U.S.A.* **58,** 1227–1234.

SCHAEFFER, G. W. (1974) By-pass mutants of tobacco resistant to aminoethylcysteine and crotylglycine. No. 71. Abstracts of Papers. 3rd International Congress Plant Tissue and Cell Culture. Univ. Leicester.

SHORT, K. C., BROWN, E. G. and STREET, H. E. (1969) Studies on the growth in culture of plant cells. VI Nucleic acid metabolism of *Acer pseudoplatanus* L. cell suspensions. *J. exp. Bot.* **20,** 579–590.

STREET, H. E. (1968) The induction of cell division in plant suspension cultures. In *Les Cultures de Tissus de Plantes.* pp 177–193. C.N.R.S., Paris.

STREET, H. E. (1973a) Cell (suspension) cultures—techniques. In *Plant Tissue and Cell Culture*. ed. STREET, H. E. pp 59–99. Blackwell Scientific, Oxford.

STREET, H. E. (1973b) Single-cell lines. In *Plant Tissue and Cell Culture*. ed. STREET, H. E. pp 191–204. Blackwell Scientific, Oxford.

STREET, H. E. (1975) Plant cell cultures: Present and projected applications for studies in genetics. In *Genetic Manipulation with Plant Materials*. ed. LEDOUX, L. pp 231–244. Plenum Press, New York.

STREET, H. E., KING, P. J. and MANSFELD, K. J. (1971) Growth control in plant cell suspension cultures. In *Les Cultures de Tissus de Plantes*. Colloq. Int. Strasbourg, 1970. pp 17–40. C.N.R.S., Paris.

STUART, R. and STREET, H. E. (1969) Studies on the growth in culture of plant cells. IV The initiation of division in suspensions of stationary phase cells of *Acer pseudoplatanus*, L. *J. exp. Bot.* **20,** 556-571.

SUNDERLAND, N. (1973) Pollen and anther culture. In *Plant Tissue and Cell Culture*. ed. STREET, H. E. pp. 205–239. Blackwell Scientific, Oxford.

VAN DEN HONERT, T. H. and HOOYMANS, J. J. M. (1955) On the absorption of nitrate by maize in water culture. *Acta. bot. neerl.* **4,** 376–384.

WIDHOLM, J. M. (1972) The use of fluorescein diacetate and phenosafranine for determining viability of cultured plant cells. *Stain Tech.* **47,** 189–194.

WIDHOLM, J. M. (1974) Selection and characteristics of biochemical mutants of cultured plant cells. In *Tissue Culture and Plant Science* 1974. ed. STREET, H. E. pp 287–300. Academic Press, London.

WILSON, G. (1971) The nutrition and differentiation of cells of *Acer pseudoplatanus*, L. in suspension cultures. Ph.D. Thesis. Univ. Birmingham.

WILSON, S. B., KING, P. J. and STREET, H. E. (1971) Studies on the growth in culture of plant cells. XII A versatile system for the large scale batch or continuous culture of plant cell suspensions. *J. exp. Bot.* **21,** 177–207.

YEOMAN, M. M. (1974) Division synchrony in cultured cells. In *Tissue Culture and Plant Science* 1974. ed. STREET, H. E. pp 1–17. Academic Press, London.

YEOMAN, M. M. and AITCHISON, P. A. (1973a) Changes in enzyme activities during the division cycle of cultured plant cells. In *The Cell Cycle in Development and Differentiation. Br. Soc. Dev. Biol. Symp.* 1 ed. BALLS, M. and BILLETT, F. S. Cambridge University Press.

YEOMAN, M. M. and AITCHISON, P. A. (1973b) Growth patterns in tissue (callus) cultures. In *Plant Tissue and Cell Culture*. ed. STREET, H. E. pp 240–268. Blackwell Scientific, Oxford.

YOUNG, M. (1973) Studies on the growth in culture of plant cells. XVI. Nitrogen assimilation during nitrogen-limited growth of *Acer pseudoplatanus*, L cells in chemostat culture. *J. exp. Bot.* **24,** 1172–1185.

ENZYMIC CONTROLS IN THE BIOSYNTHESIS OF LIGNIN AND FLAVONOIDS

V. S. Butt

Botany School, South Parks Road, University of Oxford, U.K.

Introduction

As one looks back over the changes in the study of enzymes and their role in metabolism during the last 25 years, at least three major lines of development can, I think, be recognized. There is, firstly, the spectacular success in the identification of amino acid sequences in proteins and the construction of their three-dimensional order. This has furthered our knowledge of the chemical nature of the active site, the enzyme–substrate combination and the catalytic process, but surprisingly, has done little more than add precision to concepts, such as enzyme specificity and saturation kinetics, which have been current for a very long time. Yet we have to remember that our knowledge of protein structure was vague even as recently as 1940 (see e.g. Bull, 1941) and since then, the view that all enzymes are proteins has been challenged (Binkley, 1954). A second feature has promised more direct interest to the biologist. This lies in the interaction of allosteric enzymes not only with their substrates, but also with their products and with chemically-unrelated metabolites. The number of chemical factors likely to affect enzyme activity *in vivo* has taken us away from a view that metabolism is controlled by simple kinetic considerations and chemical equilibrium (see e.g. Dixon, 1949) to one in which the overall control of any metabolic pathway is effected through the more immediate action of intermediates of that pathway and of allied pathways. This limited view of metabolism has been extended by the third development, the recognition of internal and external factors which govern the actual levels of enzymes in tissues and organs. In consequence, we now see the relative importance of many pathways to be determined as much by the changes in enzyme levels which accompany development or environmental change as by the impact of purely kinetic effects upon enzymes already present and potentially or partially active. The concepts of induction and repression have replaced such simpler ideas as overflow or shunt metabolism (see Foster, 1949).

In this paper I propose to apply these concepts of enzyme specificity, enzyme regulation and enzyme induction (and repression) to one aspect of the secondary metabolism of plants. These are the controls which determine rate and direction and dominate our view of secondary metabolism, so-called not because it is in any real sense secondary (the large mass of lignin synthesized in trees and forests or the functions of phenolics in disease resistance and pollination indicate its critical, selective importance) but because these aspects of metabolism vary in their incidence during the life of an organ under the action of internal controls and external change.

Our knowledge of phenolic synthesis has grown up with these developments in enzymology and has been determined by them, so that new approaches have not been grafted on to knowledge already existing but have been built into it. The first successful experiments were as recent as 1955 when the incorporation of shikimic acid into lignin residues in wheat was successfully demonstrated (Brown and Neish, 1955) after Davis and his group (1955) had worked out the pathway of biosynthesis of phenylalanine and tyrosine with *Escherichia coli* mutants. The pathways of lignin and flavonoid synthesis were subsequently established in detail by feeding experiments with ^{14}C-labelled precursors (Neish, 1960), but it was after this that Koukol and Conn (1961) first isolated phenylalanine ammonia-lyase (PAL), an enzyme since subjected to more detailed investigation than any other enzyme involved in secondary metabolism (see Camm and Towers, 1973). Many of the enzymes concerned in the synthesis of lignin, flavonoids and other phenolic substances have since been isolated and studied in relation to their role *in vivo*.

Lignin biosynthesis

Lignin is a polymer, produced by the mixed oxidation of three monomers, *p*-coumaryl alcohol, coniferyl alcohol and sinapyl alcohol, among which coniferyl alcohol residues occur in all woods, while those of the other alcohols have a more limited distribution

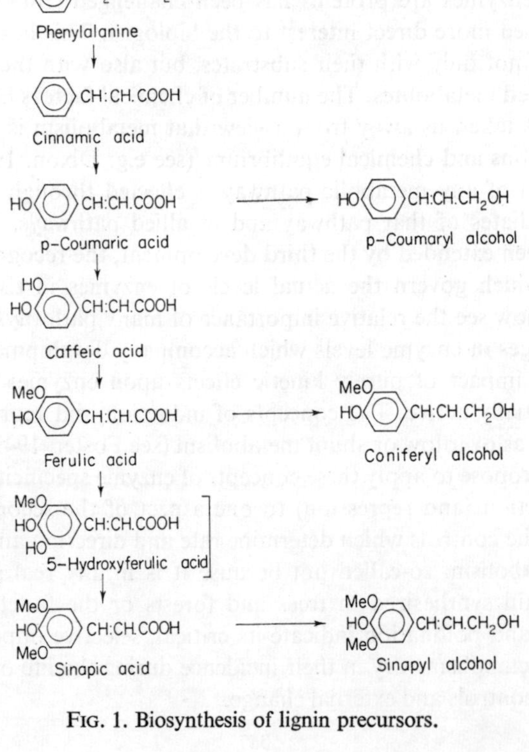

FIG. 1. Biosynthesis of lignin precursors.

TABLE 1. ENZYMES OF PHENYLPROPANOID METABOLISM

Enzyme	Reaction	K_m	Metabolite inhibition	Reference
Phenylalanine ammonia-lyase	Phenylalanine → cinnamate + NH_3	20–270 μM	Cinnamate (0.6 mM), 25 % *p*-Coumarate (0.6 mM), 64 %	Camm and Towers (1973)
Cinnamate-4-hydroxylase	Cinnamate + O_2 + NADPH + H^+ → *p*-coumarate + H_2O + $NADP^+$	20 μM	*p*-Coumarate, 30–100 μM, non-competitive, cooperative	Russell (1971)
p-Coumarate 3-hydroxylase (Phenolase)	*p*-Coumarate + O_2 + H_2A → caffeate + H_2O + A	0.64 mM	None	Vaughan (1968)
Caffeate: SAM O-methyltransferase	Caffeate + SAM → ferulate + SAH	Caffeate, 68 μM SAM, 12.5 μM	SAH, K_i = 4.4 μM competitive with SAM	Poulton (1974)
Hydroxycinnamate: CoA ligase	*p*-Coumarate + ATP + CoA → *p*-Coumaroyl-CoA + AMP + P_p	*p*-Coumarate, 25 μM CoA, 3.2 μM	*p*-Coumarate, caffeate and ferulate above 0.2 mM	Gross and Zenk (1974)

SAM = S-adenosylmethionine
SAH = S-adenosylhomocysteine
H_2A = Hydrogen or electron donor (A = oxidized species)
P_p = Pyrophosphate.

(Fig. 1). These alcohols are synthesized by the reduction of the corresponding carboxylic acids after esterification with coenzyme A (CoA) (Mansell *et al.*, 1972; Ebel and Grisebach, 1973; Stöckigt *et al.*, 1973). The acids are interrelated through a sequence of reactions involving hydroxylation and methylation from cinnamic acid, itself synthesized from phenylalanine by deamination. Synthesis of lignin is controlled by the levels of these enzymes and the regulation of their activity.

Intermediates in the biosynthesis of lignin were not immediately recognized because the free acids are found in plant cells at concentrations below 10^{-4} M, unless they accumulate to much higher concentrations as glycosides or esters, such as chlorogenic acid. When [2-^{14}C]cinnamic acid was supplied to leaf discs of spinach (Vaughan, 1968) or spinach beet (Walton, 1969), the label did not accumulate in any single intermediate but was found in small amounts almost equally in *p*-coumaric, caffeic and ferulic acids.

These observations are explained by the regulatory properties of the enzymes involved. The Michaelis constants (K_m) and inhibition constants (K_i) are shown in Table 1. K_m values ranging from 20 μM to 70 μM for four of these enzymes indicate the low concentrations at which these enzymes are effective, and the first two enzymes in the sequence are both inhibited by the accumulation of their products in low concentrations. Hydroxycinnamate:CoA ligase is inhibited by concentrations of ferulate in excess of 0.2 mM (Gross and Zenk, 1974). However, caffeic acid:S-adenosylmethionine O-methyltransferase is not inhibited by ferulic acid, but evidence for its regulation by the relative concentrations of S-adenosylhomocysteine (SAH) and S-adenosylmethionine (SAM) has been obtained, suggesting that the conversion of the former back to S-adenosylmethionine through a methylation cycle, the rate of which would be governed by the supply of methyl groups (Duerre, 1968), might limit this reaction (Poulton, 1974).

The hydroxylation of *p*-coumarate to caffeate can be catalysed by phenolases and has been studied in detail with enzymes from mushroom (Sato, 1969) and spinach-beet leaves (Vaughan and Butt, 1969), but although the use of dimethyltetrahydropteridine has been found to act as reductant with the inhibition of the further conversion of caffeate to its quinone (Vaughan and Butt, 1972), there have also been reports of a hydroxylase in *Sorghum* leaves (Stafford and Dresler, 1972) and parsley cells (Schill and Grisebach, 1973) which has no catechol oxidase activity and appears to be distinct from phenolases present in the tissues. As will be discussed later, spinach-beet leaves appear to possess a similar enzyme (Bolwell, 1974).

ENVIRONMENTAL INFLUENCE ON ENZYME LEVELS

As well as regulation of their activity by metabolites in the biosynthetic pathway, the levels of the enzymes of phenylpropanoid metabolism are especially sensitive to changes in internal and external conditions. Phenylalanine ammonia-lyase (PAL) has been most extensively investigated; its activity in different tissues has been increased by such treatments as white light, red light (far-red reversible), blue light, ethylene, gibberellin and abscisic acid and by wounding, infection or the excision of leaf discs (Camm and Towers, 1973). After the activity of cinnamate 4-hydroxylase had also been shown to increase on illumination of gherkin seedlings with blue light (Engelsma, 1965) parallel increases in this and many other enzymes were demonstrated in parsley and soya cell cultures (Hahlbrock *et al.*, 1971a, b).

Illumination of both pea seedling apices and leaves of spinach-beet seedlings has been found to induce comparable changes in the levels of PAL, cinnamate 4-hydroxylase and hydroxycinnamate: CoA ligase (Fig. 2). With *p*-coumarate 3-hydroxylase, the changes in spinach beet follow a similar pattern but appear to be superimposed upon a high (and perhaps unchanging) level of enzyme initially present. Similar changes were induced by red light treatment, reversed by subsequent far-red illumination (Table 2) but no significant change in 3,4-dihydroxyphenylalanine (DOPA) oxidase activity, used here as an index of phenolase activity, was observed. If we assume that there is no change in phenolase activity and make an arbitrary deduction from the levels of *p*-coumarate 3-hydroxylase to allow for this, we obtain data similar to those for the other three enzymes. We therefore conclude that a hydroxylase which does not possess phenolase activity is co-induced with the other enzymes and to a similar extent; the fractionation of hydroxylase activity on DEAE-cellulose and Sephadex G-200 showed peaks of activity, which had low DOPA oxidase activity and which were found only in extracts from illuminated leaves.

p-Coumarate 3-hydroxylase is not however invariably co-induced with PAL and cinnamate-4-hydroxylase. When excised leaf discs from spinach-beet seedlings were illuminated, these latter two enzymes reached maximum activity together after 24 hr,

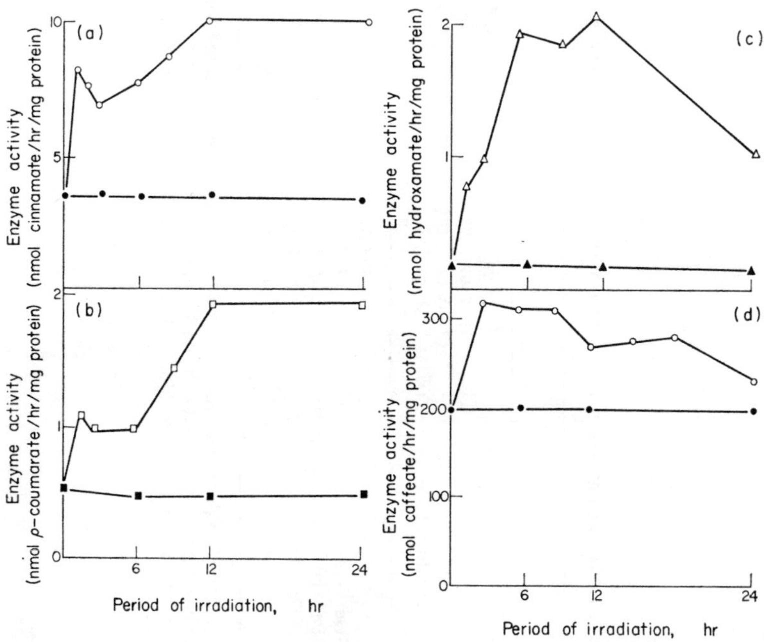

FIG. 2. Changes in activities of (a) phenylalanine ammonia-lyase, (b) cinnamate 4-hydroxylase, (c) hydroxycinnamate: CoA ligase and (d) *p*-coumarate 3-hydroxylase on illumination of 6-day etiolated spinach-beet seedlings (open symbols) and in controls without illumination (closed symbols). Data from Bolwell (1974).

TABLE 2. EFFECT OF RED/FAR-RED LIGHT TREATMENTS ON ENZYME LEVELS IN ETIOLATED SPINACH-BEET SEEDLINGS (Data from Bolwell, 1974)

Treatment	Enzyme activities: Phenylalanine ammonia-lyase (nmol hr^{-1} mg^{-1} protein)	Cinnamate 4-hydroxylase (nmol hr^{-1} mg^{-1} protein)	p-Coumarate 3-hydroxylase (nmol hr^{-1} mg^{-1} protein): 1	p-Coumarate 3-hydroxylase (nmol hr^{-1} mg^{-1} protein): 2*	DOPA oxidase (ΔE_{470} min^{-1} mg^{-1} protein)
Darkness	3.25 (1.0)	0.42 (1.0)	317 (1.0)	87 (1.0)	20.6
Red (15 min)-Darkness	9.77 (3.0)	1.12 (2.6)	481 (1.5)	251 (2.9)	21.4
Red (15 min)-Far-red (5 min)-Darkness	5.46 (1.7)	0.57 (1.4)	363 (1.1)	133 (1.5)	20.0
Far-red (5 min)-Darkness	5.52 (1.7)	0.63 (1.5)	370 (1.1)	140 (1.6)	20.4
Red	10.02 (3.1)	1.43 (3.4)	514 (1.6)	284 (3.3)	20.6

6-Day seedlings, raised in darkness, were exposed to the conditions indicated for a total period of 3 hr before assay. Figures in brackets indicate the activity of each enzyme relative to that of its dark control.

* The data in column (2) are those in column (1) after deduction of 230 nmol hr^{-1} mg^{-1} protein.

whereas *p*-coumarate hydroxylase reached a peak later after 48 hr (Fig. 3). Furthermore, this second peak was associated with a proportionately similar increase in DOPA oxidase activity. Under these conditions, it seems that increases in phenolase activity are induced but by some process of sequential induction after PAL and cinnamate 4-hydroxylase rather than by the co-induction observed with intact leaves.

These experiments have therefore not only revealed the existence of a new enzyme, already hinted at with *Sorghum* leaves (Stafford and Bliss, 1973) but also something about the physiological differences between intact and excised leaves. Illumination of intact leaves stimulates their development with increased vascularization and synthesis of lignin, whereas after excision, the production of phenols, quinones and melanins in the

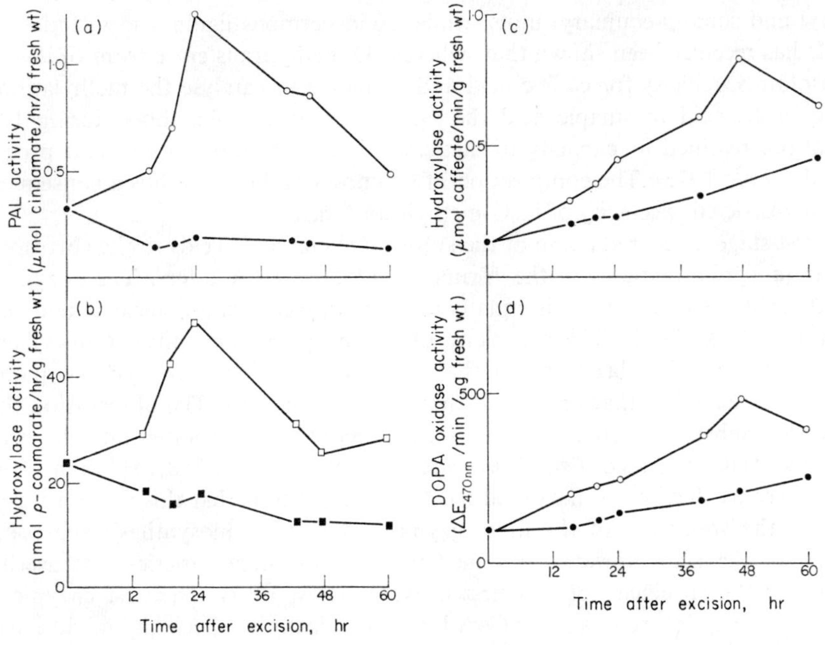

FIG. 3. Changes in activities of (a) phenylalanine ammonia-lyase, (b) cinnamate 4-hydroxylase, (c) *p*-coumarate 3-hydroxylase and (d) DOPA oxidase in excised leaf discs from 6-day etiolated spinach-beet seedlings in light (open symbols) and darkness (closed symbols). Data from Bolwell (1974).

damaged tissue is observed. In the first case, the enzymes are induced to synthesize lignin in relation to the changed physiological status of the leaf; the further oxidation of caffeate has to be discounted. Conditions after excision more closely resemble those after damage or infection, in which the chemically active quinones produced by phenolase action may help to protect the tissue against invasion by pathogens.

VARIATION IN ENZYME SPECIFICITY AND PHENOLIC PRODUCT

Plants produce not only lignin and the glycosides or esters of its biosynthetic intermediates, but also an almost bewildering variety of phenolic products. Many of these are highly specific in their distribution and have recently been used widely in chemotaxonomic studies (see e.g. Harborne, 1967). In almost all cases, they are synthesized from phenylalanine through cinnamic acid and *p*-coumaric acid and, sometimes after further hydroxylation and methylation, form their CoA esters before further conversion. The chemical differences between the products, sometimes rather small, must arise either from the production of additional enzymes or from some change in the specificities of certain enzymes involved in their synthesis. Study of the specificity characteristics of these different enzymes may give us some insight into the molecular changes associated with plant variation and micro-evolution.

A simple, but rather broad, example stems from lignin biosynthesis. The lignin from gymnosperms differs from that of angiosperms in that it is derived almost entirely from coniferyl and some *p*-coumaryl units, while angiospermous lignin also includes sinapyl units. It has recently been shown that, whereas O-methyltransferase from *Ginkgo biloba* has absolute specificity for caffeic acid and is unable to catalyse the methylation of 5-hydroxyferulic acid to sinapic acid the same enzyme from bamboo, during 100-fold purification, retained its capacity to methylate both substrates in constant proportion (Shimada *et al.*, 1973). The composition of gymnosperm lignin is thus a consequence of the more restricted specificity of its O-methyltransferase.

The first stage in the reduction of the substituted cinnamic acids to the corresponding alcohols in lignin synthesis is the formation of their CoA esters. The enzyme from cambial scrapings of *Forsythia* is equally active with *p*-coumarate, ferulate and sinapate (Gross and Zenk, 1974). The specificity of this enzyme from other sources has been examined, since it has been shown that the hydroxylation/methoxylation pattern in flavonoids is related to that in the substituted cinnamic acids. The observation that the ^{14}C-labelled substituted cinnamic acid most effectively incorporated into the B ring of the flavonoids in varieties of *Petunia hybrida* possessed the same hydroxylation/methoxylation pattern as that of the flavonoid led to the hypothesis that this pattern was determined by the specificity of the first enzyme in flavonoid biosynthesis subsequent to synthesis of the acid; variation in the flavonoids of these varieties was ascribed to variation in the specificity of this first enzyme (Hess, 1967). The first enzyme in the sequence is the hydroxycinnamate: CoA ligase, in which this specificity might have been expected to reside. However, Table 3 shows that specificities of the purified ligases from

TABLE 3. SPECIFICITIES OF HYDROXYCINNAMATE: COA LIGASE FROM DIFFERENT PLANT SPECIES

Enzyme source	Relative Activities				
	p-Coumarate	Caffeate	Ferulate	Cinnamate	3,4-Dimethoxy-cinnamate
Swede root[1]	100	68	73	0	0
Haplopappus[2]	100	76	106	0	0
Parsley[2]	100	84	118	0	0

Enzyme activities are expressed as percentage of activity with *p*-coumarate, using 0.1 mM concentrations of each substrate.
[1] Rhodes and Wooltorton (1973).
[2] Butt, unpublished results.

parsley cells, which produce apigenin, and *Haplopappus* cells, producing cyanidin, showed no differences (Butt, unpublished results), and were similar to the enzyme in swede discs, during induced synthesis of substituted cinnamic acids and lignin (Rhodes and Wooltorton, 1973). All of these ligases are similar to that from *Forsythia* and can be assumed to participate, in different tissues, in the synthesis of either lignin or flavonoids.

This does not invalidate the hypothesis. Rather, the theory has to be extended to the next enzyme, recently isolated from parsley cells, which catalyses the condensation of malonyl-CoA and *p*-coumaroyl CoA to form the flavanone or chalcone (Kreuzaler and Hahlbrock, 1972). The enzymes peculiar to flavonoid synthesis, as distinct from those also involved in lignin biosynthesis, have a different induction pattern after illumination of parsley cell cultures (Hahlbrock *et al.*, 1971a), and the purification of this first enzyme from different plants or variants is more likely to reveal any differences in specificity.

Conclusions and Prospect

The enzymes engaged in catalysing the stages of lignin biosynthesis are co-inducible (or at least co-activated—see Attridge *et al.*, 1974) and the existence of a PAL-operon, akin to similar systems in bacteria, has been suggested (Zucker, 1972). Further work however is needed to establish the co-induction of the remaining enzymes in lignin biosynthesis. This would not demonstrate the control of these enzymes by a single genetic unit, and the independent induction in excised leaf discs of PAL and cinnamate-4-hydroxylase on one hand, and phenolase on the other, without the appearance of a *p*-coumarate 3-hydroxylase, showing no catechol oxidase activity, suggests that any such units are either capable of some form of partial induction or that they are replicated within the genome, with different associations or altered components. Detailed study of these inductions is therefore likely to make available material for expanding our knowledge of the genetic organization of plant cells and perhaps for studying the molecular mechanisms which underly its expression.

The specificity studies may provide an enzymic basis for chemotaxonomy and thence the mechanisms and changes concerned in the production of chemical variants as the first stage in speciation. Determination of the amino acid sequence of proteins of relatively invariable function, such as cytochrome *c*, ferredoxins or plastocyanin has given us some insight into long-term phylogenetic change (Boulter, 1972); the structural differences between proteins with slightly different catalytic function, on the products of which phenotypic selection has operated, would give us a better view of the significant changes involved in variation and speciation at the molecular level.

The techniques acquired over recent years for the purification and study of enzymes have thus brought enzymology into the forefront of biology, furthering the fundamental knowledge which interests modern biologists. Among plant scientists, study of the enzymes of secondary metabolism, of which those engaged in the synthesis of phenolics have attracted special attention, is giving rise to new and expanding approaches in their special field of interest.

Acknowledgements: I have to thank Dr. P. F. T. Vaughan, Dr. Everild Walton, Dr. G. P. Bolwell and Dr. J. E. Poulton, who have carried out much of the experimental work quoted here, and Professor Hans Grisebach of the University of Freiburg, who suggested the specificity studies and provided me with the hospitality of his laboratory to pursue them while I was on sabbatical leave.

REFERENCES

ATTRIDGE, T. H., JOHNSON, C. B. and SMITH, H. (1974) Density-labelling evidence for the phytochrome-mediated activation of phenylalanine ammonia-lyase in mustard cotyledons. *Biochim. biophys. Acta* **343,** 440–451.

BINKLEY, F. (1954) Organization of enzymes in the synthesis of peptides. *Proc. R. Soc.* B, **142,** 170–174.

BOLWELL, G. P. (1974) The control of enzyme levels involved in the metabolism of phenolics. D.Phil. Thesis, Oxford.

BOULTER, D. (1972) The use of comparative amino acid sequence data in evolutionary studies of higher plants. In *Progress in Phytochemistry*, ed. REINHOLD, L. and LIWSCHITZ, Y. Vol. 3, pp. 199–229. Wiley–Interscience, London and New York.

BROWN, S. A. and NEISH, A. C. (1955) Shikimic acid as a precursor in lignin biosynthesis. *Nature Lond.* **175,** 688–689.

BULL, H. B. (1941) Protein structure. *Adv. Enzymol.* **1,** 1–42.

CAMM, E. L. and TOWERS, G. H. N. (1973) Phenylalanine ammonia lyase. *Phytochemistry* **12,** 961–973.

DAVIS, B. D. (1955) Biosynthesis of the aromatic amino acids. In *A Symposium on Amino Acid Metabolism*, ed. MCELROY, W. D. and GLASS, B., pp. 799–811. The Johns Hopkins Press, Baltimore.

DIXON, M. (1949) *Multi-enzyme Systems.* Cambridge University Press, Cambridge.

DUERRE, J. A. (1968) *In vivo* and *in vitro* metabolism of S-adenosyl-L-homocysteine by *Saccharomyces cerevisiae. Archs Biochem. Biophys.* **124,** 422–430.

EBEL, J. and GRISEBACH, H. (1973) Reduction of cinnamic acids to cinnamyl alcohols with an enzyme preparation from cell suspension cultures of soybean (*Glycine max*). *FEBS Lett.* **30,** 141–143.

ENGELSMA, G. (1965) Photo-induced hydroxylation of cinnamic acid in gherkin hypocotyls. *Nature Lond.* **208,** 1117–1119.

FOSTER, J. W. (1949) *Chemical Activities of Fungi*, pp. 164–169. Academic Press, New York.

GROSS, G. G. and ZENK, M. H. (1974) Isolation and properties of hydroxycinnamate: CoA ligase from lignifying tissue of *Forsythia. Eur. J. Biochem.* **42,** 453–459.

HAHLBROCK, K., EBEL, J., ORTMANN, R., SUTTER, A., WELLMANN, E. and GRISEBACH, H. (1971a) Regulation of enzyme activities related to the biosynthesis of flavone glycosides in cell suspension cultures of parsley (*Petroselinum hortense*). *Biochim. biophys. Acta* **244,** 7–15.

HAHLBROCK, K., KUHLEN, E. and LINDL, T. (1971b) Änderungen von Enzymaktivitäten während des Wachstums von Zellsuspensionkulturen von *Glycine max*: Phenylalanin Ammonia-Lyase und p-Cumarat: CoA Ligase. *Planta* **99,** 311–318.

HARBORNE, J. B. (1967) *Comparative Biochemistry of the Flavonoids*, Academic Press, London and New York.

HESS, D. (1967) Die Wirkung von Zimtsauren auf die Anthocyansynthese in isolierten Petalen von *Petunia hybrida. Z. Pfl.physiol.* **56,** 12–19.

KOUKOL, J. and CONN, E. E. (1961) The metabolism of aromatic compounds in higher plants. IV. Purification and properties of the phenylalanine deaminase of *Hordeum vulgare. J. biol. Chem.* **236,** 2692–2698.

KREUZALER, F. and HAHLBROCK, K. (1972) Enzymatic synthesis of aromatic compounds in higher plants: Formation of naringenin (5,7,4′-trihydroxyflavanone) from p-coumaroyl coenzyme A and malonyl coenzyme A. *FEBS Lett.* **28,** 69–72.

MANSELL, R. L., STÖCKIGT, J. and ZENK, M. H. (1972) Reduction of ferulic acid to coniferyl alcohol in a cell-free system from a higher plant. *Z. Pfl.physiol.* **68,** 286–288.

NEISH, A. C. (1960) Biosynthetic pathways of aromatic compounds. *A. Rev. Pl. Physiol.* **11,** 55–80.

POULTON, J. E. (1974) Enzymic methylation of caffeic acid and related reactions. D.Phil. Thesis, Oxford.

RHODES, M. J. C. and WOOLTORTON, L. S. C. (1973) Formation of CoA esters of cinnamic acid derivatives by extracts of *Brassica napo-brassica* root tissue. *Phytochemistry* **12,** 2381–2387.

RUSSELL, D. W. (1971) Metabolism of aromatic compounds in higher plants. X. Properties of cinnamic acid 4-hydroxylase of pea seedlings and some aspects of its metabolic and developmental control. *J. biol. Chem.* **246,** 3870–3878.

SATO, M. (1969) The conversion by phenolase of p-coumaric acid to caffeic acid with special reference to the role of ascorbic acid. *Phytochemistry* **8,** 353–362.

SCHILL, L. and GRISEBACH, H. (1973) Properties of a phenolase preparation from cell suspension cultures of parsley. *Hoppe Seyler's Z. physiol. Chem.* **354,** 1555–1562.

SHIMADA, M., KURODA, H. and HIGUCHI, T. (1973) Evidence for the formation of methoxyl groups of ferulic and sinapic acids in *Bambusa* by the same O-methyltransferase. *Phytochemistry* **12,** 2873–2875.

STAFFORD, H. A. and BLISS, M. (1973) The effect of greening of *Sorghum* leaves on the molecular weight of a complex containing 4-hydroxycinnamic acid hydroxylase activity. *Pl. Physiol. Lancaster* **52,** 453–458.

STAFFORD, H. A. and DRESLER, S. (1972) 4-Hydroxycinnamic acid hydroxylase and polyphenol oxidase activities in *Sorghum vulgare*. *Pl. Physiol. Lancaster* **49,** 590–595.

STÖCKIGT, J., MANSELL, R. L., GROSS, G. G. and ZENK, M. H. (1973) Enzymatic reduction of p-coumaric acid via p-coumaroyl-CoA to p-coumaryl alcohol by a cell-free system from *Forsythia* sp. *Z. Pfl.-physiol.* **70,** 305–307.

VAUGHAN, P. F. T. (1968) Aromatic hydroxylation in plants. D.Phil. Thesis, Oxford.

VAUGHAN, P. F. T. and BUTT, V. S. (1969) The hydroxylation of p-coumaric acid by an enzyme from leaves of spinach beet (*Beta vulgaris* L.) *Biochem. J.* **113,** 109–115.

VAUGHAN, P. F. T. and BUTT, V. S. (1972) The expression of catechol oxidase activity during the hydroxylation of p-coumaric acid by spinach-beet phenolase. *Biochem. J.* **127,** 641–647.

WALTON, R. E. H. (1969) Enzymes of cinnamic acid metabolism in plants. D.Phil. Thesis, Oxford.

ZUCKER, M. (1972) Light and enzymes. *A. Rev. Pl. Physiol.* **23,** 133–156.

ACCUMULATION OF IONS IN PLANT CELL VACUOLES

ENID A. C. MACROBBIE

Botany School, University of Cambridge, U.K.

INTRODUCTION

This paper deals with transport processes in plant cells at a cellular rather than a tissue level, and with activities of plant cells of various types in the regulation of their ionic or solute contents. It deals with one of the three types of ion transport found in a wide range of plant cells, but with emphasis on such transport in the particular experimental system provided by giant algal cells.

The three major transport processes in plant cells are responsible for:

1. maintenance of a high K/Na ratio within the cell, particularly in the cytoplasm; this is achieved by active extrusion of sodium, with or without an associated active influx of potassium
2. active extrusion of H^+ from the cell, presumably contributing to the maintenance of a constant pH in the cytoplasm in the face of considerable generation of H^+ in metabolism in most conditions of growth
3. net accumulation of salt to a high internal level, as one of the constituent processes involved in maintenance of a large solute-loaded central sap cavity occupying some 90 per cent of the cell volume; salt accumulation is frequently seen as KCl accumulation or as NaCl accumulation in the vacuole, but in other circumstances the vacuolar accumulation is of K^+ from the outside solution, together with an organic acid anion synthesized in the cytoplasm.

All three processes involve ion transport at the plasmalemma (at least in cells where the information is available to examine the question of the site of active transport processes). However there is likely also to be further ion regulation at the tonoplast associated with each of three processes, and, for the last two, continued operation of the transport processes at the plasmalemma may depend on further transport from cytoplasm to vacuole.

The aim in this paper is to discuss the process of net salt accumulation in plant cells; I shall present first some discussion of the general characteristics of the process, then some experimental results on chloride uptake in the giant algal cells of *Nitella translucens*, and finally some speculative discussion of the mechanisms suggested by these results.

GENERAL FEATURES OF SALT ACCUMULATION

Most work on salt accumulation in plant cells has been concerned with measurement of plasmalemma fluxes under a range of conditions, and with identification of such fluxes as passive diffusion, carrier-mediated 'facilitated' diffusion, or as active transport (defined as the movement against the electrochemical potential gradient of the ion concerned). The problems of such identification have been reviewed previously (MacRobbie, 1970a). With the identification of the processes of active transport the aim has been the characterization of each transport process in terms of its concentration dependence, its link to a specific energy-yielding metabolic sequence and the nature of the coupling between the two, and finally its response to changing salt status of the tissue. The general view has been of carrier-mediated transport, although specific carriers have not been identified, and discussion has centred on close links between salt uptake in non-green tissues and respiration (reviewed by Robertson, 1968) or protein synthesis (Steward and Sutcliffe, 1959; Sutcliffe, 1962). In green tissues most ion uptake processes seem to be light-dependent. Kinetic work is reviewed by Epstein in a recent book (1972). Problems involved in measurement of fluxes in plant cells have been considered in an earlier review (MacRobbie, 1971a).

However, Steward and Mott have recently suggested that salt accumulation by plant cells is in fact only one aspect of a much more general process, which is essential to plant cells as distinct from animal cells—an ability to secrete solutes internally in the cell vacuoles, to create new vacuolar volume (Steward and Mott, 1970; Mott and Steward, 1972). They discuss this in terms of water pumps for the creation of vacuoles, but the general points are independent of the mechanisms by which vacuoles are formed within the cell. Creation of vacuoles during growth and development of the cell, the formation of small vacuoles in the cytoplasm and their fusion into the enormous central sap cavity of mature cells, is as important for growth as is synthesis of macromolecular cytoplasmic constituents. Often the osmotic pressure in the vacuole is almost entirely contributed by inorganic salts, with both cations and anions taken up from the environment, and the process is then studied as an ion uptake process, with emphasis on ion transfer at the plasmalemma. But equally important in the overall process is the transfer of ions to vacuoles of one size or another, and thus the salt accumulation should perhaps be considered as part of the vacuolation process, involving some part of the intracellular membrane systems responsible for the creation of vacuoles. The discharge to the central vacuole of vesicles derived from both the endoplasmic reticulum and from the Golgi apparatus has been suggested, and current discussions of the so-called *endomembrane system* of the plant cell view the tonoplast as one of the end products of the processes of membrane flow and membrane differentiation taking place within the cytoplasm [see for example reviews by Buvat (1971), Morré and Mollenhauer (1974), Roberts (1974)]. It may be therefore that a process of membrane flow is involved in salt accumulation in the vacuole, and the question of the relation between the primary process of entry at the plasmalemma and subsequent transfer to the vacuole needs to be considered.

A second point was also made strongly by Steward and Mott (1970)—that we should consider *solute* relations of plant cells, rather than salt relations. They stressed the diversity of processes of solute accumulation in the vacuole during the life histories of plant cells, and suggested that the emphasis should be on secretion of *solutes* internally, with the creation of new vacuole; the fact that in some circumstances the process was one of salt accumulation was of secondary importance. In support of this view they draw attention to the changing pattern of salt accumulation during growth of cells, and to the reversible changes in pattern in mature cells resulting from changes in experimental conditions. In their work on carrot cells in a wide range of culture conditions they compared the solute relations of cells in three growth states. In the first phase after initiation of growth in a culture, the phase of rapid growth by cell division, cells have small incipient vacuoles and the osmotic pressure is largely contributed by potassium and organic acid anions; sugars, Na^+ and Cl^- contribute very little. In the subsequent stage of growth by cell expansion, in which the enormous increase in vacuolar volume takes place, Na^+, Cl^-, and organic non-electrolytes (sugars, amino acids and amides) play a much greater role. Mature cells, which are maintaining their size rather than growing, will accumulate whatever solutes are available in the culture solution, to maintain a more or less regulated internal osmotic level; if sugars are supplied then sugars are accumulated, if salts are supplied, then salts are accumulated, and reversible replacements of one for the other can easily be achieved by changing the culturing conditions. Salt accumulation in mature cells supplied with salt is the process most commonly studied experimentally, but Steward argues that this is only one facet of the more general process of secretion of solutes (of whatever kind is available) in internal vacuoles. He deprecates the emphasis on *salt* uptake and the study of non-growing systems.

However, it seems to me that if we accept Steward's thesis that salt uptake is one aspect of a much more general process in plant cells, one form of the process of vacuolation, then we ought to draw precisely the opposite conclusion about the most useful direction for our research. If we are looking at a general process of the creation of internal volume, then we may choose to study it in the easiest experimental system, that most accessible to experiment. We may then hope to be able to pose well-defined questions, and translate them into feasible experimental measurements, in a way which is simply not possible, with present experimental techniques, in the wider range of experimental material. If we are to translate these ideas into concrete molecular mechanisms we need to make a detailed study of events in the cytoplasm. Some possibility of this is offered by giant algal cells, but not by higher plant cells, whether growing or not. If we turn to the giant algal cells, we find mature cells, growing only very slowly, if at all, whose normal activities encompass only one of the facets of soluble accumulation which we have discussed—that of inorganic salt accumulation, with both anion and cation absorbed from the medium. Although in such cells the vacuolar volume is being maintained rather than increased, ability to develop vacuoles is still potentially present; non-vacuolate fragments of *Nitella*, produced by centrifugation and ligation of the cytoplasmic fragment (Hayashi, 1952), are capable of regenerating a large central vacuole. It seems likely that the mature vacuole is in fact maintained by the same processes as were responsible for its original creation, namely the addition of new solute-filled vesicles to it by processes of membrane flow within the cytoplasm. Hence in spite of the arguments for considering the more general aspects of solute accumulation, a case can be made for using the simplest experimental system available, even though its activities may be restricted to only one facet of the general process. I would therefore like to consider the current state of knowledge of chloride transport in Characean cells, with particular emphasis on *Nitella translucens*.

CHLORIDE TRANSPORT IN CHARACEAN CELLS

Ionic state: Nitella translucens

Concentration, potential and flux values for a range of giant algal cells are summarized in earlier reviews (MacRobbie, 1970a, 1974).

Nitella translucens provides cylindrical internodal cells some 10 cm long and up to about 1 mm in diameter; the cytoplasm consists of a thin lining layer about 10 μm thick just inside the cell wall, of which the outer 5–6 μm layer is stationary and contains the chloroplasts embedded in it in spiral rows, but whose inner 5–6 μm layer is free of chloroplasts and streams at about 50 $\mu m\ s^{-1}$ along the inner surface of the chloroplast array.

The cell maintains high K/Na and seems to have both active uptake of potassium and active extrusion of sodium, with a degree of linkage between these fluxes, and showing sensitivity to ouabain (MacRobbie, 1962). Existence of an active H^+ extrusion is shown by the direct observation of acidification of the medium (Spear *et al.*, 1969), although this may be masked in bicarbonate-containing solutions by the appearance of OH^- in the medium at discrete sites along the cell where bicarbonate uptake is thought to take place (Lucas and Smith, 1973). The activity of the electrogenic H^+ extrusion is also seen in the electrical behaviour of the cell. The membrane potential in light is often more negative than can be accounted for as a diffusion potential, and the electrical resistance is much lower in light than in the dark or in the presence of inhibitors (Spanswick, 1972, 1973, 1974). These observations point to the existence of an active process transferring net positive charge out of the cell, and Spanswick suggested that a voltage-dependent electrogenic H^+ extrusion pump provided the best explanation of the experimental results.

In its normal pond water, with about 1 mM chloride outside, the cell maintains a vacuolar chloride concentration of 150–170 mM, with the vacuole potential some 100–180 mV negative with respect to the outside solution. Hence a very large energy barrier exists for the accumulation of chloride in the vacuole, but an active influx of chloride of about 1–2 pmol $cm^{-2}\ s^{-1}$ is responsible for maintaining steady internal levels of chloride. The main potential is at the plasmalemma, with the tonoplast potential of only about 20 mV, the vacuole being positive with respect to the cytoplasm (Spanswick and Williams, 1964). The concentration of chloride in the cytoplasm of Characean cells is a matter of some dispute. Published figures for *Nitella translucens* are 240 mM chloride in the cytoplasm including the chloroplast layer (MacRobbie, 1964), and 65–87 mM for the flowing cytoplasm (Spanswick and Williams, 1964; Hope *et al.*, 1966). Recently Tazawa *et al.* (1974) have suggested that their previous determinations of cytoplasmic chloride in other Characean species by the same methods may be in error, and that the true values are lower than the published ones; they now suggest about 30 mM only, for cytoplasmic chloride in *Nitella flexilis* or *Nitella pulchella*. Hence the question of cytoplasmic chloride level is still uncertain. Any of these figures require active uptake of chloride at the plasmalemma, and the lower estimates of cytoplasmic chloride would require also active transfer from cytoplasm to vacuole.

Active chloride influx at the plasmalemma

The characteristics of the entry process have been extensively reviewed, and are also the subject of another paper in this volume (Raven); only a very brief discussion will be given here.

Active influx of chloride, like other processes of active ion transport in Characean cells, is strongly light-dependent. In *Nitella translucens* the evidence suggests that this is not because ATP from photophosphorylation provides the energy source for transport, but the alternative means of energy coupling remains unclear. But whereas conditions for only cyclic photophosphorylation (such as for red light) can support cation transport (MacRobbie, 1965), and the H^+ pump (Spanswick, 1974), I find marked inhibition of chloride uptake in these circumstances (MacRobbie, 1965). It seems therefore that some other product or consequence of light-driven electron flow, other than ATP, is responsible for light-dependent chloride uptake. Much more extensive studies in *Hydrodictyon africanum* by Raven (1967, 1968, 1969a, b, 1971) suggest that the same is true of that cell. By contrast there is no evidence against ATP as the energy source in *Chara corallina* (Coster and Hope, 1968; Smith and West, 1969). On the other hand there is recent evidence that *Chara* may not have cyclic phosphorylation, or that its products may not be available outside the chloroplast (Smith and Raven, 1974), since it is unable to support a wide range of light-dependent processes in this cell. If so, we have no evidence for or against ATP as the energy source in *Chara*, since we are unable to separate the possible products of the light-driven metabolism.

The active chloride influx is associated with inorganic cation influx, but the nature of the linkage is not entirely clear. There is no evidence of a large chloride effect on the membrane potential, suggesting that an electrogenic chloride pump, with electrical coupling for cation influx, is not involved, and that chemical coupling of chloride and cations, by salt entry, may be more likely.

The importance of pH, both inside and outside the cell, for the process has also been clearly established in *Chara*. High pH outside inhibits chloride influx, but pretreatment of cells at high pH can increase their subsequent ability to take up chloride in the dark from solutions at low pH; chloride influx is also stimulated by ammonium ions, or imidazole or tris buffers, treatments which are assumed to increase the internal pH (Spear *et al.*, 1969; Smith, 1970, 1972). Smith argues that Cl^- uptake in *Chara* is by a process of Cl^-/OH^- exchange, with the necessary OH^- provided in the cytoplasm by the primary secretion of H^+ by the H^+ pump. In *Nitella*, observations that Cl uptake cannot be supported in far-red light (MacRobbie, 1965), whereas the H^+ pump can function in these conditions (Spanswick, 1974), suggest the process is not a simple Cl^-/OH^- exchange, and the hypothesis would need to be considerably modified to explain this observation.

Transfer of chloride to the vacuole

Continued accumulation of salt from the outside solution demands its removal from the cytoplasm to vacuoles; measurements of the rate of the vacuolar transfer process, and of the factors controlling it, are therefore important. Attempts at such measurements can be made in giant algal cells, where samples of cytoplasm and vacuole may be obtained separately for tracer analysis, something which cannot yet be done in higher plant cells. The problem resolves itself into that of measuring the rate of appearance of tracer from the outside solution in the vacuole, and the specific activity of the cytoplasmic phase or phases from which this activity comes. The hope is that by study of such kinetics under a range of different conditions we can deduce the nature of the processes involved in transfer of ions from cytoplasm to vacuole.

The experiments therefore consist of separation of cytoplasmic and vacuolar samples after short periods of tracer influx, for measurement of the total entry of chloride to the cell (Cl_T^*, nmol cm^{-2}), and of the amount of tracer chloride in the vacuole (Cl_v^*, also expressed on an area basis, in nmol cm^{-2}). The method has been used on *Nitella translucens* and *Tolypella intricata* (MacRobbie, 1966, 1969, 1970b, 1971b, 1973), and on *Chara corallina* (Coster and Hope, 1968; Findlay *et al.*, 1971; Walker and Bostrom, 1973).

Before presenting the experimental results we may discuss the kinetic behaviour of simple models of the cell, for later comparison with the kinetic behaviour observed experimentally.

Kinetic behaviour of models of the cell

The simplest model is that of a single, small, cytoplasmic phase in series with the very large vacuolar compartment. In this system the specific activity in the cytoplasm will rise as $(1 - e^{-kt})$ during the initial stages of uptake, with the rate constant k given by the ratio of the sum of the fluxes out of the cytoplasm (to vacuole and to the solution) to the cytoplasmic content. After short times of influx the fraction of the total tracer entry which is in the cytoplasm is equal to $(1 - e^{-kt})/kt$. At very short times this cytoplasmic fraction will be approximately equal to $(1 - \frac{1}{2}kt)$, or the vacuolar fraction ($P = Q_v^*/Q_T^*$), will be equal to $\frac{1}{2}kt$. In the simplest model therefore, of a uniform cytoplasmic phase in series with the vacuole, the vacuolar fraction P will rise linearly with time, or the vacuolar tracer, Q_v^* will be proportional to the square of the time (MacRobbie, 1969).

In case this turns out to be inadequate, in view of the structural complexity of the cytoplasm and its complement of membrane-bound organelles of one type or another, we may consider the next simplest model, of two kinetically distinct cytoplasmic phases. In this model we have a rapidly exchanging bulk cytoplasmic phase (r) feeding activity both to the vacuole (v) and to a slowly exchanging cytoplasmic phase (c), which might be, for example, the chloroplasts. We then predict that after a very short initial period, in which the specific activity of the rapid phase r rises to a quasi-steady level, the vacuolar fraction will be given by: $P = P_0 + \gamma t$. The intercept P_0 is equal to the fraction $M_{rv}/(M_{rv} + M_{rc})$, where M_{rv} is the flux from phase r to the vacuole, and M_{rc} is the flux from phase r to the slow cytoplasmic phase c. The slope γ is related to the exchange of the slow cytoplasmic phase (MacRobbie, 1969, 1973). A vacuolar time course of this form suggests that two cytoplasmic phases are involved in the ion transfer to the vacuole, in one of which the specific activity rises very rapidly to a quasi-steady level, while the specific activity in the slow phase is rising linearly with time over the period of measurement. In the model suggested the specific activity in the phase r, the rapidly exchanging phase, is in fact given by $(P_0 + 2\gamma t)$, while the specific activity in the slowly exchanging phase is proportional to γt. This model applies to one possible arrangement of two cytoplasmic phases, and the time course does not, of course, define the particular arrangement of the phases; it does, however, indicate how many kinetically distinguishable cytoplasmic phases are involved.

Experimental observations

The vacuolar time course for ^{36}Cl observed experimentally is of the second form, and shows both a fast component and a component rising linearly with time. This has been found for *Nitella translucens* and *Tolypella intricata* (MacRobbie, 1969, 1971b, 1973), and for *Chara corallina* (Findlay *et al.*, 1971; Walker and Bostrom, 1973). In *Nitella* it has been shown that the fast component does not arise by gross contamination of the sap sample with cytoplasm, since it is present for ^{36}Cl and not for ^{42}K, in double labelling experiments on the same cells (MacRobbie, 1969).

Thus the experimental time course is described by two quantities, the intercept P_0 and the slope γ. From their values, and from their behaviour under different conditions, we have to interpret the results in terms of named compartments of the cytoplasm, to identify the structural sub-divisions in the cytoplasm associated with the kinetically-defined phases. A number of odd properties of the kinetics emerge from this examination, which must be accommodated in any model of the cell.

One of these oddities is that there is no discrimination between bromide and chloride within the cell, although there is a marked preference for chloride over bromide in the initial process of active influx (MacRobbie, 1971b). In double labelling experiments with ^{82}Br and ^{36}Cl, equal vacuolar fractions for the two ions are found after equal loading times; this remains true over a wide range of influx ratios, with (chloride influx)/(bromide influx) from about 0.5 to 5 (MacRobbie, unpublished observations). It is true over both long and short times of uptake, which means that both P_0 and γ must be equal for the two ions. Such equality, reflecting an inability of the vacuolar transfer to distinguish between chloride and bromide, is unexpected for a flux that is likely to be a metabolically-linked transport process. Any single-ion transfer process, any carrier system or equivalent, would be expected to show some degree of specificity. On the other hand a process involving salt transport, or the creation and transfer of new vacuolar volume, might not.

This property of non-discrimination in double labelling experiments allows the measurement of P_0 and γ on the same cell, by transfer experiments in which the cell is loaded in ^{82}Br and ^{36}Cl for different times, and thus makes the study of the behaviour of P_0 and γ a good deal easier. The results from such experiments establish a number of features of the kinetics. The first is that values of P_0 are very variable, and show a tendency to cluster in groups, having relative values in the ratio of 1:2:3, with some higher values. This clustering is the subject of dispute; it was suggested (MacRobbie, 1970b), disputed (Findlay *et al.*, 1971) and re-affirmed (MacRobbie, 1971b, 1973). I believe it is a real experimental property of the observations, although its interpretation poses difficulty.

The second property established is that P_0 and γ are independent. The wide variation in values of P_0 is not reflected in a similar wide variation in γ, which means that the fast and slow components are independent of one another, and must be in parallel rather than in series. The third feature of the kinetics to be discussed is the observation that the slope of the slow component is proportional to the influx to the cell, i.e. the transfer of chloride from cytoplasm to vacuole is proportional to the influx. In itself this provides a strong indication that the process of transfer from cytoplasm to vacuole is a metabolically-linked process. This characteristic has been established in two ways—from the comparison of values on individual cells in any given experiment, and also from the variations in the mean rate constant for exchange in the slow fraction under different experimental conditions. It is an important property of the kinetics, for which any model of the cell must provide a reasonable explanation.

Interpretation of kinetics

We have to explain the existence of the two components of the vacuolar time course, with the kinetic properties described, in terms of distinct structural compartments within the cytoplasm. There are two possible explanations of the time course, to be discussed in turn.

The first way in which such a time course could arise is that there are in fact two distinct cytoplasmic compartments, one very small, filling rapidly, and the other filling more slowly. The specific activity in the slow compartment is rising linearly with time over the period of measurement, that in the fast compartment is rising as $(P_0 + 2\gamma t)$ over the same period, the initial rise to specific activity P_0 having taken place very rapidly, over the first minute or so. The shortness of the period over which this initial rapid rise occurs, combined with the observed values of P_0 and γ, allow an upper limit to be set on the content of the rapid phase. The lowest value of this upper limit is obtained from low values of the rate of exchange in the slow fraction, at low influx values; the range set by the observed values of γ was given as 3–17 nmol cm^{-2} (MacRobbie, 1969). This is lower than the published values of the chloride content of the flowing cytoplasm (33–44 nmol cm^{-2}), although as was discussed earlier, the reliability of previous methods of cytoplasmic analysis has recently been questioned. The lower end of this range is however lower than any of the estimates for cytoplasmic content, suggesting that the rapid phase could only be some part of the flowing cytoplasm and not all of it. There are other difficulties also in identifying the rapid phase with the whole of the flowing cytoplasm. One comes from experiments in which only half the cell is in radioactive solution, and in which the fraction of activity taken up by the cell which is carried by the flow to the end not in active solution is determined. This fraction is high, and demands that the slow phase be divided between stationary and flowing layers of cytoplasm. But if the slow phase is considered to include both stationary chloroplasts and organelles in the flowing cytoplasm, then the proportionality between γ and influx is very hard to explain. γ is equal to the quantity $\frac{1}{2}P_0(1-P_0)M_{cr}/Q_c$, where M_{cr} is the flux from the slow cytoplasmic phase to the fast phase, and Q_c is the content of the slow phase. The proportionality between γ and influx, when P_0 is independent of influx, means therefore that M_{cr}/Q_c is proportional to influx. The bulk of Q_c must be in the chloroplast layer, whose content does not appear to change with influx, and hence we would require M_{cr} to be related to influx. It is very difficult to see how the flux out of a mixed population of organelles into the cytoplasm round them could change with the influx at the plasmalemma, in the absence of a change in the content of the organelles.

Further half-labelling experiments provide another reason for rejecting the identification of the fast phase with the flowing cytoplasm. In these, the time course of appearance of tracer in the inactive end was studied by the use of both ^{82}Br and ^{36}Cl, with different labelling times for the two isotopes in the same cell (MacRobbie, unpublished observations). These experiments suggest that the specific activity in the flowing cytoplasm rises linearly from zero, over a period in which the vacuolar transfer is not linear from zero, but shows the biphasic kinetics described. The conclusion is, therefore, that if the fast component of vacuolar transfer reflects a rapidly exchanging cytoplasmic phase, it is only a very small fraction of the flowing cytoplasm and contributes little to the transfer of tracer to the inactive end of the cell.

It seems that if we are to explain the kinetics in these terms we have to look for transfer from a small membrane-bound phase in the cytoplasm. Possibilities would be pinocytotic vesicles formed from the plasmalemma, or vesicles budding off from the endoplasmic reticulum for discharge to the vacuole. The clustering of values of P_0 but not of γ, and the independence of P_0 and γ, put the fast and slow components in parallel rather than in series. This would seem to rule out pinocytotic entry, but would leave an explanation in terms of transfer via the endoplasmic reticulum as a possibility. The relation between γ and influx would then require the rate of budding off of new vacuoles to be related to the influx to the cell.

However there is an alternative way in which the observed time course could arise—by action potentials on cutting the cell to extract the vacuolar sample. In this hypothesis the vacuolar transfer is primarily concerned with what appears to be the slow component, reflecting a phase in which the specific activity is proportional to time (over the period measured), and the apparent fast component arises from transfer to the vacuole during cutting, the grouping of values reflecting variation in the number of action potentials initiated. It was originally argued that this was unlikely, largely on the basis of the effects of bicarbonate on vacuolar transfer, but the argument needs to be further considered, preferably with reference to a better-defined time course in the presence of bicarbonate. An explanation based on chloride transfer during action potentials generated in cutting accounts easily for the properties of the fast fraction—its variability, its clustering of values, and its independence of the slow exchange. It is still necessary to suppose that transfer during the action potential takes place from a cytoplasmic phase that contains only a small fraction of the total cytoplasmic chloride, but a large fraction of the short-term tracer contents, and hence we must still consider a heterogeneous cytoplasm.

If the fast phase is an artifact of cutting, then we are left with the other curious features of the kinetics to explain—the non-discrimination in vacuolar transfer in spite of the link with metabolism, and the very close linkage between influx and vacuolar transfer. Both these properties are difficult to explain in terms of carriers for single ions at the tonoplast (or equivalent single-ion transfer processes). Both are easier to envisage in terms of a process of membrane flow, involving delivery of salt-filled vesicles to the main vacuole. The link between influx and vacuolar transfer might be imagined as a response to the concentration of chloride in a small membrane-bound phase from which vesicles are produced, such as the endoplasmic reticulum (e.r.), and it might be imagined that this could be related to the influx to the cell. Alternatively we could imagine an influence of pH on membrane turnover processes, with the possibility of linkage to the influx through

H^+/OH^- movements associated with the initial chloride entry of the kind that Smith (1970, 1972) has suggested.

CONCLUSIONS

At present the interpretation of the kinetics is ill-defined, but does provide indications of the type of process likely to be involved, or of the type of process which is unlikely to be involved, in the transfer of chloride from cytoplasm to vacuole. If the cell vacuole is formed from the e.r. then we might expect this membrane system to be the site of the processes transferring salt, or other solutes, to the vacuole, and a common control on all solute transfer processes might be exerted through the effects of the solute content of the e.r. cisternae on membrane flow processes. A model of this kind, in principle if not in detail, provides an easier explanation of the observed kinetic features in *Nitella*, than any model based on single-ion transfer processes across a static tonoplast membrane. It may be therefore that kinetic analysis in *Nitella* can produce some experimental evidence for the sort of mechanism discussed by Steward and Mott (1970), of salt accumulation by the creation of new vacuoles, as part of a more general process of solute transfer in vacuolation. The implication of some sort of membrane flow process might also explain a number of observations in other systems. Such a process lays stress on solute transfer rather than on salt transfer, and could account for the relations between salt and sugar accumulation in the vacuole. It might predict some kind of common control on all vacuolar transfer processes, and there is some evidence for such control. For example, Cram has evidence for non-selective anion transfer at the tonoplast in carrot and barley roots, from the equivalence of vacuolar chloride and nitrate in the regulation of chloride transfer (Cram, 1973), and in the effects, in some experiments, of chloride on the vacuolar transfer of both malate and chloride (Cram, 1974).

The question of what is, and what is not, artifact of measurement in the kinetics of vacuolar transfer in giant algal cells must obviously be settled before the details of the mechanism can be established. When this is done it should be possible to characterize the vacuolar transfer process and to identify the site involved. It is, then, likely that the present model is a stage towards the correct one, rather than the final answer. The principle of the hypothesis, the participation of membrane flow processes, rather than single-ion static membrane transport systems, is the essential feature; the details are much less important and need to be established by future work.

Acknowledgement: Support of the S.R.C., for purchase of equipment used in the experimental work described in this paper, is gratefully acknowledged.

REFERENCES

BUVAT, R. (1971) Origin and continuity of cell vacuoles. In *Results and Problems in Cell Differentiation. II. Origin and Continuity of Cell Organelles*, ed. REINERT, J. and URSPRUNG, H. Springer-Verlag, Berlin.

COSTER, H. G. L. and HOPE, A. B. (1968) Ionic relations of *Chara australis*. XI. Chloride fluxes. *Aust. J. biol. Sci.* **21**, 243–254.

CRAM, W. J. (1973) Internal factors regulating nitrate and chloride influx in plant cells. *J. exp. Bot.* **24**, 328–341.

CRAM, W. J. (1974) Effects of Cl^- on HCO_3^- and malate fluxes and CO_2 fixation in carrot and barley root cells. *J. exp. Bot.* **25**, 253–268.

EPSTEIN, E. (1972) *Mineral Nutrition of Plants: Principles and Perspectives.* Wiley, New York.

FINDLAY, G. P., HOPE, A. B. and WALKER, N. A. (1971) Quantization of a flux ratio in Charophytes? *Biochim. biophys. Acta* **233,** 155–162.

HAYASHI, T. (1952) Some aspects of behaviour of protoplasmic streaming in plant cells. *Bot. Mag. Tokyo* **65,** 765–766.

HOPE, A. B., SIMPSON, A. and WALKER, N. A. (1966) The efflux of chloride from cells of *Nitella* and *Chara. Aust. J. biol. Sci.* **19,** 355–362.

LUCAS, W. J. and SMITH, F. A. (1973) The formation of alkaline and acid regions at the surface of *Chara corallina* cells. *J. exp. Bot.* **24,** 1–14.

MACROBBIE, E. A. C. (1962) Ionic relations of *Nitella translucens. J. gen. Physiol.* **45,** 861–878.

MACROBBIE, E. A. C. (1964) Factors affecting the fluxes of potassium and chloride ions in *Nitella translucens. J. gen. Physiol.* **47,** 859–877.

MACROBBIE, E. A. C. (1965) The nature of the coupling between light energy and active ion transport in *Nitella translucens. Biochim. biophys. Acta* **94,** 64–73.

MACROBBIE, E. A. C. (1966) Metabolic effects on ion fluxes in *Nitella translucens.* II. Tonoplast fluxes. *Aust. J. biol. Sci.* **19,** 371–383.

MACROBBIE, E. A. C. (1969) Ion fluxes to the vacuole of *Nitella translucens. J. exp. Bot.* **20,** 236–256.

MACROBBIE, E. A. C. (1970a) The active transport of ions in plant cells. *Q. Rev. Biophys.* **3,** 251–294.

MACROBBIE, E. A. C. (1970b) Quantized fluxes of chloride to the vacuole of *Nitella translucens. J. exp. Bot.* **21,** 335–344.

MACROBBIE, E. A. C. (1971a) Fluxes and compartmentation in plant cells. *A. Rev. Pl. Physiol.* **22,** 75–96.

MACROBBIE, E. A. C. (1971b) Vacuolar fluxes of chloride and bromide in *Nitella translucens. J. exp. Bot.* **22,** 487–502.

MACROBBIE, E. A. C. (1973) Vacuolar ion transport in *Nitella.* In *Ion Transport in Plants,* ed. ANDERSON, W. P. Academic Press, London and New York.

MACROBBIE, E. A. C. (1974) Ion uptake. In *Algal Physiology and Biochemistry,* ed. STEWART, W. D. P. Blackwell, Oxford.

MORRÉ, J. D. and MOLLENHAUER, H. H. (1974) The endomembrane concept: a functional integration of endoplasmic reticulum and Golgi apparatus. In *Dynamic Aspects of Plant Ultrastructure,* ed. ROBARDS, A. W. McGraw-Hill, London.

MOTT, R. L. and STEWARD, F. C. (1972) Solute accumulation in plant cells. V. An aspect of nutrition and development. *Ann. Bot.* NS **36,** 915–937.

RAVEN, J. A. (1967) Light stimulation of active ion transport in *Hydrodictyon africanum. J. gen. Physiol.* **50,** 1627–1640.

RAVEN, J. A. (1968) The linkage of light-stimulated Cl influx to K and Na influxes in *Hydrodictyon africanum. J. exp. Bot.* **19,** 233–253.

RAVEN, J. A. (1969a) Effects of inhibitors on photosynthesis and the active influxes of K and Cl in *Hydrodictyon africanum. New Phytol.* **68,** 1089–1113.

RAVEN, J. A. (1969b) Action spectra for photosynthesis and light-stimulated ion transport processes in *Hydrodictyon africanum. New Phytol.* **68,** 45–62.

RAVEN, J. A. (1971) Inhibitor effects on photosynthesis, respiration and active ion transport in *Hydrodictyon africanum. J. Membrane Biol.* **6,** 89–107.

ROBERTS, K. (1974) The structural development of plant cells. In *Comprehensive Biochemistry,* ed. BULL, A. T., LAGNADO, J. R., THOMAS, J. O. and TIPTON, K. F. Longman, London.

ROBERTSON, R. N. (1968) *Protons, Electrons, Phosphorylation and Active Transport.* Cambridge University Press.

SMITH, F. A. (1970) The mechanism of chloride transport in Characean cells. *New Phytol.* **69,** 903–917.

SMITH, F. A. (1972) Stimulation of chloride transport in *Chara* by external pH changes. *New Phytol.* **71,** 595–601.

SMITH, F. A. and RAVEN, J. A. (1974) Energy-dependent processes in *Chara corallina*: absence of light stimulation when only photosystem one is operative. *New Phytol.* **73,** 1–12.

SMITH, F. A. and WEST, K. R. (1969) A comparison of the effects of metabolic inhibitors on chloride uptake and photosynthesis in *Chara corallina. Aust. J. biol. Sci.* **22,** 351–363.

SPANSWICK, R. M. (1972) Evidence for an electrogenic ion pump in *Nitella translucens.* I. The effects of pH, K^+, Na^+, light and temperature on the membrane potential and resistance. *Biochim. biophys. Acta* **288,** 73–89.

SPANSWICK, R. M. (1973) Electrogenesis in photosynthetic tissues. In *Ion Transport in Plants,* ed. ANDERSON, W. P. Academic Press, London and New York.

SPANSWICK, R. M. (1974) Evidence for an electrogenic ion pump in *Nitella translucens.* II. The control of the light-stimulated component of the membrane potential. *Biochim. biophys. Acta* **332,** 387–398.

SPANSWICK, R. M. and WILLIAMS, E. J. (1964) Electric potentials and Na, K and Cl concentrations in the vacuole and cytoplasm of *Nitella translucens*. *J. exp. Bot.* **15**, 193–200

SPEAR, D. G., BARR, J. K. and BARR, C. E. (1969) Localization of hydrogen ion and chloride ion fluxes in *Nitella*. *J. gen. Physiol.* **54**, 397–414.

STEWARD, F. C. and MOTT, R. L. (1970) Cells, solutes and growth: salt accumulation in plants re-examined. *Int. Rev. Cytol.* **28**, 275–370.

STEWARD, F. C. and SUTCLIFFE, J. F. (1959) Plants in relation to inorganic salts. In *Plant Physiology* Vol. II, ed. STEWARD, F. C. Academic Press, New York and London.

SUTCLIFFE, J. F. (1962) *Mineral Salts Absorption in Plants*, Pergamon Press, Oxford.

TAZAWA, M., KISHIMOTO, U. and KIKUYAMA, M. (1974) Potassium, sodium and chloride in the protoplasm of Characeae. *Pl. Cell Physiol. Tokyo* **15**, 103–110.

WALKER, N. A. and BOSTROM, T. E. (1973) Intercellular movement of chloride in *Chara*—a test of models for chloride influx. In *Ion Transport in Plants*, ed. ANDERSON, W. P. Academic Press, New York and London.

TRANSPORT AT ALGAL MEMBRANES

J. A. Raven

Department of Biological Sciences, University of Dundee, U.K.

Introduction

The study of ion transport at the cell level in plants has depended to a large extent on the use of giant algal cells. In order to decide if ion distribution can be accounted for, at flux equilibrium, by the passive driving forces of chemical activity differences and electrical potential differences across the membrane, the Nernst equation (eqn. (1)) is used (see Dainty, 1962):

$$E_{io} = \frac{RT}{zF} \ln \frac{a_o}{a_i} \tag{1}$$

where E_{io} = electrical potential of phase i (inside) with respect to phase o (outside)
a_o = chemical activity of the ionic species in phase o
a_i = chemical activity of the ionic species in phase i
R = Gas constant
T = Absolute temperature
z = charge on ion
F = Faraday's equivalent.

At physiological temperatures, numerical solution of this equation shows that a potential difference of 60 mV corresponds to a tenfold concentration gradient of an ion. Thus at passive equilibrium, E_{io} of -60 mV means that a_o/a_i for an anion is 0.1, and for a cation, 10. For a cell membrane (e.g. the plasmalemma or the tonoplast), the minimum information that is needed to decide if active transport occurs for a given ion is E_{oi}, a_o and a_i. The use of giant algal cells, in which physical separation of cell wall, cytoplasm and vacuole can be achieved, and the intracellular location of microelectrodes can be determined with more precision than is possible in higher plant cells, has proved of great value in investigating active ion transport at the two membranes bounding the plant cell cytoplasm (see MacRobbie, 1974; Raven, 1975a, b).

Bidirectional flux measurements are also important in both the definition of active transport, and in further investigation of active and passive transport. A *net flux* against the electrochemical potential gradient must involve active transport: methods of investigating the nature of other fluxes will be mentioned later in this paper.

The results discussed are largely those of the author using the freshwater green algal coenocyte *Hydrodictyon africanum*, and involve transport of the anions chloride and phosphate at the plasmalemma.

DIRECTION OF DRIVING FORCES

Application of the Nernst equation (eqn. (1)) to data on the values of cytoplasmic and external concentration of Cl^- and $H_2PO_4^-$, and the value of E_{co} (electrical potential of the cytoplasm with respect to the outside) (Table 1) shows that the passive driving force on both anions is directed outwards at the plasmalemma (Raven, 1967a, 1974a). The energy required to transport a mole of the anion into the cytoplasm is also given in Table 1. This value refers to flux equilibrium and to electrogenic transport of the anion (see Fig. 1(a), $A^- = Cl^-$ or $H_2PO_4^-$, and pp. 384–385). Symport (Fig. 1(c), $A^- = Cl^-$ or $H_2PO_4^-$) with a cation such as Na^+ present in the cells at a lower electrochemical potential than in the solution would decrease this energy requirement; conversely, symport with K^+, which is present at a higher electrochemical potential in the cell than outside would increase it (Raven, 1967a, 1975a, b).

TABLE 1. CYTOPLASMIC AND EXTERNAL CONCENTRATIONS OF CHLORIDE AND PHOSPHATE IN *Hydrodictyon africanum*, AND MINIMUM ENERGY REQUIREMENT FOR INFLUX

Ion	External concentration, mM	Cytoplasmic concentration, mM	Energy required per anion transported inwards, kJ mole^{-1}
Cl^-	1.3	58	21
$H_2PO_4^-$	0.1	1–2	18

Calculation of energy requirement uses Nernst equation (eqn. (1)), and assumes flux equilibrium and E_{co} of −116 mV. Data from Raven (1967a, 1974a).

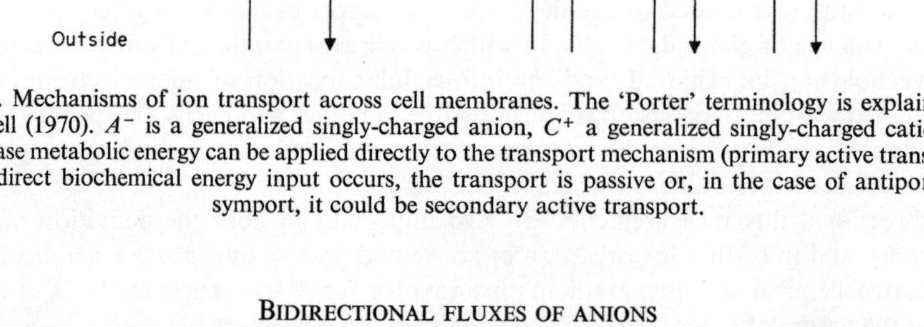

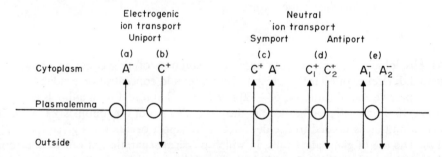

FIG. 1. Mechanisms of ion transport across cell membranes. The 'Porter' terminology is explained in Mitchell (1970). A^- is a generalized singly-charged anion, C^+ a generalized singly-charged cation. In each case metabolic energy can be applied directly to the transport mechanism (primary active transport). If no direct biochemical energy input occurs, the transport is passive or, in the case of antiport and symport, it could be secondary active transport.

BIDIRECTIONAL FLUXES OF ANIONS

Table 2 gives the results of experiments on tracer Cl^- fluxes at the plasmalemma of *Hydrodictyon africanum*. In the light in artificial pond water there is a net influx of Cl^-, while in the dark in air there is approximate flux equilibrium, and in the dark in nitrogen

TABLE 2. TRACER FLUXES OF $^{36}Cl^-$ IN *Hydrodictyon africanum* UNDER VARIOUS EXPERIMENTAL CONDITIONS

Conditions	Light, air	Dark, air	Dark, nitrogen
(1) Influx from APW (active plus exchange diffusion)	1.96 ± 0.21	0.34 ± 0.04	0.16 ± 0.02
(2) Efflux to APW (passive uniport plus exchange diffusion)	0.52 ± 0.07	0.38 ± 0.05	0.21 ± 0.03
(3) Efflux to Cl^--free (SO_4^{2-}) APW (passive uniport)	0.23 ± 0.03	0.17 ± 0.03	0.05 ± 0.03
(4) Row (2) minus Row (3) (exchange diffusion)	0.29 ± 0.08	0.21 ± 0.06	0.16 ± 0.05
(5) Row (1) minus Row (4) (active transport)	1.67 ± 0.23	0.13 ± 0.10	0.00 ± 0.06

Fluxes in pmole cm^{-2} s^{-1}. APW = Artificial pond water; NaCl 1 mM, KCl 0.1 mM, $CaCl_2$ 0.1 mM. Rows (1), (2) and (3) represent primary data; Rows (4) and (5) represent derived data. Experimental conditions as in Raven (1974c).

there is a slight net efflux. Row 3 in Table 2 gives the efflux to a solution lacking Cl^-. Since such a solution does not alter the resting potential (Table 4), a reasonable explanation for the decreased efflux in each treatment compared with the control with external Cl^- is that there is an exchange diffusion component (Fig. 1(e), $A_1^- = A_2^- = Cl^-$) of Cl^- fluxes. This assumes that chloride permeability (P_{Cl}) is independent of external chloride concentration (Cl_o). The exchange diffusion component is smaller in the dark than in the light, but is independent of anaerobiosis, while the residual efflux decreases in anaerobic conditions.

The Ussing–Teorell equation (see Dainty, 1962) describes the relationship of influx to efflux for ions acted on by passive driving forces, and for which E_{io}, a_o and a_i are known. Inserting these values (see Table 1), and the non-Cl_o-dependent (passive uniport: Fig. 1) component of Cl^- efflux, the passive uniport component of Cl^- influx can be computed. While it is clear that the Ussing–Teorell equation does not exactly describe passive ion distribution at the algal cell plasmalemma (Walker and Hope, 1969), it is very likely that passive uniport entry of Cl^- is negligible. The component of tracer Cl^- influx which cannot be accounted for by exchange diffusion (measured as the Cl^- tracer efflux dependent on external Cl^-) must be active.

Thus all the tracer Cl^- influx in dark-nitrogen can be attributed to exchange diffusion; under dark-aerobic conditions there is an active component of the influx while in the light there is a larger active component. The very low tracer loss to Cl^--free solution in the dark in the absence of oxygen suggests that P_{Cl^-} is decreased under these low energy conditions, perhaps as a means of decreasing the loss of Cl^- under conditions in which active Cl^- influx does not occur.

Table 3 gives comparable data for $H_2PO_4^-$ fluxes; similar conclusions may be drawn as for Cl^-, except that there is a net influx in the dark in air, and there is possibly an active component of the influx in dark-nitrogen (cf. Larkum and Loughman, 1969).

TABLE 3. TRACER FLUXES OF $H_2PO_4^-$ IN *Hydrodictyon africanum* UNDER VARIOUS EXPERIMENTAL CONDITIONS

Conditions	Light, air	Dark, air	Dark, nitrogen
(1) Influx from APW plus 0.1 mM P_i (active plus exchange diffusion)	0.24 ± 0.03	0.15 ± 0.02	0.031 ± 0.006
(2) Efflux to APW plus 0.1 mM P_i (passive uniport plus exchange diffusion)	0.042 ± 0.006	0.027 ± 0.004	0.022 ± 0.003
(3) Efflux to APW without P_i (passive uniport)	0.023 ± 0.003	0.014 ± 0.002	0.017 ± 0.001
(4) Row (2) minus Row (3) (exchange diffusion)	0.019 ± 0.007	0.013 ± 0.005	0.005 ± 0.003
(5) Row (1) minus Row (4) (active transport)	0.22 ± 0.03	0.14 ± 0.02	0.026 ± 0.007

APW = Artificial pond water.
Rows (1), (2) and (3) represent primary data; Rows (4) and (5) represent derived data. Experimental conditions as in Raven (1974a).

As regards the energetics of these active anion influxes, the biochemical energy source for $H_2PO_4^-$ influx appears to be ATP, supplied in the dark by oxidative phosphorylation (and, to a small degree, by fermentation), and in the light by (mainly) photophosphorylation of various kinds, including cyclic and carbon dioxide-linked non-cyclic photophosphorylation. The situation for Cl^- is much less clear; in the dark the mitochondrial electron transport chain, and in the light open-chain (non-cyclic) chloroplast electron transport, are involved. Energy transfer between these redox processes and the Cl^- pump at the plasmalemma is unlikely to be solely as ATP (Raven, 1967b, 1969a, b, 1971a, b, 1974b; Raven and Glidewell, 1975).

ELECTROGENIC VS. ELECTRONEUTRAL ANION INFLUX

In the long term, the accumulation of Cl^- and $H_2PO_4^-$ by *H. africanum* cells involves accumulation of equivalent amounts of cations, largely K^+ and Na^+ (Raven, 1975a, b). The simplest hypothesis to explain the net accumulation of cations with Cl^- and $H_2PO_4^-$ involves a mechanism at the plasmalemma which transports cations along with the anions. This could be either by electrogenic anion pumping (Fig. 1(a), $A^- = Cl^-$ or $H_2PO_4^-$) or by electroneutral symport (Fig. 1(c), $A^- = Cl^-$ or $H_2PO_4^-$). The electrogenic mechanism would involve a hyperpolarization of E_{co} as a result of net negative charge influx (as Cl^- or $H_2PO_4^-$); this would produce changes in the passive uniport of cations and anions such that net anion efflux and cation influx would increase till zero net charge transfer was regained with a new, higher steady state resting potential, and net salt accumulation. Since a hyperpolarization of the required magnitude is not found when either Cl^- replaces SO_4^{2-} in a bathing medium, or when $H_2PO_4^-$ is added to a bathing medium (Table 4), the electrogenic mechanism appears to be ruled out.

In experiments lasting 2–4 hr there is a reciprocal stimulation of the cation influxes by Cl^- or $H_2PO_4^-$ and vice versa. This is consistent with symport of cations with anions,

TABLE 4. EFFECT OF ANION ADDITION AND SUBSTITUTION ON THE VALUE OF E_{vo} (POTENTIAL OF THE VACUOLE WITH RESPECT TO THE OUTSIDE) IN *Hydrodictyon africanum*

Experimental solution	E_{vo} (mV)
1 mM NaCl, 0.1 mM KCl, 0.1 mM $CaCl_2$	-92.3 ± 7.1
1 mM Na benzenesulphonate, 0.1 mM K benzenesulphonate, 0.1 mM Ca benzenesulphonate	-93.0 ± 9.4
1 mM NaCl, 0.1 mM KCl, 0.1 mM $CaCl_2$, 5 mM MES pH 6	-97.5 ± 8.7
1 mM NaCl, 0.1 mM KCl, 0.1 mM $CaCl_2$, 5 mM MES pH 6, 0.1 mM NaH_2PO_4.	-94.1 ± 9.5

MES = N-morpholino-ethanesulphonate.
Experimental conditions as in Raven (1967a).

but does not rule out less direct interactions (see below). Cl^- differs from $H_2PO_4^-$ in three ways with respect to this 'coupling' to alkali cations (Raven, 1968, 1974a).

First, in the case of Cl^- very little cation stimulation is found in darkness, while a large effect is found with $H_2PO_4^-$ in the dark as well as in the light. Second, the V_{max} for Cl^- at saturating cation concentration is similar for K^+ and Na^+, while for $H_2PO_4^-$, Na^+ gives a much larger V_{max} than does K^+. Third, the apparent K_m for cation stimulation of the anion influx is lower for K^+ than for Na^+ in the case of Cl^- influx, while the stimulation of $H_2PO_4^-$ influx shows similar apparent K_m values for K^+ and Na^+. Thus, as with energetic linkage of their active influxes, the two anions studied show considerable differences as regards cation effects.

POSSIBLE RELATION TO H^+ OR OH^- FLUXES

Active efflux of H^+ or OH^- at the plasmalemma is an important aspect of cytoplasmic pH regulation in plant cells (Raven and Smith, 1973, 1974). These fluxes must be in some way related to fluxes of other ions. A direct linkage has been proposed by Smith (1970) for Cl^- and alkali cation transport in *Chara corallina* (Fig. 2). The primary energy transducer at the plasmalemma is the active H^+ extrusion pump. This is at least partly electrogenic (Fig. 1(b), $C^+ = H^+$) and could thus bring about K^+ and Na^+ influx either by electrogenic coupling or by cation antiport (Fig. 1(d), $C_1^+ = K^+$ or Na^+, $C_2^+ = H^+$). The inside-alkaline pH gradient set up by this pump is suggested by Smith (1970) as leading to a downhill efflux of OH^- coupled to uphill Cl^- influx (Fig. 1(e), $A_1^- = Cl^-$, $A_2^- = OH^-$: anion antiport). As well as the mechanistic requirement for an antiport system, such a pump requires that the energy available from OH^- efflux exceeds that required by Cl^- influx.

A test of this mechanism is to abolish the biochemical energy supply and to substitute an artificially imposed inside-alkaline pH gradient; active Cl^- influx should still occur.

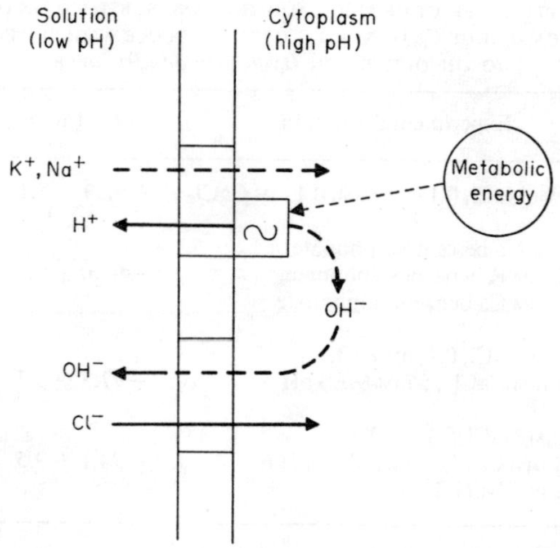

FIG. 2. Scheme for coupling Cl^- and alkali cation influx to H^+ and OH^- transport at the plasmalemma of plant cells: from Smith (1970).

Smith (1972) sought to abolish the biochemical energy supply by withholding light, and to impose a pH gradient by pretreatment at high external pH followed by influx measured at a lower external pH. After such treatment a significant increase in tracer Cl^- influx was found (Smith, 1972).

These experiments were repeated and extended by Raven (1974c) using *Hydrodictyon africanum*. It was found that *net*, as well as *tracer*, Cl^- influx was enhanced by a high–low pH transition compared with a low–high transition, or controls held at low or high pH throughout. These results are summarized in Table 5.

TABLE 5. INFLUENCE OF pH OF PRETREATMENT ON INFLUX AND EFFLUX OF $^{36}Cl^-$ IN *Hydrodictyon africanum* MEASURED AT pH 6 AND AT pH 10

pH of solution for pretreatment	pH of solution for flux measurement	Influx	Efflux
pH 6	pH 6	0.31 ± 0.04	0.37 ± 0.06
pH 6	pH 10	0.18 ± 0.03	0.33 ± 0.04
pH 10	pH 10	0.22 ± 0.03	0.26 ± 0.03
pH 10	pH 6	0.54 ± 0.05	0.35 ± 0.06

Darkness, 16°C. Pretreatment for 2 hours; flux measured for 30 minutes. Fluxes in pmoles cm^{-2} s^{-1}. Experimental conditions as in Raven (1974c).

However, results discussed earlier (pp. 382–384) suggest that there is an active, non-exchange-diffusion component of tracer Cl^- influx in *Hydrodictyon africanum* in the dark in air. The stimulation of Cl^- influx following a high pH pretreatment could have been powered directly by respiratory energy rather than by the pH gradient. In an attempt to abolish the respiratory energy component, the pH transition experiment was repeated under anaerobic conditions (Raven, 1974c). Under these conditions there appears to be no tracer Cl^- influx which cannot be attributed to exchange diffusion (pp. 382–384). While this treatment substantially inhibits the enhancement of Cl^- influx as a result of the pH transition treatment, it is possible that fermentative production of acids under anaerobic conditions decreases the pH gradient (Raven, 1974c). So this experiment does not provide an unambiguous answer to the question of energy supply to Cl^- influx after a pH transition.

There are a number of difficulties with the OH^- coupling hypothesis of Cl^- influx in algal cells, among them the possibility that the OH^- gradient is not large enough to power the Cl^- influx. Since neither the Cl^- nor the OH^- gradients are known with certainty, this is another question which awaits further investigation (MacRobbie, 1974; Raven, 1975a, b; Tazawa *et al.*, 1974).

With *Hydrodictyon africanum* a further problem is that the tracer Cl^- influx is relatively insensitive to uncouplers such as CCCP (carbonyl cyanide *m*-chlorophenyl hydrazone; Raven, 1967b, 1969b, 1971a). These compounds are known to increase membrane permeability to H^+, and thus would be expected to 'short-circuit' any secondary active transport system powered by the pH gradient. This should occur even if the pH gradient were set up by an uncoupler-insensitive (redox) mechanism. However, Lüttge *et al.* (1970) have demonstrated a net active H^+ extrusion in the light in *Atriplex* leaf slices which is *not* inhibited by a CCCP analogue at a concentration which abolishes photosynthesis. This suggests that organelle ATP synthesis can be uncoupled by proton-translocating uncouplers at concentrations which do not greatly increase the permeability of the plasmalemma to H^+.

So it is possible, but not proven, that active Cl^- influx in *Hydrodictyon africanum* can be powered by an OH^- gradient under some conditions. Thus, at high external pH values, a large enough pH gradient across the plasmalemma to power the Cl^- pump might lead to unacceptably high cytoplasmic pH. An additional energy input, e.g. primary active transport of Cl^-, would be required (Raven, 1975a, b). At all events, it is still teleologically attractive to have some mechanistic connection between pH regulation and turgor generation by accumulation of inorganic or organic salts of alkali cations (Raven and Smith, 1973, 1974).

By contrast, the results shown in Table 6 suggest that the differences between Cl^- and $H_2PO_4^-$ influxes mentioned earlier extend to responses to imposed OH^- gradients;

TABLE 6. INFLUENCE OF pH OF PRETREATMENT ON INFLUX AND EFFLUX OF $H_2{}^{32}PO_4{}^-$ IN *Hydrodictyon africanum* MEASURED AT pH 6 AND AT pH 10.

pH of solution for pretreatment	pH of solution for flux measurement	Influx	Efflux
pH 6	pH 6	0.29 ± 0.03	0.022 ± 0.003
pH 6	pH 10	0.16 ± 0.02	0.013 ± 0.003
pH 10	pH 10	0.18 ± 0.02	0.015 ± 0.003
pH 10	pH 6	0.28 ± 0.03	0.023 ± 0.003

Darkness, 16°C. Pretreatment for 2 hours; flux measured for 30 minutes. Fluxes in pmole cm^{-2} s^{-1}. Solutions as in Raven (1974a), other conditions as in Raven (1974c).

transition from high to low external pH does not stimulate $H_2PO_4^-$ influx in the dark. However, there is a relatively much larger active component of $H_2PO_4^-$ (Table 3) than of Cl^- (Table 2) influx in the dark, which would make the demonstration of an influx stimulation by an imposed pH gradient more difficult to demonstrate. In view of the evidence from mitochondria, bacteria and yeast for $H_2PO_4^-$-OH^- exchange (Mitchell, 1970; Rothstein, 1972; A. A. Eddy, personal communication) it may be premature to rule out this possibility for *Hydrodictyon africanum*.

PROSPECTS

Further work is required to elucidate the nature of the relationship between Cl^- influx, pH gradients across the plasmalemma and cell energy metabolism. The chemical and electrochemical activities of cytoplasmic H^+ can be measured by determining weak electrolyte distribution and using H^+-sensitive microelectrodes. In addition, there is a need for measurements of net H^+ fluxes under conditions which alter Cl^- influx. Further work is also needed to elucidate the relationship of the active and passive anion fluxes to the regulation of cell anion content (Raven, 1975a, b).

Acknowledgements: I should like to thank Shiela M. Glidewell and Janet I. Sprent for advice and encouragement, and D. T. Clarkson for valuable discussion.

REFERENCES

DAINTY, J. (1962) Ion transport and electrical potentials in plant cells. *A. Rev. Pl. Physiol.* **13,** 379–402.

LARKUM, A. W. D. and LOUGHMAN, B. C. (1969) Anaerobic phosphate uptake by barley plants. *J. exp. Bot.* **20,** 12–24.

LÜTTGE, U., PALLAGHY, C. K. and OSMOND, C. B. (1970) Coupling of ion transport in green cells of *Atriplex spongiosa* leaves to energy sources in the light and the dark. *J. membrane Biol.* **2,** 17–30.

MACROBBIE, E. A. C. (1974) Ion transport. In *Algal Physiology and Biochemistry*, ed. STEWART, W. D. P. Blackwell, Oxford.

MITCHELL, P. (1970) Reversible coupling between transport and chemical reactions. In *Membranes and Ion Transport*, Volume 1, ed. BITTAR, E. E. Wiley-Interscience, London.

RAVEN, J. A. (1967a) Ion transport in *Hydrodictyon africanum. J. gen. Physiol.* **50,** 1607–1625.

RAVEN, J. A. (1967b) Light stimulation of active transport in *Hydrodictyon africanum. J. gen. Physiol.* **50,** 1627–1640.

RAVEN, J. A. (1968) The linkage of light-stimulated Cl influx to K and Na influxes in *Hydrodictyon africanum. J. exp. Bot.* **19,** 233–253.

RAVEN, J. A. (1969a) Action spectra for photosynthesis and light-stimulated ion transport processes in *Hydrodictyon africanum. New Phytol.* **68,** 45–62.

RAVEN, J. A. (1969b) Effects of inhibitors on photosynthesis and the active influxes of K and Cl in *Hydrodictyon africanum. New Phytol.* **68,** 1089–1113.

RAVEN, J. A. (1971a) Inhibitor effects on photosynthesis, respiration and active ion transport in *Hydrodictyon africanum. J. membrane Biol.* **6,** 89–107.

RAVEN, J. A. (1971b) Ouabain-insensitive K influx in *Hydrodictyon africanum. Planta* **97,** 28–38.

RAVEN, J. A. (1974a) Phosphate transport in *Hydrodictyon africanum. New Phytol.* **73,** 421–432.

RAVEN, J. A. (1974b) Energetics of active phosphate influx in *Hydrodictyon africanum. J. exp. Bot.* **25,** 221–229.

RAVEN, J. A. (1974c) Time course of chloride fluxes in *Hydrodictyon africanum* in alternating light and darkness. In *Proceedings of the International Workshop on Membrane Transport in Plants and Plant Organelles*, ed. DAINTY, J. and ZIMMERMANN, U. Springer-Verlag, Berlin.

RAVEN, J. A. (1975a) Transport in algal cells and tissues. In *Encyclopaedia of Plant Physiology* (2nd edition), Volume on short-distance transport, ed. LÜTTGE, U. and PITMAN, M. G. Springer-Verlag, Berlin.

RAVEN, J. A. (1975b) Algal Cells. In *Ion Transport in Plant Cells and Tissues*, ed. BAKER, D. A. and HALL, J. L. Elsevier, Amsterdam.

RAVEN, J. A. and GLIDEWELL, S. M. (1975) Effects of CCCP on photosynthesis and on active and passive chloride transport at the plasmalemma of *Hydrodictyon africanum*. *New Phytol.* Submitted for publication.

RAVEN, J. A. and SMITH, F. A. (1973) The regulation of intracellular pH as a fundamental biological process. In *Ion Transport in Plants*, ed. ANDERSON, W. P. Academic Press, London.

RAVEN, J. A. and SMITH, F. A. (1974) Significance of hydrogen ion transport in plants. *Can. J. Bot.* **52,** 1035–1048.

ROTHSTEIN, A. (1972) Ion transport in micro-organisms. In *Metabolic Pathways*, Volume 6, 3rd edition, ed. HOKIN, L. E. Academic Press, New York.

SMITH, F. A. (1970) The mechanism of chloride transport in characean cells. *New Phytol.* **69,** 903–917.

SMITH, F. A. (1972) Stimulation of chloride transport in *Chara* by external pH changes. *New Phytol.* **71,** 595–601.

TAZAWA, M., KISHIMOTO, U. and KIKOYAMA, M. (1974) Potassium, sodium and chloride in the protoplasm of characeae. *Pl. Cell Physiol. Tokyo* **15,** 103–110.

WALKER, N. A. and HOPE, A. B. (1969) Membrane fluxes and electrical conductance in characean cells. *Aust. J. Biol. Sci.* **22,** 1179–1195.

IONIC GRADIENTS IN HIGHER PLANT TISSUES

D. J. F. Bowling

Department of Botany, University of Aberdeen, U.K.

Introduction

Plant physiologists have spent a lot of time and effort in the study of uptake of ions by plant cells and tissues. In our zeal to try to understand salt uptake we have perhaps tended to overlook the fact that there are other transport problems to be tackled. For example, ions once taken up by cells are known to move from cell to cell across tissues. This phenomenon has been called short distance transport to distinguish it from the long distance transport which occurs in the xylem and phloem.

There is increasing interest in the problem of short distance transport but hitherto relatively little work has been done in this field with the notable exception of the work of W. H. Arisz on the transport of labelled chloride in leaves of the water plant *Vallisneria* (Arisz, 1964). One reason for our lack of knowledge is the lack of suitable techniques available for studying the problem. One approach is to follow the movement of tracers as Arisz did and another is to measure gradients, either in the ionic concentrations themselves or in factors which might affect ion transport.

Historically, the main methods that have been employed to determine ionic gradients in tissues have been:

(a) histochemical methods
(b) autoradiography
(c) use of the electron probe microanalyser.

Histochemical methods and autoradiography have provided useful qualitative data but attempts to obtain quantitative information from them have met with only limited success. Quite recently the electron probe microanalyser has come on the scene and provides a technique of great promise. However, results obtained with this instrument so far, using plant tissues, have been rather disappointing. Perhaps the most useful information it has provided has been about the potassium gradients in the leaf epidermis of *Vicia faba* obtained in connection with stomatal action (Sawhney and Zelitch, 1969; Humble and Raschke, 1971). There are problems in the preparation of the tissue and in the calibration of the instrument that have to be solved before the electron probe method will reach its full potential.

Specific ion microelectrodes

The purpose of this paper is to describe a fourth method for determining ionic gradients; namely the use of the specific ion microelectrode. Specific ion electrodes which work on the same principle as the traditional pH electrode are a recent commercial development. Electrodes are now available for determining several biologically important cations and anions. They produce an electrical potential difference which is proportional to the chemical activity of the ion to which the electrode is specific. The activity or concentration of the ion can then be determined using a suitable calibration curve. It is possible to make miniature specific ion electrodes which work exactly like the larger versions but which are small enough to insert into plant cells thus making possible direct determination of the vacuolar concentration of certain ions. These specific ion microelectrodes are a development of the microelectrodes used for about thirty years by physiologists for measuring electrical potential differences in animal and plant cells. Some advantages of using specific ion microelectrodes over the other methods for measuring ion gradients are:

(a) they give results in meaningful units, i.e. chemical activity or concentration
(b) they can be used on intact tissues thereby reducing to a minimum artifacts caused by preparation of the material
(c) data obtained can be used in conjunction with potential differences measured with conventional microelectrodes to calculate driving forces on the ions concerned.

There are of course some disadvantages with specific ion microelectrodes also. The microelectrodes, usually home-made, misbehave from time to time and skilful micromanipulation is required to insert them into cells.

IONIC GRADIENTS IN ROOTS

Ions taken up by the root can move centripetally to the xylem. There has been a lot of speculation (but a dearth of reliable information) about the pathway of this transport. Through the root cortex at least, ion movement appears to occur in the cell walls but there is growing evidence that some transport also occurs from cell to cell via the cytoplasmic continuum (Jarvis and House, 1970; Ginsburg and Ginzburg, 1970). It seems reasonable to postulate that such cell to cell transport is brought about by gradients of some kind developed across the tissue. Alternatively, the ion transport itself could bring about gradients. In our laboratory we have been investigating these possibilities by using microelectrodes to determine various gradients in higher plant tissues.

We began by investigating gradients of electrical potential difference across the primary root of *Zea mays* (Dunlop and Bowling, 1971). Thick longitudinal sections were taken and mounted under the microscope. The microelectrode was then inserted into the vacuoles of the various cells in the root (Fig. 1). The potential difference (PD) was measured with a high impedance electrometer and with a reference electrode in the solution bathing the root which was either a mixture of KCl and $CaCl_2$ or a complete nutrient medium.

We could measure the PD, with respect to the bathing solution, of the vacuolar sap of every cell likely to be involved in radial ion transport. We found that there was no significant gradient in PD between the living cells of the root although all the cells were electrically negative compared to the external solution. The data are set out in Fig. 2.

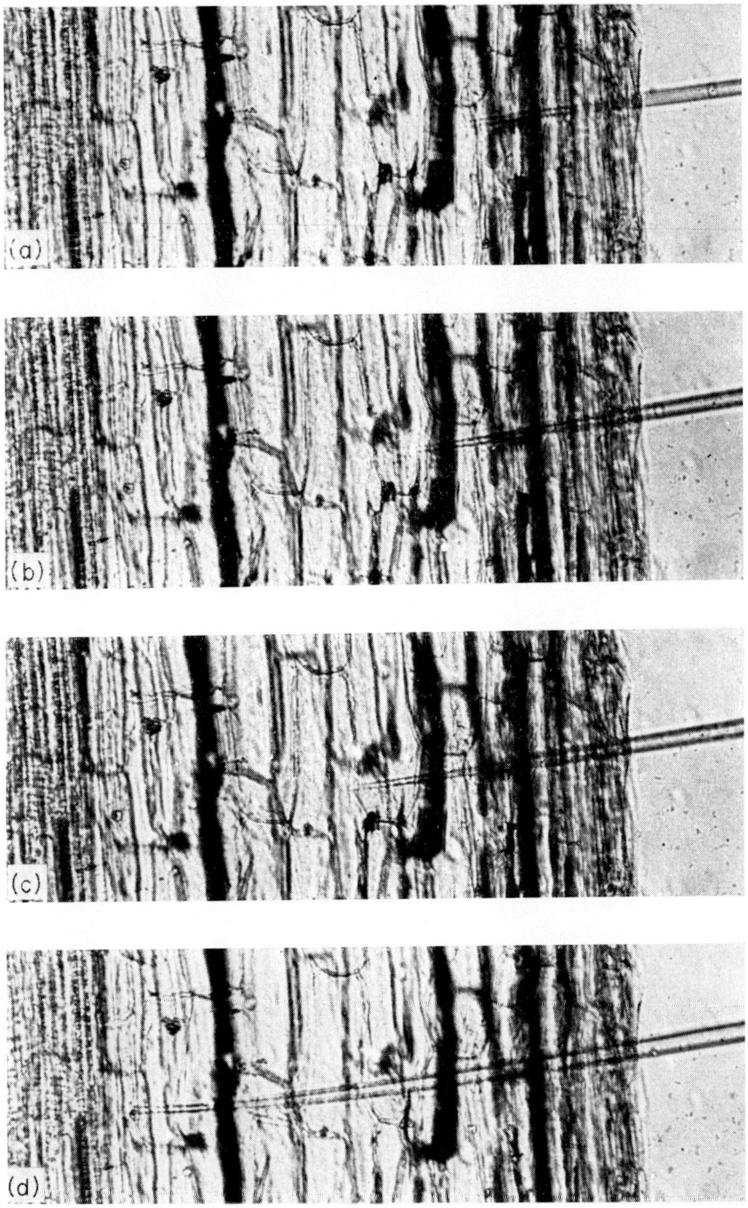

FIG. 1. A microelectrode being advanced through a longitudinal section of a root of *Zea mays*. In (a), (b) and (c), the electrode tip is respectively in the 4th, 5th and 6th cortical cells from the epidermis. In (d) it is in the endodermis. Magnification 480×. Reproduced from Dunlop and Bowling (1971).

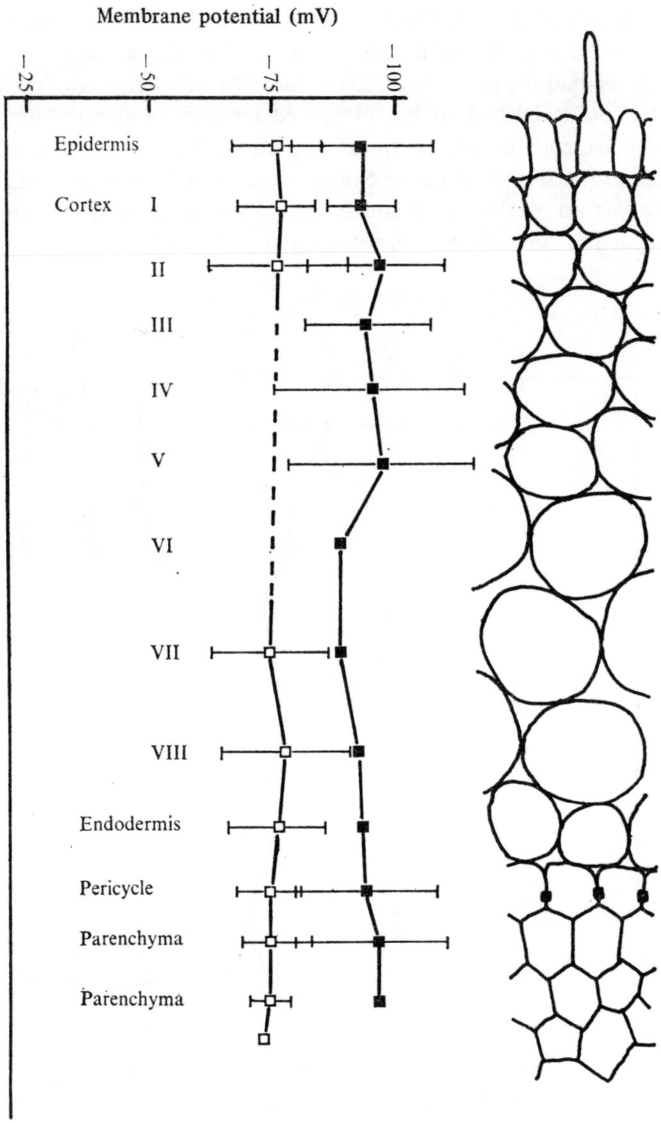

FIG. 2. Membrane potentials of various maize root cells in 1.0 mM KCl. Open symbols, tissue bathed in $CaCl_2$ solution; closed symbols, tissue bathed in nutrient medium. Horizontal bars indicate 95 per cent confidence limits. From Dunlop and Bowling (1971).

The electrical PD of plant cells is generally believed to reside mainly at the outer-cell membrane and so the cytoplasmic continuum does not appear to have an overall potential gradient which could influence cell to cell transport.

In 1968 a Soviet biophysicist, L. N. Vorobiev, described a potassium specific microelectrode made by precipitating a crystal of potassium cobaltinitrite in the tip of a conventional microelectrode. However, he only published the full technique for making this electrode later on (Vorobiev and Khitrov, 1971), so we devised our own version of it and used it to determine the profile of potassium concentration across a maize root in much the same way as we had measured the PD profile. The only major difference was that the value of the cell potential had to be subtracted from the millivolt signal reaching the potassium microelectrode before the potassium concentration of the vacuolar sap could be obtained from the calibration curve. Some of our results are shown in Fig. 3. It can be seen that there was no marked gradient in potassium concentration across the root. All the cells contained a potassium concentration of approximately 110 mM. It must be

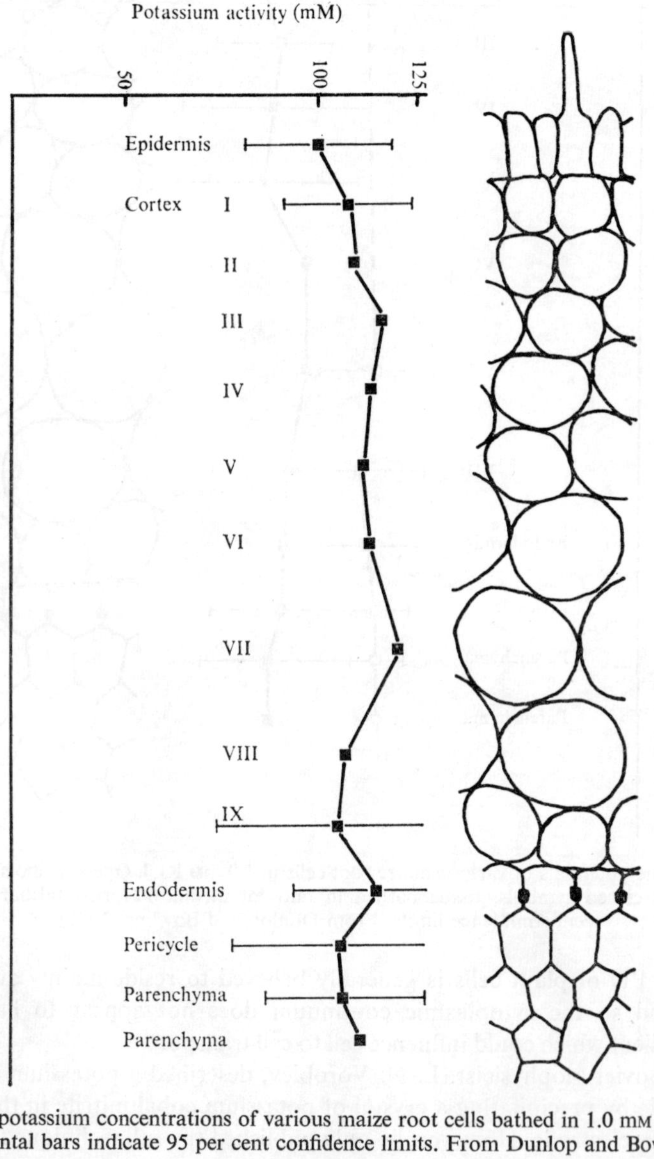

FIG. 3. Vacuolar potassium concentrations of various maize root cells bathed in 1.0 mM KCl + 0.1 mM $CaCl_2$. Horizontal bars indicate 95 per cent confidence limits. From Dunlop and Bowling (1971).

remembered that we were measuring vacuolar concentrations. The cytoplasm in most vacuolate root cells is too thin to allow insertion of a microelectrode into it with certainty. Therefore there is a faint possibility that a potassium gradient may develop in the cytoplasmic continuum and not in the vacuoles, but it seems unlikely.

It could be argued that the absence of any gradients in PD and K^+ concentration may be due to damage to the tissue on making the longitudinal sections. This is unlikely as similar results have been obtained more recently with *intact* roots of *Helianthus annuus* (Bowling, 1972) and *Lolium multiflorum* (Dunlop, 1973). All these data indicate that root cells have equal ability to accumulate potassium, and at present it is difficult to see how the ion can move centripetally in the absence of a gradient. It may be that small gradients in potassium concentration do occur but that our methods are not precise enough to detect them.

In addition to gradients in the ions themselves there may be gradients in other factors which might influence polar ion transport. Crafts and Broyer (1938) suggested that a gradient in oxygen partial pressure might occur across the root. They reasoned that as respiration is an important factor in salt accumulation, a deficiency of oxygen in the stele could cause leakage of ions into vessels from surrounding cells.

It was obviously difficult for Crafts and Broyer to measure oxygen gradients and so test their hypothesis. Recently, however, a microelectrode sensitive to oxygen partial pressure has become available. This electrode, unlike most microelectrodes, is polarographic, that is, it produces a current in the presence of oxygen rather than a potential difference. By using appropriate circuitry and otherwise employing the same techniques as were used for the other type of microelectrode, it was possible to determined the oxygen profile across the intact sunflower root (Bowling, 1973a).

The experiments were carried out on roots in culture solution saturated with air in order to provide a uniform and stable environment which would be expected to produce a steady gradient in oxygen partial pressure across the root if such a gradient was indeed developed. A definite decline in oxygen partial pressure was consistently found in a number of experiments (Fig. 4). The protoxylem appeared to have an overall oxygen level which was approximately 16 per cent less than that of the epidermis. This finding confirms Crafts and Broyer's notion of an oxygen deficit in the stele, but at the same time, whether the gradient is great enough to have an important effect on the inner cells of the root and hence on centripetal ion transport is a matter of speculation. The Crafts–Broyer hypothesis requires a sharp decline at the cortex–stele boundary, i.e. at the endodermis, but this does not occur. The steepest part of the oxygen gradient is in fact found in the outer cortex.

pH has a marked effect on many metabolic processes including salt uptake. It seems a reasonable possibility that a pH gradient could be responsible for centripetal transport. Indeed a notion of this kind was put forward by Scott and Priestley (1928) to account for root pressure. Miniaturization of the familiar pH electrode was a logical step and it has made possible the determination of the pH of the vacuolar sap of individual cells with reasonable precision.

A pH microelectrode was used to measure the pH profile across the intact sunflower root (Bowling, 1973b). In most of the experiments the root was bathed in culture solution at pH 6.4. The pH of the epidermal cells (5.7) was lower than that of the external solution

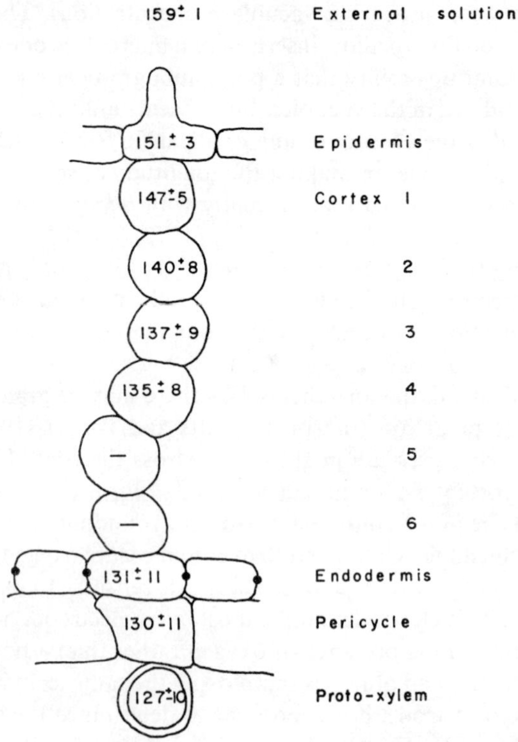

FIG. 4. Profile of oxygen partial pressure (mm Hg) across the intact root of *Helianthus annuus*. Data are means ± 95 per cent confidence limits. Bathing medium was air-saturated culture solution at 19–20°C. From Bowling (1973a).

(6.4). On moving the microelectrode inwards towards the xylem the vacuolar pH of the cells was observed to rise until the protoxylem sap was almost neutral (Fig. 5). An overall rise of 1.2 pH units was observed across the root. This is a considerable gradient and the data prompt a number of questions such as:

(a) what causes the pH gradient?
(b) is it of general occurrence in roots or does its presence depend on environmental factors or species differences?
(c) is there a parallel gradient in pH in the cytoplasmic continuum?

If there is a pH gradient developed across the root in the cytoplasmic continuum then we would have a good basis on which to build a theory to explain polar transport.

IONIC GRADIENTS IN LEAF EPIDERMIS

Another tissue which has provoked a lot of attention recently is leaf epidermis because of accumulation of potassium in the guard cells when the stomata open. Very large potassium gradients have been demonstrated by several methods including the use of the

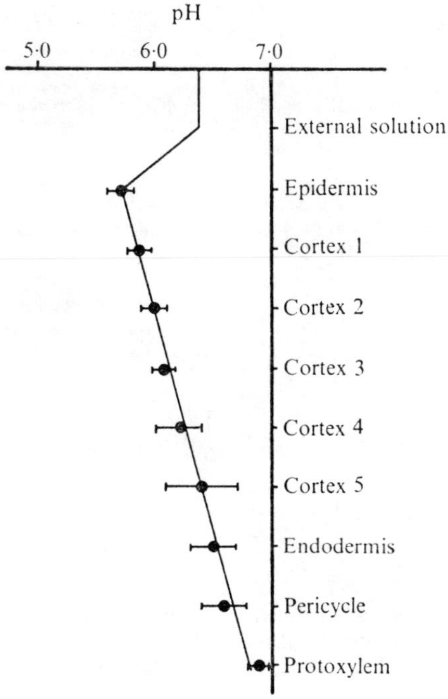

FIG. 5. Profile of vacuolar pH across the root of *Helianthus annuus*. Bars represent 95 per cent confidence limits. From Bowling (1973b).

electron probe microanalyser (Sawhney and Zelitch, 1969, Humble and Raschke, 1971). The amounts of potassium involved appear to be sufficient to produce the increase in hydrostatic pressure in the guard cells required to open the stomata.

The potassium microelectrode has proved to be a useful tool for studying potassium gradients in leaf epidermis. We have used it to investigate stomatal movement in the monocotyledonous plant *Commelina communis* (Penny and Bowling, 1974). It has been possible to measure potassium concentrations in the individual cells with a precision so far unattained by other methods. Furthermore these measurements were made on the *intact* leaf.

The stomata of *Commelina* have a number of subsidiary cells and our results indicate that they play an important role in bringing about stomatal opening and closure. When the stoma is open, as expected, we found a high potassium concentration in the guard cells (Fig. 6). There is a stepwise decline in concentration away from the guard cells through the subsidiary cells to the epidermal cells. When the stoma closes the potassium gradient is reversed (Fig. 6). At stomatal closure, potassium moves out of the guard cells and passes through the subsidiary cells to accumulate in the epidermal cells. At stomatal opening, a similar transport takes place in the opposite direction. All the cells of the epidermis therefore appear to be involved in stomatal activity.

Using conventional electrodes we failed to detect any gradients of PD in the epidermis either with the stomata open or when they were closed. Calculation of the driving forces on potassium using the Nernst equation (see Raven, this volume) indicated that the

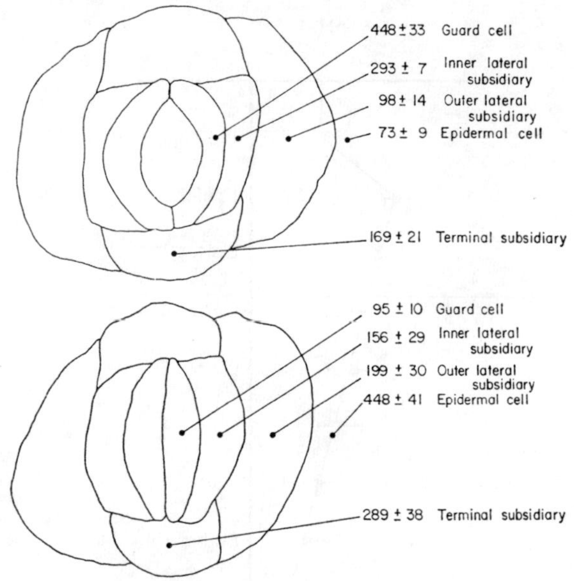

FIG. 6. Vacuolar potassium concentrations (mM) in various cells of the lower epidermis of the leaf of *Commelina communis* with the stoma open (top) and closed (bottom). Means ± standard error of the mean. From Penny and Bowling (1974).

movement of potassium both into and out of the guard cells involves active transport (Fig. 7).

Clearly the mechanism of stomatal action is intimately bound up with the processes of cell to cell transport. It seems that unless we learn more about the basis of short distance transport we are unlikely to understand fully how stomata work. For example we need to know the pathway taken by the potassium across the stomatal complex. Does it go from cell to cell via the plasmodesmata or through the cell membranes? Also what is the mechanism of the active transport and how is its direction reversed? We will need to gather much more information about ionic gradients and cell to cell transport in a wide range of tissues before these questions can be satisfactorily answered.

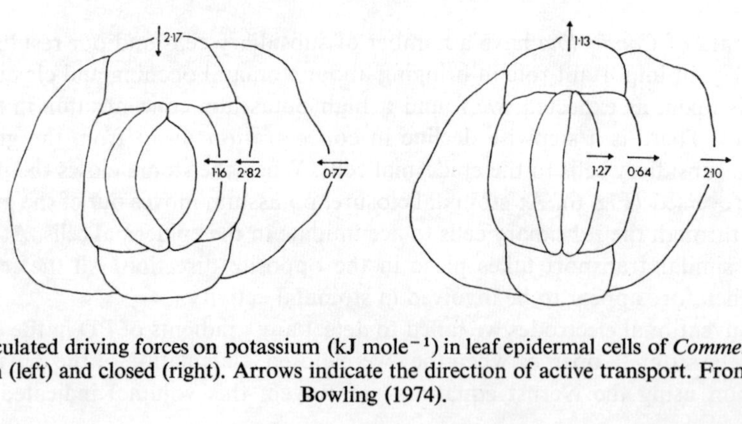

FIG. 7. Calculated driving forces on potassium (kJ mole^{-1}) in leaf epidermal cells of *Commelina* with the stoma open (left) and closed (right). Arrows indicate the direction of active transport. From Penny and Bowling (1974).

References

ARISZ, W. H. (1964) Translocation of labelled chloride ions in the symplasm of *Vallisneria* leaves. *Proc. K. ned. akad. Wet.* **C67,** 128–137.

BOWLING, D. J. F. (1972) Measurement of profiles of potassium activity and electrical potential in the intact root. *Planta* **108,** 147–151.

BOWLING, D. J. F. (1973a) Measurement of a gradient of oxygen partial pressure across the intact root. *Planta* **111,** 323–328.

BOWLING, D. J. F. (1973b) A pH gradient across the root. *J. exp. Bot.* **24,** 1041–1045.

CRAFTS, A. S. and BROYER, T. C. (1938) Migration of salts and water into the xylem of higher plants. *Am. J.Bot.* **25,** 529–535.

DUNLOP, J. (1973) The transport of potassium to the xylem exudate of rye grass. *J. exp. Bot.* **24,** 995–1002.

DUNLOP, J. and BOWLING, D. J. F. (1971) The movement of ions to the xylem exudate of maize roots. I. Profiles of membrane potential and vacuolar potassium activity across the root. *J. exp. Bot.* **22,** 434–444.

GINSBURG, H. and GINZBURG, B. Z. (1970) Radial water and solute flows in roots of *Zea Mays.* II. Ion fluxes across root cortex. *J. exp. Bot.* **21,** 593–604.

HUMBLE, G. D. and RASCHKE, K. (1971) Stomatal opening quantitatively related to potassium transport. Evidence from electron probe analysis. *Pl. Physiol. Lancaster* **48,** 447–453.

JARVIS, P. and HOUSE, C. R. (1970) Evidence for symplastic ion transport in maize roots. *J. exp. Bot.* **21,** 83–90.

PENNY, M. G. and BOWLING, D. J. F. (1974) A study of potassium gradients in the epidermis of intact leaves of *Commelina communis* L. *Planta.* **119,** 17–25.

SAWHNEY, B. L. and ZELITCH, I. (1969) Direct determination of potassium ion accumulation in guard cells in relation to stomatal opening in light. *Pl. Physiol. Lancaster* **44,** 1350–1354.

SCOTT, L. I. and PRIESTLEY, J. H. (1928) The root as an absorbing organ. I. A reconsideration of the entry of water and salt in the absorbing region. *New Phytol.* **27,** 125–140.

VOROBIEV, L. N. and KHITROV, YU-A (1971) A simple K^+ sensitive microelectrode with small effective surface. *Fiziologiya Rast.* **18,** 1054–1059.

ION TRANSPORT IN ROOT SYSTEMS

R. Scott Russell and D. T. Clarkson

A.R.C. Letcombe Laboratory, Wantage, Oxon., U.K.

Introduction

From the viewpoint of the integrated behaviour of the intact plant two of the most obvious questions concerning the performance of roots as absorbing organs are—what is the capability of different parts of the root system to absorb and translocate nutrients and how is nutrient uptake co-ordinated with plant growth? Botanists and agriculturalists were interested in both questions many decades before the Society for Experimental Biology was founded, but neither the basic physiological information nor the experimental methods necessary to resolve them were then available. The Anniversary Meeting is an appropriate occasion to review recent progress and to draw attention to major aspects which appear particularly deserving of future study.

Contribution of different components of the intact root system to the nutrition of the intact plant*

The suggestion that the absorption and translocation of water and solutes is restricted to the young apical parts of roots can be found in the literature of the 19th century (e.g. Henfrey, 1847). To the microscopist there were obvious reasons for this assumption. The delicate, thin-walled root hairs which are usually abundant close to the root tip seemed to portend special absorptive properties. Early investigations of the endodermis (Kroemer, 1903; de Lavison, 1910) suggested that this cell layer was well-designed to regulate the flow of materials between the cortex and the vascular tissue of the root. Subsequently, more detailed study showed that during later stages of their development, the walls of endodermal cells are coated with hydrophobic suberin and become greatly thickened (see Robards, this volume pp 413–422), apparently constituting an impenetrable barrier around the stele (Priestley and North, 1922; Scott, 1928).

Physiological studies reinforced the views based on anatomical characteristics; the rate of respiration in the apical region of roots considerably exceeded that in the older tissues further back from the apex. Thus, when relationships between metabolism and ion uptake were established it was natural to conclude that the young, thin-walled, metabolically active cells near the root apex had a particular capability for ion absorption. This region was often spoken of as 'the absorbing zone'. The principal obstacle in testing this

* In the description of root systems in the following discussion we have adopted the terminology of Hackett (1968) which is now widely accepted. The main cylinders of root tissue developing from seeds or nodes are respectively referred to as seminal axes and nodal axes, and the branches therefrom are called laterals.

conclusion was the technical difficulty of distinguishing between nutrients that reach the vascular tissue and the shoot from one part of the root system and those absorbed and translocated from elsewhere (e.g. Gregory and Woodford, 1939). In early attempts to resolve the limits of the absorption zone Steward and his colleagues supplied the exotic analogues of K^+ and Cl^-, Rb^+ and Br^-, to selected lengths of root axis of barley and analysed the plants by means of a spectrograph. Their results confirmed that the young tissues absorbed and translocated the ions Rb^+ and Br^- more rapidly than the older zones of the root axis (Steward *et al.*, 1942). It was clear, however, that absorption and translocation, particularly of Br^-, did occur at a distance of 4–5 cm from the root apex where the endodermis was suberized; the function of older parts of the root was not considered.

Now that radioactive tracers are available for most of the common nutrients it is much easier to investigate ion uptake by components of the intact root system. Wiebe and Kramer (1954) were among the first to take advantage of the opportunities which tracers provide for examining the performance of selected parts of the intact functioning root system. We have employed broadly the same method. Short lengths of roots on intact plants are enclosed in absorption cells made from incised plastic tubing into which the radioactive solution is introduced. The rest of the root system receives the same nutrient solution but without radioactive tracers (Russell and Sanderson, 1967; Clarkson and Sanderson, 1971b). The exposure of the entire root system to a uniform ionic environment is essential since if one zone receives an enhanced external concentration uptake elsewhere may be reduced (Drew *et al.*, 1973).

The early stages of our work with barley produced two rather unexpected results: (i) neither radial movement of phosphate (Clarkson *et al.*, 1968) and potassium (Clarkson and Sanderson, 1970; Russell and Clarkson, 1973) across the root, nor transport of these ions to shoots was confined to the absorbing zone. In barley, movement of phosphate from regions 40 cm or more from the apex was as great as within the absorbing zone

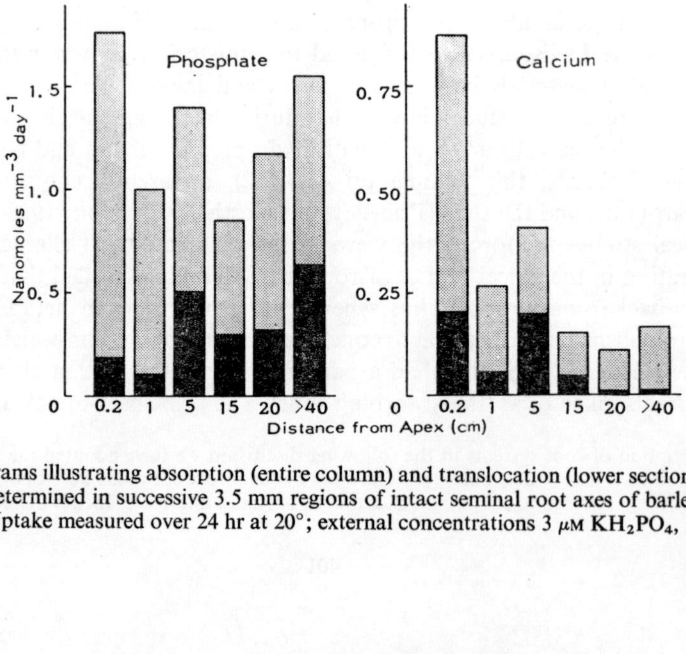

FIG. 1. Histograms illustrating absorption (entire column) and translocation (lower section) of phosphate and calcium determined in successive 3.5 mm regions of intact seminal root axes of barley plants about 3 weeks old. Uptake measured over 24 hr at 20°; external concentrations 3 μM KH_2PO_4, 1.25 μM $CaCl_2$.

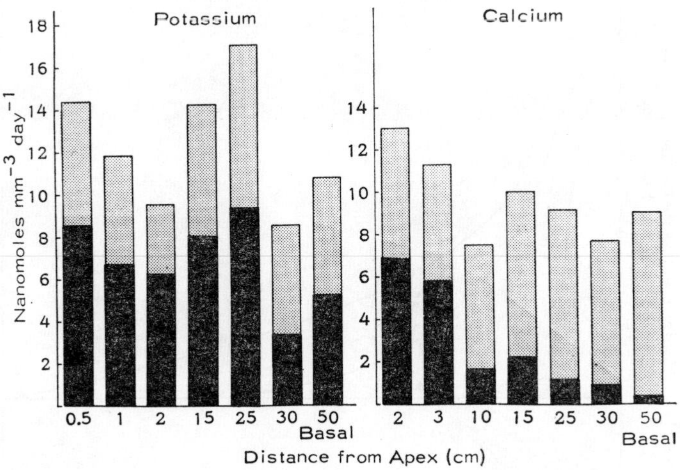

FIG. 2. Histograms illustrating absorption (entire column) and translocation (lower section) of potassium and calcium determined in successive 3.5 mm regions of seminal root axes of marrow plants (*Cucurbita pepo* cv Greenbush) 10 days old. Uptake measured over 6 hr at 20°; external concentrations 0.5 mM KCl; 0.15 mM $Ca(NO_3)_2$.

itself (Fig. 1); results with potassium were similar in *Cucurbita pepo* (Fig. 2) even in regions where secondary cambial activity had commenced (Harrison-Murray and Clarkson, 1973). Other workers using different techniques have also concluded that mature parts of the root axis both absorb and translocate significant amounts of nutrient to the shoot (e.g. Bowen, 1969, 1970; Burley *et al.*, 1970); (ii) not all ions were so readily absorbed by older regions of the root axis. In both barley and marrow (Figs. 1, 2) the movement of calcium to shoots was restricted to the relatively young parts of the root system (Clarkson and Sanderson, 1971a; Robards *et al.*, 1973). For this ion, but not for phosphate and potassium, the endodermis appeared to fulfil the role traditionally assigned to it of being a virtually impenetrable barrier once its walls began to thicken. The relationship between the development of the endodermis and ion uptake in different regions of the root could be particularly well examined in barley since it contains few, if any, passage cells (Clarkson *et al.*, 1971). Figure 3 shows that there is a very close correlation between the decrease in the transport of calcium in seminal and nodal root axes and laterals and the deposition of suberin lamellae in the endodermis (i.e. State II of endodermal development) (see Figs. 5A, B).

As it has been often assumed that the rate of ion transport is closely linked with the gross respiration rate of tissues, it is of interest to compare respiration in different parts of root axes with their ion transport. Figure 4 shows that there is no close parallel—respiration (measured as oxygen uptake per unit dry weight) declined with increasing distance from the tip whereas transport of phosphate decreases little. Since the transport of phosphate to the xylem almost certainly depends on a mechanism linked directly to metabolism (i.e. active transport) it is evident that the rate of this process is not closely linked to that of gross respiration. This is not surprising because of the manifold metabolic processes which depend on respiration, particularly in the younger tissues.

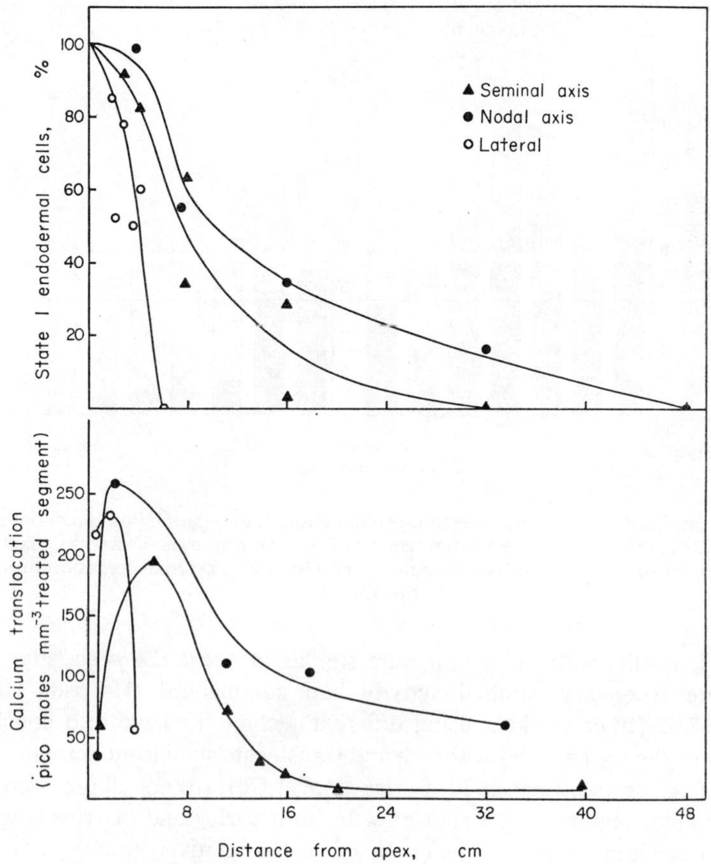

FIG. 3. A comparison of the distribution of unsuberized endodermal cells (upper graph) with the translocation of calcium by three types of root member in barley (lower graph). Symbols: ▲, seminal axis, ●, nodal axis, ○, lateral. Data from Robards *et al.* (1973).

The contrast between the patterns of uptake of calcium on the one hand and phosphate or potassium on the other can be explained on the basis that, whereas calcium moves radially across the cortex to the xylem in the apoplast (largely the cell walls), the other two ions move in the symplast (Fig. 5A). The development of the casparian bands in the endodermis restricts but does not halt the transfer of ions which cross the cortex by the apoplastic pathway; it is presumed that these ions diffuse or are transported across the plasmalemma of the endodermal cells. However, their movement is barred when suberin lamellae are laid down on the endodermal walls since the plasmalemma is then no longer directly accessible from the free space (Fig. 5B). The symplastic pathway, however, remains open because plasmodesmata penetrate the tangential endodermal walls with high frequencies in both young and old parts of the root (see Robards, this volume pp 413–422). Thus, the transfer of phosphate and potassium continues. The apparent inability of calcium to traverse the symplastic pathway is still not understood; it may

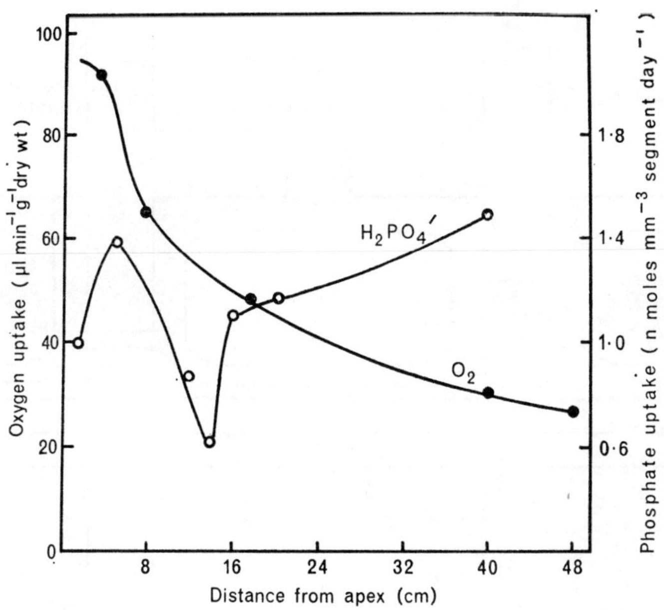

FIG. 4. Oxygen and phosphate uptake at different distances from the tip of seminal root axes of barley. Oxygen uptake measured on excised root segments over 2 hr at 20°C; phosphate uptake measured over 24 hr at 20° from a solution containing 3 μM $H_2PO_4^-$.

reflect the fact that a high proportion of the calcium in the cytoplasm is in a bound form so that the activity of calcium ions is very low.

Autoradiographic measurements of phosphate and calcium in tissues (Clarkson and Sanderson, 1971a; Sanderson and Clarkson, 1973) accord with the view that these two ions reach the stele by different pathways. When labelled nutrients are supplied to roots in which suberin lamellae have not yet been laid down, the concentration of labelled calcium at first increases more rapidly in the stele than in the cortex, but later there is little difference in the rate of increase in the inner cortex, endodermis and xylem parenchyma (Fig. 6)—a situation compatible with the view that rapid apoplastic movement of calcium takes place towards the stele and bypasses the cortical cells. The calcium transported into the cortical cells is passed on to the stelar tissues very slowly when roots are returned to unlabelled solution (Clarkson and Sanderson, 1971a; Shone *et al.*, 1973). With phosphate, however, the three tissues 'fill up' in sequence during this period as would be expected with symplastic movement (Fig. 6).

The ability of roots in regions at considerable distances from the apex to absorb and translocate nutrients to shoots shows that developing xylem vessels which still have living contents are not essential for the release of ions into them as some authors have assumed. In barley no living contents were found in xylem vessels more than 1.5 cm from the apex of root axes (see Robards, this volume pp. 413–422); yet radial transfer of ions, especially phosphate and potassium, occurred at very considerably greater distances from the tip (see also Läuchli *et al.*, 1974). Those parts of the barley root axes in which

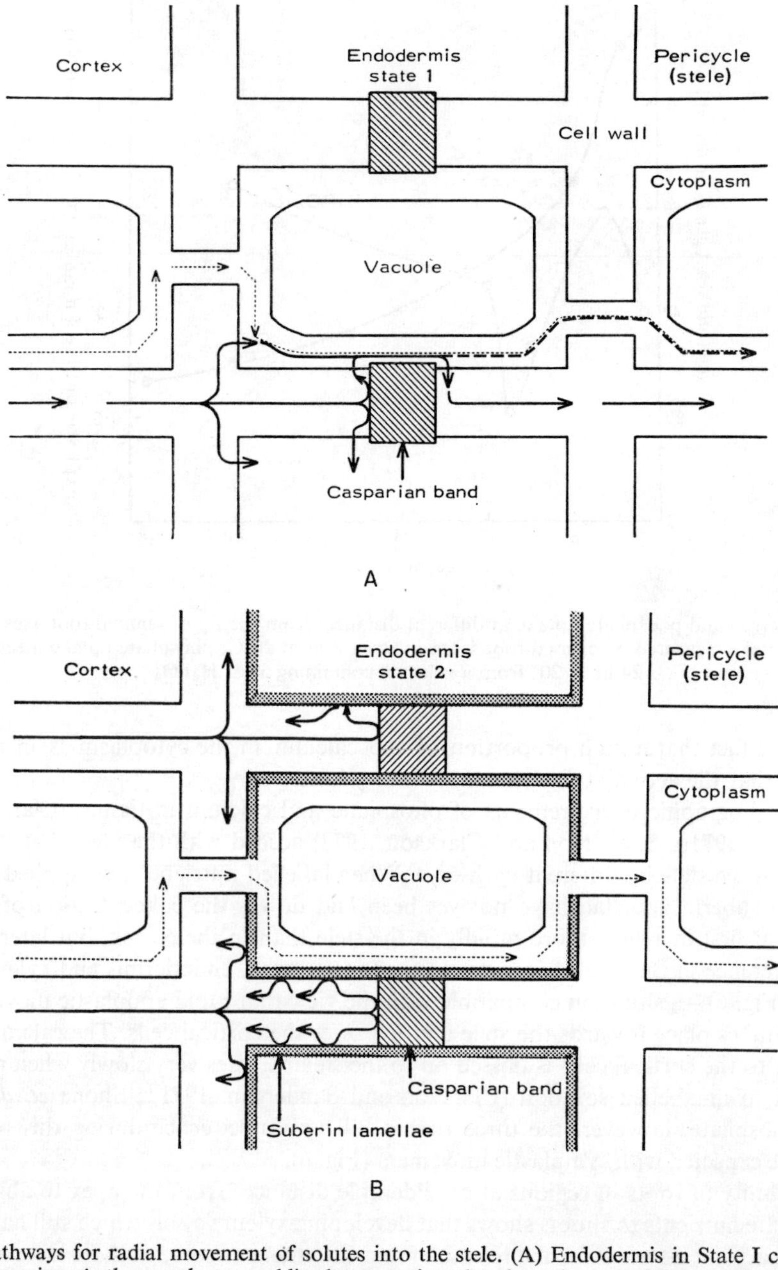

FIG. 5. Pathways for radial movement of solutes into the stele. (A) Endodermis in State I condition: materials moving *via* the apoplast are obliged to negotiate the plasmalemma of the endodermis by the casparian band. (B) Endodermis in State II condition: deposition of suberin lamellae prevents direct access to the endodermal plasmalemma from the apoplast; symplast pathway is undisturbed. ...., symplast movement; ——, apoplast movement.

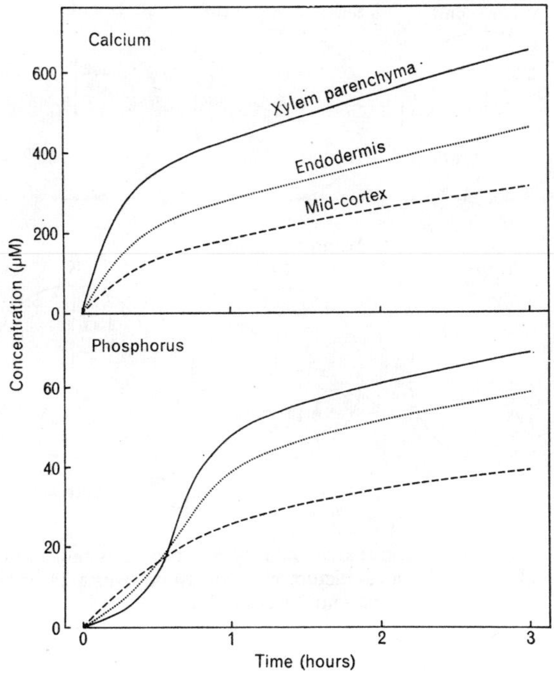

FIG. 6. Changes in concentration of labelled calcium (upper graphs) and phosphorus (lower graphs) with time in tissues of seminal root axes of barley 1 cm from the tip. Data obtained by quantitative micro-autoradiography.

there were living contents in the xylem could supply only a small fraction of the total ions which reach the shoot, unless improbably high ion fluxes into the living xylem are assumed. This supports the view that ions are released into the xylem by diffusion from, or secretion by the adjacent xylem parenchyma cells (Lundegårdh, 1950; Laties, 1969; Läuchli *et al.*, 1974). In contrast, Higinbotham *et al.* (1973) recently restated the contrary view of Hylmö (1953), that in roots of corn (*Zea*) transfer to the xylem depends on the presence of living contents in it. This was unsupported by direct observation of radial ion movement in the zones in question and is in conflict with recent observations by Ferguson (1974) that appreciable amounts of phosphate are translocated at a distance of 28 cm from the root tip of *Zea* where none of the xylem elements have living contents.

When observations of translocation are made at a sufficient number of regions along the root, it is possible to estimate the relative contribution of both axes and laterals to the nutrient supply to the shoots. The results presented in Fig. 7 relate only to conditions when the entire root system is continuously exposed to a uniform external concentration; this seldom occurs in soil. It will be seen that, in plants 3 to 4 weeks old the contributions made by axes and laterals to the uptake of phosphate and potassium are more closely related to the volume of living tissue than to their surface area or length. The uptake of calcium is also more nearly related to the volume of the roots than to the other parameters but less closely than the uptake of potassium and phosphate. This

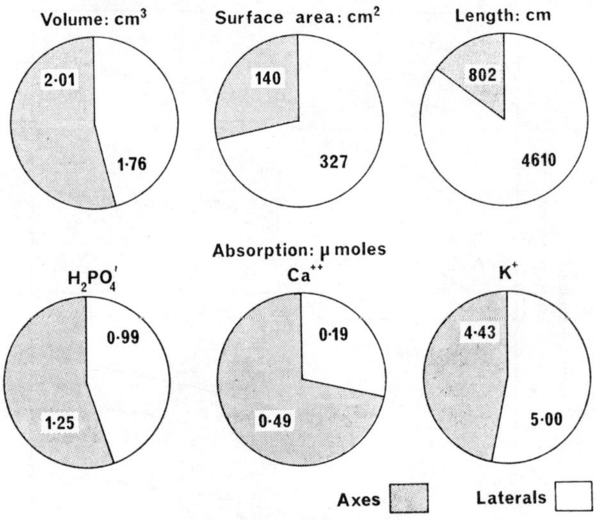

FIG. 7. Diagram illustrating volume, surface area and length of root axes and laterals, and their respective contributions to the uptake of phosphate, calcium and potassium over a 24 hr period in barley plants about 3 weeks old.

conclusion, though contrary to some earlier assumptions, is in no way incompatible with ion movement by the symplastic or apoplastic pathways here discussed.

REGULATION OF ION ABSORPTION AND TRANSLOCATION IN THE INTACT PLANT

Knowledge of the overall manner in which the uptake and transfer of ions to shoots is controlled has increased less rapidly than our understanding of relationships between the physiological anatomy of roots and their capability as absorbing organs. In short term experiments with fragments of root or slices of storage tissue only three variables which affect ion uptake need normally be considered, i.e. changes in the metabolism of the absorbing cells, changes in their ionic content, and changes in the external concentration with which they are provided. The situation is much more complex in the intact plant. One of its most obvious characteristics is that the entry of ions into roots and their movement thence to shoots can be largely controlled by metabolic utilization. Over 25 years ago Williams (1948) concluded from very meticulous pot culture experiments with cereal plants that 'the rate of uptake of phosphorus was seen to be more determined by internal factors of demand than by external factors of supply'; over a range of increasing concentration which much enhanced uptake per plant and growth, uptake per unit increment of plant weight remained virtually constant.

These results cease to be surprising if we discard the view that the movement of ions can be adequately studied in isolation from other concurrent aspects of metabolism. There is every reason to do so. The concepts of Mitchell (1966) and Robertson (1968) suggest that, in respiration or photosynthesis, charge separation brought about by the separation of protons or hydroxyl ions across the membranes of mitochondria and chloroplasts during electron transport, creates a free energy gradient which can be harnessed to a wide range of biochemical and transport processes. The protons or hydroxyl ions thus generated by metabolism may be exchanged for other cations or anions from the external medium. Thus, transport across membranes must be regarded as one of the many inter-related facets of metabolism. This view is endorsed by recent progress in animal physiology, which is enabling the cause of metabolic disorders and malignancies to be traced back to changes in membrane properties; Wallach (1972) has reviewed many aspects of this subject.

It seems, therefore, that no adequate explanation of the manner in which nutrient uptake is regulated can be expected until the processes which control metabolism as a whole are more fully understood. A number of observations support this view. In experiments in which the growth rate of plants was varied widely by altering the photoperiod, Pitman (1972) observed that changes in the net absorption of potassium into roots and its transport to shoots closely paralleled changes in shoot growth. He found it necessary to envisage that the uptake of potassium was regulated by a 'feed-back' mechanism between roots and shoots. Other observations point to the same conclusion; for example Drew *et al.* (1973) have found that if the supply of nitrate or phosphate to part of a root system is enriched, increased ion uptake by the favourably placed part of the root system causes metabolites from the shoots to be directed preferentially to it. This results in a considerable proliferation of laterals in that zone.

Evidence for the direct interaction of growth regulators with ion transport in intact plants is, however, still very limited and in the few examples discussed below relate mainly to systems simpler than the intact root system. Many findings are difficult to interpret but because general effects of hormones on growth and metabolism become confounded with their effects on transport. Most observations depend on the exogenous application of a growth regulator and it is usually assumed that any observed effect is due to its endogenous action. Recent information on the asymmetric structure and organization of membranes makes it clear that the direct interaction between a substance and a transport process may depend on which side of the membrane the substance is presented and the orientation of the transporting mechanism in the membrane. Both for this and other reasons it seems possible that an exogenous hormone may not act in quite the same way as an endogenous one.

Perhaps the best evidence for the direct involvement of a growth regulator in membrane transformation and hence on ion transport is for abscisic acid (ABA). Mansfield and Jones (1971) showed that ABA increases the passive permeability of the plasmalemma in stomatal guard cells to potassium ions; both net monovalent cation and chloride uptake and Na^+/K^+ discrimination were altered by 10^{-5} M ABA in discs of beetroot (Van Steveninck, 1972), and Collins and Kerrigan (1974) have shown that dilute solutions of ABA (10^{-8} to 10^{-9} M) rapidly increased the fluxes of exudate and potassium into the xylem of detached maize roots. In this latter study exogenous kinetin had the opposite effect. With growth regulators other than ABA the evidence is less satisfactory and frequently contradictory. Recent developments in the study of cell expansion seem to have directed attention to a search for effects of indol-3yl-acetic acid (IAA) on proton pumping and membrane potentials. Whereas persuasive evidence is cited by Marrè *et al.*

(1974) that membrane potentials and proton extension are rapidly affected by IAA in pea internode segments, Fisher and Albersheim (1974) found no effect whatever on proton extension by auxin in suspension cultured cells of *Acer pseudoplatanus*, *Lolium perenne* or *Phaseolus vulgaris*. Gibberellins may also interact directly with membrane lipids modifying their stability and permeability as has been shown recently in synthetic liposomes by Wood and Paleg (1974).

Few firm conclusions on the underlying mechanisms which control nutrient movement in the intact plant can yet be advanced but if we accept that this process is closely related to other facets of metabolism, the conclusion appears inescapable that the hormonal mechanisms which control growth and differentiation must also regulate ion movement. Animal physiologists have long recognized this. It may be reasonable for us to excuse our own ignorance on the grounds that plant physiological problems are more difficult—it would be so convenient if plants had specialized hormone secreting glands. But at the same time we would do well to remind ourselves that our understanding of the physiology of ion movement in plants will remain superficial only until ways are found to overcome this difficulty. Is there a more exciting, challenging and important aspect of ion uptake than the exploration of growth control mechanisms which integrate it with the other functions of the growing plant?

We may hope that the situation may have become clearer by the Centenary meeting of the Society. Will it then be a central topic of discussion or the cause of complacent reflection of past success?

REFERENCES

BOWEN, G. D. (1969) The uptake of orthophosphate and its incorporation into organic phosphates along roots of *Pinus radiata*. *Aust. J. biol. Sci.* **22,** 1125–1135.

BOWEN, G. D. (1970) Effects of soil temperature on root growth and phosphate uptake along roots of *Pinus radiata*. *Aust. J. Soil Res.* **8,** 31–42.

BURLEY, W. J., NWOKE, F. I. O., LEISTER, G. L. and POPHAM, R. A. (1970) The relation of xylem maturation to the absorption and translocation of ^{32}P. *Am. J. Bot.* **57,** 504–511.

CLARKSON, D. T. and SANDERSON, J. (1970) Further observations on the uptake of ions by different portions of the intact root system of barley (var. Maris Badger). Agricultural Research Council Letcombe Laboratory Report ARCRL 20, 17–19.

CLARKSON, D. T. and SANDERSON, J. (1971a) Inhibition of the uptake and long-distance transport of calcium by aluminium and other polyvalent cations. *J. exp. Bot.* **23,** 837–851.

CLARKSON, D. T. and SANDERSON, J. (1971b) Relationship between the anatomy of cereal roots and the absorption of nutrients and water. Agricultural Research Council Letcombe Laboratory Report 1970, 16–25.

CLARKSON, D. T., SANDERSON, J. and RUSSELL, R. S. (1968) Ion uptake and root age. *Nature Lond.* **220,** 805–806.

CLARKSON, D. T., ROBARDS, A. W. and SANDERSON, J. (1971) The tertiary endodermis in barley roots: fine structure in relation to radial transport of ions and water. *Planta* **96,** 292–305.

COLLINS, J. C. and KERRIGAN, A. P. (1974) The effect of kinetin and abscisic acid on water and ion transport in isolated maize roots. *New Phytol.* **73,** 309–314.

DE LAVISON, J. DE R. (1910) Du mode de pénétration de quelques sels dans la plante vivant. Rôle de l'endoderme. *Revue gén. Bot.* **22,** 225–241.

DREW, M. C., SAKER, L. R. and ASHLEY, T. W. (1973) Nutrient supply and the growth of the seminal root system in barley. I. The effect of nitrate concentration on the growth of axes and laterals. *J. exp. Bot.* **24,** 1189–1202.

FERGUSON, I. B. (1974) Ion uptake and translocation by the root system of maize. Agricultural Research Council Letcombe Laboratory Report 1973, 13–15.

FISHER, M. L. and ALBERSHEIM, P. (1974) Characterization of a H^+-efflux from suspension-cultured plant cells. *Pl. Physiol. Lancaster* **53,** 464–468.

GREGORY, F. G. and WOODFORD, H. K. (1939) An apparatus for the study of oxygen, salt and water uptake of various zones of the root, some preliminary results with *Vicia faba*. *Ann. Bot.* NS **3,** 147–154

HACKETT, C. (1968) A study of the root system of barley. I. Effects of nutrition on two varieties. *New Phytol.* **67,** 287–300.

HARRISON-MURRAY, R. S. and CLARKSON, D. T. (1973) Relationships between structural development and the absorption of ions by the root system of *Cucurbita Pepo*. *Planta* **114,** 1–16.

HENFREY, A. (1847) *Outlines of Structural and Physiological Botany*, p. 97, Van Voorst, London.

HIGINBOTHAM, N., DAVIS, R. F., MERTZ, S. M. and SHUMWAY, L. K. (1973) Some evidence that radial transport in maize roots is into living vessels. In *Ion Transport in Plants*, ed. ANDERSON, W. P. pp. 493–506, Academic Press, London.

HYLMÖ, B. (1953) Transpiration and ion absorption. *Physiologia Pl.* **6,** 333–405.

KROEMER, K. (1903) Wurzelhaut, Hypodermis und Endodermis der Angiospermenwurzel. *Biblthca bot.* Heft 59.

LATIES, G. G. (1969) Dual mechanisms of salt uptake in relation to compartmentation and long distance transport. *A. Rev. Pl. Physiol.* **20,** 89–116.

LÄUCHLI, A., KRAMER, D., PITMAN, M. G. and LÜTTGE, U. (1974) Ultrastructure of xylem parenchyma cells of barley roots in relation to ion transport to the xylem. *Planta* **119,** 85–99.

LUNDEGÅRDH, H. (1950) Translocation of salts and water through wheat roots. *Physiologia Pl.* **3,** 103–151.

MANSFIELD, T. A. and JONES, R. J. (1971) Effects of abscisic acid on potassium uptake and starch content of stomatal guard cells. *Planta* **101,** 147–158.

MARRÈ, E., FERRONI, A. and DENTI, A. B. (1974) Transmembrane potential increase induced by auxin, benzyladenine and fusicoccin. Correlation with proton extension and cell enlargement. *Pl. Sci. Lett.* **2,** 257–265.

MITCHELL, P. (1966) Chemiosmotic coupling in oxidative and photosynthetic phosphorylation. *Biol. Rev.* **44,** 445–502.

PITMAN, M. G. (1972) Uptake and transport of ions in barley seedlings. III. Correlation between transport to the shoot and relative growth rate. *Aust. J. biol. Sci.* **25,** 905–919.

PRIESTLEY, J. H. and NORTH, E. E. (1922) Physiological studies in plant anatomy. III. The structure of the endodermis in relation to its function. *New Phytol.* **21,** 111–139.

ROBARDS, A. W., JACKSON, S. M., CLARKSON, D. T. and SANDERSON, J. (1973) The structure of barley roots in relation to the transport of ions into the stele. *Protoplasma* **77,** 291–312.

ROBERTSON, R. N. (1968) *Protons, Electrons, Phosphorylation and Active Transport*, Cambridge University Press.

RUSSELL, R. S. and CLARKSON, D. T. (1973) The uptake and distribution of potassium by crop plants. In *Potassium in Biochemistry and Physiology. Proceedings of 8th Colloquium, IPI, Uppsala*, 1971. International Potash Institute.

RUSSELL, R. S. and SANDERSON, J. (1967) Nutrient uptake by different parts of the intact roots of plants. *J. exp. Bot.* **18,** 491–508.

SANDERSON, J. and CLARKSON, D. T. (1973) Quantitative microautoradiography of accumulation of phosphorus-32 in tissues from young and mature zones of the barley root. Agricultural Research Council Letcombe Laboratory Report 1972, 7–10.

SCOTT, L. I. (1928) The root as an absorbing organ. II. Delimitation of the absorbing zone. *New Phytol.* **27,** 141–174.

SHONE, M. G. T., CLARKSON, D. T., SANDERSON, J. and WOOD, A. V. (1973) A comparison of the uptake and translocation of some organic molecules and ions in higher plants. In *Ion Transport in Plants*, ed. ANDERSON, W. P. pp. 571–582. Academic Press, London.

STEWARD, F. C., PREVOT, P. and HARRISON, J. A. (1942) Absorption and accumulation of rubidium bromide by barley plants. Localization in the root of cation accumulation and of transfer to the shoot. *Pl. Physiol. Lancaster* **17,** 411–421.

VAN STEVENINCK, R. F. M. (1972) Abscisic acid stimulation of ion transport and alteration in K^+/Na^+ selectivity. *Z. Pfl-physiol.* **67,** 282–286.

WALLACH, D. F. H. (1972) *The Plasma Membrane: Dynamic Perspectives, Genetics, and Pathology*. English Universities Press, London.

WIEBE, H. H. and KRAMER, P. J. (1954) Translocation of radioactive isotopes from various regions of roots of barley seedlings. *Pl. Physiol. Lancaster* **29,** 342–348.

WILLIAMS, R. F. (1948) The effects of phosphorus supply on the rates of intake of phosphorus and nitrogen and upon certain aspects of phosphorus metabolism in gramineous plants. *Aust. J. sci. Res.* B **1,** 333–361.

WOOD, A. and PALEG, L. G. (1974) Alteration of liposomal membrane fluidity in gibberellic acid. *Aust. J. Pl. Physiol.* **1,** 31–40.

ROOT STRUCTURE AND FUNCTION—AN INTEGRATED APPROACH

A. W. Robards and S. Margaret Jackson

Department of Biology, University of York, Heslington, York, U.K.

Introduction

'Absorption is not carried out over the whole surface of a subterranean root system. It can, on the contrary, be easily shown that the functional absorbing tissue only occurs in the youngest rootlets and that it is, moreover, confined to a definite zone'. So wrote the great German botanist Haberlandt (1914) during the early years of this century. This concept of a relatively restricted absorbing zone has dominated ideas of root function right up to the present time. Modern techniques have enabled us to analyse the structure and function of a root system to an extent unthought of in Haberlandt's day. We now have a detailed picture, not only of the pattern of absorption of different ions and water *along* the length of a root, but also of the pathways of water and solute flow from cell to cell *across* the root, and of the structural elements that may regulate this flow.

Study of the function of any organ calls for active collaboration between both anatomists and physiologists. This paper should therefore be read in conjunction with that of Russell and Clarkson (this volume), in collaboration with whom much of the work described here was performed. Our paper deals specifically with the subcellular structure of roots (in particular those of barley) as determined by transmission and scanning electron microscopy.

Anatomy of a barley root

The general anatomy of a nodal root axis of barley (*Hordeum vulgare*) is illustrated in Fig. 1; the anatomy of seminal root axes and lateral roots in the same region only differs in detail (Robards *et al.*, 1973). Water and solutes taken up by the root move across the cortex, endodermis and pericycle, into the xylem ducts and are then transported to the shoot system. The concept of the absorbing zone has needed modification since it has been found that different ions are taken up into the roots at various rates, depending upon the position along the root (Robards *et al.*, 1973). It has long been considered that the endodermis acts as the major constraint upon the movement of water and solutes from cortex to stele. Therefore, a detailed examination of endodermal structure and development is well justified.

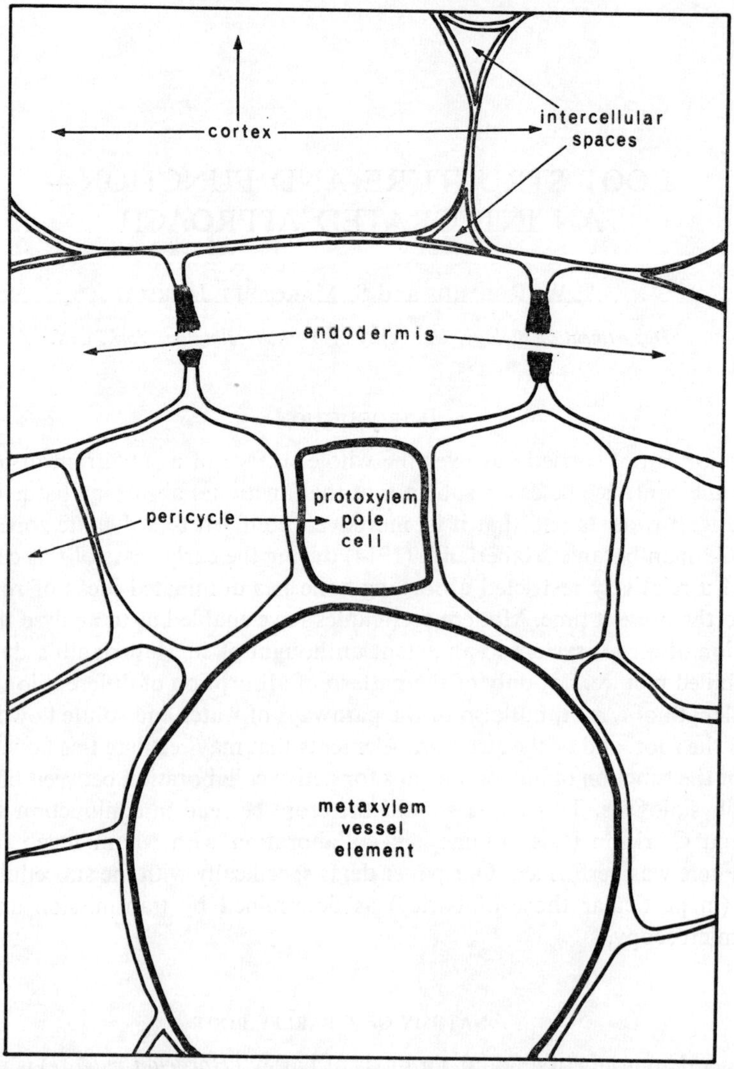

FIG. 2. Diagram to illustrate the relationships between cortical and stelar cells in a barley root. The Casparian bands completely block apoplastic movement of water and ions from cortex to stele. Water and solutes must pass through the cytoplasm of the endodermal cells: i.e., they must move into the symplasm to traverse the endodermal cells.

THE ENDODERMIS

The endodermis is a single layer of cells forming a continuous cylindrical sheath between cortex and stele (Fig. 2). A characteristic, and physiologically important, feature of endodermal cells is the specialized nature of their wall development. Four stages have been recognized in this development:

(i) *pro-endodermis*—young cells just cut off from the meristem and showing no unusual wall features;

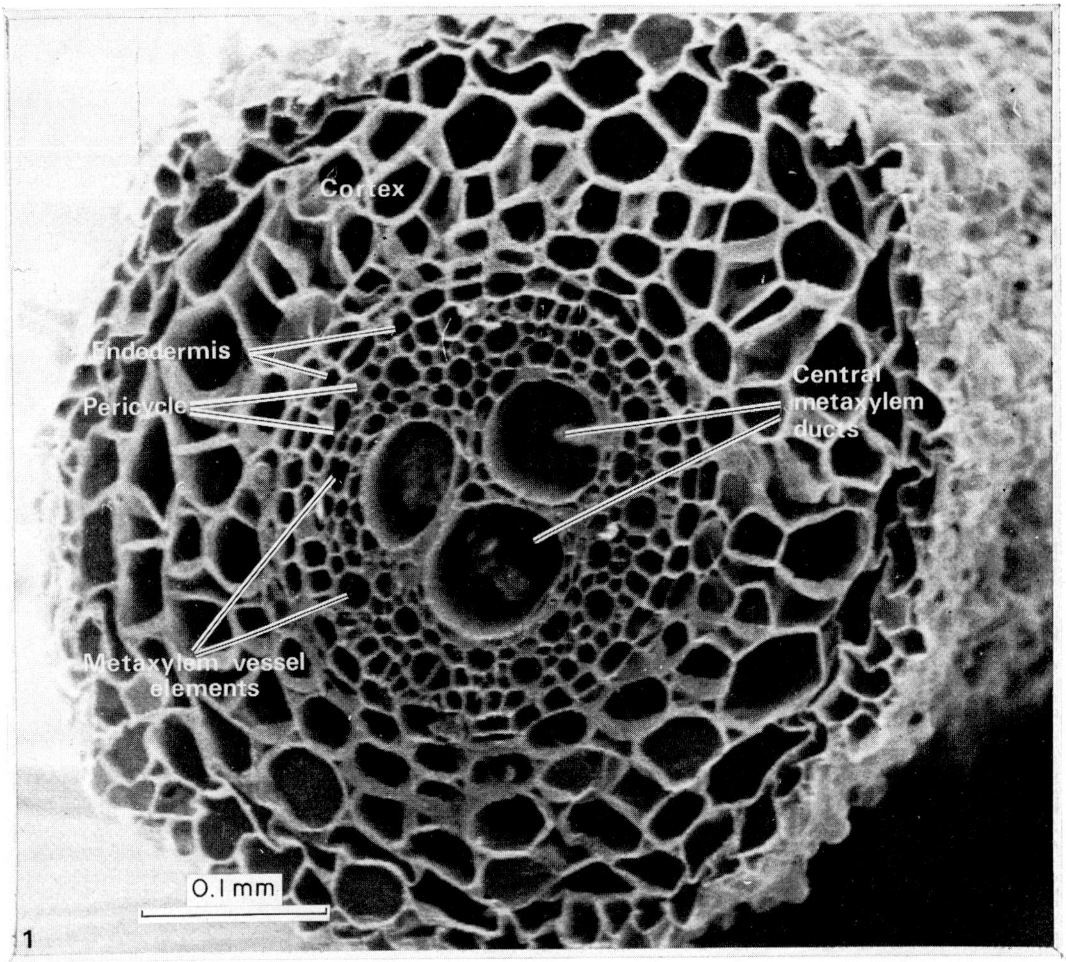

FIG. 1. Scanning electron micrograph of a transversely fractured nodal root axis of barley (*Hordeum vulgare*). Relative positions of cortex, endodermis, pericycle and metaxylem vessel elements are clearly demonstrated. Protoxylem pole cells are the small cells found in the pericycle between metaxylem elements and endodermis. This root was prepared by fixation in glutaraldehyde, dehydration in acetone, and critical point drying from liquid carbon dioxide. A thin layer of gold was sputter-coated on to the surface to minimize charging. (By courtesy of P. Crosby.) ×350.

(ii) *State I*—a heavy deposition of suberin and lignin-like material forms the Casparian band in the longitudinal and transverse radial (anticlinal) walls;

(iii) *State II*—a thin layer of suberin-containing material (suberin lamella) is deposited around the *whole* internal surface of the cell wall; and

(iv) *State III*—a heavy, lamellated, wall of cellulose is laid down internal to the suberin lamella, this may be either in the form of an even layer ('O'-type), or with far greater thickening on the inner tangential wall than the outer ('C'-type).

Although the endodermal cells of many plant roots do not progress beyond State I, others typically show the complete range, as exemplified by grasses, in which the 'C'-type State III endodermal cell wall thickening is highly characteristic. States I–III are shown diagrammatically in Fig. 3.

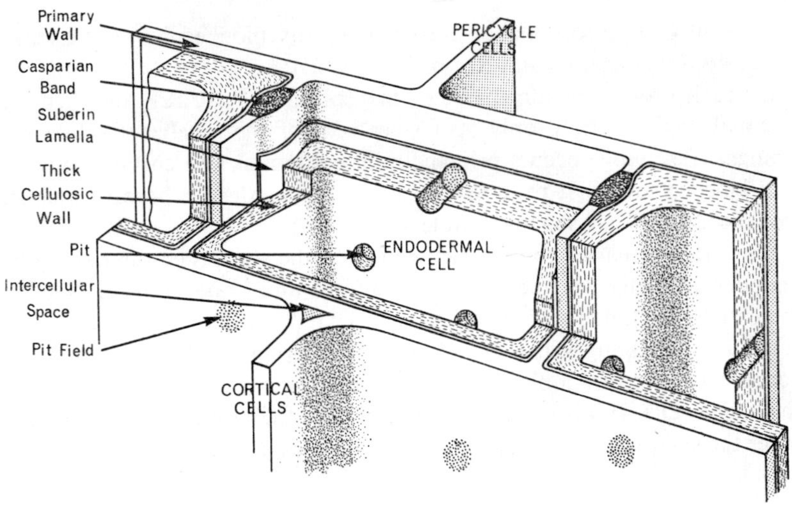

FIG 3. Diagram illustrating the structure of a State III endodermal cell of barley.

Within a very short distance of the meristem, the pro-endodermis can be recognized distinctly as the innermost rank of cells of the cortex. At a distance of 4–5 mm from the root tip the first signs of Casparian-band development may be seen. One of the most interesting features of State I is that the plasmalemma becomes extremely tightly bound to the Casparian band itself. When State II and State III development occurs, the two states are attained asynchronously, and it is not until about 30 cm or so from the root tip that all cells are found in the State III condition (Robards *et al.*, 1973).

TRANSLOCATION ACROSS THE ENDODERMIS

There are two possible pathways for the movement of water and solutes across the root: (i) via the free space of the cell walls (apoplasm) and (ii) via a cytoplasmic continuum from cell to cell (symplasm). Neither pathway need be exclusive over the whole distance from epidermis to xylem vessel elements.

Pro-endodermis

The pro-endodermis does not seem to provide any barrier to movement via the apoplasm (Fig. 4A), but whether this lack of a barrier allows gross leakage in the translocatory system is doubtful as xylem conducting elements are not fully formed at this stage (within the first millimetre or so from the root tip).

State I

The tight junction between the plasmalemma of the endodermal cell and the Casparian band apparently completely blocks movement via the apoplasm. This may be demonstrated in a number of ways:

(i) movement of electron-opaque tracers is totally blocked by the Casparian band (Figs. 4B, C) (Robards and Robb, 1972, 1974);
(ii) when barley roots are immersed first in a solution of $^{45}CaCl_2$ and then transferred to unlabelled solution, only 50–60 per cent of the labelled Ca taken up is exchanged. Exchange occurs principally in epidermal and cortical cells, indicating that labelled ions that penetrate beyond the endodermis become unavailable for free exchange (Clarkson and Sanderson, 1971);
(iii) if the cortex of barley roots is stripped from the stele, then the radial walls of the endodermal cells are the sites of fracture, and water uptake is considerably increased until the root has time to construct repair material (Clarkson and Sanderson, 1974). This suggests that, when the Casparian band is no longer available to act as a barrier to apoplastic flow, water can move unimpeded into the walls of the stelar cells. In an intact endodermal cell, therefore, *all* ions and water must enter the symplasm if they are to move into the stele.

State II

As development of the endodermis moves from State I to State II there has been found (in both barley and marrow) to be a strong correlated reduction in the amounts of water and calcium moved into the stele. Transport of both phosphorus and potassium, however, does not show such a decline (Robards *et al.*, 1973; Harrison-Murray and Clarkson, 1973; Graham *et al.*, 1974). These results are interpreted as indicating that the formation of the suberin lamella over the whole endodermal wall surface blocks movement of much of the water and most of the calcium (but not movement of phosphate and potassium). In other words, the lamella stops direct uptake into the symplasm of endodermal cells from the cortical free space. However, endodermal cell walls (including the suberin lamella) are penetrated by numerous plasmodesmata (Fig. 5) which allow ions and water to move from the *symplasm* of the cortex to that of the endodermis. Little is known about the permeability of the suberin lamella itself, but there is no evidence to suggest that it has a permeability sufficient to sustain the total observed rates of flow across endodermal walls.

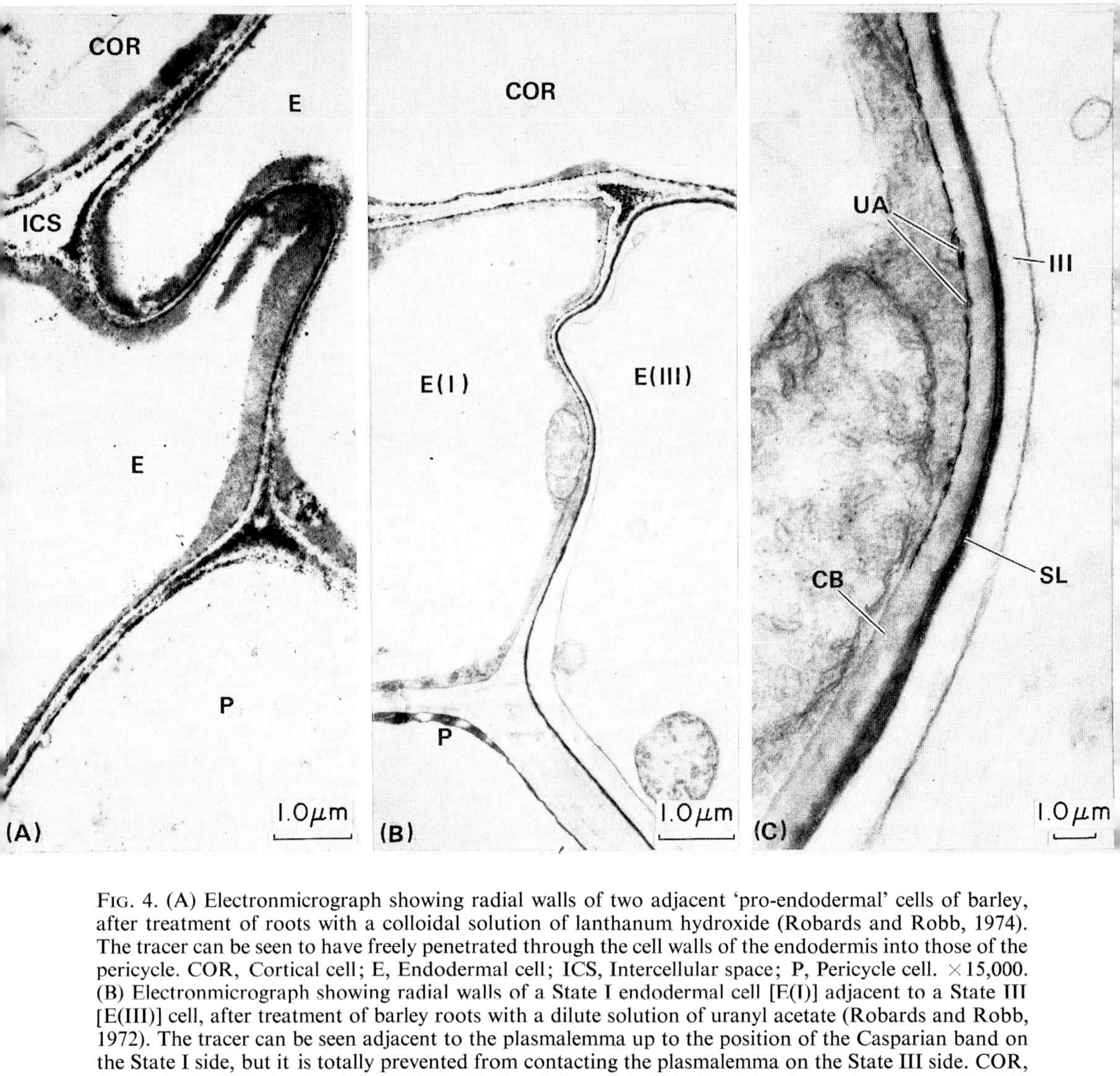

FIG. 4. (A) Electronmicrograph showing radial walls of two adjacent 'pro-endodermal' cells of barley, after treatment of roots with a colloidal solution of lanthanum hydroxide (Robards and Robb, 1974). The tracer can be seen to have freely penetrated through the cell walls of the endodermis into those of the pericycle. COR, Cortical cell; E, Endodermal cell; ICS, Intercellular space; P, Pericycle cell. ×15,000. (B) Electronmicrograph showing radial walls of a State I endodermal cell [E(I)] adjacent to a State III [E(III)] cell, after treatment of barley roots with a dilute solution of uranyl acetate (Robards and Robb, 1972). The tracer can be seen adjacent to the plasmalemma up to the position of the Casparian band on the State I side, but it is totally prevented from contacting the plasmalemma on the State III side. COR, Cortical cell; P, Pericycle cell. ×15,000. (C) Detail from (B). UA, Crystals of uranyl complex; III, Cellulosic wall of State III cell; SL, Suberin lamella; CB, Casparian band. ×90,000.

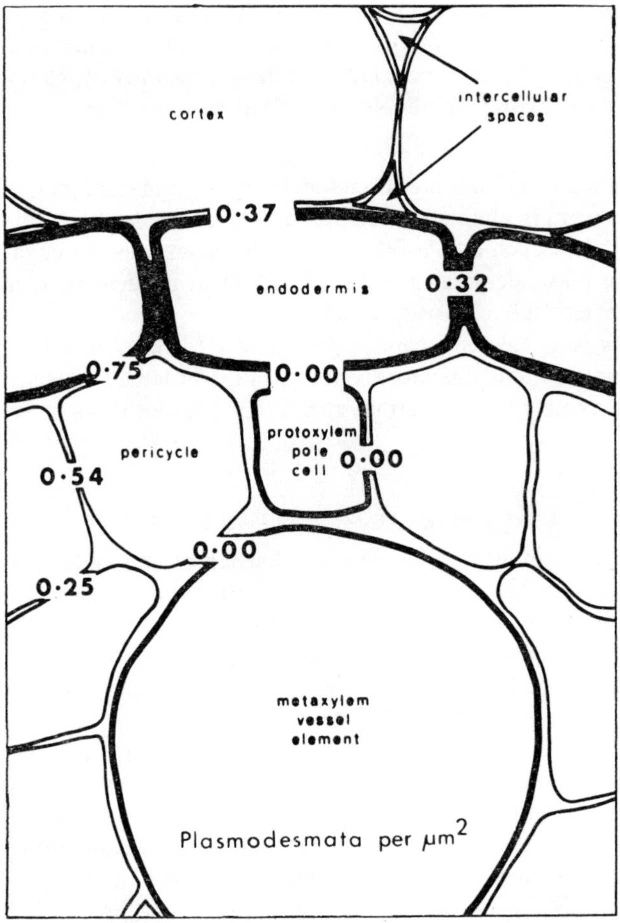

FIG. 5. Diagram illustrating plasmodesmatal frequency through endodermal and adjacent walls.

State III

Deposition of a thick cellulosic wall at this stage seems to be an irrelevance so far as translocatory function is concerned. The wall is heavily pitted, the pit membrane being perforated by plasmodesmata laid down at an earlier stage of development.

In some plants the otherwise complete sheath of State III endodermal cells is sometimes interspersed with cells in State I: these latter cells usually overlie the xylem poles, and have been referred to as passage cells. Passage cells may become a permanent feature of mature roots (as in Iridaceae) or they may simply be retarded cells (as in the case of barley, where they all eventually achieve State III structure). The question arises whether passage cells could account for the rates of flow observed across mature roots. This is unlikely since in barley roots showing total State III development, significant flow rates can be measured. In the case of phosphate it can be calculated that the total flux across the barley endodermis (assuming transport only via passage cells) is approximately 3.5×10^{-12} mole sec^{-1} mm^{-2}. This flux is three hundred times greater than fluxes observed in *Nitella translucens* and seems most improbable (Clarkson *et al.*, 1971).

Conclusions to be drawn from this section of the work are that:

(i) water and some ions (e.g. phosphate and potassium) move in appreciable amounts across the State II and State III endodermis, i.e. in mature regions of the root;
(ii) such movement is symplastic and probably takes place via plasmodesmata;
(iii) water and ions such as phosphate and potassium must be capable of movement across the plasmalemma into the cytoplasm of cortical cells, and then of symplastic movement from cortex to stele;
(iv) other ions (e.g. calcium) cannot gain access to the symplasm of the suberized endodermal cell and, therefore, once the suberin lamella has separated the outer free space from the endodermal symplasm, translocation of such ions virtually ceases.

TRANSPORT FROM ENDODERMIS TO XYLEM

Once the translocates have overcome the barrier of the endodermis, by whatever means, they are channelled into xylem elements across the pericycle. In barley, the pericycle comprises a single layer of cells forming a continuous sheath between endodermis and xylem (Fig. 1). Over each xylem pole there is, in the pericycle, a protoxylem pole cell measuring 5–6 μm in the tangential direction. In the xylem itself, there are 5–8 metaxylem elements arranged in a ring, and 1–3 central metaxylem ducts. It must be assumed that the bulk of the upward flow of translocates occurs in the larger metaxylem elements (20 μm diameter for the vessels arranged in a ring; 50 μm for the central duct(s)). There is some evidence that protoxylem pole cells may become particularly heavily loaded with translocates (Clarkson and Sanderson, 1972), suggesting that these cells play a significant role in long distance transport. As opposed to the metaxylem elements, which remain thin-walled as far as 10 mm from the root tip, the protoxylem pole cells have virtually completed their development at this distance (Fig. 6). These protoxylem pole cells vacuolate rapidly, go through a period of intense Golgi activity and of cell wall thickening, and then, apparently by hydrolytic means, produce pitted wall thickenings (Fig. 7A). The significance of this developmental sequence is that:

(i) protoxylem pole cells appear to become open, conducting elements long before the metaxylem elements;
(ii) the ratio of wall thickening to pit areas having relatively close contact between the xylem lumen and the plasmalemma of adjacent living cells is presumably important in determining the potential loading velocity of the vessel elements (Figs. 7B, C), and
(iii) in the early stages of root development there appear to be temporary anomalies in the pattern of constraints imposed upon the inward-flowing solutes.

For example, in barley, the Casparian bands do not develop until 5–7 mm from the root tip: this means that there is a zone, up to 7 mm in length, lacking a barrier to centripetal transport via the apoplasm. Whether this zone represents any major by-pass of transport

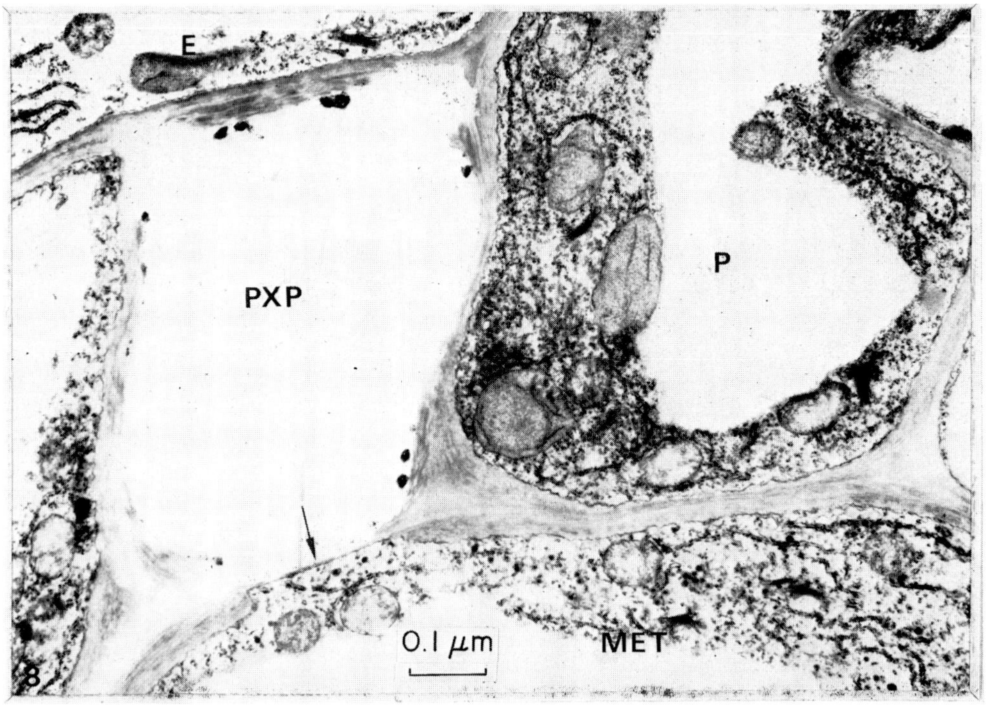

FIG. 6. Electronmicrograph of a transverse section of the pericycle of a barley root 10 mm from the tip. The protoxylem pole cell (PXP) is seen to be devoid of contents, and to have unevenly thickened walls which in some places (arrowed) show nothing but a thin pit membrane. This contrasts with earlier stages in development where the PXP wall is evenly thickened. The other pericycle cells (P) have normal cytoplasmic contents, as have the rapidly vacuolating metaxylem vessel elements (MET) which are presumed not to be fully functional at this stage. Endodermis = E. Roots were fixed in glutaraldehyde followed by osmium tetroxide; dehydrated through an acetone series (including uranyl acetate in the 70% stage); embedded in Spurr's resin; sectioned and post-stained in lead citrate. ×13,000.

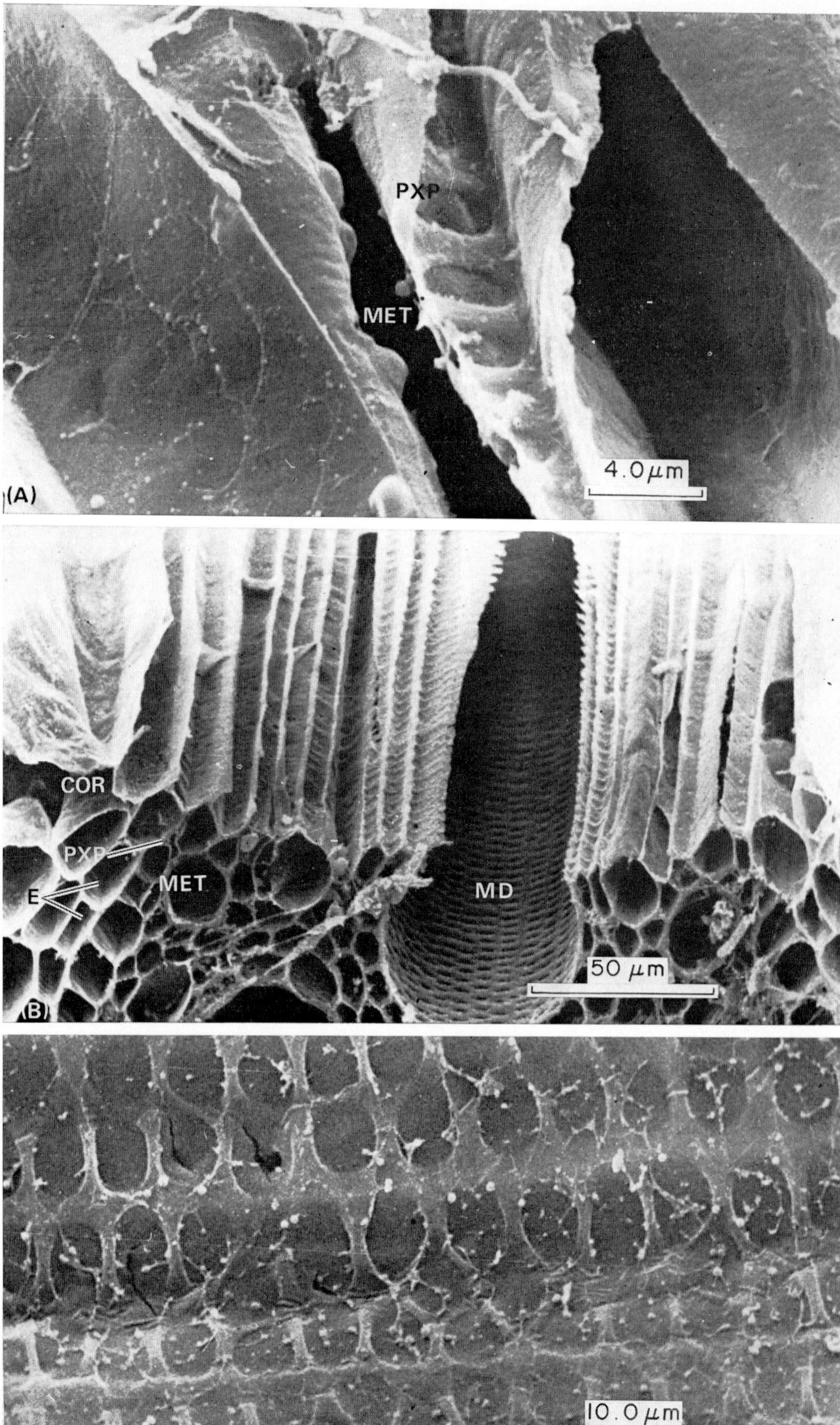

FIG. 7. (A) Scanning electron micrograph of an oblique longitudinal fracture through a protoxylem pole cell of barley, showing large pit areas in contact with underlying peripheral metaxylem vessel element (MET). Preparation as for Fig. 1. ×6250. (B) Scanning electron micrograph of an oblique fracture through a barley nodal root axis showing heavy pitting in the metaxylem central duct (MD). COR, Cortical cells; E, Endodermis; PXP, Protoxylem pole cell in pericycle; MET, Metaxylem vessel element. ×725. (C) Scanning electron micrograph showing pitting in the wall of a central metaxylem duct of barley root axis. (By courtesy of P. Crosby.) ×3100.

pathways along the main part of the root is not clear, but it seems evident that there would be no constraint upon symplastic or apoplastic movement from cortex towards differentiating xylem vessel elements. Any barrier to apoplastic flow would presumably be associated with plasmalemmata of the xylem vessel elements themselves and, in the case of protoxylem pole cells, the question arises whether the cell contents have already (totally) degenerated before endodermal Casparian bands have been deposited. Except in the youngest part of the root it appears that the xylem vessel elements are devoid of contents and, therefore, movement into such vessels must be controlled by the last living membrane across which the solutes pass.

We conclude that there is no effective barrier to the transport of water and ions across the root via the apoplasm until they arrive either at the Casparian band, or at the suberin lamella on the outer tangential wall of the endodermis adjacent to the cortex. Nor does there appear to be any major constraint upon movement from the inner tangential wall of the endodermis through to the pericycle and xylem vessel elements. Consequently, we place great emphasis upon the opportunities for transport across inner and outer tangential walls of endodermal cells via the symplasm. It should be mentioned that, in those roots where the hypodermis becomes suberized, then this barrier probably forms an important restriction upon access to the cortical symplasm, and in such cases, may be of even greater importance than the barrier of the endodermis (Ferguson, 1974).

Plasmodesmata

It has been shown that some ions can move in appreciable quantities across State II and State III endodermal cells, and we have implied that transport must occur through plasmodesmata rather than the free space of cell walls. Questions to be answered, therefore, are: are plasmodesmata present? what is their distribution? do they appear to be structurally capable of acting as conducting pores? and, finally, are they large enough and present in sufficient numbers to accommodate observed rates of flow?

The structure of plasmodesmata has been reviewed elsewhere (Robards, 1971, 1975). One controversial aspect of the plasmodesmata theory of transport is the exact nature and dimensions of the conducting tubule. In endodermal cell walls of barley roots, in common with many other situations, the plasmodesmata are plasmalemma-lined channels, approximately 60 nm in diameter. They each contain a core which is commonly supposed to be a strand of endoplasmic reticulum continuous from cell to cell. This view has been supported by high-voltage electron microscopy of thick sections (Robards and Jackson, in preparation). A striking feature of endodermal cell walls is the degree to which they are perforated by plasmodesmata (Fig. 5): frequencies of half a million per square millimeter are common in barley, while even greater numbers have been found in endodermes of marrow and maize, as well as in many other locations (Robards, 1975). By making assumptions about plasmodesmatal dimensions and by determining rates of translocation across the inner tangential wall of the endodermis, it is possible to determine approximate flow rates and hydraulic conductivities appropriate to a single plasmodesma. For example, the flux of water across the inner tangential wall of barley root endodermal cells in State II or State III has been determined as 9.1×10^{-5} cm^3 mm^{-2} hr^{-1} (Clarkson *et al.*, 1971). If the frequency of plasmodesmata through the inner

TABLE 1. EFFECT OF VARIATION IN SIZE OF THE CONDUCTING PORE THROUGH A PLASMODESMA ON VOLUME CHANGES AND FLOW VELOCITY

Assumptions	Surface area of pore mm^2	Volume of pore mm^3	Volume changes sec^{-1}	Flow velocity $\mu m\ sec^{-1}$
1. Plasmodesmatal pore (desmotubule) has radius = 5 nm (Robards, 1971)	7.9×10^{-11}	3.9×10^{-14}	954	477
2. Plasmodesmatal pore is an open channel, unrestricted by a desmotubule—radius = 10 nm	3.1×10^{-10}	1.6×10^{-13}	239	120
3. As in case (1), but assuming 50% shrinkage during processing—*in vivo* pore radius = 10 nm	3.1×10^{-10}	1.6×10^{-13}	239	120
4. As in case (2), but assuming 50% shrinkage during processing—*in vivo* pore radius = 20 nm	1.3×10^{-9}	6.3×10^{-13}	60	30

It has been experimentally determined that the flow of water through a single plasmodesma in the inner tangential wall of a barley root is $3.75 \times 10^{-11}\ mm^3\ sec^{-1}$ (see Clarkson *et al.*, 1971).

endodermal wall is 6.73×10^5 mm^{-2} then the flow through a single plasmodesma is approximately 3.75×10^{-11} mm^3 sec^{-1}. Given that the diameter of the plasmodesmatal pore is 10 nm, and the length, 0.5 μm, then its volume is 3.93×10^{-14} mm^3. Consequently the observed flow rate represents almost 1000 volume changes every second, or a velocity of about 500 μm per second. These values appear high to say the least. However, the values depend very much on the exact radius of the plasmodesmatal pore (Table 1), and with the techniques at present available, it has not been possible to determine the radius (or the structure of the pore) with a sufficient degree of accuracy. Calculations of the hydraulic conductivity of plasmodesmata compared with similar measurements from other sources (Table 2) show that the resistance to flow offered by plasmodesmata is far less than offered, for example, by cell membranes. The solute-concentration gradient required to drive the observed flow of water across the endodermis would not be unduly

TABLE 2. PERMEABILITY OF PLASMODESMATA COMPARED WITH SOME OTHER BARRIERS BETWEEN PLANT CELLS

Barrier	Hydraulic conductivity (μm s^{-1} bar^{-1})	Source
Plasmodesmata	13–1300	Clarkson *et al.* (1971)
Plasmodesmata	42	Marks (1973)
Biological membranes	0.37–270	Clarkson (1974)
'Plant cells'	0.001–0.01	Briggs (1967)

Data from Robards (1975).

high (Clarkson *et al.*, 1971). The idea that the endoplasmic reticulum is continuous from cell to cell via the plasmodesmata is an attractive one, not least because the contents of the reticulum then constitute the symplasm. However, this idea poses difficulties, because of the need for a bidirectional flow of solutes through the plasmodesmata (Clarkson and Robards, 1975).

CONCLUSIONS

The passage of water and solutes from the external solution into the upward conducting vessel elements of the xylem is primarily determined (at least in barley) by the ability of the particular molecule or ion to gain entry to the endodermal symplasm. In those plants which rapidly develop a suberized hypodermis the relative importance of the two barriers (hypodermis and endodermis) is probably inverted. Numerous plasmodesmata link endodermal cells to each other, as well as to neighbouring cortical and pericycle cells. Calculations support the contention that plasmodesmata can carry the centripetal flow of water and solutes through the symplast, but much more work remains to be done to establish the precise nature and dimensions of the conducting tubules, as well as the viscosity of the pore fluid.

Acknowledgements: We are glad to acknowledge the assistance and advice of D. T. Clarkson; P. Crosby who carried out most of the S.E.M. work; and finally Christine Johnson for general technical assistance and help in the preparation of the Figures. This work has been supported by grants from the Agricultural Research Council and Science Research Council.

References

Briggs, G. E. (1967) *Movement of Water in Plants*, Blackwell, Oxford.

Clarkson, D. T. (1974) *Ion Transport and Cell Structure in Plants*, McGraw-Hill, London.

Clarkson, D. T. and Sanderson, J. (1971) Inhibition of the uptake and long-distance transport of calcium by aluminium and other polyvalent cations. *J. exp. Bot.* **23,** 837–851.

Clarkson, D. T. and Sanderson, J. (1972) The time-course of the accumulation of calcium in the various tissues in the roots of barley seedlings. *Agric. Res. Council Letcombe Lab. Rep. 1971* 4–6.

Clarkson, D. T. and Sanderson, J. (1974) The endodermis and its development in barley roots as related to radial migration of ions and water. In *Structure and Function of Primary Root Tissues*, ed. Kolek, J., Slovak Academy of Sciences, Bratislava.

Clarkson, D. T. and Robards, A. W. (1975) The endodermis, its structural development and physiological role. In *The Development and Function of Roots*, ed. Torrey, J. G. and Clarkson, D. T., Proceedings of the 3rd Cabot Symposium, 1974, Academic Press, London.

Clarkson, D. T., Robards, A. W. and Sanderson, J. (1971) The tertiary endodermis in barley roots: fine structure in relation to radial transport of ions and water. *Planta* **96,** 292–305.

Ferguson, I. B. (1974) Ion uptake and translocation by the root system of maize. *Agric. Res. Council Letcombe Lab. Rep, 1973*, 13–15.

Graham, J., Clarkson, D. T. and Sanderson, J. (1974) Water uptake by the roots of marrow and barley plants. *Agric. Res. Council Letcombe Lab. Rep. 1973*, 9–12.

Haberlandt, G. (1914) *Physiological Plant Anatomy*, Translation from 4th German edition. Macmillan, London.

Harrison-Murray, R. S. and Clarkson, D. T. (1973) Relationships between structural development and the absorption of ions by the root system of *Cucurbita pepo*. *Planta* **114,** 1–16.

Marks, I. (1973) Ultrastructural studies of minor veins. Ph.D. thesis, Queens University, Belfast.

Robards, A. W. (1971) The ultrastructure of plasmodesmata. *Protoplasma* **72,** 315–323.

Robards, A. W. (1975) Plasmodesmata. *A. Rev. Pl. Physiol.* **26,** 13–29.

Robards, A. W. and Robb, M. E. (1972) Uptake and binding of uranyl ions by barley roots. *Science N.Y.* **178,** 980–982.

Robards, A. W. and Robb, M. E. (1974) The entry of ions and molecules into roots: an investigation using electron-opaque tracers. *Planta* **120,** 1–12.

Robards, A. W., Jackson, S. M., Clarkson, D. T. and Sanderson, J. (1973) The structure of barley roots in relation to the transport of ions into the stele. *Protoplasma* **77,** 291–312.

METABOLIC PROCESSES IN ROOTS RELATED TO ABSORPTION AND TRANSPORT OF PHOSPHATE

B. C. LOUGHMAN

Department of Agricultural Science, Parks Road, University of Oxford, U.K.

INTRODUCTION

Although some ions, notably phosphate, sulphate, nitrate and ammonium are rapidly metabolized during or after entry into the root, relatively little attention has been given to the biochemical aspects of the processes of absorption and transport of these ions. Most modern texts ignore completely the biochemical aspects of ion utilization and concentrate on the biophysical components of the systems as applied to the absorption of cations not directly involved in metabolic changes. The extent of the metabolic component can be demonstrated in the case of phosphate where, within 5 sec of entry into roots of young barley plants (12-day), a third of the radioactive inorganic phosphate fed is incorporated into nucleotides (Loughman and Russell, 1957; Miettenen and Savioja, 1958; Kursanov and Vyskrebentseva, 1960).

The normal soil environment contains inorganic orthophosphate at concentrations ranging from 1×10^{-7} M to 10^{-5} M although under conditions of heavy fertilizing of some crops, e.g. hops, levels an order of magnitude higher have been reported. However, the fact that plants normally remove phosphate from solutions approaching micromolar has particular implications in considerations of the mechanisms by which absorption and transport occur. The current interest in two systems of uptake for a number of ions, one saturated at low and the other at high concentrations is clearly of only academic interest in the study of the relationship between the plant root and the external bathing solution in the soil. Because of the low concentrations of phosphate available, and its prime importance to the economy of the plant it is perhaps to be expected that phosphate uptake is more sensitive to changes in environmental factors, e.g. concentrations of other ions, pH, oxygen availability, temperature, and the imposed effects of herbicides in the field or of metabolic inhibitors in the laboratory would have more effect on phosphate uptake than on other ions.

Over periods long enough to allow measurable transport of phosphate to the shoots of young barley plants, where it is incorporated into the photosynthetic phosphorylation system, i.e. greater than 15 min, it can be readily demonstrated that the phosphate is transported in the xylem primarily as inorganic orthophosphate. None of the labelled nucleotides, sugar nucleotides or sugar phosphates present in the root at this time are detectable in the transport system. This suggests that phosphate absorption is an integrated process involving esterification and that some organic form is hydrolysed prior to the transfer of inorganic phosphate via the xylem to the shoot.

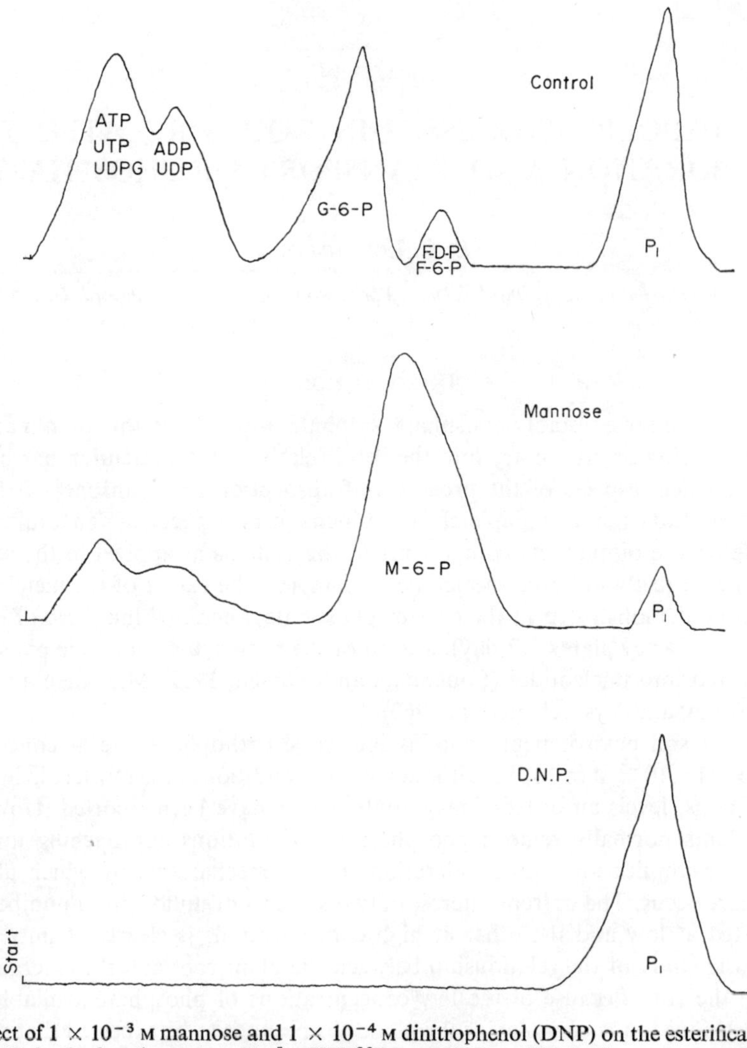

FIG. 1. Effect of 1×10^{-3} M mannose and 1×10^{-4} M dinitrophenol (DNP) on the esterification pattern in roots of 18-day barley plants. 1×10^{-5} M $KH_2{}^{32}PO_4$ at 20°C at pH 5.5. Chromatogram developed in *tert* butanol/picric acid/water (80 ml:2 g:20 ml).

A NON-METABOLIC COMPONENT IN PHOSPHATE ABSORPTION

It is not easy to establish the presence of a non-metabolic component in phosphate absorption because of the rapidity with which esterification takes place. Thus in 10-day barley plants kept at 0°C, the small amount of phosphate absorbed by the roots is rapidly incorporated into organic compounds and the pattern of incorporation is similar to that in plants kept at 20°–25°C (see Fig. 1). Nevertheless a non-metabolic component has been demonstrated by inhibiting the metabolic component with 2,4-dinitrophenol (DNP). In 18-day barley plants, for example, esterification of phosphate is severely inhibited by DNP at a concentration of 1×10^{-4} M but accumulation of phosphate in its inorganic form nevertheless takes place (Fig. 1). Moreover, transport of inorganic phosphate to the shoot is not completely stopped by DNP.

Positive evidence for a non-metabolic component has also been obtained in tissue other than roots. For example, at 0°C slices of potato tuber which have a much lower metabolic activity than barley roots accumulate inorganic phosphate irreversibly, although esterification is turned on almost simultaneously by raising the temperature to 25°C (Loughman, 1960). Results of experiments carried out with actively growing plants at low temperatures must be interpreted with particular care; such plants are clearly capable of phosphorylation at 1°C. Again, the ability of young barley plants to absorb and transport limited amounts of phosphate in the complete absence of oxygen by switching from aerobic to anaerobic metabolism also results in the incorporation of the entering phosphate into organic compounds (Larkum and Loughman, 1969). However, in plants treated with DNP (Fig. 1) the absence of esterification strongly suggests that phosphate enters the root and moves to the shoot as a result of processes not directly linked to metabolism. Although, under anaerobic conditions, the esterification process can be demonstrated at temperatures close to 0°C, the results obtained in the presence of DNP provide the best evidence available in support of non-metabolic absorption. It is therefore possible that a small but significant portion of normal absorption may occur by this passive route.

The metabolic component in phosphate absorption

The kinetics of phosphate absorption with respect to temperature are illustrated by the data of Fig. 2 for 18-day barley plants. The rate of phosphate absorption increases linearly with temperature, but between about 10° and 40°C the percentage of the absorbed phosphate incorporated into organic compounds remains almost constant. The overall process of phosphate absorption is clearly not limited by the esterification capacity of the root at normal root temperatures.

Since the metabolic component in phosphate absorption is sensitive to DNP it is concluded that the absorption process is linked directly to the process of oxidative phosphorylation (Jackson and Hagen, 1960). Absorption by barley plants is also sensitive to inhibitors of glycolytic phosphorylation such as iodoacetate (Loughman, 1974).

Both inhibitors block much of the transport of inorganic phosphate to the shoot, indicating unequivocally that the metabolic component is directly involved not only with the entry process but also with exit to the shoot irrespective of operation of a non-metabolic component. The overall metabolic process can be envisaged in three separate stages:

(i) esterification of inorganic phosphate at entry to the root;
(ii) utilization in metabolism of root tissues; and
(iii) hydrolysis of organic phosphates to inorganic phosphate before or at entry into the xylem.

The effect of DNP and iodoacetate also varies with the age of the plants. Thus, in 6-day barley plants, DNP inhibits phosphate absorption by only about 40 per cent whereas

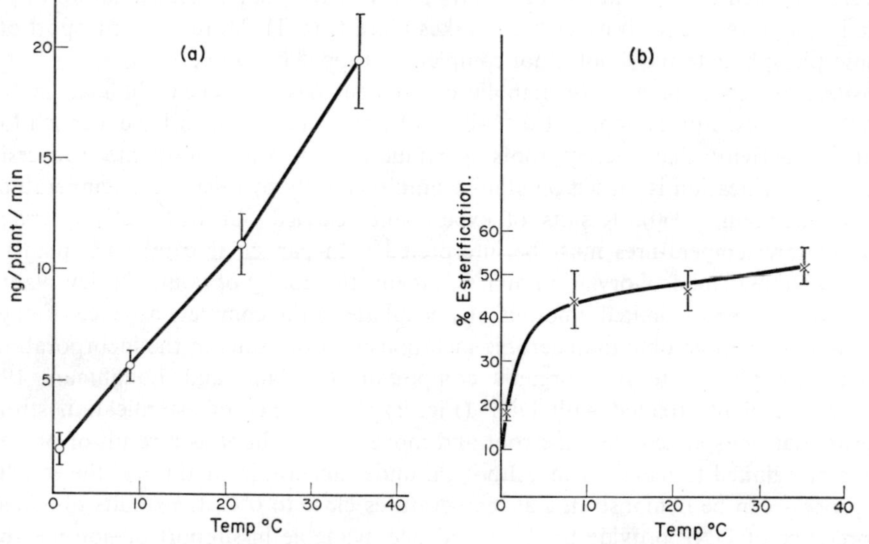

FIG. 2. Effect of temperature on (a) the rate of absorption of phosphate and (b) the percentage of phosphate esterified in 18-day barley plants. 1×10^{-5} M $KH_2{}^{32}PO_4$, pH 5.5. Mean of 4 replicates ± S.E.

in 18-day plants the inhibition reaches 95 to 98 per cent. Sensitivity to iodoacetate, on the other hand, declines with the age of the plant. These observations, incidentally, indicate that changes in the balance between glycolytic phosphorylation and oxidative phosphorylation may be important during the development of the root.

ENTRY OF PHOSPHATE INTO BARLEY ROOTS

Further evidence for the involvement of metabolic reactions in the process has come from experiments in which the energy supply is increased by means of sugars provided with the phosphate in the bathing solution. In barley, exogenous glucose increases the uptake of phosphate slightly, but has a negligible effect on the pattern of compounds into which the phosphate is incorporated. Exogenous mannose, on the other hand, although it has little effect upon the uptake of phosphate (Fig. 3) causes a marked change in the pattern of incorporation (Fig. 1). By using both labelled mannose and phosphate, it can be shown that the greater part of the label appears in mannose-6-phosphate (M-6-P) at the expense of nucleotides, glucose and fructose phosphates and inorganic phosphate. The build-up in M-6-P is due to negligible activity of the enzyme phosphomannose-isomerase.

A novel feature of the mannose effect is that the sugar must be present with the phosphate during entry into the root; pretreatment of roots with mannose has little effect upon subsequent phosphate entry (Fig. 3). This observation implies that mannose acts at an early stage of the absorption process. Whether the initial step is at the cell membrane or the mitochondrial membrane needs to be established, but it is clear that

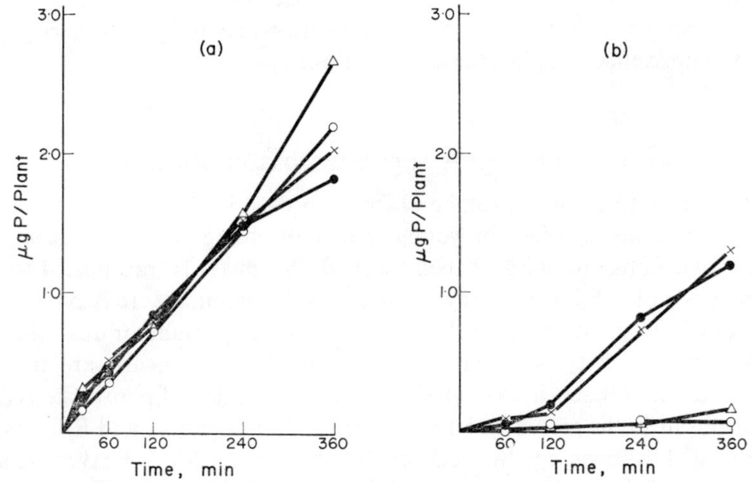

FIG. 3. Effect of mannose on (a) phosphate absorption by roots of 18-day plants of barley (*Hordeum vulgare* L var. Proctor) and (b) transport of phosphate to the shoots. Absorption from 1×10^{-5} M $KH_2{}^{32}PO_4$ at 20°C at pH 5.5. ×—× Control; ○—○ 1×10^{-3} M mannose added with the phosphate; ●—● 1×10^{-3} M mannose added 1 hr before phosphate; △—△ 1×10^{-3} M mannose added 1 hr before phosphate and with the phosphate.

the sugar must be in close proximity or in direct contact with a membrane component concerned in the transfer of phosphate from ATP to the initial sugar acceptor(s).

TRANSPORT OF PHOSPHATE FROM ROOT TO SHOOT IN BARLEY

A further and highly significant feature of the effect of mannose on phosphate absorption is that the change in metabolic pattern is accompanied by a severe inhibition of phosphate transport to the shoot (Fig. 3). Again mannose must be supplied to the plants together with the phosphate; if supplied before the phosphate, mannose has no effect on subsequent transport of phosphate at least over a 6 hr period (Fig. 3).

The effect of the exogenous mannose is rapid, and at a concentration of 1×10^{-2} M in the bathing solution, transport of phosphate to the shoot ceases within a few minutes although entry of phosphate into the root continues at a constant rate. The extreme rapidity of the mannose effect again implies that the action of the mannose is exerted at an early stage in the phosphate transfer system across membranes. These results support strongly the view already expressed that the metabolic component is a necessary prerequisite for the subsequent transport of inorganic phosphate out of the root. The mannose effect is reversible and if plants treated with mannose (in which phosphate transport has ceased) are transferred into a mannose-free solution the metabolic pattern in the roots reverts over a period of about 1 hr to that characteristic of control plants; transport to the shoot is also restored. The reversible nature of the phenomenon suggests that a re-equilibration of the interconversion of phosphorylated hexoses occurs probably as a result of flow through the previously rate-limiting phosphomannoseisomerase system. In this connection it is pertinent that in species (such as mung bean and other legumes) possessing a higher phosphomannoseisomerase activity than barley there is no significant mannose effect on phosphate absorption.

PHOSPHATE ABSORPTION AND TRANSPORT IN OTHER SPECIES

The particular features of phosphate absorption described for barley are not readily demonstrated in other species. In young plants of mung bean, for example, under comparable conditions one third of the absorbed phosphate is transported to the shoot in 6 hr but it is not inhibited to any great extent by mannose. It is possible that the glycolytic contribution to the overall process is not important or that the phosphomannoseisomerase activity is not limiting. Although mung beans are insensitive to mannose they are highly sensitive to DNP, e.g. the absorption of phosphate is reduced to less than 2 per cent in the presence of 1×10^{-4} M DNP (Loughman and Robertson, 1972).

In larger seeded legumes, e.g. the field bean (*Vicia faba* cv. Maris Beaver) no significant transport of phosphate to the shoot occurs during the first 2 weeks of growth. The seed reserves clearly provide all the phosphate required by the shoot even though the roots are absorbing and metabolizing phosphate at rates similar to those of barley plants of the same age. Removal of the cotyledons stimulates the capacity for upward transport from the root as shown in Table 1.

TABLE 1. EFFECT OF COTYLEDON REMOVAL ON TRANSPORT OF PHOSPHATE TO THE SHOOT OF FIELD BEAN (*Vicia faba* cv Maris Beaver)

Time (hr)	Per cent transport to the shoot: 8 day control plants	Per cent transport to the shoot: 8 day plants decotyledonized at 6 days
1	0.25	1.25
2.5	0.25	2.00
4.5	1.50	6.00
9	1.00	17.00
20	4.00	42.50

Treatment 1×10^{-5} M KH_2PO_4 at 20°C.
Surface-sterilized seeds germinated on moist filter paper, and raised in phosphate-free culture solution containing 1.5 mM $MgSO_4$; 1.5 mM $Ca(NO_3)_2$; 4 mM KNO_3; 1 mM K_2SO_4 and 2 mM $NaNO_3$.

Roots of intact 8-day plants of field bean absorb phosphate from 1×10^{-5} M KH_2PO_4 at a high rate but only 4 per cent is transported to the shoot in 20 hr. If the cotyledons are removed at 6 days, the plants grow subsequently at a slower rate than the control but 2 days later the transport capacity of the decotyledonized plants increases to a value of 42 per cent over the 20-hr absorption period (Robinson and Loughman, unpublished).

The inability of intact seedlings of field beans to transport phosphate from root to shoot is surprising and suggests that the requirement for externally supplied phosphate in the early stages of growth is minimal. This species often shows little response to phosphate fertilizers in the field and the timing of application of fertilizer would be crucial if maximal economy is to be achieved.

Phosphate absorption in boron-deficient roots

A further factor which affects the initial stages of phosphate entry across the root membrane is the nutritional status of the root. A comparison of phosphate uptake by roots of boron-deficient and non-deficient plants of field bean has yielded interesting results. The comparison was made on the terminal 0.5 cm of roots from 7-day plants grown in a complete nutrient solution and from plants grown in the complete medium for 5 days and then transferred to boron-free medium for a further 2 days. During the latter 2 days the roots stopped growing. As expected, the capacity for phosphate absorption was reduced in the boron-deficient roots as was that of rubidium, chloride and sulphate. However, a feature of the phosphate absorption not shared by the other ions tested was that it could be substantially restored by addition of 1×10^{-6} M boric acid to the root environment either at the time of addition of the radioactive phosphate or 30 min before. This reversal was found at all concentrations of phosphate studied (10^{-7} M–10^{-2} M). The effect of boron in restoring the absorption capacity may well prove to be specific for phosphate and probably involves a rapid reorganization of the carrier system for phosphate ions across membranes (Robertson and Loughman, 1974).

The effect of boron is clearly not due to a reduction in efflux of phosphate resulting from reduced permeability of the membranes concerned since the proportion of absorbed P released during a 1-hr efflux period was identical in control and deficient plants although less was absorbed by the deficient plants.

It is possible that boron functions as an essential component of the carrier system for transport of phosphate across membranes of either the protoplast or the mitochondria. Boron could also be essential for normal oxidative phosphorylation within the root and if, as is possible, the intake of phosphate is coupled to phosphorylation steps in the electron transport chain the absence of boron might then reduce the flux of phosphate through the system.

Other evidence suggests that phosphate absorbed by the root in the terminal 0.5 cm is less readily transferred to the rest of the root in boron-deficient roots. Boron may therefore be involved in transport of phosphate along roots as well as in the entry of this ion into roots.

Incorporation of phosphate into organic compounds is much reduced in the terminal 0.5 cm of roots of boron-deficient plants and after exposure of roots for 1 hr to labelled phosphate there is a build up of label in the hexose phosphate fraction. This build up is very similar to that found in barley plants treated with the glycolytic inhibitor iodoacetate. In field beans, however, iodoacetate produces an esterification pattern similar to that found in boron-deficient roots implying that lack of boron causes a reduction in glycolytic activity and thereby affects the level of triosephosphate dehydrogenase directly.

CONCLUSION

In discussing the metabolic events associated with the accumulation of phosphate by roots of different species the intention has been to show that each stage in the overall process can be studied separately in whole plants by biochemical methods. In addition, the information obtained at each stage can be applied to the investigation of systems in different physiological states, e.g. mineral deficiency. Furthermore it is clear that species show very considerable differences in the metabolic pattern associated with absorption and transport of phosphate, and that the stage of root development is an important factor.

The mobilization of phosphate reserves during the early stages of seedling growth is important in the control of phosphate utilization and affects the subsequent pattern of seedling development. The greater part of the reserve phosphate in cereal seeds is in the form of inositol hexaphosphate. In *Avena* this fraction represents 53 per cent of the total phosphate of the seed and inorganic orthophosphate only 12 per cent. Three days after seed-sowing, however, the inositol phosphate contains 30 per cent of the total phosphate and after 8 days the organic form has been completely converted to inorganic (Hall and Hodges, 1966). This redistribution of phosphate has a considerable effect on the metabolism of the root and it is not surprising that the pattern of this metabolism and its sensitivity to inhibitors changes rapidly with root age. As a result of such changes it appears that in cereals the capacity of the root to transport phosphate to the shoot is present at an early stage in germination and that the capacity increases with seedling age, whereas in legumes which possess large seeds onset of any significant transport capacity is delayed for a period of 2–3 weeks. The widely varying patterns of utilization of phosphate found in young seedlings of different species confirms the need for detailed studies of the individual stages of the overall absorption process. The interplay between processes of internal and external supply of phosphate, transport of phosphate across membranes, esterification within the root, transport to the xylem, incorporation of phosphate into photosynthetic intermediates and finally the effects of atmospheric and soil environment on each of these processes can best be examined by a combined physiological and biochemical approach as exemplified by the experiments described here.

REFERENCES

HALL, J. R. and HODGES, T. K. (1966) Phosphorus metabolism of germinating oat seeds. *Pl. Physiol. Lancaster* **41,** 1459–1464.

JACKSON, P. E. and HAGEN, C. E. (1960) Products of orthophosphate absorption by barley roots. *Pl. Physiol. Lancaster* **35,** 326–332.

KURSANOV, A. and VYSKREBENTSEVA, E. L. (1960) Primary inclusion of phosphate in root metabolism. *Fiziol. Rast.* **7,** 274–282.

LARKUM, A. W. D. and LOUGHMAN, B. C. (1969) Anaerobic phosphate uptake by barley plants. *J. exp. Bot.* **20,** 12–24.

LOUGHMAN, B. C. (1960) Uptake and utilization of phosphate associated with respiratory changes in potato tuber slices. *Pl. Physiol. Lancaster* **35,** 418–424.

LOUGHMAN, B. C. (1974) The metabolism of phosphate in young roots. In *Structure and Function of Primary Root Tissues.* ed., KOLEK, J., VEDA, Slovak Academy of Sciences, Bratislava.

LOUGHMAN, B. C. and ROBERTSON, G. A. (1972) Boron and phosphate metabolism in higher plants. Atti de IX Simposia Internatzionale di Agrochimica su "La fitonutrizione oligominerale". Punta Ala 2–6 Oct. 1972.

LOUGHMAN, B. C. and RUSSELL, R. S. (1957) The absorption and utilization of phosphate by young barley plants. *J. exp. Bot.* **8**, 280–293.

MIETTENEN, J. K. and SAVIOJA, T. (1958) Uptake of orthophosphate by pea plants. II. Identification of organic phosphate esters formed. *Acta. chem. scand.* **8**, 1693–1698.

ROBERTSON, G. A. and LOUGHMAN, B. C. (1974) Reversible effects of boron on the absorption and incorporation of phosphate in *Vicia faba* L. *New Phytol.* **73**, 291–298.

REGULATION OF ION TRANSPORT IN THE WHOLE PLANT

J. F. SUTCLIFFE

School of Biological Sciences, University of Sussex, Falmer, Brighton, U.K.

DEVELOPMENT OF IDEAS ON ION TRANSPORT

The period 1923–1924, in which the Society was formed, was an important watershed in the development of ideas on the mechanisms of uptake and transport of ions in plants. During the first quarter of the century there had been an increasing appreciation of the need to apply to biology the precise analytic methods of chemistry. This trend was stimulated by such workers as Jacques Loeb, Pfeffer and Overton who attempted to account for the movement of ions across cell membranes in physicochemical terms. The work of Stiles and Kidd (1919) was one of the first detailed quantitative investigations of ion absorption in plant tissues. It led in 1924 to the publication of Stiles' *New Phytologist* Reprint entitled 'Permeability', which remained the only important monograph on the subject for twenty years. In it he sought to account for ion accumulation in terms of adsorption and the establishment of Donnan equilibria.

On the other hand, Hoagland and Davis (1923) had just embarked on their classical experiments with the alga, *Nitella*, which were to lead to the establishment of links between solute absorption, growth and metabolism, and influence so profoundly the future course of research in this area.

In the 1920's, there was renewed interest too in the relationship between water and salt uptake in intact plants. The view held by Sachs (1887) that ions are taken up passively in the transpiration stream, and carried upwards by it into the leaves where they accumulate, was being increasingly questioned. This mechanism did not satisfactorily account for the selectivity of absorption that de Saussure and others had noticed and the accumulation of ions in the leaves of submerged water plants was also difficult to explain. Muenscher (1922), working at Cornell found that there was no clear quantitative relationship between transpiration and salt absorption in barley plants grown under various conditions. He showed that salt uptake was more closely related to the intensity of photosynthesis, and hence of growth, than to water absorption. Hoagland, who had already worked on the mineral nutrition of barley, returned to this problem in the 1930's, and in collaboration with Broyer (Hoagland and Broyer, 1936) carried out their well-known research, which together with the work on *Nitella* formed the basis of Hoagland's Prather Lectures at Harvard published in 1944. These lectures did more than anything else to stimulate my own interest in salt accumulation when I was a botany student in Leeds after the Second World War.

As Professor Heath has mentioned (p. 2), 1924 was the year in which Maskell left Rothamsted to take up an appointment in Trinidad. There he began his fruitful collaboration with Mason and later with Phillis which laid much of the foundation for present knowledge of phloem transport. They published a series of papers in the *Annals of Botany* from 1928 onwards in which they developed the important concept of sources and sinks. This idea was taken up enthusiastically, in a somewhat different guise, by Steward who at the SEB Symposium on Active Transport held at Bangor in 1953 described unpublished experiments of Mrs. Steward and others which showed the importance of 'growth centres' in controlling the pattern of distribution of ions in the intact plant (Steward and Miller, 1954).

Although it had been widely assumed since Malpighi's day that solutes move upwards in the xylem and downwards in the phloem, this assumption still rested on inconclusive evidence in the early 1920's as Curtis (1923) pointed out. He showed that, in some woody twigs, solutes may move upwards in the phloem if the xylem is interrupted. In one of the earliest investigations with radioactive isotopes on plants, Gustafson and Darken (1937) demonstrated some upward movement in the phloem of intact stems. In contrast, Stout and Hoagland (1939) confirmed the classical view that ions move upwards from roots to leaves in a variety of herbaceous and woody plants, mainly, if not entirely, in the xylem. They showed that the picture may be complicated by lateral exchange of solutes between xylem and phloem when the two tissues are in close proximity. Later isotope studies by Biddulph and others (e.g. Biddulph *et al.*, 1958) established that some elements, notably phosphorus and sulphur, circulate in the plant from roots to leaves and back to roots again, whereas other elements, particularly calcium, which are relatively immobile in the phloem are transported upwards from the roots in the transpiration stream to the leaves where they are deposited.

The general picture that has emerged from these various lines of research during the past fifty years is summarized in Fig. 1. Ions taken up by cells of the root cortex, if not utilized or accumulated there, are transferred laterally to the stele. From there, some move downwards in the phloem to the apical meristem of the root, while some are transported upwards, mainly in the xylem. The quantity of ions supplied to individual leaves is related to the amount of xylem sap reaching the leaf, and its concentration.

Young leaves, which behave as sinks, receive an additional supply of solutes in the phloem. As they become mature, leaves are gradually converted into sources and export the ions they receive, together with photosynthates, at an increasing rate through the sieve tubes to growing regions, such as root and shoot apices, developing fruits, seeds, rhizomes and other storage organs.

Ion distribution in germinating pea seedlings

It occurred to me some years ago that it would be interesting to examine the control of ion distribution from storage organs, such as cotyledons, during the germination of seeds. I began some experiments on peas (*Pisum sativum* L.) (Sutcliffe, 1962) because I knew that pea seedlings will grow quite well, without a supply of mineral elements, except calcium, to the roots. Therefore it was possible to study the distribution of endogenous ions in the absence of an external supply and to investigate the effects of adding specific ions to the medium on the pattern of distribution. Results of later experiments

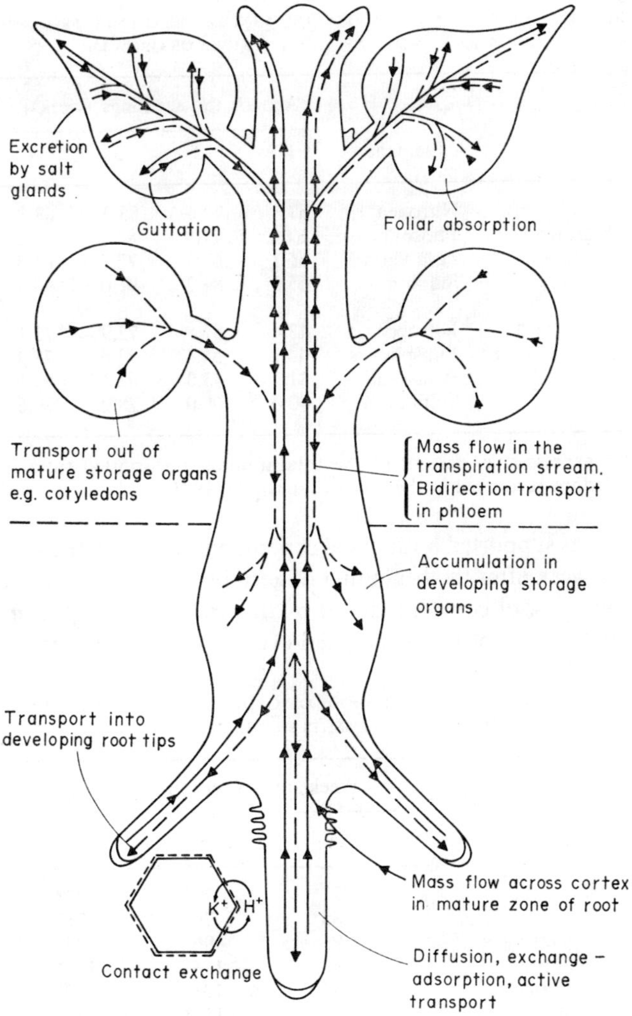

FIG. 1. Circulation of ions in plants (from Sutcliffe, 1962).

which were mainly carried out by my student J. L. Guardiola have been published elsewhere (Guardiola, 1973; Guardiola and Sutcliffe, 1971, 1972). Our main conclusions may be summarized as follows:

First, we established that the distribution of cotyledonary potassium between root and shoot of the developing seedling is closely related to the growth of each organ. Treatments, such as darkening of the plants, or application of gibberellic acid, which differentially stimulated shoot growth, increased the proportion of exported potassium and other elements transported to the shoot (Table 1). Total export from the cotyledons was increased by these treatments to such an extent that the amount going to the roots was unaffected. These results suggest that there is little or no competition between roots and shoots for the available supply, at least in the early stages of germination.

TABLE 1. ACROPETAL TRANSPORT OF MINERAL ELEMENTS FROM PEA COTYLEDONS AFTER DIFFERENT PERIODS OF GROWTH

Seedlings	Element	Age of the seedlings (weeks) 1	2	3	4
Dark-grown	Nitrogen	63.3	72.6	83.3	88.8
	Phosphorus	62.2	71.9	84.1	88.6
	Potassium	56.1	69.5	77.7	81.4
	Sulphur	65.3	84.3	91.0	93.3
Light-grown	Nitrogen	43.4	57.8	73.9	73.1
	Phosphorus	44.2	54.6	73.6	78.4
	Potassium	31.3	45.8	62.3	62.6
	Sulphur	40.0	66.0	78.9	79.6

Data expressed as percent of the total amount transported from the cotyledons recovered in the shoot (Guardiola, 1973).

This conclusion was supported by surgical experiments in which the shoot was removed from the plant one week after the beginning of germination. The movement of materials from the cotyledons was, of course, reduced by this treatment but the amount of potassium accumulating in the root was unaffected (Table 2). There was an increase of some

TABLE 2. INFLUENCE OF THE SHOOT ON GROWTH AND ACCUMULATION OF MINERAL ELEMENTS IN THE PEA ROOT

Parameter	Weeks after excision of the shoot 1	2	3
Fresh weight	77*	79*	77*
Dry weight	85*	76*	71*
Protein nitrogen	96	96	94
Total nitrogen	93	124*	126*
Phosphorus	97	133*	130*
Sulphur	101	112*	139*
Potassium	94	100	100

Plants de-shooted when 1 week old. Growth parameters of the root expressed as a percentage of those from intact plants. Values marked with an asterisk (*) differ significantly from the untreated control plants (Guardiola, 1973).

20–40 per cent in the amounts of nitrogen, sulphur and phosphorus accumulating in the root after the shoot was removed. We concluded that the increased accumulation of these elements in the de-shooted plants was due to the prevention of secondary redistribution from root to shoot, but the possibility of competition between root and shoot remains.

We found that removal of the shoot leads to a reduction in both fresh and dry weight of pea roots within 1 week (Table 2). As this had no corresponding effect on the accumulation of potassium or any of the other elements examined, it is evident that these growth parameters are not in control of the process by which materials are moved from cotyledons to the roots. The amount of protein in the root was unaffected by removal of the shoot even 3 weeks after excision, which led us to suspect that sink potential, i.e. the ability of a growing organ to attract materials towards it is governed by the intensity of protein synthesis. On the other hand we were unable to find any clear correlation between the rate of protein hydrolysis in the cotyledons and the movement of nitrogen or other elements from the cotyledons to the growing axis (Guardiola and Sutcliffe, 1971).

When pea seedlings were grown under a variety of conditions it was observed that the rate of depletion of the cotyledons, of dry matter and of individual elements, was closely related to growth of the axis. Irrespective of the rate at which depletion of the cotyledons occurred there was a linear relationship between the rate of translocation of dry matter and the individual elements examined from the first week of germination until the cotyledons were almost depleted (Fig. 2). It was proposed that the rates in which individual elements are transported is determined by the proportions in which they are released by

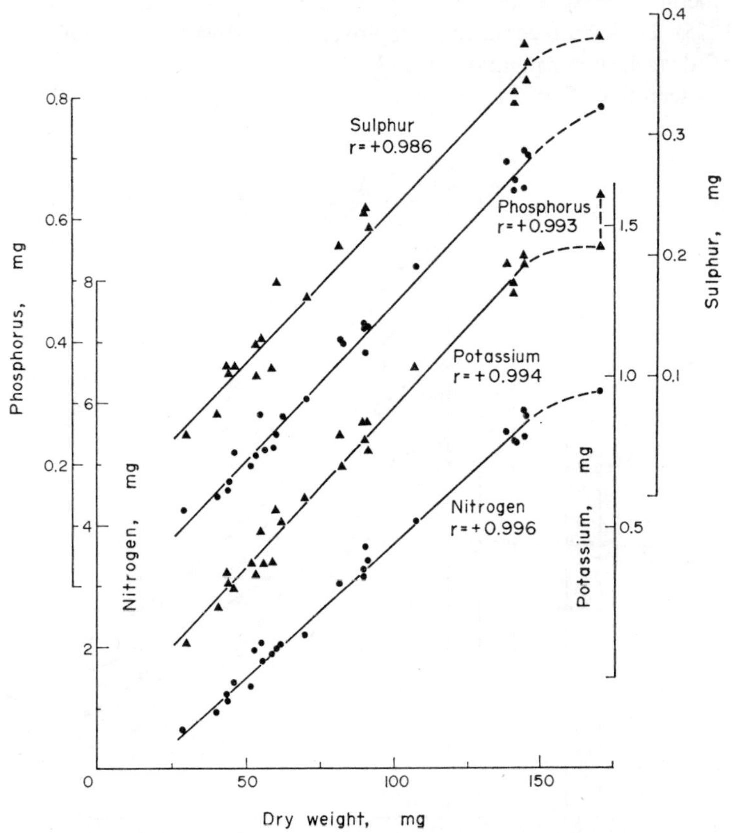

FIG. 2. Relationship between dry-matter content and amounts of individual elements in the cotyledons of *Pisum sativum* during different stages of growth over a period of 4 weeks. Results are from seedlings grown under a variety of conditions (see text). Full lines have been statistically adjusted and coefficients of correlation are given. (Data from Guardiola and Sutcliffe, 1972.)

the storage cells and utilized in the axis. Deviation from the linear relationship during the first week of germination was attributed to competition for available nutrients between the axis sinks and actively-metabolizing cells in the cotyledons. When cotyledon senescence was delayed by removing the shoot this effect was enhanced (Guardiola and Sutcliffe, 1972).

CONTROL OF TRANSLOCATION IN THE OAT SEEDLING

From the pea seedling, I turned my attention to oat, *Avena sativa* L. as an example of a plant in which the seed reserves are stored in a non-living tissue, the endosperm. In this research I was associated with Q. A. Baset, upon whose skill as an experimenter the work largely depended.

The experiments were performed on seedlings grown, from the time of germination, in darkness at 25°C in a solution of 10^{-4} M calcium chloride which was sometimes supplemented by other salts. Further details of the technique for growing the seedlings and the analytical methods used in the research have been published elsewhere (Baset and Sutcliffe, 1975). Under the conditions employed the endosperm was largely depleted of dry matter 7 days after germination (Fig. 3).

As may be seen from Fig. 3 the pattern of depletion was broadly similar for total dry

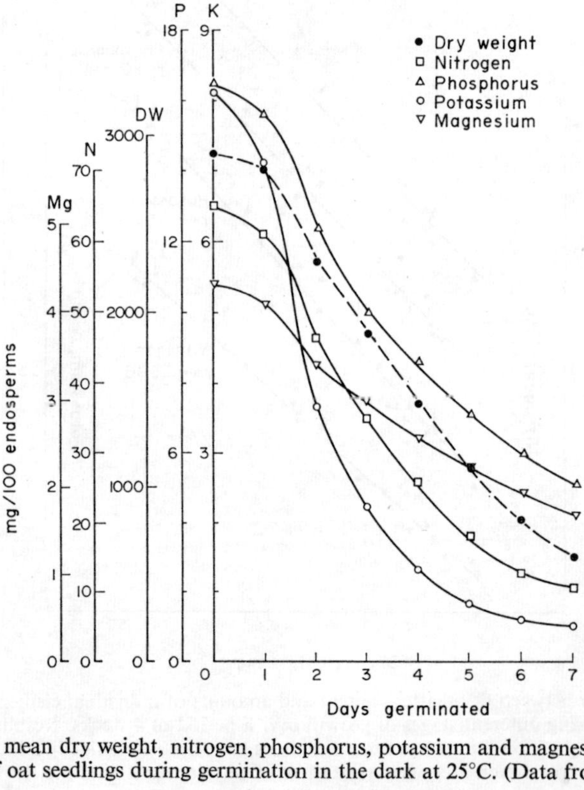

FIG. 3. Changes in mean dry weight, nitrogen, phosphorus, potassium and magnesium contents of 100 endosperms of oat seedlings during germination in the dark at 25°C. (Data from Baset, 1972.)

matter and for each of the individual element studied. There was a lag phase during an initial period of 1–2 days which built up to a maximum rate of depletion after 2–3 days, and this was followed by a phase of decline. Differences in the patterns are seen more clearly when the data are plotted as in Fig. 4. From this figure it is clear that in contrast to peas (Fig. 2), the rate of export of individual elements relative to dry matter changes markedly with time and the relationship is different for each element.

The simplest curve to analyse is that for potassium. This element is transported from the endosperm relative to dry matter most rapidly during the first day of germination and progressively more slowly thereafter. This suggests that potassium is readily available for export immediately upon hydration of the seed and it is evident that strong sinks for this element already exist in the growing axis. As sink capacity is likely to increase with time as the axis enters its exponential phase of growth, especially in the absence of an external supply of potassium to the roots, the falling rate of export of potassium from the endosperm must be related to diminishing supply. Supply is therefore likely to be the controlling factor over most, if not all, of the germination period.

The relationship between the export of nitrogen and dry matter is similar to that for potassium, but the initial rate of depletion of nitrogen relative to dry matter is lower and it falls less rapidly with time. This suggests that either the sinks for nitrogen in the axis are initially less powerful than those for potassium or that nitrogen is not so immediately available for export. We have no evidence one way or the other for the first possibility,

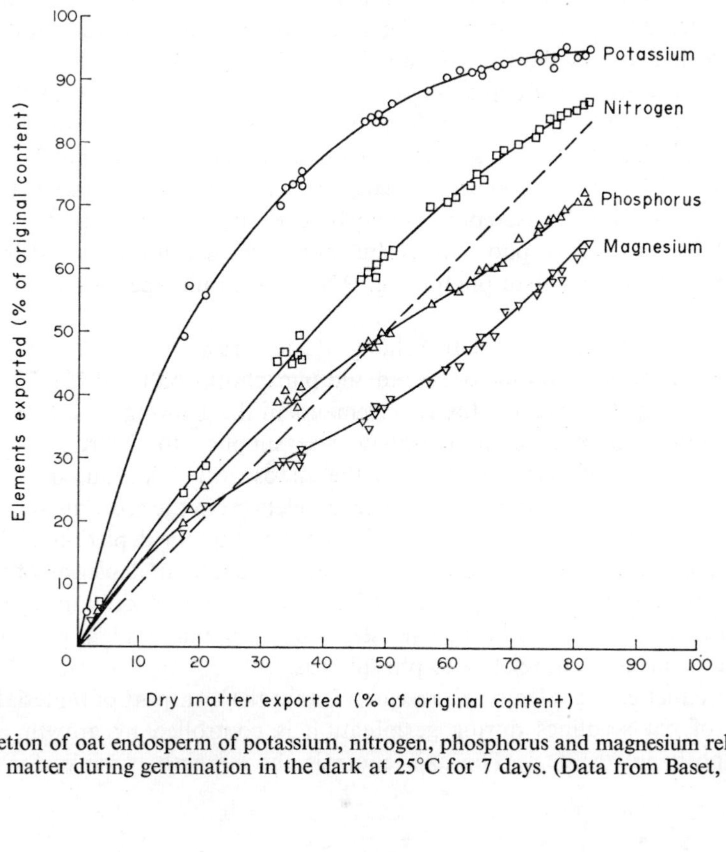

FIG. 4. Depletion of oat endosperm of potassium, nitrogen, phosphorus and magnesium relative to loss of dry matter during germination in the dark at 25°C for 7 days. (Data from Baset, 1972.)

but the latter proposition seems probable as about 88 per cent of the nitrogen in the endosperm initially is in the form of protein, which presumably must be hydrolysed before the nitrogen can be translocated. The proteolytic activity of endosperm extracts is relatively low at the beginning of germination but increases rapidly with time (Sutcliffe and Baset, 1973). We found that there is a close correlation between the *rate of increase* of protease activity in endosperm extracts and the rate of export of nitrogen. This suggests that the growing axis controls the mobilization and transport of nitrogen partly by regulating synthesis of proteolytic enzymes in the aleurone layer. A similar correlation was found between the rate of depletion of starch and the rate of increase in amylase activity in endosperm extracts. The fact that proteolytic activity develops relatively more quickly than amylase activity probably accounts for the fact that nitrogen is transported from the endosperm relatively more quickly than dry matter in the early stages of germination (Fig. 3).

It follows from these observations that only recently-acquired enzyme activity is involved in the breakdown of starch and protein. With time a progressively higher proportion of the total enzyme in endosperm extracts appears to be inactive *in vivo*. We have suggested that newly-synthesized enzyme becomes associated with individual starch or aleurone grains and that when each of these is depleted the enzyme attached to it is inactivated through lack of substrate.

The curves for depletion of magnesium and phosphorus relative to dry matter are more difficult to interpret. Both are transported from the endosperm relatively more quickly than dry matter during the first day of germination but the rate does not increase as quickly thereafter and so the relative rate falls. It increases again in the later stages of germination when the rate of depletion of the two elements falls off less quickly than that of dry matter. The similarity between the two curves may be related to the fact that both magnesium and phosphorus are stored in oat endosperm mainly in the form of phytin (magnesium and calcium myoinositol hexaphosphate) and so are mobilized together when phytin is hydrolysed. Discrepancies may be attributed to the fact that an appreciable amount (20–30 per cent) of phosphorus, but not of magnesium is stored in oat endosperm in substances other than phytin, e.g. RNA, which may be mobilized at different rates.

Alternatively, there may be a relative change during growth in the strength of the sinks in the axis for magnesium on the one hand and phosphorus on the other. This might be related to changing demands for the two elements in the growing axis. It was observed that when sodium or potassium phosphate was supplied to the roots, transport of phosphorus, but not of magnesium, from the endosperm was reduced. This is interpreted as indicating that demands for a particular element in the growing axis, if partially satisfied from elsewhere, can exert a regulating influence on the depletion of this element in the endosperm. It would be interesting to compare the rates of depletion of magnesium and phosphorus reserves in dark-grown seedlings with those of seedlings grown in light in which chlorophyll synthesis in the later stages of germination might possibly increase the demand for magnesium relative to phosphorus.

The picture that emerges from this research is that the transport of materials from the endosperm of oat seedlings during germination is controlled by growth of the axis though a 'push–pull' system. The axis controls the strength of sinks for individual elements directly through the utilization of materials in growth or (e.g. in the case of potassium) by immobilizing them in vacuoles. At the same time it regulates the mobilization of reserve materials in accordance with demand by influencing the rate of synthesis of hydrolytic enzymes in the aleurone layer by controlling the supply of gibberellins (Varner, 1964). It seems that mobilization is under more precise control in the oat seedling than in peas and transport can thus be adjusted more quickly to changing needs. Whether this is a feature of endospermous dicotyledonous seeds as well as cereal grains remains to be investigated.

Control of ion distribution—a working hypothesis

From the results of the investigations of pea and oat seedlings described above and detailed studies of ion movements between leaves during the development of bean (*Phaseolus vulgaris*) and cocklebur (*Xanthium pennsylvanicum*) (Yagi, 1972; Saeed, 1974), I have been led to the following working hypothesis to account for the control of distribution of ions and other solutes in plants.

The hypothesis is based on the assumption that, irrespective of the mechanism of phloem transport, solutes move from sources to sinks through the sieve tubes in response to gradients of electrochemical potential between the solutions bathing the two ends of the system. Transport occurs from a place where the concentration (chemical activity) of a solute is higher to a place where it is lower and each solute moves independently in response to the existing gradient. Mason *et al.* (1936) put forward a somewhat similar hypothesis based on their work on cotton and they were able to demonstrate the existence of concentration gradients between source and sink for sucrose but not for inorganic salts. The problem of course lies in the difficulty of extracting for analysis the solution at the sieve tube ends which may be different in composition from that in the bulk of the source or sink tissue.

When discussing the movement of water in plants in modern thermodynamic terms (see e.g. Sutcliffe, 1968) it is usual to fix the 'water potential' of pure water under standard conditions arbitrarily at zero. If a solute is added to this water the water potential becomes negative and the solution acquires a 'solute potential' which is numerically equal to the osmotic potential of the solution, and it is given a negative sign. To avoid confusion in what follows the same terminology will be used and solutes will be assumed to move from a region of negative 'solute potential' towards one which is less negative, i.e. from a more concentrated to a less concentrated solution.

The movement of solutes from source to sinks in both shoots and roots in the axis of a germinating seed can be described using an electrical analogy.

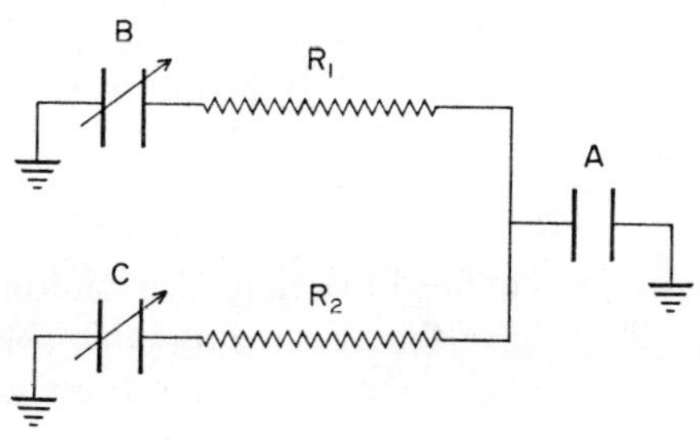

The source solution is represented by a capacitor (A) to indicate that it has both capacity (which is related to its volume) and the ability to acquire a 'charge', or potential. The solution accumulates solutes as a result of enzymic processes in the cotyledons or endosperm in which soluble particles are released from insoluble materials. The sinks (B and C) likewise have both capacity and potential. Capacity increases during growth and is therefore represented as variable. Sink potential is related to the rates at which solutes are removed from the solution to be utilized in growth or accumulated in vacuoles. The amount of an individual solute transported in unit time from the source to an individual sink will be related to the potential gradient between it and the source, and to the resistance presented by the sieve tubes (R_1 and R_2). It appears from this model that removal of one of the two sinks will not significantly affect the rate of solute movement to the other immediately as was observed in pea seedlings (p. 436). Movement of solutes to a particular sink may be affected by a change in solute potential at either source or sink or by a change in phloem resistance. It appears that in the oat seedling increased growth of the axis not only causes an increase in solute potential at the sinks, but tends to induce a reduction in solute potential at source by stimulating the hydrolysis of storage products.

As the plant grows and new leaves are produced older leaves begin to senesce and the pattern of solute distribution becomes more complex. On the basis of my hypothesis a leaf continues to act as a sink as long as the potential of the solution bathing its sieve tube terminals is sufficiently high to maintain a gradient in the right direction between it and neighbouring leaves. When growth begins to slow down solutes start to accumulate in the sink solution and the gradient is reduced until finally it is reversed. The leaf then begins to act as a source, but it does not do so necessarily for all solutes simultaneously because the rates of consumption and supply of individual substances may be different. The possibility of simultaneous bi-directional flow through the same phloem strand and even through the same sieve tube is not excluded.

It has been found (Yagi, 1972; Saeed, 1974) that, in general, the ability of a leaf to function as a sink or source is closely related to the intensity of RNA and protein synthesis. This is not surprising because the rate of synthesis of these substances must control the rate of utilization of a variety of organic and inorganic solutes. Even the accumulation of ions in vacuoles has been shown to be dependent on protein synthesis (Sutcliffe, 1962, 1973). On the other hand, net protein hydrolysis, as in senescing leaves, leads to the accumulation of a variety of solutes in the solution bathing the sieve tube ends. These solutes are released from ageing cells through an increase in the permeability of the tonoplasts and plasma membranes following reduction in protein synthesis.

It has often been observed that solutes exported from individual leaves are transported mainly to the next younger leaf in the same orthostichy. This is not surprising if the resistance of the phloem connections is a factor in the regulation of ion distribution. When a powerful sink arises, such as a developing fruit or other storage organ, solutes move into it from all over the plant, but even so movement is often greatest from the nearest mature leaves, presumably because the resistance of the transport pathway is lower.

So far this discussion has been confined to solute distribution in the phloem and it is now necessary to mention the role of the xylem. Inorganic solutes absorbed from the soil solution and organic substances synthesized in the roots are transported upwards into the leaves in the transpiration stream. The amounts of solutes reaching individual leaves in this way are related to the volume and concentration of xylem sap reaching each leaf irrespective of whether it is acting as a source or a sink. The solutes released at the vein endings diffuse in the free space to the solution bathing the sieve tube ends, and thus affect the solute potential of the bathing solution. If a greater amount of a particular solute is being added to the solution in this way than is being utilized currently in growth its solute potential is reduced and transport of this substance out of the leaf in the phloem is stimulated. This mechanism could form a basis of the regulation of ion circulation in plants.

The hypothesis outlined above appears to be consistent with the features of solute distribution that have been reported. Substantiation will require extraction and analysis of the solution from the free space which adjoins the vein endings.

References

BASET, Q. A. (1972) Mobilization and transport of food reserves in etiolated oat seedlings. D.Phil. Thesis, University of Sussex.

BASET, Q. A. and SUTCLIFFE, J. F. (1975) Regulation of the export of potassium, nitrogen, phosphorus, magnesium and dry matter from the endosperm of etiolated oat seedlings (*Avena sativa* cv Victory). *Ann. Bot* NS **39,** 31–42.

BIDDULPH, O., BIDDULPH, S. F., CORY, R. and KOONTZ, H. (1958) Circulation patterns for P^{32}, S^{35} and Ca^{45} in the bean plant. *Pl. Physiol. Lancaster* **33,** 293–300.

CURTIS, O. F. (1923) The effect of ringing a stem on the upward transfer of nitrogen and ash constituents. *Am. J. Bot.* **10,** 361–382.

GUARDIOLA. J. L. (1973) Growth and accumulation of mineral elements in the axis of young pea (*Pisum sativum* L) seedlings. *Acta bot. néerl.* **22,** 55–68.

GUARDIOLA, J. L. and SUTCLIFFE, J. F. (1971) Control of protein hydrolysis in the cotyledons of the germinating pea (*Pisum sativum* L.) seeds. *Ann. Bot.* NS **35,** 791–807.

GUARDIOLA, J. L. and SUTCLIFFE, J. F. (1972) Transport of materials from the cotyledons during germination of seeds of the garden pea (*Pisum sativum* L.) *J. exp. Bot.* **23,** 322–337.

GUSTAFSON, F. G. and DARKEN, M. (1937) Further evidence for the upward movement of minerals through the phloem of stems. *Am. J. Bot.* **24,** 615–621.

HOAGLAND, D. R. (1944) *Lectures on the Inorganic Nutrition of Plants.* Chronica Botanica. Waltham, Mass. U.S.A.

HOAGLAND, D. R. and BROYER, T. C. (1936) General nature of the process of salt accumulation by roots with description of experimental methods. *Pl. Physiol. Lancaster* **11,** 471–507.

HOAGLAND, D. R. and DAVIS, A. R. (1923) The composition of the cell sap of the plant in relation to the absorption of ions. *J. gen. Physiol.* **5,** 629–646.

MASON, T. C. and MASKELL, E. J. (1928) Studies on the transport of carbohydrates in the cotton plant. I. A study of diurnal variation in the carbohydrates of leaf bark and wood and the effects of ringing. *Ann. Bot.* OS **42,** 189–253.

MASON, T. C., MASKELL, E. J. and PHILLIS, E. (1936) Further studies on transport in the cotton plant. II. Concerning the independence of solute movement in the phloem. *Ann. Bot.* OS **50,** 23–58.

MUENSCHER, W. C. (1922) Effect of transpiration on the absorption of salts by plants. *Am. J. Bot.* **9,** 311–330.

SACHS, J. VON (1887) *Lectures on the Physiology of Plants,* translated by H. Marshall Ward. Clarendon Press, Oxford.

SAEED, A. F. H. (1974) The distribution of mineral elements in *Xanthium pennsylvanium* in relation to growth. D.Phil. thesis, University of Sussex.

STEWARD, F. C. and MILLER, F. K. (1954) Salt accumulation in plants: a reconsideration of the role of growth and metabolism. In *Active Transport and Secretion,* ed. BROWN, R. and DANIELLI, J. F. Symposium VIII of the Society for Experimental Biology held at the University of Bangor 1953. Cambridge University Press.

STILES, W. (1924) Permeability. *New Phytol.* Reprint 13, 1–296.

STILES, W. and KIDD, F. (1919) The influence of external concentration of salts on the position of equilibrium attained in the intake of salts by plant cells. *Proc. R. Soc.* B. **90**, 448–470.

STOUT, P. R. and HOAGLAND, D. R. (1939) Upward and lateral movement of salt in certain plants as indicated by radioactive isotopes of potassium, sodium and phosphorus absorbed by roots. *Am. J. Bot.* **26**, 320–324.

SUTCLIFFE, J. F. (1962) *Mineral Salts Absorption in Plants*. Pergamon Press, Oxford.

SUTCLIFFE, J. F. (1968) *Plants and Water*. Edward Arnold, London.

SUTCLIFFE, J. F. (1973) The role of protein synthesis in ion transport. In *Ion Transport in Plants*, ed. ANDERSON, W. P. pp. 399–406. Academic Press, London and New York.

SUTCLIFFE, J. F. and BASET, Q. A. (1973) Control of hydrolysis of reserve materials in the endosperm of germinating oat (*Avena sativa* L.) grains. *Pl. Sci. Lett.* **1**, 15–20.

VARNER, J. E. (1964) Gibberellic acid controlled synthesis of α-amylase in barley endosperm. *Pl. Physiol. Lancaster* **39**, 413–415.

YAGI, M. I. A. (1972) Relationship between the distribution of mineral elements and growth of bean plants. D.Phil. Thesis, University of Sussex.

ASPECTS OF LEAF WATER RELATIONS AND STOMATAL FUNCTIONING

H. MEIDNER

Biology Department, University of Stirling, U.K.

INTRODUCTION

The study of guard cell movements must take into account the anatomy and morphology of the stomatal apparatus, its metabolism and its water relations as well as the factors which affect all of these. Opening and closing of the stomatal pore results from changes in shape, chiefly of the guard cells, and to a lesser extent, of the epidermal cells. These deformations are due to changes in the pressure relations between epidermal and guard cells and they are mainly a function of the properties of the walls of the guard cells.

Two main streams in studies of the physiology of stomata can be distinguished: those concerned primarily with metabolic and biochemical processes which lead to the operative changes in osmotic and pressure potentials of guard cells and those which concern themselves primarily and more directly with these changes in pressure relations. It is, of course, not possible to consider such studies isolated one from the other because all factors, processes and forces involved in the functioning of stomata interact; however, for the purpose of this brief account I concentrate on studies of transient stomatal movements which result directly from changes in leaf water relations.

CHANGES IN LEAF THICKNESS

Before dealing specifically with stomata I wish to refer to an aspect of leaf water relations relevant to transient stomatal movements, namely, changes in leaf thickness that accompany changes in leaf water content. Figure 1 gives an example of such changes as they occur in most mesomorphic leaves. With few exceptions the epidermal tissue constitutes between 15 and 30 per cent of total thickness, the exceptions being somewhat succulent leaves in which epidermal thickness may be as low as 5 per cent of total leaf thickness. In leaves possessing in addition another water-storing tissue in the form of a hypodermis, the water-storage tissues account for a still greater percentage of total leaf thickness. A major proportion of changes in leaf thickness is due to changes in epidermal dimensions. However, the mesophyll tissue also shrinks, the palisade cells tend to become shorter and broader (seen in transverse section) and in the spongy mesophyll the air-space system reduces markedly in volume. These changes are indirectly indicated by measurements of the resistance to lateral flow of air within a leaf blade (Fig. 2). One

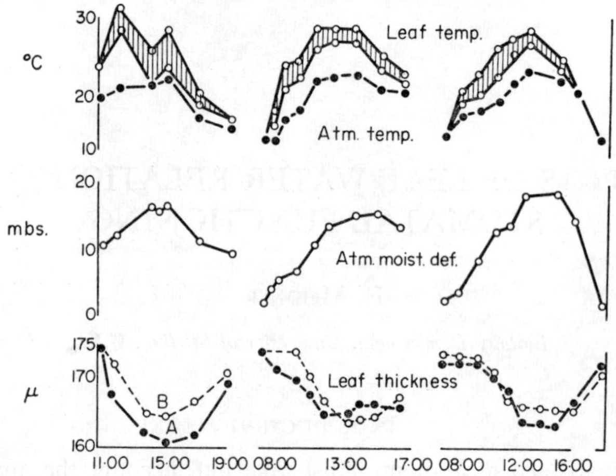

FIG. 1. Changes in atmospheric temperature and moisture deficit, and changes in temperature and thickness of leaves of *Heteromorpha involucrata*. For every 1 per cent change in leaf water content, leaf thickness changed by 4 per cent in these leaves: in *Xymalos monospora*, a shade leaf, corresponding thickness changes amounted to 7 per cent. Data from Meidner (1952).

feature of these changes is that the inner walls of epidermal cells and especially of guard cells change comparatively little in area and in the extent of their surfaces in contact with air-spaces; this will be referred to later.

TRANSIENT STOMATAL MOVEMENTS

Returning to transient stomatal movements, it may be noted that their occurrence has been known for a long time. Thus, F. Darwin (1898) commented on stomatal opening prior to wilting and Lloyd (1908) as well as Laidlaw and Knight (1916) measured such movements. Stålfelt (1929) dealt more extensively with this aspect of stomatal physiology

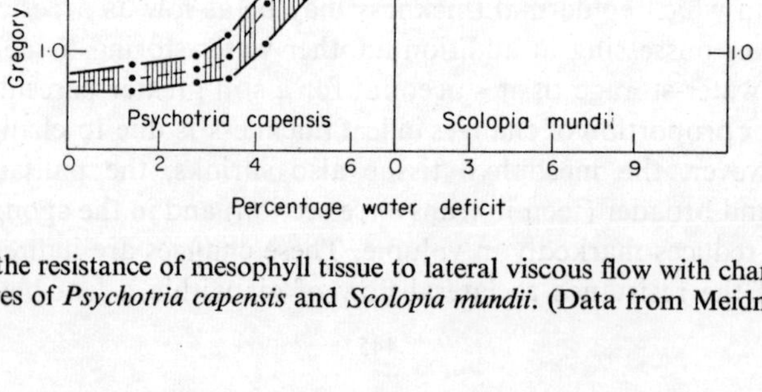

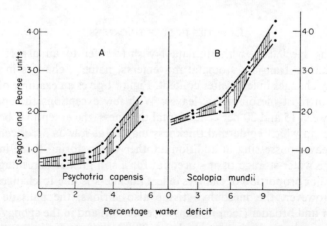

FIG. 2. Changes in the resistance of mesophyll tissue to lateral viscous flow with changes in tissue water content in leaves of *Psychotria capensis* and *Scolopia mundii*. (Data from Meidner, 1955).

for which he used the descriptive term 'passive movements', implying that they do not involve the expenditure of energy on the part of the guard cells and are indeed often the result of changes occurring in other cells (Meidner and Mansfield, 1968). In this context Stålfelt distinguished between three states of leaf water deficit:

(i) 'sub-optimal' (leaf too turgid for full stomatal opening to be possible),
(ii) 'optimal' (fullest opening possible as epidermal resistance is minimal), and
(iii) 'supra-optimal' (partial stomatal closure occurring owing to loss of turgor by guard cells).

Such stomatal closure at supra-optimal water deficits Stålfelt termed 'hydro-active' in contra-distinction to 'passive' closing movements discussed below because he thought that hydro-active closure involved events within the guard cells.

The hydro-active mechanism envisaged by Stålfelt may well be the resultant of interactions with other closing stimuli such as increases in internal carbon dioxide concentration (Heath and Meidner, 1961) or an increase in sensitivity to carbon dioxide concentration (Heath and Mansfield, 1962). Nevertheless, it is now known that in a leaf at supra-optimal water deficit, i.e. at some degree of water stress, the concentration of abscisic acid increases and that this promotes stomatal closure—in this sense, Stålfelt's term 'hydro-active closure' can be understood to distinguish these events from the truly passive movements dealt with next.

The terms 'passive' and 'active' have other connotations and since the stomatal movements considered here do not lead to steady state openings they have aptly been called transient movements. Figure 3 illustrates the time course of such events.

Hydraulic conductivities of leaf tissues

For a study of transient movements it is desirable to produce them experimentally under controlled conditions. Sheriff and Meidner (1975) used leaves of *Tradescantia*

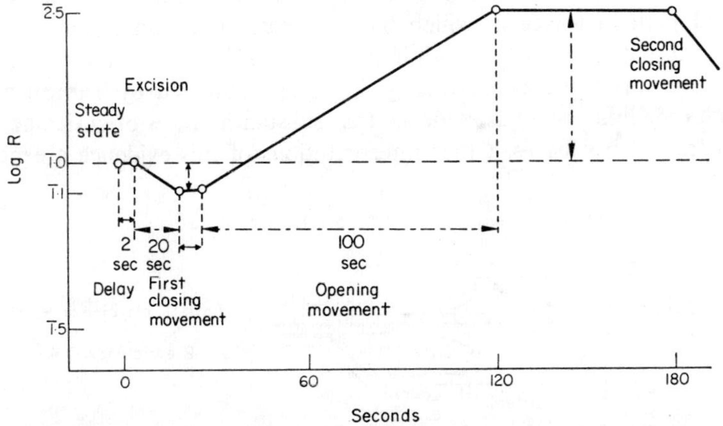

FIG. 3. Transient stomatal movements following excision of a transpiring leaflet of *Phaseolus vulgaris*. Data from Meidner (1965).

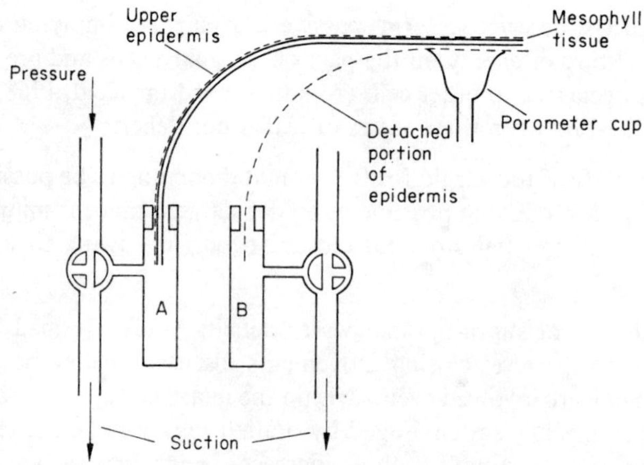

FIG. 4. Arrangement of leaf tissues for measurement of their hydraulic conductivity (from Sheriff and Meidner, 1975).

virginiana and *Hedera helix* in an arrangement shown in Fig. 4. The aim of the experiment was to obtain estimates of hydraulic conductivities of epidermal, vascular and mesophyll tissues. Transient stomatal movements were only used as indicators for the conductivity measurements. Time lags were measured between application of pressures and the occurrence of transient stomatal movements as indicated by a recording viscous flow porometer attached to the intact portion of the leaf. The pressures applied were plus or minus 0.2 bar producing on average transient changes in pore widths of 1.5 μm as we were able to estimate from the results obtained with the porometer and from the experiment reported below.

A very rough summary of the hydraulic conductivities estimated is shown in Fig. 5 where conductivities have been calculated as relative rehydration efficiencies of the three paths. The comparatively high conductivity of the epidermal path would make it the preferred path in leaves in which bundle-sheath and vein extensions are poorly developed or absent, and a major path in most leaves. The importance of this epidermal path goes some way towards explaining the results obtained by Tanton and Crowdy (1973) which they interpreted as evidence that substantial rates of cuticular vapour loss occur from transpiring leaves. Other interpretations of this evidence are possible. The

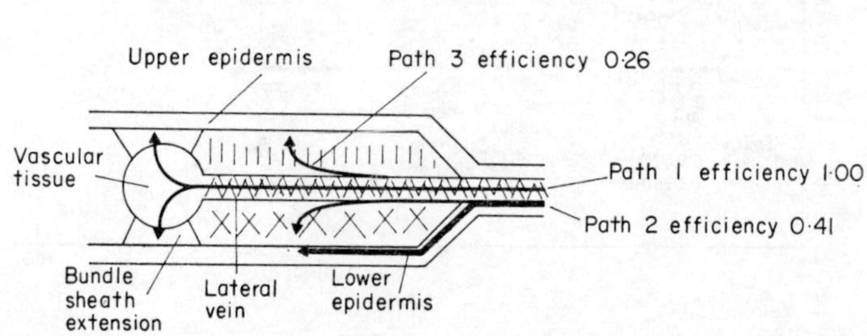

FIG. 5. Diagrammatic representation of paths of water flux in a leaf (from Sheriff and Meidner, 1975).

lead precipitates which they found within the cuticle could be considered as having arisen at least partly as a result of preferential adsorption of the lead chelate present in the water pervading the epidermal walls, so that a continuous diffusion gradient of chelate towards these sites would have prevailed during the experiment. In addition, it appears from the experimental procedures that the cuticle must have become dehydrated and subsequently rehydrated, i.e. some bulk water movement must have occurred from the epidermal walls into the cuticle. Together, adsorption, rehydration and, of course, some cuticular vapour loss could account for the deposits in the cuticle. Tanton and Crowdy's results showing a greater concentration of lead deposits in the epidermal walls than in the mesophyll walls would indeed be expected if the inner epidermal cell walls constitute a major path of leaf water supply and are therefore major sites for evaporation into the leaf air space. This is likely in view of Sheriff and Meidner's results from which it appears that this route offers only a low resistance to water movement so that water evaporated can be replenished with ease. In considering the inner epidermal walls and possibly the guard cells as major evaporation sites the fact should not be overlooked that the shortest diffusion path is between these sites and the stomatal pore, whereas the path from mesophyll walls to the entrance of the pore is considerably longer.

It is thus not surprising that, on illumination of leaves, the water content of epidermal cells declines, even if cuticular vapour loss is very low, and that the physical resistance of epidermal cells to light-induced guard cell movement falls. Once the stomata are open, under certain conditions not common in the field, a further transient opening prior to wilting may occur unless the supply of water to the epidermis proceeds at an adequate rate to replenish that lost by evaporation into the leaf air space. Such transient openings were experimentally produced by the application of a strong osmotic solution to the root systems of vigorously transpiring plants (Heath and Meidner, 1963). In contrast, if the water supply to a vigorously transpiring leaf, with its stomata at a steady state opening, is suddenly improved, this will in part suddenly replenish the water content and turgor of epidermal tissues with a consequent transient stomatal closing movement, as measured in the experiment illustrated in Fig. 3.

Manipulation of guard cell pressure potentials

The information obtained about the paths of epidermal water supply and about the magnitude of pressures which can produce transient stomatal movements was further extended in an experiment using a more direct technique of changing pressure relations between epidermal and guard cells. In essence the technique consisted of implanting microneedles in either subsidiary or guard cells and thus making it possible to apply different pressures within these cells via a special syringe. The experimental arrangement is shown in Fig. 6 and details of the experiment can be found in Meidner and Edwards (1975). As convincing evidence of transient stomatal movements Mrs. Edwards prepared a film showing repeated closing and opening of stomatal pores in response to her varying the pressure within subsidiary cells adjacent to one of the guard cells. A summary of the results obtained is shown in Fig. 7 and some explanatory notes follow.

1. Pressures quoted in Fig. 7 are total pressures prevailing inside guard cells in

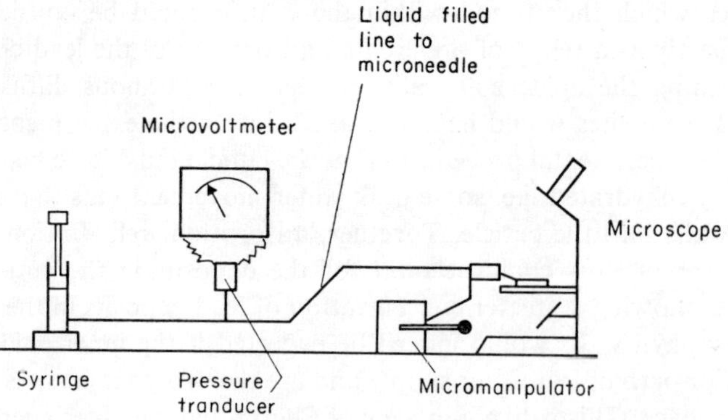

FIG. 6. Arrangement of apparatus for the measurement of pressure potentials in stomatal cells (Meidner and Edwards, 1975).

response to variation of the pressures in subsidiary cells, and not pressure differentials between epidermal and guard cells.

2. Measurements made on epidermal strips did not vary from those made on intact leaves, neither were there differences between measurements with both guard cells of a pair intact and one punctured beforehand, nor were the pressures required inside guard cells for opening of stomata different from those required in subsidiary cells for closing of stomata.

3. Completely closed stomata could not be opened with pressures up to 10 bar applied inside guard cells. This seems to support Stålfelt's concept of a *Spannungsphase* during which either greater pressures are required to cause deformations of the guard cells or during which the properties of the cell wall may change.

Two issues arise from these results.

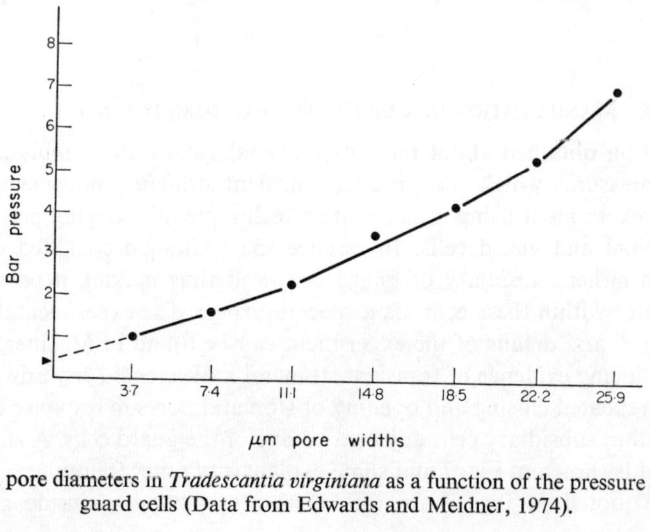

FIG. 7. Stomatal pore diameters in *Tradescantia virginiana* as a function of the pressure potentials within guard cells (Data from Edwards and Meidner, 1974).

(i) Generally low pressures of less than 10 bar prevailed in guard cells of wide open stomata. These were not specific to *Tradescantia virginiana* but similar pressures were measured in *Vicia faba*. It must be stressed once more that the quoted pressures were the total pressures prevailing in cells and not pressure differentials. Against these values plasmolytically determined osmotic potentials of saps of guard cells have been quoted which range from −10 to −40 bar. However in many cases the degree of stomatal opening was not stated (Meidner and Mansfield, 1968; Humble and Raschke, 1971). The question arises whether the plasmolytic method yields elevated values. The contrary has been argued if the time allowed for plasmolysis to occur exceeds about 2 min. In many of the investigations yielding the high values quoted this was certainly the case and therefore it is perhaps not warranted to assume that guard cells of stomata are at their maximum pressure potentials even when they have attained a steady state opening. For this situation to obtain a mechanism must exist which would prevent the attainment of maximum turgor potentials in guard cells (cf. Meidner and Edwards, 1975). It must be remembered also that as soon as the pressure potentials of guard cells exceed by a certain amount those of their subsidiary or epidermal neighbours, i.e. long before full turgor is established, the guard cells will become deformed in such a way that the pore begins to open. These excess pressure requirements may well vary considerably between different species and possibly with different leaf water contents.

(ii) The gentle curvature of the line in Fig. 7 contrasts with other much steeper curves depicting changes in turgor potentials (for instance, in the well known Höfler diagram). Up to an opening of about 11 μm a pressure change of 1 bar caused 5.5 μm of increased pore diameter. This value rose gradually until in the region of the maximum opening measured, an increase of 1 bar pressure caused an increased diameter of pore of only 2.0 μm. Attention should be drawn to the fact that pore diameters of about 10 μm–15 μm are those found under natural conditions.

These observations together with those made earlier about the existence of a *Spannungsphase* lead to the question whether the elasticity of guard cell walls is changeable. Could it be that guard cell walls offer progressively *less* resistance to expansion and that the increased pressure requirements for the attainment of extra large pore diameters are chiefly due to resistances offered by surrounding cells? A mechanism for changes in cell wall properties could be connected with the ionic movements between subsidiary and guard cells, and between vacuoles, cytoplasm and cell walls (Imamura, 1943; Fisher and Hsiao, 1968; Humble and Hsiao, 1970).

Acknowledgement: I wish to acknowledge financial support from the S.R.C. for some of the work reported in this paper.

References

Darwin, F. (1898) Observations on stomata. *Phil. Trans. R. Soc.* B **190,** 531–621.

Edwards, M. and Meidner, H. (1974) Micromanipulation of stomatal guard cells. *Nature Lond.* **253,** 114–115.

Fisher, R. and Hsiao, T. C. (1968) Stomatal opening in isolated epidermal strips of *Vicia faba*, II. *Pl. Physiol. Lancaster* **43,** 1953–1958.

HEATH, O. V. S. and MANSFIELD, T. A. (1962) A recording porometer with detachable cups operating on four separate leaves. *Proc. R. Soc.* B **156**, 1–13.

HEATH, O. V. S. and MEIDNER, H. (1961) The influence of water stress on the minimum intercellular space carbon dioxide concentration and stomatal movements in wheat leaves. *J. exp. Bot.* **12**, 226–242.

HEATH, O. V. S. and MEIDNER, H. (1963) Rapid changes in transpiration in plants. *Nature Lond.* **200**, 283–284.

HUMBLE, G. D. and HSIAO, T. C. (1970) Light-dependent influx and efflux of potassium of guard cells during stomatal opening and closing. *Pl. Physiol. Lancaster* **46**, 483–487.

HUMBLE, G. D. and RASCHKE, K. (1971) Stomatal opening quantitatively related to potassium transport. *Pl. Physiol. Lancaster* **48**, 447–453.

IMAMURA, S. (1943) Untersuchungen über den Mechanismus der Turgorschwankung der Spaltöffnungs-schliesszellen. *Jap. J. Bot.* **12**, 251–346.

LAIDLAW, C. and KNIGHT, R. C. (1916) A note on stomatal behaviour during wilting. *Ann. Bot.* **30**, 47–56.

LLOYD, F. E. (1908) The physiology of stomata. *Publ. Carnegie Inst.* **82**, 1–142.

MEIDNER, H. (1952) An instrument for the continuous determination of leaf thickness changes in the field. *J. exp. Bot.* **3**, 319–325.

MEIDNER, H. (1955) Changes in the resistance of the mesophyll tissue with changes in the leaf water content. *J. exp. Bot.* **6**, 94–99.

MEIDNER, H. (1965) Stomatal control of transpirational water loss. In *The State and Movement of Water in Living Organisms* ed. FOGG, G. E. *Proceedings of the XIXth Symposium of the Society for Experimental Biology*. Cambridge Univ. Press.

MEIDNER, H. and EDWARDS, M. (1975) Direct measurements of turgor pressure potentials of stomatal guard cells. *J. exp. Bot.* **26**, in press.

MEIDNER, H. and MANSFIELD, T. A. (1968) *The Physiology of Stomata*. McGraw-Hill, London.

SHERIFF, D. W. and MEIDNER, H. (1975) Water pathways in leaves of *Hedera helix* and *Tradescantia virginiana* L. *J. exp. Bot.* **28**, 1147–1156.

STÅLFELT, M. G. (1929) Die Abhängigkeit der Spaltöffnungsreaktionen von der Wasserbilanz. *Planta* **8**, 287–301.

TANTON, T. W. and CROWDY, S. H. (1972) Water pathways in higher plants, II. *J. exp. Bot.* **23**, 619–626.

MECHANISMS INVOLVED IN TURGOR CHANGES OF GUARD CELLS

T. A. MANSFIELD

Department of Biological Sciences, University of Lancaster, Bailrigg, Lancaster, U.K.

INTRODUCTION

The past few years have seen major advances in our understanding of the physiology of stomatal guard cells. These advances fall essentially into two main areas. First, there is the discovery of the massive movements of potassium to and from guard cells as stomata open and close, and secondly there is the hormonal control of guard cell activity, in which abscisic acid (ABA) plays a principal part.

ION TRANSPORT AND STOMATAL MOVEMENTS

Fujino (1967) reported that stomatal opening on isolated epidermis was stimulated in solutions of potassium salts, and he postulated that active transport of K^+ ions to and from the guard cells was responsible for the changes in their turgor associated with stomatal movements. Independent work by Fischer (1968) showed that stomatal opening in *Vicia faba* probably depended on active K^+ transport, and it was later found by use of labelled Rb^+ ions that the changes in solute potential resulting from the movement of monovalent cations and associated anions would be sufficient to account for the turgor changes (Fischer, 1971).

Movements of K^+ are easily demonstrated qualitatively by histochemical means (Willmer and Mansfield, 1970; Mansfield and Jones, 1971), and have been estimated quantitatively by X-ray microanalysis (Humble and Raschke, 1971). Potassium levels in epidermis with open and closed stomata, as revealed by simple histochemistry, are shown in Fig. 1. Associated movements of anions have proved rather more elusive and even the microanalysis technique has failed to show changes in Cl, P or other elements that would indicate anion transfer to and from the guard cells in *Vicia faba*. This contrasts with *Zea mays* in which chloride ions seem to move to and from subsidiary cells alongside potassium (Raschke and Fellows, 1971). The well-established breakdown of starch in guard cell chloroplasts in light, to which turgor changes were at one time attributed (Meidner and Mansfield, 1968), could be the missing link in *Vicia faba*. Organic anions might be generated from the starch, and indeed malate levels in guard cells do rise markedly when stomata open (Allaway, 1973).

The most detailed picture of K^+ movements as stomata open and close has come from recent determinations made with the potassium microelectrode (Penny and Bowling, 1974). There is an active movement not only into and out of the guard cells, but also along a series of neighbouring cells.

The demonstration that massive, active ion movements are occurring raises the question of how energy requirements are met. This is an obvious problem awaiting solution and no doubt research in the near future will be much concerned with it. Some studies have, indeed, already been carried out and I am somewhat dubious about the methods of approach that have been adopted. What I think is basically wrong is the tacit assumption that light is the principal driving force behind stomatal opening.

Are stomatal movements controlled by light?

It is generally true that, in mesophytes, stomata open during the day and close with the onset of night, and because of this there is a tendency to assume that they have been caused to open by light. However, in most plants opening can begin before dawn if the night is long enough. This is the result of an endogenous circadian rhythm, which is revealed if the plant is kept in the dark for several days, and which in natural conditions brings it into a state of preparedness to open in the light (Mansfield and Heath, 1963; Martin and Meidner, 1972). If light is not forthcoming and if other environmental factors are unchanged, the degree of opening achieved in the dark is small (Fig. 2). But if there is an increase in temperature at a time roughly equivalent to that of the dawn, substantial opening can occur in the dark. This temperature-induced opening is reinforced in CO_2-free air conditions, and the aperture then reached is little short of that occurring in light (Fig. 3). The magnitude of the temperature response is greatly dependent on the phase of the circadian rhythm. The similarity of temperature and light responses of rhythms has been demonstrated previously (Wilkins, 1962) and it appears that stomata are merely exhibiting a pattern of behaviour that is common to several physiological processes.

During the day a leaf experiences both an increase in temperature and a reduction in internal CO_2 concentration brought about by photosynthesis. These two factors are obviously normally linked to light, but the data above clearly demonstrate both that the signal to open can be given by each factor independently of light, and that the energy

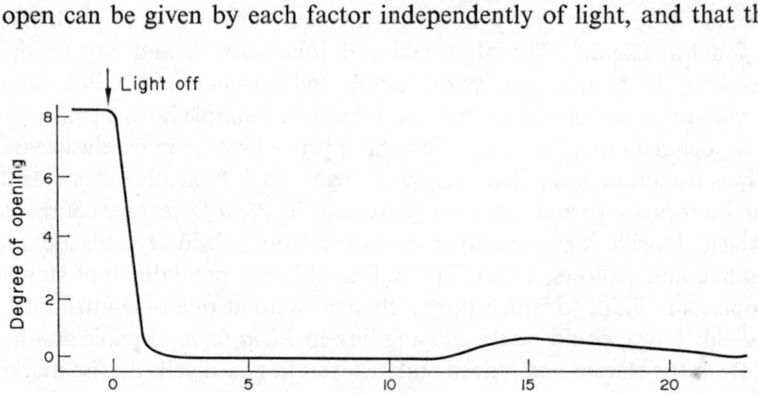

FIG. 2. The small amount of opening that occurs during prolonged darkness, at about the time when light would have been received. Darkness commenced at the time indicated by the arrow, and prior to this the light intensity was 15 K lx. Data obtained for *Xanthium strumarium* by means of an automated porometer. The ordinate is in arbitrary units.

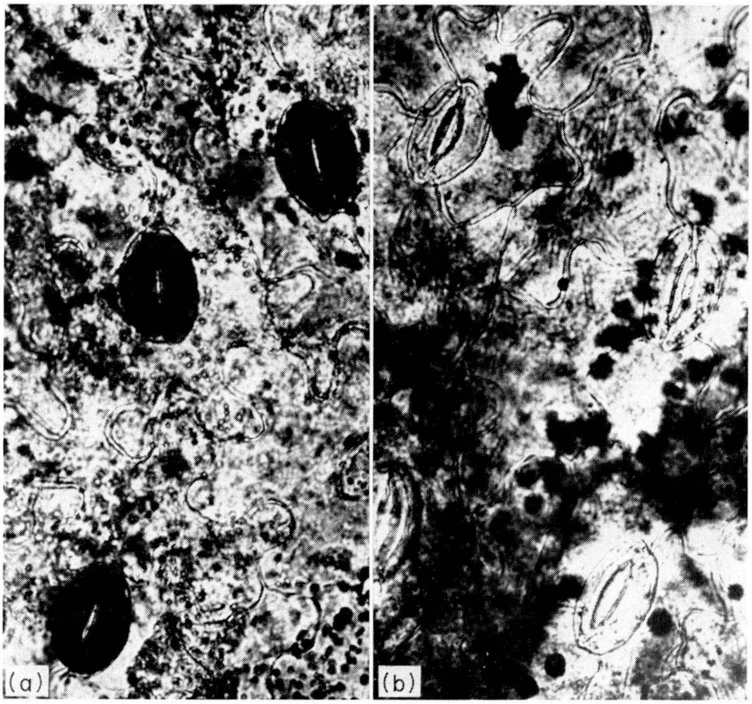

FIG. 1. Potassium distribution in epidermis of *Vicia faba* with open stomata in light (a) and closed stomata in darkness (b). The histochemical procedure (reagent: sodium cobaltinitrite) caused the stomata in (a) to close.

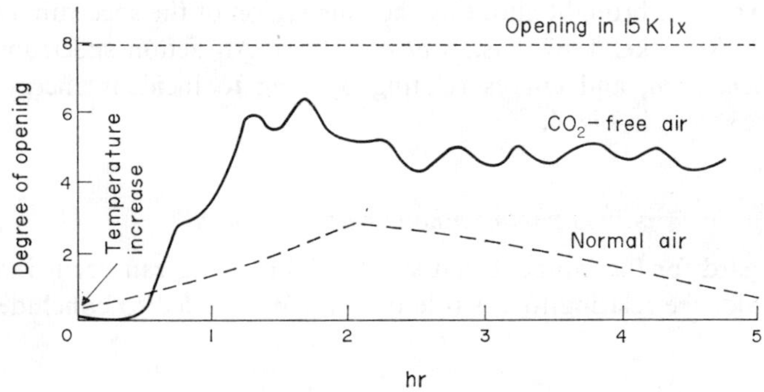

FIG. 3. The small degree of opening exhibited by *Xanthium strumarium* in darkness (see Fig. 2) approaches that in light if the temperature is increased from 27° to 36°C and the leaf is flushed with CO_2-free air. The continuous trace from the porometer shows short-period fluctuations such as are often found in light. Also shown are the opening (from measurements at 30-min intervals) attributed to the temperature increase alone in normal air, and the maximum opening achieved by leaves of comparable age in light of 15 K lx. See also Mansfield (1965).

needed for opening can be provided in the absence of light. This last point is important because it conflicts with some conclusions reached in experiments with metabolic inhibitors of photophosphorylation.

The role of photophosphorylation

Zelitch (1963) first suggested that stomatal opening might utilize ATP from non-cyclic photophosphorylation. He envisaged more energy being available from this source when the demands made by CO_2 fixation were less, i.e. in a CO_2-depleted atmosphere, and thus offered an explanation of the increased opening of stomata in low CO_2 concentrations. More recently, Humble and Hsiao (1970) concluded from responses of *Vicia faba* to metabolic inhibitors, and an action spectrum for opening, that 'photosynthetic cyclic electron flow can be sufficient and possibly necessary for K^+ uptake and stomatal opening in light'. Raghavendra and Das (1972) attached importance to ATP production by cyclic photophosphorylation, regarding this as a requirement for light-stimulated opening in *Commelina benghalensis*. This is not unlike the conclusion reached by Willmer and Mansfield (1970) for *Commelina communis*, although in this species substantial opening can occur in the dark, especially in the absence of CO_2. More data on the amount of opening that can take place under favourable conditions in the dark are obviously necessary before any conclusions about the *essential* role of photophosphorylation can be reached.

The CO_2-independent light response

The opening that can be achieved in darkness is never as great as that reached in light, if we compare the two in CO_2-free air. As far as I know this is one of the few generalizations that can be made for all mesophytes that have been investigated, and it is equally true whether stomata are observed on intact leaves or on isolated epidermis. This CO_2-independent light reaction was first recognized by Heath and Russell (1954) and subsequently shown to be brought about by the blue region of the spectrum (Mansfield and Meidner, 1966; Raschke, 1967; Hsiao *et al.*, 1973). An action spectrum for stomatal opening in *Vicia faba*, and curves relating opening to incident energy in *Xanthium strumarium*, are shown in Fig. 4.

Is photosynthetic energy required?

Let us disregard for the moment that substantial opening can occur in the dark, and consider the evidence relating to the role of light. We are led to conclude, from a sub-

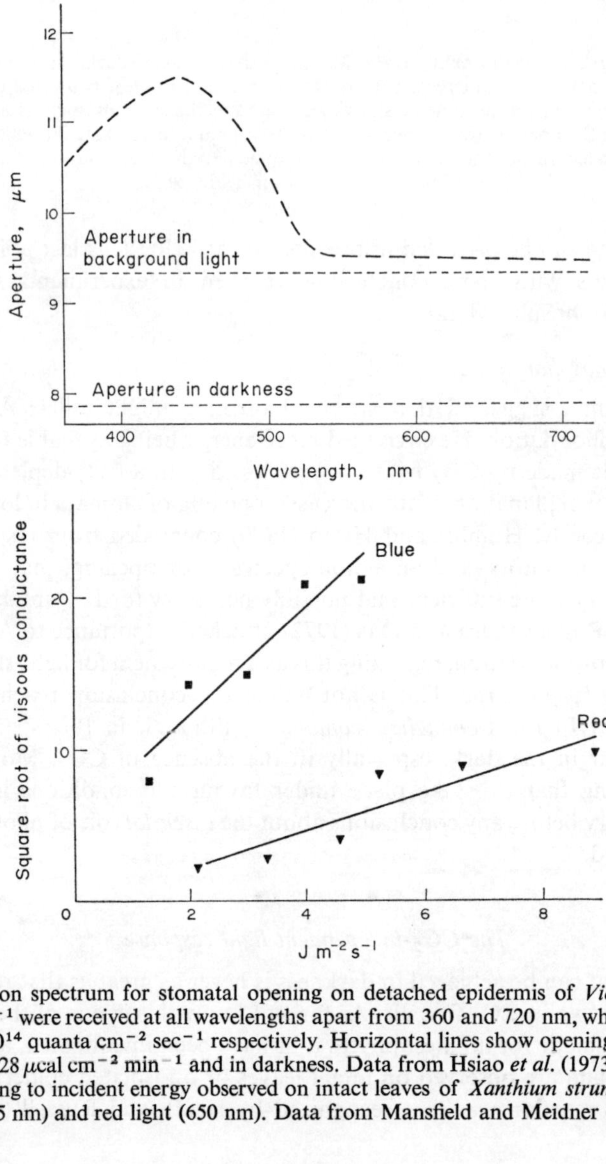

FIG. 4. Top. Action spectrum for stomatal opening on detached epidermis of *Vicia faba*. 7.8 × 10^{14} quanta cm^{-2} sec^{-1} were received at all wavelengths apart from 360 and 720 nm, where there were 5.9 × 10^{14} and 6.6 × 10^{14} quanta cm^{-2} sec^{-1} respectively. Horizontal lines show opening in low background light of less than 28 μcal cm^{-2} min^{-1} and in darkness. Data from Hsiao *et al.* (1973). Bottom. Relation of stomatal opening to incident energy observed on intact leaves of *Xanthium strumarium* in blue light (475 nm) and red light (650 nm). Data from Mansfield and Meidner (1966).

stantial body of evidence in each case (a) that the light-stimulated component of opening is dependent on cyclic electron flow or photophosphorylation and (b) that the blue region of the spectrum achieves this. The first of these conclusions is not difficult to accept, for active cation transport in algae can be supported by ATP synthesized during cyclic electron flow about photosystem I (Raven, 1967). However, in moving on to the second conclusion we immediately encounter problems, because blue light does not specially favour photosystem I. Furthermore, far-red light of 700–720 nm which does activate photosystem I but not photosystem II is not effective in producing stomatal opening, at a quantum flux sufficient to produce wide opening in blue light (Hsiao *et al.*, 1973; cf. Fig. 4). It is also difficult to accommodate the finding that wide opening can occur in blue light at low quantum fluxes, even at around the light compensation point (Mansfield and Meidner, 1966; Raschke, 1967; Hsiao *et al.*, 1973). While it might be argued that at these low quantum fluxes there is still sufficient energy input to provide for active ion transport, this involves assuming that most if not all photosynthetically available energy is used for this purpose and not for CO_2 fixation by the guard cell chloroplasts. If blue light were providing energy via either photophosphorylation or electron flow in the chloroplasts, competition for available energy would be expected according to the supply of CO_2. Near the light compensation point there is no severe CO_2 limitation of photosynthesis, but there is still wide opening of stomata in low intensity blue light.

The situation could probably be clarified if the experiments with photosynthetic inhibitors and uncouplers were repeated in blue light, of a low intensity but sufficient to produce measurable opening over and above that found in CO_2-free air in the dark. In the preceding discussion it has been necessary to assume that results obtained with inhibitors in high intensity white light would also apply in low intensity blue light. It may be that this assumption is incorrect, and that this is the cause of the major dilemma that emerges from these apparently conflicting data.

Mouravieff (1958, 1973) has pursued an independent line of inquiry, and has obtained evidence that both blue light and absence of CO_2 cause hydrolysis of starch in guard cell chloroplasts. While the precise role of starch breakdown and reformation remains uncertain, his observations serve as a useful reminder that these may be events which play a central role in the physiology of guard cells.

The role of the subsidiary cells

Penny and Bowling (1974) have presented a detailed picture of the movement of potassium not only in and out of the guard cells of *Commelina communis*, but also across the complex of subsidiary cells into the unspecialized epidermal cells. Potassium gradients were maintained against the electrochemical gradients between guard and subsidiary cells, between inner and outer subsidiary cells, and also between the latter and the neighbouring epidermal cells. There is thus active transport along a series of cells.

These observations are pertinent here because they show that we greatly over-simplify the problem if we suppose that it is necessary only to consider energy expenditure by the guard cells to account for stomatal movements. In *Commelina communis* at least, the surrounding cells play an equally important part, and these cells do not contain chloroplasts. Hence their contribution cannot come from photosynthetic phosphorylation or electron flow.

Squire and Mansfield (1972) developed a method for killing the protoplasts of subsidiary cells and epidermal cells, but which left the guard cells unharmed. Such 'isolated stomata' behaved differently from those on intact epidermis, and we deduced that the subsidiary cells played an important regulatory role.

Experiments with strips of epidermis are made easier if the subsidiary cells are killed by one means or another. If they are intact there is much greater variability in the data. Stomata of *Commelina communis* were found to behave most predictably in media containing salt concentrations high enough to lead to incipient plasmolysis of all cells except the guard cells (Fujino, 1967; Willmer and Mansfield, 1969). For similar reasons, in work with *Vicia faba* (in which there are no morphologically distinct subsidiary cells), strips containing a low proportion of intact epidermal cells have been selected for use (e.g. Pallaghy and Fischer, 1974). Thus most of the studies of the energetics of stomatal movements have been performed on material in which an essential part of the K^+ transport system had been removed or rendered inactive. Conclusions from studies with metabolic inhibitors using 'isolated' guard cells are clearly unlikely to be applicable to the intact leaf.

Now that we have a technique for 'isolating' guard cells without apparently harming them we should, however, be much better able to estimate the role of the subsidiary cells. We are at present engaged in a comparison of the effects of metabolic inhibitors on stomatal movements when the guard cells are 'isolated' and when all the cells in the epidermis are intact. Preliminary experiments have shown that the uncoupler of oxidative phosphorylation, dinitrophenol (DNP), prevents stomatal closure if the subsidiary cells are intact, but has no effect if their protoplasts have been destroyed. It would have been difficult to explain this finding without the detailed picture of potassium levels in epidermis with open and closed stomata as presented by Penny and Bowling (1974). They have shown that transport of K^+ across the subsidiary cells as stomata close is an uphill process requiring expenditure of energy. The prevention of closure by DNP suggests that the source of energy is ATP from oxidative phosphorylation. If, however, the subsidiary cells are not present the 'uphill route' for K^+ is replaced by a 'downhill route', i.e. leakage into the surrounding medium. It is also clear that in the presence of the subsidiary cells the K^+ cannot take this 'downhill route', which suggests that in the intact epidermis it does not move into the free space. Electron micrographs have, however, revealed few if any plasmodesmata between guard cells and their neighbours once the cells are mature (e.g. Allaway and Setterfield, 1972).

Energy sources

Perhaps only one thing emerges clearly from the foregoing discussion, which is that we still know very little about energy sources for stomatal movements. Some essential points have, however, been established. It is clear that the subsidiary and epidermal cells play a part in active ion movements which is perhaps equal in importance to that of the guard cells. This finding casts doubt on conclusions that energy from photosynthetic sources plays the major part, since these other cells do not contain chloroplasts. The fact that substantial opening occurs in darkness in response to environmental stimuli also indicates that energy from non-photosynthetic sources can be utilized. Though there is a special role for blue light, there is no evidence yet that blue light provides the energy for the extra opening that it stimulates.

Although I have argued against the conclusion that there is an *essential* role for energy coming from the guard cell chloroplasts, I do not want to give the impression that such energy is never utilized in stomatal movements. It would be surprising if it were not put to use in some circumstances. For example, in light when photosynthesis is CO_2-limited, surplus energy might be available from the chloroplasts. We might, therefore, expect that this would be utilized for the final step in the sequence of ion movements from cell to cell, i.e. movement into the guard cells themselves. When stomata of *Commelina communis* are opening in light in the absence of CO_2, the chloroplasts can be seen in close association with the plasmalemma adjacent to the subsidiary cells (Fig. 5). They are not usually found in this position under other treatments, and it is tempting to suppose that when they have ready energy which is surplus to other requirements, it is made available at the plasmalemma to support active transport.

Hormonal control of stomatal movements

Small effects of auxins and cytokinins had been reported on several occasions since 1954, but the true extent of hormonal control of stomata only became known in the period 1968–1970. Little and Eidt (1968) made the first observation that applied abscisic acid (ABA) markedly reduced transpiration, from which closure of stomata could be implied. The true physiological significance of the response to ABA was, however, revealed by the work of Wright (1969). He showed that ABA levels rose appreciably when wheat leaves were subjected to water stress. For many years it had been known that stomata were unable to open fully after plants had suffered water stress. The inhibition persisted for days or even weeks after re-watering, and the return of full turgor. Allaway and Mansfield (1970) were unable to suggest any explanation apart from the effect of some chemical regulator, and there now seems little doubt that ABA is the principal agent involved. Simultaneous work by Tal and Imber (1970, 1971) showed that the lack of transpiration control in the *flacca* mutant of tomato could be restored by supplying it with ABA, thus suggesting a role for the hormone in normal stomatal functioning.

At present there is little evidence that other hormones apart from ABA are involved. Effects of the major growth promoting substances are, surprisingly, lacking (Tucker and Mansfield, 1971) though some responses to cytokinins have been noted in grasses (Cooper *et al.*, 1972). Chlorogenic acid, which is also formed by wilting plants, can induce stomatal closure, but not at the very low concentrations at which ABA is effective (Ogunkanmi *et al.*, 1973).

Stomata respond to ABA over a very wide range of concentrations, and there is a detectable response even at 10^{-10} M (Fig. 6). It is clear that endogenous changes in ABA levels could provide sensitive regulation of stomatal opening, and there seems little doubt that it functions as a stress hormone, reducing transpirational water loss at times when environmental conditions make this necessary. This physiological role of ABA is much better established than its involvement in growth regulation.

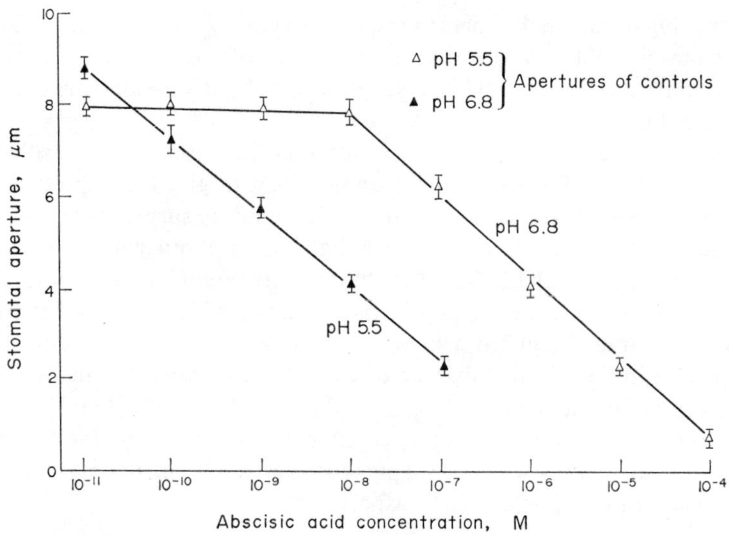

FIG. 6. Relationship between stomatal aperture and concentration of abscisic acid on isolated epidermis of *Commelina communis*. The response depends on the pH of the incubation medium. Data from Ogunkanmi *et al.* (1973).

The response of stomata on isolated epidermis to ABA provides a sensitive bioassay for the hormone. It is also possible to use the assay in the search for other endogenous compounds capable of regulating stomatal aperture, and some success has been achieved recently in an examination of extracts from water-stressed sorghum plants (Ogunkanmi *et al.*, 1974). The role of these other compounds in relation to that of ABA will be an objective of our own future studies.

ABA and its derivatives cause stomatal closure when applied to leaf surfaces, and the effects remain for a week or more (Jones and Mansfield, 1970, 1972). This finding has aroused new interest in the possibility of imposing artificial controls on the stomata of field crops, thus regulating transpirational water loss. This could be of great benefit to agriculture in drier parts of the world, particularly if used in conjunction with irrigation schemes. More work is needed, however, to determine whether what has been achieved in the laboratory is feasible on a large scale in the field.

Acknowledgement: I acknowledge the support of the Science Research Council for my current research on this subject, some of which is described in this paper.

REFERENCES

ALLAWAY, W. G. (1973) Accumulation of malate in guard cells of *Vicia faba* during stomatal opening. *Planta* **110,** 63–70.

ALLAWAY, W. G. and MANSFIELD, T. A. (1970) Experiments and observations on the after effect of wilting on stomata of *Rumex sanguineus*. *Can. J. Bot.* **48,** 513–521.

ALLAWAY, W. G. and SETTERFIELD, G. (1972) Ultrastructural observations on guard cells of *Vicia faba* and *Allium porrum*. *Can. J. Bot.* **50,** 1405–1413.

COOPER, M. J., DIGBY, J. and COOPER, P. J. (1972) Effects of plant hormones on the stomata of barley: a study of the interaction between abscisic acid and kinetin. *Planta* **105,** 43–49.

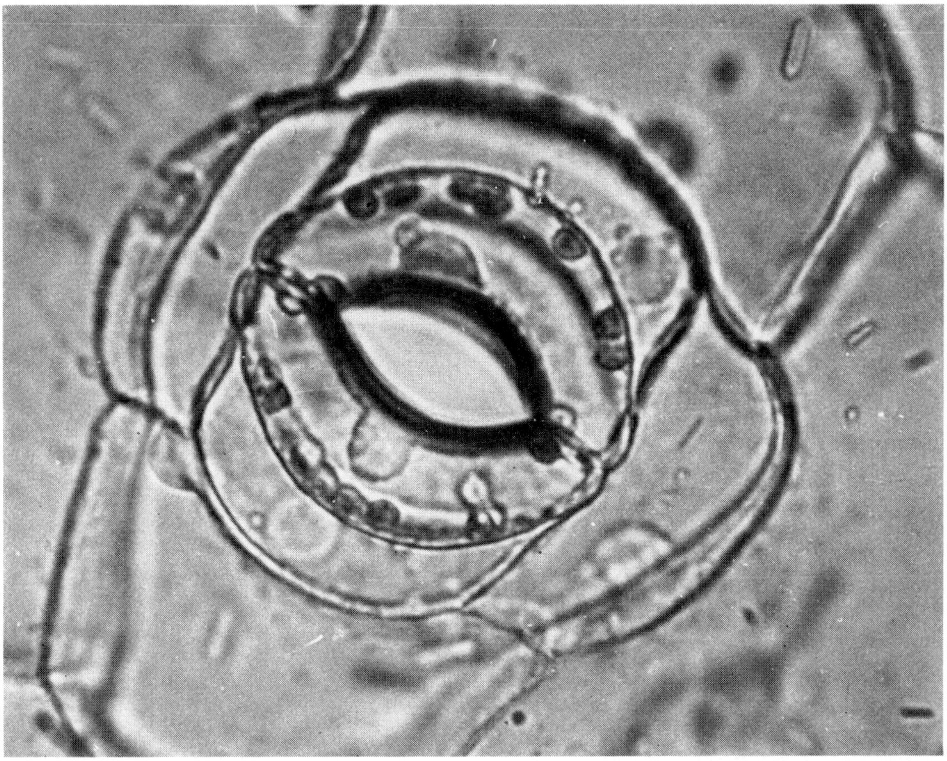

FIG. 5. Position of chloroplasts adjacent to the plasmalemma nearest the subsidiary cells. Epidermis was removed from leaves of *Commelina communis* and photographed within 2 min. Stomata were opening, but had achieved only half their full aperture, in CO_2-free air.

FISCHER, R. A. (1968) Stomatal opening: role of potassium uptake by guard cells. *Science N.Y.* **160,** 784–785.

FISCHER, R. A. (1971) Role of potassium in stomatal opening in the leaf of *Vicia faba. Pl. Physiol. Lancaster* **47,** 555–558.

FUJINO, M. (1967) Role of adenosinetriphosphate and adenosinetriphosphatase in stomatal movement. *Sci. Bull. Fac. Educ. Nagasaki Univ.* **18,** 1–47.

HEATH, O. V. S. and RUSSELL, J. (1954) An investigation of the light responses of wheat stomata with the attempted elimination of control by the mesophyll. Part II. *J. exp. Bot.* **5,** 269–292.

HSIAO, T. C., ALLAWAY, W. G. and EVANS, L. T. (1973) Action spectra for guard cell Rb^+ uptake and stomatal opening in *Vicia faba. Pl. Physiol. Lancaster* **51,** 82–88.

HUMBLE, G. D. and HSIAO, T. C. (1970) Light-dependent influx and efflux of potassium of guard cells during stomatal opening and closing. *Pl. Physiol. Lancaster* **46,** 483–487.

HUMBLE, G. D. and RASCHKE, K. (1971) Stomatal opening quantitatively related to potassium transport. *Pl. Physiol. Lancaster* **48,** 447–453.

JONES, R. J. and MANSFIELD, T. A. (1970) Suppression of stomatal opening in leaves treated with abscisic acid. *J. exp. Bot.* **21,** 714–719.

JONES, R. J. and MANSFIELD, T. A. (1972) Effects of abscisic acid and its esters on stomatal aperture and the transpiration ratio. *Physiologia Pl.* **26,** 321–327.

LITTLE, C. H. A. and EIDT, D. C. (1968) Effect of abscisic acid on budbreak and transpiration of woody species. *Nature Lond.* **220,** 498–499.

MANSFIELD, T. A. (1965) Studies in stomatal behaviour. XII. Opening in high temperature in darkness. *J. exp. Bot.* **16,** 721–731.

MANSFIELD, T. A. and HEATH, O. V. S. (1963) Studies in stomatal behaviour IX. Photoperiodic effects on rhythmic phenomena in *Xanthium pennsylvanicum. J. exp. Bot.* **14,** 334–352.

MANSFIELD, T. A. and JONES, R. J. (1971) Effects of abscisic acid on potassium uptake and starch content of stomatal guard cells. *Planta* **101,** 147–158.

MANSFIELD, T. A. and MEIDNER, H. (1966) Stomatal opening in light of different wavelengths: effects of blue light independent of carbon dioxide concentration. *J. exp. Bot.* **17,** 510–521.

MARTIN, E. S. and MEIDNER, H. (1972) The phase response of the dark stomatal rhythm in *Tradescantia virginiana* to light and dark treatments. *New Phytol.* **71,** 1045–1054.

MEIDNER, H. and MANSFIELD, T. A. (1968) *Physiology of Stomata*, McGraw-Hill, London.

MOURAVIEFF, I. (1958) Action de la lumière sur la cellule végétale. *Bull. Soc. bot. Fr.* **105,** 467–475.

MOURAVIEFF, I. (1973) Microphotométrie des fluctuations de la teneur en amidon des stomates éclairés par la lumière de 436 nm et 665 nm en absence ou en présence de gaz carbonique. *Anns. Sci. nat. Bot.* **14,** 377–383.

OGUNKANMI, A. B., TUCKER, D. J. and MANSFIELD, T. A. (1973) An improved bio-assay for abscisic acid and other antitranspirants. *New Phytol.* **72,** 277–282.

OGUNKANMI, A. B., WELLBURN, A. R. and MANSFIELD, T. A. (1974) Detection and preliminary identification of endogenous antitranspirants in water-stressed *Sorghum* plants. *Planta* **117,** 293–302.

PALLAGHY, C. K. and FISCHER, R. A. (1974) Metabolic aspects of stomatal opening and ion accumulation by guard cells in *Vicia faba. Z. Pfl.physiol.* **71,** 332–344.

PENNY, M. G. and BOWLING, D. J. F. (1974) A study of potassium gradients in the epidermis of intact leaves of *Commelina communis* L. in relation to stomatal opening. *Planta* **119,** 17–25.

RAGHAVENDRA, A. S. and DAS, V. S. R. (1972) Control of stomatal opening by cyclic photophosphorylation. *Curr. Sci.* **41,** 150–151.

RASCHKE, K. (1967) Der Einfluss von Rot und Blau Licht auf die Öffnungs und Schliess Geschwindigkeit der Stomata von *Zea mays. Naturwissenschaften* **54,** 73.

RASCHKE, K. and FELLOWS, M. P. (1971) Stomatal movement in *Zea mays*: Shuttle of potassium and chloride between guard cells and subsidiary cells. *Planta* **101,** 296–316.

RAVEN, J. A. (1967) Light stimulation of active transport in *Hydrodictyon africanum. J. gen. Physiol.* **50,** 1627–1640.

SQUIRE, G. R. and MANSFIELD, T. A. (1972) A simple method of isolating stomata on detached epidermis by low pH treatment: observations of the importance of the subsidiary cells. *New Phytol.* **71,** 1033–1043.

TAL, M. and IMBER, D. (1970) Abnormal stomatal behaviour and hormonal imbalance in *flacca*, a wilty mutant of tomato. II. Auxin and abscisic acid-like activity. *Pl. Physiol. Lancaster* **46,** 373–376.

TAL, M. and IMBER, D. (1971) Abnormal stomatal behaviour and hormonal imbalance in *flacca*, a wilty mutant of tomato. III. Hormonal effects on the water status of the plant. *Pl. Physiol. Lancaster* **47,** 849–850.

Tucker, D. J. and Mansfield, T. A. (1971) A simple bioassay for detecting 'antitranspirant' activity of naturally occurring compounds such as abscisic acid. *Planta* **98,** 157–163.

Willmer, C. M. and Mansfield, T. A. (1969) A critical examination of the use of detached epidermis in studies of stomatal physiology. *New Phytol.* **68,** 363–375.

Willmer, C. M. and Mansfield, T. A. (1970) Effects of some metabolic inhibitors and temperature on ion-stimulated stomatal opening in detached epidermis. *New Phytol.* **69,** 983–992.

Wilkins, M. B. (1962) An endogenous rhythm in the rate of carbon dioxide output of *Bryophyllum.* III. The effects of temperature changes on the phase and period of the rhythm. *J. exp. Bot.* **11,** 269–288.

Wright, S. T. C. (1969) An increase in the inhibitor-β content of detached wheat leaves following a period of wilting. *Planta* **86,** 10–20.

Zelitch, I. (1963) Chapter 2 of *Stomata and water relations in plants.* Connecticut Agricultural Experiment Station Bulletin no 664.

ENZYME SECRETION AND DIGEST UPTAKE IN CARNIVOROUS PLANTS

YOLANDE HESLOP-HARRISON

Cell Physiology Laboratory, Royal Botanic Gardens, Kew, London, U.K.

INTRODUCTION

Carnivory in flowering plants depends upon the adaptation of leaves to capture living prey, digest it and resorb the products. Carnivorous (or more narrowly, insectivorous) plants have intrigued botanists and zoologists for upwards of 200 years, but it is particularly appropriate to review the topic now, since 1974 marks the centenary of the paper given by J. D. Hooker to a joint meeting of Sections D and K of the British Association in Belfast, in which he surveyed much of the then current knowledge of insectivory. Hooker concentrated principally on the pitcher plants; by prior arrangement with Darwin, he touched little upon the physiology of insectivory, leaving this to Darwin himself, whose remarkable book, *Insectivorous Plants*, was published the following year. Hooker and Darwin had, of course, collaborated for more than a decade in the study of insectivorous species: indeed it was probably Hooker's initial interest in the phenomenon that had enthused Darwin, whose experimental work was made possible by a steady supply of plants from the Royal Botanic Gardens, Kew.

For much of the century that has elapsed since the publications of Hooker and Darwin, studies on insectivorous species have been concerned with taxonomy, distribution and general phenomenology, and these aspects were well summarized by Lloyd up to 1942. In the last 10 years or so, however, new light has been cast upon the physiology and biochemistry of insectivory, notably through the application of electron microscopy and cytochemistry.

One must at the outset dispel one of the myths that haunts the literature of insectivorous plants, namely the suggestion that the plant itself does not carry out the digestion but that this is due to associated bacteria. It is now established beyond doubt for most classes of insectivorous plants that certain glands of the leaf surface *do* synthesize and secrete enzymes, and that these are the principal agents of digestion. Secondary infection occurs in the liquid of the pitcher plants, and the flora there may contribute to the digestion of some constituents of the prey; but this is probably relatively unimportant, and no such commensalism occurs in other classes.

The importance of carnivory as a means of supplementing mineral nutrition is also not in any doubt. The feeding experiments carried out by Darwin and his son, Francis (1878), and by Büsgen (1883) proved that *Drosera* benefited from supplementary nutrition. In general, it seems that the habit is most important in supplementing nitrogen supply, but the tests of Pringsheim and Pringsheim (1962, 1967) and Harder (1963) on *Utricularia*, and of Harder and Zemlin (1967, 1968) on *Pinguicula* grown in axenic culture, show that insectivory may contribute to the supply of other mineral elements also.

On the basis of their morphological and physiological adaptations, carnivorous species may be grouped in four classes. In the first, typified by the pitcher plants such as *Nepenthes*, the glandular surfaces are immersed in the pitcher fluid, which is secreted during the early growth of the leaf. The fluid contains the digestive enzymes, and from it the products of digestion are withdrawn continuously during the life of the leaf, which is of several months duration. At the end of a season, a pitcher may contain large numbers of insect remains in various stages of decomposition, representing the catch for the life of the pitcher. In the second class, microscopic prey, principally in the form of free swimming copepods, are drawn into small but elaborate traps by suction (*Utricularia*) or water currents (*Genlisea*), and the glands probably do not release their enzymes until stimulation. After digestion and absorption the undigested detritus is ejected or remains in the utricle (*Utricularia*), or is passed into a bladder (*Genlisea*) for temporary storage before the whole organ is shed. In the third class the secretion product accumulates on the surface of the individual gland head forming the familiar globules seen on the leaves of *Drosera* (Fig. 1). Digestion begins when the globules are brought into contact with prey, and again the products are resorbed, seemingly once more through the glands. In the fourth class the glands responsible for digestion remain dry until stimulated by capture of the prey, as in *Dionaea* (Figs. 2A, B). The genera *Pinguicula* and *Drosophyllum* occupy a somewhat intermediate position between *Drosera* and *Dionaea* since their leaves possess stalked glands (the capturing glands), bearing permanent charges of mucilage, as well as sessile glands that do not secret until stimulated (Figs. 2C, D, 3A, B).

In the second two classes—the leaf-trapping insectivores as opposed to the pitcher plants and the small utricle types—a local pool of secretion product is formed on the leaf, and digestion takes place in this. Resorption then follows, and after this phase the undigested remains of the carcase dry out on the exposed leaf surface and often drop off. In all carnivorous plants but those with pitchers, the leaves are relatively ephemeral and new trapping organs are produced throughout the growing season. In *Pinguicula*, where the secretion–absorption cycle takes place only once, a new leaf is produced every 5 days or so (Heslop-Harrison and Knox, 1971). Where the leaves have a rather longer life, as in *Dionaea*, the cycle may occur repeatedly.

Most of the recent fine-structural and physiological work has been carried out on the leaf trapping species of *Drosera*, *Dionaea*, *Drosophyllum* and *Pinguicula*, and it is these genera that are mainly referred to in the following account, although results from other genera are mentioned where they are pertinent.

GLAND ARCHITECTURE

A reasonably comprehensive anatomical picture of carnivorous plant glands was built up in the late 19th and early 20th centuries through the work of Goebel (1891–1893), Fenner (1904) and Macfarlane (1908), and this was later supplemented by Lloyd (1933), who provided an excellent review of the field (Lloyd, 1942). In the last few years, transmission and scanning electron microscopy have provided further understanding both of

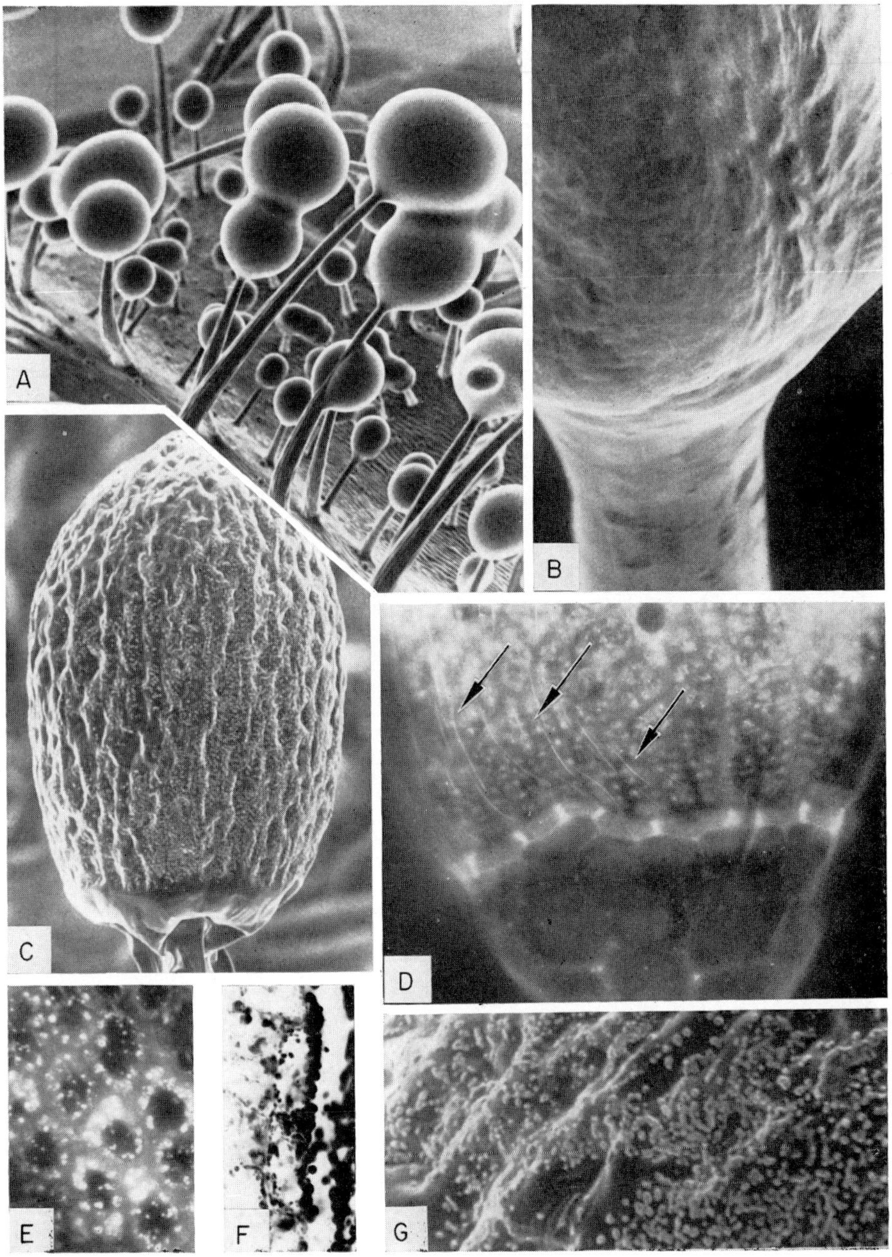

FIG. 1. *Drosera capensis* (A) Scanning electron micrograph (SEM) of the upper surface of an unstimulated leaf; fresh uncoated preparation. The leaf margin (foreground) bears large tentacles which curve inwards on stimulation; stalked glands of the centre of the leaf are smaller and are not mobile. On the right the glutinous nature of the gland secretions may be seen where three heads have come into contact. × c. 40. (B, C and G) SEMs of portions of a gland head. In (B) the surface secretion has been removed by blotting and light rinsing and the cell outlines are partly visible. In (C) all of the surface secretion has been removed after preliminary fixation. The upper part of the gland head is covered with an irregular coating of cutin particles, while the cuticle of the neck area and the gland stalk is continuous and without particles. The endodermal layer, with thickened radial walls has supported the upper part of the gland head, although the neck and stalk have collapsed somewhat. (G) details showing cutin particles. (B), × c. 550; (C), × c. 300; (G), × c. 1,500. (D) Fluorescence micrograph of a portion of the head and neck area of a gland head; benzpyrene staining after removal of surface lipids. The thin cuticle over the secretory cells has fallen into pleats (arrows). The endodermal layer where it comes to the surface at the junction of the secretory head cells and neck cells forms a conspicuous girdle with heavy cutinization along the radial walls corresponding to the regions of the Casparian strips. The polygonal cells of the neck may be compared with the elongated rectangular secretory cells bearing the cutinized particles seen in the SEMs of (C) and (G). × c. 550. (E) Detail of the secretory gland cell surface showing the cutinized particles; fluorescence micrograph, benzpyrene staining. × c. 800. (F). Longitudinal section of the two outer layers of secretory cells (left) and the endodermal layer (right). The endodermal cells are vertically elongated. Glutaraldehyde fixation, osmium tetroxide staining. The endodermal cells contain many osmiophilic bodies. × c. 400.

Facing p. 464

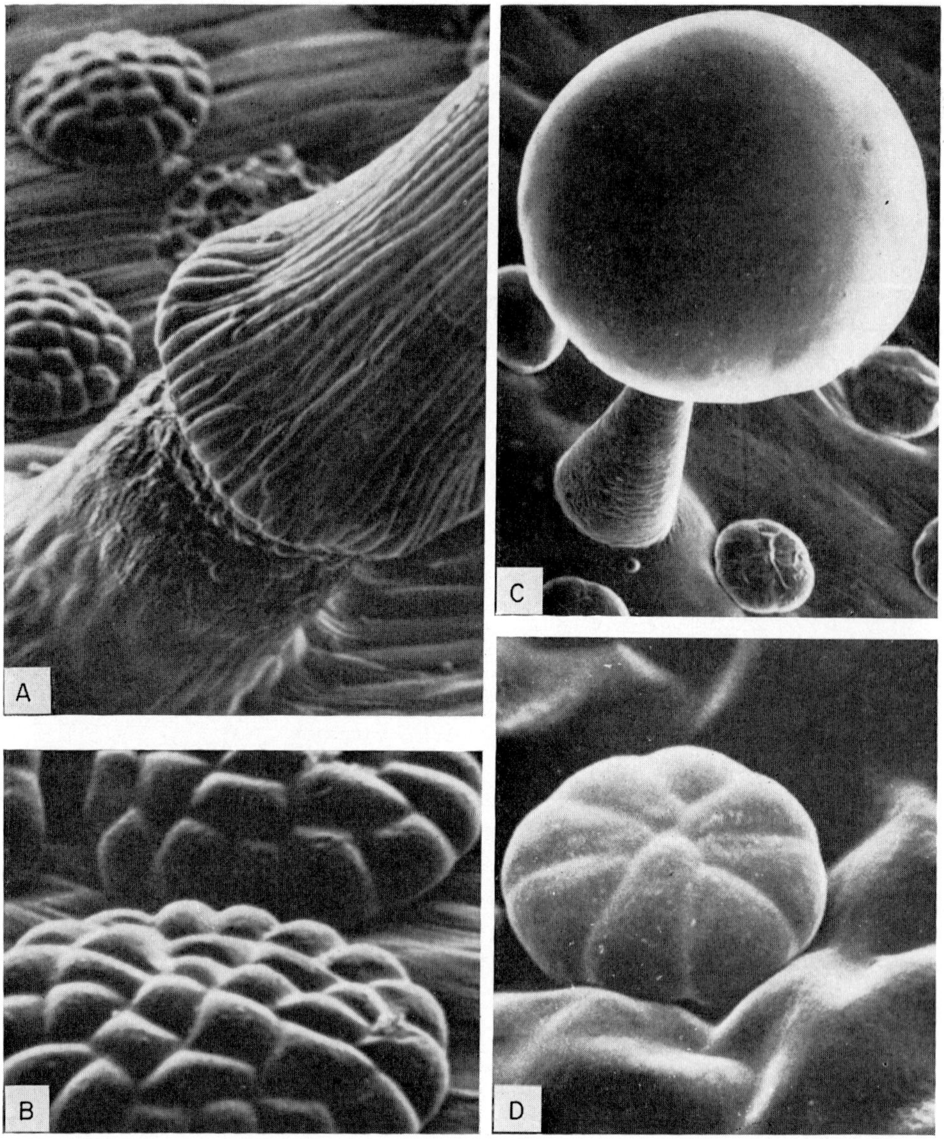

FIG. 2. (A, B) *Dionaea muscipula.* SEMs of the upper leaf surface, fresh uncoated preparations. In (A), foreground, the irritable base (podium) of a trigger hair may be seen and passing towards the top right a portion of the rigid shaft. The leaf closes when these hairs are stimulated. × c. 250. (B) Two digestive glands, × c. 500. That in the foreground consists of a head of thirty outer secretory cells which remain dry until stimulated. A second layer of secretory cells underlies these, in contact with the endodermal layer.

(C, D) *Pinguicula grandiflora.* SEMs of the unstimulated upper leaf surface, fresh and uncoated. (C) shows a stalked gland. The tapering single-celled stalk surmounts a slightly raised large reservoir cell, and the gland head bears a large mucilage droplet. Sessile glands may be seen below. × c. 300. (D) shows a sessile digestive gland. The secretory head consists of eight cells; the gland is unstimulated, and carries no secretion. The surrounding epidermal cells form a sump around it (cf. Fig. 4a). × c. 800.

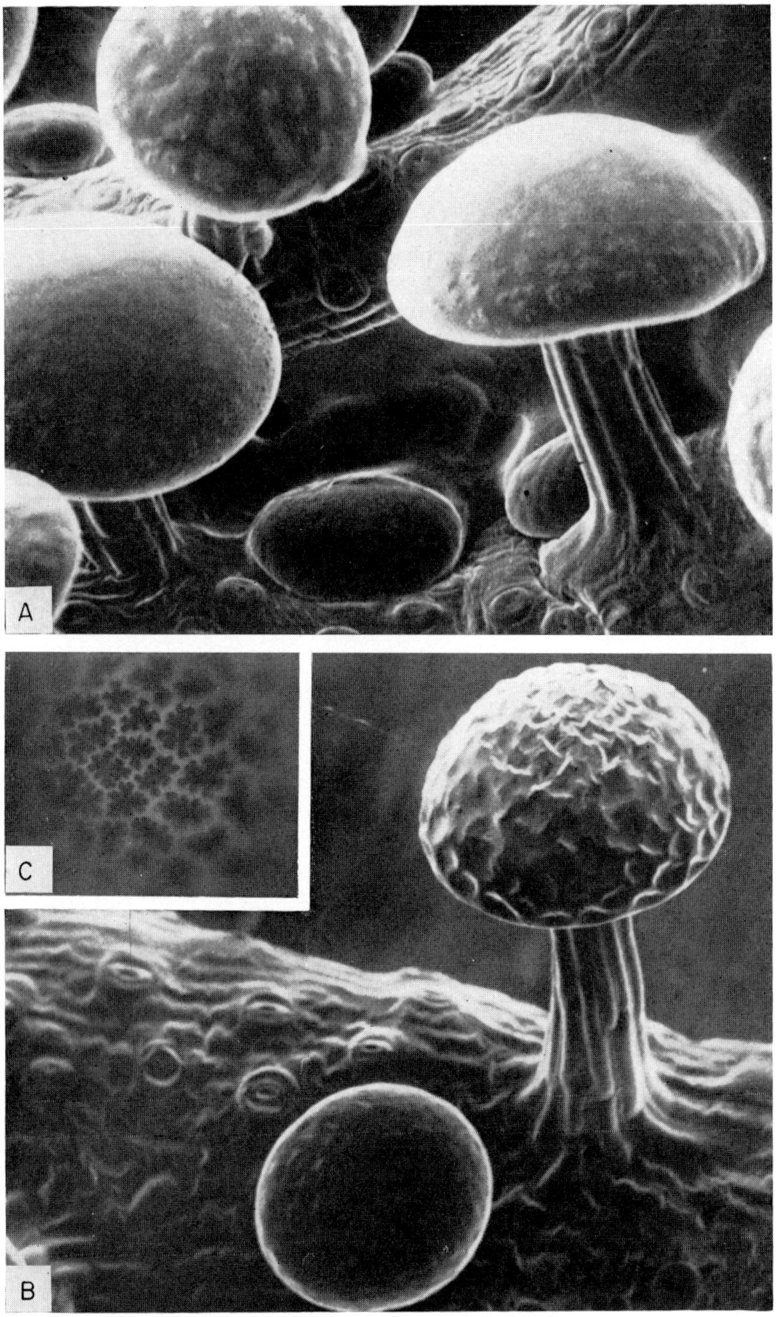

FIG. 3. (A, B) *Drosophyllum lusitanicum*. SEMs of the upper leaf surface, uncoated, × c. 200. (A) shows the stalked glands with the surface charge of secretion. The ribbed stalks are multicellular; numerous stomata occur on the leaf surface. The leaf has been slightly stimulated and the sessile glands have begun to secrete. In (B) the secretion has been drawn off the stalked gland (a single strand may be seen to the left). The outer walls of the secretory cells have collapsed somewhat so that their outlines have become distinguishable. (C). Optical micrograph of gland surface of *Drosophyllum lusitanicum*, secretion removed. The cell sap contains anthocyanins and the crenellated radial walls appear pale in contrast. × c. 400.

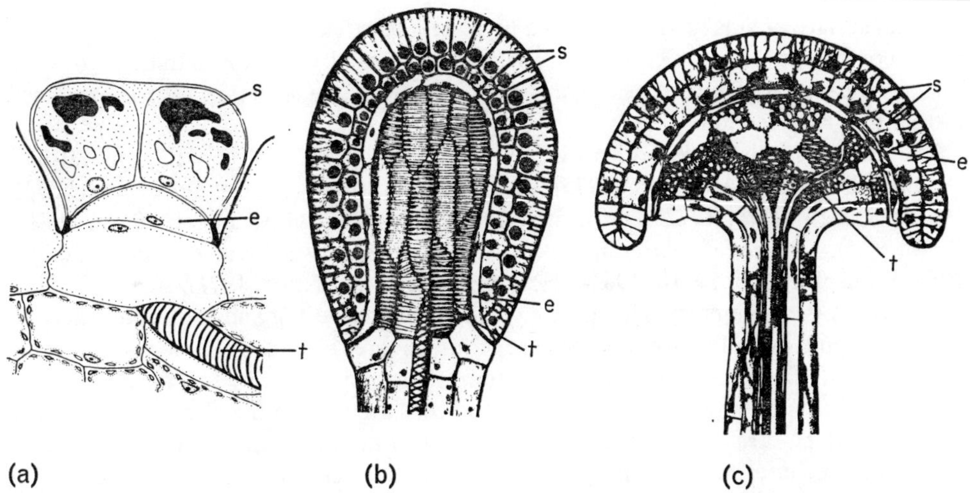

FIG. 4. A comparison of the anatomical features of digestive glands of (a) *Pinguicula* (sessile gland), (b) *Drosera* (stalked gland) and (c) *Drosophyllum* (stalked gland), seen in longitudinal section. s, Secretory cells; e, endodermal cell or cells; t, tracheid or tracheids. In (a) the tracheid does not abut directly on the endodermal cell and there is an intermediate reservoir cell; in (b) and (c) the swollen tracheidal cores of the gland heads are in direct communication with the endodermal cells and taper into the vascular cores of the gland stalk. (a) × c. 600; (b) × c. 200; (c) × c. 160.

the structure of the traps and the morphology of the secretory glands (Heslop-Harrison, 1970, 1975; Heslop-Harrison and Knox, 1971).

Notwithstanding differences in architectural detail, the gland heads of all classes of insectivorous plants consist essentially of but a small number of components, considered from a functional point of view (Fig. 4 and Heslop-Harrison, 1975). These components may be classed as 'secretory', 'endodermal', and 'communicatory'. This is true for the glands made up of relatively massive tissues, such as those of *Drosera*, *Drosophyllum* and *Nepenthes*, as well as for the much simpler ones of *Pinguicula*, *Genlisea* and *Utricularia*. The appearance of the glands does, of course, vary a good deal between the different genera, the more so since the gland heads may be either sunken in the leaf tissue, lie flush with the leaf surface, or be raised upon stalks of varying length and complexity.

The following survey refers only to glands associated with enzyme secretion and so excludes, for example, alluring glands found in pitcher plants, sessile glands on the outer trap surfaces of *Genlisea* and *Utricularia*, and sessile glands on thc upper leaf surface, and even tentacle stalks, of *Drosera*.

The secretory cells

These form a group of a few or several cells in a single layer (*Pinguicula*, *Dionaea*, *Genlisea* and *Utricularia*), a double layer (*Drosera* and *Drosophyllum*) or a glomerulus (*Nepenthes*). Whatever their number or disposition, each secretory cell is characterized by possessing one or more walls with a spongy inner surface of the transfer cell type. In *Pinguicula* (Figs. 6D–F) and *Genlisea* (Figs. 6H, I), the surface ramifications are present on the radial walls and to a lesser extent on the outer walls. In *Drosera* and *Drosophyllum*, the total surface area is further increased by plate-like extensions, visible with the optical microscope (Figs. 1D, E, 3C, 6A, B), themselves covered with labyrinthine ingrowths. The significance of the presence of this type of wall in the sercetory cells is examined further in a later section.

The secretory cells possess other distinctive features apart from the wall ramifications, reviewed by Heslop-Harrison (1975) which summarizes work on *Drosera* (Schnepf, 1961b; Ragetli *et al.*, 1972; Dexheimer, 1972), *Drosophyllum* (Schnepf, 1960a, 1961a, 1963a, b, c), *Dionaea* (Scala *et al.*, 1968; Schwab *et al.*, 1969), and *Pinguicula* (Vogel, 1960; Schnepf, 1960b, 1961b, 1963a; Heslop-Harrison and Knox, 1971).

As with other secretory glands, e.g., nectaries and salt glands, the pathways of secretion—and, as will be seen, in the case of insectivores also of resorption—are related in an intimate fashion to the cuticularization of the cell surfaces. In *Pinguicula* (Figs. 7A, C–E) the outer walls of the secretory head cells are seen in the electron microscope to bear a thin, discontinuous cuticle compared with the much thicker, continuous cuticle of the outer walls of neighbouring epidermal cells. In *Drosera* the cuticle is penetrated by regular pores (Heslop-Harrison, 1975; Williams and Pickard, 1969, 1974; Figs. 8B, C); these lie mostly over the anticlinal walls of the gland head cells, and they must provide the principal pathway for the passage of secretion and resorption products. The cuticle, although appearing more continuous than in *Pinguicula*, is nevertheless quite delicate; it readily falls into tucks and folds when the cells collapse (Fig. 1D) and does not hold the cells together after cellulase digestion of the underlying wall. The surface is spangled with larger particles, also identifiable as cutin from their cytochemical reactions (Figs. 1B–E, G).

Endodermal layer

This consists of a single layer of cells separating the secretory cells from the underlying tissues. In *Drosera* the layer is thimble-shaped; in *Nepenthes* it is sub-spherical; and in *Drosophyllum* it takes the form of a very shallow cup. In *Pinguicula*, *Genlisea* and *Utricularia* there is only a single endodermal cell, and in *Dionaea*, two.

The heavy cutinization of the Casparian strips of the endodermal cells is readily demonstrated cytochemically (Fig. 1D), and the thickening is also very conspicuous optical microscopically through its higher refractive index, and electron microscopically because of its electron density after osmication (Fig. 7D). Continuity of the Casparian strip with the cuticle of the epidermal cells on the one hand and the thin cuticle of the secretory head cells on the other may be traced from Fig. 7D, of *Pinguicula*. The significance of this relationship is discussed further below in connection with the secretion mechanism.

In *Pinguicula* the nucleus of the endodermal cell, which is usually seen to contain massive protein bodies of a distinct fibrillar organization (Schnepf, 1960b, 1961b; Heslop-Harrison, 1975), usually lies against the upper wall, adjacent to the glandular head cells. In certain genera, notably *Drosera*, the endodermal cell contains many osmiophilic bodies (Fig. 1F), but in others such as *Pinguicula* these are not conspicuous. Invariably the mitochondria are large, and with well developed cristae.

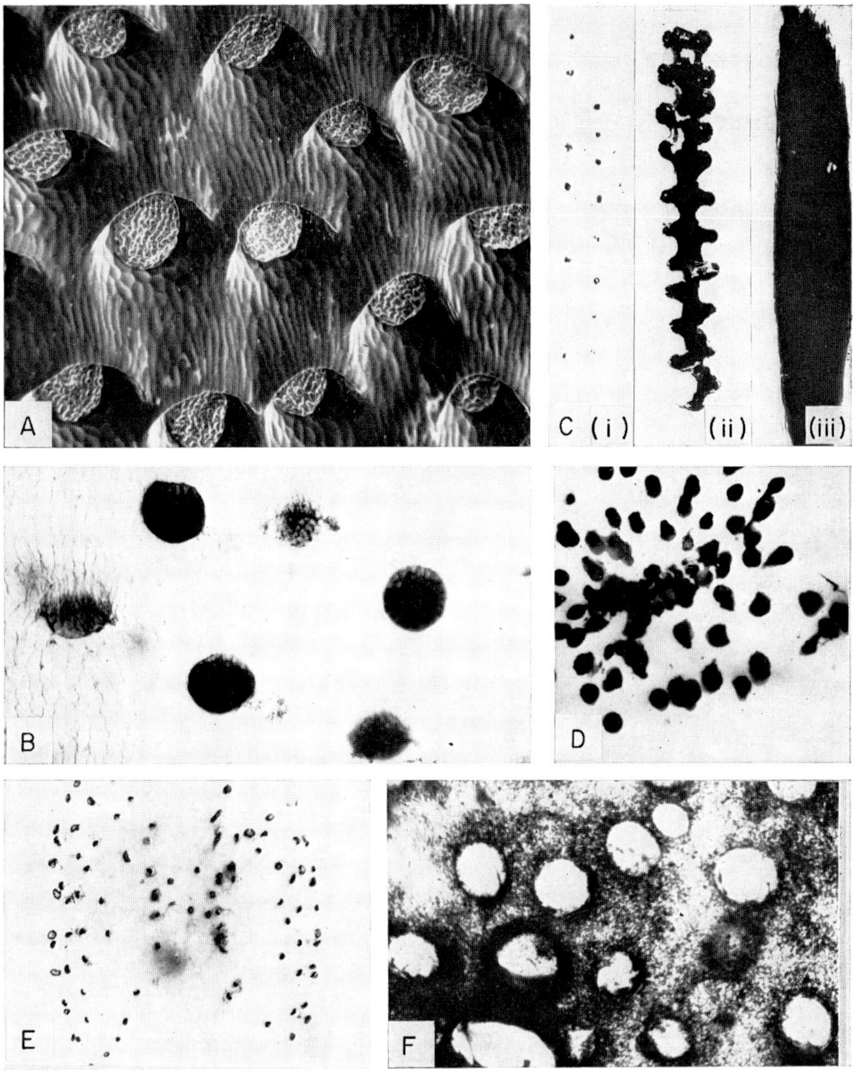

FIG. 5. (A, B) *Nepenthes rafflesiana*. (A) SEM of the inner surface of the lower part of the pitcher showing the sunken digestive glands partly protected by epidermal flaps. × c. 100. (B) Acid phosphatase reaction (ASBI-phosphate substrate; hexazotized pararosanilin coupling reagent) in digestive gland cell walls. Some of the glands have secreted, and give little reaction; the wall localization is clearest in the gland to the right. (C) *Drosophyllum lusitanicum*. Progressive digestion of the gelatin layer of processed colour film by proteolytic enzymes secreted by the glands after application of leaf strips to the dry emulsion surface for periods of (i) ½ hr, (ii) 1½ hr and (iii) 3 hr. The first holes in (i) have been formed over the sites of the stalked glands; later digestion begins over the central part of the leaf, as in (ii) and (iii) as the sessile glands begin to secrete. × c.6. (D, E) *Drosera capensis*. Esterase activity in the gland secretion of an unstimulated leaf. The secretions were printed onto 5 per cent gelatin and then bathed immediately with the reaction mixture (α-naphthyl acetate substrate, fast blue B salt coupling agent). × c. 3. (E). Imprint of gland secretions from unstimulated leaves on glass; proteins fixed and stained with Coomassie Blue stain-fixing medium. × c. 1. (F) *Pinguicula grandiflora*. Digestion of a photographic emulsion by proteolytic enzymes secreted by the sessile glands on the upper leaf surface. × c. 140.

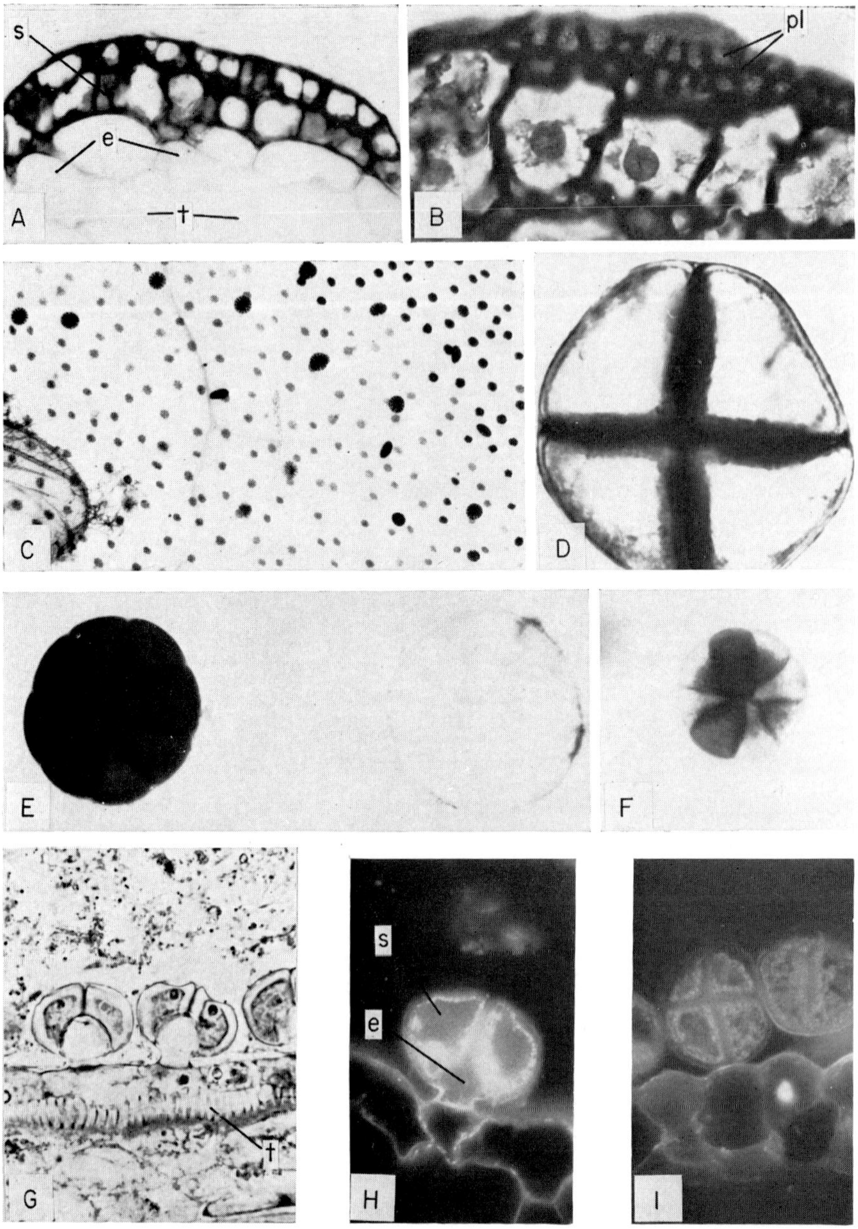

FIG. 6. (A, B). *Drosophyllum lusitanicum*. Freeze-sectioned gland heads showing enzyme storage in the walls of the secretory cells; acid phosphatase reaction medium, α-naphthyl phosphate substrate, hexazotized pararosanilin coupling agent. In (A) the section is longitudinal and the two layers of secretory cells (s) may be distinguished, each with wall-held enzymes. The endodermal cells (e) and the tracheidal cells (t) show no phosphatase activity. × c. 350. In (B) the section plane is almost paradermal. The plate-like in-growths of the anticlinal cells (pl) seen in a living gland in Fig. 3C, bear a heavy enzyme load. The large secretory cell nuclei have been stained lightly with safranin. × c. 500. (C–F) *Pinguicola grandiflora*. Acid phosphatase activity. (C) Whole mount of a portion of a leaf bearing a captured insect, portion of wing visible to the left. The zone of stimulated sessile glands (pale grey, small gland heads) corresponding to the secretion pool, extends around the insect. The limits of the pool are visible along a diagonal line at the top right, and beyond this the sessile glands are unstimulated and still carry the full enzyme load. The larger stalked glands are stimulated (pale grey) in the deeper central part of the pool, and unstimulated (black) at the margins where the heads stand above the level of the secretion pool. × c. 25. (D) Shows a partly stimulated gland. The enzyme has been leached from the outer walls, but activity remains in the spongy radial walls. × c. 1000. In (E) the left hand gland has not discharged, and enzyme activity remains in the outer and radial walls and also in the vacuoles; in that on the right the enzymes have been leached out of all of the storage sites and only a trace of activity remains associated with the gland surface on the right. That two adjacent glands of the same leaf should behave differently in this manner shows that they respond independently to the stimulus eliciting discharge. × c. 500. In the partially stimulated sessile gland in (F), six head cells have almost completely discharged and two are yet to do so. Independent discharge is rather rare; generally the head cells of a stimulated gland secrete simultaneously, as would be expected from a mechanism of the kind described in the text. × c. 400. (G–I) *Genlisea africana*. (G) Phase contrast image of a longitudinal section through the utricle region showing three glands, each abutting the same tracheid (t). The gland heads are surrounded by partly digested debris in the tubular leaf cavity. × c. 400. (H) and (I) are of freeze-sectioned digestive glands, 1-anilino-naphthylsulphonic acid stained for total protein. The protein is mostly associated with the thickened spongy radial walls (compare (I) with (D) of the head of a sessile gland of *Pinguicula grandiflora*).

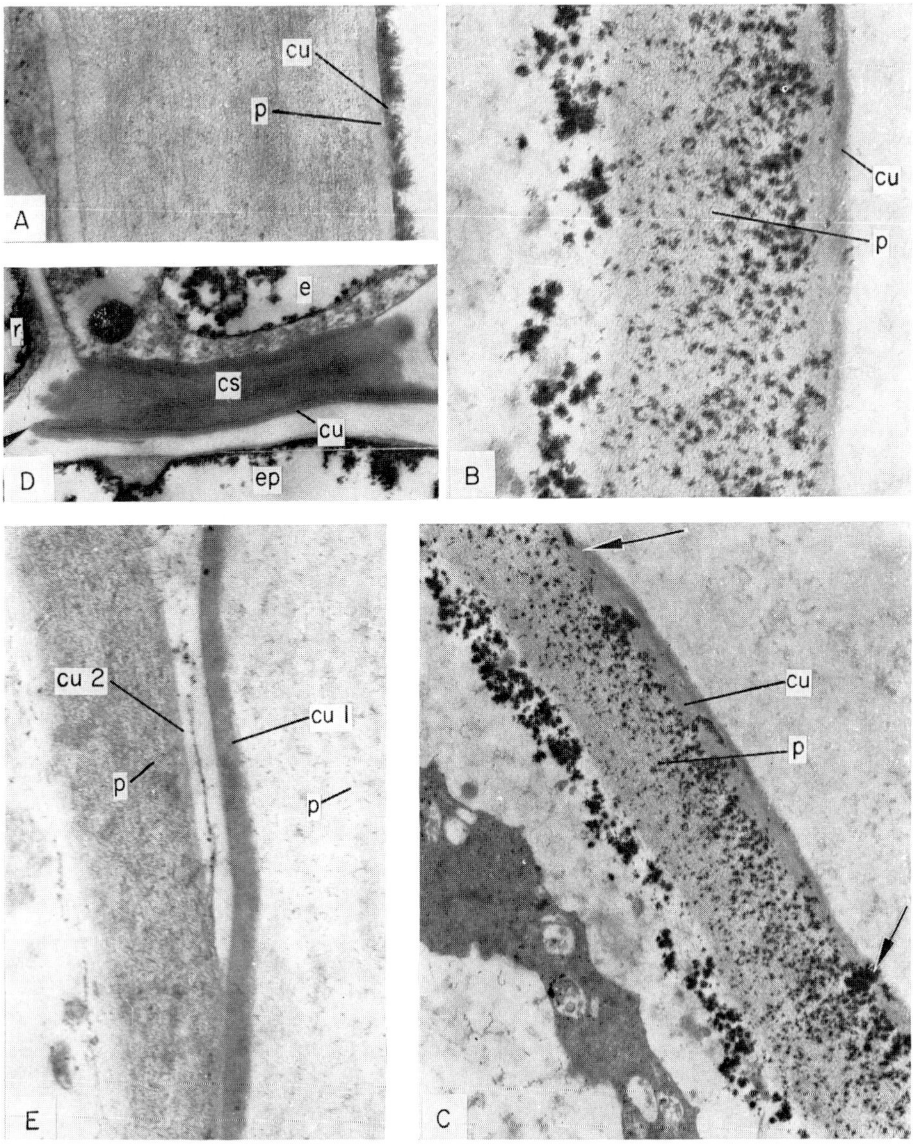

FIG. 7. *Pinguicula grandiflora*. (A–C) Electron micrographs of cuticularized regions. (A), outer wall of an epidermal cell showing the retention of colloidal lanthanum nitrate added to the secretion pool at the cuticle (cu); none has penetrated the underlying pectocellulosic wall (p). (B) (× c. 50,000) and (C) (× c. 25,000) are of parts of the outer wall of a secretory cell from a sessile gland, colloidal lanthanum nitrate added to the secretion pool during the resorption phase. The tracer has passed through the discontinuities in the thin cuticle (arrow); much is still present in the pectocellulosic wall, but the main accumulations at this time are in the spongy inner wall. (D) Morphology of the Casparian strip of the endodermal cell of *Pinguicula grandiflora*; e, endodermal cell; cs. Casparian strip; ep, adjacent epidermal cell; r, reservoir cell; cu, cuticle of epidermal cell. The cuticle of the epidermal cell is continuous with the Casparian strip (arrow, left), ensuring a complete seal. The plasmalemma of the endodermal cell is tightly appressed against the Casparian strip. × c. 15,000. (E) Adjacent walls of an epidermal cell (right) and a sessile gland head cell (left) in *P. grandiflora*. The cuticle of the epidermal cell is thick and continuous (cu 1); that of the gland cell is barely distinguishable with this preparation (cu 2). Glutaraldehyde-osmium tetroxide fixation; uranyl acetate post-staining. × c. 20,000.

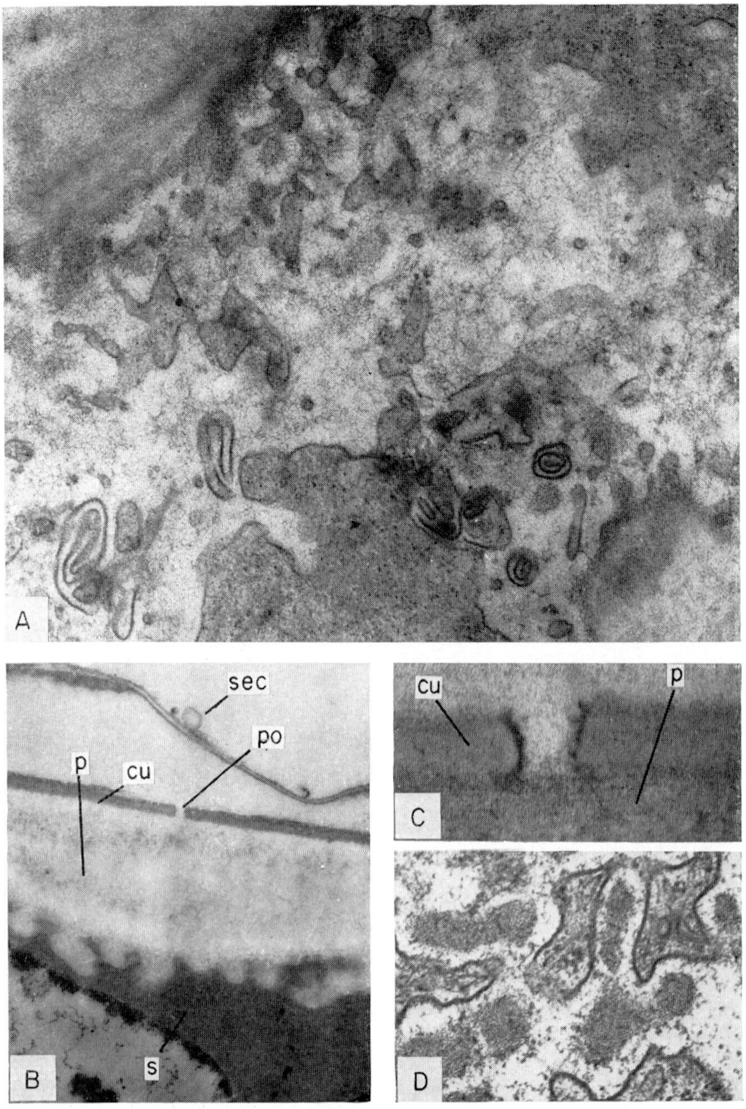

FIG. 8. (A–C) *Drosera capensis*. Electron micrographs of secretory cells of gland heads. (A) is of the inner labyrinthine part of the wall. The ramifications are invested in evaginations of the plasmalemma, but some of the configurations suggest that portions have been cut off into the wall complete with associated cytoplasm. (B) illustrates the outer wall in true longitudinal section. The cuticle is well-defined (cu), and is penetrated by a pore (po). The labyrinthine wall underlies the pectocellulosic wall (p). A film of secretion is retained near the surface (sec). × c. 25,000. (C) shows the well-formed profile of a cuticular pore. × c. 90,000. (D) *Pinguicula grandiflora*. Section of the spongy wall of a secretory cell. The evaginations of cytoplasm are invested in a continuous plasmalemma. The wall ramifications show a denser granular core with a more electron-transparent outer zone, characteristic of enzyme-secretory cells which do not release their contents until stimulation. × c. 65,000.

Communicatory layers

The secretory glands are always in more or less direct communication with the vascular supply of the leaf. In the more complex glands such as those of *Drosera* and *Drosophyllum* which discharge large amounts of fluid on stimulation, the centre of the gland head itself consists of a core of tracheids abutting directly on the endodermal layer; this core connects with the vessels of the stalk of the gland, which in turn tap into those of the vascular system of the leaf. In the simpler glands of *Pinguicula, Genlisea* and *Utricularia* there is no direct connection between the vascular system and the endodermal cells, but the latter are never more than two or three cells removed from vessel terminations, and sometimes specialized 'reservoir' cells serve as intermediate links (Heslop-Harrison, 1975).

Enzyme storage and secretion

Physical and chemical properties of the secretions

In the pitcher plants the bulk of the secretion accumulates as the pitcher matures and before it opens for the capture of prey. In some species of *Nepenthes* the secretion is a glutinous fluid amounting to as much as a litre in a mature pitcher. In other pitcher-plant genera the secretion is very much less. There is no evidence of a stimulation of secretion flow following upon the capture of prey.

As we have already noted, the amounts of secretion held by the glands of the leaf-trapping species varies with the genus. All *Drosera* leaf digestive glands and the stalked glands of *Drosophyllum* and *Pinguicula* hold permanent globules of secretion, while the sessile glands of *Drosophyllum* and *Pinguicula* and all the glands of *Dionaea* are dry on the unstimulated leaf. After the capture of prey, however, all of the digestive glands of the leaf-trapping species are stimulated into further secretion, to add to the existing globules in *Drosera*, and to form secretion pools in *Drosophyllum, Dionaea* and *Pinguicula.* The size of the secretion pool is more or less related to the amount of stimulation; generally it is restricted to the area stimulated, which in *Pinguicula* becomes slightly dished, but occasionally the pool may become so large as to run off the leaf, as noted by Darwin.

A fair amount of analytical work has been carried out on the fluid of the pitcher plants but regrettably many of the results—including some of very recent date—are meaningless physiologically because fluid has been derived from open pitchers contaminated with bacteria and decaying prey, a hazard noted by Hepburn *et al.* as far back as 1919. The difficulties of separating, purifying and characterizing the pitcher enzymes are increased by the presence of carbohydrates in the secretion itself (Lüttge, 1964a,b; Steckelburg *et al.*, 1967; Jentsch, 1970, 1972; Nakayama and Amagase, 1968; Amagase *et al.*, 1972; Amagase, 1972; Tökes *et al.*, 1974). In the species with leaf surface secretions it is not easy to acquire large enough samples for biochemical study. Various micromethods have been used for the secretions of *Drosera* (Tate, 1875; Holter and Linderstrøm-Lang, 1932), and some studies have been made on crude extracts of whole leaves (Dernby, 1917; Amagase, 1972; Amagase *et al.*, 1972). The results from extracts naturally need not give a true picture of what is in the secretion itself, since they necessarily include all the enzymes of intermediate metabolism from within the leaf cells. In *Dionaea* the leaf has to be stimulated by feeding before secretions for analysis can be obtained (Lüttge, 1964a).

The permanent secretion globules of leaf-trapping genera invariably contain mucopolysaccharides. This fraction, which has been characterized for *Drosophyllum* by Schnepf (1961b), is effective in capturing and entrapping the prey and also in stabilizing the secretion on the leaf surface and around the prey after its release from the gland heads. The secretion has marked detergent properties, so that droplets from different glands readily coalesce, the pool formed flowing over the whole of the insect exoskeleton, including the normally non-wettable wing surfaces even when these are not actually immersed (Heslop-Harrison and Knox, 1971). The secretion globules on unstimulated glands are more acid than cell sap, the pH being in the range 2.5–3.0 in *Drosophyllum* and 3.0–5.0 in *Drosera*.

Using high-resolution substrate film and transfer methods developed to characterize emissions from the walls of pollen grains (Heslop-Harrison *et al.*, 1973, 1974) unequivocal evidence for the presence of various physiologically significant constituents in the permanent secretion droplets has been obtained. Whilst there are obvious similarities between the secretion droplets of *Drosera*, *Drosophyllum* and *Pinguicula*, there are also notable differences. Thus the droplets of *Drosophyllum* are very rich in peroxidases, whilst those of *Drosera*, though lacking marked peroxidase activity, are remarkably rich in esterase (Fig. 5D). The total protein content is also high (Fig. 5E). Some indication of the quantitative differences, so far as these can be judged from cytochemical evidence, is given in Table 1. In *Pinguicula*, the stalked 'mucilage' glands are concerned principally with the capture of the prey, and it is significant therefore that little enzyme activity can

TABLE 1. ENZYMES OCCURRING IN THE DIGESTIVE GLANDS OF CARNIVOROUS PLANTS, AS DETERMINED BY (1) HISTOCHEMICAL REACTIONS OR (2) SUBSTRATE FILM METHODS.

Genus	Oxidases 1.11.1.7 Peroxidase	Transferases 2.7.7.16 Ribonuclease	Hydrolases 3.1.1 Esterase	3.1.3.2 Acid phosphatase	3.2.1.1 Amylase	3.4.4 Proteases
Nepenthaceae						
Nepenthes	+ ?(2)	N/T	+(1)	++(1,2)	N/T	N/T
Droseraceae						
Drosera	+ (2)	N/T	++(1,2)	+(1,2)	N/T	+(2)
Drosophyllum	++ (2)	N/T	+(1,2)	+(1,2)	N/T	++(2)
Dionaea	N/T	N/T	+(1,2)	+(1,2)	N/T	++(2)
Lentibulariaceae						
Pinguicula						
stalked glands	− (2)	+(1)	+(1)	+(1)	+(2)	+(2)
sessile glands	− (2)	++(1)	++(1)	++(1)	−(2)	++(2)
Genlisea	N/T	N/T	+(1)	+(1)	N/T	
Utricularia	N/T	N/T	+(1)	+(1)	N/T	

+ denotes presence; ++ denotes heavy reaction; − absence; N/T not tested. All results except for those of *Pinguicula* (Heslop-Harrison and Knox, 1971) from Heslop-Harrison (1975).

be detected in the secretions before stimulation. The stimulation of prey is required for enzyme discharge, just as with the sessile, dry digestive glands, mentioned further below.

Substrate film and transfer methods can also be used to follow the kinetics of enzyme release. When a leaf of *Drosophyllum* is left in contact with processed colour film, some digestion of the surface gelatine layers occurs over the stalked glands within 30 min. However, in a further 2 hr a secretion pool builds up along the whole length of the leaf in contact with the film, and this contains enzymes which quickly remove all of the gelatine (Figs. 5C(i–iii)). This new secretion is mostly from the sessile glands, but it is not excluded that stalked glands contribute further as a result of stimulation. With *Dionaea*, a spectacular demonstration of the effectiveness of the secreted proteases may be obtained by feeding the leaf small fragments of exposed and processed colour film: the trap closes, the glands are stimulated, and by the time the leaf reopens only the transparent film base remains. Figure 5F illustrates the puncturing of autoradiographic liquid emulsion over the sessile glands of *Pinguicula* by secreted proteases.

Various other techniques are available for following enzyme discharge. The build up of esterase activity in the secretion pool on the leaf of *Pinguicula* has been timed using fluorescein diacetate which releases fluorescein on cleavage by non-specific esterase, detectable by its fluorescence in trace amounts (Heslop-Harrison and Knox, 1971). The emission of proteins from the head cells of the glands of *Pinguicula* after stimulation can even be observed directly using a Coomassie Blue stain-fixing medium (Heslop-Harrison, 1975).

Enzyme storage sites and discharge patterns

The precise intracellular localization of the enzyme storage sites was established first for *Pinguicula* (Heslop-Harrison and Knox, 1971) by means of cytochemical techniques developed for work on pollen (Knox and Heslop-Harrison, 1969, 1971; Heslop-Harrison and Heslop-Harrison, 1970). The methods have now been extended to several other genera (Heslop-Harrison, 1975). Some of the results are summarized in Table 1. In all species so far studied the principal storage sites are associated with the ramifying wall zones of the secretory cells (Figs. 5A, B, 6A–I). In certain species, notably *Pinguicula*, acid phosphatase and esterase are also stored in the vacuoles. It is to be noted that the digestive enzymes so far located intracellularly are those for which suitable cytochemical methods exist; proteases and amylases, readily detected in the leaf secretions, have not been localized within the gland cells because of the lack of an appropriate technique. There seems little doubt, however, that they will prove to be held in the same sites as the other acid hydrolases.

On stimulation the enzymes leave the storage sites in the gland head cells in a regular manner. The sequence has been traced in detail in *Pinguicula* (Heslop-Harrison and Knox, 1971). The secretion pool accumulates gradually around a captured insect, so that a transect from the prey at the centre to the pool margin represents a time sequence, with the glands in contact with the prey discharged first and those on the margin, last (Fig. 6C). Following the sequence across such a pool, it may be seen that the enzymes are leached first from the outer walls and vacuoles of the sessile gland head cells, and then more gradually from the anticlinal walls. In Fig. 6D, the mottled appearance of the reaction product in a partially stimulated gland shows the close association of the enzyme with the spongy inner wall surface, but two interpretations are possible; the enzyme may at this stage be held within the wall processes, or alternatively in the space between the plasmalemma and the wall. As yet there has been no adequate electron microscopic localization, but techniques are available which should allow a decision. The enzymes are ultimately entirely leached from the anticlinal walls, and the glands then lack any indication of activity (Fig. 6E). Occasionally on a stimulated leaf of *Pinguicula* a gland will fail to secrete; Fig. 6E shows this situation in a leaf exposed to a reaction for acid phosphatase; of the two glands seen, one is unstimulated, and the other wholly discharged.

The pattern of discharge from the stalked glands of *Pinguicula* is much the same as with the sessile glands, although the former show much less vacuolar storage. The stalked glands provide a clear demonstration that stimulation by the advancing secretion pool is needed before discharge begins; in a shallow pool, or along pool margins, the heads of the stalked glands above the surface may remain undischarged, even though the enzymes from the storage sites of neighbouring sessile glands are entirely lost (Fig. 6C).

In *Nepenthes* acid hydrolases can be detected in the gland heads sunken in pits on the pitcher walls (Figs. 5A, B). Again there is a clear association with the walls of the secretory cells, but there is no regular pattern of discharge such as that seen in the secretion pool of the leaf surface of *Pinguicula*; the release into the pitcher fluid occurs irregularly among the different glands in any one area of the pitcher wall (Fig. 5B). However, as in *Pinguicula*, it is usual for all of the secretory cells of any one gland to discharge simultaneously. Just occasionally do some cells of a gland secrete independently (Fig. 6F), and the fact that this can happen suggests that the stimuli are not necessarily transferred immediately between neighbouring cells, notwithstanding the plasmodesmatal continuity.

Although the techniques available for intracellular enzyme localizations are not particularly well adapted for quantitative studies, they do permit some broad conclusions about the relative amounts held in different classes of glands. In *Pinguicula*, there is no doubt that acid phosphatase and esterase activity is more pronounced in the sessile than in the stalked glands; on the other hand, substrate film methods suggest that amylases are mostly associated with the stalked glands (Heslop-Harrison and Knox, 1971).

Fine structural features associated with secretion

The investigations of Schnepf (1961b, 1963a, b, c) have given a reasonably convincing picture of the processes whereby polysaccharide fractions of permanent secretion globules of *Drosophyllum*, *Drosera* and *Pinguicula* are secreted. According to Schnepf, the synthesis is in dictyosomes, which release vesicles that discharge at the plasmalemma. The contents then pass into the wall, and are emitted from the surface through pores or other discontinuities in the cuticle. The wall ramifications, by increasing the area of interface between protoplast and wall and so decreasing the impedance, are presumably important for this process, which certainly involves a massive fluid flow. The passage is undoubtedly mainly through the spongy walls, and correlated with this is the distribution of the cuticular pores, which as we have seen are concentrated over the anticlinal walls in *Drosera*. Lloyd (1942) and later Schnepf (1961b) showed that when the droplet is removed from a gland head, secretion begins again, but mainly over these anticlinal walls. In *Pinguicula*, where there is no special system of pores, the discharge probably takes place irregularly over the whole outer surface of the gland head.

As yet there is no full picture of the mechanism of transfer of the secretion proteins into the wall, and indeed it is possible that slightly different processes are involved in the different genera. In the young sessile glands of *Pinguicula*, the developing wall processes are enclosed within evaginations of the plasmalemma, and at the time of enzyme synthesis there is an abundant peripheral ribosomal endoplasmic reticulum. The large, amoeboid plastids are completely ensheathed in extensions of the endoplasmic reticulum (Vogel, 1960), suggesting that they may play some part in the synthesis. Later there is some indication that embayments of cytoplasm may be passed completely into the wall even before stimulation (Fig. 8D), but this is uncertain. The observations of Schnepf (1963c) and of Schwab *et al.* (1969) on *Dionaea* suggest that after stimulation and with the onset of the secretion phase the embayments of cytoplasm are cut off, the plasmalemma withdrawing from the wall and becoming continuous, a process discussed further below. In *Drosera*, there seems little doubt that a cutting off of peripheral evaginations of the cytoplasm continues throughout the life of the gland, so allowing for a continuous passage of cytoplasmically synthesized proteins into the wall. In consequence of this process, the spongy zone of the wall is at all times seen to contain membrane fragments, sometimes in wildly convoluted forms (Fig. 8A).

Gland secretion and absorption

The secretion phase

A partial scheme purporting to explain the hydrodynamics of secretion has been proposed for *Pinguicula* by Heslop-Harrison and Knox (1971). This can now be extended to all of those species where gland secretion is not continuous but depends upon stimulation by captured prey. As already noted, such stimulation results in a remarkably rapid release of fluid, the size of the secretion pool produced being related more or less to the degree of stimulation. In *Pinguicula*, although the stalked capturing glands do contribute to the enzymes present in the secretion pool, the bulk of the fluid passes onto the leaf surface from the sessile glands, which do not bear any secretion before stimulation (Fig. 2D). The anatomy of the leaf in the neighbourhood of the sessile glands suggests that much of the secreted water is derived at first from the adjacent reservoir and epidermal cells. The reservoir cells are always associated with vessel terminations in the mesophyll, and thus can tap directly the vascular system of the plant. The passage of water into the gland head cells is controlled by the protoplast of the endodermal (basal) cell, since the cutinized periclinal walls forming the Casparian strips preclude permeation through the free space of the wall. Such enforced symplastic flow appears to be common for the basal cells of many plant glands (Lüttge, 1971).

The pathways of water flow in the stimulated gland may therefore be set out as in Fig. 9, which can be taken as referring either to the paucicellular or multicellular type of gland. If it be assumed that water follows a diffusion pressure gradient, then the initial

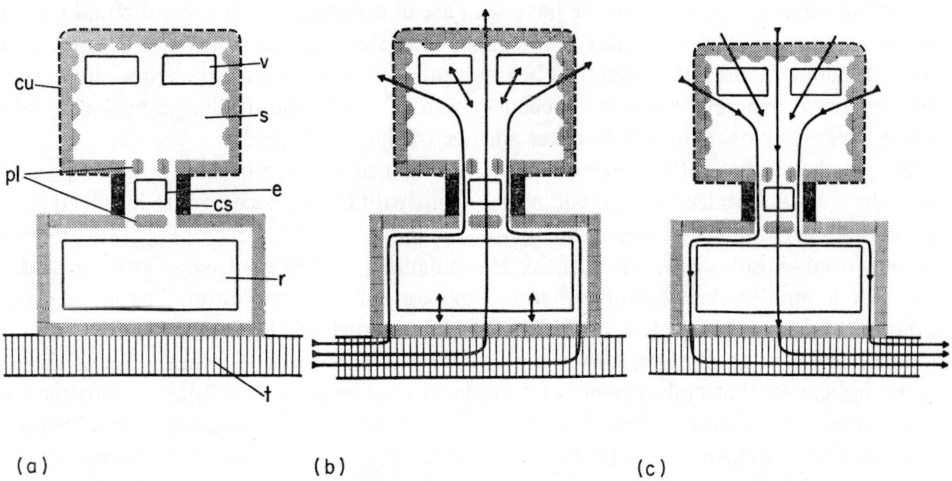

FIG. 9. Scheme for the secretion–resorption cycle in *Pinguicula*; s, secretory cell with vacuoles, v, and thin cuticular layer on the outer surface of the wall, cu; e, endodermal cell with Casparian strip, cs, on the radial walls and plasmodesmatal connections, pl, with the reservoir cell, r, which adjoins a tracheid, t. (a) Resting phase; (b) secretory phase; (c) resorption phase. For explanation see text.

movement into the head cells must result from a rise of osmotic potential there. It can be seen that if stimulation were to bring about an increase in the concentration of osmoticum in the head cells or in the apoplast beyond the Casparian strip, exosmosis of water from the basal cell and then the adjacent reservoir cell would result. The tension would then be transmitted to the tracheid associated with the latter, and water would be abstracted from the general vascular system of the leaf.

In *Pinguicula*, the vacuoles in the gland head cells decrease in size during the secretion phase, and this suggests that the osmotic drive is not developed by an increase of osmoticum in the vacuoles themselves. Rather does it seem that changes in the wall are involved, and a clue as to how this occurs is now available from fine-structural observations. According to Schwab *et al.* (1969), the cell wall protuberances progressively diminish during the secretion phase in *Dionaea* and the space between the plasmalemma and the primary wall becomes filled with filamentous material. The same kind of changes have been noted in *Pinguicula* (Heslop-Harrison, 1975), and it is significant that they also apparently occur in nectar-secreting cells (Schnepf, 1964). The protuberances of the inner wall face are presumably polysaccharide, and their break down would be expected to release sugars; this then could be the source of the osmoticum responsible for the withdrawal of water from the head cells in the first phase of secretion. The rest of the sequence then suggests itself. With the passage of water into the walls of the head cells, leakage to the outside would begin through the discontinuities in the cuticle, and the wall-held enzymes would be leached out. As long as the protoplast of the endodermal cell with its intimate contact with the bounding Casparian strip continued to act as a compound semipermeable membrane between the head cells and the reservoir cells beneath, water would be drawn through it, and tensions transmitted along the vascular system would ensure that eventually even the water supply from the root would be tapped. That this must happen in *Pinguicula* is adequately demonstrated by the fact that the secretion released by a single well-stimulated leaf may ultimately exceed the volume of the leaf itself.

For the proposed system to operate, return pathways for the secretion pool building up on the leaf surface must necessarily be blocked (Fig. 7A). This is indeed ensured by the continuity of the cuticle of the epidermal cells, which, as may be seen from Fig. 7D, links without a break with the Casparian strip.

The build up of secretion in the pitcher plants, notably in *Nepenthes*, does not depend on prey stimulation, but takes place long before the pitcher opens (Heslop-Harrison, 1975). Clearly, then, the mechanism cannot be the same as that outlined above, which is one appropriate to a rapid discharge of stored enzymes immediately upon the capture of living prey. The mechanism is more likely to be akin to that producing the permanent secretion droplets of the stalked glands of *Pinguicula*, or of the glands of *Drosera*. The latter, as we have seen, do secrete enzymes before the stimulation of the leaf and may be provoked into further secretion by the capture of prey, but they do not generate a flow comparable with that of the sessile glands of *Pinguicula*.

The resorption phase

For the capture and digestion of prey to be of any nutritional value to the plant, the digestion products (or the break-down products resulting from secondary infection in the case of some of the pitcher plants) must be absorbed into the leaf and translocated to areas of active metabolism. Studies of the resorption of digestion products have been made for several genera, and there is now no doubt that all of the species examined have adequate methods of recovering products from the secretion pool (*Aldrovanda* and *Drosera*, Fabian-Galan and Sălăgeanu, 1968; *Nepenthes* and *Dionaea*, Lüttge, 1963, 1964b, 1965a, b; *Sarracenia*, Hepburn *et al.*, 1920, 1927; Plummer and Kethley, 1964; *Pinguicula*, Heslop-Harrison and Knox, 1971; Heslop-Harrison, 1975).

Microautoradiography has been used to follow the uptake of digestion products following the feeding of *Pinguicula* leaves with ^{14}C-labelled *Chlorella* protein (Heslop-Harrison and Knox, 1971). The principal entry points into the leaf were found to be the sessile glands. Within 4 hr, radioactive digestion products (amino acids and peptides) could be detected in the head cells, and within 12 hr they had moved into the vascular system and were passing out of the leaf. The indications are, in fact, that the digest is absorbed into the leaf through the same pathways as those taken by the secretion. Although there is a possibility of some leakage of the secretion pool into the leaf via the stomata, this was not observed in *Pinguicula*. In *Drosera* and *Drosophyllum*, passage through the stomata would be even less probable since the heads of the glands are held at some distance from the leaf surface. In *Drosera*, some return passage through the scattered sessile glands is not however excluded.

The resorption pathway in *Pinguicula* has been traced electron microscopically (Heslop-Harrison, 1975). Fresh leaves were stimulated, and then small amounts of colloidal lanthanum nitrate were added to the secretion pool. After the beginning of the resorption phase, leaves were fixed and prepared for electron microscopy by standard methods. The cuticles of the epidermal cells offered a complete barrier to the penetration of the tracer (Fig. 7A), but it was taken up into the walls of the gland head cells through the discontinuities (Fig. 7C). Temporary accumulations appeared within the wall in the space occupied by the residue of the spongy wall (Figs. 7B, C). In these experiments, the penetration of the tracer has been followed as far as the endodermal cell. In *Pinguicula* the secretion pool passes back into the leaf almost as quickly as it is originally secreted. It is hardly conceivable that the glands could permit a simultaneous two-way traffic, so it must be supposed that, in each, the phase of secretion gives place quite rapidly to that of absorption. To account for this, it is only necessary to assume that the protoplast of the endodermal cell—the effective osmotic barrier in the first phase—progressively becomes more permeable, until a reverse mass flow becomes possible through it in response to water tensions in the xylem (Fig. 9). The gland would then become a channel of passage for secretion fluid from the leaf surface, a fluid containing digestion products, residual digestive enzymes and presumably also the sugars responsible for establishing the original outward flow (Heslop-Harrison and Knox, 1971). This type of resorption by mass flow depending essentially on the loss of membrane control at the sites of uptake would explain very neatly the rapid disappearance of the secretion pool in *Drosophyllum* and *Drosera* as well as in *Pinguicula*. Interestingly, it has not so far been possible to follow the movement of colloidal lanthanum nitrate back into the leaf during the resorption phase further than the endodermal cell, suggesting that although its protoplast certainly becomes permeable at this time to sugars and amino acids it continues to present an appreciable impedance to the passage of particles of colloidal size.

It is obvious that no such system as that just described could account for the preferential uptake of nutrients from the pitcher fluid of *Nepenthes*. The routes of entry into the pitcher wall nevertheless appear to be through the glands (Lüttge, 1965a), and it seems likely that active mechanisms are responsible for the passage of ions across intact membranes in the gland head cells. The process is probably much more like the uptake of nutrients by roots from their bathing media, without any mass flow component. The case of *Dionaea* is interesting; here the secretion–resorption cycle is much slower than in other leaf-trapping species, and according to Schwab *et al.* (1969) detectable amounts of proteolytic enzymes do not appear until after 36 hr from trap closure. Schwab *et al.* suggest that the delay in secretion is due to the fact that the enzymes are held in the gland cell vacuoles, their activation and release being quite slow. However, whatever the mechanism of discharge, certain of the digestive enzymes of *Dionaea* have now been shown to be held in the gland cell walls before stimulation, much as in *Pinguicula*, *Drosera* and *Drosophyllum* (Heslop-Harrison, 1975). The delayed release must therefore reflect a slower operation of the discharge mechanism. It may be noted also that the resorption system in *Dionaea* is not likely to be the same as that found in *Pinguicula*. In the latter, the glands do not recover after one secretion–resorption sequence, and it is probable that the gland head and endodermal cells are destroyed during the resorption phase when the membranes lose control of water movement. In *Dionaea*, in contrast, the glands do recover, and the cycle can be repeated. The wall ramifications disappear during this point of the cycle, but they then reappear during the recovery phase. There must therefore be no irreversible change in the gland tissues (Schwab *et al.*, 1969). This indicates that there is no mass-flow phase during resorption. The long period for the completion of the cycle suggests a slow, controlled withdrawal through intact membranes perhaps akin in some sense with that found in the pitcher plants.

References

AMAGASE, S. (1972) Digestive enzymes in insectivorous plants. III. Acid proteases in the genus *Nepenthes* and *Drosera peltata*. *J. Biochem. Tokyo* **72,** 73–81.

AMAGASE, S., MORI, M. and NAKAYAMA, S. (1972) Digestive enzymes in insectivorous plants. IV. Enzymatic digestion of insects by *Nepenthes* secretion and *Drosera peltata* extract: proteolytic and chitinolytic activities. *J. Biochem. Tokyo* **72,** 765–767.

BÜSGEN, M. (1883) Die Bedeutung des Insektfanges für *Drosera rotundifolia*. *Bot. Ztg* **41,** 569–577; 585–594.

DARWIN, C. (1875) *Insectivorous Plants*, Murray, London.

DARWIN, F. (1878) Experiments on the nutrition of *Drosera rotundifolia*. *J. Linn. Soc.* (*Bot.*) **17,** 17–32.

DERNBY, K. G. (1917) Notiz betreffend die proteolytischen Enzyme der *Drosera rotundifolia*. *Biochem. Z.* **78,** 197.

DEXHEIMER, J. (1972) Quelques aspects ultrastructuraux de la sécretion de mucilage par les glandes digestives de *Drosera rotundifolia* L. *C.r. hebd. Séanc. Acad. Sci. Paris* **275,** 1983–1986.

FABIAN-GALAN, G. and SĂLĂGEANU, N. (1968) Considerations on the nutrition of certain carnivorous plants (*Drosera capensis* and *Aldrovanda vesiculosa*). *Revue roum. Biol. Bot.* **13,** 275–280.

FENNER, C. A. (1904) Beiträge zur kenntnis der Anatomie Entwicklungsgeschichte und biologie der Laublätter und Drüsen einiger Insektivoren. *Flora Jena* **93,** 335–434.

GOEBEL, K. (1891–1893) Insektivoren. *Pflbiol. Schilderungen Marburg* **2,** 51–214.

HARDER, R. (1963) Blutenbildung durch Tierische Zusatznahnung und andere Faktoren bei *Utricularia exoleta*. *Planta* **59,** 459–471.

HARDER, R. and ZEMLIN, I. (1967) The development of flowering of *Pinguicula lusitanica* in axenic culture. *Planta* **73,** 181–193.

HARDER, R. and ZEMLIN, I. (1968) Blutenbildung von *Pinguicula lusitanica* in vitro durch Fütterung mit Pollen. *Planta* **78,** 72–78.

HEPBURN, J. S., ST. JOHN, E. Q. and JONES, F. M. (1919) Biochemical studies of insectivorous plants. *Contr. bot. Lab. Univ. Pa* **4,** 419–463.

HEPBURN, J. S., ST. JOHN, E. Q. and JONES, F. M. (1920) The absorption of nutrients and allied phenomena in the pitchers of the Sarraceniaceae. *J. Franklin Inst.* **189,** 147–184.

HEPBURN, J. S., JONES, F. M. and ST. JOHN, E. Q. (1927) Biochemical studies of the North American Sarraceniaeae. *Trans. Wagner free Inst. Sci. Philad.* **11,** 1–95.

HESLOP-HARRISON, J. and HESLOP-HARRISON, Y. (1970) Evaluation of pollen viability by enzymically induced fluorescence: intracellular hydrolysis of fluorescein diacetate. *Stain Technol.* **45,** 115–120.

HESLOP-HARRISON, J., HESLOP-HARRISON, Y., KNOX, R. B. and HOWLETT, B. (1973) Pollen wall proteins: 'gametophytic' and 'sporophytic' fractions in the pollen walls of the Malvaceae. *Ann. Bot.* **37,** 403–412.

HESLOP-HARRISON, J., KNOX, R. B. and HESLOP-HARRISON, Y. (1974) Pollen-wall proteins: exine-held fractions associated with the incompatibility response in Cruciferae. *Theoret. appl. Genet.* **44,** 133–137.

HESLOP-HARRISON, Y. (1970) Scanning electron microscopy of fresh leaves of *Pinguicula*. *Science N.Y.* **167,** 172–174.

HESLOP-HARRISON, Y. (1975) Enzyme release in carnivorous plants. In *Lysosomes in Biology and Pathology*, ed. DINGLE, J. T. and DEAN, R. Volume 4, A.S.P. Biological and Medical Press, B.V. (North Holland Division) Amsterdam, pp. 525–576.

HESLOP-HARRISON, Y. and KNOX, R. B. (1971) A cytochemical study of the leaf-gland enzymes of insectivorous plants of the genus *Pinguicula*. *Planta* **96,** 183–211.

HOLTER, H. and LINDERSTRØM-LANG, K. (1932) Beitrage zur enzymatischen Histochemie. III. Über die Proteinasen von *Drosera rotundifolia*. *Hoppe-Seyler's Z. physiol. Chem.* **214,** 223–240.

HOOKER, J. D. (1874) Presidential address to the Botanical and Zoological sessions: on the carnivorous habits of some of our brother organisms-plants. Report of the British Association for the Advancement of Science meeting held at Belfast, 1874. 102–116.

JENTSCH, J. (1970) Probleme bei der Reinigung von *Nepenthes*-kannensaft Enzymen. *Ber. dt. bot. Ges.* **83,** 171–176.

JENTSCH, J. (1972) Enzymes from carnivorous plants (*Nepenthes*): isolation of the protease nepenthacin. *FEBS Lett.* **21**, 273–276.

KNOX, R. B. and HESLOP-HARRISON, J. (1969) Cytochemical localisation of enzymes in the wall of the pollen grain. *Nature Lond.* **223**, 92–94.

KNOX, R. B. and HESLOP-HARRISON, J. (1971) Pollen-wall proteins: localisation and enzymic activity. *J. Cell Sci.* **6**, 1–27.

LLOYD, F. E. (1933) Carnivorous plants—a review with contributions. *Trans. R. Soc. Can.* III **27**, 35–101.

LLOYD, F. E. (1942) *The Carnivorous Plants*, Chronica Botanica Co., Massachusetts.

LÜTTGE, U. (1963) Die Bedeutung des chemischen Reizes bei der Carnivoren Resorption von ^{14}C-Glutaminsäure, $^{35}SO_4^{--}$ und $^{45}Ca^{++}$ durch *Dionaea muscipula, Naturwissenschaften* **50**, 4–22.

LÜTTGE, U. (1964a) Untersuchungen zur Physiologie der Carnivoren-Drüsen. I. Die an den Verdauungsgängen beteiligten Enzyme. *Planta* **63**, 103–117.

LÜTTGE, U. (1964b) Untersuchungen zur Physiologie der Carnivoren-Drüsen. *Ber. dt. bot. Ges.* **77**, 181–187.

LÜTTGE, U. (1965a) Untersuchungen zur Physiologie der Carnivoren-Drüsen. II. Über die Resorption verschiedener Substanzen. *Planta* **66**, 331–344.

LÜTTGE, U. (1965b) Untersuchungen zur Physiologie der Carnivoren-Drüsen. III. Der Stoffwechsel der resorbierten Substanzen. *Flora Jena* **155**, 228–236.

LÜTTGE, U. (1971) Structure and function of plant glands. *A. Rev. Pl. Physiol.* **22**, 23–44.

MACFARLANE, J. M. (1908) Accounts of Sarraceniaceae and Nepenthaceae. In *Das Pflanzenreich* ed. ENGLER, A. Volume 4, Engelmann, Leipzig.

NAKAYAMA, S. and AMAGASE, S. (1968) Acid protease in *Nepenthes*. Partial purification and properties of the enzyme. *Proc. Jap. Acad.* **44**, 358–362.

PLUMMER, G. L. and KETHLEY, J. B. (1964) Foliar absorption of amino acids, peptides and other nutrients by the pitcher plant *Sarracenia flava*. *Bot. Gaz.* **125**, 245–260.

PRINGSHEIM, E. and PRINGSHEIM, O. (1962). Axenic culture of *Utricularia*. *Am. J. Bot.* **49**, 898–901.

PRINGSHEIM, E. and PRINGSHEIM, O. (1967) Small contributions to the physiology of *Utricularia*. *Z. Pflphysiol.* **57**, 1–10.

RAGETLI, H. W. J., WEINTRAUB, M. and LO, E. (1972) Characteristics of *Drosera* tentacles. I. Anatomical and cytological details. *Can. J. Bot.* **50**, 159–168.

SCALA, J., SCHWAB, D. and SIMMONS, E. (1968) The fine structure of the digestive gland of Venus's flytrap. *Am. J. Bot.* **55**, 649–657.

SCHNEPF, E. (1960a) Zur Feinstruktur der Drüsen *Drosophyllum lusitanicum*. *Planta* **54**, 641–674.

SCHNEPF, E. (1960b) Kernstrukturen bei *Pinguicula*. *Ber. dt. bot. Ges.* **73**, 243–245.

SCHNEPF, E. (1961a) Über Veränderunge der plasmatischen Feinstrukturen während des Welkens. *Planta* **57**, 156–175.

SCHNEPF, E. (1961b) Licht- und elektronenmikroskopische Beobachtungen an Insektivoren-Drüsen über die Sekretion des Fangschleimes. *Flora Jena* **151**, 73–87.

SCHNEPF, E. (1963a) Zur Cytologie und Physiologie pflanzlicher Drüsen. I. Über den Fangschleim der Insektivoren. *Flora Jena* **153**, 1–22.

SCHNEPF, E. (1963b) Zur Cytologie und Physiologie pflanzlicher Drüsen. II. Über die Wirkung von Sauerstoffentzug und von Atmungsinhibitoren auf die Sekretion des Fangschleimes von *Drosophyllum* und auf die Feinstruktur der Drüsenzellen. *Flora Jena* **153**, 23–48.

SCHNEPF, E. (1963c) Zur Cytologie und Physiologie pflanzlicher Drüsen. III. Cytologische veränderungen in den Drüsen von *Drosophyllum* während der verdauung. *Planta* **59**, 351–379.

SCHNEPF, E. (1964) Zur Cytologie und Physiologie pflanzlicher Drüsen. IV. Licht- und elektronenmikroscopische Untersuchungen an Septalnektarien. *Protoplasma* **58**, 137–171.

SCHWAB, D. W., SIMMONS, E. and SCALA, J. (1969) Fine structure changes during function of the digestive gland of Venus's flytrap. *Am. J. Bot.* **56**, 88–100.

STECKELBURG, R., LÜTTGE, U. and WEIGL, J. (1967) Reinigung der Proteinase aus *Nepenthes*-kannensaft. *Planta* **76**, 238–241.

TATE, L. (1875) Insectivorous plants. *Nature Lond.* **12**, 251.

TÖKES, Z. A., WOON, W. C. and CHAMBERS, S. M. (1974) Digestive enzymes secreted by the carnivorous plant *Nepenthes macfarlanei* L. *Planta* **119**, 39–46.

VOGEL, A. (1960) Feinstructur der Drüsen von *Pinguicula*. *Beih. Zn. schweiz. Forstver.* **30**, 113–122.

WILLIAMS, S. E. and PICKARD, B. G. (1969) Secretion, absorption and cuticle structure in *Drosera* tentacles. (Abstract). *Pl. Physiol. Lancaster* **44**, Supplement, 5.

WILLIAMS, S. E. and PICKARD, B. G. (1974) Connections and barriers between cells of *Drosera* tentacles in relation to their electrophysiology. *Planta* **116**, 1–16.

ABSORPTION AND TRANSPORT OF CALCIUM IN THE STALKED GLANDS OF *DROSERA CAPENSIS* L.

B. E. JUNIPER and A. J. GILCHRIST

School of Botany, University of Oxford, South Parks Road, Oxford, U.K.

INTRODUCTION

Stalked glands on the leaves of the sundew, *Drosera capensis*, provide a model system for the experimental study of relationships between cell structure and physiological processes. The glands are thought to secrete enzymes which digest prey caught on the leaves, digestion products being subsequently absorbed into the gland and transported down the stalk into the leaf. Recent X-ray-analytical procedures allow the precise tracing of pathways of transport of specific substances placed on the glands. In this paper we describe (i) some ultrastructural features (some previously unreported) of gland and stalk cells both before and after gland-stimulation by cow's milk and (ii) the transport pathway of milk-calcium into the leaf traced with the EMMA-4 analysing microscope.

STRUCTURAL FEATURES

The stalked (as opposed to the sessile) glands in *Drosera capensis* are composed of three layers of cells covering a central tracheid mass (Fig. 1). The gland is borne on a stalk through which runs a single xylem strand connecting with the tracheid mass. Surrounding this tracheid mass is a layer of long tubular cells, suberized or cutinized like a root endodermis and for this reason referred to as the endodermis of the gland. On top of the endodermal cells are two or occasionally three layers of glandular cells. In the outer layer is synthesized the mucilage (produced by the golgi) and possibly, also, the digestive enzymes. The mucilage passes out to the gland exterior and entraps prey on the leaf. The whole of the gland and stalk is covered by a cuticle which is continuous with that of the leaf. The cuticle of the gland is thinner than that of the stalk; moreover, unlike the cuticle of the stalk, the cuticle of the gland is a discontinuous layer interrupted by many pores up to 300 nm in diameter (Williams and Pickard, 1969; Ragetli *et al.*, 1972; Chafe and Wardrop, 1973). Pores are more frequent in the upper half of the gland where they account for 12 per cent of the total surface area, compared with only 2 per cent in the lower half.

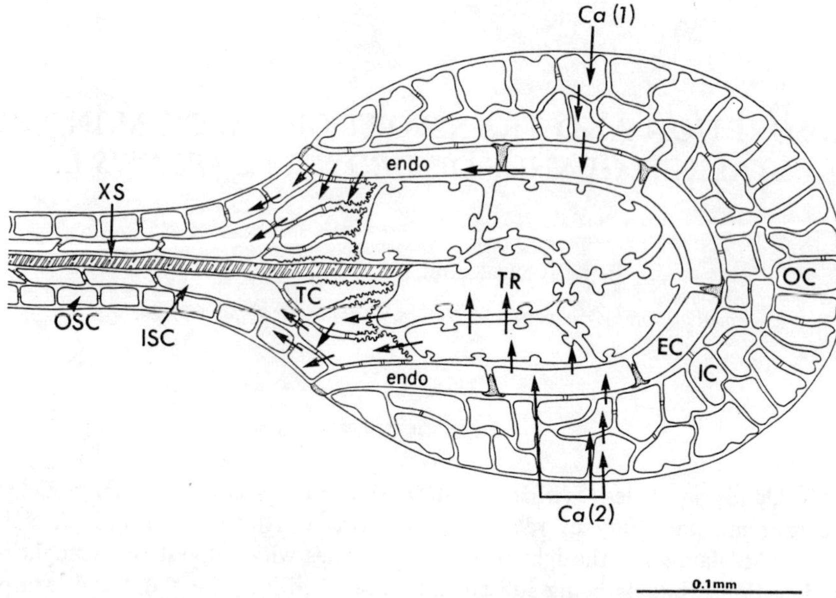

FIG. 1. Diagram of a stalked gland of *Drosera capensis*. TR, tracheid mass; EC, endodermal cells; OC and IC, outer and inner gland cells; TC, transfer cells; XS, xylem strand; OSC and ISC, outer and inner stalk cells. Ca(1) and Ca(2) refer to the two proposed transport pathways.

The Stalk

The stalk is composed of two concentric layers of cells surrounding the central xylem strand. The epidermal cells are considerably shorter than the subepidermal and both are rich in well defined plasmodesmata, grouped in large pitfields along adjacent transverse walls. At the top of the stalk, where the stalk joins with both the central tracheid mass and the basal endodermal cells, are the cells referred to by Fenner (1904) as the 'Halskranz' or neck cells (TC, Fig. 1). In the electron microscope, the walls of the uppermost inner stalk cells where they join the central tracheids, can be seen to bear labyrinthine wall projcctions (Fig. 2). Some elaboration of the wall can also be seen where the innermost cells join with the central xylem strand, but not where the outermost cells join with the basal endodermal cells. The projcetions are confined to the upper part of the neck cells and appear nowhere else in the stalk.

The labyrinthine wall projections are similar in profile and electron density to those described by Gunning and Pate (1969) in transfer cells of minor leaf veins, and in legume root nodules, where the overall morphology of the nodule is similar to that of the stalked glands of *Drosera capensis*. There is also sometimes a space in the neck cells between plasmalemma and wall projection, as in transfer cells of minor veins. We propose, therefore, to refer to those neck cells bearing wall elaborations as *stalk transfer cells*.

It was the discovery of the labyrinthine wall projections which suggested to us that they may be associated with transport of materials from tracheids to stalk. The transfer

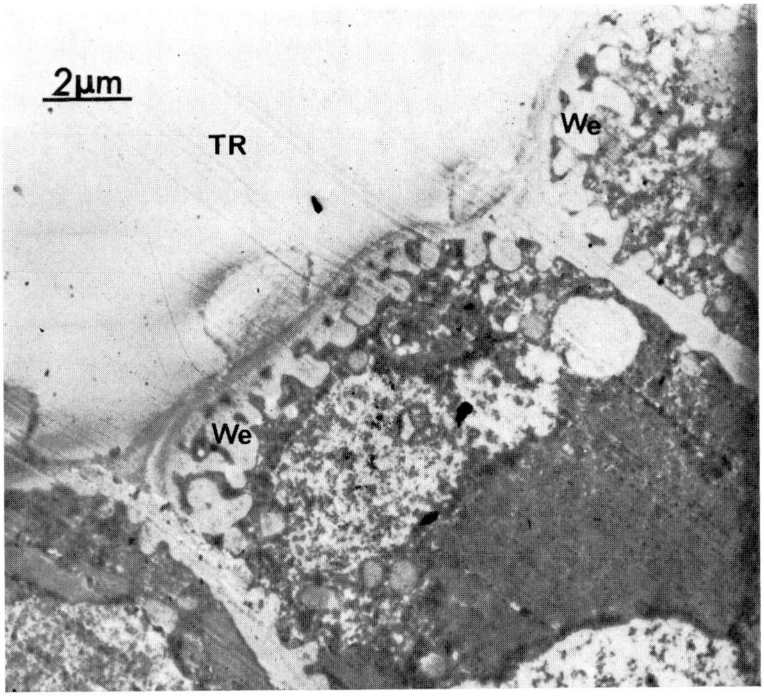

FIG. 2. Electron micrograph of a longitudinal section of a stalk transfer cell in *Drosera capensis*. The transfer apparatus or wall elaborations (We) are visible on the wall adjacent to the tracheids (TR)

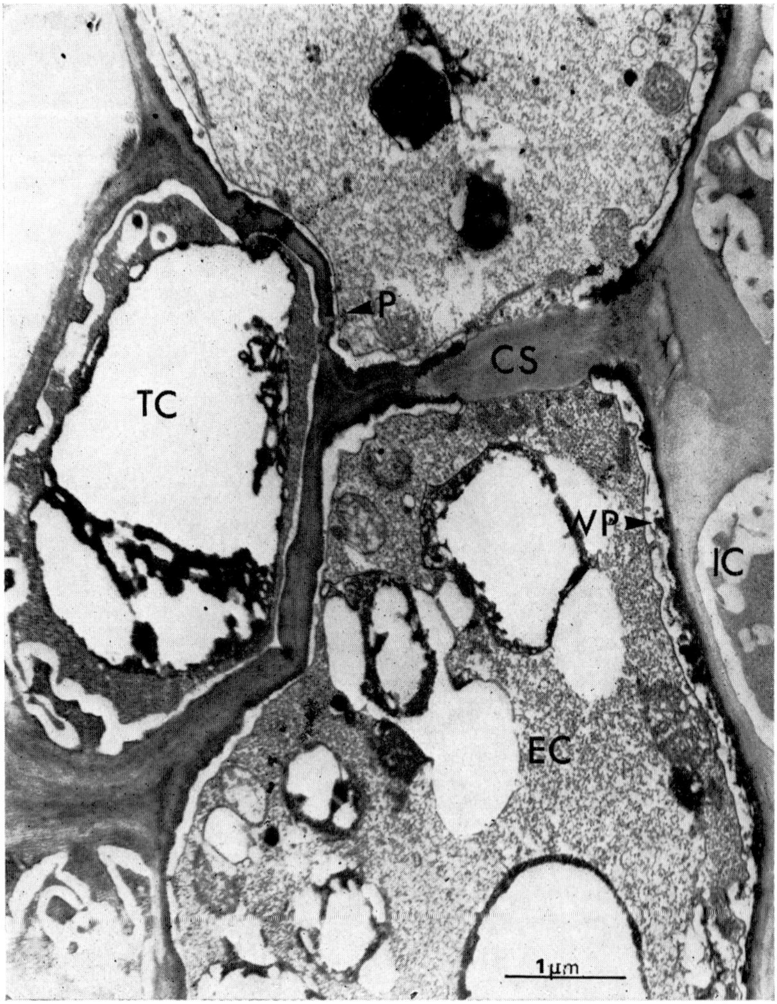

FIG. 3. Electron micrograph of a section in the stalked gland of *Drosera capensis* of two adjoining endodermal cells (EC) showing part of an inner gland cell (IC) and an adjacent transfer cell (TC). The Casparian strip between the two endodermal cells is clearly visible (CS). Wall projections (WP) are visible on the outer endodermal wall, and plasmodesmata (P) can be seen in the wall between one endodermal cell and a transfer cell.

cells protrude up into the tracheid mass and thus increase the area of contact between the two tissues; the area of contact is additionally increased by the wall elaborations.

It may be emphasized that the epidermal cells of the stalk are not in direct contact with the central tracheid mass; the epidermal cells join with the basal endodermal cells, which are themselves of epidermal origin (Homes, 1927).

The endodermis

The endodermis has been a subject of interest to all who have worked with *Drosera*. Few workers dispute that the endodermal cells possess a Casparian strip (Lloyd, 1942; Williams and Pickard, 1974) similar to that in a root endodermis (see Robards this volume). The strip (CS, Fig. 3) has a different electron-density when compared to other cell-wall material, and is thus easily identified in the electron microscope. The strip is continuous around each cell in the functionally radial walls including the wall that reaches the outside of the gland; it covers only the outer two-thirds of these walls and spreads out slightly where it contacts the inner glandular layer. At the surface of the gland the Casparian strip is a thin band under the cuticle covering the whole of the exposed end of the endodermal cell.

The Casparian strip presumably provides an effective barrier to apoplastic flow. As in roots (see Robards, pp. 413–422) the plasmalemma of the endodermal cells is always tightly appressed to that part of the wall adjacent to the Casparian strip, though loosely associated elsewhere (Fig. 3). This appression of the plasmalemma in conjunction with the suberized barrier completely seals off the apoplast so that transport of any material across the endodermis into the central tracheid mass must be confined solely to the symplast. Apoplastic flow of materials from gland cells directly to stalk cells is also prevented by the suberization of the end-wall where each basal endodermal cell comes to the surface of the gland. Schnepf (1961) has suggested that the Casparian strip ensures a unidirectional flow of mucilage from the outer gland layer to the exterior and this might apply to the digestive enzymes also.

In the electron microscope, plasmodesmata (probably a key feature of symplastic flow) are seen to traverse the unsuberized parts of the endodermal walls both in the radial and tangential directions. Between endodermal and inner gland cells the plasmodesmata are usually ill defined and occur singly or in small groups of two to four (Schnepf, 1961; Ragetli *et al.*, 1972). On the other hand, between the basal endodermal cells and the stalk transfer cells, and between the endodermal cells and the adjoining stalk epidermal cells, plasmodesmata are numerous and well defined, and almost invariably associated with tubular endoplasmic reticulum in both cells.

Finally, the outer tangential walls of the endodermal cells, but not the inner, bear a series of projections (Fig. 3). These are smaller than those present in the stalk transfer cells and are also more electron-dense.

Gland cells

Gland cell structure has been described extensively elsewhere (Schnepf, 1961, 1966; Ragetli *et al.*, 1972). The walls of the gland cells bear small projections not unlike those present on the outer tangential wall of the endodermal cells.

CHANGES ASSOCIATED WITH GLAND-STIMULATION

Several changes take place in the gland after stimulation by food. Drops of whole cow's milk held on the tip of a needle and placed on the gland cause a strong and rapid initiation of curvature of the gland-stalks towards the centre of the leaf. This initial response is rapidly succeeded by release of accumulated mucilage from the outer gland cells, probably by reverse pinocytosis or exocytosis; the golgi also become more active in the production of mucilage (Schnepf, 1961). In the endodermis, 15 minutes after application of the first drops of milk to the leaf, the normally smooth plasmalemma adjacent to the outer tangential wall becomes highly invaginated (Fig. 4). Sometimes thc invaginations are long and finger-like, reminiscent of invaginations associated early in the formation of pinocytotic vesicles, but we have not observed such vesicles in the endodermis still attached to the invaginated membrane nor suspended free in the cytoplasm.

At the same time in the endodermis, small evaginations develop from the plasmalemma adjacent to the inner tangential wall (Figs. 5, 6A, B). Frequently these evaginations are seen to be associated with tubular ER. Sometimes there appears to be a 'septum' across the base of an evagination (Fig. 6B) but such 'septa' may be artefacts arising from the smallness of the evaginations relative to the thickness of the sections, i.e. images of a smooth region of the plasmalemma either above or below the evaginations and normal to the plane of the section may appear as one septum-like image.

Evaginations of the endodermal plasmalemma take the form of microvilli protruding outwards into the extracytoplasmic space between the plasmalemma and cell wall. The evaginations may represent a form of exocytosis but we have failed to observe anything resembling rupture or discharge of materials into the extracytoplasmic space. Whenever the membrane is active, groups of small bodies arc sccn in the extracytoplasmic space, and give the appearance of free vesicles (Fig. 5), but without resorting to detailed serial sections it is difficult to know if the 'vesicles' are truly detached from the membrane.

Twenty-four hours after the first application of milk to the gland, the endodermal plasmalemma adjacent to the outer tangential wall is no longer greatly folded or invaginated but some evaginations are usually still visible in the plasmalemma adjacent to the inner tangential wall.

In one chase-type experiment, in which the glands were give a 1:1 mixture of ferritin and milk for 2 hr, then left for a further 24 hr without milk before fixation, some of the cvaginations appeared to contain ferritin particles (Fig. 6B). This suggested that the invaginations and evaginations of the endodermal plasmalemma are associated with transport of materials across the endodermis presumably via the symplast. We, therefore, turned to X-ray microanalysis to try and trace the transport pathway of the particles. Ferritin however, proved to be troublesome as a tracer and was eventually abandoned in favour of calcium present in the milk.

TRANSPORT OF CALCIUM IN STALKED GLANDS OF *Drosera capensis*

Milk was fed both to leaves and isolated glands, and fixations were made in glutaraldehyde at intervals after commencement of feeding. Thin sections were examined in the EMMA-4 microscope to localize the calcium. Some degree of quantitative analysis was

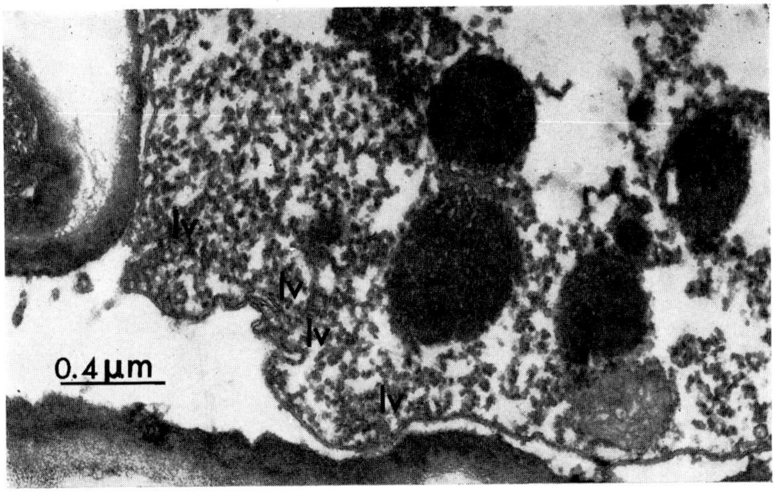

FIG. 4. Electron micrograph of part of the endodermal plasmalemma adjacent to the inner gland layer in *Drosera capensis*, 15 min after application of milk to the gland. Several long invaginations (Iv) of the membrane are visible.

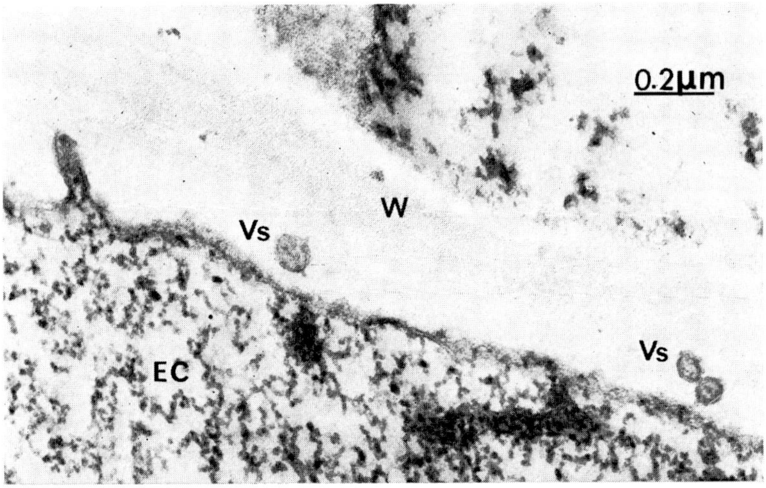

FIG. 5. Electron micrograph of part of the plasmalemma of the endodermis (EC) adjacent to the central tracheid mass in the stalked gland of *Drosera capensis*. An evagination of the plasmalemma is visible as are several possible vesicles (Vs) in the extracytoplasmic space between the plasmalemma and the wall (W).

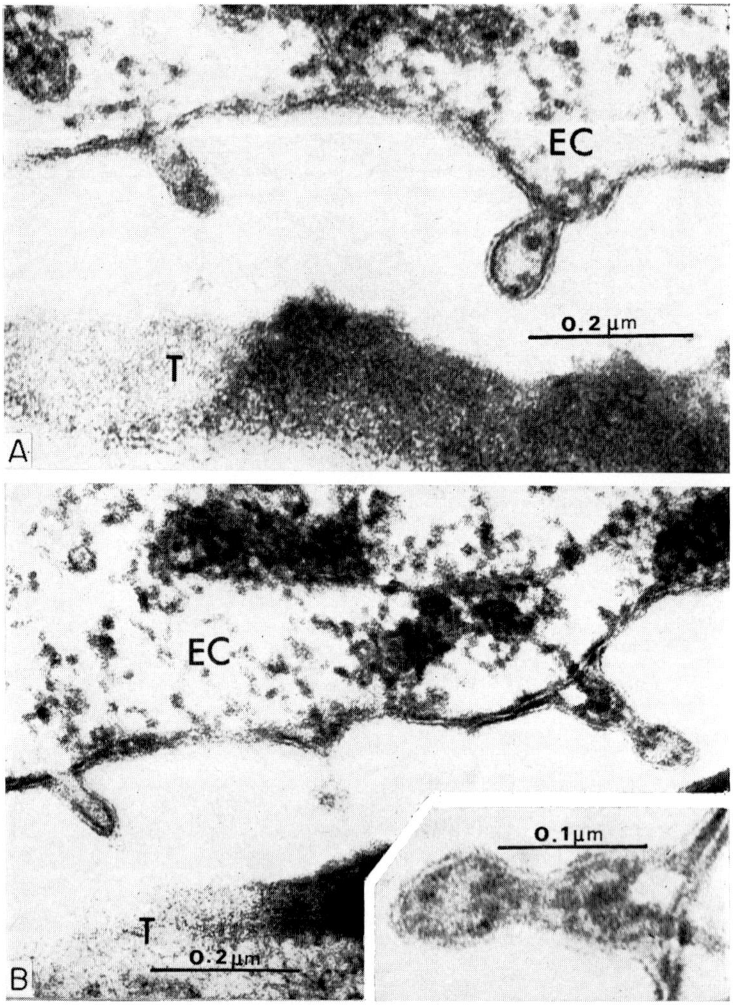

FIG. 6. Electron micrographs of the plasmalemma of the endodermis (EC) in the stalked gland of *Drosera capensis*, adjacent to the central tracheid mass. Evaginations of the membrane are visible protruding into the extracytoplasmic space between the plasmalemma and the tracheid wall (T). Inset: higher magnification of one evagination of the plasmalemma.

possible with the instrument according to the number of counts recorded at each site, but we shall restrict discussion here mainly to the route taken by calcium during its passage through the gland into the stalk.

Similar results were obtained on isolated glands (plus stalk) and glands still attached to the leaf. Controls treated with distilled water instead of milk gave no significant recordings of calcium. In timing the speed of calcium uptake into the gland some allowance needs to be made for the speed of fixation by glutaraldehyde. However, penetration of this fixative is likely to be as rapid or more so than that of the milk components.

Entry and transport of calcium into the gland

Our experiments indicate that calcium enters the gland rapidly. Within 1 min after the first application of milk to either the entire leaf or individual glands, calcium is detectable in both the cytoplasm and apoplast of the outer gland cells, and to a lesser extent in cytoplasm and apoplast of the inner gland cells. As expected, there is no calcium in the apoplast of the central tracheid mass. Nor is calcium detectable in the cytoplasm of the endodermal cells after 1 min, although certain areas of the plasmalemma adjacent to the outer tangential walls give strong counts in the analysing microscope. However, 1 min later, calcium is indeed present in the cytoplasm of the endodermal cells and is associated strongly with the folds and invaginations of the plasmalemma adjacent to the outer tangential wall. Also after 2 min, calcium is more concentrated in the cytoplasm and apoplast of the two glandular layers, more so in the outer layer. By sequential analysis we found that the concentration of calcium in the apoplast of the gland layers declines steadily, progressing inwards, until the Casparian strip is reached. Calcium is never found in the Casparian strip itself; immediately adjacent to the strip, however, the calcium count increases dramatically to a level approaching that in the apoplast of the outer gland layer (Fig. 7).

At 15 min, calcium is detected in the evaginations of the endodermal plasmalemma adjacent to the central tracheid mass. These sites of 'exocytosis' give high counts relative to the cytoplasm of the endodermal cells. Calcium is also detectable in the cytoplasm and apoplast of both gland layers although at this stage fewer sites in the apoplast give appreciable counts (about 25 per cent of those examined).

At 60 min, counts for calcium in the apoplast of the outer gland layer are about double those in the apoplast of the inner layer. There is, however, no difference in counts for the cytoplasms of the two layers. The calcium count for the cytoplasm of the gland layers is about three times that of the cytoplasm of the endodermal cells, but in the invaginated and evaginated areas of the endodermal plasmalemma even higher counts may be recorded. At 60 min we also found calcium in the apoplast of the central tracheid mass.

There seems little doubt that calcium, and probably other components of the milk, are absorbed by the gland cells and transported via the apoplast at least as far as the endodermis. The calcium then traverses the endodermis in the symplast, transport being mediated in some way by the plasma-membrane. The calcium is then discharged into the apoplast of the central tracheid mass. Corroborative evidence for this pathway has come from chase-type experiments in which glands are fed for 2 hr then left for a further 24 hr without milk before fixation. In these circumstances, calcium can only be detected

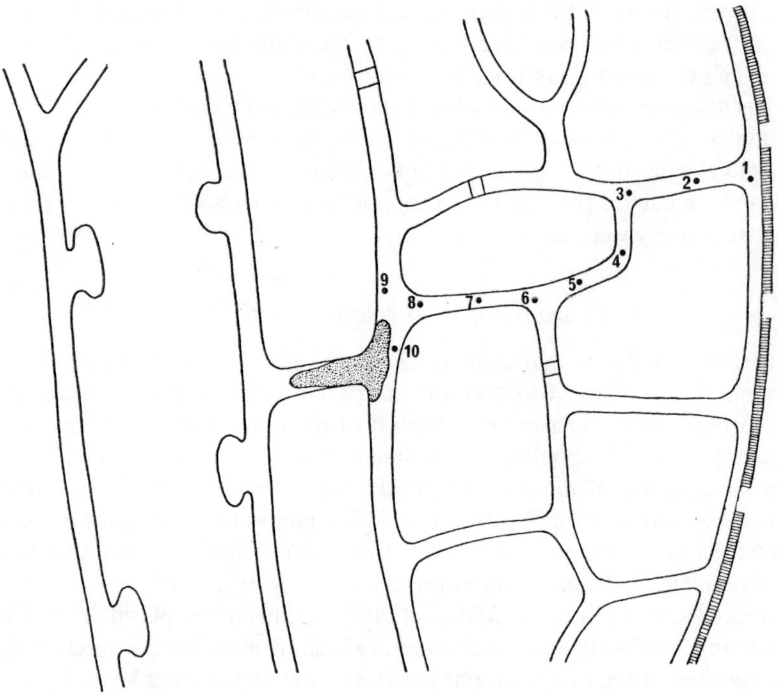

Table of relative peaks for calcium obtained at the sites numbered 1–10 above:

Position:	1	2	3	4	5	6	7	8	9	10
Relative peak value:	145	145	150	92	55	55	45	35	15	85

FIG. 7. Diagram (not to scale) showing the sites analysed in the glandular apoplast of *Drosera capensis*, 2 min after application of whole milk to the gland. The table gives relative values of calcium recorded at each site.

in the apoplast of the tracheids; it is completely absent from the gland layers and present only in traces in the endodermal cells where plasmalemmal activity is then minimal.

Transport from gland to stalk

Calcium is not detected in the stalk cells until 15 min after the first application of milk to the glands. Epidermal cells appear rich in calcium whereas the subepidermal cells seem to contain little or none. In both cell layers the calcium is sited mainly in the pit-fields situated in the transverse walls, the pitfields of the epidermal layer giving higher counts than those of the subepidermal layer. This suggests that both layers may be involved in transport down the stalk and that transport occurs mainly in the symplast. If so, downward transport in the subepidermal layer must be more rapid than it is in the epidermal layer. Ragetli *et al.* (1972) have suggested that the subepidermal layer is the most likely pathway of transport down the stalk. However, the higher concentration in the epidermis could indicate that this is the main pathway of flow.

With regard to the stalk transfer cells, counts have been recorded occasionally at various sites, but there is no regularly detectable calcium. This again suggests that if calcium is transported via the stalk transfer cells the rate of transport must be very high.

CONCLUSIONS

Hypotheses relating to pathways of transport in the stalked glands of *Drosera* have been influenced by the apparent conflict between secretion of digestive enzymes and mucilage to the exterior of the gland, and the simultaneous resorption of digestive products in the opposite direction. Referring to this conflict in *Pinguicula*, Heslop-Harrison and Knox (1971) thought it unlikely that materials could move in opposite directions simultaneously. Certainly, it is intellectually simpler to think in terms of separate pathways of secretion and resorption which do not require two-way traffic.

In order to obviate the need for a two-way traffic process, Williams and Pickard (1969) have suggested that resorption is restricted to the exposed ends of the basal endodermal cells, thus leaving the gland with a purely secretory function. There are several reasons for rejecting this view.

1. The exposed wall of the basal endodermal cells is covered by a cuticle (Ragetli *et al.*, 1972) and unless pores are present in this region the cuticle will slow the uptake of at least complex molecules. Beneath the cuticle is the suberin or cutin of the Casparian strip.
2. Molecules entering the exposed end of the basal endodermal cell are presumably envisaged as moving directly into the adjacent epidermal and subepidermal cells of the stalk and thence downwards into the leaf, thus bypassing the rest of the endodermis and the gland layers. On structural grounds, however, there is no *à priori* reason why molecules entering the exposed end of the basal endodermal cell should not be transmitted throughout the rest of the endodermis via the symplast, and thence in both directions either outwards or inwards.
3. Calcium in milk is detectable in the cells of the gland layer within 1 min, and moves into the endodermis within 15 min, and this occurs against an outward moving stream of mucilage. For calcium to reach the endodermal barrier in 2 min, as it appears to do, indicates a rate of movement through the gland apoplasm of approximately 1 mm/hr or 0.017 mm/min. We cannot say at present in what form the calcium is taken up by the gland-layer cells. It probably moves in organic forms rather than as ions, and is probably bound to partially-degraded proteins or to amino acids. Hence, its ability to remain in the tissues during fixation and dehydration procedures.

We think it highly probable therefore that resorption of digestive products can occur through the outer glandular layer against the inflow of mucilage and digestive enzymes. Huie (1899) for example, found that calcium phosphate will stimulate the glands to react in a manner similar to that of milk, and Arisz (1942) much later demonstrated that inorganic phosphates applied in aqueous solutions to the glands is transported from the gland into the leaf. We can however be more specific about the precise pathways taken. Two proposals may be considered.

First proposal

The steps envisaged are (see Fig. 1):

(1) diffusion of digestion products (as exemplified by calcium) through the pores of the gland cuticle into the apoplast of the outer layers of glandular cells;
(2) uptake of digestion products (or some of them such as calcium) from the apoplast into the cytoplasm of the outer glandular cells, either by a pinocytotic mechanism or by active transport across the plasmalemma;
(3) continued passage of products via the symplast into the endodermal cells;
(4) transport of products via the symplast from one endodermal cell to another;
(5) transport (again via the symplast) from the basal endodermal cells directly into the stalk cells; and
(6) downward transport (still in the symplast) from stalk cell to stalk cell.

In the first proposal we envisage a completely symplastic pathway of transport from the inner glandular layer into the leaf. Spanswick (1974) points out that a purely symplastic pathway has the advantage that it avoids the expenditure of energy in uptake and discharge of materials and avoids possible loss from the tissue.

This proposal, however, does not account for all the observed facts, and in particular the accumulation of calcium at the Casparian barrier, the presence of calcium in association with activity of the endodermal plasmalemma, and finally the appearance of calcium in the apoplast of the central tracheid mass.

Second proposal

In this pathway steps 1–3, as enumerated above, are again included (see Fig. 1), but an additional pathway via the apoplast of the two gland layers is proposed (Fig. 1; cf. Fig. 7). Digestion products are then seen as being transported across the endodermis via the symplast, as well as from one endodermal cell to another; transport across the endodermis is effected by invaginations of the plasmalemma:

(4) discharge of digestion products via the evaginated plasmalemma of the endodermal cells into the extracytoplasmic space adjacent to the central tracheids;
(5) movement into the apoplast of the central tracheid mass;
(6) movement from the apoplast of the tracheids into that of the stalk transfer cells and thence down the stalk into the leaf.

The second pathway clearly requires further investigation, in particular the mechanism involved in the invagination and evagination processes of the endodermal plasmalemma. Pinocytosis and exocytosis may be involved.

Phosphotungstic acid (PTA) is thought to be a selective stain for areas of plasmalemma concerned with pinocytosis (Mayo and Cocking, 1969) and which may represent areas of high protein concentration such as have been detected in excited animal membranes by freeze-etching techniques (Satir *et al.*, 1972). In its most active state, however, the endodermal plasmalemma of the *Drosera* glands does not stain with PTA in the direction of supposed inflow, but does stain densely on the opposite side of the cell in the direction of outflow. This dense staining of the evaginated areas of membrane of microvilli might be an indication of membrane reorganization of the 'exocytotic' type. However, we are inclined to the view that the tubular ER associated with the evaginations may be directly concerned with transport and that the evaginations and microvilli are purely devices which increase membrane area at specific sites to which calcium is bound and then released by local changes in membrane permeability.

The second proposal suggests a pumping of calcium into the vascular tissue, resulting in the diffusion of calcium throughout the tracheid mass. The calcium is then absorbed by the stalk transfer cells and passes into the inner and outer cells of the stalk, and thence down the stalk to the leaf. This pathway might be expected to bypass the bottleneck which has been suggested in a purely symplastic transport system, where calcium and carbohydrate, moving in opposite directions, would have to pass through the very limited area of connection between the stalk and endodermal cells.

Acknowledgements: Our thanks are due to Mrs. Anita Fursey and Miss Margaret Chapman of the National Physical Laboratory, Teddington for the EMMA-4 analyses which form a major part of this work and to the SRC for financial support.

References

ARISZ, W. H. (1942) Absorption and transport by *Drosera capensis* I. Active transport of asparagine in the parenchyma cells of the tentacles. *Proc. Sect. Sci. K. ned. Akad. Wet.* **45,** 2–8.

CHAFE, S. C. and WARDROP. A. B. (1973) Fine structural observations on the epidermis. II. The cuticle. *Planta* **109,** 39–48.

FENNER, C. A. (1904) Beiträge zur Kenntnis der Anatomie, Entwickelungsgeschichte und Biologie der Laubblätter und Drüsen einiger Insectivoren. *Flora Jena* **93,** 335–434.

GUNNING, B. E. S. and PATE, J. S. (1969) Transfer cells—plants cells with wall ingrowths, specialized in relation to short distance transport of solutes—Occurrence, structure and development, *Protoplasma* **68,** 107–133.

HESLOP-HARRISON, Y. and KNOX, R. B. (1971) A cytochemical study of the leaf-gland enzymes of insectivorous plants of the genus *Pinguicula. Planta* **96,** 183–211.

HOMÈS, M. (1927) Evolution du vacuome au cours de la differenciation des tissus chez *Drosera intermedia* Hayne. *Bull Acad. r. Belg. Cl. Sci.* **13,** 731–746.

HUIE, L. (1899) Further studies of cytological changes produced in *Drosera. Q. Jl. microsc. Sci.* **42,** 203–222.

LLOYD, F. E. (1942) *The Carnivorous Plants*, Chronica Botanica, Waltham, Mass.

MAYO, M. A. and COCKING, E. C. (1969) Detection of pinocytotic activity using selective staining with phosphotungstic acid. *Protoplasma* **68,** 231–236.

RAGETLI, H. W. J., WEINTRAUB, M. and LO, E. (1972) Characteristics of *Drosera* tentacles. I. Anatomical and cytological detail. *Can. J. Bot.* **50,** 159–168.

SATIR, P., SATIR, B. and SCHOOLEY, C. (1972) Membrane reorganization during secretion in *Tetrahymena. Nature Lond.* **235,** 53–54.

SCHNEPF, E. (1961) Licht und elektron mikroskopische Beobachtungen on Insektivoren-Drüsen über die Sekretion des Fangschleimes. *Flora Jena* **151,** 73–87.

SCHNEPF, E. (1966) Feinbau und Funktion pflanzlicher Drüsen. *Umshau* **16,** 522–527.

SPANSWICK, R. M. (1974) Symplastic transport in plants. In *Transport at the Cellular Level*, ed. SLEIGH, M. A. and JENNINGS, D. H., pp. 127–137, Proceedings of the 28th Symposium of the Society for Experimental Biology, London, 1973. Cambridge University Press, Cambridge.

WILLIAMS, S. E. and PICKARD, B. G. (1969) Secretion, absorption and cuticle structure in *Drosera* tentacles. *Pl. Physiol. Lancaster* **44** Suppl. 5, No. 22.

WILLIAMS, S. E. and PICKARD, B. G. (1974) Connections and barriers between cells of *Drosera* tentacles in relation to their electrophysiology. *Planta* **116**, 1–16.

SECRETORY PROCESSES IN SEAWEEDS

L. V. Evans and Maureen E. Callow

Department of Plant Sciences, University of Leeds, Leeds, U.K.

Introduction

Although there has been a lot of work on the chemistry of many of the seaweed polysaccharides (see Percival and McDowell, 1967), there has been relatively little work on their intracellular sites of synthesis and the biosynthetic pathways involved, or on their movement and subsequent location within the plant. Seaweed polysaccharides have been identified by physical methods e.g. X-ray diffraction in conjunction with extraction procedures but this method does not permit precise localization in intact material. Polarized light has been used on sectioned material for localization of polysaccharides showing birefringence e.g. cellulose and alginic acid, but neither specific identification nor localization is easy with this method. However, various histochemical methods are available which enable more precise localization of seaweed polysaccharides and the present communication is concerned with aspects of recent work in this field and with aspects of polysaccharide secretion and biosynthesis.

Chemistry of seaweed polysaccharides

Brown algae

Alginic acid is found in all brown algae and amounts to 14–40 per cent of their dry weight. It is composed of D-mannuronic and L-guluronic acids in varying proportions. The two acids are arranged along the molecule in three types of block subunits: mannuronic acid (M—M blocks), guluronic acid (G—G blocks) or alternating mannuronic and guluronic acid (M—G blocks) (Haug *et al.*, 1967). X-ray diffraction studies by Frei and Preston (1962) indicated that polymannuronic acid-rich alginic acid is characteristic of the cell walls of young tissue and/or intercellular regions whereas polyguluronic acid-rich alginic acid is located in the cell wall proper (see also Mackie and Preston,

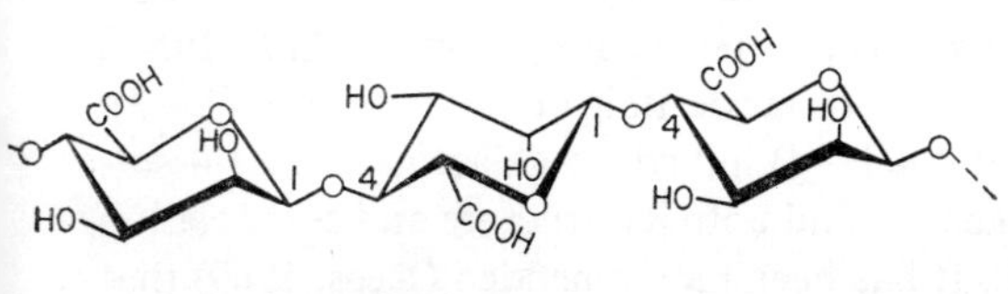

Segment of polymannuronic acid chain

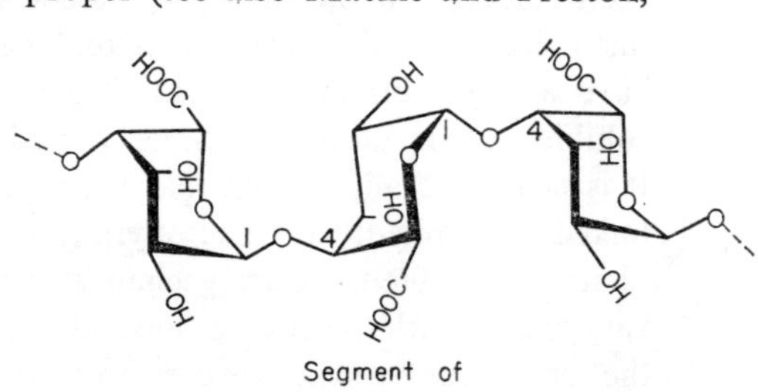

Segment of polyguluronic acid chain

1974). Chemical analyses support this finding; alginate from young tissues is rich in M—M blocks while that from older tissues is richer in G—G blocks and M—G blocks (Haug *et al.*, 1969). Alginate is supposed to occur within the plant as a mixed salt, the amount of bound calcium being mainly responsible for its gel strength. Since guluronic acid-rich alginates have a much higher affinity for calcium ions than mannuronic acid-rich alginates (Haug and Smidsrød, 1967; Smidsrød and Haug, 1968) it would appear that change in composition between young and old tissues would be advantageous to the plant.

In addition to alginic acid all brown seaweeds contain a wide range of sulphated polysaccharides. These are typified by the predominance of fucose and sulphate, although other constituents such as galactose, xylose, glucuronic acid and protein are often present. Fucoidan is the most well-known of a family of glucuronoxylofucans (Percival and McDowell, 1967; Bidwell *et al.*, 1972). The proposed structure of fucoidan is as follows (Percival and Mian, 1973):

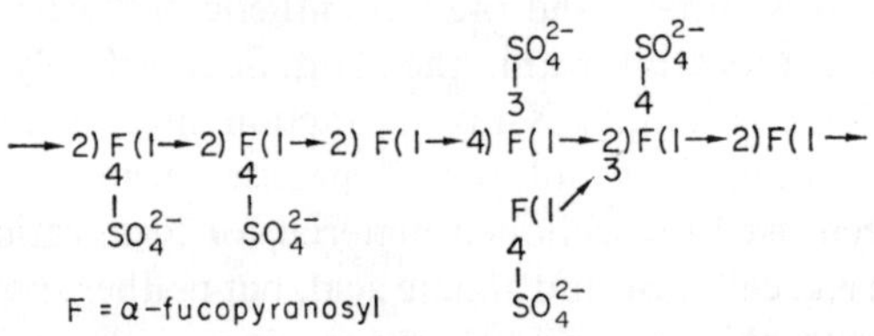

As may be seen, α 1-2 linked fucopyranose units are the most common (with sulphate on C4) but α 1-4 and α 1-3 linkages (with sulphate on C3 and C4 respectively) are also present.

Red algae

The most important polysaccharides of red algae are the galactans of which there are three main types, agaroids, carrageenans and porphyrans. These are all essentially linear molecules consisting of alternating 1,3- and 1,4-linked galactose units which may be masked by modification or substitution. The main differences between the types are in (1) the proportions of D- and L-galactose, (2) the extent to which the galactose is modified with the 3,6-anhydroderivative, (3) the amount and position of sulphation and methylation of the individual sugar residues and (4) the presence of other constituents such as xylose, uronic acid, glycerol and pyruvic acid (Percival, 1972). Since studies so far have been confined to the biology of those red seaweeds which produce carrageenan, only this galactan need concern us here. The carrageenans are all based on D-galactose and have a high sulphate content. They have been traditionally fractionated into KCl-soluble and -insoluble fractions, λ- and κ-carrageenan respectively (Smith *et al.*, 1954). It is now generally agreed that the repeating disaccharide unit of the latter is α-(1→3)-linked 3,6-anhydro-D-galactopyranose and β-(1→4)-linked D-galactopyranose-4-sulphate (Rees, 1969). λ-Carrageenan is heterogenous and both its structure and composition vary widely with the source (Percival, 1972). It has been recommended (Rees, 1969) that the term λ-carrageenan be used to represent a disaccharide of α(1→3)-linked D-galactopyranose-2-disulphate and β-(1→4)-linked D-galactopyranose-2,6-sulphate

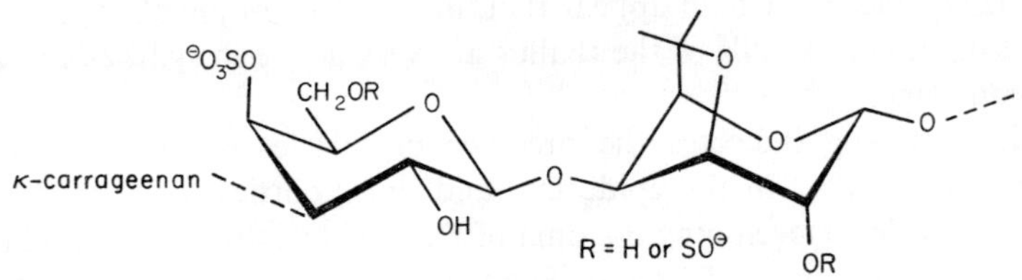

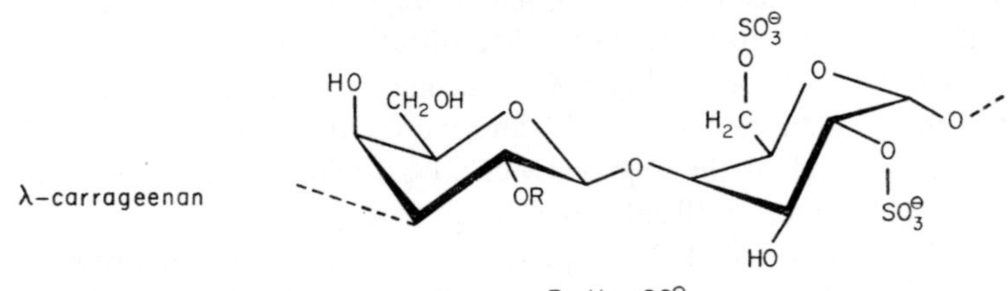

Another KCl-soluble carrageenan termed μ-carrageenan has been isolated from *Chondrus crispus*.

Localization of seaweed polysaccharides

Brown algae

Localization of alginic acid and sulphated polysaccharides within brown algal tissues can be carried out by means of a variety of light and electron microscope histochemical techniques. A range of staining reactions e.g. toluidine blue at pH 6.8 (O'Brien *et al.*, 1964), alcian blue/alcian yellow (Parker and Diboll, 1966) and the critical electrolyte alcian blue method (Scott and Dorling, 1965), and periodic acid Schiff (PAS) (Feder and O'Brien, 1968) all indicate that alginic acid is a major component of immediate thallus cell wall layers as e.g. in *Fucus* (McCully, 1966, 1968a) and *Dictyota* (Evans and Holligan, 1972), whereas the outer cell wall layer is predominantly sulphated.

Sulphated polysaccharides also occur as a layer on the thallus surface, and the intercellular mucilage between cortex and medulla cells of e.g. *Fucus* (McCully, 1965, 1966, 1968a), *Pelvetia* and *Himanthalia* is composed largely of sulphated materials although some alginic acid is also present. More sensitive labelled antibody techniques confirm the presence of alginic acid in the inner cell walls and in the intercellular matrix (especially in older regions of the medulla) of *Fucus* (Vreeland, 1970, 1972). An investigation of the sites of synthesis of sulphated polysaccharides by means of labelled inorganic sulphate ($^{35}SO_4^{2-}$) and light microscope autoradiography (Evans *et al.*, 1973) shows that in *Pelvetia* labelled for 2 hr and then transferred for a short period to unlabelled medium, silver grains during the chase are at first located mainly around the nuclei of epidermal and outer cortical cells (Fig. 1), an area which electron microscope examination reveals to be rich in Golgi bodies. With increased chase-time in unlabelled medium, the label moves mainly into the external mucilage with some going into the intercellular mucilage

(Fig. 2). Since similar observations have been made on *Fucus spiralis*, *Ascophyllum*, *Halidrys* and *Himanthalia* it would appear that, in the blade of members of the Fucales, epidermal and outer cortex cells of the thallus are very active in synthesis and secretion of sulphated mucilage.

Histochemical staining indicates the presence of vesicles of both sulphated polysaccharide and alginic acid in the epidermal and outer cortical cells of the thallus of members of the Fucales. Developing oogonia of *Fucus* (McCully, 1968b) and developing oogonia and tetrasporangia of *Dictyota* (Evans and Holligan, 1972) also contain both types of vesicle. In young oogonia, alginic acid-containing vesicles predominate but during maturation those containing sulphated polysaccharide become more numerous. Following release, and within 15 min of fertilization an alginic acid-rich wall forms around the zygote. After about 12 hr there is a sharp increase in the intracellular content of sulphated polysaccharide, followed by an asymmetric secretion of this material at the prospective rhizoid end of the zygote (McCully, 1970). During the period of rhizoid initiation (10–16 hr after fertilization) $^{35}SO_4^{2-}$ is rapidly incorporated into fucoidan (Quatrano and Crayton, 1973). Unsulphated fucoidan is however present immediately after fertilization and since this undergoes little metabolic activity or turnover in the subsequent 24 hr, it was concluded that incorporation of $^{35}SO_4^{2-}$ involves the sulphation of a pre-existing unsulphated fucose polymer (Quatrano and Crayton, 1973). Eggs of *Fucus serratus* have been used to obtain Golgi-rich preparations (Fig. 6) (Matthews and Evans, 1975) which are being used to investigate further the role of the Golgi in the secretion of sulphated polysaccharides.

In the blade of members of the Laminariales both the cell walls and intercellular mucilage consist mainly of alginic acid, and sulphated material is confined mainly to the thallus surface and to internal canals (Fig. 3) from which it is secreted onto the surface. Labelling with $^{35}SO_4^{2-}$ for 2 hr results in heavy accumulations of silver grains associated with the specialized secretory cells (Fig. 4) lining the canals, and after a chase period in

---

FIG. 1. Light microscope autoradiograph of a longitudinal section of epidermal cells of *Pelvetia*, 2 hr incubation in $^{35}SO_4^{2-}$ followed by a 4 hr chase. Silver grains are seen around the nuclei (arrowed) and basal regions of epidermal cells and in outer cortical cells. No labelling of the outer mucilage. (Embedded in glycol methacrylate, 1 μm section, toluidine blue stained.) × 1500.

FIG. 2. Light microscope autoradiograph of a longitudinal section of the apical region of *Pelvetia*, 2 hr incubation in $^{35}SO_4^{2-}$ followed by an 18 hr chase. Some silver grains are associated with the intercellular mucilage but the outer layer of mucilage is heavily labelled. (Section preparation as in Fig. 1) × 250.

FIG. 3. Light micrograph (interference contrast) of a transverse section of the thallus of *Laminaria saccharina* showing canal (c) with associated secretory cells. Chloroplasts (arrowed) are visible in surrounding thallus cells. (Epon embedded, 1 μm.) × 375.

FIG. 4. Light microscope autoradiograph of a transverse section of the thallus of *Laminaria hyperborea* incubated in $^{35}SO_4^{2-}$ for 8 hr, showing heavy labelling of the secretory cells. No labelling of the secretory canal (c) or surrounding thallus cells. (Section preparation as in Fig. 1.) × 375.

FIG. 5. Electron micrograph of a transverse section of part of a secretory cell of *Laminaria hyperborea* showing part of the large central nucleus (n) surrounded by Golgi bodies (g). Also visible are Golgi-derived vesicles (v), extensive areas of distended endoplasmic reticulum (er) and mitochondria (m). ×20,000.

FIG. 6. Electron micrograph of isolated Golgi-rich fraction from *Fucus serratus* eggs. Also present is contaminating vesicular material and a mitochondrion (m). Homogenization and fractionation as described by Matthews and Evans (1975). × 30,000.

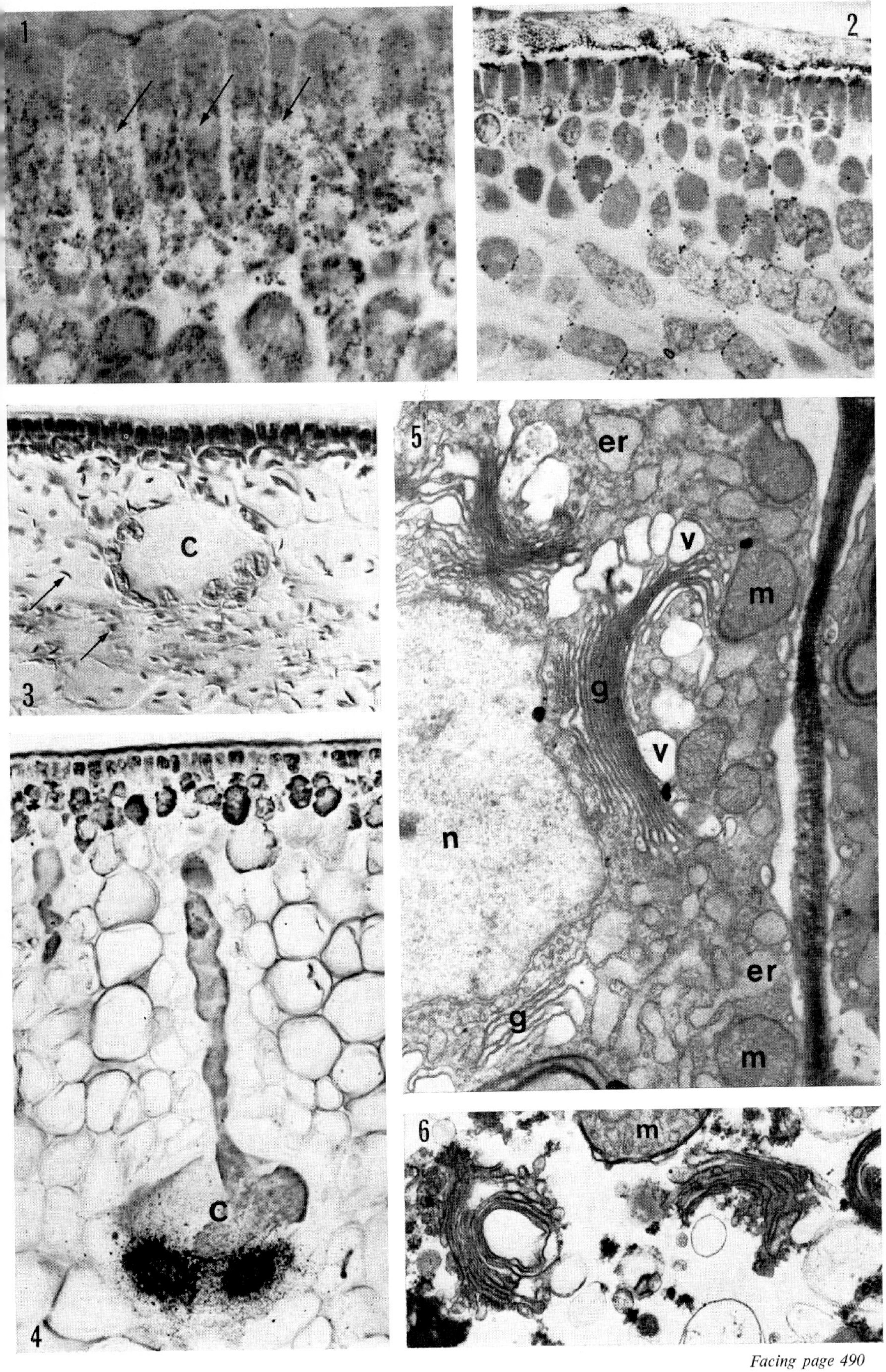

Facing page 490

non-radioactive medium, label is seen within the secretory canals and later on the thallus surface. Electron microscope (e.m.) examination of *Laminaria* secretory cells (Fig. 5) shows many large perinuclear Golgi bodies and in tissues incubated in $^{35}SO_4^{2-}$ for 2 hr, e.m. autoradiography shows that the Golgi bodies are the sites of polysaccharide sulphation in this plant (Evans and Callow, 1974). Although histochemical tests indicate a predominance of sulphated polysaccharide in secretory canals, differential removal of these in *Macrocystis* reveals some material with staining properties of alginic acid (Vreeland, personal communication). However, in spite of the extensive system of distended endoplasmic reticulum (e.r.) in secretory cells (Fig. 5) protein has not been detected in the canals, and it is possible that the e.r. functions in input of new Golgi membranes.

In young haptera of *Laminaria hyperborea*, epidermal cells synthesize sulphated polysaccharide which is secreted onto the surface. However, in older haptera a canal system develops and this serves to transport sulphated polysaccharides synthesized within adjacent secretory cells (Davies *et al.*, 1973; Johnston, personal communication).

Since histochemical staining reactions reveal the presence of vesicles of alginic acid within cells of all types of vegetative and reproductive tissues of brown algae, it is generally assumed that all cells have the potential for synthesis. However, there is no direct experimental evidence for this or for the inferred involvement of the Golgi bodies in synthesis. As there is no specific precursor molecule of alginic acid, results from autoradiographic experiments are difficult to interpret. However, preliminary experiments in which pieces of young *Himanthalia* receptacles were incubated in [^{3}H]mannose show preferential labelling of the outer regions of epidermal cells and the immediate cell walls of epidermal and cortical cells. [^{14}C]mannose has also been shown to be taken up and incorporated into the cell walls of *Fucus* zygotes (Forman, personal communication).

Red algae

Involvement of the Golgi in sulphation of polysaccharide has been shown in the unicellular red alga *Rhodella*. Although without walls the cells are enveloped in mucilage (Fig. 7) and this comprises 50 per cent carbohydrate, 16 per cent protein and 10 per cent sulphate (Evans *et al.*, 1974). In the light, mucilage is continuously synthesized and solubilized into the medium by *Rhodella*, a situation similar to that which occurs in the related genus *Porphyridium* (Ramus, 1973). Distended cisternae containing finely fibrillar material become detached from the maturing faces of the Golgi bodies and round off to give membrane-bounded vesicles with a fibrillar finger-print like content closely resembling the external mucilage (Fig. 10). These move to the cell periphery and discharge their contents through the plasma membrane by reverse pinocytosis. An autoradiographic investigation of the intracellular formation of mucilage shows that in cells labelled with $^{35}SO_4^{2-}$ for 30 min silver grains are preferentially associated with Golgi bodies (Fig. 11) which therefore have a very high activity compared with other organelles (Table 1). After a short chase in unlabelled medium, activity of the Golgi is much reduced (Table 1) and silver grains are associated with detached cisternal vesicles (Fig. 12), giving increased activity in this category (Table 1). With increased chase time, heavy labelling occurs in the extracellular mucilage envelope (Fig. 13); the Golgi is no more active than any other organelle and activity in the vesicles is reduced (Table 1).

TABLE 1. DISTRIBUTION OF SILVER GRAINS ON AUTORADIOGRAPHS AFTER INCUBATION OF *Rhodella* IN $^{35}SO_4^{2-}$

Organelle	30 min pulse		30 min pulse/ 90 min chase		30 min pulse/ 180 min chase	
	% Total grains	Relative activity	% Total grains	Relative activity	% Total grains	Relative activity
Cytoplasm	8.0	1.0	8.0	1.0	7.5	1.0
Chloroplast	22.0	0.7	23.0	0.8	22.0	0.7
Pyrenoid	2.7	0.6	4.6	0.6	8.5	0.8
Starch grains	6.8	0.5	5.0	0.3	8.5	0.5
Nucleus	3.4	1.3	3.4	0.6	4.3	0.6
Vacuole	15.0	0.6	11.0	0.4	14.2	0.3
Mitochondrion	1.0	1.6	0.5	1.0	—	—
Golgi	21.0	17.0	11.0	9.0	4.3	1.3
Vesicles	3.4	1.0	9.0	4.0	4.9	2.5
Miscellaneous	5.5	2.0	6.0	0.8	3.6	1.2
Mucilage A	10.0	0.5	16.0	1.0	16.8	1.0
Mucilage B	1.0	0.1	1.0	0.2	3.7	0.3
Mucilage C	0.5	0.1	2.0	0.3	1.0	0.1
Total no. grains	147		237		162	

(Data from Evans *et al.*, 1974). Relative activity was calculated after analysis of autoradiographs as described by Callow and Evans (1974); all categories are expressed relative to the cytoplasm which is designated as 1.0 in all treatments.

Chemical analysis (Evans *et al.*, 1974; Fareed and Percival, 1975) of the mucilage of *Rhodella* shows that the major sugar, xylose, is present as long chains of 1,4- and 1,3-linked units, together with smaller quantities of glucuronic acid, galactose, glucose, rhamnose and an unknown reducing substance. The position and conformation of the sulphate residue is yet to be determined but appears to be different from that in other sulphated algal polysaccharides. In addition to the sulphated polysaccharide, solubilized mucilage contains 16 per cent by weight of protein. Beneath the cell plasma membrane is an anastomosing network of the main cytoplasmic e.r. system (Fig. 9) and small ducts from the subplasmalemmal network approach and fuse with the plasma membrane (Fig.

FIGS. 7–13: *Rhodella.* FIG. 7. Light micrograph of living cells in a weak suspension of Indian ink, showing the extent of the encapsulating mucilage. × 500.

FIG. 8. Electron micrograph showing part of the lobed chloroplast and smooth endoplasmic reticulum with ducts (arrowed) projecting towards the plasmalemma. × 40,000.

FIG. 9. Electron micrograph showing the sub-plasmalemmal endoplasmic reticulum (arrowheads) in continuity (arrowed) with the endoplasmic reticulum (*er*) found in association with the forming face of the Golgi body (*g*). Also visible are lobes of the chloroplast (*c*) and a mitochondrion (*m*). × 25,000.

FIG. 10. Electron micrograph through a Golgi body showing a developmental sequence of vesicle formation towards the maturing face. The Golgi cisternae, vesicle (v_1) newly-detached from the maturing face and vesicle (v_2) in close proximity to the plasmalemma contain fibrillar material similar in appearance to the external mucilage (*mu*). × 45,000.

FIG. 11. Electron microscope autoradiograph of part of a cell incubated in $^{35}SO_4^{2-}$ for 0.5 hr showing heavy labelling of the Golgi body (*g*). Chloroplast (*c*) and starch grains (*sg*) are also visible. × 30,000.

FIG. 12. Electron microscope autoradiograph of part of a cell incubated in $^{35}SO_4^{2-}$ for 0.5 hr and chased for 1.5 hr showing silver grains associated with a Golgi body (*g*) and detached vesicles (arrowed). × 30,000.

FIG. 13. Light microscope autoradiograph of cells incubated in $^{35}SO_4^{2-}$ for 0.5 hr and chased for 5 hr showing heavy labelling around the cell peripheries. (Section preparation as in Fig. 1). × 800.

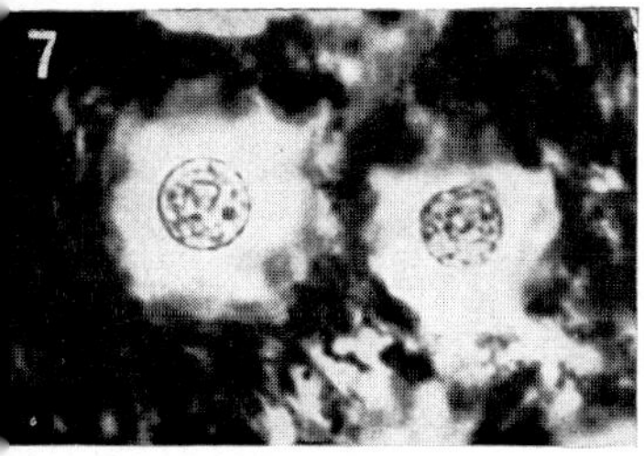
7

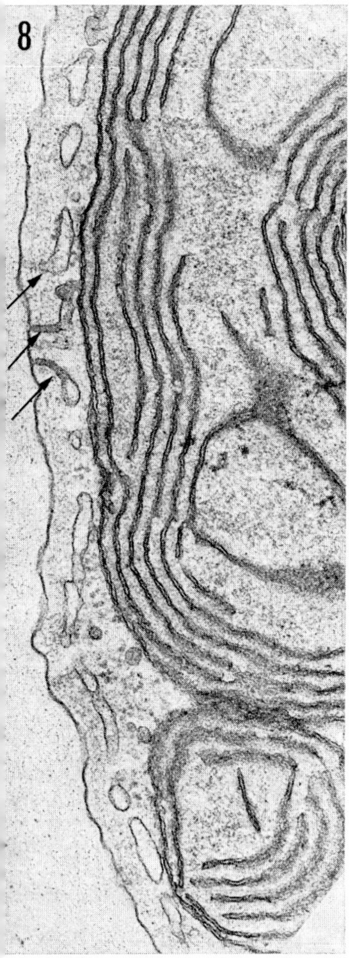
8

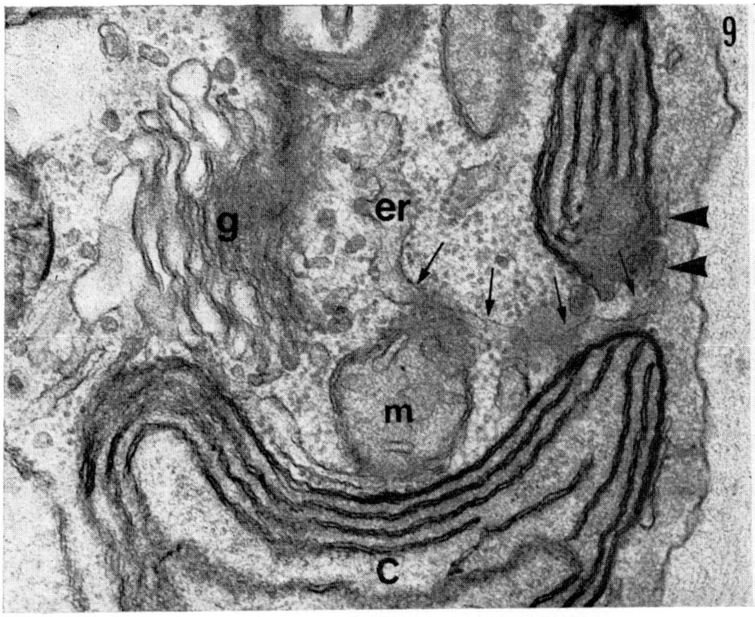
9
g
er
m
c

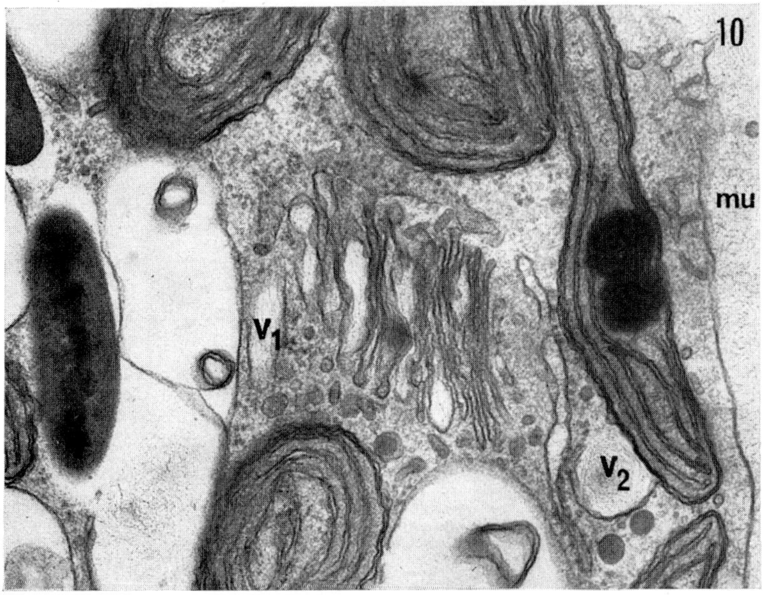
10
mu
v_1
v_2

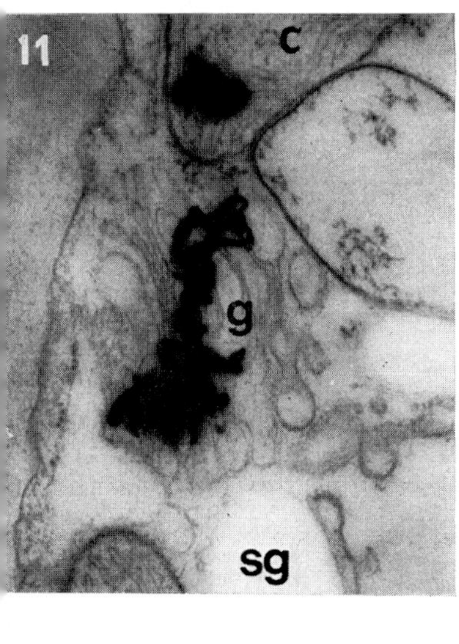
11
c
g
sg

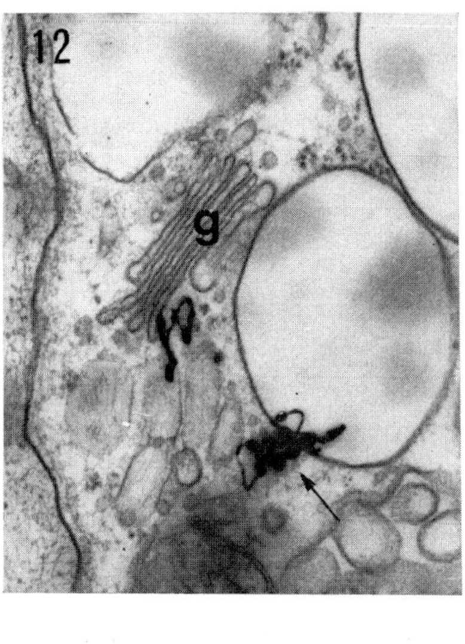
12
g

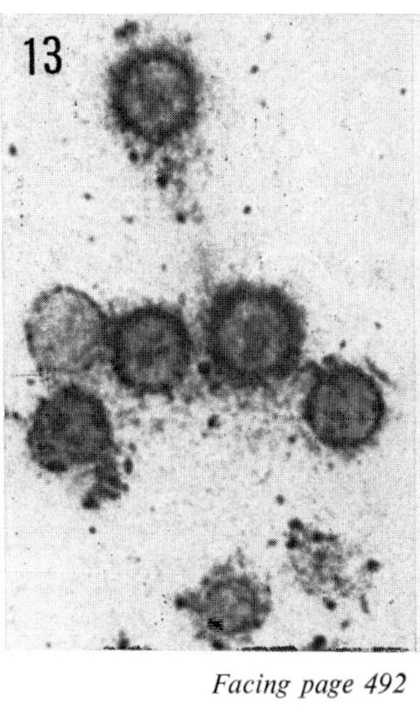
13

8), thus providing a channel from the interior of the e.r. to the outside and a possible route for protein passing out of the cell. Electron microscope autoradiography is being used to gain further information on this stage of the secretion process. Synthesis of the protein component has been studied by labelling cells with $^{35}SO_4^{2-}$ or [^{14}C]arginine and then fractionating the secreted mucilage by polyacrylamide gel electrophoresis (Callow and Evans, 1975a). In gels obtained from cells labelled with $^{35}SO_4^{2-}$, radioactivity coincides with the polysaccharide stained bands, whilst in those obtained from [^{14}C]-arginine labelled cells, radioactivity coincides with the three protein stained bands, indicating that protein is not linked to polysaccharide. Secretion of amino acids, polypeptides and protein is known to occur in a wide range of algae (Newell *et al.*, 1972). Protein is also present in cell walls (Thompson and Preston, 1967) and is the major constituent of the thin cuticle found in many members of the green, brown and red seaweeds (Hanic and Craigie, 1969). Little is known of the biological significance or of the process of secretion of the chemically resistant cuticle.

Amongst the larger red seaweeds, secreted mucilages take on the major structural role in the plant. It has been known for many years that the gelling properties of carrageenan extracted from members of the Gigartinaceae are extremely variable. Chemical analysis and infra-red spectroscopy have shown that sporophytic (tetrasporic) plants of *Chondrus crispus* contain λ-carrageenan whilst gametophytic (male and female) plants contain κ-carrageenan, together with lesser amounts ($\sim$25 per cent) of μ-carrageenan and a third carrageenan, as yet unidentified, but shown not to be λ-carrageenan (McCandless *et al.*, 1973). Labelling of vegetative tissue with specific fluorescent antibodies confirms these results. Moreover, cell walls of tetrasporangia stain with anti-κ antibody and cell walls of carposporophytes stain with anti-λ antibody, indicating that the carrageenans associated with each stage of the life cycle are synthesized by the first cells of each stage (Gordon-Mills and McCandless, 1975). Analysis of a number of *Gigartina* species shows similar differences in carrageenans between the two isomorphic nuclear phases (Chen *et al.*, 1973; Pickmere *et al.*, 1973; McCandless, 1975).

Role of the Golgi in Secretion

It has been inferred from electron micrographs that synthesis of cell wall materials and other secreted mucilages occurs in the Golgi bodies e.g. *Chondrus crispus* (Gordon and McCandless, 1973), *Ceramium* (Chamberlain and Evans, 1973), *Ptilota* (Scott and Dixon, 1973), *Griffithsia* (Peryière, 1970), *Smithora* (McBride and Cole, 1971) and *Levringiella* (Kugrens and West, 1972). Autoradiographic techniques have confirmed that the Golgi has a key role in the production of sulphated polysaccharide in *Rhodella* and in several brown algae, as has been found in animal cells which synthesize a wide range of sulphated compounds (Berg and Young, 1971; Young, 1973). However, only the transfer of activated sulphate to an acceptor molecule can definitely be said to take place in the Golgi, since soluble low molecular weight precursor compounds are lost during preparation for electron microscopy.

A commonly observed phenomenon is some seaweeds where secretory activities have been investigated is the close relationship between Golgi bodies and e.r. Whilst it is likely that the e.r. is providing a source for input of new membranes into either the forming face of the Golgi or into cisternal peripheries in many organisms, in others, material passes from the e.r. into the cisternae of the Golgi. In the motile zoospores of the green seaweed *Enteromorpha* Golgi bodies are always partially enclosed by an arc of rough e.r. which cuts off small vesicles from its smooth inner face next to the forming face of the Golgi (Fig. 14) (Evans and Christie, 1970). The Golgi itself cuts off large vesicles from the maturing face and by use of specific enzymes (Christie *et al.*, 1970) and e.m. cytochemical techniques, in particular periodic acid/thiosemicarbazide/silver proteinate (Thiéry, 1967, 1969), the contents of the detached vesicles have been shown to be glycoprotein in nature (Figs. 15, 16) (Callow and Evans, 1974). Autoradiography with [^{3}H]leucine shows the involvement of the Golgi e.r. in the production of the glycoprotein (Callow and Evans, 1974). After a 30 min pulse the label appears predominantly in the cytoplasmic e.r., Golgi e.r. and Golgi (Fig. 17; Table 2), but after a chase period in unlabelled medium, labelling declines in both cytoplasmic e.r. and the Golgi e.r. and increases in the Golgi and detached vesicles. Material synthesized in the Golgi e.r. moves via transfer vesicles into the Golgi cisternae and from there into the distended cisternal ends at the maturing face. Thus, functional continuity exists between the e.r. and the Golgi apparatus, a situation often found in animal tissues (Beams and Kessel, 1968; Morré *et al.*, 1971). Complete cisternae, each with a distended end containing electron dense material (Fig. 16), then become detached from the Golgi stack and e.m. cytochemistry shows that polymerization of the carbohydrate moiety is completed after detachment has occurred. However, inosine diphosphatase (IDP-ase) activity (Dauwalder *et al.*, 1969) is associated with cisternae in the Golgi stack at this stage (Fig. 18) and not with detached cisternae (Callow and Evans, 1975b). Although the function of IDP-ase is not clearly understood, it is thought to be involved in polysaccharide synthesis and is generally associated with Golgi bodies, hence its use as a Golgi marker enzyme. From the cytochemical results IDP-ase activity would be expected in detached cisternae also. Clearly a very complex situation exists and this is being investigated. During settlement of a zoospore the vesicles discharge their glycoprotein content through the plasma

---

FIGS. 14–18: Swimming zoospores of *Enteromorpha*. FIG. 14. Electron micrograph of the anterior region, showing Golgi body (*g*) abstricting an adhesive vesicle (arrow) and free vesicles (*v*) in the cytoplasm, the membrane of one forming a tube-like extension (arrowheads). Profiles of endoplasmic reticulum (*er*) are seen in the cytoplasm and in association with the forming face of the Golgi bodies. (Fixed in glutaraldehyde with post-osmication, and stained in uranyl acetate and lead citrate.) ×34,000.

FIG. 15. Electron micrograph showing adhesive vesicles from the anterior region of a spore. (Fixed glutaraldehyde/unstained.) ×54,000.

FIG. 16. Electron micrograph of part of the anterior region of a spore, showing the staining reactions characteristic of adhesive vesicles (*v*), vesicle membranes and the tube-like extensions of the latter (arrowed); Golgi bodies (*g*) and endoplasmic reticulum (*er*) are visible but are not stained. (Fixed in glutaraldehyde, sections treated with 1 per cent periodic acid for 20 min, then 1 per cent thiosemicarbazide in 10 per cent acetic acid for 24 hr and 1 per cent silver proteinate for 35 min.) ×54,000.

FIG. 17. Electron microscope autoradiograph of a median L.S. of a spore after 30 min incubation in [^{3}H]leucine showing association of silver grains with the chloroplast (*c*), mitochondrion (*m*), Golgi bodies (*g*) and vesicles (*v*). (Specimen preparation as in Fig. 14.) ×20,000.

FIG. 18. Electron micrograph showing IDPase activity associated with cisternae of the Golgi body (*g*). Also visible but unstained are vesicles (*v*), part of the chloroplast (*c*), mitochondrion (*m*) and nucleus (*n*). (Fixed in glutaraldehyde, IDP-ase incubation after Dauwalder *et al.*, 1969, post-stained in uranyl acetate/lead citrate.) ×45,000.

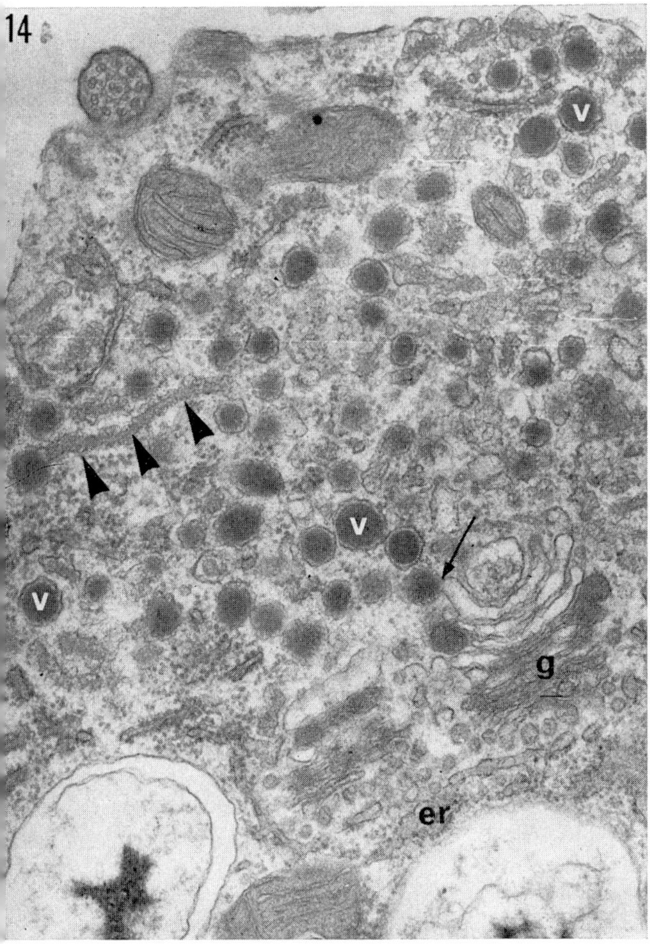
14
v
v
v
g
er

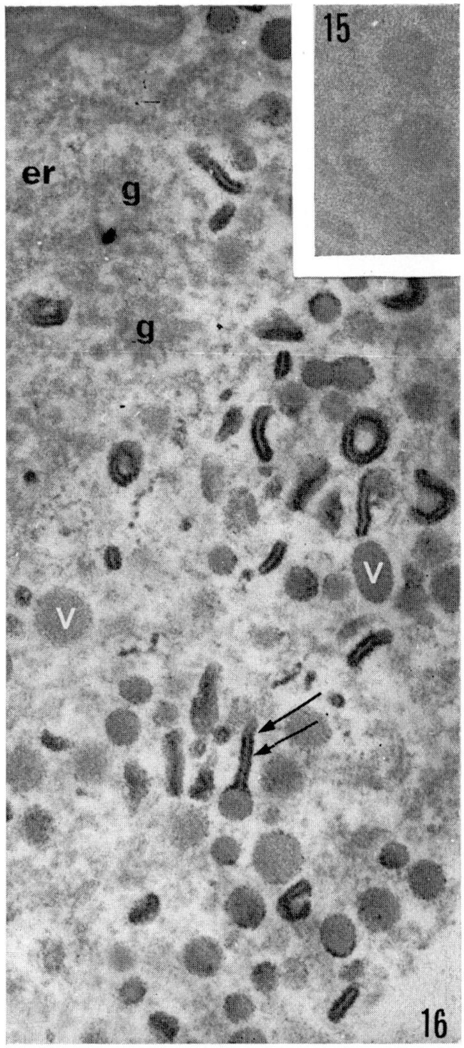
15
er
g
g
v
v
16

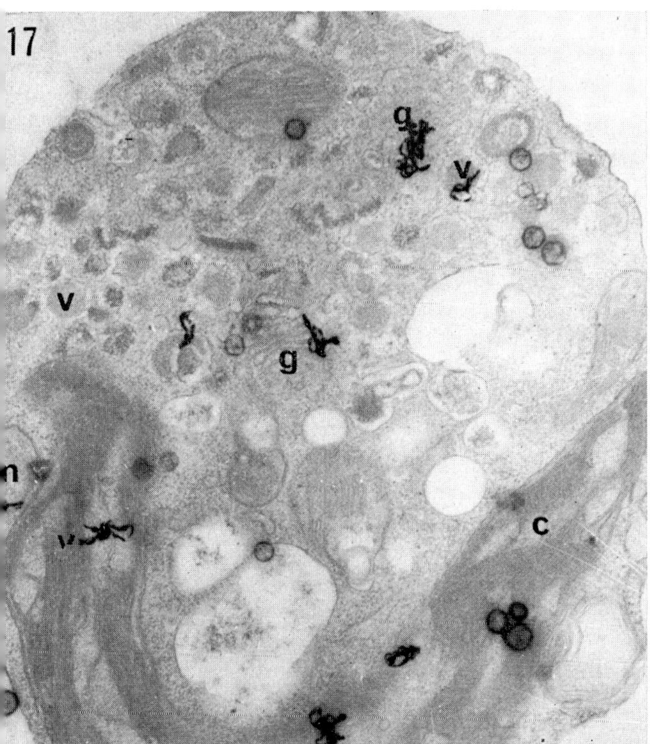
17
g
v
v
g
n
c

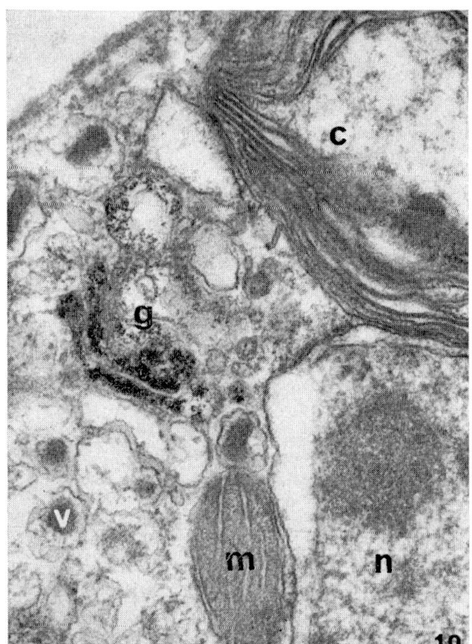
c
g
v
m
n
18

TABLE 2. DISTRIBUTION OF SILVER GRAINS ON AUTORADIOGRAPHS AFTER INCUBATION OF *Enteromorpha* ZOOSPORES IN [^{3}H]-L-LEUCINE

	30 min pulse		30 min pulse/ 30 min chase		30 min pulse/ 180 min settled	
Organelle	% Total grains	Relative activity	% Total grains	Relative activity	% Total grains	Relative activity
Cytoplasm	20.5	1.0	18.0	1.0	1.5	1.0
Chloroplast	22.5	1.2	25.0	1.5	43.0	2.8
Nucleus	4.0	1.0	4.0	1.3	1.0	1.0
Mitochondrion	5.5	1.5	6.0	1.7	1.0	1.6
Golgi	6.5	7.8	9.0	11.4	1.0	1.0
Endoplasmic reticulum (e.r.)	9.5	12.2	2.0	4.0	0	—
Golgi e.r.	7.0	12.6	2.0	2.8	0	—
Vesicle	20.5	4.6	28.0	6.0	—	—
Adhesive	—	—	—	—	54.5	5.0
Storage vesicle	4.5	0.8	6.0	0.6	1.5	1.7
Miscellaneous	0	—	0	—	2.0	2.0
Total no. grains	201		183		441	

(Data from Callow and Evans, 1974). Relative activity was calculated after analysis of autoradiographs as described by Callow and Evans (1974); all categories are expressed relative to the cytoplasm which is designated as 1.0 in all treatments.

membrane by reverse pinocytosis as a fibrillar adhesive material. At this stage the function of the Golgi changes; adhesive synthesis ceases and the Golgi becomes involved in the production of wall material.

Carbohydrate has been detected cytochemically in Golgi-produced adhesive vesicles of other seaweeds, e.g. *Ectocarpus* (Baker and Evans, 1973; Clitheroe, personal communication) and *Ceramium* (Chamberlain and Evans, 1973, 1975). In the latter, Golgi bodies in developing spores produce large numbers of dense-cored adhesive vesicles, the content of which are composed of protein and polysaccharide. Enzymic and chemical treatments however, show that a protein/polypeptide component is unlikely to be essential in spore adhesion and that the functional glue of *Ceramium* spores is a polysaccharide.

BIOSYNTHESIS

The biosynthesis of polysaccharides involves the formation of the appropriate precursor molecules, polymerization and in some cases, modification of the polysaccharide by for example substitution or sulphation.

In brown algae, the biosynthetic pathway leading to incorporation of mannuronic acid into a polymer of alginic acid is known (Lin and Hassid, 1966), but the way in which mannuronic acid is converted to guluronic acid still remains to be elucidated. Since Lin and Hassid (1966) detected small quantities of GDP-L-guluronic acid in extracts of *Fucus* it seemed likely that such nucleotides would be directly incorporated into alginic acid chains. However, work on the extracellular production of alginate by the bacterium *Azotobacter vinelandii* indicates that a C-5 epimerization reaction takes place at the polymer level, converting 'in-chain' D-mannuronic acid units to L-guluronic acid units (Haug and Larsen, 1974). There is as yet no direct evidence for this reaction in brown algae. However, when plants of *Laminaria* were kept in darkness after labelling with $NaH^{14}CO_3$ in the light, an increase in the radioactivity in guluronic acid occurred although there was very little net synthesis of alginate (Hellebust and Haug, 1972). These results could be explained by the epimerization of D-mannuronic acid to L-guluronic acid at the polymer level and would therefore represent modification of the polysaccharide in the direction of higher gel strength.

A variety of sugar nucleotides has been isolated from various red algae (see Haug and Larsen, 1974), although the biosynthetic pathways involved in the formation of the polymer do not appear to be well characterized. As stated earlier, many of the algal polysaccharides are commonly substituted. In animal cells there is evidence that sulphation takes place at the polymer level (Stoolmiller and Dorfman, 1969). The rapidity of sulphate incorporation in those seaweeds so far studied e.g. *Chondrus crispus* (Loewus *et al.*, 1971), *Porphyridium aerugineum* (Ramus and Groves, 1972) and *Rhodella maculata* (Evans *et al.*, 1974) suggests that sulphation of a preformed polymer is occurring. However, the pathway of sulphate metabolism in algae is little understood. The starting point for a variety of reactions involving sulphate is the formation of a high energy phosphorylated compound adenosine 3′-phosphate 5′-phosphosulphate (PAPS) (Schiff and Hodson, 1973) which would act as the sulphate donor for the polymer backbone. Alternatively, sugar monomers (probably as nucleotides) may become sulphated first (with PAPS as donor), and the resulting sugar sulphates then polymerized to give polysaccharides.

Modification at the polymer level may also occur in red algae, e.g. treatment of carrageenan with an enzyme isolated from *Gigartina stellata* resulted in elimination of sulphate with concomitant formation of 3,6-anhydro-L-galactose residues (Lawson and Rees, 1970). As in the case of the epimerase reaction in alginic acid formation in brown algae, higher gel strength will result from the conversion of 4-linked galactose-6-sulphate units to 3,6-anhydro-galactose residues.

LOOKING AHEAD

In looking ahead to future developments in this field it is necessary first to take stock of the present state of knowledge. Localization of seaweed polysaccharides can now be carried out with some precision, providing an array of light and electron microscope techniques are used including highly sensitive immunocytochemical methods. Valuable analytical information can also be obtained in some cases by means of X-ray microanalytical techniques. However, with advances in techniques and instrument technology, improved methods of fixing, embedding and hence specific staining of seaweed polysaccharides should be possible. The further development of microanalytical techniques to quantify and extend conventional histochemical studies should also enable significant advances to be made, in particular in enzyme histochemical studies. The localization of soluble substances should also be facilitated by the development of improved preparative procedures for soluble compound autoradiography.

Such techniques of experimental microscopy must however, be integrated with chemical work on seaweed polysaccharides and with the use of current animal and higher plant biochemical methods. It is clear that the Golgi plays a key role in the synthesis and sulphation of seaweed polysaccharides and it is now necessary to investigate in detail the biosynthetic pathways involved in their formation, synthesis and polymerization, to locate the sites of these and to gain greater understanding of the mechanisms of regulation and control.

References

BAKER, J. R. J. and EVANS, L. V. (1973) The ship fouling alga *Ectocarpus*. I. Ultrastructure and cytochemistry of plurilocular reproductive stages. *Protoplasma* **77,** 1–13.

BEAMS, H. W. and KESSEL, R. G. (1968) The Golgi apparatus, structure and function. *Int. Rev. Cytol.* **23,** 209–276.

BERG, N. B. and YOUNG, R. W. (1971) Sulfate metabolism in pancreatic acinar cells. *J. Cell Biol.* **50,** 469–483.

BIDWELL, R. G. S., PERCIVAL, E. and SMESTAD, B. (1972) Photosynthesis and metabolism of marine algae. VIII. Incorporation of ^{14}C into the polysaccharides metabolized by *Fucus vesiculosus* during pulse labelling experiments. *Can. J. Bot.* **50,** 191–197.

CALLOW, M. E. and EVANS, L. V. (1974) Studies on the ship-fouling alga *Enteromorpha*. III. Cytochemistry and autoradiography of adhesive production. *Protoplasma* **80,** 15–27.

CALLOW, M. E. and EVANS, L. V. (1975a) Studies on the synthesis and secretion of mucilage in the unicellular red alga *Rhodella*. In *Proceedings of the VIIIth International Seaweed Symposium Bangor, 1974*. In press.

CALLOW, M. E. and EVANS, L. V. (1975b) Studies on the ship-fouling alga *Enteromorpha*. IV. Polysaccharide and nucleoside diphosphatase localization. In preparation.

CHAMBERLAIN, A. H. L. and EVANS, L. V. (1973) Aspects of spore production in the red alga *Ceramium*. *Protoplasma* **76,** 139–159.

CHAMBERLAIN, A. H. L. and EVANS, L. V. (1975) Chemical and histochemical studies on the spore adhesive of *Ceramium*. In *Proceedings of the VIIIth International Seaweed Symposium Bangor, 1974*. In press.

CHEN, L. C.-M., MCLACHLAN, J., NEISH, A. C. and SHACKLOCK, P. F. (1973) The ratio of kappa- to lambda-carrageenan in nuclear phases of the rhodophycean algae, *Chondrus crispus* and *Gigartina stellata*. *J. mar. biol. Ass. U.K.* **53,** 11–16.

CHRISTIE, A. O., EVANS, L. V. and SHAW, M. (1970) Studies on the ship-fouling alga *Enteromorpha*. II. The effect of certain enzymes on the adhesion of zoospores. *Ann. Bot.* NS **34,** 467–482.

DAUWALDER, M., WHALEY, W. G. and KEPHART, J. E. (1969) Phosphatases and differentiation of the Golgi apparatus. *J. Cell Sci.* **4,** 455–497.

DAVIES, J. M., FERRIER, N. C. and JOHNSTON, C. S. (1973) The ultrastructure of the meristoderm cells of the hapteron of *Laminaria*. *J. mar. biol. Ass. U.K.* **53,** 237–246.

EVANS, L. V. and CALLOW, M. E. (1974) Polysaccharide sulphation in *Laminaria*. *Planta* **117,** 93–95.

EVANS, L. V. and CHRISTIE, A. O. (1970) Studies on the ship-fouling alga *Enteromorpha* I. Aspects of the fine-structure and biochemistry of swimming and newly settled zoospores. *Ann. Bot.* NS **34,** 451–466.

EVANS, L. V. and HOLLIGAN, M. S. (1972) Correlated light and electron microscope studies on brown algae. I. Localization of alginic acid and sulphated polysaccharides in *Dictyota*. *New Phytol.* **71,** 1161–1172.

EVANS, L. V., SIMPSON, M. and CALLOW, M. E. (1973) Sulphated polysaccharide synthesis in brown algae. *Planta* **110,** 237–252.

EVANS, L. V., CALLOW, M. E., PERCIVAL, E. and FAREED, V. (1974) Studies on the synthesis and composition of extracellular mucilage in the unicellular red alga *Rhodella*. *J. Cell. Sci.* **16,** 1–21.

FAREED, V. and PERCIVAL, E. (1975) Extracellular mucilage from *Rhodella maculata*. In *Proceedings of the VIIIth International Seaweed Symposium, Bangor, 1974*. In press.

FEDER, N. and O'BRIEN, T. P. (1968) Plant microtechnique: some principles and new methods. *Am. J. Bot.* **55,** 123–142.

FREI, E. and PRESTON, R. D. (1962) Configuration of alginic acid in marine brown algae. *Nature Lond.* **196,** 130–134.

Gordon-Mills, E. M. and McCandless, E. L. (1975) Studies on carrageenan in the cell walls of *Chondrus crispus*. In *Proceedings of the VIIIth International Seaweed Symposium Bangor, 1974*. In press.

Hanic, L. A. and Craigie, J. S. (1969) Studies on the algal cuticle. *J. Phycol.* **5,** 89–102.

Haug, A. and Larsen, B. (1974) Biosynthesis of algal polysaccharides. In *Plant Carbohydrate Biochemistry*. ed. Pridham, J. B. *Proceedings of the Phytochemical Society Symposium* 1973. Academic Press, New York.

Haug, A. and Smidsrød, O. (1967) Strontium, calcium and magnesium in brown algae. *Nature Lond.* **215,** 1167–1168.

Haug, A., Larsen, B. and Smidsrød, O. (1967) Studies on the sequence of uronic acid residues in alginic acid. *Acta chem. scand.* **21,** 691–703.

Haug, A., Larsen, B. and Baardseth, E. (1969) Comparison of the constituents of alginate from different sources. In *Proceedings of the VIth International Seaweed Symposium, Madrid, 1968*. pp. 443–451.

Hellebust, J. H. and Haug, A. (1972) *In situ* studies on alginic acid synthesis and other aspects of the metabolism of *Laminaria digitata*. *Can. J. Bot.* **50,** 177–184.

Kugrens, P. and West, J. A. (1972) Ultrastructure of tetrasporogenesis in the parasitic red alga *Levringiella gardneri* (Setchell) Kylin. *J. Phycol.* **8,** 370–383.

Lawson, C. J. and Rees, D. A. (1970) An enzyme for the metabolic control of polysaccharide conformation and function. *Nature, Lond.* **227,** 392–393.

Lin, T.-Y. and Hassid, W. Z. (1966) Pathway of alginic acid synthesis in the marine brown alga *Fucus gardneri* Silva. *J. biol. Chem.* **241,** 5284–5297.

Loewus, F., Wagner, G., Schiff, J. A. and Weistrop, J. 1971) The incorporation of ^{35}S-labelled sulfate in carrageenan in *Chondrus crispus*. *Pl. Physiol. Lancaster* **48,** 373–375.

McBride, D. L. and Cole, K. (1971) Electron microscopic observations on the differentiation and release of monospores in the marine red alga *Smithora naiadum*. *Phycologia* **10,** 49–61.

McCandless, E. L. (1975) Biological control of the structure of carrageenans. In *Proceedings of the VIIIth International Seaweed Symposium, Bangor, 1974*. In press.

McCandless, E. L., Craigie, J. S. and Walter, J. A. (1973) Carrageenans in the gametophytic and sporophytic stages of *Chondrus crispus*. *Planta* **112,** 201–212.

McCully, M. E. (1965) A note on the structure of the cell walls of the brown alga *Fucus*. *Can. J. Bot.* **43,** 1101–1104.

McCully, M. E. (1966) Histological studies on the genus *Fucus*. I. Light microscopy of the vegetative plant. *Protoplasma* **62,** 287–305.

McCully, M. E. (1968a) Histological studies on the genus *Fucus*. III. Fine structure and possible functions of the epidermal cells of the vegetative thallus. *J. Cell Sci.* **3,** 1–16.

McCully, M. E. (1968b) Histological studies on the genus *Fucus*. II. Histology of the reproductive tissues. *Protoplasma* **66,** 205–230.

McCully, M. E. (1970) The histological localization of the structural polysaccharides of seaweeds. *Ann. N.Y. Acad. Sci.* **175,** 702–711.

Mackie, W. and Preston, R. D. (1974) Cell wall and intercellular region polysaccharides. In *Algal Physiology and Biochemistry*, ed. Stewart, W. D. P. pp. 40–85. Blackwell Scientific Publications, Oxford.

Matthews, R. A. and Evans, L. V. (1975) Extraction of Golgi apparatus from larger brown seaweeds. In *Proceedings of the VIIIth International Seaweed Symposium, Bangor, 1974*. In press.

Morré, D. J., Mollenhauer, H. H. and Bracker, C. E. (1971) Origin and continuity of Golgi apparatus. In *Origin and Continuity of Cell Organelles*, ed. Reinert, J. and Ursprung, Z., pp. 82–126. Springer-Verlag, Berlin.

Newell, B. S., Dalpont, G. and Grant, B. R. (1972) The excretion of organic nitrogen by marine algae in batch and continuous culture. *Can. J. Bot.* **50,** 2605–2611.

O'Brien, T. P., Feder, N. and McCully, M. E. (1964) Polychromatic staining of plant cell walls by toluidine blue O. *Protoplasma* **59,** 368–373.

Parker, B. C. and Diboll, A. G. (1966) Alcian stains for histochemical localization of acid and sulfated polysaccharides in algae. *Phycologia* **6,** 37–46.

Percival, E. (1972) Chemistry of agaroids, carrageenans and furcellarans. *J. Sci. Fd Agric.* **23,** 933–940.

Percival, E. and McDowell, R. H. (1967) *Chemistry and Enzymology of Marine Algal Polysaccharides*. Academic Press, London.

Percival, E. and Mian, A. J. (1973) Fucose-containing polysaccharides in brown seaweeds. *Proceedings of the VIIth International Seaweed Symposium, Sapporo, Japan, 1971*. pp. 443–446.

Perytére, M. (1970) Evolution de l'appareil de Golgi au cours de la tetrasporogenèse de *Griffithsia flosculosa* (Rhodophycée). *C. r. hebd. Séanc. Acad. Sci. Paris* **270,** 2071–2074.

PICKMERE, S. E., PARSONS, M. J. and BAILEY, R. W. (1973) Composition of *Gigartina* carrageenan in relation to sporophyte and gametophyte stages of the life cycle. *Phytochemistry* **12,** 2441–2444.

QUATRANO, R. S. and CRAYTON, M. A. (1973) Sulfation of fucoidan in *Fucus* embryos. I. Possible role in localization. *Devl Biol.* **30,** 29–41.

RAMUS, J. (1973) Cell surface polysaccharides of the red alga *Porphyridium.* In *Biogenesis of Plant Cell Wall Polysaccharides* ed. LOEWUS, F. pp. 333–359. Academic Press, New York.

RAMUS, J. and GROVES, S. T. (1972) Incorporation of sulfate into the capsular polysaccharide of the red alga *Porphyridium. J. Cell Biol.* **54,** 399–407.

REES, D. A. (1969) Structure, conformation and mechanism in the formation of polysaccharide gels and networks. *Adv. Carbohyd. Chem. Biochem.* **24,** 267–332.

SCHIFF, J. A. and HODSON, R. C. (1973) The metabolism of sulfate. *A. Rev. Pl. Physiol.* **24,** 381–414.

SCOTT, J. L. and DIXON, P. S. (1973) Ultrastructure of tetrasporogenesis in the marine red alga *Ptilota hypnoides. J. Phycol.* **9,** 29–46.

SCOTT, J. E. and DORLING, J. (1965) Differential staining of acid glycosaminoglycans (mucopolysaccharides) by alcian blue in salt solutions. *Histochemie* **5,** 221–233.

SMIDSRØD, O. and HAUG, A. (1968) Dependence upon uronic acid composition of some ion-exchange properties of alginates. *Acta chem. scand.* **22,** 1989–1997.

SMITH, D. B., COOK, W. H. and NEAL, J. L. (1954) Physical studies on carrageenin and carrageenin fractions. *Archs Biochem. Biophys.* **53,** 192–204.

STOOLMILLER, A. C. and DORFMAN, A. (1969) The metabolism of glycosaminoglycans. In *Comprehensive Biochemistry* ed. FLORKIN, M. and STOTZ, E. H., vol. 17, pp. 241–275. Elsevier, Amsterdam.

THIÉRY, J. P. (1967) Mise en évidence des polysaccharides sur coupes fines en microscopie électronique. *J. Microsc.* **6,** 987–1017.

THIÉRY, J. P. (1969) Rôle de l'appareil de Golgi dans la synthèse des mucopolysaccharides. Etude cytochemique. I. Mise en évidence des mucopolysaccharides dans les vésicules de transition entre l'ergastoplasme et l'appareil de Golgi. *J. Microsc.* **8,** 689–708.

THOMPSON, E. W. and PRESTON, R. D. (1967) Proteins in the cell walls of some green algae. *Nature Lond.* **213,** 684–685.

VREELAND, V. (1970) Localization of a cell wall polysaccharide in a brown alga with labeled antibody. *J. Histochem. Cytochem.* **18,** 371–373.

VREELAND, V. (1972) Immunocytochemical localization of the extracellular polysaccharide alginic acid in the brown seaweed, *Fucus distichus. J. Histochem. Cytochem.* **20,** 358–367.

YOUNG, R. W. (1973) The role of the Golgi complex in sulfate metabolism. *J. Cell Biol.* **57,** 175–189.

AUTHOR INDEX

SUBJECT INDEX